DYNAMICS
SI-VERSION

Also by J. L. Meriam

STATICS
SI-VERSION

DYNAMICS

SECOND EDITION
SI-VERSION

J. L. Meriam

PROFESSOR OF MECHANICAL ENGINEERING
CALIFORNIA POLYTECHNIC STATE UNIVERSITY

John Wiley & Sons, Inc.

NEW YORK LONDON SYDNEY TORONTO

Library of Congress Cataloging in Publication Data:

Meriam, James L
Dynamics.

Companion volume: Statics. These two works are also published together as Statics and dynamics.

1. Dynamics. I. Title.

TA352.M45 1975 620.1'04 74-30017

ISBN 0-471-59607-8

Printed in the United States of America

10 9 8

PREFACE SI-VERSION

Widespread adoption of the International System of Units (abbreviated SI after the French "Le Système International d'Unités") by technology throughout most of the world is an accomplished fact. This modernized metric system, which inherently is a simple system and which greatly facilitates the international practice of engineering, is rapidly replacing the British, MKS, CGS, and other systems of units which have been used for technological and scientific work in various countries over the years. In consequence of this change it is imperative that engineering students be thoroughly familiar with the SI system, since most of their professional work following graduation will be conducted with the use of SI units.

Many instructors and students will find it preferable in their respective teaching and study of mechanics to use the new system of units exclusively rather than be burdened with two sets of units. Once the principles and procedures of mechanics are mastered, conversion to other units becomes a routine procedure. To satisfy this need the Second Editions of *Dynamics* and its companion volume *Statics* have been converted entirely to the new metric system in the *SI-Editions*. Tables and charts for conversion between British-U.S. units and SI units have been included on the inside front and back covers of the books for convenient reference. The rules which govern the correct usage of SI units have been followed consistently throughout, and the student will find it to his advantage to observe these details carefully so as to form proper habits of accepted practice from the beginning.

Many of the numerical problems in the SI-Editions have been converted from the Second Editions by retaining dimensional similarity, thus making it possible to convert the results to SI units by the principles of similitude. The work of conversion has been skillfully accomplished by an expert, Dr. A. L. Hale of the Bell Telephone Laboratories, whose contributions to the author's previous books are a matter of record. The conversions have been checked, and there is every reason to believe that the SI-Editions are remarkably free of error.

It is my express hope that this SI-Edition with its modernized metric units will help to promote the study of mechanics as preparation for the practice of engineering in a modern metric world.

Santa Barbara, California
March 1975

J. L. Meriam

PREFACE

To the Student

The challenge and responsibility of modern engineering practice demand a high level of creative activity which, in turn, requires the support of strong analytical capability. The subject of engineering mechanics, which includes statics and dynamics, constitutes one of the cornerstones of analytical capability, and all engineers should have a basic background in this field of study.

Today's student of engineering becomes tomorrow's practicing engineer who must, through the exercise of his creative imagination and his professional knowledge, successfully combine theory and practice in the development of new structures, machines, devices, and processes which provide benefit to man. This process of modern creative design depends on the ability to visualize new configurations in terms of real materials and processes and the physical laws which govern them. Maximum progress to support the development of this design capability will be made when engineering theory is learned within the context of engineering reality so that the significance of theory can be perceived as it is being studied. This book is written with the foregoing view in mind, and it is hoped that the student will find interest and stimulation in the many problems which are taken from a wide variety of contemporary engineering situations to provide realistic and significant applications of the theory.

The purpose of the study of mechanics is to predict through calculation the behavior of engineering components and systems involving force and motion. Successful prediction in engineering design requires the careful formulation of problems with the aid of a dual thought process of physical understanding and mathematical reasoning. The process of formulating a problem is one of constructing a mathematical model which incorporates appropriate physical assumptions and mathematical approximations and approaches the actual situation with sufficient accuracy for the purpose at hand. Indeed, this process of matching the symbolic model to its physical prototype is, undoubtedly, one of the most valuable experiences of engineering study, and the problems which are included are intended to provide a comprehensive opportunity to develop this ability.

Success in analysis depends to a surprisingly large degree on a well-disciplined method of attack from hypothesis to conclusion where a straight path of rigorous application of principles has been followed. The student is urged to develop ability to represent his work in a clear, logical, and neat manner. The basic training in mechanics is a most excellent place for early development of this disciplined approach which is so necessary in most engineering work which follows.

More material is contained in this text than is covered in the usual first course in engineering mechanics, so that the book also includes an introduction to more advanced topics in mechanics and can serve as a future reference for basic principles. The more advanced topics are identified along the margins of the pages to alert the reader. To aid the student in his initial study of each topic this second edition of the text includes an expanded collection of introductory problems and problems of intermediate difficulty. Problems that offer special challenge because of their difficulty are placed at the end of each set and are identified by a black triangle and a red number.

I extend encouragement to all students of mechanics, and I hope that this book will provide substantial assistance to them in acquiring a strong and meaningful engineering background.

July 1971

J. L. Meriam

PREFACE

To the Instructor

In recent times the strong trend to increase the analytical capability in engineering has resulted in increased emphasis on the mathematical generalities in mechanics. When adequate emphasis on physical understanding and engineering application is preserved, then the trend is of great benefit in extending capabilities for the analytical description of difficult problems. On the other hand, when primary attention is focused on the mathematical framework of mechanics with secondary attention to physical reality and engineering application, then the trend is of questionable benefit. Instruction in engineering mechanics has as its basic purpose the development of capacity to predict the effects of force and motion as an aid in carrying out the creative design process of engineering. Therefore the primary focus should be on the engineering significance of physical quantities with the mathematical structure acting in a supporting role. When this basic purpose is kept in mind, a proper balance between theory and application can be realized.

In this same connection there is often the temptation for the instructor of mechanics who has reached a high level of theoretical ability to forget the frame of reference of his students and present the subject with an overemphasis on generalization. There is considerable danger in this approach for the first basic course, since students lack the background necessary to cope with excessive early generality and they are also deprived of experiencing some of the historical and natural development of the subject.

A further consideration of philosophy is the strong need to provide an environment of challenging engineering reality as a means of developing the motives for learning mechanics. The importance of a solid background of analytical capability can be established in no better way than by creating a genuine interest and a compelling engineering need for the effective use of theory.

Dynamics is an engineering text and is written with these views in mind. Effort has been made to present the theory rigorously, concisely, and with a generality commensurate with the background of basic calculus assumed of the reader. For students who do not have a background in vector analysis, the necessary concepts and explanation are introduced in the text as needed. A more formal introduction and summary of the algebra and calculus of vectors as used in mechanics are included in Appendix B for convenient reference. Usually, I have found that facility with vector analysis is developed best within the context of its meaningful application in mechanics, and this view has guided the treatment of vectors in this book. For two-dimensional analysis the scalar-geometric method is frequently used as the simplest and most direct description. For three-dimensional problems, vector notation is employed as the most direct and appropriate description. Matrix and tensor methods are used where it becomes necessary to make transformations from one coordinate system to a rotated coordinate system in three dimensions. The exclusive use of scalar notation, of vector notation, or of tensor index notation is rejected in favor of the choice of the mathematical tool which is most appropriate for the situation at hand. In my view it is far more important in the basic course in mechanics to preserve and strengthen dependence on geometrical visualization and physical understanding than it is to emphasize the extensive or exclusive use of a tensor notation that reduces geometry essentially to a notational manipulation. The creative ideas that find greatest use in those branches of engineering which are supported by mechanics are born and developed more through the visualization of geometrical configurations than through the manipulation of notation in analysis.

The presentation and the problems in *Dynamics* have not been intentionally structured to provide exercises in the use of the computer. If such were the case, the role of the computer would be largely artificial. However, the student who has access to computer facilities should be encouraged to use them for the solution of occasional problems where a machine solution offers a distinct advantage.

Dynamics is intended for use by engineering students in their second year and beyond. The treatment is arranged under the three main headings of particles, rigid bodies, and nonrigid systems, each with its appropriate coverage of kinematics and kinetics. This arrangement is particularly convenient, since it permits a logical choice of topics to match a variety of course outlines. If primary emphasis on particle dynamics is desired, rigid-body motion is easily minimized, or vice versa. It is also possible to follow an outline based on a two-dimensional treatment with coverage of three-dimensional motion reserved for later study. Included are introductions to more advanced topics such as the space motion of rigid bodies, the general equations for variable mass, wave motion, and Lagrange's equations. These topics will serve as a basis for further study and future reference.

The problem sets include numerous examples taken from the subject of space mechanics and other contemporary developments. The problems are arranged approximately in order of increasing difficulty. The assessment of difficulty, of course, depends not only on the recognition of applicable theory but on the obstacles encountered in constructing the idealized model and in carrying out the required mathematics, and these factors vary considerably among individuals.

A number of significant changes have been incorporated into the second edition of *Dynamics* to enhance the usefulness of the book. The collection of problems has been greatly enlarged by the addition of 400 new and modified problems mainly at the introductory and intermediate levels of difficulty. This addition should prove helpful to most students in gaining early experience and initial confidence in their study of dynamics. The explanatory material in the basic sections of the book has been expanded in most of the articles to aid the student in his self-study efforts. The advanced material remains more concentrated and is included primarily for those who wish to pursue their study of dynamics somewhat beyond the basic introductory topics.

A further change that should prove helpful is the identification of the more advanced sections of the book by a gray band along the outer margins of the pages. Most of the three-dimensional topics, for instance, have been placed in this category. This identification, coupled with certain rearrangements of material, will make it easier to choose assignments within a desired level of treatment and with the distinction between two- and three-dimensional analysis in mind.

In the first edition of *Dynamics* effort was made to unify various sections of the subject through certain consolidations. Curvilinear motion of a particle with its several coordinate descriptions and the plane motion of a rigid body with its three categories of motion are examples where this consolidation took place with attention focused within a single article on the unifying ideas of each respective topic. To ease the learning difficulty associated with consolidation without at the same time losing the advantage of a unified treatment, each of these broad topics, and a number of others, has been broken down within its respective article into more identifiable sections. The first few examples which are included with the problem set for each of these broad articles are identified as to the coordinate system or as to the type of motion with the balance of the problems left for the student to identify. This arrangement should approach an optimum experience for the student by providing him with initial guidance followed by the opportunity to develop his own judgment.

The emphasis placed in the first edition on clear and detailed illustrations in an effort to establish a sense of engineering reality has been continued in the second edition. This effort is consistent with my firm belief that experience with the formulation of problems which incorporate a high degree of reality, including a choice of the approach to their solution, is perhaps the most important aspect of the study of mechanics. With this approach, theory takes on a significance that it cannot possibly have when the student encounters primarily idealized, and hence preformulated, problems. Special attention has been given to the format of the second edition through the identification of optional topics and difficult problems. Color has been introduced which greatly facilitates the recognition of external forces and motion vectors and which provides a new dimension to the function and appearance of the book.

I am pleased to give continued recognition to Dr. A. L. Hale of Bell Telephone Laboratories for his valuable assistance in reviewing the manuscript and in offering numerous helpful suggestions for the second edition. Dr. Hale rendered similar assistance in my previous books on mechanics, and it is a genuine pleasure to have his continued interest and contribution in this new volume. Acknowledgment is also given to the critical reviews and numerous helpful suggestions of Professor Paul Jones of the University of Illinois and Professor Andrew Pytel of The Pennsylvania State University during the preparation of the present edition. Also, I am grateful for the high standards and professional contributions of the staff of John Wiley & Sons in the planning and production of this book. The support of Duke University in arranging for a sabbatical leave of absence, during which time the present edition was largely prepared, is gratefully acknowledged. Finally, I acknowledge continued encouragement, patience, and assistance from my wife, Julia, during the many hours required to prepare this manuscript.

Durham, North Carolina
July 1971

J. L. Meriam

CONTENTS

* Symbol ■ indicates that the article contains topics of a somewhat more advanced or specialized nature.

PART II DYNAMICS OF RIGID BODIES

5 PLANE KINEMATICS OF RIGID BODIES

6 PLANE KINETICS OF RIGID BODIES

7 SPACE KINEMATICS OF RIGID BODIES

8 SPACE KINETICS OF RIGID BODIES

9 VIBRATION AND TIME RESPONSE

PART III DYNAMICS OF NONRIGID SYSTEMS

10 DYNAMICS OF NONRIGID SYSTEMS

GUIDE TO THE USE OF DYNAMICS

1 *Principal equations* are identified by a red triangle to the left and a red equation number to the right, such as

▶ $$\mathbf{F} = m\mathbf{a} \tag{1}$$

2 *Advanced and specialized topics* included in the text for optional study are preceded by a row of triangles

▼ ▼ ▼ ▼ ▼

and are identified by a gray band along the outer margin of the page.

3 **Sample Problems**

are set off from the remainder of the text for ready identification by horizontal red rules and by a vertical red rule along the outer margin of the page.

4 *Problems* in the problem sets are
numbered consecutively by chapter,
arranged generally in order of increasing difficulty,
identified by a black triangle and red number (◀**2/159** for example) when they incorporate special challenge or difficulty.

5 *Force vectors* are represented on the diagrams by heavy red arrows to focus attention on their unique significance and to distinguish them from other lines or arrows. *Motion vectors* are represented by red lines of lighter weight.

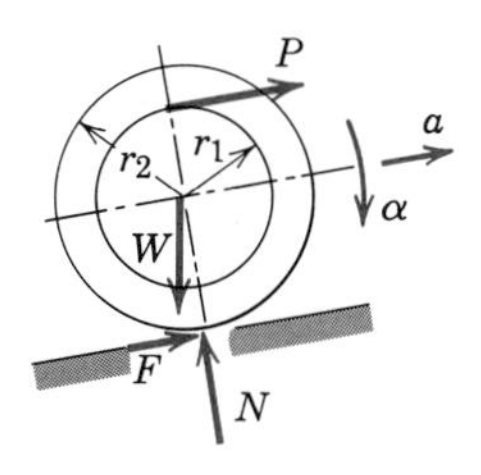

Color is also used selectively to highlight or clarify other geometric elements in the figures.

1 PRINCIPLES OF DYNAMICS

1 **Introduction.** Dynamics is that branch of mechanics which deals with the motion of bodies under the action of forces. The study of dynamics in engineering usually follows the study of statics which deals with the action of forces on bodies at rest. Dynamics has two distinct parts—*kinematics,* which is the study of motion without reference to the forces which cause motion, and *kinetics,* which relates the action of forces on bodies to their resulting motions. The student of engineering will find that a thorough comprehension of dynamics will provide him with one of his most useful and powerful tools for analysis in engineering.

Historically, dynamics is a relatively recent subject compared with statics. The beginning of a rational understanding of dynamics is credited to Galileo (1564–1642) who made careful observations concerning bodies in free fall, motion on an inclined plane, and motion of the pendulum. He was largely responsible for bringing a scientific approach to the investigation of physical problems. Galileo was continually under severe criticism for refusing to accept the established beliefs of his day, such as the philosophies of Aristotle which held, for example, that heavy bodies fall more rapidly than light bodies. The lack of accurate means for the measurement of time was a severe handicap to Galileo, and further significant development in dynamics awaited the invention of the pendulum clock by Huygens in 1657. Newton (1642–1727), guided by Galilieo's work, was able to make an accurate formulation of the laws of motion and, hence, to place dynamics on a sound basis. Newton's famous work was published in the first edition of his *Principia,** which is generally recognized as one of the greatest of all recorded contributions to knowledge. In addition to stating the laws governing the motion of a particle, Newton was the first to formulate correctly the law of universal gravitation. Although his mathematical description was accurate, he felt that the concept of remote transmission of gravitational force without a supporting medium was an absurd notion. Following Newton's time, important contributions to mechanics were made by Euler, D'Alembert, Lagrange, Laplace, Poinsot, Coriolis, Einstein, and others.

In terms of engineering application, dynamics is an even more recent science. Only since machines and structures have operated with high speeds and appreciable accelerations has it been necessary to make calculations based on the principles of dynamics rather than on the principles of statics. The rapid technological developments of the present day require increasing

* The original formulations of Sir Isaac Newton may be found in the translation of his *Principia* (1687), revised by F. Cajori, University of California Press, 1934.

application of the principles of mechanics, particularly dynamics. These principles are basic to the analysis and design of moving structures, to fixed structures subject to shock loads, to high-speed computer mechanisms, to automatic control systems, to rockets, missiles, and spacecraft, to ground and air transportation vehicles, to electron ballistics of electrical devices, and to machinery of all types such as turbines, pumps, reciprocating engines, hoists, machine tools, etc. The student whose interests lead him into one or more of these and many other activities will find a constant need for applying his basic knowledge of dynamics.

2 Basic Concepts. Certain concepts and definitions are basic to the study of dynamics, and they should be thoroughly understood at the outset.

Space is the geometric region in which events take place. In this book the word space will be used to refer to a three-dimensional region. It is not uncommon, however, to refer to motion along a straight line or in a plane as occurring in one- or two-dimensional space, respectively. The concept of n-dimensional space is an abstract device for describing relations among n quantities.

Reference Frame. Position in space is determined relative to some geometric reference system by means of linear and angular measurements. The basic frame of reference for the laws of Newtonian mechanics is the *primary inertial system* or *astronomical frame of reference* which is an imaginary set of rectangular axes assumed to have no translation or rotation in space. Measurements show that the laws of Newtonian mechanics are valid for this reference system as long as any velocities involved are negligible compared with the speed of light.* Measurements made with respect to this reference are said to be *absolute,* and this reference system is considered to be "fixed" in space. A reference frame attached to the surface of the earth has a somewhat complicated motion in the primary system, and a correction to the basic equations of mechanics must be applied for measurements made relative to the earth's reference frame. In the calculation of rocket and space flight trajectories, for example, the absolute motion of the earth becomes an important parameter. For most engineering problems of machines and structures which remain on the earth's surface, the corrections are extremely small and may be neglected. For these problems the laws of mechanics may be applied directly for measurements made relative to the earth, and in a practical sense such measurements will be referred to as *absolute.*

Time is a measure of the succession of events and is considered an absolute quantity in Newtonian mechanics. The unit of time is the second, which is a convenient fraction of the 24-hour day.

Force is the action of one body on another. A force tends to move a body in the direction of its action upon it.

Matter is substance which occupies space. A *body* is matter bounded by a closed surface.

*For velocities of the same order as the speed of light, 300 000 km/s, the theory of relativity must be applied. See Art. 16 for a brief discussion of this theory and a numerical example of its effect.

Inertia is the property of matter causing a resistance to change in motion.

Mass is the quantitative measure of inertia. Mass is also a property of every body which is always accompanied by mutual attraction to other bodies.

Particle. A body of negligible dimensions is called a particle. In the mathematical sense a particle is a body whose dimensions approach zero so that it may be analyzed as a point mass. Frequently a particle is chosen as a differential element of a body. Also, when the dimensions of a body are irrelevant to the description of its position or its motion, the body may be treated as a particle.

Rigid body. A body is said to be rigid when it has no relative deformation between its parts. This is an ideal limiting condition, since all real bodies undergo some change in shape when they are subjected to forces. When such changes in shape are negligible compared with the overall dimensions of the body or with the changes of position of the body as a whole, the assumption of rigidity is permissible. For a rigid body, then, the difference in configurations between the initial and the deformed conditions is neglected. As an example of the assumption of rigidity, for an airplane flying through turbulent air, the small flexural movement of its wing tip in relation to the body of the aircraft is clearly of no consequence to the average distribution of aerodynamic forces on its wings or to the specification of the motion of the airplane as a whole in its flight path. The treatment, then, of the airplane as a rigid body offers no complication. On the other hand, if the problem is one of examining the internal stresses in the wing structure due to changing dynamics loads, then the deformable characteristics of the structure would have to be examined, and for this purpose the airplane could no longer be considered a rigid body.

Scalar. A quantity with which a magnitude only is associated is known as a scalar. Examples of scalars are time, volume, density, speed, energy, and mass.

Vector. A quantity with which a direction as well as a magnitude is associated and which obeys the parallelogram law of addition is a vector. Examples of vectors are displacement, velocity, acceleration, force, moment, and momentum.

In *Dynamics* boldface type is used for vectors and lightface type is used for scalars. Thus $\mathbf{V} = \mathbf{V}_1 + \mathbf{V}_2$ represents the vector sum of two vectors whereas $S = S_1 + S_2$ represents the scalar sum of two scalars. The magnitude of a vector $\mathbf{V}$ is written as V, and its direction may be indicated by a unit vector, say $\mathbf{n}$, which has a magnitude of unity and the direction of $\mathbf{V}$. Thus $\mathbf{V} = V\mathbf{n}$. Vectors appearing in diagrams are labeled either with boldface or lightface symbols. When the magnitude only of the vector is indicated on the diagram, a lightface symbol is used. Scalar multiplication of vector quantities is indicated by the dot, $\mathbf{V}_1 \cdot \mathbf{V}_2$, and vector multiplication by the cross, $\mathbf{V}_1 \times \mathbf{V}_2$. It is essential that the student make a clear and consistent distinction between vectors and scalars in his written work. If this distinction is not preserved in his notation, serious difficulty can arise. It is recommended that a distinguishing mark for each vector quantity, such as an underline $\underline{V}$, be used in all handwritten work to take the place of the boldface type in print.

It is assumed that the reader is familiar with the physical properties of force vectors and the mathematics of vector algebra through previous study of statics and mathematics. Reference may be made to Appendix B for an introduction and summary of the algebra and calculus of vectors as used in mechanics for those who may need additional assistance.

Dynamics involves the frequent use of time derivatives of both vectors and scalars. As a notational shorthand a dot over a quantity will frequently be used to indicate a derivative with respect to time. Thus $\dot{x}$ means dx/dt and $\ddot{x}$ stands for d^2x/dt^2.

3 Newton's Laws. Sir Isaac Newton was the first to state correctly the basic laws governing the motion of a particle and to demonstrate their validity. Slightly reworded to use modern terminology, these laws are as follows:

Law I. A particle remains at rest or continues to move in a straight line with a uniform velocity if there is no unbalanced force acting on it.

Law II. The acceleration of a particle is proportional to the resultant force acting on it and is in the direction of this force.*

Law III. The forces of action and reaction between interacting bodies are equal in magnitude, opposite in direction, and collinear.

The correctness of these laws has been verified by innumerable accurate physical measurements. The first two laws hold for measurements made in an absolute frame of reference but are subject to some correction when the motion is measured relative to a reference system having acceleration, such as one attached to the earth's surface.

Newton's second law forms the basis for most of the analysis in mechanics. As applied to a particle of mass m it may be stated as

$$\mathbf{F} = m\mathbf{a} \tag{1}$$

where $\mathbf{F}$ is the resultant force acting on the particle and $\mathbf{a}$ is the resulting acceleration. Newton's first law is a consequence of the second since there is no acceleration when the force is zero, and the particle either is at rest or moves with a constant velocity. The first law adds nothing new to the description of motion but is included since it was a part of Newton's classical statements.

The third law is basic to our understanding of force. It states that forces always occur in pairs of equal and opposite forces. This principle holds for all forces, variable or constant, regardless of their source and holds at every instant of time during which the forces act. Lack of careful attention to this basic law is the cause of frequent error by the beginner. In analyzing bodies under the action of forces, it is absolutely necessary to be clear as to which

* To some it is preferable to interpret Newton's second law as meaning that the resultant force acting on a particle is proportional to the time rate of change of momentum of the particle and that this change is in the direction of the force. Both formulations are equally correct when applied to a particle of constant mass.

of the pair of forces is being considered. This clarification comes by *isolating* the body in question from other contacting bodies and considering only the one force of the pair which acts *on* the body isolated.

4 Units. Throughout the years several systems of units have been used to express the magnitudes of the several quantities in mechanics and other subject areas. In very recent times virtually all countries throughout the world have adopted the International System of Units, abbreviated SI (from the French, Système International d'Unités) for all engineering and scientific work. The following table summarizes those units which form the basis for calculations made in mechanics.

SI System of Units

Quantity	Dimensional Symbol	Basic SI Unit (symbol)
Length	L	metre* (m)
Time	T	second (s)
Mass	M	kilogram (kg)
Force	F	newton (N)

The kilogram (1000 grams) and not the gram (g) is selected as the basic unit of mass. The first three quantities are the base units in the SI system, and the fourth quantity, force, is derived from the base units by Newton's second law of motion. By definition one newton is that force which will give a one-kilogram mass an acceleration of one metre per second squared. Thus, from $F = ma$ the equivalence between the units is

$$(1 \text{ N}) = (1 \text{ kg})(1 \text{ m/s}^2) \text{ or } \text{N} = \text{kg}\cdot\text{m/s}^2$$

In the English-speaking countries the SI system replaces the U.S.–British system of units used in engineering in which length in feet (ft), time in seconds (s), and force in pounds (lbf) are the base units and mass in slugs ($\text{lbf}\cdot\text{s}^2/\text{ft}$) is derived from Newton's second law. The pound is also used as a unit of mass (lbm) in which case 32.2 lbm are equivalent to one slug. Conversion tables between U.S.–British units and SI units are given on the inside covers of the book.

Detailed guidelines have been recommended for the consistent use of SI units, and these guidelines have been followed throughout this book. Only three of them will be mentioned here.

(a) *Unit prefixes.* Numerical values should generally be kept between 0.1 and 1000. To this end the following prefixes are the ones most commonly used:

Amount	Multiple	Prefix	Symbol
1 000 000 000	10^9	giga	G
1 000 000	10^6	mega	M
1 000	10^3	kilo	k
0.001	10^{-3}	milli	m
0.000 001	10^{-6}	micro	μ
0.000 000 001	10^{-9}	nano	n

*Spelled alternatively *meter*.

For example, a length of 4245 m is written 4.245 km*; a mass of 0.0326 kg is written 32.6 g; and a force of 0.0068 N is written 6.8 mN.

(b) *Unit designations.* To avoid confusion a dot will be used to separate units which are multiplied together. Thus the unit for the moment of a force, newton-metre, is written N·m whereas mN would mean millinewton. Also, when compound units are formed by division, such as pressure, the unit should be written N/m^2 or $N \cdot m^{-2}$ and not in the ambiguous form N/m/m. Accepted form further requires that prefixed units not be used in the denominator. Thus 25 N/mm^2 should be written 25 MN/m^2.

(c) *Number grouping.* The use of a comma to separate groups of numbers in powers of 10^3 is ruled out in favor of a space. Thus correct practice is illustrated by the number 4 607 321.048 72. Note that the space is used on both sides of the decimal point. The space may be omitted for numbers of four digits only, such as 4296 or 0.0476.

5 Gravitation. The *law of gravitation,* first formulated by Newton, governs the mutual attraction between bodies and is expressed by the equation

$$F = K\frac{m_1 m_2}{r^2} \tag{2}$$

where F = the mutual force of attraction between two particles
K = a universal constant known as the constant of gravitation
m_1, m_2 = the masses of the two particles
r = the distance between the centers of the particles

The mutual forces F obey the law of action and reaction since they are equal and opposite and are directed along the line joining the centers of the particles. Experimental data yield the value $K = 6.673(10^{-11})\ m^3/(kg \cdot s^2)$ for the gravitational constant. Gravitational forces exist between every pair of bodies. On the surface of the earth the only gravitational force of appreciable magnitude is the force due to the earth's attraction. Thus each of two iron spheres 100 mm in diameter is attracted to the earth with a gravitational force of 37.9 N which is called its *weight.* On the other hand, the force of mutual attraction between them if they are just touching is 0.000 000 099 4 N. This force is clearly negligible compared with the earth's attraction of 37.9 N, and, consequently, the gravitational attraction of the earth is the only gravitational force of any appreciable magnitude which need be considered for experiments conducted on the earth's surface.

The force of gravitational attraction of the earth on a body depends on the position of the body relative to the earth. If the earth is regarded as a perfect sphere, a body with a mass of exactly 1 kg would be attracted to the earth by a force of 9.824 N on the earth's surface, 9.821 N at an altitude of 1 km, 9.523 N at an altitude of 100 km, 7.340 N at an altitude of 1000 km, and 2.456 N at an altitude equal to the mean radius 6371 km of the earth. It is

*The accepted pronunciation of kilometre (km) is "kilo-metre" and not "kil-ometre". As in metre an accepted alternative spelling is *kilometer*.

at once apparent that the variation in gravitational attraction of high-altitude rockets and spacecraft becomes a major consideration.

The gravitational attraction of the earth on a body is known as the "weight" of the body. This force exists whether the body is at rest or in motion. Since this attraction is a force, strictly speaking the weight of a body should be expressed in newtons (N) under the SI system of units. Unfortunately in common practice the mass unit kilogram (kg) has been used extensively as a measure of weight. When expressed in kilograms the word "weight" technically means mass. To avoid confusion the word "weight" in this book shall be restricted to mean the force of gravitational attraction, and it will always be expressed in newtons.

Every object which is allowed to fall in a vacuum at a given position on the earth's surface will have the same acceleration g as can be seen by combining Eqs. 1 and 2 and canceling the term representing the mass of the falling object. This combination gives

$$g = \frac{Km_0}{r^2}$$

where m_0 is the mass of the earth and r is the radius of the earth.* The mass m_0 and the mean radius r of the earth have been found through experimental measurements to be $5.976(10^{24})$ kg and $6.371(10^6)$ m, respectively. These values, together with the value of K already cited, when substituted into the expression for g, give

$$g = 9.824 \text{ m/s}^2$$

The acceleration due to gravity as determined from the gravitational law is the acceleration which would be measured from a set of reference axes with origin at the center of the earth but not rotating with the earth. With respect to these "fixed" axes, then, this value may be termed the absolute value of g. Because of the fact that the earth rotates, the acceleration of a freely falling body as measured from a position attached to the earth's surface is slightly less than the absolute value. Accurate values of the gravitational acceleration as measured relative to the earth's surface account for the fact that the earth is a rotating oblate spheroid with flattening at the poles. These values are given by the International Gravity Formula which is

$$g = 9.780\,49(1 + 0.005\,288\,4 \sin^2 \gamma - 0.000\,005\,9 \sin^2 2\gamma)$$

where γ is the latitude and g is expressed in metres per second squared. The constants account for the deviation of the earth's shape from that of a sphere and also for the effect of the earth's rotation. The absolute acceleration due to gravity as determined for a nonrotating earth may be computed from the relative values to a close approximation by adding $3.382(10^{-2}) \cos^2 \gamma$ m/s^2, which removes the effect of the earth's rotation. The variation of both the

*It can be proved that the earth, when taken as a sphere with a symmetrical distribution of mass about its center, may be considered a particle with its entire mass concentrated at its center.

absolute and the relative values of g with latitude is shown in Fig. 1 for sea-level conditions.* The standard value adopted internationally for the gravitational acceleration relative to the rotating earth at sea level and at a latitude of 45° is 9.806 65 m/s².

The proximity of large land masses and the variations in the density of the earth's crust also influence the local value of g by a small but detectable amount. In almost all engineering problems where measurements are made on the surface of the earth the difference between the absolute and relative values of the gravitational acceleration and the effect of local variations are neglected, and 9.81 m/s² (32.2 ft/s² in U.S.-British units) is used for the sea-level value of g.

The variation of g with altitude is easily determined by the gravitational law. If g_0 represents the absolute acceleration due to gravity at sea level, the absolute value at an altitude h is

$$g = g_0 \frac{r^2}{(r+h)^2}$$

where r is the radius of the earth.

The earth's gravitational attraction on a body may be calculated from the results of the simple gravitational experiment. If the gravitational force of attraction or true weight of a body is $\mathbf{W}$, then, since the body will fall with an absolute acceleration $\mathbf{g}$ in a vacuum, Eq. 1 gives

$$\blacktriangleright \qquad \mathbf{W} = m\mathbf{g} \qquad \textbf{(3)}$$

The apparent weight of a body as determined by a spring balance, calibrated to read the correct force and attached to the surface of the earth, will be

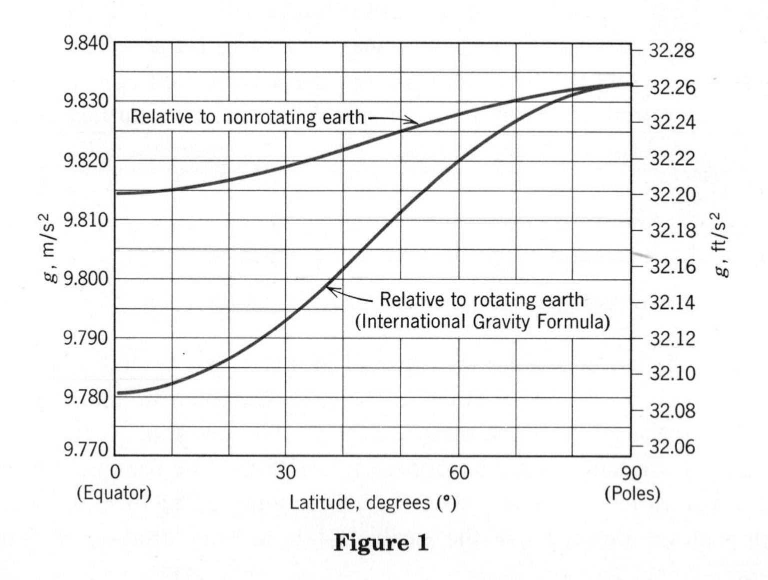

Figure 1

* The student will be able to derive these relations for a spherical earth following his study of relative motion in Chapter 3.

slightly less than its true weight. The difference is due to the rotation of the earth. The ratio of the apparent weight to the apparent or relative acceleration due to gravity still gives the correct value of mass. The apparent weight and the relative acceleration due to gravity are, of course, the quantities which are measured in experiments conducted on the surface of the earth.

6 Dimensions. A given dimension such as length can be expressed in a number of different units such as metres, millimetres, or kilometres. Thus the word *dimension* is distinguished from the word *unit*. Physical relations must always be dimensionally homogeneous, that is, the dimensions of all terms in an equation must be the same. It is customary to use the symbols L, M, T, and F, to stand for length, mass, time, and force, respectively. In the SI system the units of force are derived. From Eq. 1 force has the dimensions of mass times acceleration or

$$F = ML/T^2$$

One important use of the theory of dimensions is found in checking the dimensional correctness of some derived physical relation. The following expression for the velocity v of a body of mass m which is moved from rest a horizontal distance x by a force F may be derived:

$$Fx = \tfrac{1}{2}mv^2$$

where the $\frac{1}{2}$ is a dimensionless coefficient resulting from integration. This equation is dimensionally correct since substitution of L, M, and T gives

$$[MLT^{-2}][L] = [M][LT^{-1}]^2$$

Dimensional homogeneity is a necessary condition for correctness, but it is not sufficient since the correctness of dimensionless coefficients cannot be checked in this way.

A second important use for dimensional theory is in the prediction of full-scale performance from the results of experiments on models. There are many problems, such as the flow resistance of ships and aircraft and the behavior of loaded structures of complex shape, where a mathematical solution is not feasible by reason of the great complexity involved. The form of the relation which describes a physical problem certainly does not depend on the size of the units employed, and therefore a physical relation should describe equally well the behavior of a model or its prototype. A full discussion of this use of dimensional analysis is beyond the scope of this book,* and only one simple example of the procedure followed is given here.

Let it be desired to determine the expression for the period τ of vibration for a simple pendulum consisting of a small mass m suspended by a cord of length l. On the basis of observation, the period is assumed to be a function of the length l, the acceleration of gravity g, and the mass m. Next, it is

*The classical reference in this subject is *Dimensional Analysis* by P. W. Bridgman, Yale University Press, 1932.

assumed that this functional relationship is given by the products of these quantities raised to unknown powers α, β, γ, or

$$\tau = kl^{\alpha}g^{\beta}m^{\gamma}$$

where k is a dimensionless constant to account for the units used. Expressing this relation in dimensional symbols gives

$$\begin{aligned}[T] &= [L]^{\alpha}[LT^{-2}]^{\beta}[M]^{\gamma} \\ &= [L^{\alpha+\beta}][T^{-2\beta}][M^{\gamma}]\end{aligned}$$

In order that the equation be dimensionally homogeneous, it is necessary for the exponents of each of the three fundamental dimensions to be identical on the two sides of the equation. Equating the exponents of T, L, and M in that order gives

$$\begin{aligned}1 &= -2\beta \\ 0 &= \alpha + \beta \\ 0 &= \gamma\end{aligned}$$

The solutions are clearly $\gamma = 0$, $\beta = -\frac{1}{2}$, $\alpha = \frac{1}{2}$, and the assumed relation becomes

$$\tau = kl^{1/2}g^{-1/2} = k\sqrt{l/g}$$

Dimensional considerations disclose that the period does not depend on the mass m. One carefully executed experiment for small amplitudes of vibration will give measurements for τ and l. Substitution of these measured values along with the known value of g will give $k = 6.283$. Therefore the equation

$$\tau = 6.283\sqrt{l/g}$$

may be used to describe the period for *any* similar pendulum of a different size as long as a consistent set of units is used. In the case of the simple pendulum, direct solution will disclose the fact that $k = 2\pi$ for small amplitudes.

7 Accuracy, Limits, and Approximations. The number of significant figures shown in a numerical calculation should be no greater than the number of figures which can be justified by the accuracy of the given data. Hence the cross-sectional area of a square bar whose side, 24 mm, say, was measured to the nearest half millimetre should be written as 580 mm^2 and not as 576 mm^2, as would be indicated if the numbers were multiplied out.

When calculations involve small differences in large quantities, greater accuracy must be achieved. Thus it is necessary to know the numbers 4.2503 and 4.2391 to an accuracy of five significant figures in order that their difference 0.0112 be expressed to three-figure accuracy. It is often difficult in somewhat lengthy computations to know at the outset the number of significant figures needed in the original data to ensure a certain accuracy in the answer.

Accuracy to three significant figures is considered satisfactory for the majority of engineering calculations. The decimal point should be located by a rough longhand approximation which also serves as a check against large computational error.

The *order* of differential quantities is the subject of frequent misunderstanding. Higher-order differentials may always be neglected compared with lower-order differentials. As an example the element of volume ΔV of a right circular cone of altitude h and base radius r may be taken to be a circular slice a distance x from the vertex and of thickness Δx. It can be verified that the exact expression for the volume of the element may be written as

$$\Delta V = \frac{\pi r^2}{h^2}[x^2\,\Delta x + x(\Delta x)^2 + \tfrac{1}{3}(\Delta x)^3]$$

It should be recognized that, when passing to the limit in going from ΔV to dV and from Δx to dx, the terms in $(\Delta x)^2$ and $(\Delta x)^3$ drop out, leaving merely

$$dV = \frac{\pi r^2}{h^2}x^2\,dx$$

which gives an exact expression when integrated.

When trigonometric functions of differential quantities are used, it is well to recall the following relations which are true in the mathematical limit:

$$\sin d\theta = \tan d\theta = d\theta$$
$$\cos d\theta = 1$$

The angle $d\theta$ is, of course, expressed in radian measure. When small but finite angles are used, it is often convenient to replace the sine by the tangent or either function by the angle itself. These approximations, $\sin\theta = \theta$ and $\tan\theta = \theta$, amount to retaining only the first term in the series expansions for the sine and tangent. If a closer approximation is desired, the first two terms may be used which give $\sin\theta = \theta - \theta^3/6$ and $\tan\theta = \theta + \theta^3/3$. As an example of the first approximation, for an angle of 1°,

$$\sin 1° = 0.017\,452\,4 \qquad \text{and} \qquad 1° \text{ is } 0.017\,453\,3 \text{ radian}$$

The error in replacing the sine by the angle for 1° is only 0.005 per cent. For 5° the error is 0.13 per cent, and for 10° the error is still only 0.51 per cent. Similarly, for small angles the cosine may be approximated by the first two terms in its series expansion which gives $\cos\theta = 1 - \theta^2/2$.

A few of the mathematical relations which are useful in mechanics are listed in Table C3, Appendix C.

8 Description of Dynamics Problems. The study of dynamics is directed toward the understanding and description of the various quantities involved in the motions of bodies. This description, which is largely mathematical, enables predictions to be made of dynamical behavior. A dual thought process is necessary in formulating this description. It is necessary to think both in terms of the physical situation and in terms of the corresponding mathematical description. Analysis of every problem will require this repeated transition of thought between the physical and the mathematical.

Without question, one of the greatest difficulties encountered by the student is the inability to make this transition of thought freely. He should recognize that the mathematical formulation of a physical problem represents an ideal and limiting description, or model, which approximates but never quite matches the actual physical situation.

In the course of constructing the idealized mathematical model for any given engineering problem, certain approximations will always be involved. Some of these approximations may be mathematical, whereas others will be physical. For instance, it is often necessary to neglect small distances, angles, or forces compared with large distances, angles, or forces. If the change in velocity of a body with time is nearly uniform, then an assumption of constant acceleration may be justified. An interval of motion which cannot be easily described in its entirety is often divided into small increments each of which can be approximated. The retarding effect of bearing friction on the motion acquired by a machine as the result of applied forces or moments may often be neglected if the friction forces are small. However, these same friction forces cannot be neglected if the purpose of the inquiry is a determination of the drop in efficiency of the machine due to the friction process. Thus the degree of assumption involved depends on what information is desired and on the accuracy required. The student should be constantly alert to the various assumptions called for in the formulation of real problems. The ability to understand and make use of the appropriate assumptions in the course of the formulation and solution of engineering problems is, certainly, one of the most important characteristics of a successful engineer. Along with the development of the principles and analytical tools needed for modern dynamics, one of the major aims of this book is to provide a maximum of opportunity to develop ability in formulating good mathematical models. Strong emphasis is placed on a wide range of practical problems which not only require the full exercise of theory but also force consideration of the decisions which must be made concerning relevant assumptions.

An effective method of attack on dynamics problems, as in all engineering problems, is essential. The development of good habits in formulating problems and in representing their solutions will prove to be an invaluable asset. Each solution should proceed with a logical sequence of steps from hypothesis to conclusion, and its representation should include a clear statement of the following parts, each clearly identified:

1. Given data
2. Results desired
3. Necessary diagrams
4. Calculations
5. Answers and conclusions

In addition it is well to incorporate a series of checks on the calculations at intermediate points in the solution. The reasonableness of numerical magnitudes should be observed, and the accuracy and dimensional homogeneity of terms should be frequently checked. It is also important that the arrangement of work be neat and orderly. Careless solutions which cannot be easily read by others are of little or no value. It will be found that the discipline involved in adherence to good form will in itself be an invaluable aid to the development of the abilities for formulation and analysis. Many problems which at first may seem difficult and complicated become clear and straightforward once they are begun with a logical and disciplined method of attack.

The subject of dynamics is based upon a surprisingly few fundamental concepts and principles which, however, are extended and applied over an exceedingly wide range of conditions. One of the most valuable aspects of the study of dynamics is the experience afforded in reasoning from fundamentals. This experience cannot be obtained merely by memorizing the kinematical and dynamical equations which describe various motions. It must be obtained through exposure to a wide variety of problem situations which force the choice, use, and extension of basic principles to meet the given conditions.

In describing the relations between forces and the motions they produce, it is essential that the system to which a principle is applied be clearly defined. At times a single particle or a rigid body is the system to be isolated, whereas at other times two or more bodies taken together constitute the system. The definition of the system to be analyzed is made clear by constructing its *free-body diagram.* This diagram consists of a closed outline of the external boundary of the system defined. All bodies which contact and exert forces on the system but are not a part of it are removed and replaced by vectors representing the forces they exert *on* the system isolated. In this way a clear distinction is made between the action and reaction of each force, and account is taken of *all* forces on and external to the system. It is assumed that the student is familiar with the technique of drawing free-body diagrams from his prior work in statics.

In applying the laws of dynamics, numerical values of the quantities may be used directly in proceeding toward the solution, or algebraic symbols may be used to represent the quantities involved and the answer left as a formula. With numerical substitution the magnitudes of all quantities expressed in their particular units are evident at each stage of the calculation. This approach offers advantage when the practical significance of the magnitude of each term is important. The symbolic solution, however, has several advantages over the numerical solution. First, the abbreviation achieved by the use of symbols aids in focusing attention on the interconnection between the physical situation and its related mathematical description. Second, a symbolic solution permits a dimensional check to be made at every step, whereas dimensional homogeneity may not be checked when numerical values are used. Third, a symbolic solution may be used repeatedly for obtaining answers to the same problem when different sets and sizes of units are used. Facility with both forms of solution is essential, and ample practice with each should be sought in the problem work.

The student will find that solutions to the various equations of dynamics may be obtained in one of three ways. First, a direct mathematical solution by hand calculation may be carried out with answers appearing either as algebraic symbols or as numerical results. The large majority of the problems come under this category. Second, certain problems are readily handled by graphical solutions, such as with the determination of velocities and accelerations in two-dimensional relative motion of rigid bodies. Third, there are a number of problems in *Dynamics* which are appropriate for computer solution, and students who have access to digital computation facilities may wish to use them for some of the problems. The choice of the most expedient method of solution is an important aspect of the experience to be gained from the problem work.

I DYNAMICS OF PARTICLES

2 KINEMATICS OF PARTICLES

9 Description of Motion. Kinematics deals with position in space as a function of time and is often referred to as the "geometry of motion." The calculation of flight trajectories for aircraft, rockets, and spacecraft and the design of cams, gears, and linkages to control or produce certain desired motions are examples of kinematical problems. Kinematics, which deals only with motion, is a necessary introduction to kinetics, since ability to describe motion is prerequisite to an understanding of the relations between forces and their accompanying motions. The subject of kinematics is developed by first studying in this chapter the motions of points or particles. The kinematics of rigid bodies is developed in Chapter 5 in Part II of the book on rigid-body dynamics.

The motion of particles may be described through the specification of both linear and angular coordinates and their time derivatives. Particle motion on straight lines is termed *rectilinear motion,* whereas motion on curved paths is called *curvilinear motion.* The curved path may be two- or three-dimensional. The kinematics of particles will be developed progressively by discussing motion with one, then two, and finally three space coordinates. Although a large share of motion problems in engineering are two-dimensional, it is becoming increasingly important to develop ability to analyze three-dimensional motion. Rocket and spacecraft flight control, a precessing gyroscope, and the coupling in a rotating universal joint are examples of three-dimensional motion.

The motion of particles and rigid bodies may be described by using coordinates measured from fixed axes (*absolute-motion* analysis) or by using coordinates measured from moving axes (*relative-motion* analysis). Both types of description are developed in articles which follow.

10 Rectilinear Motion of a Particle. Consider a particle P moving along a straight line in the direction of s, Fig. 2. The position of P at any instant of time t may be specified by its *displacement* s from some convenient reference point O fixed on the line. If the particle moves a distance Δs from P to P' during time Δt, its *average velocity* for that interval is $v_{av} = \Delta s/\Delta t$. As the time interval becomes smaller and approaches zero, the average velocity approaches the *instantaneous velocity* of P which is $v = \lim_{\Delta t \to 0} \Delta s/\Delta t$ or

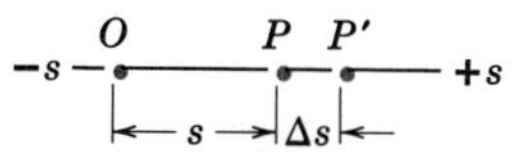

$-s$ — O — P — P' — $+s$; s, Δs

Figure 2

$$v = \frac{ds}{dt} = \dot{s} \tag{4}$$

If the instantaneous velocity of the particle changes from v at P to $v + \Delta v$ at P', the *average acceleration* during the corresponding time interval Δt is $a_{av} = \Delta v/\Delta t$ and will be positive or negative depending on whether the velocity is increasing or decreasing. The instantaneous acceleration a of the particle is the instantaneous time rate of change of the velocity, $a = \lim_{\Delta t \to 0} \Delta v/\Delta t$, which is

$$a = \frac{dv}{dt} = \dot{v} \quad \text{or} \quad a = \frac{d^2s}{dt^2} = \ddot{s} \tag{5}$$

By eliminating the time dt between Eq. 4 and the first of Eqs. 5 a differential equation relating displacement, velocity, and acceleration results which is

$$v\,dv = a\,ds \quad \text{or} \quad \dot{s}\,d\dot{s} = \ddot{s}\,ds \tag{6}$$

Equations 4, 5, and 6 are the differential equations for the rectilinear motion of a particle. Problems in rectilinear motion involving finite changes in the displacements and velocities are solved by integration of these basic differential relations. Displacement s, velocity v, and acceleration a are algebraic quantities, so that their signs may be positive or negative.

In the solution of many problems it is useful to represent the relations among s, v, a, and t graphically. Figure 3a represents any arbitrary variation

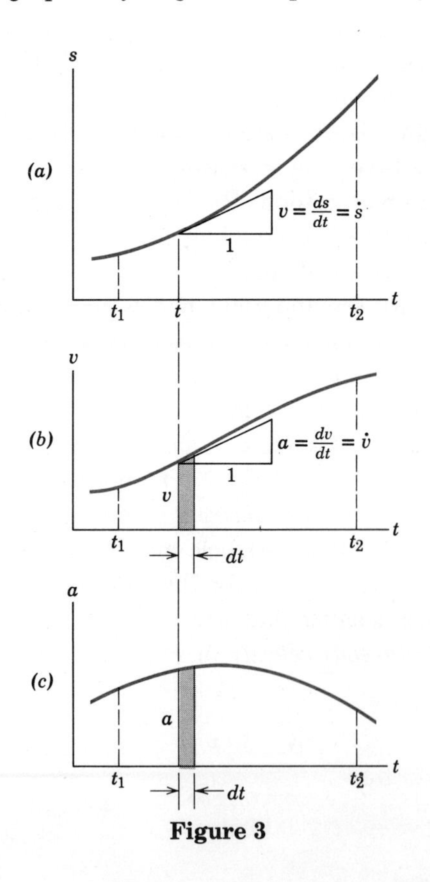

Figure 3

of s as a function of t during an interval from t_1 to t_2. The slope of the curve at any time t is the velocity $v = \dot{s}$ at that instant. Figure 3*b* is a schematic plot of these slopes which gives the variation of the velocity v with the time t. The slope of the velocity curve at any time is the acceleration $a = \dot{v}$ at this instant. Figure 3*c* is a plot of these slopes which gives the variation of acceleration a with the time t.

It is readily seen from Fig. 3*b* that the area under the v-t curve during time dt is $v\,dt$ which from Eq. 4 is the displacement ds. Consequently the net displacement of the particle during the interval from t_1 to t_2 is the corresponding area under the curve which is

$$\int_{s_1}^{s_2} ds = \int_{t_1}^{t_2} v\,dt \qquad \text{or} \qquad s_2 - s_1 = (\text{area under } v\text{-}t \text{ curve})$$

Similarly, from Fig. 3*c* it is seen that the area under the a-t curve during time dt is $a\,dt$ which, from the first of Eqs. 5, is dv. Thus the net change in velocity between t_1 and t_2 is the corresponding area under the curve which is

$$\int_{v_1}^{v_2} dv = \int_{t_1}^{t_2} a\,dt \qquad \text{or} \qquad v_2 - v_1 = (\text{area under } a\text{-}t \text{ curve})$$

Two additional graphical relations are noted. When the acceleration a is plotted as a function of the displacement s, Fig. 4*a*, the area under the curve during a displacement ds is $a\,ds$ which, from Eq. 6, is $v\,dv = d(v^2/2)$. Thus the net area under the curve between displacements s_1 and s_2 is

$$\int_{v_1}^{v_2} v\,dv = \int_{s_1}^{s_2} a\,ds \qquad \text{or} \qquad \tfrac{1}{2}(v_2^2 - v_1^2) = (\text{area under } a\text{-}s \text{ curve})$$

When the velocity v is plotted as a function of the displacement s, Fig. 4*b*, the slope of the curve at any point A is dv/ds. By constructing the normal AB to the curve at this point, it is seen from the similar triangles that

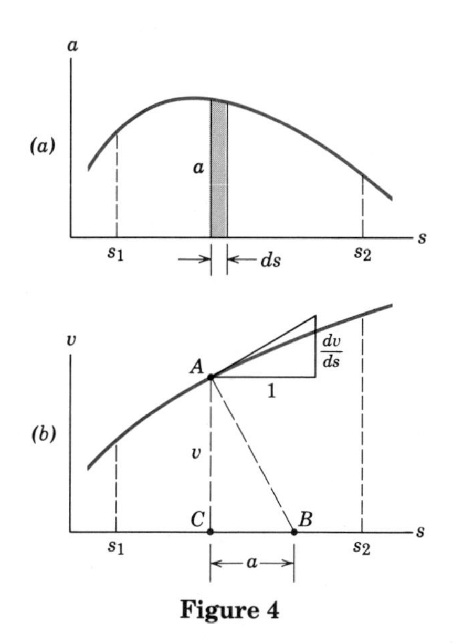

Figure 4

$\overline{CB}/v = dv/ds$. Thus, from Eq. 6, $\overline{CB} = v(dv/ds) = a$, the acceleration. It is necessary that the velocity and displacement axes have the same numerical scale so that the acceleration read on the displacement scale in metres, say, will represent the actual acceleration in the units metres per second squared.

The graphical representations described are useful not only in visualizing the relationships between the several motion quantities but also in approximating results by graphical integration or differentiation when a lack of knowledge of the mathematical relationship prevents its expression as an explicit mathematical function. Experimental data are frequently analyzed graphically.

The quantities displacement, velocity, and acceleration may be written alternatively in vector notation. If the x-axis replaces the s-direction and if $\mathbf{i}$ represents a unit vector in the positive x-direction, then the motion terms may be written as displacement $\mathbf{s} = \mathbf{i}x$, velocity $\mathbf{v} = \mathbf{i}\dot{x}$, and acceleration $\mathbf{a} = \mathbf{i}\dot{v} = \mathbf{i}\ddot{x}$. For rectilinear motion with the direction of motion known along a specified straight line, the vector notation is unnecessary, and scalar algebraic notation is entirely adequate. The sense of the quantity is specified by its algebraic sign.

If the displacement s is known for all values of the time t, then successive mathematical or graphical differentiation with respect to t gives the velocity v and acceleration a. In many problems, however, the functional relationship between displacement and time is unknown and must be determined by successive integration from the acceleration. Acceleration is determined by the forces and moments which act on moving bodies and is computed from the equations of kinetics discussed in subsequent chapters. Depending on the nature of the forces and moments, the acceleration may be specified as a function of time, velocity, displacement, or as a combined function of these quantities. The procedure for integrating the differential equation in each case is indicated as follows:

Constant acceleration. When a is constant, the first of Eqs. 5 and Eq. 6 may be integrated directly. For simplicity with $s = 0$, $v = v_1$, and $t = 0$ designated at the beginning of the interval, then for a lapse of time t the integrated equations become

$$\int_{v_1}^{v} dv = a\int_{0}^{t} dt \qquad \text{or} \qquad v = v_1 + at$$

$$\int_{v_1}^{v} v\,dv = a\int_{0}^{s} ds \qquad \text{or} \qquad v^2 = v_1^2 + 2as$$

Substitution of the integrated expression for v into Eq. 4 and integrating with respect to t gives

$$\int_{0}^{s} ds = \int_{0}^{t} (v_1 + at)\,dt \qquad \text{or} \qquad s = v_1 t + \tfrac{1}{2}at^2$$

These relations are necessarily restricted to the special case where the acceleration is constant. The integration limits depend on the initial and final conditions and for a given problem may be different from those used here. It may be more convenient, for instance, to begin the integration at some specified time t_1 or initial displacement s_1.

Acceleration given as a function of time, $a = f(t)$. Substitution of the function into the first of Eqs. 5 gives $f(t) = dv/dt$, so that the velocity is obtained by direct integration in the form

$$\int_{v_1}^{v} dv = \int_0^t f(t)\, dt \qquad \text{or} \qquad v = v_1 + \int_0^t f(t)\, dt$$

From this integrated expression for v as a function of t, the displacement s is obtained by integrating Eq. 4 which, in form, would be

$$s = \int_0^s ds = \int_0^t v\, dt$$

If the indefinite integral is employed, the end conditions are used to establish the constants of integration with results which are identical with those obtained by using the definite integral.

If desired, the displacement s may be obtained by a direct solution of the second-order differential equation $\ddot{s} = f(t)$ obtained by substitution of $f(t)$ into the second of Eqs. 5.

Acceleration given as a function of velocity, $a = f(v)$. Substitution of the function into the first of Eqs. 5 gives $f(v) = dv/dt$ which would be integrated in the form

$$t = \int_0^t dt = \int_{v_1}^{v} \frac{dv}{f(v)}$$

This result would give v in terms of t. Then, as before, integration of v with respect to t gives the displacement s.

The acceleration may also be substituted into Eq. 6 and integrated to obtain

$$\int_{v_1}^{v} \frac{v\, dv}{f(v)} = \int_0^s ds = s$$

Acceleration given as a function of displacement, $a = f(s)$. Substitution of the function into Eq. 6 and integrating gives the form

$$\int_{v_1}^{v} v\, dv = \int_0^s f(s)\, ds \qquad \text{or} \qquad v^2 = v_1^2 + 2\int_0^s f(s)\, ds$$

Substitution of ds/dt for v into this integrated equation yields an equation which may be integrated to obtain s as a function of t.

In each of the foregoing cases when the acceleration varies according to some functional relationship, the ability to solve the equations by direct mathematical integration will depend on the form of the function. In cases where the integration is excessively awkward or difficult, integration by graphical, numerical, or computer methods may be utilized.

Sample Problems

2/1 The displacement of a particle which is confined to move along a straight line is given by $s = 2t^3 - 24t + 6$ where s is measured in metres from a convenient origin and where t is in seconds. Determine (*a*) the time required for the particle to reach a velocity of 72 m/s from its initial condition at $t = 0$, (*b*) the acceleration of the particle when $v = 30$ m/s, (*c*) the net displacement of the particle during the interval from $t = 1$ s to $t = 4$ s, and (*d*) the total distance traveled during the interval from $t = 1$ s to $t = 4$ s irrespective of the direction along the path.

Solution. The velocity and acceleration are obtained by successive differentiation of s with respect to the time. Thus

$$[v = \dot{s}] \qquad v = 6t^2 - 24 \text{ m/s}$$

$$[a = \dot{v}] \qquad a = 12t \text{ m/s}^2$$

(*a*) Substituting $v = 72$ m/s into the expression for v gives $72 = 6t^2 - 24$ from which $t = \pm 4$ s. The negative root describes a mathematical solution for t before the initiation of motion, so is of no physical interest. Thus the desired result is

$$t = 4 \text{ s} \qquad \textit{Ans.}$$

(*b*) Substituting $v = 30$ m/s into the expression for v gives $30 = 6t^2 - 24$ from which the positive root is $t = 3$ s, and the corresponding acceleration is

$$a = 12(3) = 36 \text{ m/s}^2 \qquad \textit{Ans.}$$

(*c*) The net displacement during the specified interval is

$$\Delta s = s_4 - s_1 \qquad \text{or}$$

$$\Delta s = [2(4^3) - 24(4) + 6] - [2(1^3) - 24(1) + 6]$$

$$= 54 \text{ m} \qquad \textit{Ans.}$$

which represents the net advancement of the particle along the s-axis from the position it occupied at $t = 1$ s to its position at $t = 4$ s.

(*d*) The total distance traveled during the interval from $t = 1$ s to $t = 4$ s is the numerical sum of the corresponding areas under the v-t curve which is plotted as shown. It is seen from the graph that from $t = 1$ s to $t = 2$ s the particle has a negative displacement of

$$\left[\Delta s = \int v\, dt\right] \qquad \Delta s_{1-2} = \int_1^2 (6t^2 - 24)\, dt = -10 \text{ m}$$

which represents a 10-m travel in the negative s-direction. From $t = 2$ s to $t = 4$ s

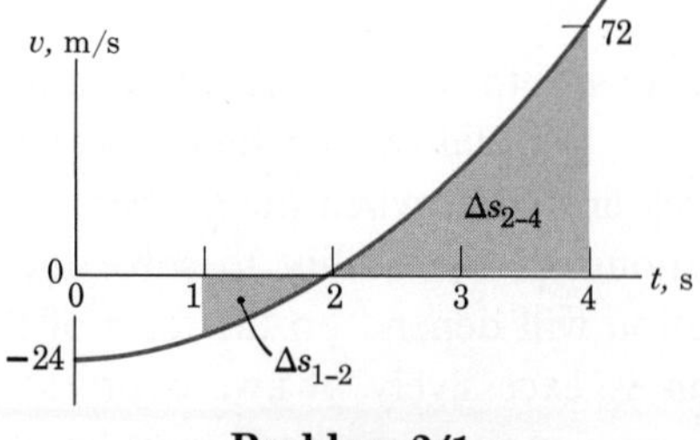

Problem 2/1

the particle travels a distance equal to its displacement of

$$\Delta s_{2-4} = \int_2^4 (6t^2 - 24)\, dt = 64 \text{ m}$$

Thus the total distance traveled during the two intervals is

$$S = |-10| + 64 = 74 \text{ m} \qquad \textit{Ans.}$$

2/2 A particle moves along the x-axis with an initial velocity $v_x = 50$ m/s at the origin when $t = 0$. For the first 4 s it has no acceleration, and thereafter it is acted upon by a retarding force which gives it a constant acceleration $a_x = -10$ m/s². Calculate the velocity and the x-coordinate of the particle for the conditions of $t = 8$ s and $t = 12$ s and find the maximum positive x-coordinate reached by the particle.

Solution. The velocity of the particle after $t = 4$ s is computed from

$$\left[\int dv = \int a\, dt\right] \qquad \int_{50}^{v_x} dv_x = -10 \int_4^t dt, \qquad v_x = 90 - 10t \text{ m/s}$$

and is plotted as shown. At the specified times the velocities are

$$t = 8 \text{ s}, \qquad v_x = 90 - 10(8) = 10 \text{ m/s}$$

$$t = 12 \text{ s}, \qquad v_x = 90 - 10(12) = -30 \text{ m/s} \qquad \textit{Ans.}$$

The x-coordinate of the particle at any time greater than 4 s is the distance traveled during the first 4 s plus the distance traveled after the discontinuity in acceleration occurred. Thus

$$\left[\int ds = \int v\, dt\right] \qquad x = 50(4) + \int_4^t (90 - 10t)\, dt = -5t^2 + 90t - 80 \text{ m}$$

For the two specified times

$$t = 8 \text{ s}, \qquad x = -5(8^2) + 90(8) - 80 = 320 \text{ m}$$

$$t = 12 \text{ s}, \qquad x = -5(12^2) + 90(12) - 80 = 280 \text{ m} \qquad \textit{Ans.}$$

The x-coordinate for $t = 12$ s is less than that for $t = 8$ s since the motion is in the negative x-direction after $t = 9$ s. The maximum positive x-coordinate is, then, the value of x for $t = 9$ s which is

$$x_{max} = -5(9^2) + 90(9) - 80 = 325 \text{ m} \qquad \textit{Ans.}$$

These displacements are seen to be the net positive areas under the v-t graph up to the value of t in question.

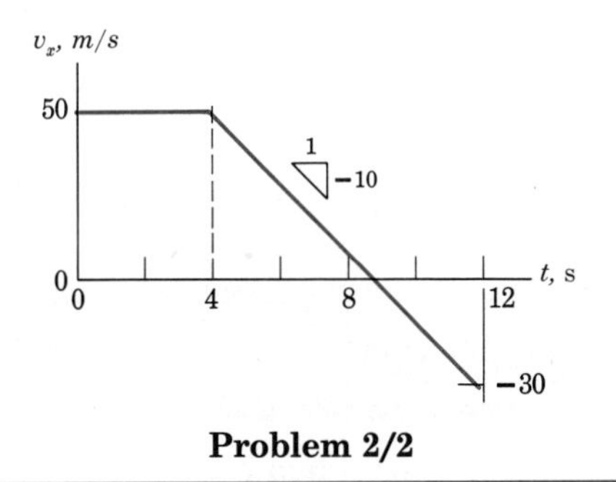

Problem 2/2

2/3 The spring-mounted slider moves in the horizontal guide with negligible friction and has a velocity v_0 in the s-direction as it crosses the mid-position where $s = 0$ and $t = 0$. The two springs together exert a retarding force to the motion of the slider which gives it an acceleration proportional to the displacement but oppositely directed and equal to $a = -k^2s$ where k is constant. (The constant is arbitrarily squared for later convenience in the form of the expressions.) Determine the expressions for the displacement s and velocity v as functions of the time t. Also determine the period τ which is the time for one complete oscillation of the slider.

Solution I. Since the acceleration is specified in terms of the displacement, the differential relation $v\,dv = a\,ds$ may be integrated. Thus

$$\int v\,dv = \int -k^2 s\,ds + C_1 \quad \text{a constant}$$

or

$$\frac{v^2}{2} = -\frac{k^2s^2}{2} + C_1$$

When $s = 0$, $v = v_0$, so that $C_1 = v_0^2/2$, and the velocity becomes

$$v = +\sqrt{v_0^2 - k^2s^2}$$

The plus sign of the radical is taken when v is positive in the plus s-direction. This last expression may be integrated by substituting $v = ds/dt$. Thus,

$$\int \frac{ds}{\sqrt{v_0^2 - k^2s^2}} = \int dt + C_2 \quad \text{a constant}$$

or

$$\frac{1}{k}\sin^{-1}\frac{ks}{v_0} = t + C_2$$

With the requirement of $t = 0$ when $s = 0$, the constant of integration becomes $C_2 = 0$, and the equation may be solved for s so that

$$s = \frac{v_0}{k}\sin kt \qquad \textit{Ans.}$$

The velocity is $v = \dot{s}$ which gives

$$v = v_0 \cos kt \qquad \textit{Ans.}$$

It is noted that both s and v are periodic in the time. The period τ is the time to complete one entire oscillation during which the argument of the cosine increases 2π. Thus $k(t + \tau) = kt + 2\pi$, and

$$\tau = 2\pi/k \qquad \textit{Ans.}$$

The frequency f of the motion is the number of oscillations or complete cycles per unit time and is $f = 1/\tau = k/2\pi$.

This motion is called *simple harmonic motion* and is characteristic of all oscillations

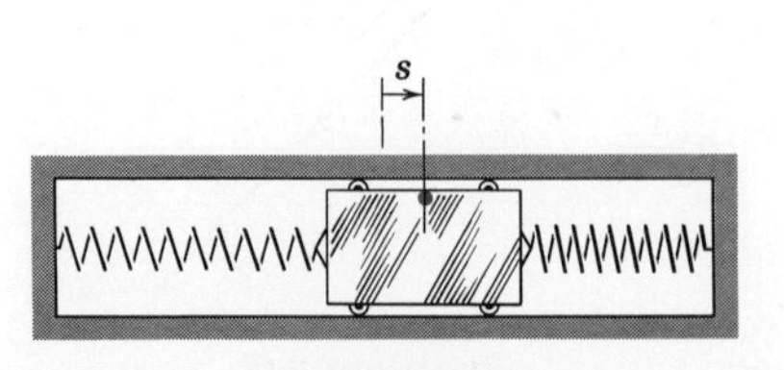

Problem 2/3

where the restoring force, and hence the acceleration, is proportional to the displacement but opposite in sign.

Solution II. Since $a = \ddot{s}$, the given relation may be written at once as

$$\ddot{s} + k^2 s = 0$$

This is an ordinary linear differential equation of second order for which the solution is well known and is

$$s = A \sin Kt + B \cos Kt$$

where A, B, and K are constants. Substitution of this expression into the differential equation shows that it satisfies the equation provided that $K = k$. The velocity is $v = \dot{s}$ which becomes

$$v = Ak \cos kt - Bk \sin kt$$

The boundary condition $v = v_0$ when $t = 0$ requires that $A = v_0/k$, and the condition $s = 0$ when $t = 0$ gives $B = 0$. Thus the solution is

$$s = \frac{v_0}{k} \sin kt \qquad \text{and} \qquad v = v_0 \cos kt \qquad \textit{Ans.}$$

Problems

2/4 A jet aircraft with a landing speed of 200 km/h has a maximum of 600 m of available runway after touch-down in which to reduce its ground speed to 30 km/h. Compute the average acceleration a required of the aircraft during braking.
Ans. $a = -2.51 \text{ m/s}^2$

2/5 In the final stages of a moon landing the lunar module descends under retro-thrust of its descent engine to within $h = 5$ m of the lunar surface where it has a downward velocity of 4 m/s. If the descent engine is cut off abruptly at this point, compute the impact velocity of the landing gear with the moon. Lunar gravity is $\frac{1}{6}$ of the earth's gravity.

Problem 2/5

2/6 A projectile is fired vertically with an initial velocity of 200 m/s. Calculate the maximum altitude h reached by the projectile and the time t after firing for it to return to the ground. Neglect air resistance and take the gravitational acceleration to be constant at 9.81 m/s^2.
Ans. $h = 2040$ m, $t = 40.8$ s

2/7 A particle is given an initial upward velocity of 30 m/s in an evacuated vertical tube. Calculate the net rise or fall of the particle from the end of the first second to the beginning of the fourth second and find its velocity at the end of the fourth second after release.

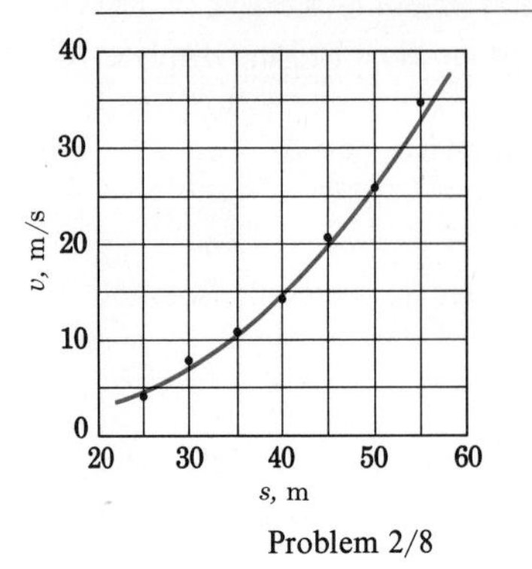

Problem 2/8

2/8 Experimental data for the motion of a particle along a straight line yield measured values of the velocity v for various displacements s. A smooth curve is drawn through the points as shown in the graph. Determine the acceleration of the particle when $s = 40$ m.

Ans. $a = 15\text{ m/s}^2$

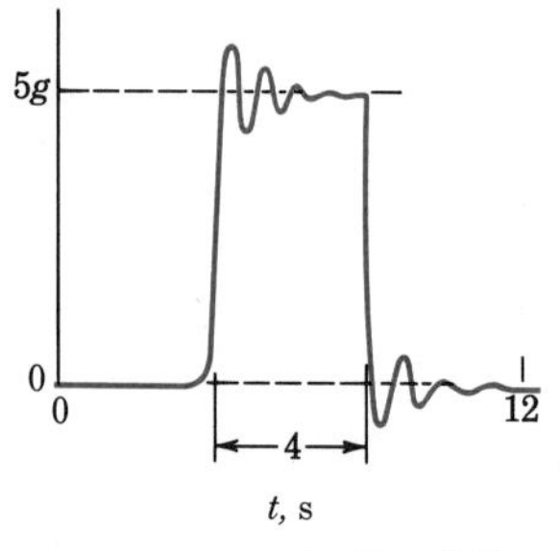

Problem 2/9

2/9 An impulsive retarding force of 4-s duration acts on a particle moving initially with a velocity of 100 m/s. The oscilloscope record of the deceleration is shown. Approximate the velocity of the particle at $t = 12$ s.

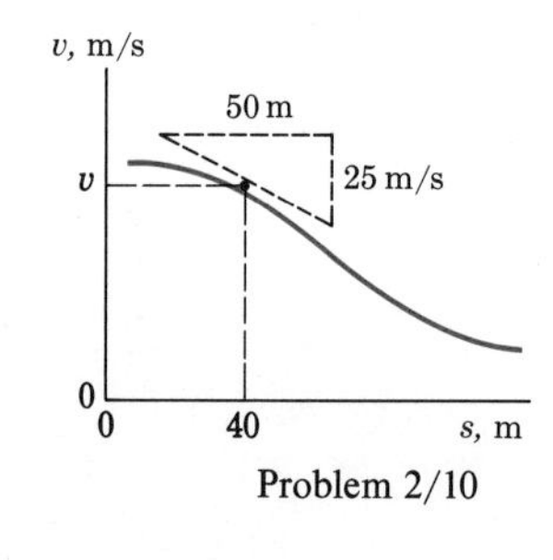

Problem 2/10

2/10 The velocity v of a particle varies with the displacement s as shown in the graph for an interval of its motion. Find the velocity v of the particle for the instant when $s = 40$ m if its velocity is decreasing 30 m/s per second at this position.

2/11 A particle which moves along a straight line has a velocity in millimetres per second given by $v = 300 - 75t^2$ where t is in seconds. Calculate the total distance D covered during the interval from $t = 0$ to $t = 3$ s and find the net displacement s of the particle during this same interval.

Ans. $D = 575$ mm, $s = 225$ mm

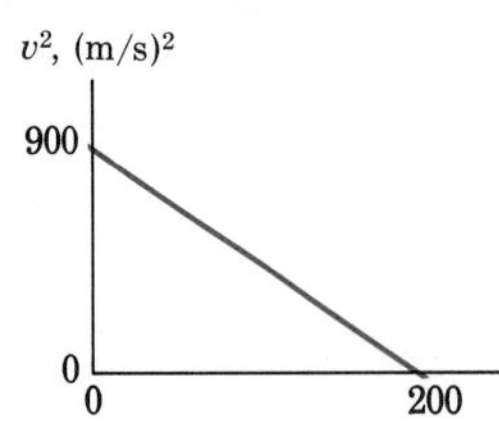

Problem 2/12

2/12 A body moves in a straight line with a velocity whose square decreases linearly with its displacement s as shown. Determine the time t required for the body to travel the 200 m and find the distance traveled during the last 3 s before it comes to rest.

2/13 A low-velocity bullet is fired horizontally with an initial velocity v_0 into an energy-absorbing medium which offers resistance proportional to the velocity of the bullet. The acceleration in the direction of the motion is written $a = -Kv$ where K is a constant which depends on the properties of the medium and the shape of the bullet. Write an expression for the distance S traveled by the bullet in the medium before coming to rest. Also find the time t for the bullet to reach one half of its initial velocity after it enters the medium.

Ans. $S = v_0/K$, $t = 0.693/K$

2/14 A projectile is fired horizontally into a resisting medium with a velocity v_1, and the resulting deceleration is equal to cv^n, where c and n are constants and v is the velocity within the medium. Find the expression for the velocity v of the projectile in terms of the time t of penetration.

2/15 A projectile is fired horizontally into a fluid medium as shown. Measurements of its horizontal displacement x in metres versus the corresponding time t in milliseconds are made and tabulated. Approximate the velocity v and acceleration a in the x-direction at $t = 3$ ms. Solve graphically by drawing a smooth curve through the plotted points.

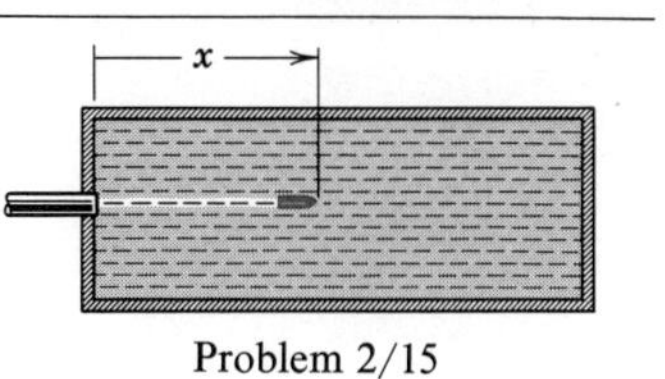

Problem 2/15

x, m	t, ms	x, m	t, ms
0	0	3.4	3.5
1.0	0.5	3.6	4
1.5	1	3.7	4.5
2.2	1.5	4.1	5
2.5	2	4.2	5.5
2.7	2.5	4.3	6
3.2	3		

Ans. $v = 550$ m/s
$a = -140$ km/s^2

2/16 Small steel balls drop from rest through an opening, one after the other, at the steady frequency n. Neglect air resistance and determine the vertical separation h between any two balls in terms of the time t during which the upper one has dropped.

2/17 A vacuum-propelled capsule for a high-speed tube transportation system of the future is being designed for operation between two stations 8 km apart. If maximum acceleration and deceleration are to have a limiting magnitude of $0.7g$ and if velocities are to be limited to 400 km/h, determine the minimum time t for the capsule to make the 8-km trip.

Ans. $t = 1.47$ min

2/18 The magnitude of the acceleration and deceleration of an express elevator is limited to $0.4g$, and maximum vertical speed is 400 m/min. Calculate the minimum time t required for the elevator to go from rest at the 10th floor to a stop at the 30th floor, a distance of 100 m.

2/19 The steel ball A of diameter D slides freely on the horizontal rod which leads to the pole face of the electromagnet. The force of attraction obeys an inverse-square law, and the resulting acceleration of the ball is $a = K/(L - x)^2$ where K is a measure of the strength of the magnetic field. If the ball is released from rest at $x = 0$, determine the velocity v with which it strikes the pole face.

Ans. $v = 2\sqrt{\dfrac{K(L - D/2)}{LD}}$

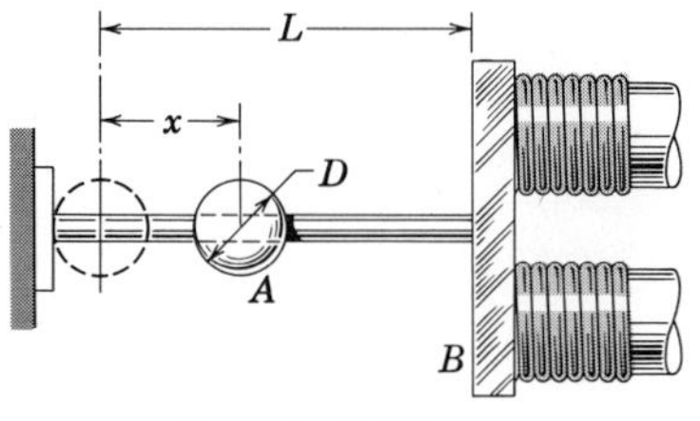

Problem 2/19

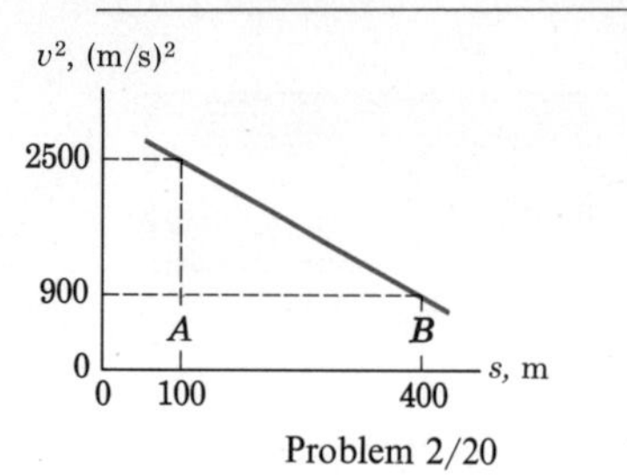

Problem 2/20

2/20 A body moves in a straight line with a velocity whose square decreases linearly with the displacement between two points A and B which are 300 m apart as shown. Determine the displacement Δs of the body during the last 2 s before arrival at B. *Ans.* $\Delta s = 65.3$ m

2/21 A rocket of initial total mass m_0 is fired vertically up from rest. Fuel is consumed at the constant rate m', so that after t seconds the mass of the rocket is $m_0 - m't$. With constant thrust T the upward acceleration becomes $a = \dfrac{T}{m_0 - m't} - g$ where atmospheric resistance has been neglected and where the gravitational acceleration g is assumed constant for low-altitude flight. Derive an expression for the upward velocity v of the rocket as a function of the time t of flight prior to fuel burn-out.

2/22 An object moves with constant acceleration along a straight path. When $t = 0$, the displacement is $+4$ m, and it is -6 m after 10 s. Also the object comes to rest momentarily when $t = 6$ s. Find the initial velocity v at $t = 0$.
Ans. $v = -6$ m/s

2/23 A particle which moves along the x-axis with constant acceleration has a velocity of 3 m/s in the negative x-direction at time $t = 0$ when its x-coordinate is 2.25 m. Three seconds later the particle passes the origin going in the positive x-direction. How far to the negative side of the origin does the particle travel?

2/24 A motorcycle starts from rest at point A and travels 300 m along a straight horizontal track to point B where it comes to a stop. If the acceleration of the motorcycle is limited to $0.7g$ and its deceleration is limited to $0.6g$, calculate the least possible time t to cover the distance. What maximum velocity v is reached?
Ans. $t = 13.76$ s, $v = 157.0$ km/h

2/25 A stone is thrown vertically up with a velocity of 15 m/s from the top of a well. If the sound of the stone hitting the bottom of the well is heard 5 s later, calculate the depth h of the well. The velocity of sound in air may be taken as 335 m/s.

2/26 To test the effects of "weightlessness" for short periods of time a test facility is designed which accelerates a test package vertically up from A to B by means of a gas-activated piston and allows it to ascend and descend from B to C to B under free-fall conditions. The test chamber consists of a deep well and is evacuated to eliminate any appreciable air resistance. If a constant acceleration of $40g$ from A to B is provided by the piston and if the total test time for the "weightless" condition from B to C to B is 10 s, calculate the required working height h of the chamber. Upon returning to B the test package is recovered in a basket filled with polystyrene pellets and inserted in the line of fall. *Ans.* $h = 125.7$ m

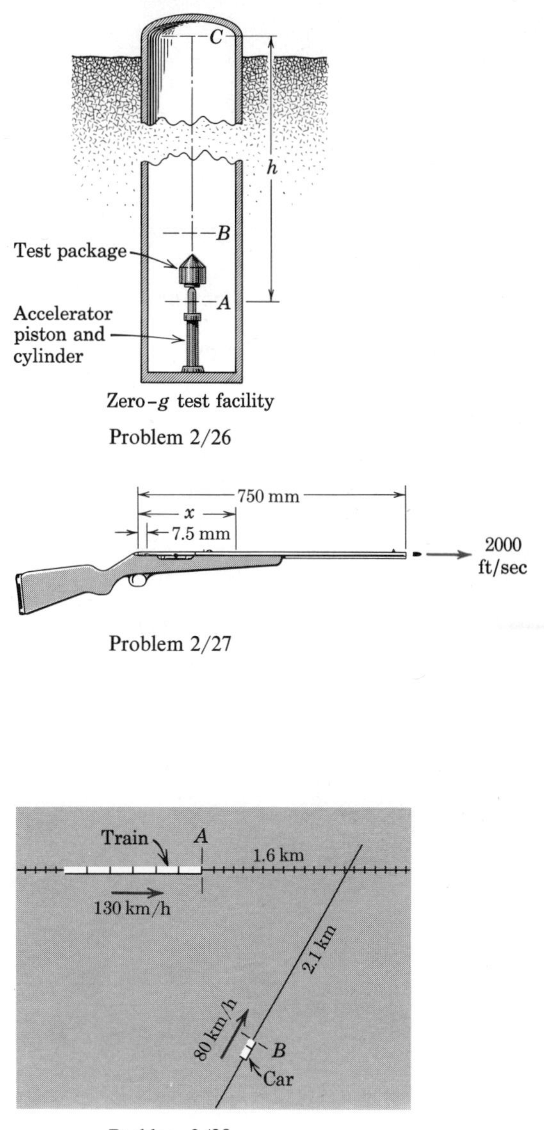

Problem 2/26

Problem 2/27

Problem 2/28

2/27 To a close approximation the pressure behind a rifle bullet varies inversely with the distance traveled by the bullet along the barrel. Thus the acceleration of the bullet may be written as $a = k/x$ where k is a constant. If the bullet starts from rest at $x = 7.5$ mm and if the muzzle velocity of the bullet is 600 m/s at the end of the 750-mm barrel, compute the acceleration of the bullet as it passes the midpoint of the barrel at $x = 375$ mm.

2/28 A train which is traveling at 130 km/h applies its brakes as it reaches point A and slows down with a constant deceleration. Its decreased velocity is observed to be 100 km/h as it passes a point 0.8 km beyond A. A car moving at 80 km/h passes point B at the same instant that the train reaches point A. In an unwise effort to beat the train to the crossing the driver "steps on the gas." Calculate the constant acceleration a which the car must have in order to beat the train to the crossing by 4 s, and find the velocity v of the car as it reaches the crossing.

Ans. $a = 0.481$ m/s^2, $v = 180.5$ km/h

2/29 A retarding force is applied to a body moving in a straight line so that, during an interval of its motion, its speed v decreases with increased displacement s according to the relation $v^2 = k/s$, where k is a constant. If the body has a forward speed of 50 mm/s and a displacement of 225 mm at time $t = 0$, determine the speed v at $t = 3$ s.

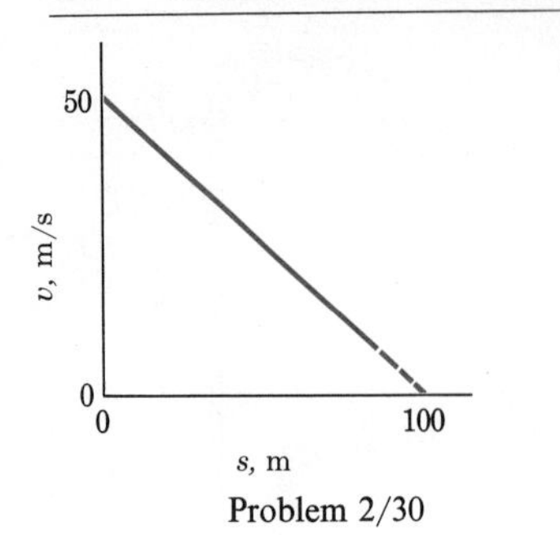

Problem 2/30

2/30 If the velocity v of a particle moving along a straight line decreases linearly with its displacement s from 50 m/s to a value approaching zero at $s = 100$ m, show that the particle never reaches the 100-m displacement. What is the acceleration a of the particle when $s = 60$ m?
Ans. $a = -10 \text{ m/s}^2$

2/31 In a test drop in still air, a small steel ball is released from rest at a considerable height. Its initial acceleration g is reduced by the increment kv^2, where k is a constant and v is the downward velocity. Determine the maximum velocity attainable by the ball and find the vertical drop y in terms of the time t from release.

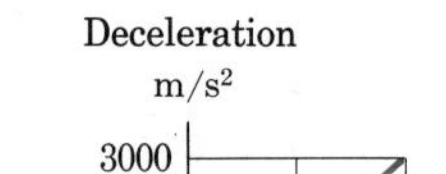

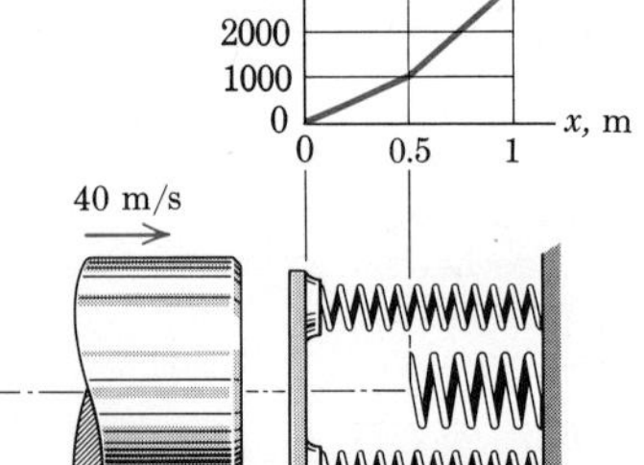

Problem 2/32

2/32 A bumper, consisting of a nest of three springs, is used to arrest the horizontal motion of a large mass which is traveling at 40 m/s as it contacts the bumper. The two outer springs cause a deceleration proportional to the spring deformation. The center spring increases the deceleration rate when the compression exceeds 0.5 m as shown on the graph. Determine the maximum compression x of the outer springs.
Ans. $x = 0.831$ m

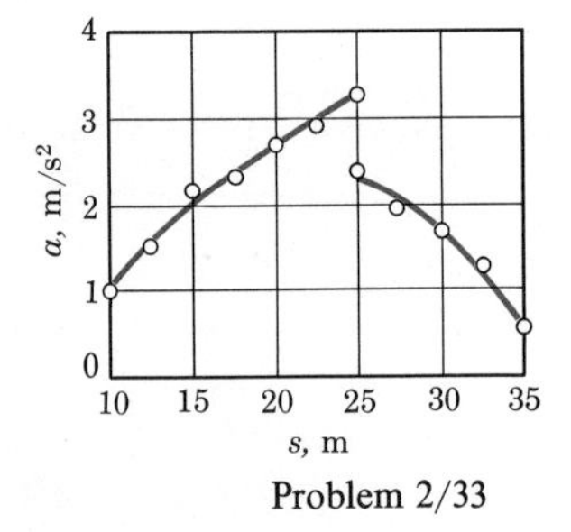

Problem 2/33

2/33 The forward acceleration a of a certain test vehicle is measured experimentally during an interval of its motion and plotted against its displacement as shown. At the 25-m position, the driving mechanism shifts abruptly, and a discontinuity in the acceleration occurs. If the vehicle has a velocity of 7.2 km/h at $s = 10$ m, plot the velocity during the measured motion. Find the velocity at $s = 35$ m.
Ans. $v_{35} = 10.2$ m/s

2/34 A certain lake is proposed as a landing area for large jet aircraft. The touchdown speed of 160 km/h upon contact with the water is to be reduced to 30 km/h in a distance of 400 m. If the deceleration is proportional to the square of the velocity of the aircraft through the water, $a = -Kv^2$, find the value of the design parameter K, which would be a measure of the size and shape of the landing gear vanes which plow through the water. Also find the time t elapsed during the specified interval.

2/35 A ship with a total displacement of 16 kt starts from rest in still water under a constant propeller thrust $T = 250$ kN. The ship develops a total resistance to motion through the water given by $R = 4.50v^2$ where R is in kilonewtons and v is in metres per second. The acceleration of the ship is $a = (T - R)/m$ where m equals the mass of the ship. Compute the distance s in nautical miles which the ship must go to reach a speed of 14 knots. (1 nautical mile = 1.852 km; 1 knot = 1 nautical mile per hour.)

Ans. $s = 2.60$ mi (nautical)

2/36 A particle which is constrained to move in a straight line is subjected to an accelerating force which increases with time and a retarding force which increases directly with the displacement. The resulting acceleration is $a = Kt - k^2x$ where K and k are positive constants and where both x and $\dot{x}$ are zero when $t = 0$. Determine x as a function of t.

Ans. $x = \dfrac{K}{k^3}(kt - \sin kt)$

2/37 The record of acceleration measurements made on an experimental vehicle during its rectilinear motion is shown in the full line. The vehicle starts from rest at $t = 0$. Use the dotted approximation and draw the v-t curve for the 9-s interval and determine the total distance s traveled.

Ans. $s = 3.30$ km

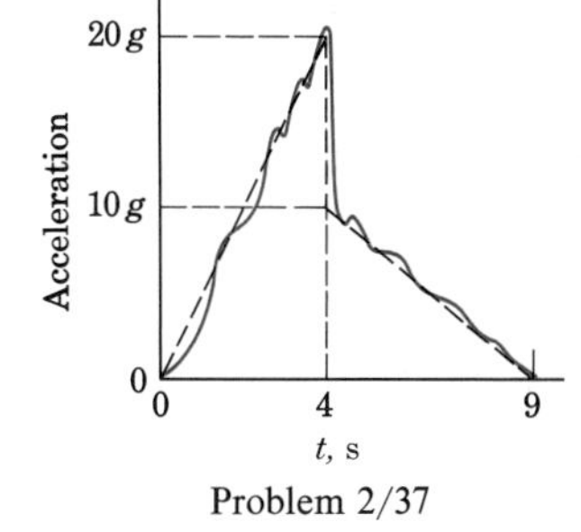

Problem 2/37

2/38 The vertical acceleration of a certain solid-fuel rocket is given by $a = ke^{-bt} - cv - g$, where k, b, and c are constants, v is the vertical velocity acquired, and g is the gravitational acceleration, essentially constant for atmospheric flight. The exponential term represents the effect of a decaying thrust as fuel is burned, and the term $-cv$ approximates the retardation due to atmospheric resistance. Determine the expression for the vertical velocity of the rocket t seconds after firing.

◀ **2/39** A spacecraft is designed for Martian impact on a course heading directly toward the center of the planet. If the craft is 10 000 km from the planet and is traveling at a relative speed of 20 000 km/h, compute the impact velocity v. The gravitational acceleration at the surface of Mars is 3.73 m/s^2, and the planet has a diameter of approximately 6400 km. Neglect any atmospheric resistance.

Ans. $v = 25\,200$ km/h

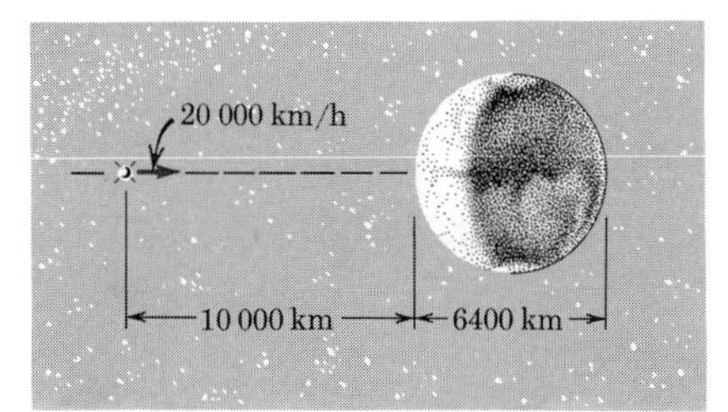

Problem 2/39

◀ **2/40** A rocket propels a space probe for measuring micrometeorite density vertically upward from the north pole and separates from the probe at burn-out conditions. The probe continues to move upward subject to the influence of the earth's gravity and has an upward velocity of 16 000 km/h at an altitude of 160 km above the earth's surface at the pole. Calculate the maximum distance h from the earth's surface reached by the probe. The polar radius of the earth is 6360 km, and the value of g at the pole is 9.833 m/s^2. *Ans.* $h = 1420$ km

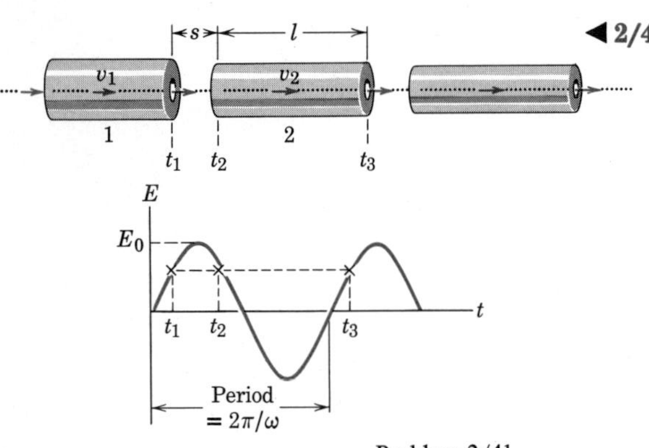

Problem 2/41

◀ **2/41** The acceleration of atomic particles to large velocities in a drift-tube linear accelerator is accomplished by subjecting them to electrical forces due to an alternating radio-frequency voltage. The machine is designed so that a particle which is moving in the shielded tube 1 at a constant velocity v_1 enters the gap between tubes 1 and 2 at time t_1 and leaves the gap at time t_2. During this interval the particle is subjected to the time-varying force $eE = eE_0 \sin \omega t$ where e is the charge on the particle, E_0 is the peak voltage gradient across the gap, and ω is the constant circular frequency. The acceleration during the interval is, then, $(e/m)E_0 \sin \omega t$, where m is the mass of the particle (constant for low velocities). The particle travels with a new constant velocity through tube 2, which shields it from the negative electrical force while the field is reversed, and emerges into the next gap at time t_3. If t_1 is $\frac{1}{8}$ of the period as shown and t_2 and t_3 are as indicated, calculate the necessary length l of tube 2 and the gap length s.

$$Ans.\ l = \frac{3\pi}{2\omega}\left(v_1 + \frac{eE_0\sqrt{2}}{m\omega}\right)$$

$$s = \frac{\pi}{2\omega}\left(v_1 + \frac{eE_0}{m\omega\sqrt{2}}\right)$$

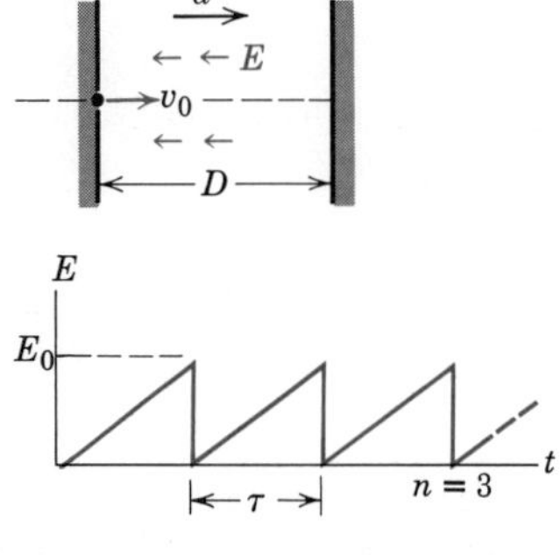

Problem 2/42

◀ **2/42** An electron moving with a velocity v_0 enters an electric field E which gives it an acceleration eE/m in the direction of its velocity, where e is the electron charge and m is its mass. If the field E (voltage gradient) is energized with saw-tooth impulses as shown, plot the v-t curve for the electron during the first three impulses and write the expression for v at the end of n impulses. Also find the separation D of the charged plates if the electron crosses the gap in the time required for n impulses.

$$Ans.\ v_n = v_0 + \frac{neE_0\tau}{2m}$$

$$D = nv_0\tau + \frac{n(3n-1)}{12}\frac{eE_0\tau^2}{m}$$

11 Angular Motion of a Line. Rectilinear displacement and its time derivatives were discussed in Art. 10. A second quantity used in kinematics is angular displacement. In the present article the angular motion of a line will be described for the case where the motion lies in a fixed plane. Figure 5 shows two examples where position in a plane is described by angular measurement. In Fig. 5*a* the position of a particle P moving along a fixed path in the x-y plane may be described by the angle θ made by the line OA with the x-axis. In Fig. 5*b* the angle θ from the reference axis to the line AB attached to the face of the rolling disk may be used to describe the angular position of the disk. It is noted that the angular position of a line does not require that there be a fixed point on the line about which rotation occurs. In general, then, for the line AB of Fig. 5*c*, which is confined to move in the plane of the figure, the *angular displacement* of the line with respect to a convenient fixed reference direction is the angle θ. For the case illustrated the angular displacement is positive in a counterclockwise sense and negative in a clockwise sense. The choice of reference axis and of the sense for positive measurements is quite arbitrary and is wholly a matter of convenience.

The angular motion of a line depends only on its angular coordinate and its time derivatives. The center line AB of the link in Fig. 6, for example, has no angular motion during the rotation of the equal links O_1A and O_2B since there is no change in the angle made by AB with respect to a fixed reference axis such as O_1O_2. Also, it should be noted that a point or particle can, of itself, have no angular motion since this concept is associated only with the angular motion of a line.

The *angular velocity* ω and *angular acceleration* α of a line are, respectively, the first and second time derivatives of its angular displacement θ. These definitions give

$$\omega = \frac{d\theta}{dt} = \dot{\theta}$$

$$\alpha = \frac{d\omega}{dt} = \dot{\omega} \quad \text{or} \quad \alpha = \frac{d^2\theta}{dt^2} = \ddot{\theta} \tag{7}$$

$$\omega\, d\omega = \alpha\, d\theta \quad \text{or} \quad \dot{\theta}\, d\dot{\theta} = \ddot{\theta}\, d\theta$$

The third relation is obtained by eliminating dt from the first two. In each of these relations the positive direction for ω and α, clockwise or counter-

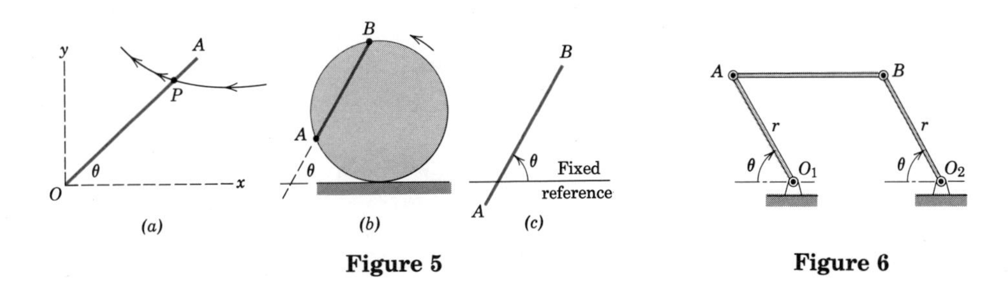

Figure 5

Figure 6

clockwise, is the same as that chosen for θ. Equations 7 should be recognized as analogous to the defining equations for the rectilinear motion of a particle expressed by Eqs. 4, 5, and 6. In fact all relations which were described for rectilinear motion in Art. 10 apply to the case of rotation in a plane if the linear quantities s, v, and a are replaced by their respective equivalent angular quantities θ, ω, and α.

The integration of Eqs. 7 is completely analogous to the integration of the corresponding linear motion relations, Eqs. 4, 5, and 6, which was described and illustrated in Art. 10. Discussion of this integration will not be repeated here since the student need only interchange the symbols θ for s, ω for v, and α for a.

The graphical relationships described for s, v, a, and t in Figs. 3 and 4 may be used for θ, ω, and α merely by the substitution of corresponding symbols. The student should sketch these graphical relations for plane rotation. The mathematical procedures for obtaining velocity and displacement from acceleration described for rectilinear motion may be applied to rotation by merely replacing the linear quantities by their corresponding angular quantities.

Angular measurements may be expressed in degrees, revolutions, or radians. Inasmuch as the relation between arc length and angle occurs so frequently, the radian is the simplest unit to use. The need to change units from revolutions per minute to radians per second, for example, occurs frequently, and the student should be prepared to make this and similar changes readily.

Angular motion may be described in vector notation. Thus the radial line OA of the disk in Fig. 7, which is rotating about the z-axis normal to its plane, will have an angular velocity and an angular acceleration given, respectively, by

$$\boldsymbol{\omega} = \mathbf{k}\dot{\theta} = \mathbf{k}\omega \qquad \text{and} \qquad \boldsymbol{\alpha} = \mathbf{k}\dot{\omega} = \mathbf{k}\alpha$$

where $\mathbf{k}$ is a unit vector in the z-direction, normal to the plane of rotation. This direction specifies uniquely the plane of motion. The positive sense for the rotation vectors is arbitrarily chosen to agree with the right-hand rule. As long as rotation is confined to the single plane, the vectors will remain unchanged in direction, and scalar notation with its algebraic sign to keep track of the sense of the rotation is entirely adequate to describe the motion.

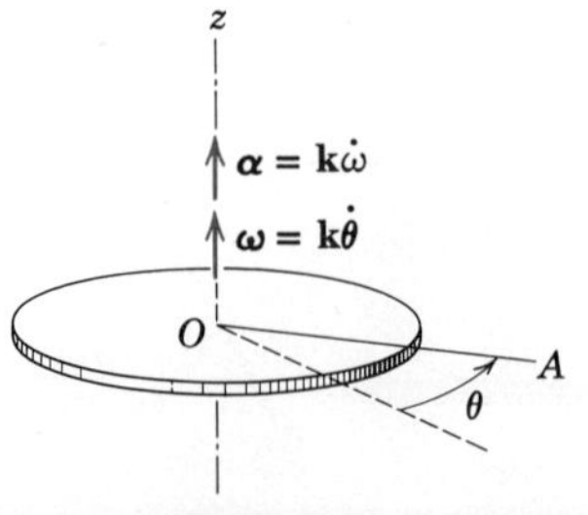

Figure 7

Sample Problems

2/43 The rotational velocity in revolutions per second of a radial line on a rotating gear is given by $\omega = -200 + 8t^2$ where t is in seconds. Calculate (a) the angle θ in radians through which the gear rotates during the third second of its motion after $t = 0$, (b) the angular acceleration α of the disk when its angular velocity is momentarily zero, and (c) the time t required for the gear to make as many revolutions clockwise as counterclockwise after $t = 0$.

Solution. (a) The angular displacement θ in revolutions during the third second of motion starting from time $t = 0$ is

$$\left[\Delta\theta = \int \omega\, dt\right] \qquad \theta = \int_2^3 (-200 + 8t^2)\, dt = -149.3 \text{ rev}$$

$$\text{or} \qquad \theta = -149.3(2\pi) = -938.3 \text{ rad} \qquad \textit{Ans.}$$

The minus sign indicates that the disk revolved in the sense opposite to that for the positive sense of ω.

(b) The angular velocity is momentarily zero when $0 = -200 + 8t^2$ or $t = 5$ s. Thus the acceleration at this time is

$$[\alpha = \dot{\omega}] \qquad \alpha = 16t = 16(5) = 80 \text{ rev/s}^2 \qquad \textit{Ans.}$$

(c) The angular displacement in terms of t is

$$\left[\Delta\theta = \int \omega\, dt\right] \qquad \Delta\theta = \int_0^t (-200 + 8t^2)\, dt = -200t - \tfrac{8}{3}t^3$$

The disk has made as many revolutions in the clockwise as in the counterclockwise sense when $\Delta\theta = 0$. Thus

$$0 = -200t + \tfrac{8}{3}t^3 \qquad t = \sqrt{\frac{600}{8}} = 5\sqrt{3} = 8.66 \text{ s} \qquad \textit{Ans.}$$

2/44 An aircraft tracking device is trained on an airplane A flying horizontally with a constant velocity v at an altitude h. Compute the angular velocity ω and the angular acceleration α of the line of sight OA for any angle θ.

Solution. The angular velocity and acceleration of the line of sight will be the first and second time derivatives, respectively, of its angular coordinate θ. Thus an expression for θ must be obtained first and then differentiated. From the right triangle, $x = h \tan\theta$,

$$[v = \dot{x}] \qquad v = h\dot{\theta}\sec^2\theta = h\omega\sec^2\theta$$

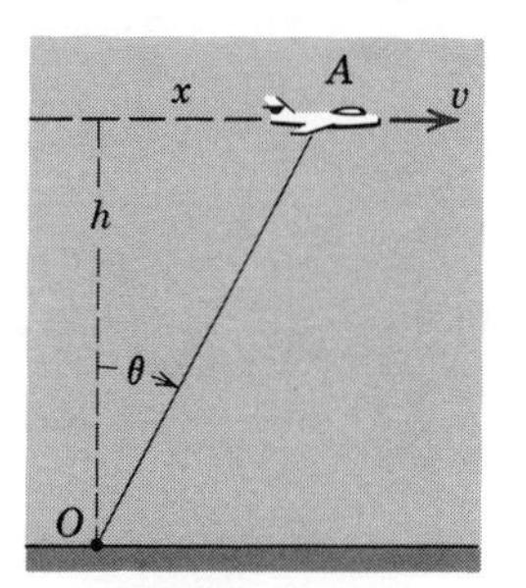

Problem 2/44

Thus

$$\omega = \frac{v}{h}\cos^2\theta \qquad \textit{Ans.}$$

The angular acceleration is

$$[\alpha = \dot{\omega}] \qquad \alpha = -2\frac{v}{h}\dot{\theta}\cos\theta\sin\theta$$

$$= -\frac{v^2}{h^2}\sin 2\theta\cos^2\theta \qquad \textit{Ans.}$$

If the velocity v were not constant, its derivative would have to be evaluated when differentiating both x and ω.

Problems

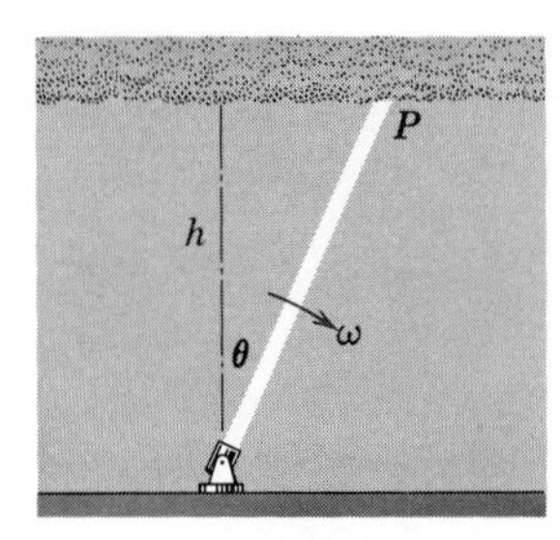

Problem 2/45

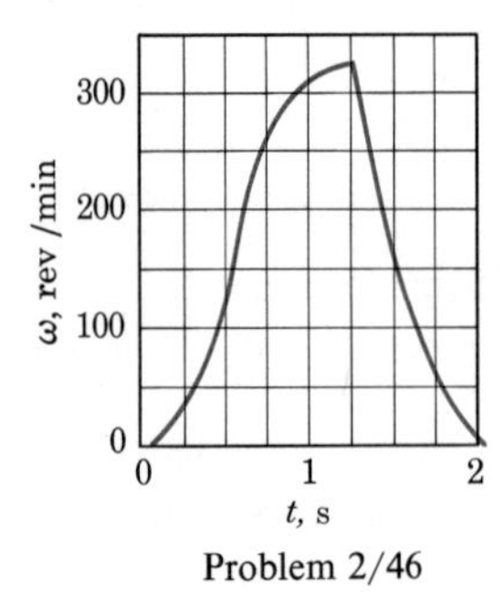

Problem 2/46

2/45 A search light puts a spot of light P on the under side of a horizontal layer of clouds at an altitude h. If the beam of light is revolved in a vertical plane by turning the light about a fixed horizontal axis at a constant angular velocity ω, determine the linear acceleration a of the spot in terms of the angle θ.

2/46 The change in the rotational velocity of a certain pulley is shown. Determine the number of revolutions N through which the pulley turned during the 2 s. *Ans.* $N = 5.4$ rev

2/47 After the fuel is turned off, the rotor of a jet engine decelerates under the action of air friction, which depends on the square of its speed, and constant bearing friction. Thus the deceleration is written as $b + c\omega^2$, where b and c are constants and ω is the angular velocity of the rotor. Determine the time t required for the rotor to come to rest from a speed ω_0.

2/48 The angular velocity ω of a rotating disk, expressed in radians per second, varies during an interval of its motion according to $\omega = 20(1 + 2t/3 - t^2/3)$ where the time t is in seconds. Compute the number of revolutions N through which the disk revolves during the interval from $t = 2$ s to $t = 3$ s. Also find the angular acceleration α when $t = 3$ s.
Ans. $N = 1.77$ rev, $\alpha = -26.7\ \text{rad/s}^2$

2/49 A rotating pulley increases its speed at the constant rate of 30 rad/min each second in a clockwise sense. If the pulley was rotating at the clockwise speed of 4 rev/s at time $t = 0$, determine the number of revolutions N through which the pulley rotated during the first 12 s of motion starting with $t = 0$.

2/50 The square of the angular velocity ω of a certain wheel increases linearly with the angular displacement during 100 rev of the wheel's motion as shown. Compute the time t required for the increase. *Ans.* $t = 17.95$ s

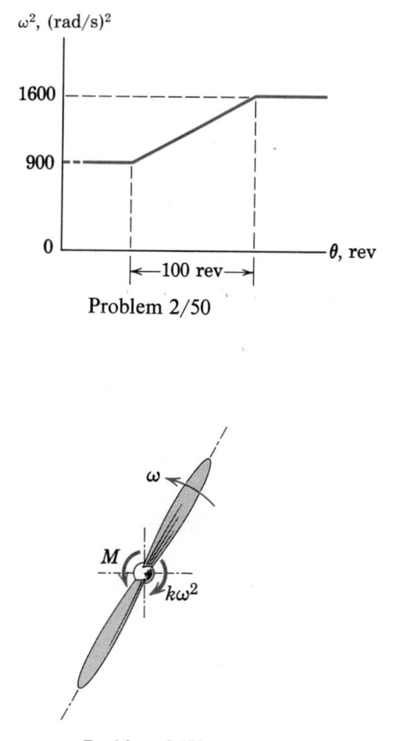

Problem 2/50

Problem 2/52

2/51 The angular motion of a rotating element in a mechanism is programmed so that the rate of change of angular velocity ω with angular displacement θ is a constant k. If the angular velocity is ω_0 when both θ and the time t are zero, determine θ, ω, and the angular acceleration α as functions of t.

2/52 In the test of a propeller a constant torque M is applied to it starting from rest. For low rotational speeds the air resistance causes a resisting torque which is proportional to the square of the angular velocity. The resulting angular acceleration of the propeller may be written as $\alpha = (M - k\omega^2)/I$ where I is the constant moment of inertia representing the fixed radial distribution of mass of the propeller and k is a constant which depends on propeller shape and air conditions. Write the expression for the angular velocity ω as a function of the time t from the start of motion.

2/53 A disk which is rotating freely at a clockwise angular speed ω_1 is suddenly subjected to a torque which produces a constant counterclockwise acceleration of 150 rev/min per second. The torque is applied for 6 s and changes the speed of the disk to 300 rev/min in the counterclockwise direction. Determine ω_1 and find the total number of revolutions N (clockwise plus counterclockwise) through which the disk turns during the 6-s interval. *Ans.* $N = 25$ rev

2/54 The rotational velocity of a rotating disk decreases linearly with angular displacement from 120 rev/min to 60 rev/min during 10 rev of the disk as shown. Compute the time t during which this change takes place.

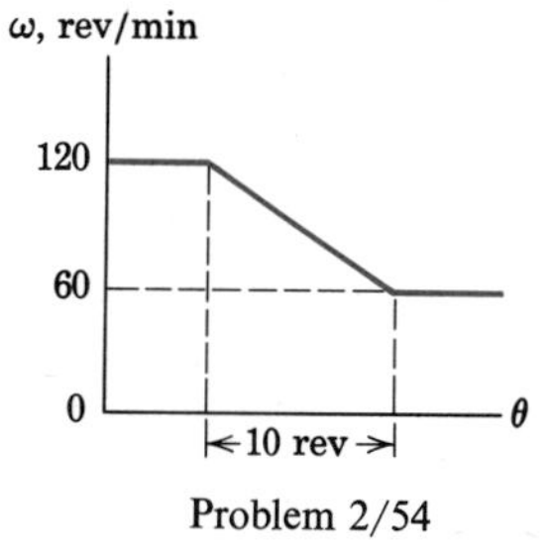

Problem 2/54

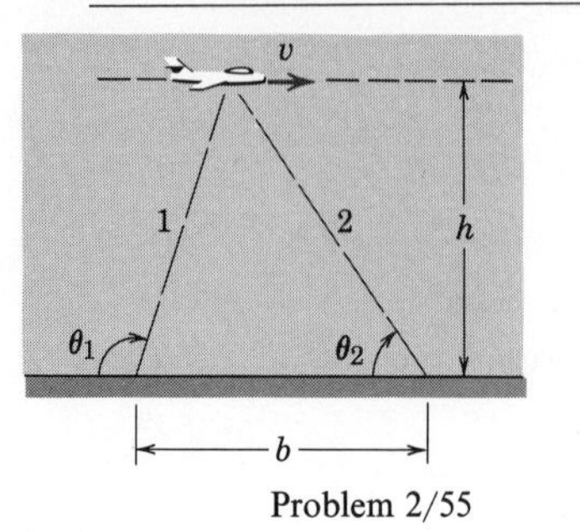

Problem 2/55

2/55 An aircraft tracking device employs two directional beams 1 and 2 emanating from two ground stations a distance b apart. The angular coordinates and their time derivatives are fed into a computer which calculates the position and motion of the aircraft. For the simple case of horizontal flight of the target in the vertical plane containing the two stations, determine the expressions for altitude h and velocity v of the aircraft in terms of the angles θ_1 and θ_2 and the angular velocity ω_1 of beam 1.

Ans. $h = \dfrac{b}{\text{ctn}\,\theta_2 - \text{ctn}\,\theta_1}, \quad v = \dfrac{b\omega_1 \csc^2\theta_1}{\text{ctn}\,\theta_2 - \text{ctn}\,\theta_1}$

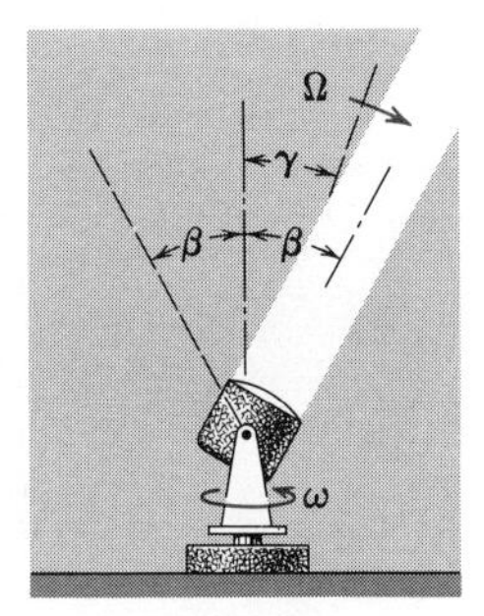

Problem 2/56

2/56 A search light with its beam fixed at a constant angle β with the vertical revolves around its vertical axis with a constant angular velocity ω. From a distance the pencil of light appears to oscillate in a vertical plane through a total angle 2β. Derive an expression for the angular velocity Ω of the projection of the beam on a vertical plane for any angle γ between the vertical axis and the projection of the beam on the vertical plane. The angular velocity is $\Omega = \dot{\gamma}$.

Ans. $\Omega = \omega \cos^2\gamma\sqrt{\tan^2\beta - \tan^2\gamma}$

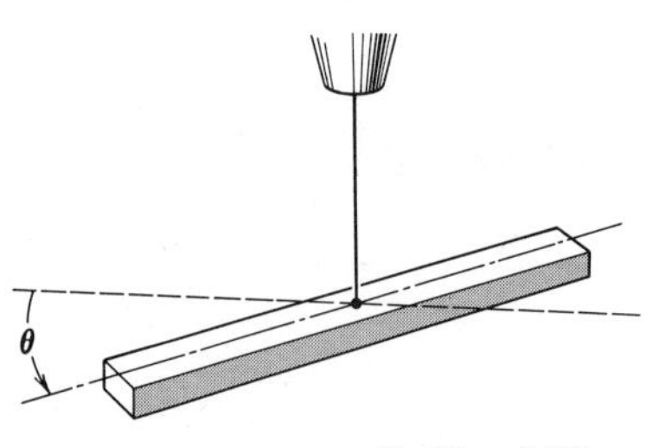

Problem 2/57

2/57 A uniform bar is suspended by a wire attached to its center. The bar is given an initial angular displacement $\theta = \theta_0$ from the equilibrium position and then released from rest at this position at time $t = 0$. The angular acceleration of the horizontal center line of the bar about the vertical wire is proportional to the angular displacement and is oppositely directed, so that $\alpha = -K\theta$. Derive the expression for θ as a function of t during the subsequent oscillation of the bar and find the magnitude of the maximum angular velocity ω_{max} of the bar.

Ans. $\theta = \theta_0 \cos\sqrt{K}\,t, \quad |\omega|_{max} = \theta_0\sqrt{K}$

2/58 A special control operates to increase the torque acting on a rotating disk so that its angular acceleration α increases linearly with the angular displacement θ measured from the rest position. The relation $I\alpha = M_0 + k\theta$ may be written, where I is the constant moment of inertia of the disk, M_0 is the fixed starting torque at $\theta = 0$, and k is a positive constant. Determine the expression for the angular displacement θ as a function of the elapsed time t from the start of motion.

Ans. $\theta = \dfrac{M_0}{k}\left(\cosh\sqrt{\dfrac{k}{I}}\,t - 1\right)$

◀**2/59** A ground station O is tracking a jet aircraft P flying at a constant velocity v in a horizontal direction at an altitude h. The beam OP revolves in a plane defined by the line of flight and the point O. Use the x-y-z coordinate system and express the angular velocity of the beam OP in vector notation in terms of the angle θ. The y-axis through O is chosen parallel to the direction of flight, and the z-axis is vertical.

Ans. $\omega = \dfrac{-v\cos^2\theta}{b^2 + h^2}(\mathbf{i}h + \mathbf{k}b)$

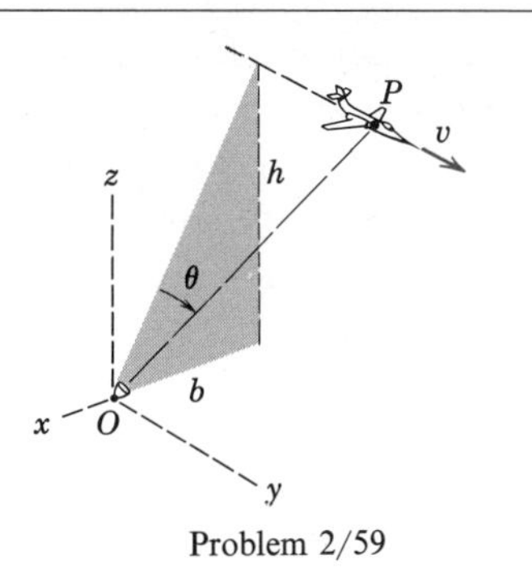

Problem 2/59

12 Plane Curvilinear Motion. The motion of a particle along a curved path which lies in a single plane is called plane curvilinear motion. The vast majority of the motions of points or particles encountered in engineering practice are of this type. Consider a particle moving along the plane curved path shown in Fig. 8. At position A the particle is located by its *position vector* $\mathbf{r}$, measured from a convenient fixed origin O, and at A' by the vector $\mathbf{r} + \Delta\mathbf{r}$. The vector change of position is called the *displacement* $\Delta\mathbf{r}$ and is clearly independent of the choice of origin O. The *distance* actually traveled is the scalar length Δs measured along the path. As the interval approaches zero, the distance Δs may be written as a differential ds which equals in the limit the magnitude of the corresponding differential displacement $d\mathbf{r}$.

The *average velocity* of the particle in going from A to A' in time Δt is the vector $\Delta\mathbf{r}/\Delta t$. As the interval becomes smaller, the direction of $\Delta\mathbf{r}$ approaches that of the tangent to the path, and the *velocity* of the particle at A is defined as

$$\mathbf{v} = \lim_{\Delta t \to 0} \frac{\Delta\mathbf{r}}{\Delta t} = \frac{d\mathbf{r}}{dt} = \dot{\mathbf{r}}$$

It should be noted carefully that the vector $\dot{\mathbf{r}}$ is the velocity and is tangent to the path, whereas $\dot{r}$ is a scalar and represents the rate at which the length or magnitude of $\mathbf{r}$ changes with time. The magnitude of the velocity is often referred to as the *speed* of the particle and is $\dot{s} = |\mathbf{v}|$.

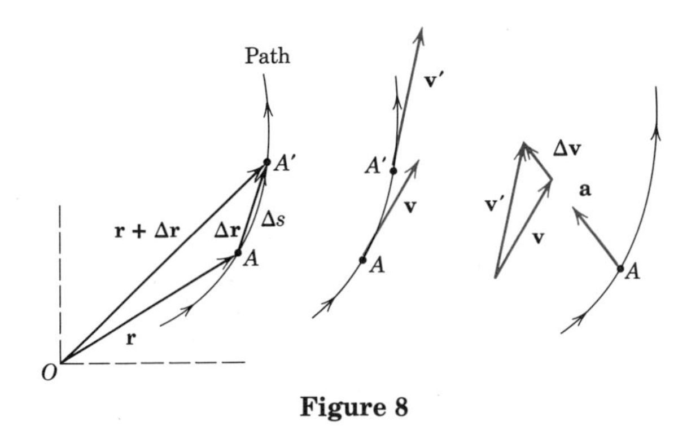

Figure 8

In the motion from A to A' the velocity changes from $\mathbf{v}$ to $\mathbf{v}'$ by the increment $\Delta\mathbf{v}$ as shown in Fig. 8. The *average acceleration* of the particle between A and A' is $\Delta\mathbf{v}/\Delta t$. In general, $\Delta\mathbf{v}$ is neither tangent nor normal to the path. As the interval becomes smaller, $\Delta\mathbf{v}$ approaches the direction of the instantaneous *acceleration* $\mathbf{a}$ which is defined as

$$\mathbf{a} = \lim_{\Delta t \to 0} \frac{\Delta\mathbf{v}}{\Delta t} = \frac{d\mathbf{v}}{dt} = \dot{\mathbf{v}} = \ddot{\mathbf{r}}$$

A further approach to the visualization of acceleration is shown in Fig. 9, where the position vectors to three arbitrary positions on the path of the particle are shown for illustrative purpose. There is a velocity vector tangent to the path corresponding to each position vector, and the relation is $\mathbf{v} = \dot{\mathbf{r}}$. Now if these velocity vectors are plotted from some arbitrary point C, then a curve, known as the *hodograph*, is formed. The derivatives of these velocity vectors will be the acceleration vectors $\mathbf{a} = \dot{\mathbf{v}}$ which are tangent to the hodograph. It is seen that the acceleration bears the same relation to the velocity that the velocity bears to the position vector.

The geometric portrayal of the derivatives of the position vector $\mathbf{r}$ and velocity vector $\mathbf{v}$ in Fig. 8 can be used to describe the derivative of any vector quantity with respect to t or with respect to any other scalar variable. With the introduction of the derivative of a vector in the definitions of velocity and acceleration it is important at this point to establish the rules under which the differentiation of vector quantities may be carried out. These rules are the same as for the differentiation of scalar quantities except for the case of the cross product where the order of the terms must be preserved. These rules are developed in Art. B6 of Appendix B and should be reviewed at this point.

For curvilinear motion of a particle in a plane there are three different coordinate systems which are in common use to describe this motion. An important lesson to be learned from the study of these coordinate systems is the proper choice of reference system for a given problem. This choice is usually revealed by the manner in which the motion is generated or by the form in which the data are specified. Each of the three coordinate systems will now be developed and illustrated.

Part A: Rectangular Coordinates (x-y)

This reference system is particularly useful for motions where the x- and y-components of acceleration are independently generated or determined.

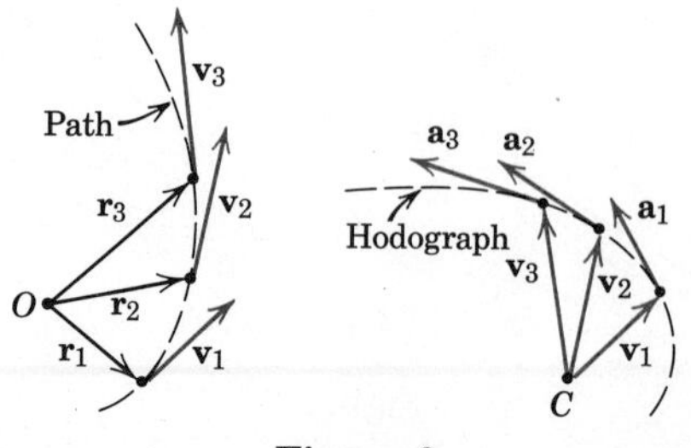

Figure 9

The resulting curvilinear motion is, then, obtained by vector combination of the respective motion components.

The position vector **r**, the velocity **v**, and acceleration **a** of the particle in Fig. 8 may be written in terms of their x- and y-components with the aid of the unit vectors **i** and **j**. Thus

$$\begin{aligned} \mathbf{r} &= \mathbf{i}x + \mathbf{j}y \\ \mathbf{v} = \dot{\mathbf{r}} &= \mathbf{i}\dot{x} + \mathbf{j}\dot{y} \\ \mathbf{a} = \dot{\mathbf{v}} = \ddot{\mathbf{r}} &= \mathbf{i}\ddot{x} + \mathbf{j}\ddot{y} \end{aligned} \tag{8}$$

as shown in Fig. 10, where the magnitudes of the velocity and acceleration components are $v_x = \dot{x}$, $v_y = \dot{y}$ and $a_x = \dot{v}_x = \ddot{x}$, $a_y = \dot{v}_y = \ddot{y}$. (As drawn in this particular figure, a_x is to the left, so that $\ddot{x}$ would be a negative number.) The direction of the velocity is always along the path, and it is clear from the figure that

$$v^2 = v_x^2 + v_y^2 \qquad \text{and} \qquad \tan\theta = \frac{v_y}{v_x}$$

The unit vectors **i** and **j** have no time derivatives since both their directions and magnitudes are constant.

Integration of each acceleration component twice with respect to the time gives the corresponding coordinate as a function of the time, $x = f_1(t)$ and $y = f_2(t)$. Elimination of t between these two parametric equations gives the equation of the path $y = f(x)$.

From the foregoing discussion it should be recognized that the rectangular-coordinate representation of curvilinear motion is merely the superposition of the coordinates of two simultaneous rectilinear motions in the x- and y-directions.

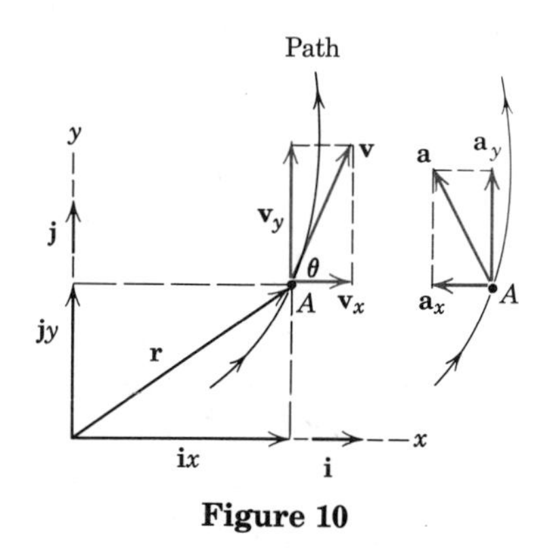

Figure 10

Sample Problem

2/60 A rocket has expended all its fuel when it reaches position A where it has a velocity u at an angle θ with respect to the horizontal. It then begins unpowered flight and attains a maximum added height h at position B after traveling a horizontal distance s from A. Determine the expressions for h and s, the time t of flight from A to B, and the equation of the path. For the interval concerned, assume a flat earth with a constant gravitational acceleration g and neglect any atmospheric resistance. If the rocket encountered appreciable atmospheric resistance from A to B, indicate qualitatively what modifications would have to be made in the solution.

Solution. Since all motion components are directly expressible in terms of horizontal and vertical coordinates, a rectangular set of axes x-y will be employed. With the neglect of atmospheric resistance, $a_x = 0$ and $a_y = -g$, and the resulting motion is a direct superposition of two rectilinear motions with constant acceleration. Thus

$$[dx = v_x\,dt] \qquad x = \int_0^t u \cos\theta\,dt \qquad x = ut\cos\theta$$

$$[dv_y = a_y\,dt] \qquad \int_{u\sin\theta}^{v_y} dv_y = \int_0^t (-g)\,dt \qquad v_y = u\sin\theta - gt$$

$$[dy = v_y\,dt] \qquad y = \int_0^t (u\sin\theta - gt)\,dt \qquad y = ut\sin\theta - \tfrac{1}{2}gt^2$$

Position B is reached when $v_y = 0$, which occurs for $0 = u\sin\theta - gt$ or

$$t = (u\sin\theta)/g \qquad \textit{Ans.}$$

Substitution of this value for the time into the expression for y gives the maximum added altitude

$$h = u\left(\frac{u\sin\theta}{g}\right)\sin\theta - \frac{1}{2}g\left(\frac{u\sin\theta}{g}\right)^2 \qquad h = \frac{u^2\sin^2\theta}{2g} \qquad \textit{Ans.}$$

The horizontal distance is seen to be

$$s = u\left(\frac{u\sin\theta}{g}\right)\cos\theta \qquad s = \frac{u^2\sin 2\theta}{2g} \qquad \textit{Ans.}$$

which is clearly a maximum when $\theta = 45°$. The equation of the path is obtained by eliminating t from the expressions for x and y, which gives

$$y = x\tan\theta - \frac{gx^2}{2u^2}\sec^2\theta \qquad \textit{Ans.}$$

This equation describes a vertical parabola as indicated in the figure.

With appreciable atmospheric resistance, the forces and, hence, the accelerations in both the x- and y-directions are no longer constant. The dependency of the acceleration components on velocity would have to be established before an integration of the acceleration equations could be carried out.

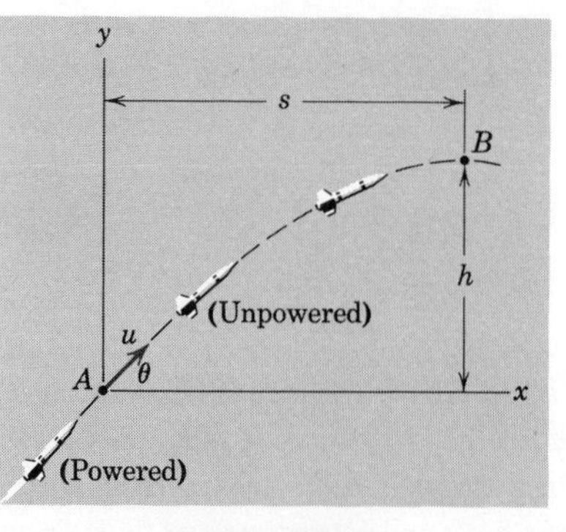

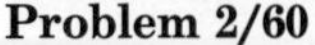
Problem 2/60

Part B: Normal and Tangential Coordinates (n-t)

Components of the motion vectors which are along the tangent t and the normal n to the curved path at the instantaneous position of the particle, Fig. 11*a*, provide the most common and useful description of curvilinear motion. The positive direction for n is taken toward the center of curvature O of the path.

The velocity $\mathbf{v}$ of the particle along its path has a magnitude

$$v = \dot{s} = \frac{ds}{dt} = \frac{\rho\, d\theta}{dt} = \rho\dot{\theta} \tag{9}$$

where ρ is the radius of curvature of the path at the position considered. Between A and A' the vector change in velocity, shown in Fig. 11*b*, is

$$d\mathbf{v} = d\mathbf{v}_n + d\mathbf{v}_t \tag{10}$$

where the n-component is due to the change in the direction of $\mathbf{v}$ and has the magnitude $|d\mathbf{v}_n| = v\, d\theta$, and where the t-component is due to the change in the magnitude of $\mathbf{v}$ and is $|d\mathbf{v}_t| = d(\rho\dot{\theta})$. It should be noted carefully that the magnitude $|d\mathbf{v}|$ of the differential change in the vector $\mathbf{v}$ is *not* the same as the differential change dv in the magnitude of $\mathbf{v}$. In other words, for a vector which changes its direction, the magnitude of the derivative is not the same as the derivative of the magnitude.

The acceleration $\mathbf{a}$ is obtained in the limit by dividing Eq. 10 by dt and has two components, Fig. 11*c*,

$$\mathbf{a} = \mathbf{a}_n + \mathbf{a}_t \tag{11}$$

The n-component is due to the change in the direction of $\mathbf{v}$ and has the magnitude

$$a_n = \frac{|d\mathbf{v}_n|}{dt} = \frac{v\, d\theta}{dt} = v\dot{\theta} = \rho\dot{\theta}^2 = \frac{v^2}{\rho} \tag{12}$$

The last two alternative expressions for a_n come from the substitution of Eq. 9 and should be learned thoroughly. It is essential to observe that the n-com-

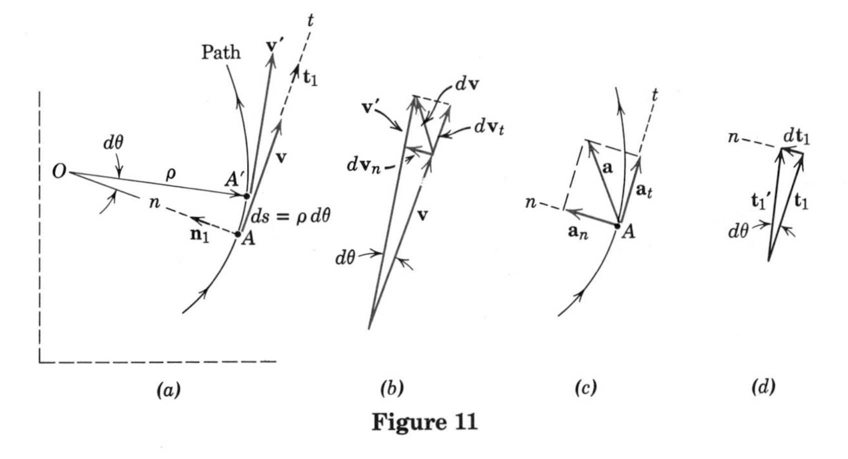

Figure 11

ponent of acceleration is *always* directed *toward* the center of curvature, as disclosed in Fig. 11*b* and *c*.

The t-component of the acceleration is due to the change in the magnitude of $\mathbf{v}$ and has the value

$$a_t = \frac{|d\mathbf{v}_t|}{dt} = \dot{v} = \ddot{s} \tag{13}$$

where s is the distance measured along the path. Both the n- and t-components of the acceleration are shown in Fig. 11*c*. It should be understood that, if the speed of the particle were diminishing rather than increasing, a_t would be in the direction opposite to that shown.

The foregoing analysis of acceleration provides a physical and geometrical picture of the vectors and their changes and is a necessary background for a clear understanding of the direct differentiation of the vector equation for velocity to obtain acceleration. With this alternative approach, unit vectors $\mathbf{t}_1$ and $\mathbf{n}_1$ are introduced along the tangent and normal directions to the path at the point in question, as shown in Fig. 11*a*. Therefore the velocity may be written in vector form as

$$\mathbf{v} = \mathbf{t}_1 v$$

where v is the magnitude of $\mathbf{v}$, and $\mathbf{t}_1$ gives the direction of $\mathbf{v}$. Differentiation of the product with respect to time gives

$$\mathbf{a} = \dot{\mathbf{v}} = \dot{\mathbf{t}}_1 v + \mathbf{t}_1 \dot{v}$$

where the unit vector has a time derivative since its direction changes even though its magnitude remains unity. The vector change in $\mathbf{t}_1$ during the interval between A and A' is shown in Fig. 11*d*. The magnitude of this change during time dt is $|\mathbf{t}_1|\, d\theta = d\theta$, and the direction is specified by the unit vector $\mathbf{n}_1$. Thus

$$\dot{\mathbf{t}}_1 = \frac{d\mathbf{t}_1}{dt} = \frac{\mathbf{n}_1\, d\theta}{dt} = \mathbf{n}_1 \dot{\theta} \tag{14}$$

Therefore the acceleration becomes

▶ $$\mathbf{a} = \mathbf{n}_1 v\dot{\theta} + \mathbf{t}_1 \dot{v} \tag{15}$$

and the two terms are the same as the components expressed in Eqs. 12 and 13.

Circular motion is a special case of plane curvilinear motion where the radius of curvature ρ becomes the constant radius r of the circle. If the symbol ω is used for the angular velocity $\dot{\theta}$ and if α is used for the angular acceleration $\ddot{\theta}$ of the radius vector, the velocity and the acceleration components for simple circular motion are

▶ $$\begin{aligned} v &= r\omega \\ a_n &= r\omega^2 = \frac{v^2}{r} = v\omega \\ a_t &= r\alpha \end{aligned} \tag{16}$$

Equations 9 through 16 are among the most frequently used relations in dynamics and, consequently, should be thoroughly understood.

Sample Problem

2/61 A certain 3rd-stage rocket maintains a horizontal attitude of its axis during the powered phase of its flight at a high altitude. The thrust imparts a horizontal acceleration of 12 m/s^2 and the downward acceleration due to gravity is 9 m/s^2. At the instant represented, the rocket has a velocity $v = 25\,000$ km/h tangent to the path of its center of mass G in the 15° direction shown, and the radius of curvature ρ of the path is increasing at the rate of 60 km/s. The vertical component of the velocity is due to the upward motion of the 2nd-stage booster prior to release of the 3rd-stage rocket. For the instant considered determine the radius of curvature ρ of the flight trajectory and the angular acceleration α of the radius-of-curvature vector.

Solution. The radius of curvature appears in the expression for the component of acceleration normal to the curve, so that n- and t-components will be used to describe the motion of G. The n- and t-components of acceleration are obtained by resolving the horizontal and vertical acceleration components and are seen from the figure to be

$$a_n = 9 \cos 15° + 12 \sin 15° = 11.80 \text{ m/s}^2$$

$$a_t = 12 \cos 15° - 9 \sin 15° = 9.26 \text{ m/s}^2$$

Therefore the radius of curvature becomes*

$$\left[a_n = \frac{v^2}{\rho}\right] \qquad \rho = \frac{(25\,000)^2/(3.6)^2}{11.80} = 4.09(10^6) \text{ m} \qquad \textit{Ans.}$$

The angular acceleration of the radius-of-curvature vector is obtained from

$$\left[a_t = \frac{dv}{dt}\right] \qquad a_t = \frac{d}{dt}(\rho\omega) = \rho\dot{\omega} + \dot{\rho}\omega$$

$$9.26 = 4.09(10^6)\dot{\omega} + 60\,000\,\frac{25\,000/3.6}{4.09(10^6)}$$

$$\alpha = \dot{\omega} = -22.7(10^{-6}) \text{ rad/s}^2 \qquad \textit{Ans.}$$

The negative sign for angular acceleration indicates that it is of opposite sign to the angular velocity $\omega = v/\rho$.

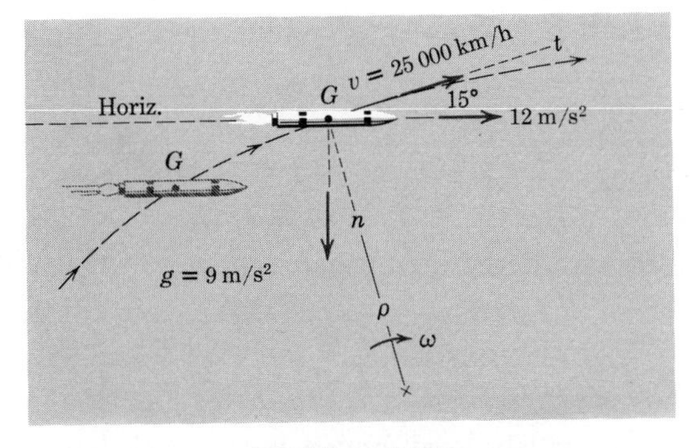

Problem 2/61

*Note that the conversion from km/h to m/s involves multiplication by $1000/(60)^2 = 1/3.6$ which is a simple conversion worth remembering.

Part C: Polar Coordinates (r-θ)

There are many instances where the curvilinear motion of a particle is determined or specified by polar coordinates r and θ. In Fig. 12*a* the differential displacement $d\mathbf{r}$ of the particle between A and A' is represented by its two polar components $d\mathbf{r} = d\mathbf{r}_r + d\mathbf{r}_\theta$. The radial component $d\mathbf{r}_r$ has a magnitude dr equal to the differential change in length of $\mathbf{r}$. The transverse or θ-component $d\mathbf{r}_\theta$ has a magnitude equal to the length of the equivalent differential arc $r\,d\theta$. Division by the time dt gives the magnitude of the velocity components, so that the velocity may be written as

$$\mathbf{v} = \mathbf{v}_r + \mathbf{v}_\theta \tag{17}$$

where

$$v_r = \frac{dr}{dt} = \dot{r}$$

$$v_\theta = r\frac{d\theta}{dt} = r\dot{\theta}.$$

As the particle moves from A to A' in Fig. 12*b*, it is seen that each of the velocity components $\mathbf{v}_r$ and $\mathbf{v}_\theta$ will undergo a vector change, as shown in Fig. 12*c*. The components of $d\mathbf{v}_r$ have the magnitudes $|(d\mathbf{v}_r)_r| = dv_r = d\dot{r}$ in the r-direction due to the change in the magnitude of $\dot{r}$, and $|(d\mathbf{v}_r)_\theta| = v_r\,d\theta = \dot{r}\,d\theta$ in the θ-direction due to the change in direction of $\mathbf{v}_r$. The acceleration components due to $d\mathbf{v}_r$ are obtained by dividing its two velocity increments by dt and are $d\dot{r}/dt = \ddot{r}$ in the r-direction and $\dot{r}\,d\theta/dt = \dot{r}\dot{\theta}$ in the θ-direction.

Similarly the components of $d\mathbf{v}_\theta$ have the magnitudes $|(d\mathbf{v}_\theta)_\theta| = dv_\theta = d(r\dot{\theta})$ in the θ-direction due to the change in magnitude of $\mathbf{v}_\theta$ and $|(d\mathbf{v}_\theta)_r| = v_\theta\,d\theta = r\dot{\theta}\,d\theta$ in the negative r-direction due to the change in direction of $\mathbf{v}_\theta$. The acceleration components due to $d\mathbf{v}_\theta$ are obtained by dividing its two velocity increments by dt and are $d(r\dot{\theta})/dt = r\ddot{\theta} + \dot{r}\dot{\theta}$ in the θ-direction and $r\dot{\theta}\,d\theta/dt = r\dot{\theta}^2$ in the negative r-direction.

Collecting the r- and θ-terms gives the polar-coordinate expressions for the acceleration components in curvilinear motion. Thus the acceleration, indicated in Fig. 12*d*, may be written as

$$\mathbf{a} = \mathbf{a}_r + \mathbf{a}_\theta \tag{18}$$

where

$$a_r = \ddot{r} - r\dot{\theta}^2$$

$$a_\theta = r\ddot{\theta} + 2\dot{r}\dot{\theta}$$

It is easily verified by differentiation that a_θ may be written alternatively as

$$a_\theta = \frac{1}{r}\frac{d}{dt}(r^2\dot{\theta})$$

Equations 18 are widely used, and it is essential that the student learn these relations and be able to identify the physical and geometrical basis of each of the terms.

An alternative derivation for $\mathbf{a}$ in polar coordinates will now be developed by direct differentiation of the vector equation for velocity to obtain acceleration. This approach requires the use of unit vectors $\mathbf{r}_1$ and $\boldsymbol{\theta}_1$, shown in Fig. 12*a*, and their time derivatives. In the previous section on *n*- and *t*-components, the expression for the time derivative of the unit vector $\mathbf{t}_1$ was developed. The time derivatives of $\mathbf{r}_1$ and $\boldsymbol{\theta}_1$ due to their rotation with time are obtained by exactly the same procedure used to obtain $\dot{\mathbf{t}}_1$, and the sketches and proof will be left to the student. The results are

$$\dot{\mathbf{r}}_1 = \boldsymbol{\theta}_1\dot{\theta} \qquad \text{and} \qquad \dot{\boldsymbol{\theta}}_1 = -\mathbf{r}_1\dot{\theta}$$

where the minus sign indicates that the sense of $\dot{\boldsymbol{\theta}}_1$ is in the negative *r*-direction. These expressions will be held in readiness for substitution into the equations which follow.

The position vector to the particle is written as $\mathbf{r} = \mathbf{r}_1 r$ so as to disclose both the magnitude and direction of $\mathbf{r}$ separately. The first time derivative gives

$$\mathbf{v} = \dot{\mathbf{r}} = \mathbf{r}_1\dot{r} + \dot{\mathbf{r}}_1 r = \mathbf{r}_1\dot{r} + \boldsymbol{\theta}_1 r\dot{\theta}$$

where $\mathbf{v}_r = \mathbf{r}_1\dot{r}$ and $\mathbf{v}_\theta = \boldsymbol{\theta}_1 r\dot{\theta}$. This result is equivalent to Eq. 17.

The acceleration is obtained by direct differentiation of $\mathbf{v}$. Thus

$$\mathbf{a} = \dot{\mathbf{v}} = \ddot{\mathbf{r}} = \dot{\mathbf{r}}_1\dot{r} + \mathbf{r}_1\ddot{r} + \dot{\boldsymbol{\theta}}_1 r\dot{\theta} + \boldsymbol{\theta}_1\dot{r}\dot{\theta} + \boldsymbol{\theta}_1 r\ddot{\theta}$$

Substituting the expressions for the derivatives of the unit vectors and collecting terms give

$$\mathbf{a} = (\ddot{r} - r\dot{\theta}^2)\mathbf{r}_1 + (r\ddot{\theta} + 2\dot{r}\dot{\theta})\boldsymbol{\theta}_1$$

which agrees with the expressions obtained in Eqs. 18. The student should identify each term in the foregoing equation with the corresponding velocity

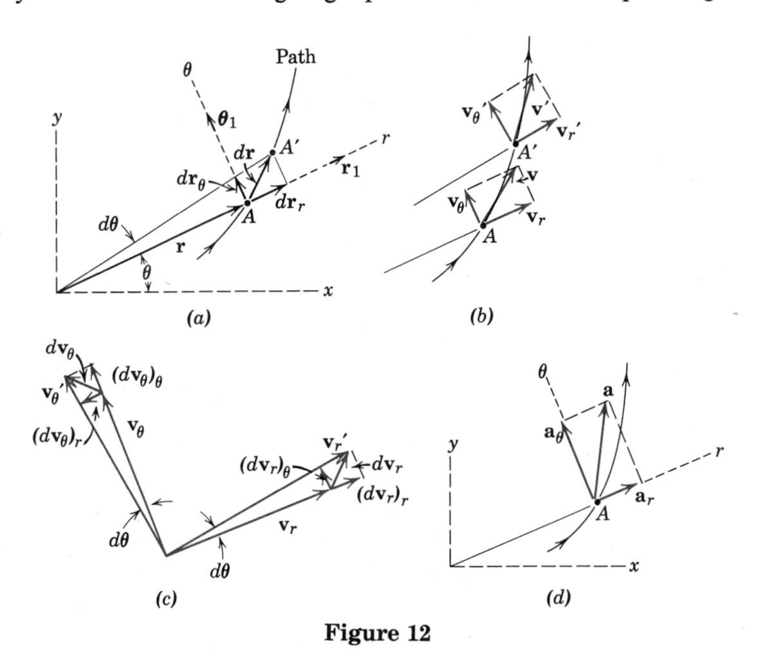

Figure 12

increment pictured in Fig. 12*c*. It should also be observed that, when r is constant, the particle moves in a circle, and the expressions reduce to those developed in the previous section on n- and t-components for the case of circular motion.

The expressions for a_r and a_θ in scalar form may also be obtained by direct differentiation of the coordinate relations $x = r\cos\theta$ and $y = r\sin\theta$ to get $a_x = \ddot{x}$ and $a_y = \ddot{y}$. Each of these rectangular components of acceleration may then be resolved into r- and θ-components which, when combined, will yield the expressions of Eq. 18. This approach is straight-forward but has the disadvantage of somewhat more algebraic manipulation than either of the foregoing proofs.

Sample Problem

2/62 A missile is launched at point O and follows a trajectory in a vertical plane as shown. A tracking device stationed at O records the coordinates r and θ as functions of time, and it is found that the functions may be approximated closely by the relations $r = 2t - t^2/40$ and $\theta^2 = 1600 - t^2$, where r is in kilometres, θ is in degrees, and t is in seconds. Compute the velocity and acceleration of the missile at the instant when $t = 30$ s.

Solution. The terms which appear in Eqs. 17 and 18 for velocity and acceleration in polar coordinates are determined as functions of the time and then evaluated for $t = 30$ s as follows:

$$r = 2t - \frac{t^2}{40} \qquad r_{30} = 2(30) - \frac{(30)^2}{40} = 37.5 \text{ km}$$

$$\dot{r} = 2 - \frac{t}{20} \qquad \dot{r}_{30} = 2 - \tfrac{30}{20} = 0.5 \text{ km/s}$$

$$\ddot{r} = -\tfrac{1}{20} \qquad \ddot{r}_{30} = -0.05 \text{ km/s}^2$$

$$\theta^2 = 1600 - t^2 \qquad \theta_{30} = \sqrt{1600 - (30)^2} = 26.5^\circ$$

$$2\theta\dot{\theta} = -2t, \quad \dot{\theta} = -\frac{t}{\theta} \qquad \dot{\theta}_{30} = -\left(\frac{30}{26.5}\right)\frac{\pi}{180} = -0.01975 \text{ rad/s}$$

$$\dot{\theta}^2 + \theta\ddot{\theta} = -1, \quad \ddot{\theta} = -\frac{1 + \dot{\theta}^2}{\theta} \qquad \ddot{\theta}_{30} = -\frac{1 + (1.134)^2}{26.5}\frac{\pi}{180} = -0.00151 \text{ rad/s}^2$$

Thus, from Eqs. 17 and 18, the components of velocity and acceleration of the missile at $t = 30$ s are

$[v_r = \dot{r}]$ $\qquad v_r = 0.5$ km/s

$[v_\theta = r\dot{\theta}]$ $\qquad v_\theta = (37.5)(-0.01975) = -0.742$ km/s

Hence, $\qquad v = \sqrt{(0.5)^2 + (0.742)^2} = 0.894$ km/s $\qquad$ *Ans.*

$[a_r = \ddot{r} - r\dot{\theta}^2]$ $\qquad a_r = -0.05 - (37.5)(-0.01975)^2 = -0.0646$ km/s^2

$[a_\theta = r\ddot{\theta} + 2\dot{r}\dot{\theta}]$ $\qquad a_\theta = (37.5)(-0.00151) + (2)(0.5)(-0.01975)$

$\qquad = -0.0764$ km/s^2

Thus, $\qquad a = \sqrt{(0.0646)^2 + (0.0764)^2} = 0.100$ km/s^2 $\qquad$ *Ans.*

The figure shows the trajectory plotted from the values of r and θ which correspond to given values of the time. Point P corresponds to the position when $t = 30$ s where $r = 37.5$ km and $\theta = 26.5°$. The velocity **v** and acceleration **a** together with their components are also represented in the figure for this particular instant. Observe that the signs of the calculated values of v_θ, a_r, and a_θ are negative. It should be noted in the solution that the numerical coefficients in the equations have dimensions, and that the coefficient of unity before the t^2 in the expression for θ^2 also has dimensions. The conversion from degrees to radians is necessary when computing $r\dot{\theta}$ and $r\ddot{\theta}$.

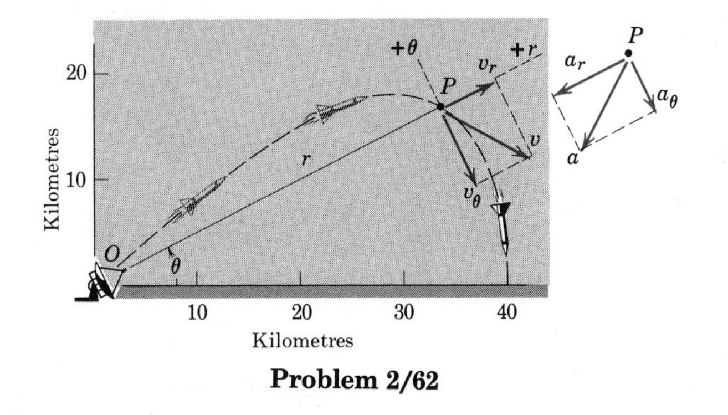

Problem 2/62

Problems

(The first group of introductory problems is arranged by coordinate category. The second group of general problems is not arranged by coordinate category, and, in that case, the student must exercise his own choice of reference system. Unless otherwise indicated air resistance is to be neglected in problems involving projectile motion.)

Rectangular Coordinates (x-y)

2/63 The x- and y-coordinates of a particle moving with plane curvilinear motion are given by $x = 2t^2 + 3t$ and $y = t^3/3 - 8$ where x and y are in metres and t is in seconds. Determine the magnitudes of the velocity **v** and acceleration **a** and the angles which the vectors make with the x-axis when $t = 3$ s.

Ans. $v = 17.49$ m/s with $\theta_x = 31.0°$
$a = 7.21$ m/s^2 with $\theta_x = 56.3°$

2/64 Calculate the minimum possible muzzle velocity u which a projectile must have when fired from point A to reach a target B on the same horizontal plane 10 km away.

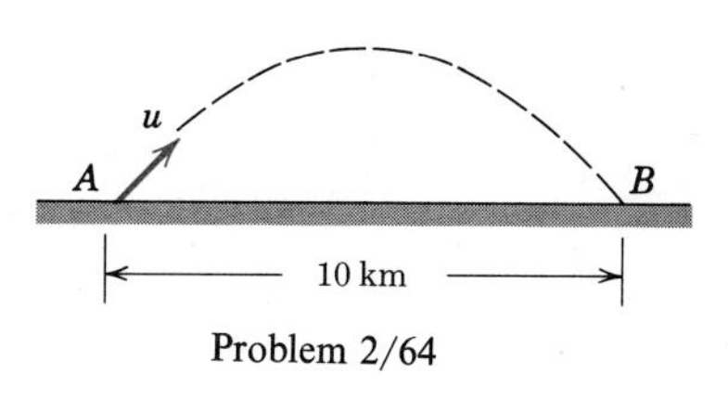

Problem 2/64

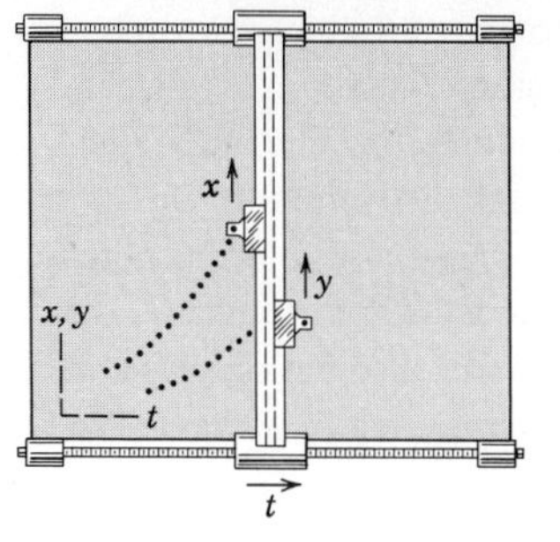

Problem 2/65

2/65 The coordinates of a particle which moves in the X-Y plane are recorded every second, and the signals are fed through a computer to the automatic plotter which draws simultaneous graphs of both coordinates against the time. Each millimetre of x-motion plotted on the graph represents 0.25 m of actual X-movement, and each millimetre of y-motion on the graph represents 1.25 m of actual Y-movement. On the horizontal time scale t each millimetre on the graph represents 0.04 s of real time. The resulting graphs with all measurements in millimetres are closely approximated by the relations $x = 50 + 0.1t^2$ and $y = 50 + 0.0015t^3/3 - 0.04t^2/2$. Determine the magnitude of the actual acceleration of the particle for the condition corresponding to $t = 2$ s and find the angle θ_X made by the velocity vector of the particle with the X-direction at this same instant.

Ans. $a = 91.4 \text{ m/s}^2$, $\theta_X = 41.2°$

2/66 The position vector of a point which moves in the x-y plane is given by

$$\mathbf{r} = \left(\frac{2}{3}t^3 - \frac{3}{2}t^2\right)\mathbf{i} + \frac{t^4}{12}\mathbf{j}$$

where $\mathbf{r}$ is in metres and t is in seconds. Determine the velocity $\mathbf{v}$ and the acceleration $\mathbf{a}$ when $t = 3$ s. What do the results tell about the curvature of the path at this instant?

2/67 A particle moves along the positive branch of the curve $y = 1 + x^2/10$ with its x-coordinate controlled as a function of time according to $x = 2t^3/3$ where x and y are in metres and t is in seconds. Compute the magnitude of the acceleration $\mathbf{a}$ of the particle at its position when $t = 2$ s.

Ans. $a = 22.8 \text{ m/s}^2$

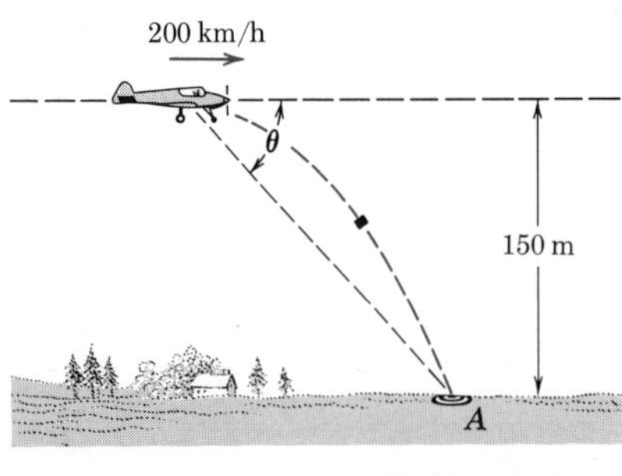

Problem 2/68

2/68 The pilot of an airplane carrying a package of mail to a remote outpost wishes to release the package at the right moment to hit the recovery location A. What angle θ with the horizontal should the pilot's line of sight to the target make at the instant of release? The airplane is flying horizontally at an altitude of 150 m with a velocity of 200 km/h.

Ans. $\theta = 26.0°$

2/69 A point moves on a circular path in the x-y plane with coordinates given by $x = r\cos\dfrac{\alpha t^2}{2}$ and $y = r\sin\dfrac{\alpha t^2}{2}$ where r and α are constants. Interpret the meaning of r and α and determine the x- and y-components of the acceleration of the particle for the position when (*a*) $t = 0$ and (*b*) $\alpha t^2/2 = \pi$.

2/70 A particle moves in the *x-y* plane with a *y*-component of velocity in metres per second given by $v_y = 8t$ with t in seconds. The acceleration of the particle in the *x*-direction in metres per second squared is given by $a_x = 4t$ with t in seconds. When $t = 0$, $y = 2$ m, $x = 0$, and $v_x = 0$. Find the equation of the path of the particle and calculate the magnitude of the velocity **v** of the particle for the instant when its *x*-coordinate reaches 18 m.

Ans. $(y - 2)^3 = 144x^2$, $v = 30$ m/s

Normal and Tangential Coordinates (n-t)

2/71 Consider the polar axis of the earth to be fixed in space and compute the acceleration a of a point P on the earth's surface at latitude 40° north. The mean diameter of the earth is 12 742 km and its angular velocity is $0.729(10^{-4})$ rad/s. *Ans.* $a = 0.0259$ m/s^2

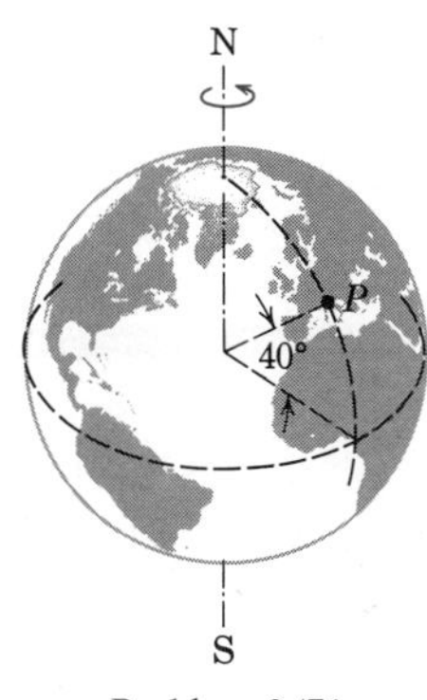

Problem 2/71

2/72 A point on the rim of a flywheel has a peripheral speed of 10 m/s at an instant when this speed is decreasing at the rate of 60 m/s^2. If the total acceleration of the point at this instant is 100 m/s^2, find the radius r of the flywheel.

2/73 At the top of its trajectory at an altitude of 800 m above sea level a certain bullet has a horizontal velocity of 620 m/s. Calculate the radius of curvature ρ of its trajectory at this point.

Ans. $\rho = 39.2$ km

2/74 To simulate a condition of "weightlessness" in its cabin, a jet transport traveling at 720 km/h moves on a sustained vertical curve as shown. At what rate $\dot{\theta}$ in degrees per second should the pilot drop his longitudinal line of sight to effect the desired condition? The maneuver takes place at a mean altitude of 8 km, and the gravitational acceleration may be taken as 9.81 m/s^2.

Ans. $\dot{\theta} = 2.81$°/s

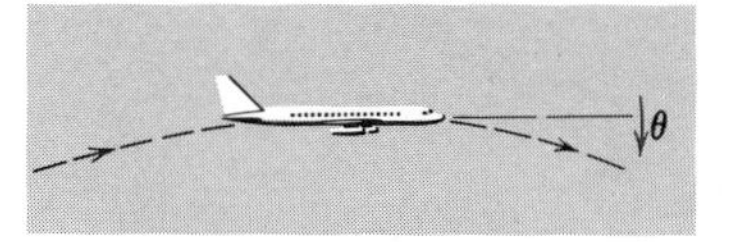

Problem 2/74

2/75 The direction of motion of a flat tape in a numerical-control device is changed by the two pulleys shown. If pulley A increases its speed at a constant rate with time from 20 rev/min to 120 rev/min in 5 revolutions, compute the acceleration of a point P on the tape in contact with pulley B at the instant when A has a speed of 30 rev/min. Assume that no slipping of the tape on the pulleys occurs.

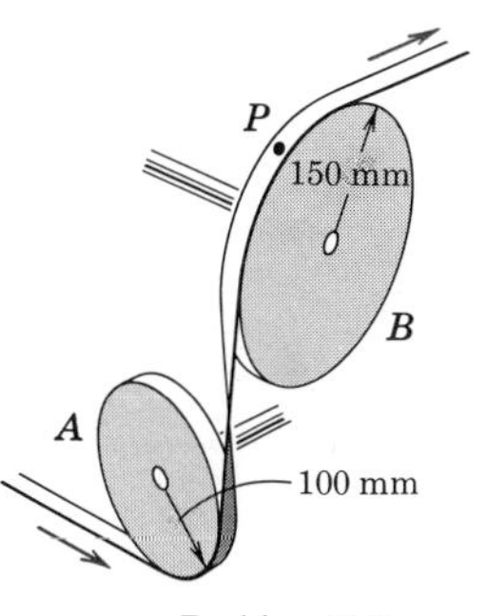

Problem 2/75

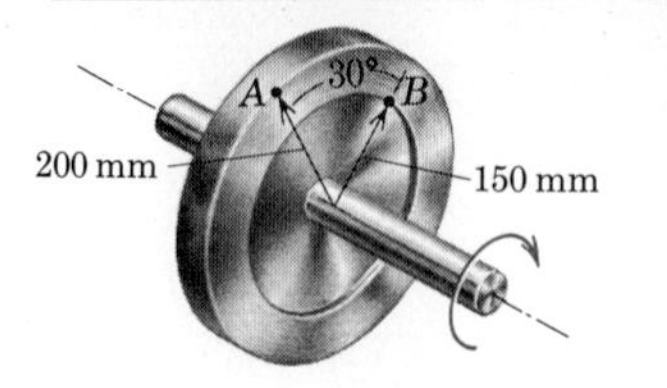

Problem 2/76

2/76 The flywheel is revolving with a changing angular velocity. At a certain instant point A has a component of acceleration tangent to its path of $1\ \text{m/s}^2$ and point B has a component of acceleration normal to its path of $0.6\ \text{m/s}^2$. For this instant compute the rim speed of point A and the total acceleration of point B.

Ans. $v_A = 0.4\ \text{m/s}$, $a_B = 0.960\ \text{m/s}^2$

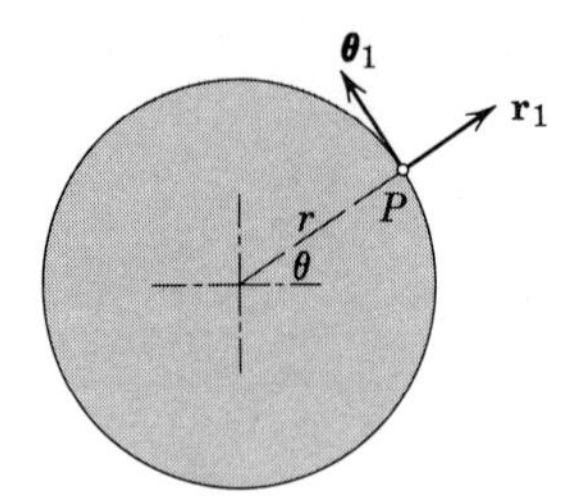

Problem 2/77

2/77 A particle P moves with a constant speed v around the circular path of radius r. Start with the expression $\mathbf{r} = r\mathbf{r}_1$ for the position vector of P and derive the expressions for the velocity $\mathbf{v}$ and acceleration $\mathbf{a}$ of P. The radial and transverse unit vectors are $\mathbf{r}_1$ and $\boldsymbol{\theta}_1$, respectively.

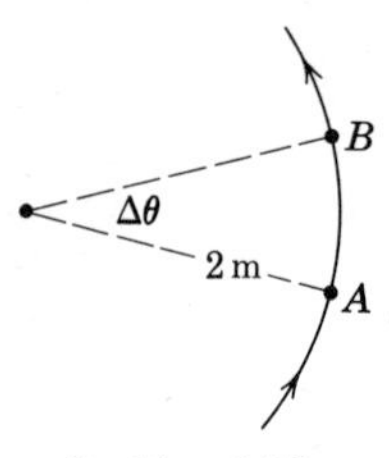

Problem 2/78

2/78 A particle moves on the circular path of 2-m radius with a constant speed of 10 m/s. During the interval from A to B the velocity undergoes a vector change. Divide this change by the time interval between the two points to obtain the average normal acceleration for (*a*) $\Delta\theta = 30°$, (*b*) $\Delta\theta = 15°$, and (*c*) $\Delta\theta = 5°$. Compare the results with the instantaneous normal acceleration.

Polar Coordinates $(r\text{-}\theta)$

2/79 Prove in detail that the time derivatives of the unit vectors $\mathbf{r}_1$ and $\boldsymbol{\theta}_1$ are $\dot{\mathbf{r}}_1 = \boldsymbol{\theta}_1\dot{\theta}$ and $\dot{\boldsymbol{\theta}}_1 = -\mathbf{r}_1\dot{\theta}$.

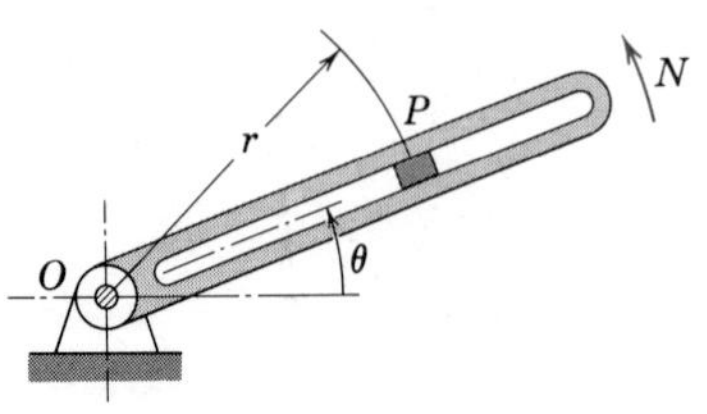

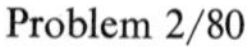

Problem 2/80

2/80 At the instant represented the slotted arm is rotating at a speed $N = 80$ rev/min and is slowing down at the rate of 280 rev/min per second. Also at the same time the radial distance r to the slider P is 250 mm and is decreasing at the constant rate of 300 mm/s. For this instant determine the acceleration $\mathbf{a}$ of P.

Ans. $\mathbf{a} = -17.55\mathbf{r}_1 - 12.36\boldsymbol{\theta}_1\ \text{m/s}^2$

2/81 For the configuration of Prob. 2/80, if $\dot{\theta} = 4\ \text{rad/s}$, $\ddot{\theta} = 16\ \text{rad/s}^2$, and $r = 250$ mm at a certain instant of time, what would be the corresponding time derivatives of r if the slider P had zero acceleration at this instant?

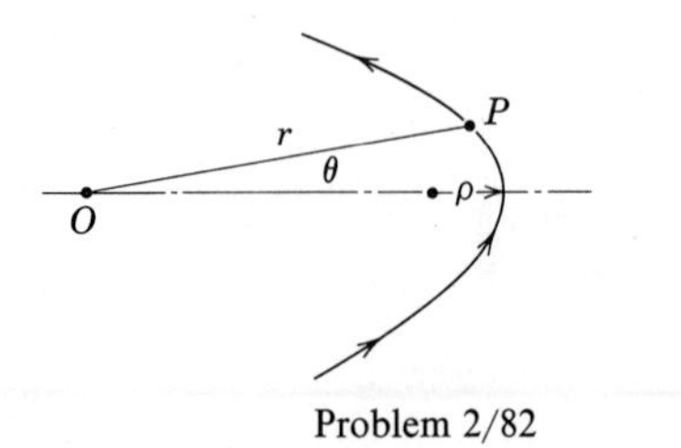

Problem 2/82

2/82 A particle P moves along a path, given by $r = f(\theta)$, which is symmetrical about the line $\theta = 0$. As the particle passes the position $\theta = 0$ where the radius of curvature of the path is ρ, the velocity of P is v. Derive an expression for $\ddot{r}$ in terms of v, r, and ρ for the motion of the particle at this point.

2/83 The motion of a point in the vertical plane is given by its polar coordinates $r = 2t^2$ and $\theta = 0.4 \sin \pi t/3$, where r is in metres, θ is in radians, and t is in seconds. Determine the velocity $\mathbf{v}$ and acceleration $\mathbf{a}$ of the particle when $t = 2$ s.

Ans. $\mathbf{v} = 8\mathbf{r}_1 - 1.675\boldsymbol{\theta}_1$ m/s
$\mathbf{a} = 3.65\mathbf{r}_1 - 6.39\boldsymbol{\theta}_1$ m/s^2

2/84 A particle moving along a plane curve has a position vector $\mathbf{r}$, a velocity $\mathbf{v}$, and an acceleration $\mathbf{a}$. Unit vectors in the r- and θ-directions are $\mathbf{r}_1$ and $\boldsymbol{\theta}_1$, and both r and θ are changing with time. Explain why each of the following statements is wrong.

$$\dot{\mathbf{r}} \neq v \qquad \ddot{\mathbf{r}} \neq a \qquad \dot{\mathbf{r}} \neq \dot{r}\mathbf{r}_1$$
$$\dot{r} \neq v \qquad \ddot{r} \neq a \qquad \ddot{\mathbf{r}} \neq \ddot{r}\mathbf{r}_1$$
$$\dot{r} \neq \mathbf{v} \qquad \ddot{r} \neq \mathbf{a} \qquad \dot{\mathbf{r}} \neq r\dot{\theta}\boldsymbol{\theta}_1$$

2/85 For a certain curvilinear motion of a particle expressed in polar coordinates the product $r^2\dot{\theta}$ in m^2/s varies with the time t in seconds measured over a short period as shown. Approximate the θ-component of the acceleration of the particle at the instant when $t = 7$ s at which time $r = 0.5$ m. *Ans.* $a_\theta = 4$ m/s^2

2/86 The cam has a shape such that the center of the roller A which follows the contour moves on a limacon defined by $r = b - c \cos \theta$, where $b > c$. If the cam does not rotate, determine the acceleration a of A in terms of θ if the slotted arm revolves with a constant counterclockwise angular velocity ω.

Ans. $a = \omega^2 \sqrt{4c^2 - 4bc \cos \theta + b^2}$

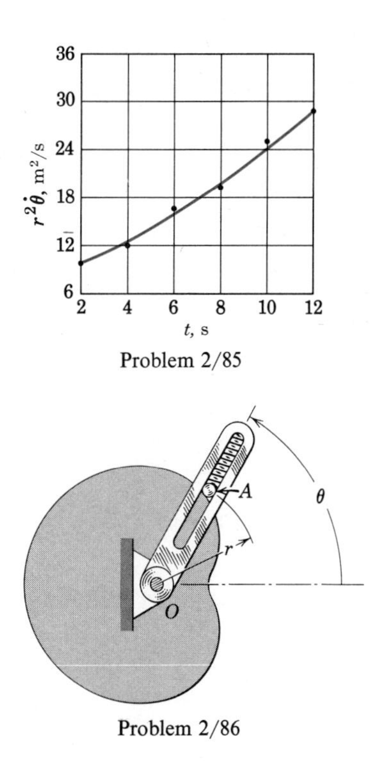

Problem 2/85

Problem 2/86

General

2/87 A 40-caliber rifle bullet has a horizontal muzzle velocity of 720 m/s. Air resistance causes an initial deceleration of 5.4 m/s^2 in the horizontal direction. Determine the radius of curvature ρ of the trajectory of the bullet an instant after it leaves the gun.

2/88 The muzzle velocity for a certain rifle is 600 m/s. If the rifle is pointed vertically upward and fired from an automobile moving horizontally at a speed of 72 km/h, determine the radius of curvature ρ of the path of the bullet at its maximum altitude. Neglect air resistance.

Ans. $\rho = 40.8$ m

Problem 2/88

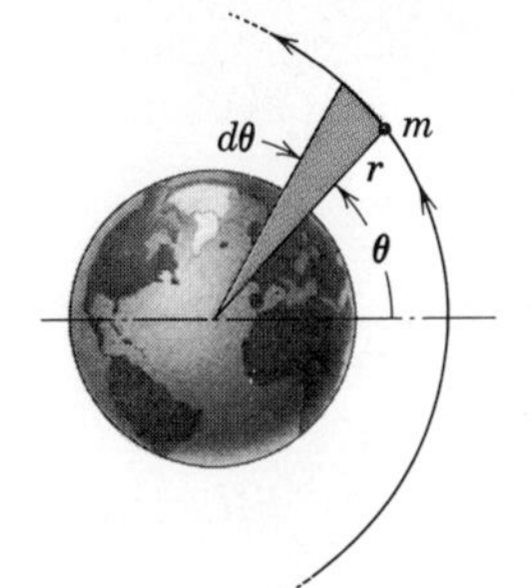

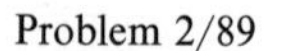
Problem 2/89

2/89 A satellite m moves in an elliptical orbit around the earth. There is no force on the satellite in the θ-direction, so that $a_\theta = 0$. Prove Kepler's second law of planetary motion which says that the radial line r sweeps through equal areas in equal times. The area dA swept by the radial line during time dt is shaded in the figure.

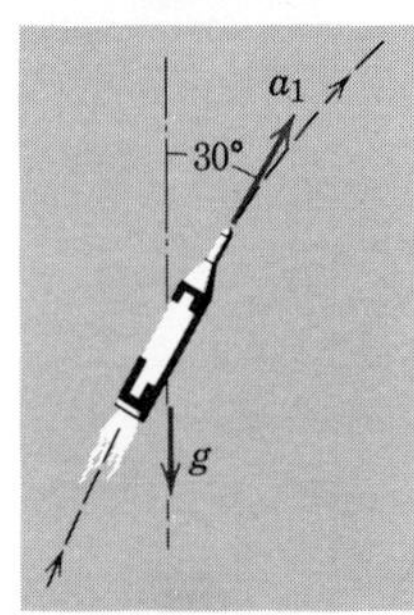

Problem 2/90

2/90 A rocket traveling above the atmosphere at an altitude of 1000 km would have a free-fall acceleration $g = 7.32$ m/s^2 in the absence of forces other than gravitational attraction. Because of thrust, however, the rocket has an additional acceleration component a_1 of 7.62 m/s^2 tangent to its trajectory, which makes an angle of 30° with the vertical at the instant considered. If the velocity v of the rocket is 40 000 km/h at this position, compute the radius of curvature ρ of the trajectory and the rate at which v is changing with time. *Ans.* $\rho = 33\,700$ km
$\dot{v} = 1.28$ m/s^2

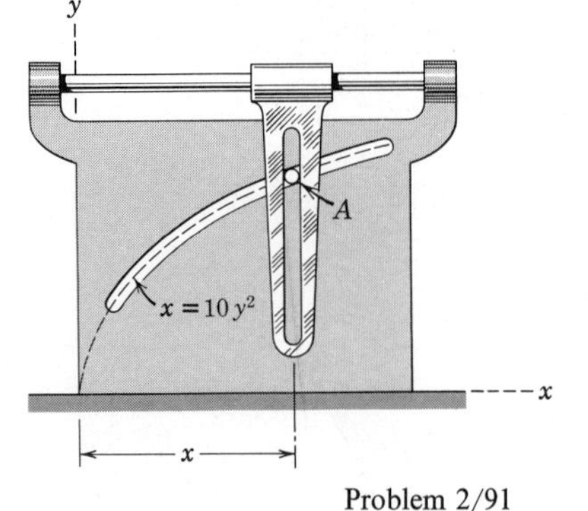

Problem 2/91

2/91 The pin A is confined to move in the parabolic slot in the fixed plate. The pin is also guided by the vertical slot, which is given a constant velocity to the right of 0.1 m/s for an interval of its motion. Determine the velocity **v** and acceleration **a** of pin A for the position $x = 0.1$ m.

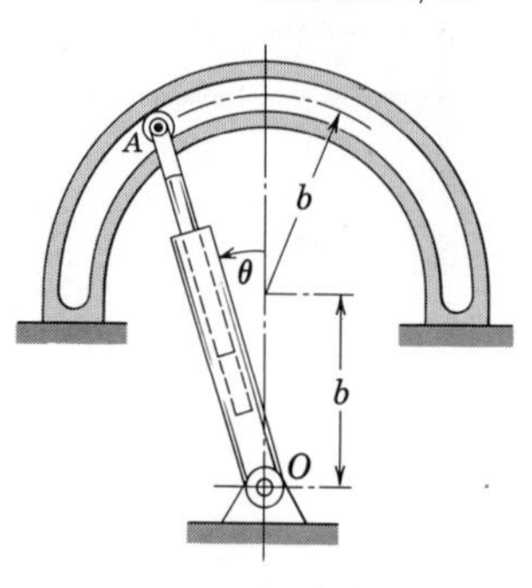

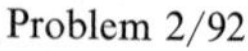
Problem 2/92

2/92 The motion of the roller A in the fixed circular slot is controlled by the arm OA, the upper part of which is free to slide in the lower part to accommodate to the changing distance from A to O as θ changes. If the arm has a constant counterclockwise angular velocity $\dot{\theta} = K$ during an interval of its motion, determine the acceleration a of point A for any position in the interval.
Ans. $a = 4K^2b$, independent of θ

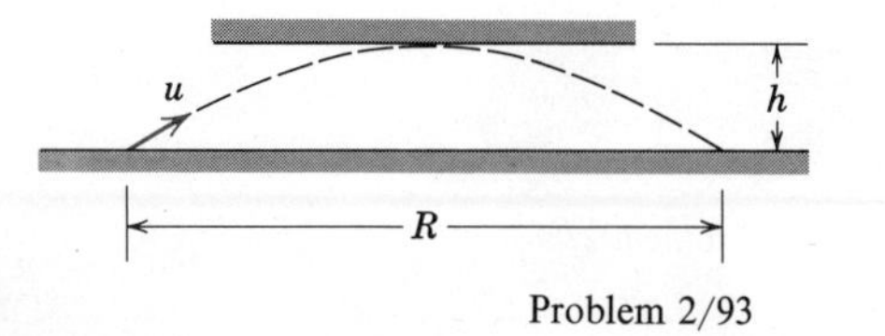

Problem 2/93

2/93 For a given muzzle velocity u find the maximum range R on a horizontal plane which can be achieved by the projectile without exceeding an altitude h.

2/94 The path of a fluid particle P in a certain centrifugal pump with radial vanes is to be approximated by the spiral $r = r_0e^{n\theta}$, where n is a dimensionless constant. If the pump turns at a constant rate $\dot{\theta} = K$, determine the expression for the total acceleration of the particle just prior to leaving the vane in terms of R, K, and n.

Ans. $a = RK^2(n^2 + 1)$

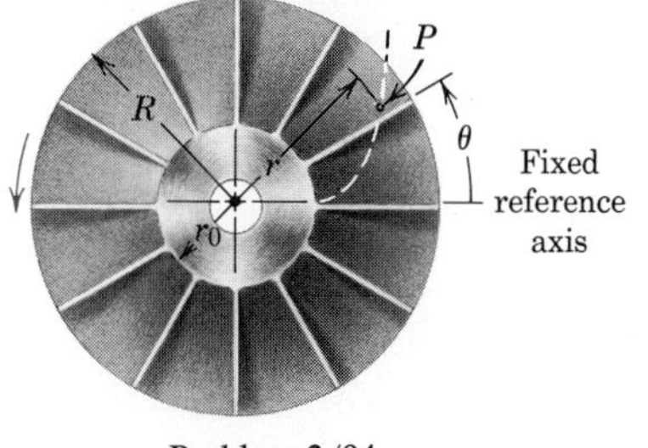

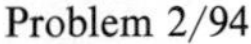
Problem 2/94

2/95 A bomber flying with a horizontal velocity of 500 km/h at an altitude of 5.4 km is to score a direct hit on a train moving with a constant velocity of 150 km/h in the same direction and in the same vertical plane. Calculate the angle θ between the line of sight to the target and the horizontal at the instant the bomb should be released.

Ans. $\theta = 59.2°$

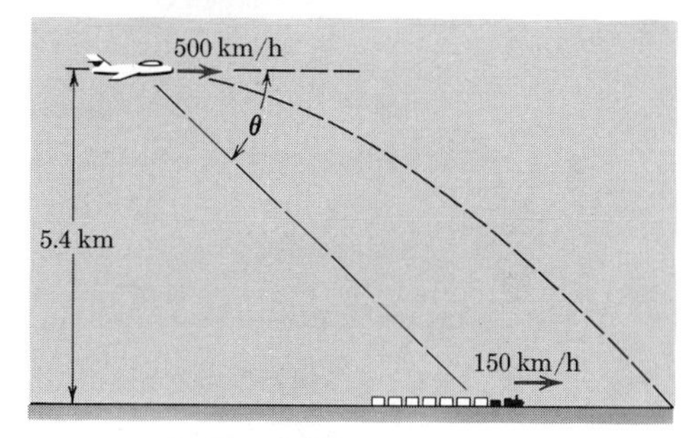

Problem 2/95

2/96 By what angle δ should the pilot of a dive bomber lead his target if he is to release his bomb at an altitude $h = 1.8$ km at a speed $u = 1000$ km/h while diving at the angle $\theta = 45°$ shown?

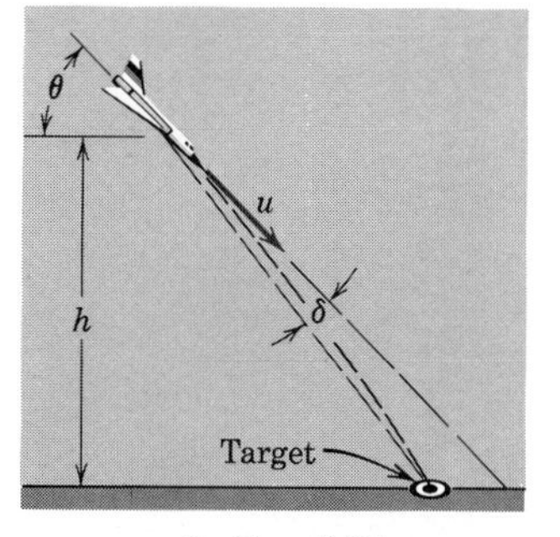

Problem 2/96

2/97 A scientific satellite is injected into an elliptical orbit about the earth and has a velocity of 32 000 km/h at its point of nearest approach (perigee) 400 km above the earth. Calculate the radius of curvature ρ of the orbit at this point. The absolute gravitational acceleration at the surface of the earth is 9.821 m/s^2, and the mean diameter of the earth is 12 742 km.

Ans. $\rho = 9090$ km

2/98 The command module of a lunar mission is orbiting the moon in a circular path at an altitude of 160 km above the moon's surface. Consult Table C2, Appendix C, for information about the moon as is needed and compute the orbital velocity v of the module with respect to the moon.

2/99 An object which is released from rest relative to the earth at A at a distance h above a horizontal surface will appear not to fall straight down by virtue of the effect of the earth's rotation. It may be shown that the object has an eastward acceleration relative to the horizontal surface on the earth equal to $2v_y\omega \cos \gamma$, where v_y is the free-fall downward velocity, ω is the angular velocity of the earth, and γ is the north latitude. Determine the expression for the distance δ to the east of the vertical line through A at which the object strikes the ground.

Ans. $\delta = \dfrac{2\sqrt{2}}{3}\omega h \sqrt{\dfrac{h}{g}} \cos \gamma$

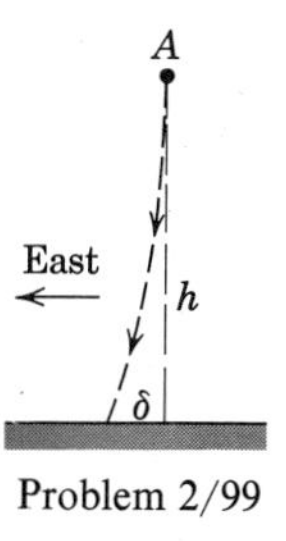

Problem 2/99

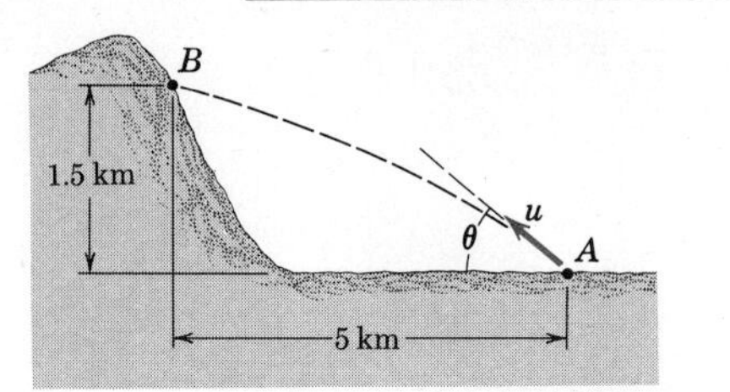

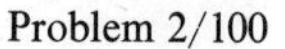
Problem 2/100

2/100 The muzzle velocity of a long-range rifle at A is $u = 400$ m/s. Determine the two angles of elevation θ which will permit the projectile to hit the mountain target B.

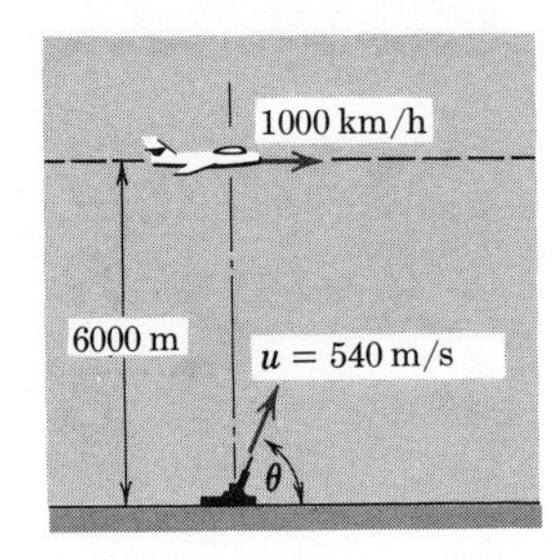

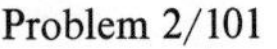
Problem 2/101

2/101 Calculate the firing angle θ of the anti-aircraft gun with a muzzle velocity of 540 m/s if a direct hit is to be scored on an aircraft flying horizontally at 1000 km/h at an altitude of 6000 m. The gun is fired at the instant when the aircraft is directly overhead. Find the time t required for the shell to reach the aircraft.

Ans. $\theta = 59.0°, \quad t = 15.50$ s

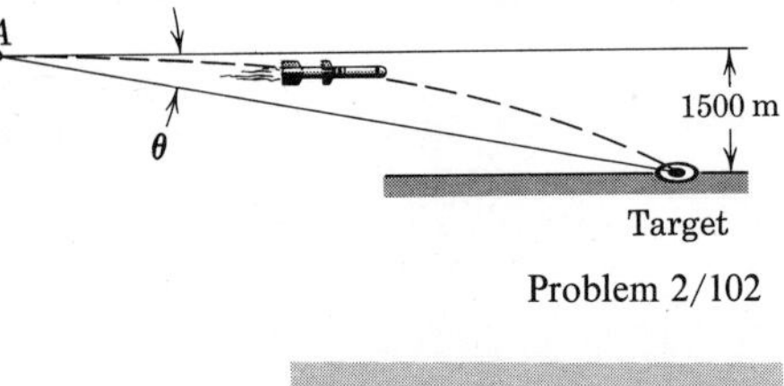

Problem 2/102

2/102 A rocket is released at point A from a jet aircraft flying horizontally at 1000 km/h at a 1500-m altitude. If the rocket thrust gives it a constant horizontal acceleration of $0.5g$, determine the angle θ between the horizontal and the line of sight to the target for a direct hit.

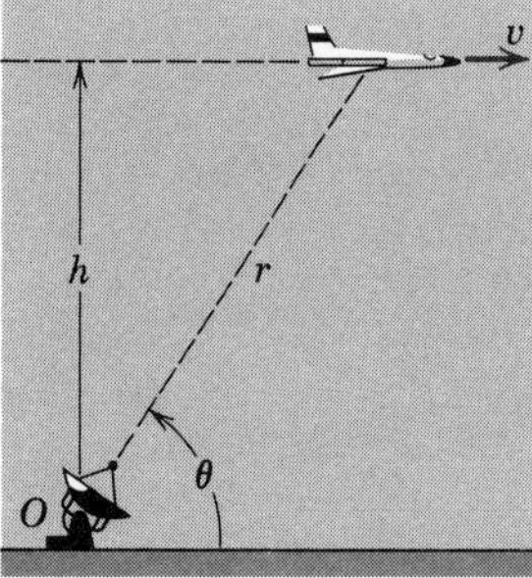

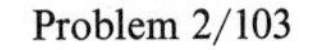
Problem 2/103

2/103 A jet plane flying at a constant velocity v at an altitude $h = 8$ km is being tracked by radar located at O directly below the line of flight. If the angle θ is decreasing at the rate of 0.025 rad/s when $\theta = 60°$, determine the value of $\ddot{r}$ at this instant and the velocity v of the plane.

Ans. $\ddot{r} = 5.77$ m/s^2, $\quad v = 960$ km/h

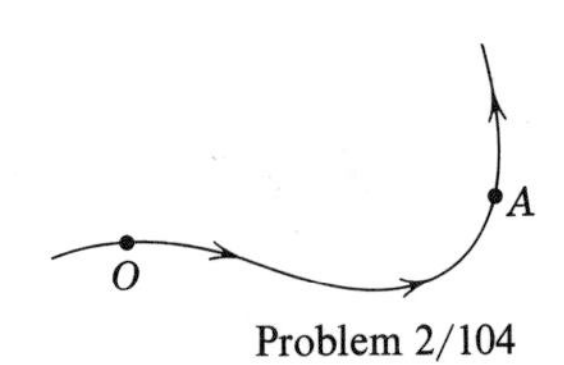

Problem 2/104

2/104 A particle which moves along the curved path shown passes point O with a speed of 12 m/s and slows down to 6 m/s at point A in a distance of 18 m, measured along the curve from O. The deceleration measured along the curve is proportional to the distance from O. If the total acceleration of the particle is 10 m/s^2 as it passes A, determine the radius of curvature ρ of the path at A.

2/105 A particle P moves along a plane curve with a speed v, measured in metres per second, given by $v = 2 + 0.3t^2$ where t is the time in seconds after P passes a certain fixed point on the curve. If the total acceleration of P is 2.4 m/s^2 when $t = 2$ s, compute the radius of curvature ρ of the curve for the position of the particle at this instant.

Ans. $\rho = 4.93$ m

2/106 A particle moving along a plane curve has a velocity v of 6 m/s and a total acceleration a of 30 m/s^2 as it passes point A. At this position the radius of curvature ρ of the path is 2 m, and the angle θ between the velocity vector and a fixed reference line has the second time derivative $\ddot{\theta} = 3$ rad/s^2. Determine the time rate of change of the radius of curvature ρ of the path to the particle as it passes A. *Ans.* $\dot{\rho} = 6$ m/s

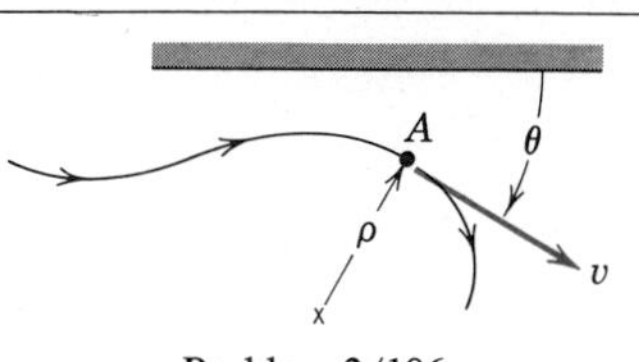

Problem 2/106

2/107 The slotted arm OA forces the small pin to move in the fixed spiral guide defined by $r = K\theta$. Arm OA starts from rest at $\theta = \pi/4$ and has a constant counterclockwise angular acceleration α. Determine the acceleration a of the pin when $\theta = 3\pi/4$.

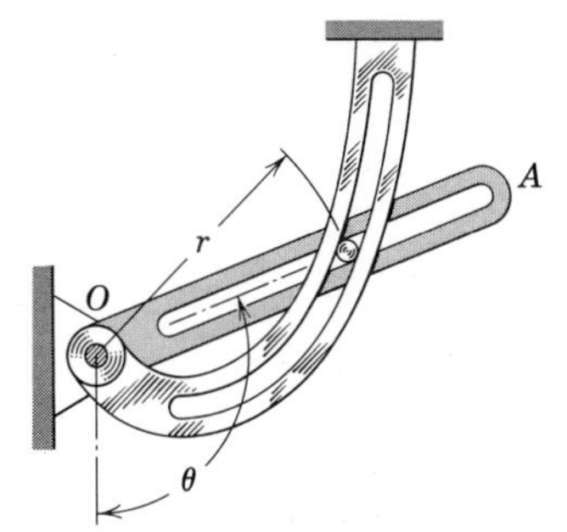

Problem 2/107

2/108 A small object is thrown down the slope as shown. Determine the magnitude u of the initial velocity. *Ans.* $u = 20.6$ m/s

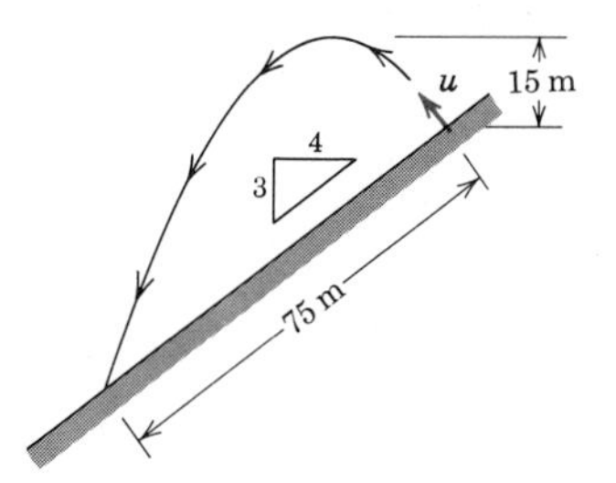

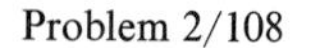
Problem 2/108

2/109 The circular disk rotates about its center O with a constant angular velocity $\omega = \dot{\theta}$ and carries the two spring-loaded plungers shown. The distance b which each plunger protrudes from the rim of the disk varies according to $b = b_0 \sin 2\pi nt$ where b_0 is the maximum protrusion, n is the constant frequency of oscillation of the plungers in the radial slots, and t is the time. Determine the maximum magnitudes of the r- and θ-components of the acceleration of the ends A of the plungers during their motion.

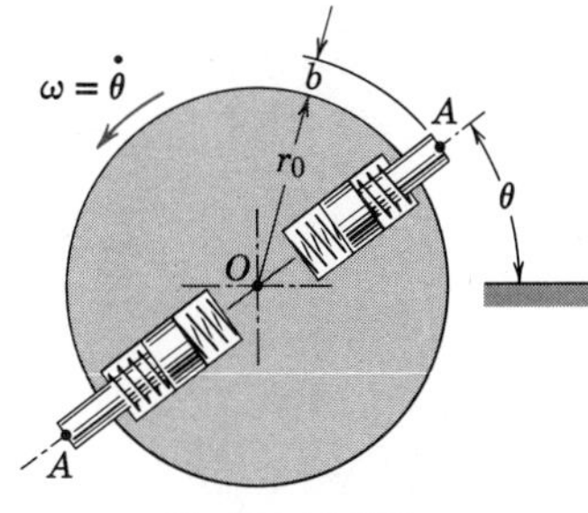

Problem 2/109

2/110 A particle moves in the x-y plane according to $x = a \cos \omega t$ and $y = b \sin \omega t$ where a, b, and ω are constants. Determine the equation of the path and find the maximum magnitude of the acceleration in each of the x- and y-directions. Show that the acceleration of the particle is always directed toward the origin of coordinates for this particular motion.

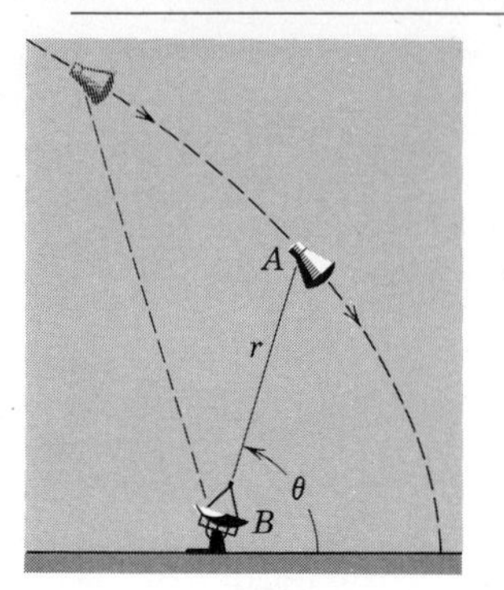

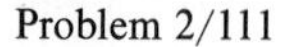
Problem 2/111

2/111 During re-entry a space capsule A is tracked by radar station B located in the vertical plane of the trajectory. The values of r and θ are read against time and are recorded in the following table. Determine the velocity v of the capsule when $t = 40$ s. Explain how the acceleration of the capsule could be found from the given data.

t, s	r, km	θ, °
0	36.4	110.5
5	29.9	100.0
10	26.2	91.0
15	24.1	83.7
20	22.7	77.7
30	20.9	67.7
40	20.1	58.6
50	19.3	52.0
60	19.0	45.0
70	18.8	38.5
80	18.7	32.8
90	18.7	27.0
100	18.7	21.6
110	18.8	16.8
120	19.0	12.0

Ans. $v = 1020$ km/h

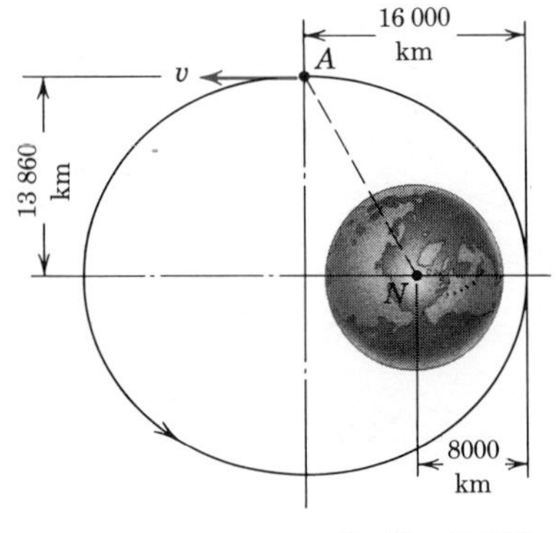

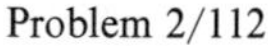
Problem 2/112

2/112 An earth satellite which moves in the elliptical equatorial orbit shown has a velocity v in space of 17 970 km/h when it passes the end of the semi-minor axis at A. The earth has an absolute surface value of g of 9.821 m/s^2, a radius of 6370 km, and has an angular velocity ω of $0.729(10^{-4})$ rad/s. Determine the radius of curvature ρ of the orbit at A. *Ans.* $\rho = 18\,480$ km

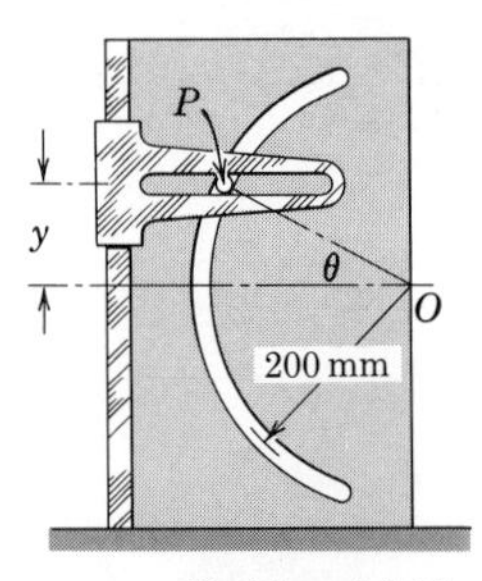

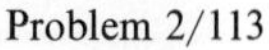
Problem 2/113

2/113 The guide with the horizontal slot is made to move up the vertical edge of the fixed plate at the constant rate $\dot{y} = 2$ m/s before reversing the direction of its motion at $y = 175$ mm. Pin P is constrained to move in both the horizontal and circular slots. Calculate the angular acceleration $\ddot{\theta}$ of line OP for the instant when $y = 100$ mm.

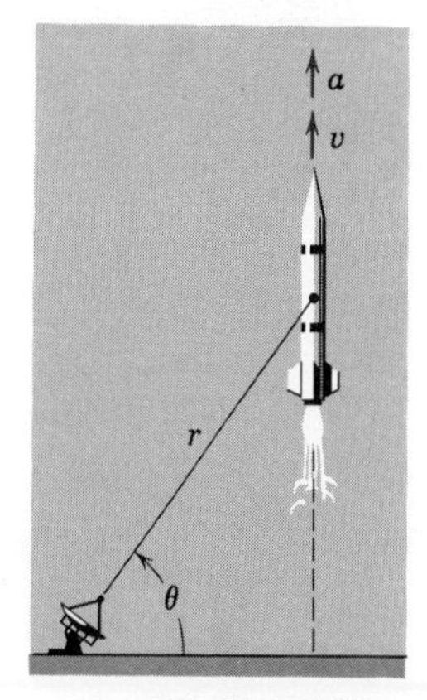

Problem 2/114

2/114 A rocket is fired vertically and tracked by the radar antenna shown. At the instant when $\theta = 60°$, measurements give $\dot{\theta} = 0.03$ rad/s and $r = 7500$ m, and the vertical acceleration of the rocket is found to be $a = 20$ m/s^2. For this instant determine the values of $\ddot{r}$ and $\ddot{\theta}$.

Ans. $\ddot{r} = 24.1$ m/s^2, $\ddot{\theta} = -1.784$ mrad/s^2

2/115 The slotted arm AB is pivoted at O and carries the slider C. The position of C in the slot is governed by the cord which is fastened at D and remains taut. The arm has a constant counterclockwise angular velocity $\dot{\theta} = 4$ rad/s during an interval of its rotation, and $r = 0$ when $\theta = 0$. Determine the acceleration a of the slider at the position for which $\theta = 30°$. The distance R is 400 mm. *Ans.* $a = 13.04$ m/s^2

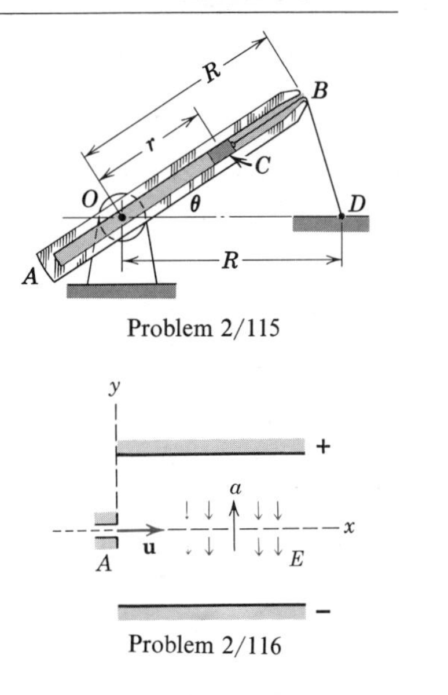

Problem 2/115

2/116 At time $t = 0$ an electron is emitted at A with a velocity u in the x-direction into an electric field $E = E_0 \sin pt$ at right angles to $\mathbf{u}$. The electron has an acceleration in the direction of E equal to eE/m, where e is the electron charge and m is its mass. Find the x- and y-coordinates of the electron at the end of the first complete cycle of E. The field frequency $f = p/2\pi$ is constant.

Problem 2/116

2/117 The third stage of a rocket is injected by its booster with a velocity u of 15 000 km/h at A into an unpowered coasting flight to B. At B its rocket motor is ignited when the trajectory makes an angle of 20° with the horizontal. Operation is effectively above the atmosphere, and the gravitational acceleration during this interval may be taken as 9 m/s^2, constant in magnitude and direction. Determine the time t to go from A to B. (This quantity is needed in the design of the ignition control system.) Also determine the corresponding increase h in altitude.

Ans. $t = 3$ min 28 s, $h = 418$ km

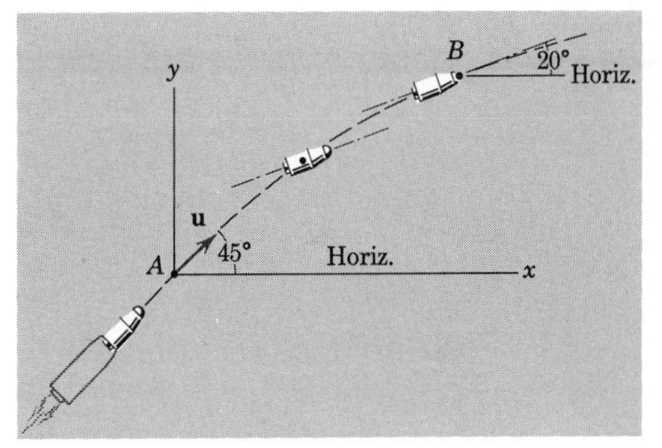

Problem 2/117

2/118 Determine the equation for the envelope a of the parabolic trajectories of a projectile fired at any angle but with the same muzzle velocity u. (*Hint:* Substitute $m = \tan\theta$, where θ is the firing angle with the horizontal, into the equation of the trajectory. The two roots m_1 and m_2 of the equation written as a quadratic in m give the two firing angles for the two trajectories shown such that the shells pass through the same point A. Point A will approach the envelope a as the two roots approach equality.) Neglect air resistance and assume g is constant.

Ans. $y = \dfrac{u^2}{2g} - \dfrac{gx^2}{2u^2}$

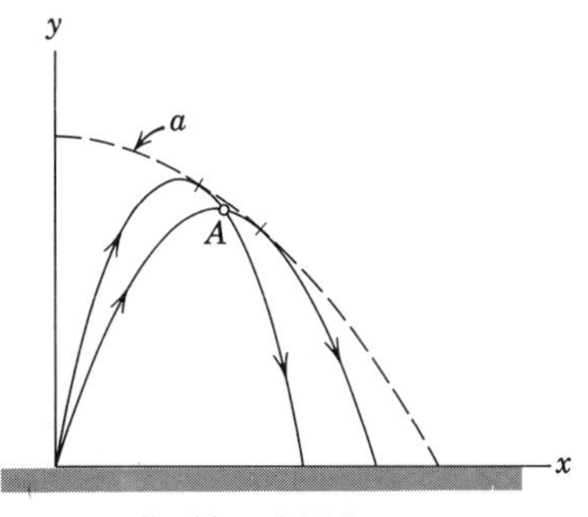

Problem 2/118

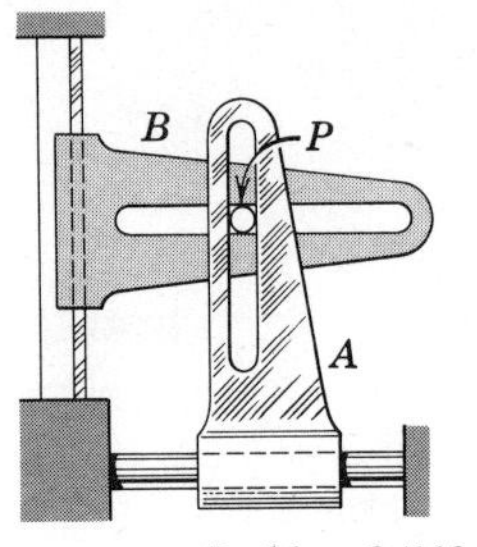

Problem 2/119

◀ **2/119** The pin P is constrained to move in the slotted guides which move at right angles to one another. At the instant represented, A has a velocity to the right of 0.2 m/s which is decreasing at the rate of 0.75 m/s each second. At the same time B is moving down with a velocity of 0.15 m/s which is decreasing at the rate of 0.5 m/s each second. For this instant determine the radius of curvature ρ of the path followed by P. Is it possible to determine also the time rate of change of ρ? *Ans.* $\rho = 1.25$ m

◀ **2/120** A centrifugal pump with radial vanes, similar to that illustrated with Prob. 2/94, rotates at the constant rate $\dot{\theta} = K$. An element of the fluid being pumped will be considered here as a smooth particle P which is introduced at the radius r_0 without radial velocity and which moves outward along the vane without friction. With no force on the particle in the radial direction, its acceleration in this direction is zero. Under these conditions, determine the equation of the path of the particle if the time t is zero for $r = r_0$. *Ans.* $r = r_0 \cosh Kt$

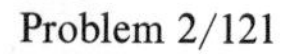
Problem 2/121

◀ **2/121** The slotted arm is pivoted at O and revolves counterclockwise with a constant angular velocity ω about the eccentrically mounted circular cam which is fixed and does not rotate. Determine the expression for the velocity v and acceleration a of the pin A for the position $\theta = \pi/2$. The pin has negligible diameter and always contacts the cam.

Ans. $v = b\omega, \quad a = \dfrac{b^2\omega^2}{\sqrt{b^2 - e^2}}$

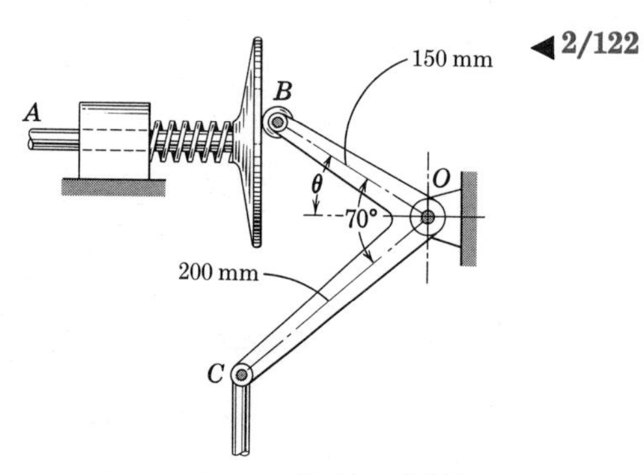

Problem 2/122

◀ **2/122** The horizontal plunger A, which operates the 70-deg bell crank BOC, has a velocity to the right of 75 mm/s and is speeding up at the rate of 100 mm/s per second at the position for which $\theta = 30°$. Compute the angular acceleration $\ddot{\theta}$ of the bell crank at this instant.

Ans. $\ddot{\theta} = -0.399$ rad/s^2

◀ **2/123** If the slotted arm (Prob. 2/86) is revolving counterclockwise at the constant rate of 40 rev/min and the cam is revolving clockwise at the constant rate of 30 rev/min, determine the acceleration a of the center of the roller A when the cam and arm are in the relative position for which $\theta = 30°$. The limacon has the dimensions $b = 100$ mm and $c = 75$ mm. (*Caution:* Redefine the coordinates as necessary after noting that the θ in the expression $r = b - c\cos\theta$ is not the absolute angle appearing in Eq. 18.)

Ans. $a = 3.68\ \text{m/s}^2$

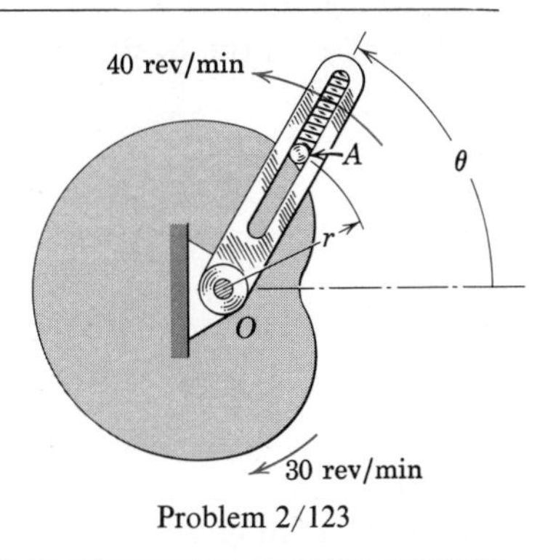

Problem 2/123

13 Relative Motion in a Plane.

13 Relative Motion in a Plane. There are many engineering problems for which the analysis of motion is simplified by using measurements made with respect to a moving coordinate system. These measurements, when combined with the observed motion of the moving coordinate system, permit the determination of the absolute motion in question. This approach is known as a *relative motion* analysis.

The motion of the moving coordinate system is specified with respect to a fixed coordinate system. Strictly speaking, this fixed system in Newtonian mechanics is the primary inertial system which is assumed to have no motion in space. From an engineering point of view the fixed system may be taken as any system whose absolute motion is negligible for the problem at hand. For most earth-bound engineering problems it is sufficiently precise to take for the fixed reference system a set of axes attached to the earth where the motion of the earth is neglected. For problems involving the motion of satellites around the earth, a nonrotating coordinate system with origin on the earth's axis of rotation is a convenient reference. For a description of interplanetary travel, a nonrotating coordinate system fixed to the sun would be appropriate. Hence the choice of the fixed system depends on the type of problem involved.

Attention in this article will now be confined to curvilinear motion which takes place in a given plane. This motion will be described by observations made with respect to a set of axes which moves in the plane. A translating set of reference axes will be described in the first part of the article, and a set of axes which rotates as well as translates will be described in the second part of the article.

Part A: Translating Reference Axes

Consider the plane curvilinear motion of two particles or points, Fig. 13. The motion of A will be observed from a translating frame of reference x-y with origin attached to B. The position vector of A relative to the frame of reference of B is $\mathbf{r}_{A/B} = \mathbf{i}x + \mathbf{j}y$ where $\mathbf{i}$ and $\mathbf{j}$ are unit vectors along the x- and y-axes. The position of B, in turn, is measured by its vector $\mathbf{r}_B$ located from the fixed reference axes X-Y. The absolute position, velocity, and acceleration of A in the fixed system X-Y are given by

$$\mathbf{r}_A = \mathbf{r}_B + \mathbf{r}_{A/B}$$

$$\blacktriangleright \quad \dot{\mathbf{r}}_A = \dot{\mathbf{r}}_B + \dot{\mathbf{r}}_{A/B} \quad \text{or} \quad \mathbf{v}_A = \mathbf{v}_B + \mathbf{v}_{A/B} \tag{19}$$

$$\blacktriangleright \quad \ddot{\mathbf{r}}_A = \ddot{\mathbf{r}}_B + \ddot{\mathbf{r}}_{A/B} \quad \text{or} \quad \mathbf{a}_A = \mathbf{a}_B + \mathbf{a}_{A/B} \tag{20}$$

In these relative motion equations the velocity of A measured relative to x-y is $\dot{\mathbf{r}}_{A/B} = \mathbf{v}_{A/B} = \mathbf{i}\dot{x} + \mathbf{j}\dot{y}$, and the acceleration of A measured relative to x-y is $\ddot{\mathbf{r}}_{A/B} = \dot{\mathbf{v}}_{A/B} = \mathbf{a}_{A/B} = \mathbf{i}\ddot{x} + \mathbf{j}\ddot{y}$. It should be noted in these differentiations that the unit vectors have no derivatives because their directions as well as their magnitudes remain unchanged.

In words, Eq. 19 (or 20) states that the absolute velocity (or acceleration) of A equals the absolute velocity (or acceleration) of B plus, vectorially, the velocity (or acceleration) of A relative to B. The relative term is the velocity (or acceleration) measurement which an observer attached to the moving coordinate system x-y would make. The relative motion terms may be expressed in whatever coordinate system is convenient—rectangular, normal and tangential, or polar—and the formulations in the preceding article may be used for this purpose. The appropriate fixed system of the previous article becomes the moving system in the present article.

The selection of the moving point B for attachment of the reference coordinate system is arbitrary. Point A could be used just as well for the attachment of the moving system, in which case the three corresponding relative motion equations for position, velocity, and acceleration are

$$\mathbf{r}_B = \mathbf{r}_A + \mathbf{r}_{B/A} \qquad \mathbf{v}_B = \mathbf{v}_A + \mathbf{v}_{B/A} \qquad \mathbf{a}_B = \mathbf{a}_A + \mathbf{a}_{B/A}$$

It is seen, therefore, that $\mathbf{r}_{B/A} = -\mathbf{r}_{A/B}$, $\mathbf{v}_{B/A} = -\mathbf{v}_{A/B}$, and $\mathbf{a}_{B/A} = -\mathbf{a}_{A/B}$.

The relative velocity equations for the motions of three particles A, B, and C may be written

$$\mathbf{v}_A = \mathbf{v}_B + \mathbf{v}_{A/B} \qquad \mathbf{v}_A = \mathbf{v}_C + \mathbf{v}_{A/C} \qquad \mathbf{v}_C = \mathbf{v}_B + \mathbf{v}_{C/B}$$

Eliminating $\mathbf{v}_C$ from the last two equations and combining with the first equation to eliminate $\mathbf{v}_A$ and $\mathbf{v}_B$ give

$$\mathbf{v}_{A/B} = \mathbf{v}_{A/C} + \mathbf{v}_{C/B}$$

Similarly, the equivalent relative acceleration equation is

$$\mathbf{a}_{A/B} = \mathbf{a}_{A/C} + \mathbf{a}_{C/B}$$

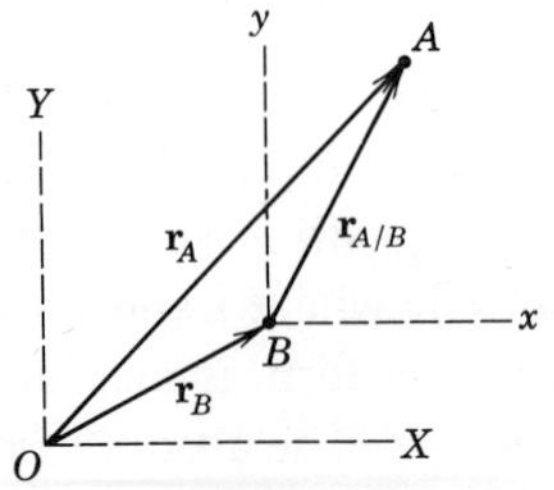

Figure 13

These relationships among the relative motion terms are intuitive and may often be used to advantage when dealing with the related motions of three particles.

An important observation to be made in relative motion analysis is that the acceleration of a particle as observed in a translating system x-y is the same as that observed in a fixed system X-Y if the moving system has a constant velocity. This conclusion broadens the application of Newton's second law of motion, treated in Chapter 3.

Sample Problem

2/124 Two ships A and B are at the positions shown at the same time. Ship A is moving at the constant speed of 16 knots (1 knot = 1.852 km/h) in a circular arc of 3-km radius. The speed of ship B is 8 knots in the direction shown, but its captain is reducing speed at the rate of 2 knots per minute to avoid risk of collision with A.

(a) Determine the velocity which A appears to have from an observation position attached to B, and from this result determine the values of $\dot{r}$ and $\dot{\theta}$ as seen from B.

(b) Determine the acceleration which A appears to have with respect to B, and from this result and the results found in part (a) calculate the values of $\ddot{r}$ and $\ddot{\theta}$.

Solution. The coordinates r and θ are measured in the reference system x-y moving with B. Thus the motion of A relative to B will be described here with a moving polar coordinate system.

(a) From the relative-velocity equation $\mathbf{v}_A = \mathbf{v}_B + \mathbf{v}_{A/B}$ the relative velocity and its direction are seen from the b-part of the figure to be

$$v_{A/B} = \sqrt{8^2 + 16^2} = 17.89 \text{ knots and } \beta = \tan^{-1}\tfrac{8}{16} = 26.57° \qquad \textit{Ans.}$$

The path which A appears to follow with respect to B is along the direction of $\mathbf{v}_{A/B}$. In addition $\alpha = 45° - (26.57°) = 18.43°$. Thus from the r- and θ-components of $\mathbf{v}_{A/B}$ in the b-part of the figure there results*

$$[v_r = \dot{r}] \qquad -17.89(\cos 18.43°)\frac{1.852}{3.6} = \dot{r}, \qquad \dot{r} = -8.73 \text{ m/s}$$

$$[v_\theta = r\dot{\theta}] \qquad -17.89(\sin 18.43°)\frac{1.852}{3.6} = 3000\sqrt{2}\dot{\theta}, \quad \dot{\theta} = -0.686 \text{ mrad/s} \qquad \textit{Ans.}$$

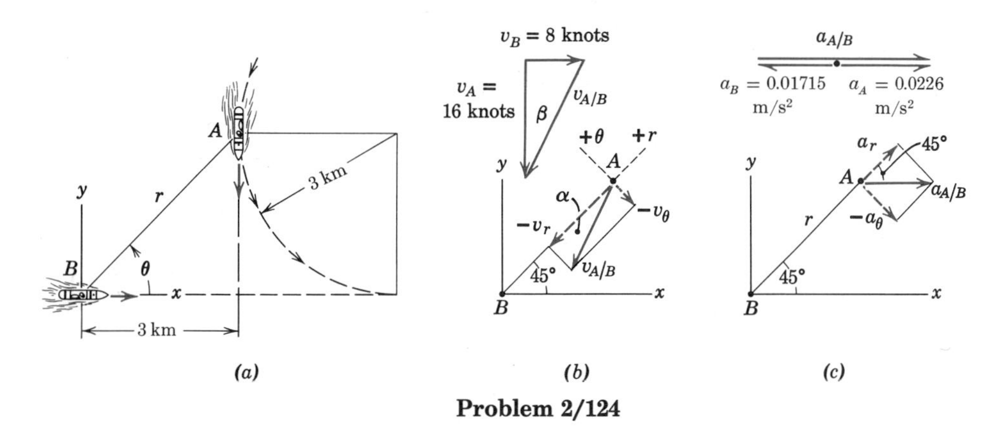

Problem 2/124

*Recall that kilometres per hour divided by 3.6 gives metres per second.

(*b*) The absolute accelerations of A and B are

$$\left[a_n = \frac{v^2}{\rho}\right] \qquad a_A = \frac{[16(1.852/3.6)]^2}{3000} = 0.0226 \text{ m/s}^2$$

$$[a = \dot{v}] \qquad a_B = \frac{2(1.852/3.6)}{60} = 0.01715 \text{ m/s}^2$$

From the relative acceleration equation $\mathbf{a}_A = \mathbf{a}_B + \mathbf{a}_{A/B}$ the relative acceleration term is seen from the *c*-part of the figure to be a vector to the right with a magnitude

$$a_{A/B} = 0.0397 \text{ m/s}^2 \qquad \textit{Ans.}$$

The polar-coordinate components of $\mathbf{a}_{A/B}$ are seen in the *c*-part of the figure and are $a_r = 0.0397 \cos 45° = 0.0281 \text{ m/s}^2$ and $-a_\theta = 0.0397 \sin 45° = 0.0281 \text{ m/s}^2$. The polar-coordinate acceleration components for motion relative to the x-y system now give

$$[a_r = \ddot{r} - r\dot{\theta}^2] \qquad 0.0281 = \ddot{r} - 3000\sqrt{2}(-0.000686)^2$$

$$\ddot{r} = 0.0301 \text{ m/s}^2 \qquad \textit{Ans.}$$

$$[a_\theta = r\ddot{\theta} + 2\dot{r}\dot{\theta}] \qquad -0.0281 = 3000\sqrt{2}\ddot{\theta} + 2(-8.73)(-0.000686)$$

$$\ddot{\theta} = -9.44\ \mu\text{rad/s}^2 \qquad \textit{Ans.}$$

Note that the velocity and acceleration components are measured positive in the direction of increasing r and θ according to the derivations of Eqs. 17 and 18.

Part B: Rotating Reference Axes

The use of a rotating reference system greatly facilitates the solution of many problems in kinematics where motion is generated within or observed from a system which itself is rotating. An example of such a motion would be the path of a fluid particle moving along the curved vane of a centrifugal pump where the path relative to the vanes of the impeller becomes an important design consideration.

The curvilinear plane motion of two particles A and B in the fixed X-Y plane, Fig. 14*a*, will be considered. The motion of A will be observed from a moving reference frame x-y with origin attached to B and which rotates with an angular velocity $\omega = \dot{\theta}$. This angular velocity may be written as the vector $\boldsymbol{\omega} = \mathbf{k}\omega = \mathbf{k}\dot{\theta}$ where the vector is normal to the plane of motion and where

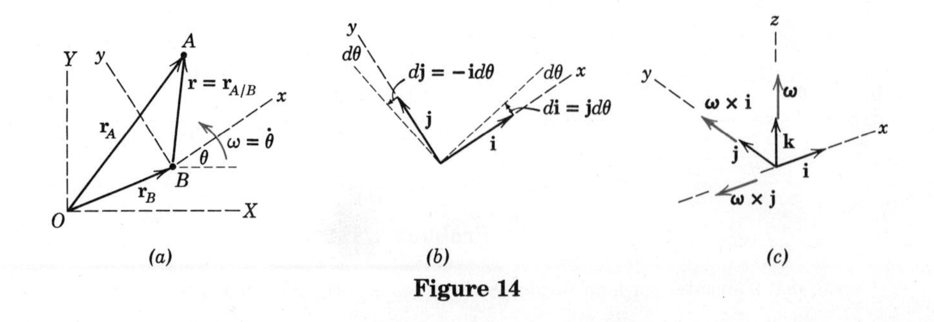

Figure 14

its positive sense is in the positive Z-direction (out from the paper) as established by the right-hand rule. The absolute position of A is given by

$$\mathbf{r}_A = \mathbf{r}_B + \mathbf{r} = \mathbf{r}_B + (\mathbf{i}x + \mathbf{j}y)$$

where it is understood that $\mathbf{r}$ stands for $\mathbf{r}_{A/B}$, the position vector of A with respect to B. The velocity and acceleration equations require differentiation of the position-vector equation. In contrast to the case of translating axes treated in Part A, the unit vectors $\mathbf{i}$ and $\mathbf{j}$ are now rotating and, hence, have time derivatives which must be evaluated. These derivatives may be seen from Fig. 14*b* which shows the infinitesimal change in each unit vector during time dt when the reference axes rotate through an angle $d\theta = \omega\, dt$. The differential change in $\mathbf{i}$ is $d\mathbf{i}$, and it has the direction of $\mathbf{j}$ and a magnitude equal to the angle $d\theta$ times the length of the vector $\mathbf{i}$, which is unity. Thus $d\mathbf{i} = \mathbf{j}\, d\theta$. Similarly the unit vector $\mathbf{j}$ has an infinitesimal change $d\mathbf{j}$ which points in the negative x-direction, so that $d\mathbf{j} = -\mathbf{i}\, d\theta$. Dividing by dt and replacing $d\mathbf{i}/dt$ by $\dot{\mathbf{i}}$, $d\mathbf{j}/dt$ by $\dot{\mathbf{j}}$, and $d\theta/dt$ by $\dot{\theta} = \omega$ give

$$\dot{\mathbf{i}} = \mathbf{j}\omega \qquad \text{and} \qquad \dot{\mathbf{j}} = -\mathbf{i}\omega$$

When the cross product is introduced, it is noted from Fig. 14*c* that $\boldsymbol{\omega} \times \mathbf{i} = \mathbf{j}\omega$ and $\boldsymbol{\omega} \times \mathbf{j} = -\mathbf{i}\omega$. Therefore the time derivatives of the unit vectors may be written as

$$\blacktriangleright \qquad \dot{\mathbf{i}} = \boldsymbol{\omega} \times \mathbf{i} \qquad \text{and} \qquad \dot{\mathbf{j}} = \boldsymbol{\omega} \times \mathbf{j} \qquad \textbf{(21)}$$

With these expressions for the time derivatives of the unit vectors, the position-vector equation for A and B may now be differentiated to obtain the relative velocity equation, which becomes

$$\dot{\mathbf{r}}_A = \dot{\mathbf{r}}_B + (\dot{\mathbf{i}}x + \dot{\mathbf{j}}y) + (\mathbf{i}\dot{x} + \mathbf{j}\dot{y})$$

But $\dot{\mathbf{i}}x + \dot{\mathbf{j}}y = \boldsymbol{\omega} \times \mathbf{i}x + \boldsymbol{\omega} \times \mathbf{j}y = \boldsymbol{\omega} \times \mathbf{r}$, and $\mathbf{i}\dot{x} + \mathbf{j}\dot{y} = \mathbf{v}_{\text{rel}}$ which is the velocity which would be measured by an observer attached to the x-y frame of reference. Thus the relative velocity equation becomes

$$\blacktriangleright \qquad \mathbf{v}_A = \mathbf{v}_B + \boldsymbol{\omega} \times \mathbf{r} + \mathbf{v}_{\text{rel}} \qquad \textbf{(22)}$$

Comparison of Eq. 22 with Eq. 19 for nonrotating reference axes shows that $\mathbf{v}_{A/B} = \boldsymbol{\omega} \times \mathbf{r} + \mathbf{v}_{\text{rel}}$ from which it is concluded that the term $\boldsymbol{\omega} \times \mathbf{r}$ is the difference between the relative velocity as measured from nonrotating and rotating axes.

To illustrate further the meaning of the last two terms in Eq. 22, the motion of particle A relative to the x-y plane is shown in Fig. 15 as taking place in a curved slot in a plate which represents the rotating x-y reference system. The velocity of A as measured relative to the plate, $\mathbf{v}_{\text{rel}}$, would be tangent to the path fixed in the x-y plate and would have a magnitude $\dot{s}$. This relative velocity may also be viewed as the velocity $\mathbf{v}_{A/P}$ relative to a point P attached to the plate and coincident with A at the instant under consideration. The term $\boldsymbol{\omega} \times \mathbf{r}$ has a magnitude $r\dot{\theta}$ and a direction normal to **r** and is the velocity relative to B of point P as seen from nonrotating axes attached to B.

The following comparison will help to establish the equivalence and clarify the difference between the relative velocity equations written for rotating and nonrotating reference axes.

$$\begin{aligned}
\mathbf{v}_A &= \mathbf{v}_B + \boldsymbol{\omega} \times \mathbf{r} + \mathbf{v}_{\text{rel}} \\
\mathbf{v}_A &= \underbrace{\mathbf{v}_B + \mathbf{v}_{P/B}} + \mathbf{v}_{A/P} \\
\mathbf{v}_A &= \mathbf{v}_P \qquad \underbrace{\qquad + \mathbf{v}_{A/P}} \\
\mathbf{v}_A &= \mathbf{v}_B + \qquad \mathbf{v}_{A/B}
\end{aligned} \tag{22a}$$

In the second equation the term $\mathbf{v}_{P/B}$ is measured from a nonrotating position, otherwise it would be zero. The term $\mathbf{v}_{A/P}$ is the same as $\mathbf{v}_{\text{rel}}$ and is the velocity of A as measured in the x-y frame. In the third equation $\mathbf{v}_P$ is the absolute velocity of P and represents the effect of the moving coordinate system, both translational and rotational. The fourth equation is the same as that developed for nonrotating axes, Eq. 19, and it is seen that $\mathbf{v}_{A/B} = \mathbf{v}_{P/B} + \mathbf{v}_{A/P} = \boldsymbol{\omega} \times \mathbf{r} + \mathbf{v}_{\text{rel}}$.

The relative acceleration equation may be obtained by differentiating the relative velocity relation, Eq. 22. Thus

$$\mathbf{a}_A = \mathbf{a}_B + \dot{\boldsymbol{\omega}} \times \mathbf{r} + \boldsymbol{\omega} \times \dot{\mathbf{r}} + \dot{\mathbf{v}}_{\text{rel}}$$

From the derivation of Eq. 22 it is seen that the third term on the right of the acceleration equation becomes

$$\boldsymbol{\omega} \times \dot{\mathbf{r}} = \boldsymbol{\omega} \times \frac{d}{dt}(\mathbf{i}x + \mathbf{j}y) = \boldsymbol{\omega} \times (\boldsymbol{\omega} \times \mathbf{r}) + \boldsymbol{\omega} \times \mathbf{v}_{\text{rel}}$$

The last term on the right of the equation for $\mathbf{a}_A$ is

$$\begin{aligned}
\dot{\mathbf{v}}_{\text{rel}} &= \frac{d}{dt}(\mathbf{i}\dot{x} + \mathbf{j}\dot{y}) = (\dot{\mathbf{i}}\dot{x} + \dot{\mathbf{j}}\dot{y}) + (\mathbf{i}\ddot{x} + \mathbf{j}\ddot{y}) \\
&= \boldsymbol{\omega} \times (\mathbf{i}\dot{x} + \mathbf{j}\dot{y}) + (\mathbf{i}\ddot{x} + \mathbf{j}\ddot{y}) \\
&= \boldsymbol{\omega} \times \mathbf{v}_{\text{rel}} + \mathbf{a}_{\text{rel}}
\end{aligned}$$

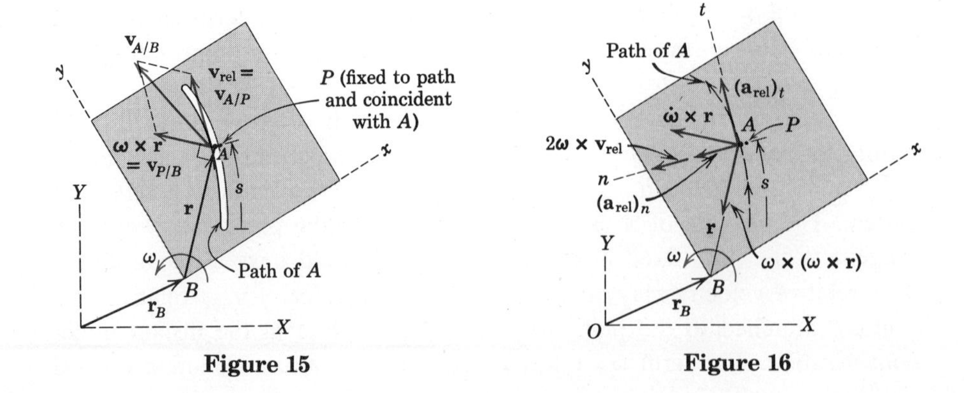

Figure 15 Figure 16

Substitution into the expression for $\mathbf{a}_A$ and collection of terms give

$$\mathbf{a}_A = \mathbf{a}_B + \dot{\boldsymbol{\omega}} \times \mathbf{r} + \boldsymbol{\omega} \times (\boldsymbol{\omega} \times \mathbf{r}) + 2\boldsymbol{\omega} \times \mathbf{v}_{\text{rel}} + \mathbf{a}_{\text{rel}} \tag{23}$$

Equation 23 is the general vector expression for the absolute acceleration of a particle A in terms of its acceleration $\mathbf{a}_{\text{rel}}$ measured relative to a moving coordinate system which rotates with an angular velocity $\boldsymbol{\omega}$. The terms $\dot{\boldsymbol{\omega}} \times \mathbf{r}$ and $\boldsymbol{\omega} \times (\boldsymbol{\omega} \times \mathbf{r})$ are shown in Fig. 16. They represent, respectively, the t- and n-components of the acceleration $\mathbf{a}_{P/B}$ of the coincident point P in its circular motion with respect to B. This motion would be observed from a set of nonrotating axes moving with B. The magnitude of $\dot{\boldsymbol{\omega}} \times \mathbf{r}$ is $r\ddot{\theta}$ and its direction is tangent to the circle. The magnitude of $\boldsymbol{\omega} \times (\boldsymbol{\omega} \times \mathbf{r})$ is $r\omega^2$ and its direction is from P to B along the normal to the circle.

The term $2\boldsymbol{\omega} \times \mathbf{v}_{\text{rel}}$, shown in Fig. 16, is known as the *Coriolis* acceleration* and represents the difference between the acceleration of A relative to P as measured from nonrotating axes and from rotating axes. The direction is always normal to the term $\mathbf{v}_{\text{rel}}$, and the sense is established by the right-hand rule for the cross product. The acceleration of A relative to the path, $\mathbf{a}_{\text{rel}}$, may be expressed in rectangular, normal and tangential, or polar coordinates in the rotating system. Generally n- and t-components are most frequently used, and these components are depicted in Fig. 16. The tangential component would have the magnitude $(a_{\text{rel}})_t = \ddot{s}$ where s is the distance measured along the path to A. The normal component would have the magnitude $v_{\text{rel}}{}^2/\rho$ where ρ is the radius of curvature of the path as measured in x-y. The sense of the vector would always be toward the center of curvature.

The following comparison will help to establish the equivalence and clarify the difference between the relative acceleration equations written for rotating and nonrotating reference axes.

$$\begin{aligned}
\mathbf{a}_A &= \mathbf{a}_B + \underbrace{\dot{\boldsymbol{\omega}} \times \mathbf{r} + \boldsymbol{\omega} \times (\boldsymbol{\omega} \times r)} + \underbrace{2\boldsymbol{\omega} \times \mathbf{v}_{\text{rel}} + \mathbf{a}_{\text{rel}}} \\
\mathbf{a}_A &= \underbrace{\mathbf{a}_B + \qquad \mathbf{a}_{P/B}} \qquad + \qquad \mathbf{a}_{A/P} \\
\mathbf{a}_A &= \qquad \mathbf{a}_P \qquad \underbrace{\qquad + \qquad \mathbf{a}_{A/P}} \\
\mathbf{a}_A &= \mathbf{a}_B + \qquad \mathbf{a}_{A/B}
\end{aligned} \tag{23a}$$

The equivalence of $\mathbf{a}_{P/B}$ and $\dot{\boldsymbol{\omega}} \times \mathbf{r} + \boldsymbol{\omega} \times (\boldsymbol{\omega} \times \mathbf{r})$, as shown in the second equation, has already been described. From the third equation where $\mathbf{a}_B + \mathbf{a}_{P/B}$ has been combined to give $\mathbf{a}_P$, it is seen that the relative acceleration term $\mathbf{a}_{A/P}$, unlike the corresponding relative velocity term, is not equal to the relative acceleration $\mathbf{a}_{\text{rel}}$ measured from the rotating x-y frame of reference. The Coriolis term is, therefore, the difference between the acceleration $\mathbf{a}_{A/P}$ of A relative to P as measured in a nonrotating system and the acceleration $\mathbf{a}_{\text{rel}}$ of A relative to P as measured in a rotating system. From the fourth equation it is seen that the acceleration $\mathbf{a}_{A/B}$ of A with respect to

*Named after a French military engineer G. Coriolis (1792–1843) who was the first to call attention to this term.

B as measured in a nonrotating system, Eq. 20, is a combination of the last four terms in the first equation for the rotating system.

In the analysis of acceleration using a rotating frame of reference it is frequently convenient to take the origin of the reference coordinates at the point P, coincident with the position of the particle A at the instant under observation. This choice eliminates the terms $\dot{\boldsymbol{\omega}} \times \mathbf{r}$ and $\boldsymbol{\omega} \times (\boldsymbol{\omega} \times \mathbf{r})$ since the vector $\mathbf{r}$ vanishes. These terms become submerged in the calculation of the acceleration of P. Hence, from the foregoing discussion and comparisons, it is seen that the relative acceleration equation may be written simply as

$$\mathbf{a}_A = \mathbf{a}_P + 2\boldsymbol{\omega} \times \mathbf{v}_{\text{rel}} + \mathbf{a}_{\text{rel}} \tag{23b}$$

When this form is used, it is noted that the point P may not be picked at random since it is the one point in the rotating reference system coincident with the particle A at the instant of analysis.

It is important to observe that the vector notation employed depends on the consistent use of a *right-handed* set of coordinate axes. Before the student endeavors to use Eqs. 22 or 23 it is important that he study the derivations carefully and understand the physical interpretation of each of the terms prior to attempting the problem work. It is appropriate to mention that Eqs. 22 and 23, developed here for plane motion, also hold for space motion, and this extension of generality will be covered in Art. 15 on relative motion in space.

Sample Problems

2/125 Particle A moves in the circular groove of 80-mm radius at the same time that the grooved plate rotates about its corner O at the rate $\omega = \dot{\theta}$. Determine the absolute velocity of A in the position for which $\theta = 45°$ and $\beta = 45°$ if at this instant $\dot{\theta} = 3$ rad/s and $\dot{\beta} = 5$ rad/s.

Solution. Axes x-y, attached to the plate with origin at B, constitute the rotating reference system. The relative velocity relation, Eq. 22, is $\mathbf{v}_A = \mathbf{v}_B + \boldsymbol{\omega} \times \mathbf{r} + \mathbf{v}_{\text{rel}}$. The terms, shown in the b-part of the figure, are evaluated as follows.

Point B moves in a circular arc around O, so its velocity has the magnitude

$$|\mathbf{v}_B| = r_B\omega = 0.10\sqrt{2}(3) = 0.424 \text{ m/s}$$

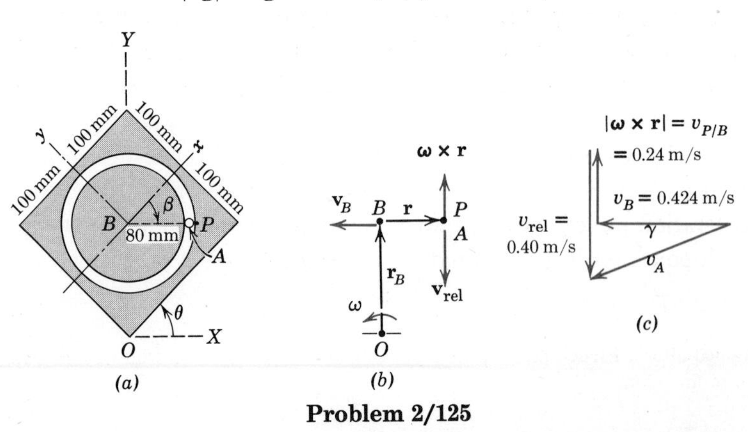

Problem 2/125

The term $\omega \times \mathbf{r}$ is the velocity relative to B of point P on the plate and coincident with A. The line PB rotates with the angular velocity $\omega = \dot{\theta}$ so that

$$|\omega \times \mathbf{r}| = |\mathbf{v}_{P/B}| = r\omega = 0.08(3) = 0.24 \text{ m/s}$$

and its direction is up as shown. The relative velocity $\mathbf{v}_{\text{rel}}$ of A with respect to the plate depends on $\dot{\beta}$ and has the magnitude

$$|\mathbf{v}_{\text{rel}}| = r\dot{\beta} = 0.08(5) = 0.40 \text{ m/s}$$

and its direction is clearly down as shown in the figure. Adding the three vectors, shown in the *c*-part of the figure, gives

$$v_A = \sqrt{(0.424)^2 + (0.40 - 0.24)^2} = 0.453 \text{ m/s} \qquad \textit{Ans.}$$

$$\gamma = \tan^{-1}\frac{0.40 - 0.24}{0.424} = 20.7°$$

The result could be expressed in vector notation if desired but is probably more useful in the form given.

2/126 For the conditions of Prob. 2/125 determine the absolute acceleration of particle A if, in addition to the given data, $\ddot{\theta} = 7$ rad/s² and $\ddot{\beta} = 12$ rad/s² for the specified position.

Solution. The relative acceleration relation, Eq. 23, is $\mathbf{a}_A = \mathbf{a}_B + \dot{\omega} \times \mathbf{r} + \omega \times (\omega \times \mathbf{r}) + 2\omega \times \mathbf{v}_{\text{rel}} + \mathbf{a}_{\text{rel}}$ and will be used to find the acceleration of the particle A. For the simplified geometry of this problem it is unnecessary to express each of the vectors in terms of its **i**- and **j**-components and to carry out the corresponding vector operations. Instead the vectors are shown in the *a*-part of the figure with this problem and their magnitudes are found as follows.

The term $\mathbf{a}_B$ is the acceleration which B has in its circular motion about O. Its components are

$$(a_B)_n = r_B\dot{\theta}^2 = 0.10\sqrt{2}(3^2) = 1.273 \text{ m/s}^2$$

$$(a_B)_t = r_B\ddot{\theta} = 0.10\sqrt{2}(7) = 0.990 \text{ m/s}^2$$

The acceleration with respect to B of point P, which is attached to the plate and is momentarily coincident with A, is due to the rotation of the plate (x-y axes) and has the components

$$|\dot{\omega} \times \mathbf{r}| = (a_{P/B})_t = r\ddot{\theta} = 0.08(7) = 0.56 \text{ m/s}^2$$

$$|\omega \times (\omega \times \mathbf{r})| = (a_{P/B})_n = r\dot{\theta}^2 = 0.08(3^2) = 0.72 \text{ m/s}^2$$

The Coriolis acceleration is in the positive X-direction as determined by the cross product $\omega \times \mathbf{v}_{\text{rel}}$ and has the magnitude

$$|2\omega \times \mathbf{v}_{\text{rel}}| = 2(3)(0.40) = 2.40 \text{ m/s}^2$$

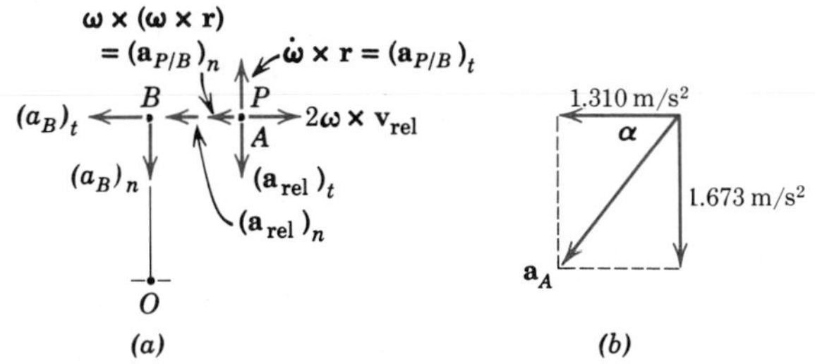

Problem 2/126

The acceleration $\mathbf{a}_{\text{rel}}$ of A relative to the rotating plate is due to $\dot{\beta}$ and $\ddot{\beta}$ and has the components

$$(a_{\text{rel}})_n = r\dot{\beta}^2 = 0.08(5^2) = 2.00 \text{ m/s}^2$$

$$(a_{\text{rel}})_t = r\ddot{\beta} = 0.08(12) = 0.96 \text{ m/s}^2$$

Collecting the X- and Y-components of the terms in the figure gives

$$(a_A)_X = -0.990 - 0.72 - 2.00 + 2.40 = -1.310 \text{ m/s}^2$$

$$(a_A)_Y = -1.273 + 0.56 - 0.96 = -1.673 \text{ m/s}^2$$

The absolute acceleration of A and its direction α now become

$$a_A = \sqrt{(1.310)^2 + (1.673)^2} = 2.12 \text{ m/s}^2$$

$$\alpha = \tan^{-1}\frac{1.673}{1.310} = \tan^{-1} 1.2770 = 51.9°$$

Ans.

Again, as with Prob. 2/125, the result could be written in vector notation but is probably more useful in the form presented. When the vectors are not as simply oriented as in the case of this problem, addition by **i**- and **j**-components or by graphical combination would be called for.

Problems

Translating Reference Axes

2/127 Aircraft A is flying west with a velocity $v_A = 900$ km/h, while aircraft B is flying north with a velocity $v_B = 600$ km/h at approximately the same altitude. Determine the magnitude and direction of the velocity which A appears to have to a passenger riding in B.

Ans. $v_{A/B} = 1082$ km/h directed 33.7° south of west

Problem 2/127

2/128 In Prob. 2/127 if aircraft B is accelerating in its northward direction at the rate of 4.5 km/h each second while aircraft A is slowing down at the rate of 3 km/h each second in its westward direction of flight, determine the acceleration in m/s^2 which B appears to have to an observer in A.

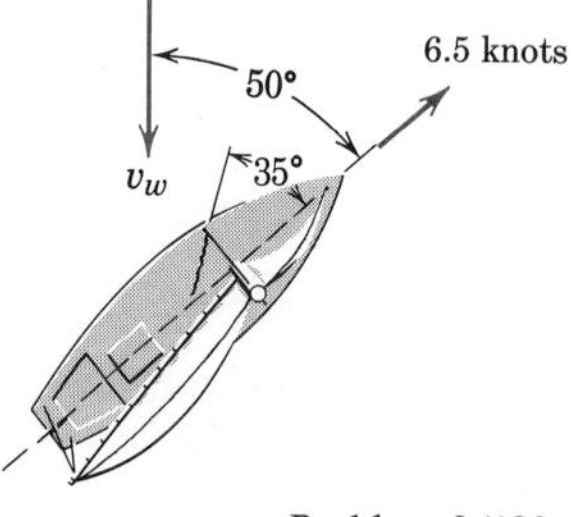

Problem 2/129

2/129 A sailboat moving in the direction shown is tacking to windward against a north wind. The log registers a hull speed of 6.5 knots. A "tell-tale" (light string tied to the rigging) indicates that the direction of the apparent wind is 35° from the center line of the boat. What is the true wind velocity v_w? *Ans.* $v_w = 14.4$ knots

2/130 An earth satellite is put into a circular polar orbit at an altitude of 240 km, which requires an orbital velocity of 27 940 km/h with respect to the center of the earth considered fixed in space. In going from south to north, when the satellite passes over an observer on the equator, in what direction does the satellite appear to be moving? The equatorial radius of the earth is 6378 km, and the angular velocity of the earth is $0.729(10^{-4})$ rad/s.

2/131 Three ships, A, B, and C, are cruising on straight courses in the vicinity of one another. With the x-direction as east and the y-direction as north the relative velocities in knots of A with respect to B and of C with respect to B are given by $\mathbf{v}_{A/B} = 6.8\mathbf{i} - 3.6\mathbf{j}$ and $\mathbf{v}_{C/B} = -8.2\mathbf{i} - 10.6\mathbf{j}$. Find both the magnitude and the direction of the velocity that A appears to have to an observer on C. Express the direction in terms of the clockwise angle β measured from the north.

Ans. $v_{A/C} = 16.55$ knots, $\beta = 65.0°$

2/132 The destroyer moves at 30 knots (1 knot = 1.852 km/h) and fires a rocket at an angle which trails the line of sight to the fixed target by the angle α. The launching velocity is 75 m/s relative to the ship and has an angle of elevation of 30° above the horizontal. If the missile continues to move in the same vertical plane as that determined by its absolute velocity at launching, determine α for $\theta = 60°$. *Ans.* $\alpha = 11.88°$

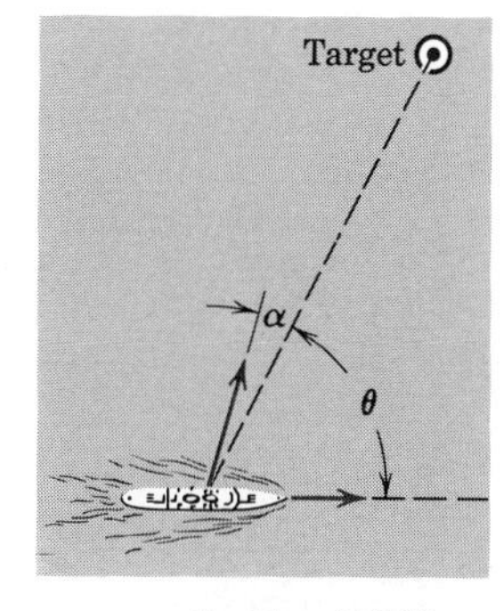

Problem 2/132

2/133 Airplane A is flying north with a constant horizontal velocity of 500 km/h. Airplane B is flying southwest at the same altitude with a velocity of 500 km/h. From the frame of reference of A determine the magnitude v_r of the apparent or relative velocity of B. Also find the magnitude of the apparent velocity v_n with which B appears to be moving sideways or normal to its center line. Would the results be different if the two airplanes were flying at different but constant altitudes?

Ans. $v_r = 924$ km/h, $v_n = 354$ km/h

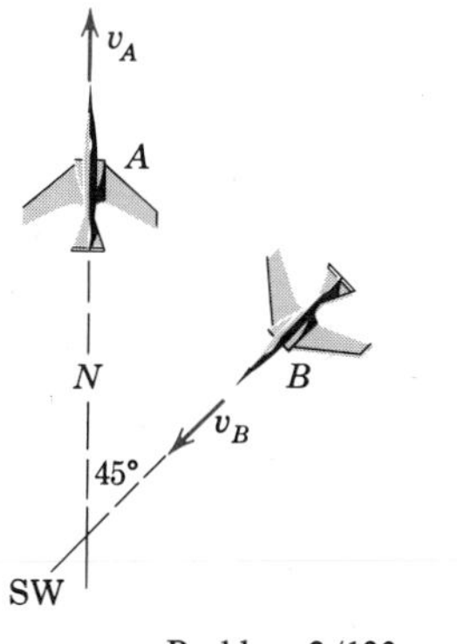

Problem 2/133

2/134 For the airplanes shown with Prob. 2/133, if A has a forward acceleration of 45 km/h per second and if B has a forward acceleration of 30 km/h per second, determine the expression in vector notation for the acceleration which A appears to have from a reference frame attached to B. Each airplane is in horizontal flight. Take the x-direction north and the y-direction west.

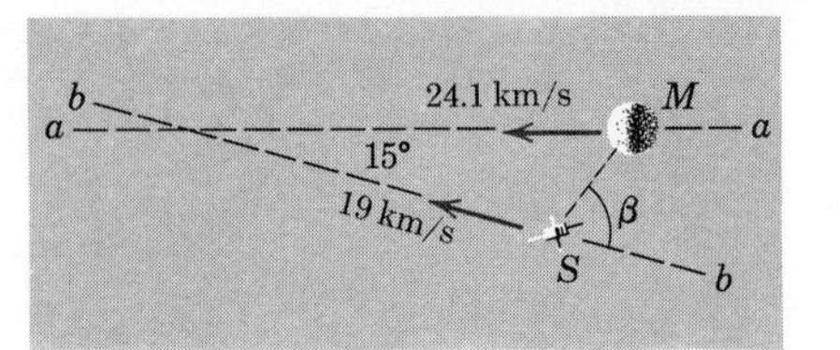

Problem 2/135

2/135 The spacecraft S approaches the planet Mars along a trajectory b-b in the orbital plane of Mars with an absolute velocity of 19 km/s. Mars has a velocity of 24.1 km/s along its trajectory a-a. Determine the angle β between the line of sight S-M and the trajectory b-b when Mars appears from the spacecraft to be approaching it head on. *Ans.* $\beta = 55.6°$

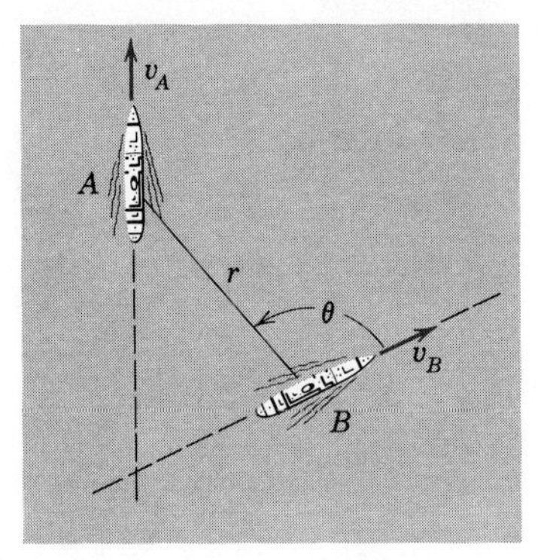

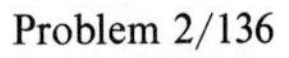
Problem 2/136

2/136 Two ships A and B are moving with constant speeds v_A and v_B, respectively, along straight intersecting courses. The navigator of ship B notes the time rates of change of the separation distance r between the ships and the bearing angle θ. Show that $\ddot{\theta} = -2\dot{r}\dot{\theta}/r$ and $\ddot{r} = r\dot{\theta}^2$.

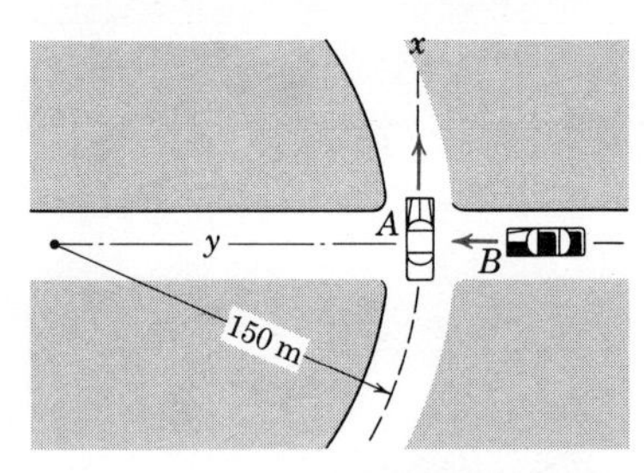

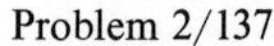
Problem 2/137

2/137 Car A rounds a curve of 150-m radius at a constant speed of 54 km/h. At the instant represented, car B is moving at 81 km/h but is slowing down at the rate of 3 m/s². Determine the velocity and acceleration of car A as observed from car B.

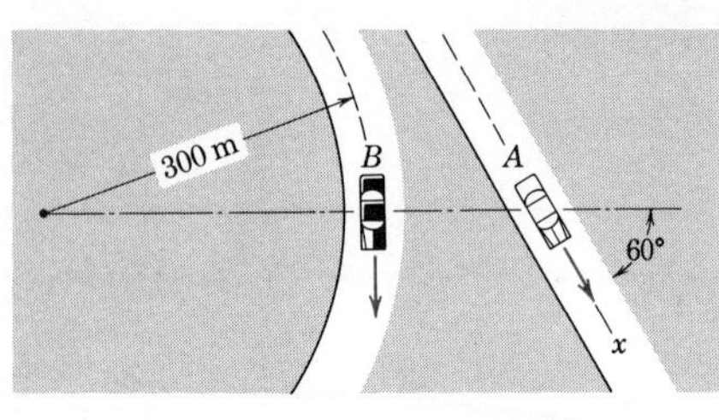

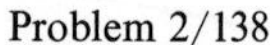
Problem 2/138

2/138 Car A travels at a constant speed of 90 km/h along a straight road, and car B travels at the same constant speed on a road which makes a circular arc of 300-m radius. For the instant when the cars are in the relative positions shown, determine the x-component of the acceleration which car A appears to have as seen by a nonrotating observer in car B.

Ans. $(a_{A/B})_x = 1.042$ m/s²

2/139 Work Prob. 2/138 for the conditions where car B is speeding up at the rate of 12 km/h per second at the instant represented. All other conditions remain the same as in Prob. 2/138.

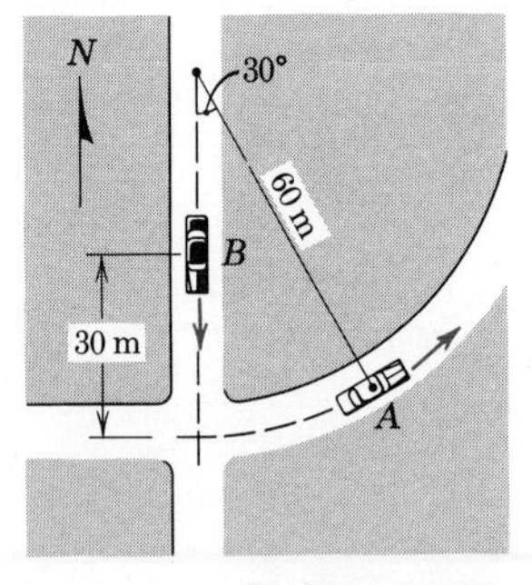

Problem 2/140

2/140 Car A negotiates a curve of 60-m radius at a constant speed of 50 km/h. When A passes the position shown, car B is 30 m from the intersection and is accelerating south toward the intersection at the rate of 1.5 m/s². Determine the acceleration which A appears to have when observed by an occupant of B at this instant.

Ans. $a_{A/B} = 4.58$ m/s², 20.6° west of north

2/141 The aircraft A with radar detection equipment is flying horizontally at 12 km and is increasing its speed at the rate of 1.2 m/s each second. Its radar locks onto an aircraft flying in the same direction and in the same vertical plane at an altitude of 18 km. If A has a speed of 1000 km/h at the instant that $\theta = 30°$, determine the values of $\ddot{r}$ and $\ddot{\theta}$ at this same instant if B has a constant speed of 1500 km/h. *Ans.* $\ddot{r} = -0.637$ m/s^2
$\ddot{\theta} = 0.1660$ mrad/s^2

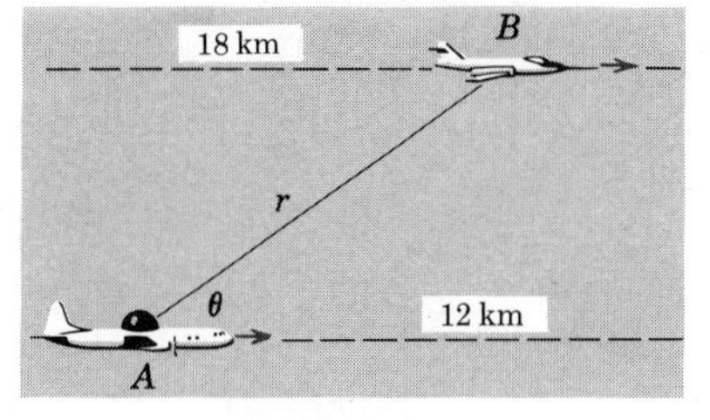

Problem 2/141

2/142 A jet aircraft is launched from an aircraft carrier along its angled deck catapult in a distance of 150 m with constant acceleration relative to the carrier. If the required air speed measured in the direction of the center line of the aircraft is 180 knots and if the carrier is moving at a constant speed of 35 knots against a head wind of 25 knots, determine the minimum acceleration $\ddot{x}$ needed to make the aircraft airborne at the end of the runway. (*Suggestion:* Superpose the two relative velocity equations expressing the velocity of the aircraft relative to the carrier and the velocity of the aircraft relative to the wind and note that the airborne speed is the component along the center line of the aircraft of its velocity relative to the wind.) (1 knot = 1.852 km/h.)
Ans. $\ddot{x} = 13.33$ m/s^2

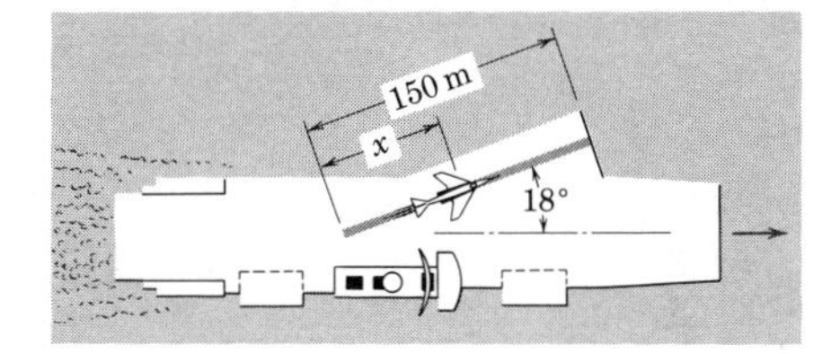

Problem 2/142

Rotating Reference Axes

2/143 The guide rail for an experimental rocket sled runs east and west and is situated at a north latitude of 37°. If the sled attains a ground speed of 2500 km/h in a westerly direction, compute the Coriolis component a_c of the acceleration of the sled due to the rotation of the earth. The angular velocity of the earth is $0.729(10^{-4})$ rad/s.
Ans. $a_c = 0.1012$ m/s^2 away from earth normal to polar axis

2/144 The slider P moves out along the rod which rotates about its vertical shaft with an angular speed ω. Assume $\dot{r} \neq 0$, $\ddot{r} \neq 0$, and $\omega \neq 0$, and show the equivalence of the determination of the acceleration of P by Eqs. 23 and 18.

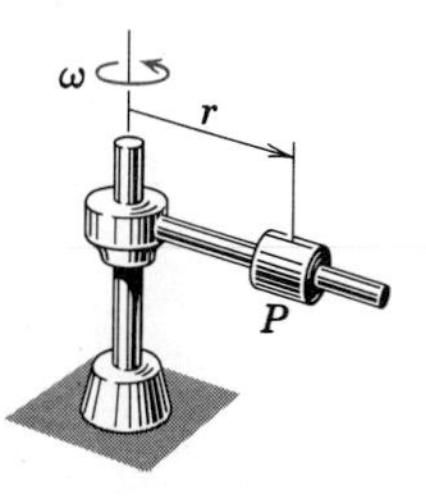

Problem 2/144

2/145 For the cars described in Prob. 2/137, determine the velocity which car B appears to have to an observer riding in and turning with car A. Is this apparent velocity the negative of the velocity that A appears to have to an observer riding in car B? The distance separating the cars at the instant considered is 30 m.

Ans. $\mathbf{v}_{rel} = -18\mathbf{i} + 22.5\mathbf{j}$ m/s; No

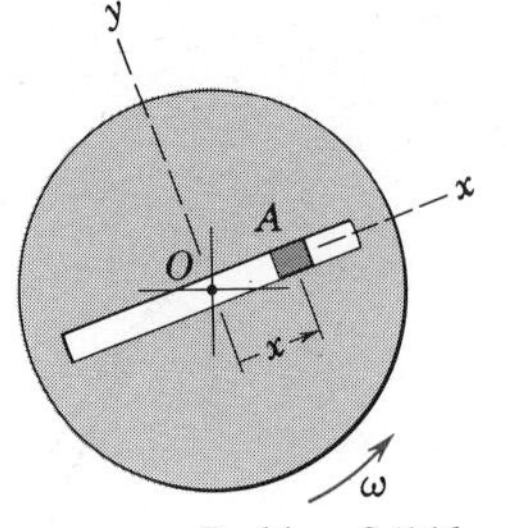

Problem 2/146

2/146 The disk with the radial slot rotates about O with an angular acceleration $\dot{\omega} = 15$ rad/s² while the slider A moves with a constant speed $\dot{x} =$ 100 mm/s relative to the slot during a certain interval of its motion. If the angular velocity of the disk is $\omega = 12$ rad/s at the instant when the slider crosses the center O of rotation of the disk, determine the acceleration of the slider at this instant.

2/147 A particle is dropped down a vertical 30-m pipe situated on the equator. What are the magnitude and direction of the particle's absolute acceleration component normal to the pipe axis after it has fallen 25 m from rest? Assume frictionless vertical fall. The angular velocity of the earth is $0.729(10^{-4})$ rad/s and the gravitational acceleration relative to the earth at the equator is 9.781 m/s². (See Fig. 1.)

Ans. $a_n = 3.22$ mm/s² west

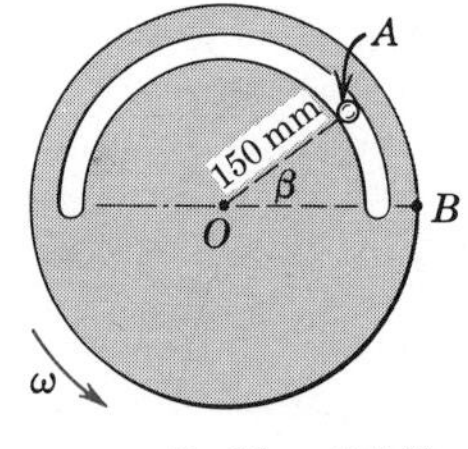

Problem 2/148

2/148 The disk with the circular slot rotates about O with a constant counterclockwise angular velocity $\omega = 10$ rad/s. The pin A moves in the slot so that its radial line AO rotates relative to the line OB, fixed to the disk, with a constant rate $\dot{\beta} = 5$ rad/s for a certain interval of its motion. Solve for the acceleration of A by both Eqs. 18 and 23.

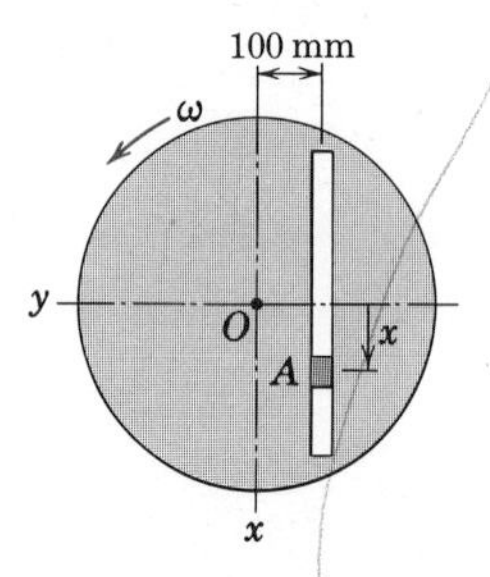

Problem 2/149

2/149 The slider A moves in the slot at the same time that the disk rotates about its center O with an angular speed ω positive in the counterclockwise sense. Determine the x- and y-components of the absolute acceleration of A if, at the instant represented, $\omega = 5$ rad/s, $\dot{\omega} = -10$ rad/s², $x =$ 100 mm, $\dot{x} = 150$ mm/s, and $\ddot{x} = 500$ mm/s².

Ans. $a = -1$ m/s², $a_y = 2$ m/s²

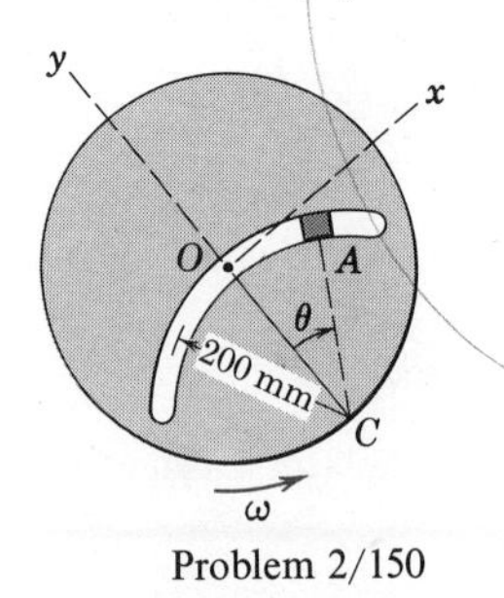

Problem 2/150

2/150 The disk with the circular slot of 200-mm radius rotates about O with a constant angular velocity $\omega = 15$ rad/s. Determine the acceleration of the slider A at the instant when it passes the center of the disk if, at that moment, $\dot{\theta} = 12$ rad/s and $\ddot{\theta} = 0$.

2/151 Ship A is proceeding north at a constant speed of 12 knots, while ship B is making a constant speed of 10 knots while turning to port (left) at the constant rate of 10°/min. When the ships are separated by a distance of 2 nautical miles in the relative positions shown, with B momentarily heading west and A crossing its bow, the navigator of B measures the apparent velocity of A. Find this velocity and specify the clockwise angle β which it makes with the north direction.

Ans. $|\mathbf{v}_{rel}| = 34.4$ knots, $\beta = 16.89°$

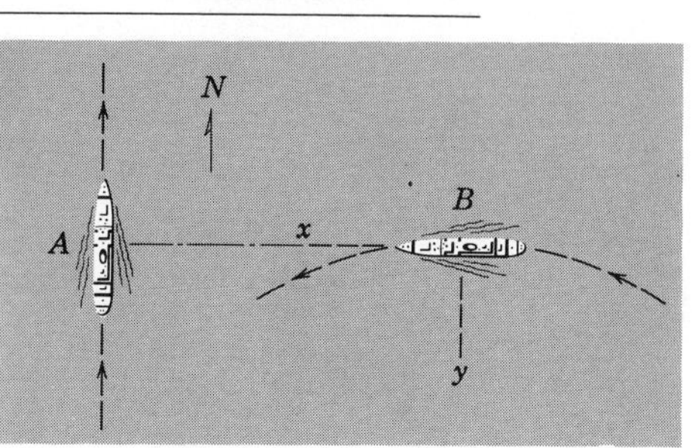

Problem 2/151

2/152 For the two ships described in Prob. 2/151 and for their relative positions shown where the distance between them is 2 nautical miles, determine the acceleration which A appears to have to an observer moving with ship B. Express the result in **i**- and **j**-components where the x-y axes are attached to B as indicated.

2/153 Cars A and B are both moving with a constant speed of 50 km/h in the directions shown. Calculate the acceleration which A appears to have to an observer riding in B and turning with it as it rounds the curve of 150-m radius.

Ans. $a_{rel} = 2.57$ m/s^2 from B to A

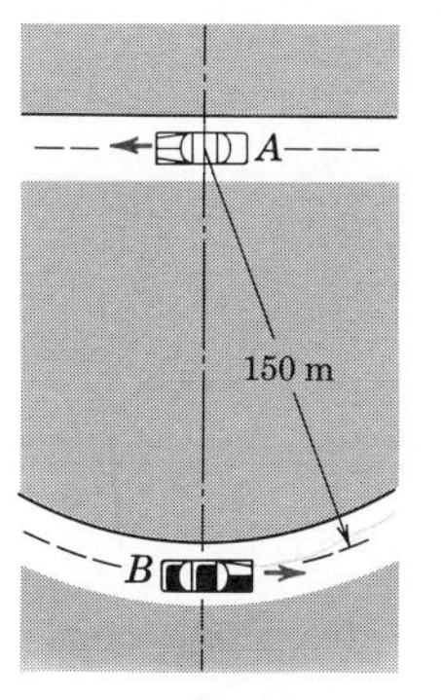
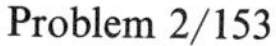

Problem 2/153

2/154 Two boys A and B are sitting on opposite sides of a horizontal turntable which rotates at a constant counterclockwise angular velocity ω as seen from above. Boy A throws a ball toward B by giving it a horizontal velocity $\mathbf{u}$ relative to the turntable toward B. Assume that the ball has no horizontal acceleration once released and write an expression for the acceleration $\mathbf{a}_{rel}$ which B would observe the ball to have in the plane of the turntable just after it is thrown. Sketch the path of the ball on the turntable as observed by B.

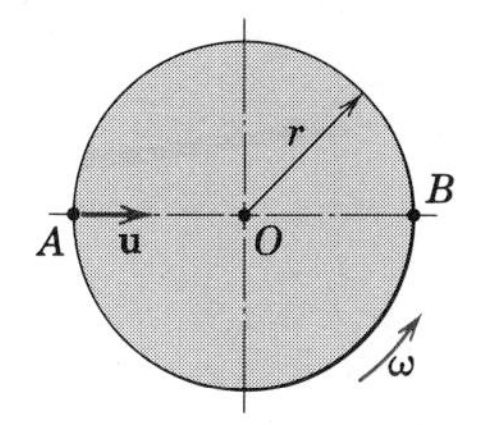

Problem 2/154

2/155 The slider A oscillates in the slot about the neutral position O with a frequency of 2 cycles per second and an amplitude x_{max} of 50 mm so that its displacement in millimetres may be written $x = 50 \sin 4\pi t$ where t is the time in seconds. The disk, in turn, is set into angular oscillation about O with a frequency of 4 cycles per second and an amplitude $\theta_{max} = 0.20$ rad. The angular displacement is thus given by $\theta = 0.20 \sin 8\pi t$. Calculate the acceleration of A for the positions (*a*) $x = 0$ with $\dot{x}$ positive and (*b*) $x = 50$ mm.

Ans. (*a*) $\mathbf{a}_A = 6.32\mathbf{j}$ m/s^2
(*b*) $\mathbf{a}_A = -9.16\mathbf{i}$ m/s^2

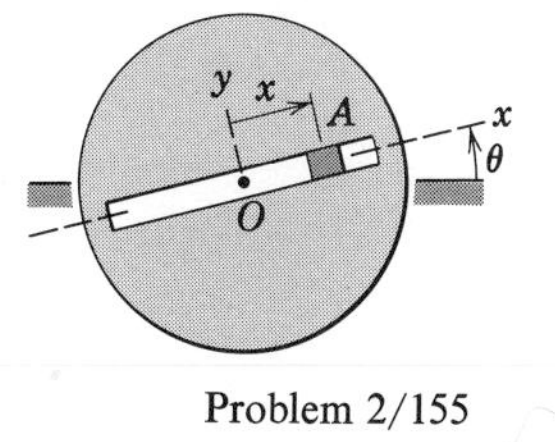

Problem 2/155

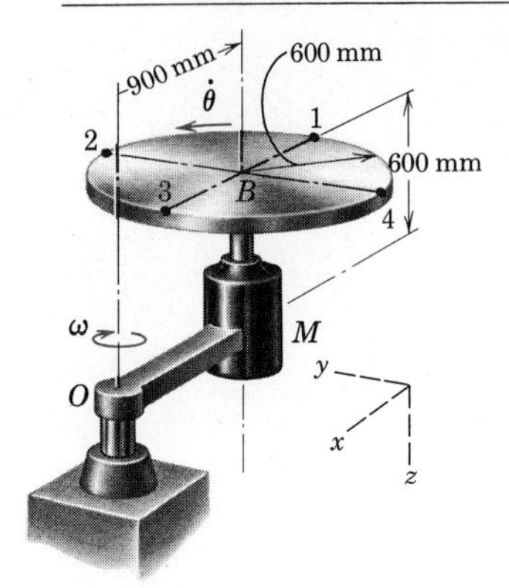

Problem 2/156

2/156 The shaft and attached disk of the motor M turn counterclockwise when viewed from above at a constant rate $\dot{\theta} = 3$ rad/s *relative* to the motor housing and attached arm OM. Simultaneously the arm is set into clockwise rotation with a constant angular speed $\omega = 2$ rad/s. Determine the absolute acceleration of each of the four points on the disk when in the position shown.

Ans. $\mathbf{a}_1 = 4.2\mathbf{i}$ m/s^2, $\mathbf{a}_2 = 3.6\mathbf{i} - 0.6\mathbf{j}$ m/s^2
$\mathbf{a}_3 = 3.0\mathbf{i}$ m/s^2, $\mathbf{a}_4 = 3.6\mathbf{i} + 0.6\mathbf{j}$ m/s^2

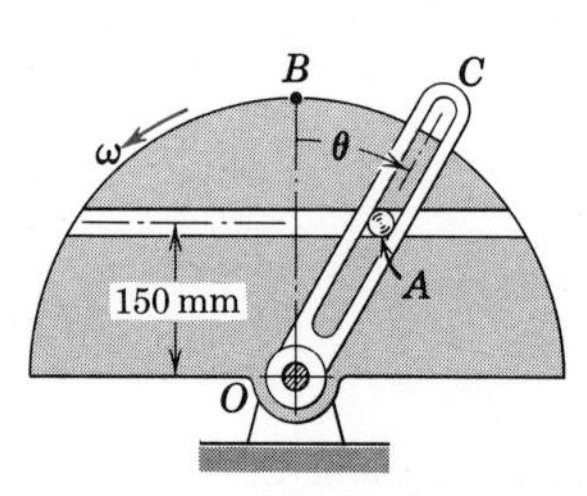

Problem 2/157

2/157 The slotted disk sector rotates with a constant counterclockwise angular velocity $\omega = 3$ rad/s. Simultaneously the slotted arm OC oscillates about the line OB (fixed to the disk) so that θ changes at the constant rate of 2 rad/s except at the extremities of the oscillation during reversal of direction. Determine the total acceleration of the pin A when $\theta = 30°$ and $\dot{\theta}$ is positive (clockwise). *Ans.* $a = 1.060$ m/s^2

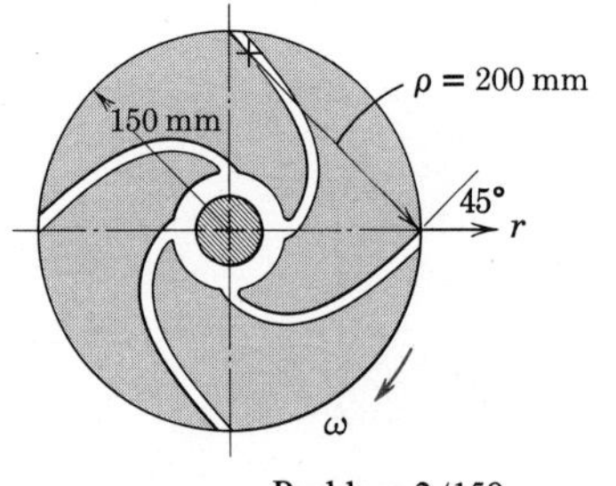

Problem 2/158

2/158 The figure shows the vanes of a centrifugal pump impeller which turns with a constant clockwise speed of 200 rev/min. The fluid particles are observed to have an absolute velocity whose component in the r-direction is 3 m/s at discharge from the vane. Furthermore, the magnitude of the velocity of the particles measured relative to the vane is increasing at the rate of 24 m/s^2 just before they leave the vane. Determine the magnitude of the total acceleration of a fluid particle an instant before it leaves the impeller. The radius of curvature ρ of the vane at its end is 200 mm. *Ans.* $a = 46.9$ m/s^2

2/159 Use the notation of Fig. 15 and work out a geometric proof for Eq. 23*b*. Proceed by evaluating the change during time dt in the relative velocity $\mathbf{v}_{A/P}$ in order to obtain the term $\mathbf{a}_{A/P}$ in the basic equation $\mathbf{a}_A = \mathbf{a}_P + \mathbf{a}_{A/P}$. The point P is fixed to the rotating path and is momentarily coincident with the moving particle A. (*Guideline:* In the *a*-part of the problem figure is shown the relative velocity $v_{rel} = v_{A/P} = \dot{s}$ for the particle A on the path at time t and the relative velocity at time $t + dt$ where the path has changed orientation. The center of curvature of the path is C, and the radius of curvature is ρ. Note that the new velocity of A relative to P' is given by $\mathbf{v}_{A/P'} = \mathbf{v}_{A/Q} + \mathbf{v}_{Q/P'}$ where Q is the point on the path coinciding with A at time $t + dt$. The vector difference between $\mathbf{v}_{A/P'}$ and $\mathbf{v}_{A/P}$ divided by dt gives in the limit $\mathbf{a}_{A/P}$. Evaluate this increment in terms of its n- and t-components with the aid of the *b*-part of the figure. Identify the vector changes which make up the Coriolis acceleration term.)

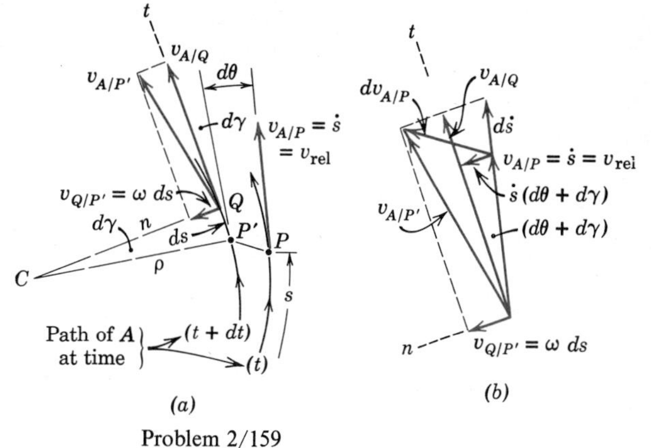

Problem 2/159

14 Space Curvilinear Motion. The motion of a particle along a curved path in space, Fig. 17, is known as space curvilinear motion. At position A the motion may be considered as taking place in the plane P which contains the curve at this position. This plane, often called the *osculating plane,* may be defined by point A and two adjacent points on the curve, one on either side of A. As these points are brought closer to A the plane containing the three points approaches the limiting plane P. The velocity $\mathbf{v}$, which is tangent to the curve, lies in the osculating plane. The acceleration $\mathbf{a}$ of the particle also lies in the osculating plane, and, as in the case of plane motion, may be represented in terms of its component $a_t = \dot{v}$ tangent to the path due to the change in the magnitude of the velocity and its component $a_n = v^2/\rho$ normal to the curve due to the change in direction of the velocity. As before ρ is the radius of curvature of the path at the position considered, and the normal acceleration component is directed toward the center of curvature C in the osculating plane. The description of the osculating plane

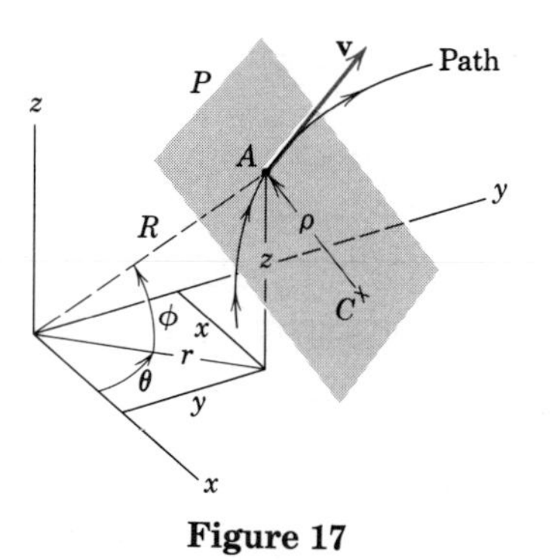

Figure 17

is generally inconvenient for three-dimensional motion, and coordinates other than the path coordinates are more frequently used. Three such coordinate systems are described in Part A, and the transformations expressed in matrix algebra between coordinate systems are covered in Part B of the article.

Part A: Coordinate Systems

Rectangular coordinates (x-y-z): The extension of Eqs. 8 cited in Art. 12 for plane curvilinear motion requires merely the addition of the z-coordinate to describe curvilinear motion in space. Consequently, the position vector, the velocity, and the acceleration of a particle moving along a space curve are

$$\begin{aligned} \mathbf{R} &= \mathbf{i}x + \mathbf{j}y + \mathbf{k}z \\ \mathbf{v} = \dot{\mathbf{R}} &= \mathbf{i}\dot{x} + \mathbf{j}\dot{y} + \mathbf{k}\dot{z} \\ \mathbf{a} = \dot{\mathbf{v}} = \ddot{\mathbf{R}} &= \mathbf{i}\ddot{x} + \mathbf{j}\ddot{y} + \mathbf{k}\ddot{z} \end{aligned} \tag{24}$$

where x, y, and z are the coordinates of the particle, and $\mathbf{i}$, $\mathbf{j}$, and $\mathbf{k}$ are the unit vectors in these respective coordinate directions.

Cylindrical coordinates (r-θ-z): Description of space motion in cylindrical coordinates calls merely for the addition of the z-coordinate, as indicated in Fig. 17, to the polar coordinate expressions in the x-y plane described in Art. 12. Thus Eq. 17, which describes the velocity for plane motion, becomes

$$\mathbf{v} = \mathbf{v}_r + \mathbf{v}_\theta + \mathbf{v}_z \tag{25}$$

where

$$\begin{aligned} v_r &= \dot{r} \\ v_\theta &= r\dot{\theta} \\ v_z &= \dot{z} \end{aligned}$$

Also, Eq. 18 for the acceleration is extended to

$$\mathbf{a} = \mathbf{a}_r + \mathbf{a}_\theta + \mathbf{a}_z \tag{26}$$

where

$$\begin{aligned} a_r &= \ddot{r} - r\dot{\theta}^2 \\ a_\theta &= r\ddot{\theta} + 2\dot{r}\dot{\theta} \\ a_z &= \ddot{z} \end{aligned}$$

Equations 25 and 26 may be written in vector form by using the unit vectors $\mathbf{r}_1$, $\boldsymbol{\theta}_1$, and $\mathbf{k}$ in the r-, θ-, and z-directions, respectively. It should be noted that, whereas $\mathbf{r}_1$ and $\boldsymbol{\theta}_1$ have time derivatives as described in Art. 12, the unit vector $\mathbf{k}$ in the z-direction remains fixed in direction and therefore has no derivative with time.

Spherical coordinates (R-θ-ϕ): The location of the particle at A may also be described by spherical coordinates as shown in Fig. 17. The velocity of the particle may be expressed in terms of its three components, Fig. 18,

$$\mathbf{v} = \mathbf{v}_R + \mathbf{v}_\theta + \mathbf{v}_\phi \tag{27}$$

where

$$v_R = \dot{R}$$
$$v_\theta = R\dot{\theta}\cos\phi$$
$$v_\phi = R\dot{\phi}$$

The R-component is due to the change in the length of $\mathbf{R}$. The θ-component comes from the change in direction of the x-y component of $\mathbf{R}$ due to the swinging of $\mathbf{R}$ with θ. The ϕ-component is due to the swinging of $\mathbf{R}$ with ϕ.

The acceleration of the particle is determined from the vector changes in the three velocity components, Fig. 18, as these components change in magnitude and direction. During time dt, the component $\mathbf{v}_R$ has the changes dv_R, $v_R \cos\phi\, d\theta$, and $v_R\, d\phi$ in the R-, θ-, and ϕ-directions respectively. The θ-component is due only to the x-y projection of $\mathbf{v}_R$. During the same interval $\mathbf{v}_\theta$ has the changes dv_θ in the θ-direction and $v_\theta\, d\theta$ parallel to the x-y plane. This last change has two components, $v_\theta\, d\theta \cos\phi$ in the negative R-direction and $v_\theta\, d\theta \sin\phi$ in the positive ϕ-direction. Lastly, $\mathbf{v}_\phi$ has the vector changes dv_ϕ, $v_\phi \sin\phi\, d\theta$, and $v_\phi\, d\phi$ in the ϕ-, minus-θ, and minus-R directions, respectively. The θ-component is due only to the swinging of the x-y projection of $\mathbf{v}_\phi$. By collecting terms, the three components of the vector change in velocity are found to be

$$|d\mathbf{v}|_R = dv_R - v_\theta \cos\phi\, d\theta - v_\phi\, d\phi$$
$$|d\mathbf{v}|_\theta = v_R \cos\phi\, d\theta + dv_\theta - v_\phi \sin\phi\, d\theta$$
$$|d\mathbf{v}|_\phi = v_R\, d\phi + v_\theta \sin\phi\, d\theta + dv_\phi$$

Upon substitution of the expressions for v_R, v_θ, and v_ϕ, dividing by dt, and rearranging terms, the acceleration and its components become

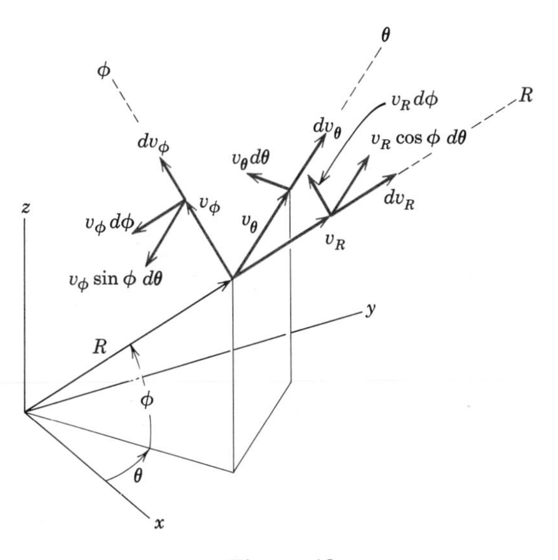

Figure 18

$$\mathbf{a} = \mathbf{a}_R + \mathbf{a}_\theta + \mathbf{a}_\phi \tag{28}$$

where

$$a_R = \ddot{R} - R\dot{\phi}^2 - R\dot{\theta}^2 \cos^2 \phi$$

$$a_\theta = \frac{\cos\phi}{R}\frac{d}{dt}(R^2\dot{\theta}) - 2R\dot{\theta}\dot{\phi}\sin\phi$$

$$a_\phi = \frac{1}{R}\frac{d}{dt}(R^2\dot{\phi}) + R\dot{\theta}^2 \sin\phi\cos\phi$$

Evaluation of the foregoing acceleration components is facilitated by careful study of Fig. 18 where it should be noted that each of the velocity increments, with the exception of the one labeled $v_\theta\, d\theta$, is parallel to a coordinate direction. This one increment is parallel to the x-y plane.

An alternative derivation of the expressions in Eqs. 27 and 28 may be employed by introducing unit vectors $\mathbf{R}_1$, $\boldsymbol{\theta}_1$, and $\boldsymbol{\phi}_1$ in the three coordinate directions and differentiating the vector equation $\mathbf{R} = R\mathbf{R}_1$ directly. To carry out this process requires expressions for $\dot{\mathbf{R}}_1$, $\dot{\boldsymbol{\theta}}_1$, and $\dot{\boldsymbol{\phi}}_1$ due to the rotation of the unit vectors with θ and ϕ. This procedure will be left to the student to develop.

The choice of x-y-z, r-θ-z, or R-θ-ϕ coordinates for a particular space problem will depend upon the manner in which the measurements are made or the motion generated. This choice is usually fairly evident.

Sample Problem

2/160 The disk A rotates with a constant speed $\omega = \dot{\theta} = \pi/3$ rad/s. Simultaneously the hinged arm OB is elevated at the constant rate $\dot{\phi} = 2\pi/3$ rad/s. At time $t = 0$, both $\theta = 0$ and $\phi = 0$. The angle θ is measured from the fixed reference x-axis. The sphere P slides out along the rod according to $R = 50 + 200t^2$, where R is in millimetres and t is in seconds. Determine the magnitude of the total acceleration $\mathbf{a}$ of P when $t = \frac{1}{2}$ s.

Solution. Spherical coordinates will be used since the motion of P is generated by these coordinates. The terms appearing in Eq. 28 are:

$R = 50 + 200(\tfrac{1}{2})^2 = 100$ mm $\quad \dot{R} = 400t = 400(\tfrac{1}{2}) = 200$ mm/s $\quad \ddot{R} = 400$ mm/s^2

$$\theta = \left(\frac{\pi}{3}\right)\left(\frac{1}{2}\right) = \frac{\pi}{6} \text{ rad} \qquad \dot{\theta} = \frac{\pi}{3} \text{ rad/s} \qquad \ddot{\theta} = 0$$

$$\phi = \left(\frac{2\pi}{3}\right)\left(\frac{1}{2}\right) = \frac{\pi}{3} \text{ rad} \qquad \dot{\phi} = \frac{2\pi}{3} \text{ rad/s} \qquad \ddot{\phi} = 0$$

$$\sin\theta = \frac{1}{2} \qquad \cos\theta = \frac{\sqrt{3}}{2} \qquad \sin\phi = \frac{\sqrt{3}}{2} \qquad \cos\phi = \frac{1}{2}$$

$$\frac{d}{dt}(R^2\dot{\theta}) = 2R\dot{R}\dot{\theta} + R^2\ddot{\theta} = 2(0.1)(0.2)\frac{\pi}{3} + 0 = \frac{0.04\pi}{3} \ (\text{m/s})^2$$

$$\frac{d}{dt}(R^2\dot{\phi}) = 2R\dot{R}\dot{\phi} + R^2\ddot{\phi} = 2(0.1)(0.2)\frac{2\pi}{3} + 0 = \frac{0.08\pi}{3} \ (\text{m/s})^2$$

Thus the components of $\mathbf{a}$ from Eq. 28 are

$$a_R = 0.40 - 0.10\left(\frac{2\pi}{3}\right)^2 - 0.10\left(\frac{\pi}{3}\right)^2\left(\frac{1}{2}\right)^2 = -0.0661 \text{ m/s}^2$$

$$a_\theta = \frac{\frac{1}{2}}{0.10}\frac{0.04\pi}{3} - 2(0.10)\left(\frac{\pi}{3}\right)\left(\frac{2\pi}{3}\right)\left(\frac{\sqrt{3}}{2}\right) = -0.1704 \text{ m/s}^2$$

$$a_\phi = \frac{1}{0.10}\frac{0.08\pi}{3} + 0.10\left(\frac{\pi}{3}\right)^2\left(\frac{\sqrt{3}}{2}\right)\left(\frac{1}{2}\right) = 0.885 \text{ m/s}^2$$

The magnitude of the acceleration is, then,

$$|\mathbf{a}| = a = \sqrt{(-0.0661)^2 + (-0.1704)^2 + (0.885)^2} = 0.904 \text{ m/s}^2 \qquad \textit{Ans.}$$

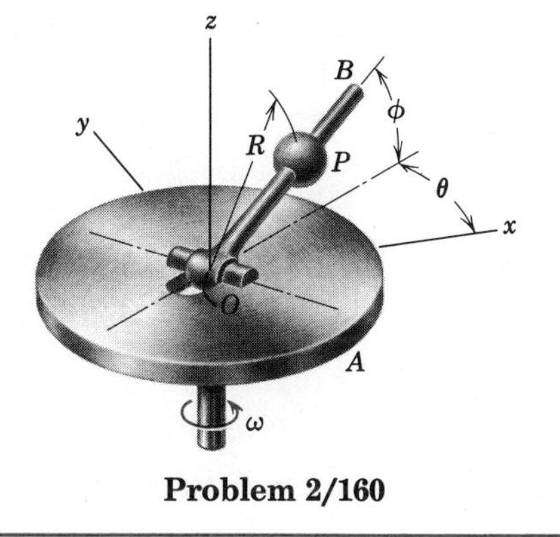

Problem 2/160

Part B: Coordinate Transformations

It is frequently necessary, particularly in three-dimensional analysis, to transform vector quantities from one set of coordinates to another. Whereas this transformation may be accomplished by writing the correct relations directly from the spatial geometry involved, the problem lends itself to routine treatment with the aid of matrix algebra since the transformation equations are linear. Consider a vector quantity **V** with components V_x, V_y, V_z in rectangular coordinates, V_r, V_θ, V_z in cylindrical coordinates, and V_R, V_θ, V_ϕ in spherical coordinates. This quantity might be the velocity or acceleration of a particle. It could be its momentum or merely its position vector. Consider the three transformation equations for **V** when changing from rectangular to cylindrical coordinates. The vector quantity **V** is represented in Fig. 19*a*. Since its *z*-component is identical in both systems, the

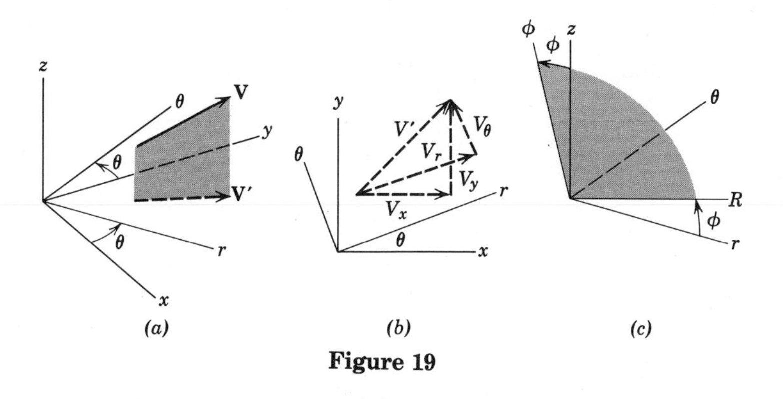

Figure 19

transformations are readily perceived from Fig. 19*b* which shows the projection V' of $\mathbf{V}$ on the x-y and r-θ planes. The relations are seen to be

$$\begin{aligned} V_r &= \cos\theta(V_x) + \sin\theta(V_y) + 0(V_z) \\ V_\theta &= -\sin\theta(V_x) + \cos\theta(V_y) + 0(V_z) \\ V_z &= 0(V_x) + 0(V_y) + 1(V_z) \end{aligned}$$

where the zero terms have been added merely to preserve the symmetry of the three equations. Note that the terms are arranged in the same order, V_x, V_y, V_z, in all three equations. In matrix notation* these equations may be written

$$\begin{Bmatrix} V_r \\ V_\theta \\ V_z \end{Bmatrix} = \begin{bmatrix} \cos\theta & \sin\theta & 0 \\ -\sin\theta & \cos\theta & 0 \\ 0 & 0 & 1 \end{bmatrix} \begin{Bmatrix} V_x \\ V_y \\ V_z \end{Bmatrix} \tag{29}$$

where the order of combination is that required to produce the given equations. The first equation, for example, gives V_r and is obtained by multiplying each element in the column matrix on the right by its corresponding element in the first row of the square matrix and adding the results together. This combination is equated to V_r. This matrix equation may also be written as

$$\{V_{r\theta z}\} = [T_\theta]\{V_{xyz}\} \tag{30}$$

where $[T_\theta]$ stands for the square matrix in θ and is called the transfer matrix from x-y-z to r-θ-z. Since the transfer involves merely a rotation of the axes through the angle θ, the transfer matrix will contain θ only.

The transformation from cylindrical to rectangular components becomes

$$\{V_{xyz}\} = [T_\theta]^{-1}\{V_{r\theta z}\} \tag{31}$$

where

$$[T_\theta]^{-1} = \begin{bmatrix} \cos\theta & -\sin\theta & 0 \\ \sin\theta & \cos\theta & 0 \\ 0 & 0 & 1 \end{bmatrix}$$

The inverse matrix $[T_\theta]^{-1}$ becomes the transfer matrix from cylindrical to rectangular coordinates, and its validity is easily verified by expansion and comparison with the geometry of Fig. 19 when V_x and V_y are written in terms of V_r and V_θ. The two transfer matrices obey the rule

$$[T_\theta][T_\theta]^{-1} = [1]$$

where the unit matrix is

$$[1] = \begin{bmatrix} 1 & 0 & 0 \\ 0 & 1 & 0 \\ 0 & 0 & 1 \end{bmatrix}$$

* See Appendix B for a brief summary of matrix algebra.

It may be shown that the inverse matrix is obtained by interchanging the corresponding elements on either side of the main diagonal.

The change from cylindrical to spherical coordinates is accomplished by a single rotation ϕ of the axes around the θ-axis, Fig. 19*c*. With a little thought the transfer matrix can be written directly from Eq. 29 where the rotation ϕ occurs in the R-ϕ plane rather than in the r-θ plane. Thus

$$\begin{Bmatrix} V_R \\ V_\theta \\ V_\phi \end{Bmatrix} = \begin{bmatrix} \cos\phi & 0 & \sin\phi \\ 0 & 1 & 0 \\ -\sin\phi & 0 & \cos\phi \end{bmatrix} \begin{Bmatrix} V_r \\ V_\theta \\ V_z \end{Bmatrix} \tag{32}$$

Again this matrix equation may be written in abbreviated form as

$$\{V_{R\theta\phi}\} = [T_\phi]\{V_{r\theta z}\} \tag{33}$$

from which it follows that

$$\{V_{r\theta z}\} = [T_\phi]^{-1}\{V_{R\theta\phi}\} \tag{34}$$

where the inverse transfer matrix is

$$[T_\phi]^{-1} = \begin{bmatrix} \cos\phi & 0 & -\sin\phi \\ 0 & 1 & 0 \\ \sin\phi & 0 & \cos\phi \end{bmatrix}$$

Direct transfer from rectangular to spherical coordinates may be accomplished by combining Eqs. 30 and 33 to give

$$\{V_{R\theta\phi}\} = [T_\phi][T_\theta]\{V_{xyz}\} \tag{35}$$

where the transfer matrix, upon expansion, becomes

$$[T_\phi][T_\theta] = \begin{bmatrix} \cos\phi\cos\theta & \cos\phi\sin\theta & \sin\phi \\ -\sin\theta & \cos\theta & 0 \\ -\sin\phi\cos\theta & -\sin\phi\sin\theta & \cos\phi \end{bmatrix}$$

The first of the transfer equations from the matrix, for example, becomes

$$V_R = V_x \cos\phi\cos\theta + V_y \cos\phi\sin\theta + V_z \sin\phi$$

Matrix notation provides a convenient shorthand for handling the linear transformation equations for solution by routine computer programs.

Sample Problem

2/161 From the results of Prob. 2/160 for the components of the acceleration of the spherical slider P expressed in spherical coordinates, determine the corresponding x-, y-, and z-components of this acceleration by direct coordinate transformation.

Solution. The x-, y-, and z-components of **a** may be obtained from the R-, θ-, ϕ-

components by direct coordinate transformations using Eqs. 31 and 34 which give

$$\{a_{xyz}\} = [T_\theta]^{-1}[T_\phi]^{-1}\{a_{R\theta\phi}\}$$

Expansion of the transfer matrix produces

$$[T_\theta]^{-1}[T_\phi]^{-1} = \begin{bmatrix} \cos\theta & -\sin\theta & 0 \\ \sin\theta & \cos\theta & 0 \\ 0 & 0 & 1 \end{bmatrix} \begin{bmatrix} \cos\phi & 0 & -\sin\phi \\ 0 & 1 & 0 \\ \sin\phi & 0 & \cos\phi \end{bmatrix}$$

$$= \begin{bmatrix} \cos\theta\cos\phi & -\sin\theta & -\cos\theta\sin\phi \\ \sin\theta\cos\phi & \cos\theta & -\sin\theta\sin\phi \\ \sin\phi & 0 & \cos\phi \end{bmatrix}$$

Expansion gives

$$a_x = a_R \cos\theta\cos\phi - a_\theta \sin\theta - a_\phi \cos\theta\sin\phi$$
$$a_y = a_R \sin\theta\cos\phi + a_\theta \cos\theta - a_\phi \sin\theta\sin\phi$$
$$a_z = a_R \sin\phi \quad + 0 \quad + a_\phi \cos\phi$$

Substitution of the R-, θ-, and ϕ-components of the acceleration of P from Prob. 2/160 gives

$$a_x = (-0.0661)\left(\frac{\sqrt{3}}{2}\right)\left(\frac{1}{2}\right) - (-0.1704)\left(\frac{1}{2}\right) - (0.885)\left(\frac{\sqrt{3}}{2}\right)\left(\frac{\sqrt{3}}{2}\right) = -0.607 \text{ m/s}^2$$

$$a_y = (-0.0661)\left(\frac{1}{2}\right)\left(\frac{1}{2}\right) + (-0.1704)\left(\frac{\sqrt{3}}{2}\right) - (0.885)\left(\frac{1}{2}\right)\left(\frac{\sqrt{3}}{2}\right) = -0.547 \text{ m/s}^2$$

$$a_z = (-0.0661)\left(\frac{\sqrt{3}}{2}\right) + 0 + (0.885)\left(\frac{1}{2}\right) = 0.385 \text{ m/s}^2$$ *Ans.*

Problems

Coordinate Systems

2/162 At a certain instant the velocity and acceleration of a particle are given by $\mathbf{v} = 3\mathbf{i} + 2\mathbf{j} - 9\mathbf{k}$ m/s and $\mathbf{a} = 4\mathbf{i} + 3\mathbf{j} + 2\mathbf{k}$ m/s² respectively. Show that **v** and **a** are perpendicular at this instant and describe the characteristic of this motion as viewed from the osculating plane. (Recall that two vectors are perpendicular if their scalar product is zero.)

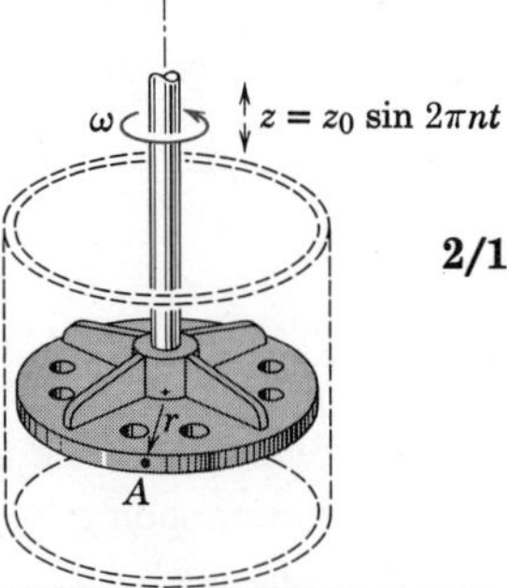

Problem 2/163

2/163 The rotating element in a mixing chamber is given a periodic axial movement $z = z_0 \sin 2\pi nt$ while it is rotating at the constant angular velocity $\dot{\theta} = \omega$. Determine the expression for the maximum magnitude of the acceleration of a point A on the rim of radius r. The frequency n of vertical oscillation is constant.

Ans. $a_{max} = \sqrt{r^2\omega^4 + 16n^4\pi^4 z_0^2}$

2/164 Assign unit vectors $\mathbf{R}_1$, $\boldsymbol{\theta}_1$, and $\boldsymbol{\phi}_1$ along the spherical coordinate directions, Fig. 18, and determine $\dot{\mathbf{R}}_1$, $\dot{\boldsymbol{\theta}}_1$, and $\dot{\boldsymbol{\phi}}_1$.

Ans. $\dot{\mathbf{R}}_1 = \boldsymbol{\phi}_1\dot{\phi} + \boldsymbol{\theta}_1\dot{\theta}\cos\phi$

$\dot{\boldsymbol{\theta}}_1 = -\mathbf{R}_1\dot{\theta}\cos\phi + \boldsymbol{\phi}_1\dot{\theta}\sin\phi$

$\dot{\boldsymbol{\phi}}_1 = -\mathbf{R}_1\dot{\phi} - \boldsymbol{\theta}_1\dot{\theta}\sin\phi$

2/165 Use the results of Prob. 2/164 to obtain the acceleration components in spherical coordinates, Eqs. 28, by a direct differentiation of the position vector $\mathbf{R} = R\mathbf{R}_1$.

2/166 Small objects are released from rest at A and slide with negligible friction down the cylindrical spiral chute of constant helix angle $\gamma = \tan^{-1}(h/2\pi r)$. The component of acceleration measured tangent to the path is $g \sin \gamma$. Determine the radial component of acceleration a_r for each object as it passes B after one complete turn.

Ans. $a_r = -2\pi g \sin 2\gamma$

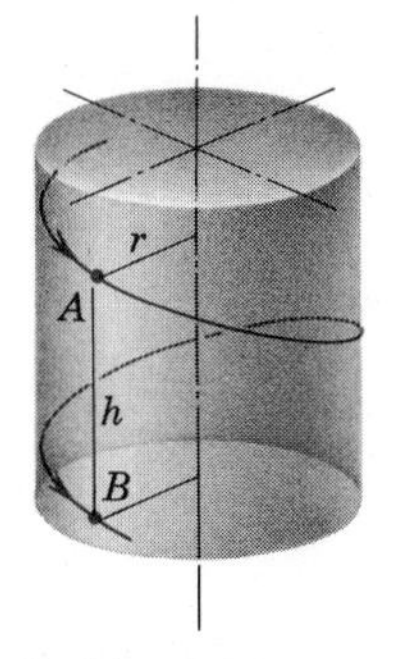

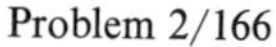
Problem 2/166

2/167 The revolving crane has a boom of length $OP = 24$ m and is turning about its vertical axis at a constant rate of 2 rev/min. At the same time, the boom is being lowered at the constant rate $\dot{\beta} = 0.10$ rad/s. Calculate the velocity and acceleration of the end P of the boom at the instant when the position $\beta = 30°$ is passed.

Ans. $v = 3.48$ m/s, $a = 1.104$ m/s^2

2/168 The rod OA is held at the constant angle $\beta = 30°$ and rotates about the vertical with a constant angular velocity $\omega = 120$ rev/min. Simultaneously the slider P oscillates along the rod with a variable distance from the fixed pivot O given in millimetres by $R = 200 + 50 \sin 2\pi nt$ where the frequency n of oscillation along the rod is a constant 2 cycles per second and where the time t is in seconds. Calculate the acceleration of the slider for an instant when its velocity $\dot{R}$ along the rod is a maximum.

Ans. $a = 17.66$ m/s^2

2/169 Solve Prob. 2/168 for the acceleration of the slider P in the position where its acceleration component $\ddot{R}$ along the rod has a maximum magnitude and is directed toward O.

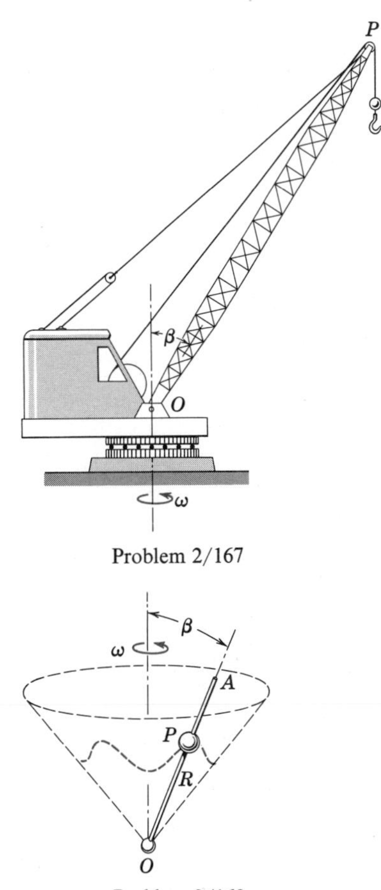

Problem 2/167

Problem 2/168

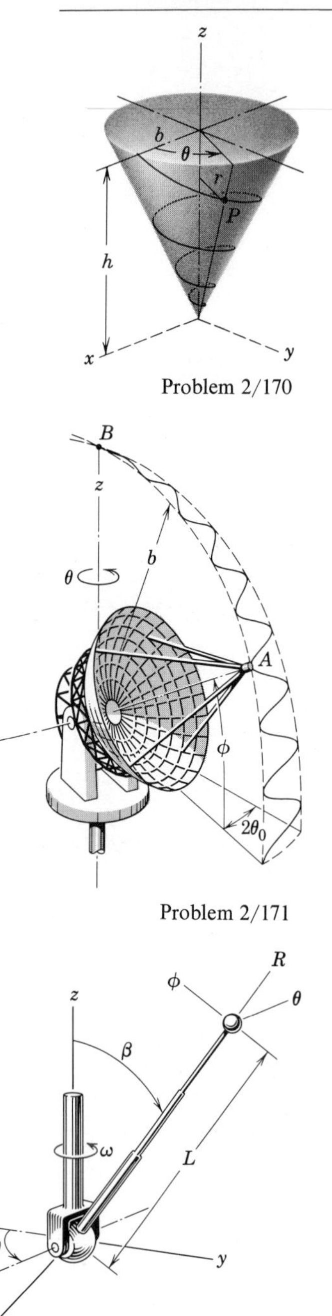

Problem 2/170

Problem 2/171

Problem 2/172

2/170 The particle P moves down the spiral path which is wrapped around the surface of a right circular cone of base radius b and altitude h. The angle γ between the tangent to the curve at any point and a horizontal tangent to the cone at this point is constant. Also the motion of the particle is controlled so that $\dot{\theta}$ is constant. Determine the expression for the radial acceleration a_r of the particle for any value of θ.

Ans. $a_r = b\dot{\theta}^2(\tan^2\gamma \sin^2\beta - 1)e^{-\theta \tan\gamma \sin\beta}$
where $\beta = \tan^{-1}(b/h)$

2/171 The radar tracking antenna oscillates about its vertical axis according to $\theta = \theta_0 \cos pt$ where p is the constant circular frequency and $2\theta_0$ is the double amplitude of oscillation. Simultaneously, the angle of elevation ϕ is increasing at the constant rate $\dot{\phi} = K$. Determine the expression for the magnitude a of the acceleration of the signal horn (a) as it passes position A and (b) as it passes the top position B, assuming that $\theta = 0$ at this instant.

Ans. (a) $a = b\sqrt{K^4 + p^4\theta_0^2 \cos^2\phi}$
(b) $a = bK\sqrt{K^2 + 4p^2\theta_0^2}$

2/172 In a test of the actuating mechanism for a telescoping antenna on a spacecraft the supporting shaft rotates about the fixed z-axis with an angular velocity ω. Determine the R-, θ-, and ϕ-components of the acceleration $\mathbf{a}$ of the end of the antenna at the instant when $L = 1.2$ m and $\beta = 45°$ if the rates $\omega = 2$ rad/s, $\dot{\beta} = \frac{3}{2}$ rad/s, and $\dot{L} = 0.9$ m/s are constant during the motion.

Ans. $a_R = -5.10$ m/s^2
$a_\theta = 7.64$ m/s^2
$a_\phi = -0.3$ m/s^2

2/173 As a further test of the operation of the control mechanism for the telescoping antenna of Prob. 2/172, the vertical shaft is made to oscillate about the fixed z-axis according to $\theta = \pi/4 + 0.12 \sin 4\pi t$, where θ is in radians and t is in seconds. During the oscillation β is increasing at the rate of $\frac{3}{2}$ rad/s, and L is increasing at the rate of 0.9 m/s, both constant. Determine the magnitude of the acceleration $\mathbf{a}$ of the tip of the antenna when $\beta = 60°$ if this occurs at the position $\theta = \pi/4$ and when $L = 1.2$ m.

Ans. $a = 7.11$ m/s^2

Coordinate Transformations

2/174 The displacement of a particle in moving from A to B along a straight line is $\mathbf{s}$. If the x-, y-, z-

coordinates of A and B in metres are (2, 2, 3) and (7, 4, 6), respectively, determine the r-θ-z components of $\mathbf{s}$.

Ans. $s_r = \sqrt{29}$ m, $s_\theta = 0$, $s_z = 3$ m

2/175 The components of velocity $\mathbf{v}$ of a certain particle expressed in spherical coordinates are $v_R = 3$ m/s, $v_\theta = -4$ m/s, and $v_\phi = 2$ m/s at the instant when $\theta = 60°$ and $\phi = 30°$. Determine the corresponding x-, y-, z-components of $\mathbf{v}$ by a direct coordinate transformation.

2/176 The radar antenna at P tracks the jet aircraft A which is flying horizontally at a speed u and an altitude h above the level of P. Determine the expressions for the components of the velocity in the spherical coordinates of the antenna motion.

Ans. $v_R = u \cos\phi \cos\theta$
$v_\theta = -u \sin\theta$
$v_\phi = -u \sin\phi \cos\theta$

2/177 A projectile is fired with an initial velocity $\mathbf{u}$ at an angle β above the horizontal. Neglect air resistance and express the azimuth velocity $\dot{\theta}$ and the elevation velocity $\dot{\phi}$ for the radar antenna P which tracks the projectile in terms of the spherical coordinates and t, the time of flight of the projectile from its starting point O.

2/178 Use the results of Prob. 2/172 and determine the x-, y-, and z-components of acceleration of the tip of the telescoping antenna under the conditions specified and for the position $\theta = 45°$.

Ans. $a_x = -7.80$ m/s^2
$a_y = 3.00$ m/s^2
$a_z = -3.82$ m/s^2

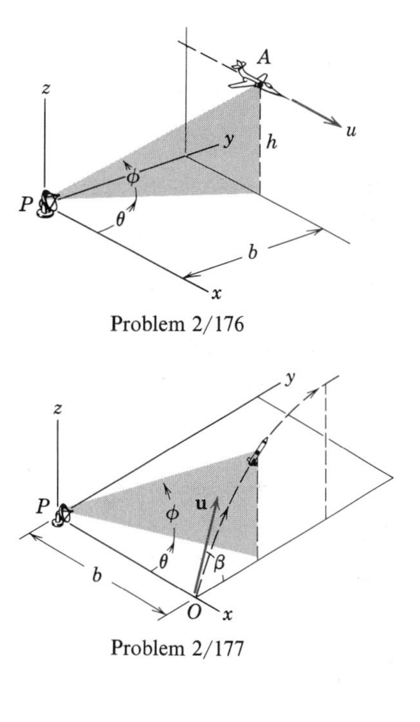

Problem 2/176

Problem 2/177

---

15 Relative Motion in Space. The analysis of the plane motion of a particle referred to translating and rotating coordinate axes, discussed in Art. 13, will now be extended to cover the space motion of a particle. If the analysis of relative motion in a plane is understood, then the extension to three-dimensional motion is easily accomplished by merely adding the third dimension.

Part A: Translating Reference Axes

Consider now the curvilinear motion of two particles A and B in space. The motion of A will be observed first from a translating frame of reference x-y-z moving with B, Fig. 20. The position vector of A relative to B is $\mathbf{r}_{A/B} = \mathbf{i}x + \mathbf{j}y + \mathbf{k}z$ where $\mathbf{i}$, $\mathbf{j}$, and $\mathbf{k}$ are the unit vectors in the moving x-y-z sys-

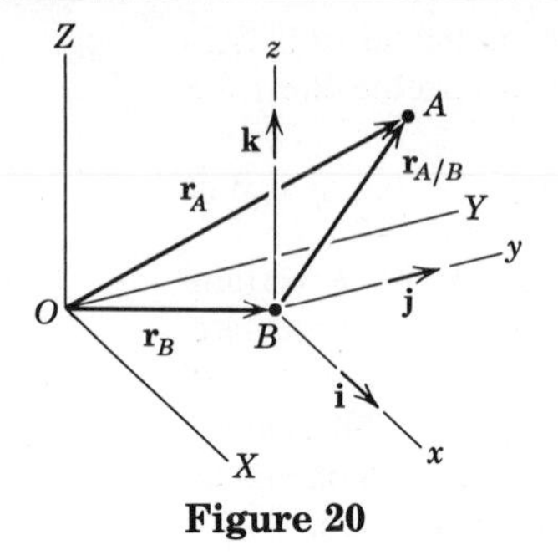

Figure 20

tem. The absolute position, velocity, and acceleration of A are, then,

$$\mathbf{r}_A = \mathbf{r}_B + \mathbf{r}_{A/B}$$

$$\dot{\mathbf{r}}_A = \dot{\mathbf{r}}_B + \dot{\mathbf{r}}_{A/B} \quad \text{or} \quad \mathbf{v}_A = \mathbf{v}_B + \mathbf{v}_{A/B} \tag{36}$$

$$\ddot{\mathbf{r}}_A = \ddot{\mathbf{r}}_B + \ddot{\mathbf{r}}_{A/B} \quad \text{or} \quad \mathbf{a}_A = \mathbf{a}_B + \mathbf{a}_{A/B} \tag{37}$$

where the velocity measured relative to x-y-z is $\mathbf{v}_{A/B} = \dot{\mathbf{r}}_{A/B} = \mathbf{i}\dot{x} + \mathbf{j}\dot{y} + \mathbf{k}\dot{z}$ and where the acceleration measured relative to x-y-z is $\mathbf{a}_{A/B} = \dot{\mathbf{v}}_{A/B} = \mathbf{i}\ddot{x} + \mathbf{j}\ddot{y} + \mathbf{k}\ddot{z}$. For the translating axes x-y-z the unit vectors have no time derivatives since their directions remain unchanged.

The relative motion terms may be expressed in whatever translating coordinate system is convenient—rectangular, cylindrical, or spherical. As noted previously in Art. 13, the choice is dependent on how the motion is generated or how the data are given.

The reciprocal use of A rather than B as a reference point and the equation relating the relative motions of three particles A, B, and C follow precisely the discussion developed in Art. 13 for these same situations in plane motion, and further discussion should be unnecessary.

Part B: Rotating Reference Axes

There are many problems in three-dimensional motion where the description is greatly facilitated by making observations relative to a rotating reference system. The relative motion equation for the three-dimensional case can be derived by adding the third dimension to the analysis for plane motion, described in Art. 13, and it is recommended that the student carry out this process in detail. A somewhat more general approach will be developed here, however, which establishes the relationship between the time derivative of any vector in a fixed and in a rotating reference system.

Whether one proceeds with this more general approach or with a direct extension of the analysis leading to Eq. 23, it is first necessary to establish the spatial concept of the angular velocity ω of a rotating frame of reference. In Fig. 21a are shown axes x-y-z which are in the process of rotating as a rigid unit about O. Point O may be considered fixed insofar as the description of rotation is concerned. During time dt the reference frame rotates about the x-axis through an angle $d\theta_x$, and the y-axis tilts to y', the z-axis

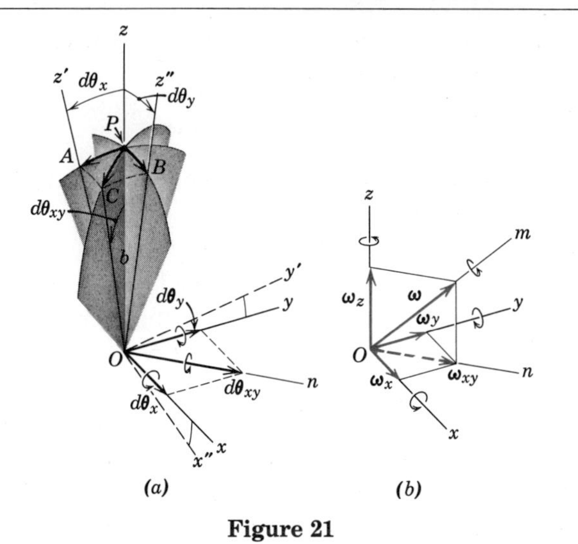

Figure 21

tilts to z', and point P on the z-axis, a distance b from O, moves to A through the distance $b\ d\theta_x$. The rotation is specified by the vector $d\boldsymbol{\theta}_x$, where the direction of the vector coincides with the axis of rotation and the sense of the vector is determined by the right-hand rule. The magnitude of the vector is the angle $d\theta_x$ expressed in radians. Similarly, an independent rotation of the reference frame through an angle $d\theta_y$ about the y-axis causes the x-axis to tilt to x'', the z-axis to tilt to z'', and point P to move to B through the distance $b\ d\theta_y$. This rotation is specified by the vector $d\boldsymbol{\theta}_y$. If both rotations occur simultaneously, point P will move to C through the arc length $b\ d\theta_{xy}$. This resultant movement may be considered due to a rotation $d\theta_{xy}$ of the reference frame about the axis O-n which lies in the x-y plane and is perpendicular to the rotation plane OPC. Since in the limit PC is the diagonal of an elemental rectangle of sides PA and PB, it follows that $(d\theta_{xy})^2 = (d\theta_x)^2 + (d\theta_y)^2$, and a similar elemental rectangle is formed by the rotation vectors. Furthermore, it is observed that the order of combination of differential rotations makes no difference, as P would end up at C regardless of which rotation occurs first. Hence *differential* rotation vectors obey the established rules for vector quantities and may be so treated. *Finite* rotations, on the other hand, do not obey the commutative law as the final configuration depends on the order of addition. This fact is easily demonstrated by noting the different final positions of P for, say, two 90-deg rotations, one about the x-axis and one about the y-axis, depending on the order in which they are taken.

If the reference frame in Fig. 21*a* is given an additional infinitesimal rotation $d\boldsymbol{\theta}_z$ about the z-axis, it may be proved in like manner by taking a point on either the x- or the y-axis, that this rotation may be combined vectorially with $d\boldsymbol{\theta}_x$ and $d\boldsymbol{\theta}_y$. Dividing the differential rotations by dt gives the angular velocity components $\omega_x = \dot{\theta}_x$, $\omega_y = \dot{\theta}_y$, $\omega_z = \dot{\theta}_z$ for the reference frame. Thus it is proved that the angular velocity of the reference axes *x-y-z* may be represented by the vector

$$\omega = \mathbf{i}\omega_x + \mathbf{j}\omega_y + \mathbf{k}\omega_z$$

as shown in Fig. 21*b*. At the instant shown the reference frame is, therefore, rotating about the axis *Om*.

In Art. 13 the time derivatives of the unit vectors **i** and **j** due to rotation of the *x-y* axes about the *z*-axis were established. Before proceeding further it is necessary to evaluate these derivatives due to the more general rotation ω. In Fig. 22 is shown the effect of the rotation of axes on the unit vector **i** during time *dt*. Because of the rotation $d\theta_y$ the tip of the vector **i** moves through an arc length $|\mathbf{i}|\, d\theta_y$ in the negative *z*-direction and has a vector change $-\mathbf{k}\, d\theta_y$ since $|\mathbf{i}|$ is unity. Similarly, because of $d\theta_z$ the tip of **i** moves through an arc length $|\mathbf{i}|\, d\theta_z$, so that the vector change is $\mathbf{j}\, d\theta_z$. Clearly no change in **i** occurs as a result of rotation about the *x*-axis. Thus the vector change in **i** due to an infinitesimal rotation $\omega\, dt$ is $d\mathbf{i} = \mathbf{j}\, d\theta_z - \mathbf{k}\, d\theta_y$. Dividing by *dt* in passing to the limit and substituting the angular velocity components give $\dot{\mathbf{i}} = \mathbf{j}\omega_z - \mathbf{k}\omega_y$. This expression is identical with the cross product $\omega \times \mathbf{i}$ as is easily verified by carrying out the cross-product expansion. Similar changes occur for the effect of the rotation components on **j** and **k**, which the reader should examine. The time derivatives of the unit vectors for the reference system rotating with an angular velocity ω may, therefore, be written

$$\dot{\mathbf{i}} = \omega \times \mathbf{i} \qquad \dot{\mathbf{j}} = \omega \times \mathbf{j} \qquad \dot{\mathbf{k}} = \omega \times \mathbf{k} \tag{38}$$

With the meaning established for the angular velocity ω of a rotating reference frame and the time derivatives of its unit vectors, attention is now turned to the meaning of the time derivative of any vector quantity $\mathbf{V} = \mathbf{i}V_x + \mathbf{j}V_y + \mathbf{k}V_z$ in the rotating system, Fig. 23. The derivative of **V** with respect to time as measured in the fixed frame *X-Y-Z* is

$$\begin{aligned}\left(\frac{d\mathbf{V}}{dt}\right)_{XYZ} &= \frac{d}{dt}(\mathbf{i}V_x + \mathbf{j}V_y + \mathbf{k}V_z) \\ &= (\dot{\mathbf{i}}V_x + \dot{\mathbf{j}}V_y + \dot{\mathbf{k}}V_z) + (\mathbf{i}\dot{V}_x + \mathbf{j}\dot{V}_y + \mathbf{k}\dot{V}_z)\end{aligned}$$

With the substitution of Eqs. 38 the terms in the first parentheses become $\omega \times \mathbf{V}$. The terms in the second parentheses represent the components of

Figure 22

Figure 23

the time derivative $(d\mathbf{V}/dt)_{xyz}$ as measured relative to the moving x-y-z reference axes. Thus

$$\left(\frac{d\mathbf{V}}{dt}\right)_{XYZ} = \omega \times \mathbf{V} + \left(\frac{d\mathbf{V}}{dt}\right)_{xyz} \tag{39}$$

Equation 39 establishes the relation between the time derivative of a vector quantity in a fixed system and the time derivative of the vector as observed in the rotating system.

With Eq. 39 in readiness, consider now the space motion of a particle A, Fig. 24, as observed both from a rotating system x-y-z and a fixed system X-Y-Z. The origin of the rotating system coincides with the position of a second reference particle B, and the system has an angular velocity ω. The position-vector equation is $\mathbf{r}_A = \mathbf{r}_B + \mathbf{r}$ where $\mathbf{r}$ stands for $\mathbf{r}_{A/B}$. The time derivative of this relation gives $\mathbf{v}_A = \mathbf{v}_B + \dot{\mathbf{r}}$. From Eq. 39

$$\dot{\mathbf{r}} = \left(\frac{d\mathbf{r}}{dt}\right)_{XYZ} = \omega \times \mathbf{r} + \mathbf{v}_{\text{rel}}$$

where the velocity measured in x-y-z is $\mathbf{v}_{\text{rel}} = (d\mathbf{r}/dt)_{xyz} = \mathbf{i}\dot{x} + \mathbf{j}\dot{y} + \mathbf{k}\dot{z}$. Thus the relative velocity equation becomes

$$\mathbf{v}_A = \mathbf{v}_B + \omega \times \mathbf{r} + \mathbf{v}_{\text{rel}} \tag{40}$$

which is identical with Eq. 22 derived for plane relative motion.

The relative acceleration equation is the time derivative of Eq. 40 which gives $\mathbf{a}_A = \mathbf{a}_B + \dot{\omega} \times \mathbf{r} + \omega \times \dot{\mathbf{r}} + \dot{\mathbf{v}}_{\text{rel}}$. From Eq. 39 the last term becomes

$$\dot{\mathbf{v}}_{\text{rel}} = \left(\frac{d\mathbf{v}_{\text{rel}}}{dt}\right)_{XYZ} = \omega \times \mathbf{v}_{\text{rel}} + \mathbf{a}_{\text{rel}}$$

where $\mathbf{a}_{\text{rel}} = (d\mathbf{v}_{\text{rel}}/dt)_{xyz} = \mathbf{i}\ddot{x} + \mathbf{j}\ddot{y} + \mathbf{k}\ddot{z}$. This result together with the expression for $\dot{\mathbf{r}}$ previously obtained gives, upon collection of terms,

$$\mathbf{a}_A = \mathbf{a}_B + \dot{\omega} \times \mathbf{r} + \omega \times (\omega \times \mathbf{r}) + 2\omega \times \mathbf{v}_{\text{rel}} + \mathbf{a}_{\text{rel}} \tag{41}$$

This expression is identical with Eq. 23 derived for plane relative motion in Art. 13.

As was pointed out in Eq. 22*a* the terms $\omega \times \mathbf{r} + \mathbf{v}_{\text{rel}}$ which appear in Eqs. 22 and 40 constitute the velocity of A relative to B as measured from nonrotating axes. Therefore $\omega \times \mathbf{r}$ is the difference between the relative veloc-

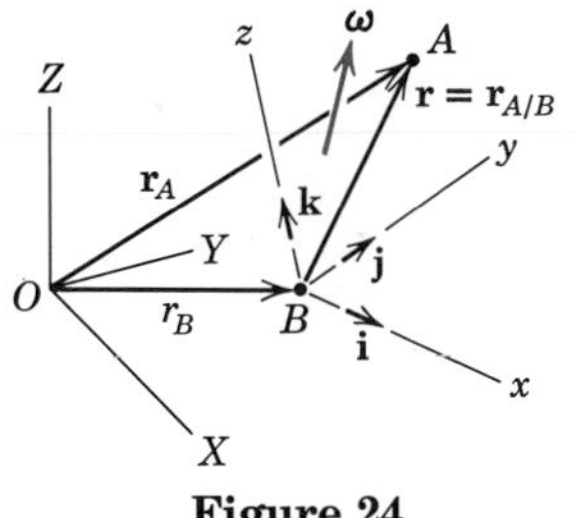

Figure 24

ity as measured from nonrotating axes and as measured from rotating axes. In the acceleration relation it was pointed out in Eq. 23*a* that the terms $2\boldsymbol{\omega} \times \mathbf{v}_{\text{rel}} + \mathbf{a}_{\text{rel}}$ together constitute the acceleration of A with respect to a point P fixed in *x-y-z* and coincident with the moving particle at the instant under observation. The terms $\dot{\boldsymbol{\omega}} \times \mathbf{r} + \boldsymbol{\omega} \times (\boldsymbol{\omega} \times \mathbf{r})$ constitute the acceleration of P with respect to B. The Coriolis acceleration $2\boldsymbol{\omega} \times \mathbf{v}_{\text{rel}}$ is the difference between the acceleration of A relative to P measured from nonrotating axes and that measured from rotating axes. It is strongly recommended that the equivalences expressed by Eqs. 22*a* and 23*a* be restudied, as they apply equally well to space motion as to plane motion.

Sample Problems

2/179 The circular disk of radius r is mounted in bearings in a yoke and spins with a constant angular velocity $\dot{\beta} = p$ about its y-axis. Simultaneously the yoke revolves about the z-axis with an angular velocity ω. Determine the acceleration of a point A on the rim of the disk in terms of the angle β from the vertical z-axis and indicate the values of the acceleration as A passes the points for which $\beta = 0$ and $\beta = 90°$. The *x-y-z* axes are attached to the yoke.

Solution. The point A is observed to have a simple circular motion with the constant angular rate p relative to the *x-y-z* axes which rotate with the constant angular velocity $\boldsymbol{\omega} = \mathbf{k}\omega$. With the origin B at rest the acceleration of A from Eq. 41 is

$$\mathbf{a}_A = \dot{\boldsymbol{\omega}} \times \mathbf{r} + \boldsymbol{\omega} \times (\boldsymbol{\omega} \times \mathbf{r}) + 2\boldsymbol{\omega} \times \mathbf{v}_{\text{rel}} + \mathbf{a}_{\text{rel}}$$

where $\dot{\boldsymbol{\omega}} = 0$ since $\boldsymbol{\omega}$ is constant. The terms in the equation are

$$\boldsymbol{\omega} \times \mathbf{r} = \mathbf{k}\omega \times r(\mathbf{i} \sin \beta + \mathbf{k} \cos \beta) = (r\omega \sin \beta)\mathbf{j}$$

$$\boldsymbol{\omega} \times (\boldsymbol{\omega} \times \mathbf{r}) = \mathbf{k}\omega \times (r\omega \sin \beta)\mathbf{j} = -(r\omega^2 \sin \beta)\mathbf{i}$$

$$2\boldsymbol{\omega} \times \mathbf{v}_{\text{rel}} = 2\omega\mathbf{k} \times rp(\mathbf{i} \cos \beta - \mathbf{k} \sin \beta) = (2\omega rp \cos \beta)\mathbf{j}$$

$$\mathbf{a}_{\text{rel}} = rp^2(-\mathbf{i} \sin \beta - \mathbf{k} \cos \beta)$$

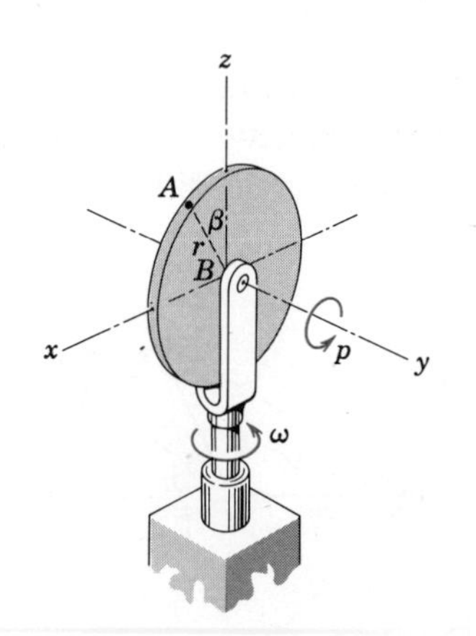

Problem 2/179

The acceleration of point A is

$$\mathbf{a}_A = \mathbf{0} - (r\omega^2 \sin\beta)\mathbf{i} + (2\omega rp\cos\beta)\mathbf{j} + rp^2(-\mathbf{i}\sin\beta - \mathbf{k}\cos\beta)$$

$$= -\mathbf{i}r(\omega^2 + p^2)\sin\beta + (2\omega rp\cos\beta)\mathbf{j} - (rp^2\cos\beta)\mathbf{k} \qquad \textit{Ans.}$$

When $\beta = 0$, $\quad \mathbf{a}_A = rp(2\omega\mathbf{j} - p\mathbf{k}) \qquad$ *Ans.*

When $\beta = 90°$ $\quad \mathbf{a}_A = -\mathbf{i}r(\omega^2 + p^2) \qquad$ *Ans.*

2/180 The disk of radius $r = 100$ mm rotates about its shaft OC with an angular speed p in the direction indicated, and at the same time the shaft, at the fixed angle $\phi = 30°$, rotates about the vertical Z-axis with an angular speed ω. Determine the velocity $\mathbf{v}$ and the acceleration $\mathbf{a}$ of point A on the rim of the disk when the position $\beta = 0$ is passed if, at this instant, $\omega = \frac{3}{2}$ rad/s, $\dot{\omega} = 2$ rad/s^2, $p = 4$ rad/s, $\dot{p} = -3$ rad/s^2, and $\overline{OC} = R = 150$ mm.

Solution. To take full advantage of the relative motion equations, a set of x-y-z axes is chosen which will disclose the rotation p of the disk and which rotates at the angular velocity $\boldsymbol{\omega}$ with the system. The angular speed of the disk measured in or relative to the x-y-z frame is $p = \dot{\beta}$, and, therefore, $p/2\pi$ represents the frequency with which A crosses the y-axis. As a vector, $\mathbf{p}$ is along the z-axis. The frame x-y-z rotates about the Z-axis with the angular velocity $\boldsymbol{\omega}$ which may be expressed in component form as $\boldsymbol{\omega} = (3\sqrt{3}/4)\mathbf{j} + (\frac{3}{4})\mathbf{k}$ rad/s. The component ω_x is zero since ϕ is constant. Note that C has circular motion about Z and that the x-axis is tangent to this path.

The velocity of A is given by Eq. 40, which is

$$\mathbf{v}_A = \mathbf{v}_C + \boldsymbol{\omega} \times \mathbf{r} + \mathbf{v}_{\text{rel}}$$

For the position $\beta = 0$, the terms are

$$\mathbf{v}_C = (R\cos\phi)\,\omega\mathbf{i} = \left(\frac{150\sqrt{3}}{2}\right)\frac{3}{2}\mathbf{i} = \frac{225\sqrt{3}}{2}\mathbf{i} \text{ mm/s}$$

$$\boldsymbol{\omega} \times \mathbf{r} = \left(\frac{3\sqrt{3}}{4}\mathbf{j} + \frac{3}{4}\mathbf{k}\right) \times 100\mathbf{j} = -75\mathbf{i} \text{ mm/s}$$

$$\mathbf{v}_{\text{rel}} = -rp\mathbf{i} = -100(4)\mathbf{i} = -400\mathbf{i} \text{ mm/s}$$

Thus the velocity of A is

$$\mathbf{v}_A = \left(\frac{225\sqrt{3}}{2} - 75 - 400\right)\mathbf{i} = -280\mathbf{i} \text{ mm/s} \qquad \textit{Ans.}$$

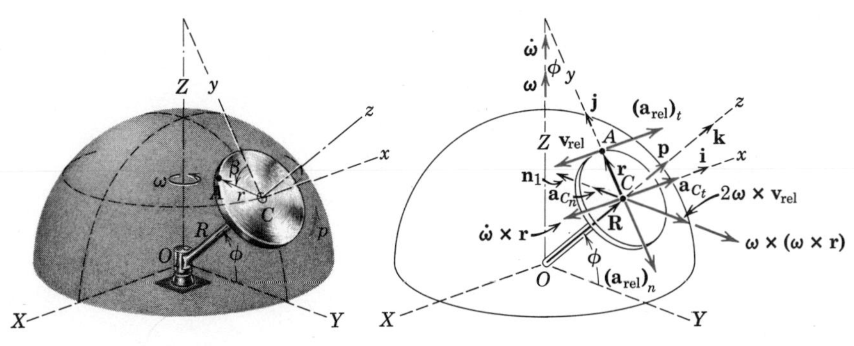

Problem 2/180

The acceleration of A is given by Eq. 41, which is

$$\mathbf{a}_A = \mathbf{a}_C + \dot{\omega} \times \mathbf{r} + \omega \times (\omega \times \mathbf{r}) + 2\omega \times \mathbf{v}_{rel} + \mathbf{a}_{rel}$$

For the position $\beta = 0$, the terms are evaluated as follows:

The tangential and normal components of the acceleration of C in its circular motion of radius $R \cos \phi$ are

$$\mathbf{a}_{C_t} = (R \cos \phi)\, \dot{\omega}\mathbf{i} = 150\frac{\sqrt{3}}{2}2\mathbf{i} = 150\sqrt{3}\mathbf{i} \text{ mm/s}^2$$

$$\mathbf{a}_{C_n} = (R \cos \phi)\, \omega^2\mathbf{n}_1 = 150\frac{\sqrt{3}}{2}\left(\frac{3}{2}\right)^2\mathbf{n}_1 = \frac{675\sqrt{3}}{8}\mathbf{j} - \frac{2025}{8}\mathbf{k} \text{ mm/s}^2$$

The acceleration components of a point P with respect to C, where P is attached to the y-axis and coincides momentarily with A, are

$$\dot{\omega} \times \mathbf{r} = (\sqrt{3}\mathbf{j} + \mathbf{k}) \times 100\mathbf{j} = -100\mathbf{i} \text{ mm/s}^2$$

$$\omega \times (\omega \times \mathbf{r}) = \left(\frac{3\sqrt{3}}{4}\mathbf{j} + \frac{3}{4}\mathbf{k}\right) \times \left[\left(\frac{3\sqrt{3}}{4}\mathbf{j} + \frac{3}{4}\mathbf{k}\right) \times 100\mathbf{j}\right] = -\frac{225}{4}\mathbf{j} + \frac{225\sqrt{3}}{4}\mathbf{k} \text{ mm/s}^2$$

The Coriolis acceleration is

$$2\omega \times \mathbf{v}_{rel} = 2\left(\frac{3\sqrt{3}}{4}\mathbf{j} + \frac{3}{4}\mathbf{k}\right) \times (-400\mathbf{i}) = -600\mathbf{j} + 600\sqrt{3}\mathbf{k} \text{ mm/s}^2$$

The relative acceleration of A as seen by an observer attached to x-y-z has n- and t-components due to p and $\dot{p}$, respectively, which are

$$(\mathbf{a}_{rel})_n = -rp^2\mathbf{j} = -100(4)^2\mathbf{j} = -1600\mathbf{j} \text{ mm/s}^2$$

$$(\mathbf{a}_{rel})_t = -r\dot{p}\mathbf{i} = -100(-3)\mathbf{i} = 300\mathbf{i} \text{ mm/s}^2$$

Addition of the **i**-, **j**-, and **k**-components of all acceleration terms gives

$$\mathbf{a}_A = 459\mathbf{i} - 2110\mathbf{j} + 884\mathbf{k} \text{ mm/s}^2 \qquad \textit{Ans.}$$

The magnitude of $\mathbf{a}_A$ is

$$a_A = \sqrt{(459)^2 + (2110)^2 + (884)^2} = 2330 \text{ mm/s}^2 \text{ or } 2.33 \text{ m/s}^2$$

Each of the acceleration terms should be studied so that its physical and geometrical significance is clear. The student should also examine other possible choices for the rotating coordinate system.

Problems

Translating Reference Axes

2/181 The rocket rises vertically with an acceleration in metres per second squared given by $a = 9.81(0.1 + 0.05t)$ where t is the time in seconds after lift-off. To stabilize the rocket it is set into rotation about its longitudinal axis by small tangential jets, and its angular velocity ω in radians per second for the early part of its ascent increases according to $\omega = 0.4t$. The x-y-z axes are attached to point C on the center line of the rocket but do not rotate with it. The point A is located on the shell of the rocket a distance of 0.9 m from the center line. For the instant when $t = 2$ s, determine the magnitude of the acceleration of A with respect to C as viewed from the translating axes and the acceleration of A with respect to the ground point O.

Ans. $a_{A/C} = 0.679 \text{ m/s}^2 \quad a_{A/O} = 2.08 \text{ m/s}^2$

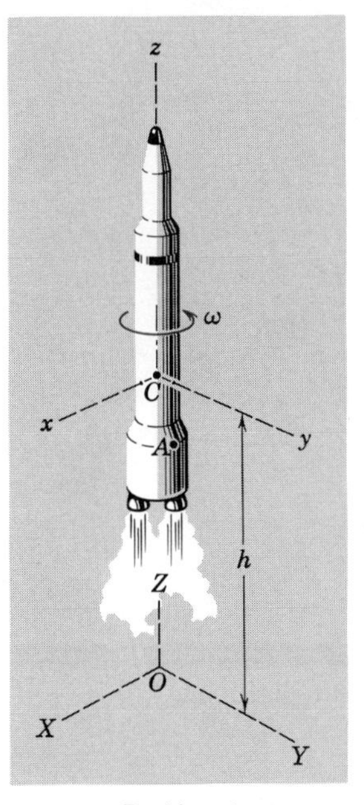

Problem 2/181

2/182 The spotter plane A is flying north with a constant speed of 150 km/h at an altitude $h = 4.5$ km and is 4.5 km from crossing over the course of the aircraft carrier B, which is heading west with a constant speed of 35 knots (1 knot = 1.852 km/h). For the instant described compute the quantity $\dot{R}$ as measured from the carrier.

2/183 For the conditions of Prob. 2/182 determine the values of $\ddot{r}$ and $\ddot{\theta}$ as measured from the carrier.

Ans. $\ddot{r} = 877 \text{ km/h}^2, \quad \ddot{\theta} = -409 \text{ rad/h}^2$

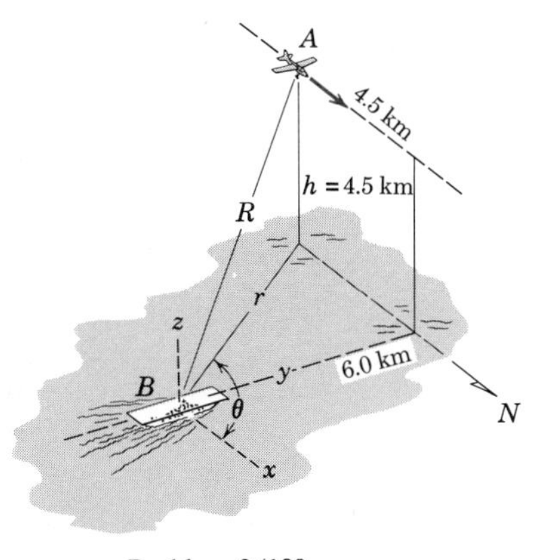

Problem 2/182

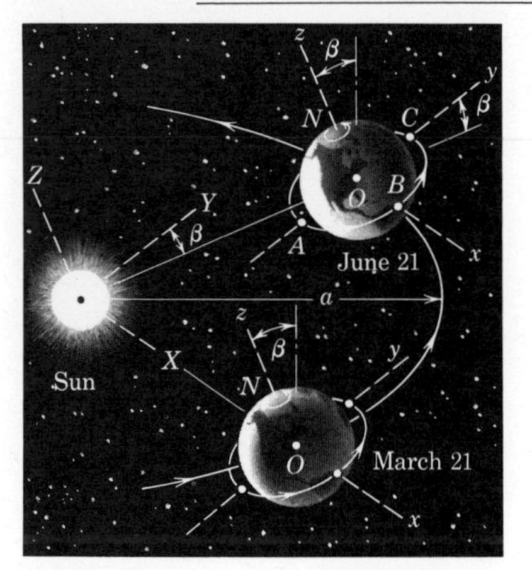

Problem 2/184

2/184 One coordinate system used to decribe space motion is a "fixed" set of X-Y-Z axes with origin at the sun, with X-axis in the plane of the ecliptic (earth's orbital plane), and with Z-axis parallel to the earth's polar axis. The direction of the polar axis is essentially constant, and the earth's equatorial plane makes an angle $\beta = 23.45°$ with the plane of the ecliptic. The moving but nonrotating coordinates x-y-z have their origin at the center of the earth with the z-axis coinciding with the polar axis and the x-axis parallel to the X-axis and in the ecliptic plane. The figure shows the earth on March 21 (zero declination) and on June 21 (declination β). If an earth satellite makes a circular equatorial orbit about the earth from west to east at an altitude of 480 km, determine the absolute velocity of the satellite in X-Y-Z when at A, B, and C on June 21. The mean radius of the earth is 6370 km, and the period of the satellite for one complete revolution relative to x-y-z is 1.57 h. The mean radius a of the near-circular orbit of the earth is $a = 149.6(10^6)$ km, and the earth's period about the sun is $365\frac{1}{4}$ days.

Ans. $\mathbf{v}_A = -79\,800\mathbf{i}$ km/h
$\mathbf{v}_B = -107\,200\mathbf{i} + 27\,400\mathbf{j}$ km/h
$\mathbf{v}_C = -134\,600\mathbf{i}$ km/h

2/185 For noon on March 21 determine the latitude of a point on the earth's surface which has no absolute acceleration in the primary X-Y-Z system. Use the data given in Prob. 2/184. The angular velocity of the earth is $\omega = 0.729(10^{-4})$ rad/s.

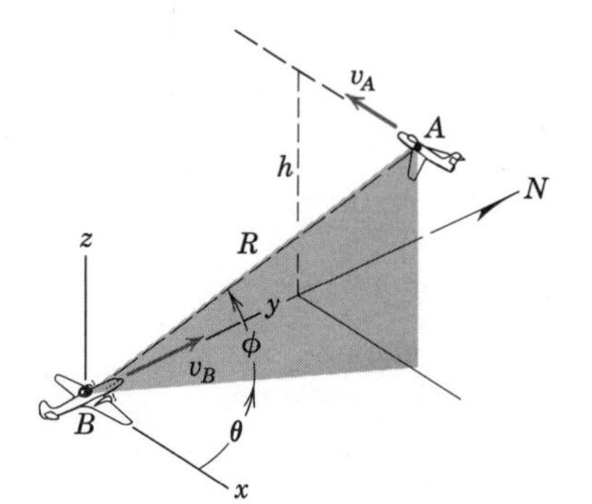

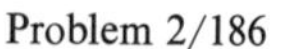
Problem 2/186

2/186 Aircraft A, flying horizontally due west with a speed of 1200 km/h, is tracked by radar from aircraft B, flying horizontally due north with a speed of 900 km/h at an altitude h below A. Determine the angle θ for the radar beam at which the distance R between the two aircraft is momentarily neither increasing nor decreasing.

Ans. $\theta = 126.9°$

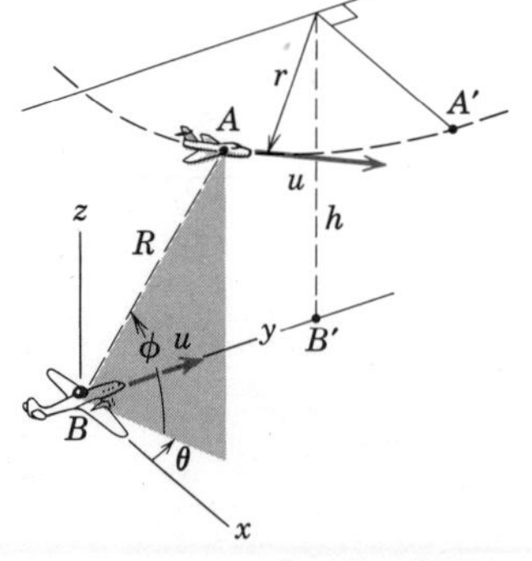

Problem 2/187

2/187 Aircraft A is flying in a horizontal circle of radius r at a constant speed u. Aircraft B with radar tracking is flying in a horizontal line with the same constant speed u at an altitude h below the center of A's circle. If A and B arrive simultaneously at the primed positions shown, write expressions for the values of $\ddot{R}$ and $\ddot{\phi}$ at this particular position.

◀**2/188** An airplane begins its landing approach from an altitude of 1 km above point A. The airport runway B is 18 km due north of A. If the airspeed during the approach is 240 km/h and if a 50-km/h wind is blowing from the south-east, determine the proper direction which the airplane should be headed by specifying the angles θ and ϕ. *Ans.* $\theta = 8.49°$, $\phi = 3.61°$

◀**2/189** A guided missile A is launched from point O and pursues the hostile aircraft B. The guidance system maintains the direction of flight always toward B. At the instant represented the missile has a velocity $v = 3750$ km/h, and the aircraft has a horizontal northerly velocity $u = 1200$ km/h. Determine the rate at which the distance r between the aircraft and the missile is decreasing for this position if the coordinates of A are $R = 45$ km, $\theta = 60°$, and $\phi = 12°$.

Ans. $-\dot{r} = 2590$ km/h

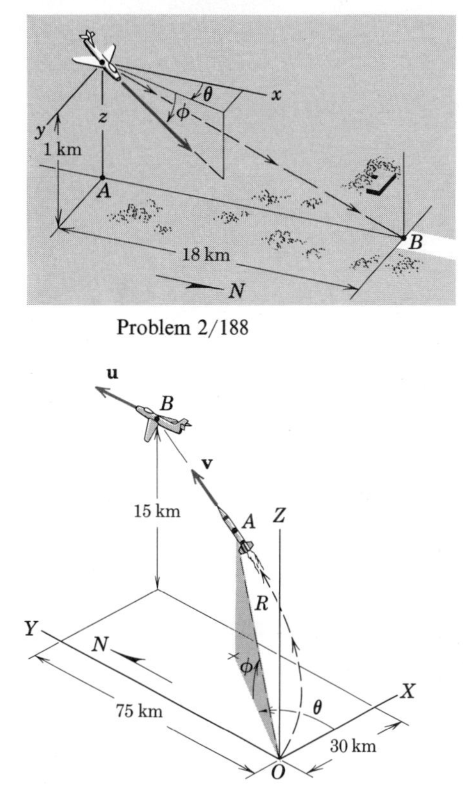

Problem 2/188

Problem 2/189

Rotating Reference Axes

2/190 In making a landing approach, aircraft B, flying at 210 km/h, is reducing the angle θ between its flight path and the ground at the constant rate of 1.0°/s. At the instant represented, when $\theta = 5°$, an aircraft A, flying horizontally at 300 km/h, crosses the path of B at right angles to it and 9 km from B. Determine the velocity which A appears to have to the pilot of B in the x-y-z frame of reference attached to B.

Ans. $\mathbf{v}_{rel} = -300\mathbf{i} - 210\mathbf{j} - 565\mathbf{k}$ km/h

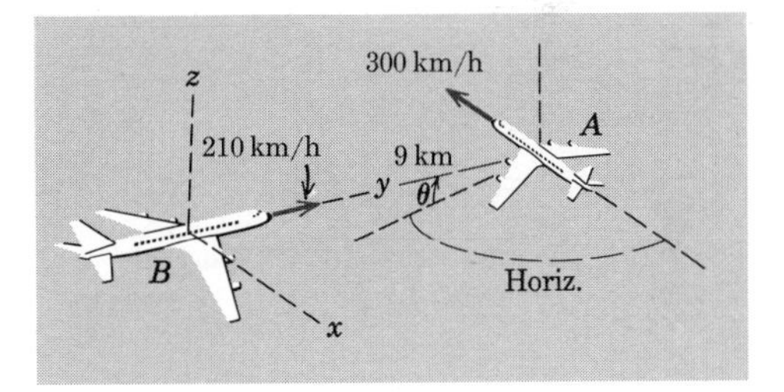

Problem 2/190

2/191 A car is traveling due south on a straight level road at a speed of 100 km/h. The latitude is 40° north. Consider the center of the earth fixed and compute the absolute easterly or westerly component of acceleration of the car. The angular velocity of the earth is $\omega = 0.729(10^{-4})$ rad/s.

2/192 Derive the formula for the second time derivative, referred to a fixed coordinate reference system X-Y-Z, of a vector $\mathbf{P}$ which is fixed in a moving system x-y-z that has an angular velocity $\boldsymbol{\omega}$.

$$\textit{Ans.}\quad \left(\frac{d^2\mathbf{P}}{dt^2}\right)_{XYZ} = \dot{\boldsymbol{\omega}} \times \mathbf{P} + \boldsymbol{\omega} \times (\boldsymbol{\omega} \times \mathbf{P})$$

2/193 The angular velocity of an x-y-z reference frame is given in radians per second by $\omega = 2t\mathbf{i} + 3t^2\mathbf{j} + 4t^3\mathbf{k}$ where t is the time in seconds. If a momentum vector $\mathbf{H}$ has x-, y-, z-components of 6, 3, 5 kg·m²/s which are invariant in x-y-z, determine the time derivative of $\mathbf{H}$ in a fixed coordinate reference frame X-Y-Z at the instant when $t = 2$ s.

2/194 Apply Eq. 39 as a vector operator to itself to obtain the relation between the second time derivatives of a vector quantity $\mathbf{V}$ as observed both in fixed coordinates X-Y-Z and in coordinates x-y-z which rotate with an angular velocity ω.

Ans. $$\left(\frac{d^2\mathbf{V}}{dt^2}\right)_{XYZ} = \dot{\omega} \times \mathbf{V} + \omega \times (\omega \times \mathbf{V}) + 2\omega \times \left(\frac{d\mathbf{V}}{dt}\right)_{xyz} + \left(\frac{d^2\mathbf{V}}{dt^2}\right)_{xyz}$$

2/195 Determine the second derivative with respect to time of the vector $\mathbf{H}$ of Prob. 2/193 referred to the fixed coordinates X-Y-Z at the instant when $t = 2$ s.

Ans. $$\left(\frac{d^2\mathbf{H}}{dt^2}\right)_{XYZ} = -6308\mathbf{i} - 634\mathbf{j} + 1054\mathbf{k} \text{ kg·m}^2/\text{s}^3$$

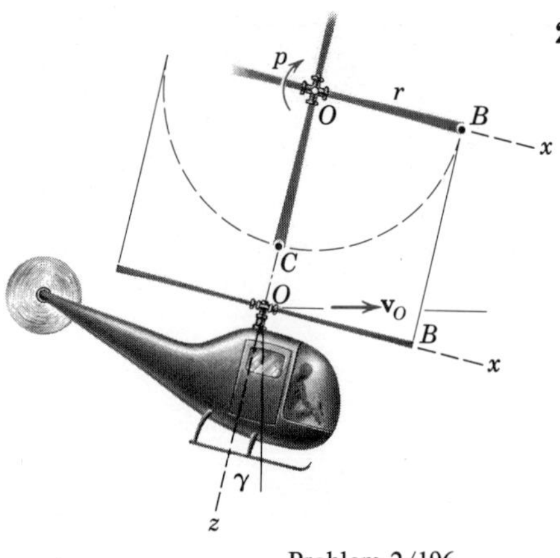

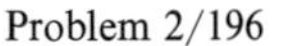
Problem 2/196

2/196 A helicopter, which is flying with a horizontal velocity $\mathbf{v}_O$ with its rotor axis tilted an angle γ from the vertical as shown, prepares for a vertical ascent by decreasing γ at the constant rate ω rad/s. At the same time, the velocity of O is decreasing at the rate a_O. Write the expressions for the velocity and acceleration of point B on the tip of the blade as it crosses the x-axis in the position shown. The x- and z-axes are fixed to the helicopter body in the vertical plane, and the z-axis coincides with the rotor axis. The constant angular speed of the rotor is p, clockwise as viewed from above.

Ans.
$$\mathbf{v}_B = \mathbf{i}v_O \cos\gamma + \mathbf{j}rp - \mathbf{k}(v_O \sin\gamma + r\omega)$$
$$\mathbf{a}_B = -\mathbf{i}(a_O \cos\gamma + r\omega^2 + rp^2) + \mathbf{k}a_O \sin\gamma$$

2/197 Write the expressions for the velocity and the acceleration of point C on the tip of the helicopter blade at the instant shown for the conditions of Prob. 2/196.

2/198 While the disk rotates about the z-axis with a constant angular velocity ω, the slider A oscillates in its slot with a displacement given by $s = s_0 \sin 2\pi nt$ where n is the frequency of oscillation and t is the time. Determine the acceleration $\mathbf{a}$ of the slider (a) when it reaches the extreme position $s = s_0$ with $\ddot{s}$ negative and (b) when it passes the position $s = 0$ with $\dot{s}$ positive. The x-y-z axes are attached to the disk.

Ans. (a) $\mathbf{a}_A = s_0[-\mathbf{j}(\omega^2 + 4\pi^2 n^2)\cos\beta + \mathbf{k}(4\pi^2 n^2 \sin\beta)]$
for $s = s_0$ and $\ddot{s}$ negative
(b) $\mathbf{a}_A = -(4\pi n\omega s_0 \cos\beta)\mathbf{i}$
for $s = 0$ and $\dot{s}$ positive

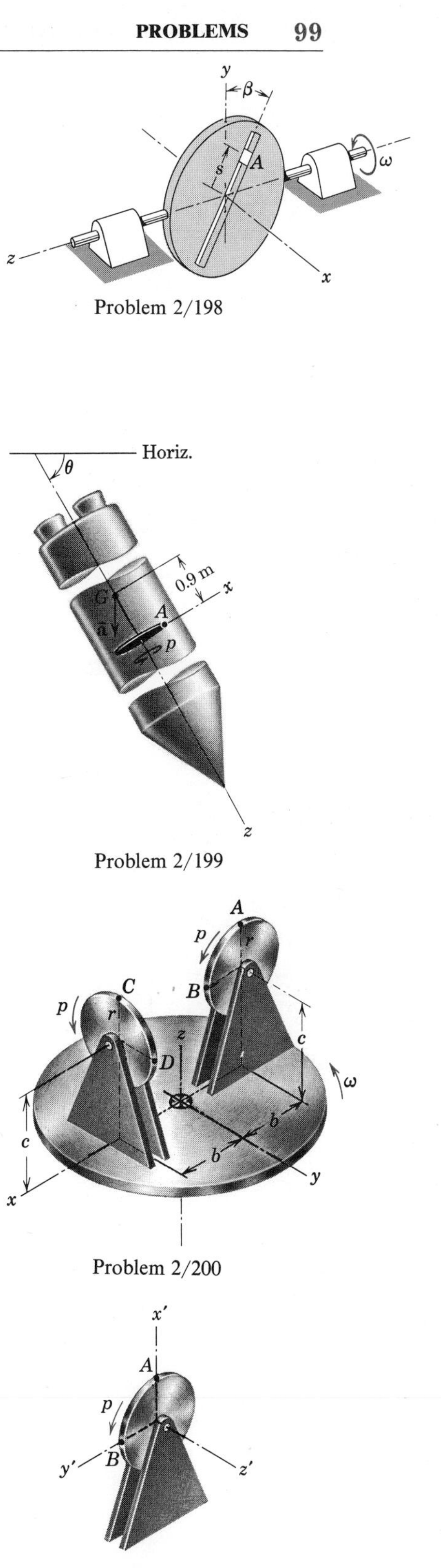

Problem 2/198

Problem 2/199

Problem 2/200

Problem 2/202

2/199 At the top of a vertical trajectory a certain rocket undergoes a rapid angular maneuver to point its nose downward for return to earth. At the instant shown, $\theta = 60°$, $\dot{\theta} = 3$ rad/s, and $\ddot{\theta}$ is negligibly small. Also the acceleration $\bar{\mathbf{a}}$ of the mass center G is 9 m/s² downward. For this instant, determine the acceleration of a point A on the uppermost part of the rim of a 150-mm-diameter disk rotating about the rocket axis with a constant speed $p = 12$ rad/s relative to the rocket, which itself has no spin about its own axis. The x-z axes are attached to the rocket and lie in the vertical plane.

Ans. $\mathbf{a}_A = -(15.98\mathbf{i} + 0.306\mathbf{k})$ m/s²

2/200 The large disk turns about its central vertical axis with a constant angular speed ω in the direction indicated. The small disks in turn spin about their horizontal axes at the constant angular speed p relative to the large disk in the directions shown. Determine the expressions for the accelerations of points A and B at the instant represented.

Ans. $\mathbf{a}_A = b\omega^2\mathbf{i} + 2rp\omega\mathbf{j} - rp^2\mathbf{k}$
$\mathbf{a}_B = [(b - r)\omega^2 - rp^2]\mathbf{i}$

2/201 Determine the expressions for the accelerations of points C and D at the instant represented for Prob. 2/200.

2/202 Introduce the coordinate axes shown with this problem and solve Prob. 2/200 modified in that $\dot{\omega} \neq 0$. The x'- and y'-axes remain vertical and horizontal, respectively, and remain in the plane of the small disk.

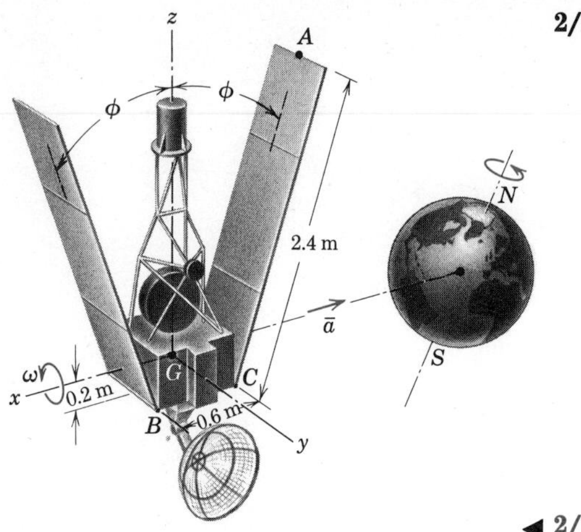

Problem 2/203

2/203 The spacecraft with attached x-y-z axes is tumbling about the x-axis at the constant rate $\omega = 2$ rad/s. Determine the acceleration of point A on the tip of the solar panel as the x-axis comes in line with the center of the earth if the panels are unfolding at the rate $\dot{\phi} = 4$ rad/s with $\ddot{\phi} = 6$ rad/s^2 for the position $\phi = 30°$. The mass center G of the spacecraft has an acceleration $\bar{a} = 6$ m/s^2 toward the earth's center, which is here considered fixed in space. The hinge axes B and C for the panels are 0.6 m apart and 0.2 m "below" G.

Ans. $\mathbf{a}_A = 0.729\mathbf{i} + 19.2\mathbf{j} - 48.0\mathbf{k}$ m/s^2

◀ **2/204** A certain rocket test sled reaches a velocity of 600 m/s in a north-easterly direction on a horizontal track at 30° north latitude. Compute the horizontal component of the acceleration of the sled normal to the rails due to the rotation of the earth. The mean radius and angular velocity of the earth are 6370 km and $0.729(10^{-4})$ rad/s, respectively.

Ans. $a_n = 0.0541$ m/s^2 (north-west)

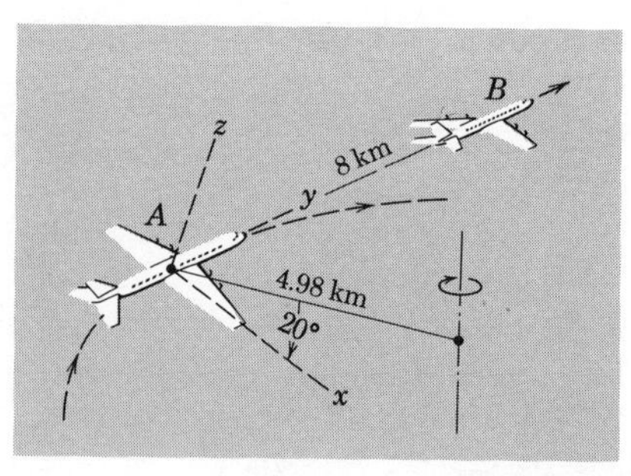

Problem 2/205

◀ **2/205** Aircraft A is flying in a horizontal circle of 4.98-km radius at a constant speed of 480 km/h and is banked at the correct angle which is 20°. The velocity of a second aircraft B flying horizontally in a straight line at a constant speed of 640 km/h momentarily becomes collinear with the velocity of A when the two aircraft are 8 km apart in the positions shown. Use the x-y-z frame of reference attached to A and determine the velocity and acceleration of B as observed from and relative to x-y-z at this instant.

Ans. $\mathbf{v}_{rel} = -725\mathbf{i} + 160\mathbf{j} - 264\mathbf{k}$ km/h

$\mathbf{a}_{rel} = -5.59\mathbf{i} - 5.73\mathbf{j} - 2.03\mathbf{k}$ m/s^2

2/206 The electric motor with attached disk is mounted on the base S which swivels about the vertical Z-axis with the angular speed ω. The motor axis makes a fixed $30°$ angle with the horizontal. The x-y-z axes are fixed to the motor frame with the y-axis always in a vertical plane through the shaft z-axis. The disk rotates at the speed p relative to x-y-z so that a point P on the rim crosses the y-axis with the frequency $p/2\pi$. Determine the vector expressions for the velocity and acceleration of P as it crosses the position $\theta = 60°$ under the conditions that $\omega = 10$ rad/s and $p = 20$ rad/s, both constant.

Ans. $\mathbf{v}_P = -1.907\mathbf{i} + 1.562\mathbf{j} - 0.541\mathbf{k}$ m/s
$\mathbf{a}_P = -43.8\mathbf{i} - 63.7\mathbf{j} + 35.3\mathbf{k}$ m/s^2

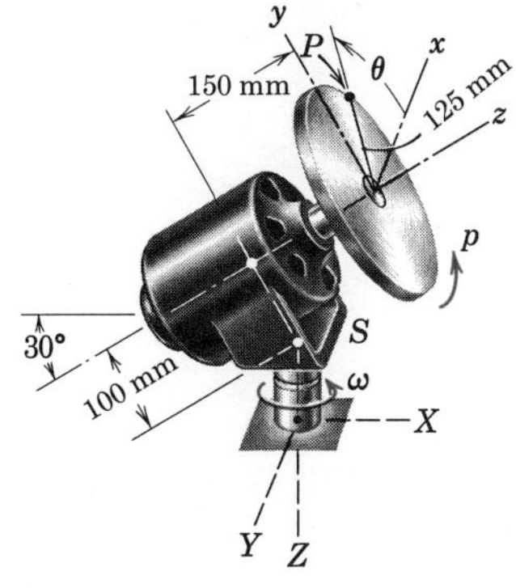

Problem 2/206

2/207 An earth satellite, which makes a circular polar orbit of altitude $h = 640$ km above the earth, travels at a speed of 27 100 km/h with respect to nonrotating axes through the earth's center considered fixed. Calculate the velocity and acceleration of the satellite parallel to the plane of the earth's surface with an earth observer A at $30°$ north latitude would measure as the satellite passed overhead. Specify the direction in terms of the angle θ measured clockwise from the north. The radius R and angular velocity ω of the earth are 6370 km and $0.729(10^{-4})$ rad/s, respectively.

Ans. $v_{\text{rel}} = 27\,150$ km/h, $\theta = 356.6°$
$(a_{\text{rel}})_{xy} = 0.549$ m/s^2, $\theta = 88.3°$

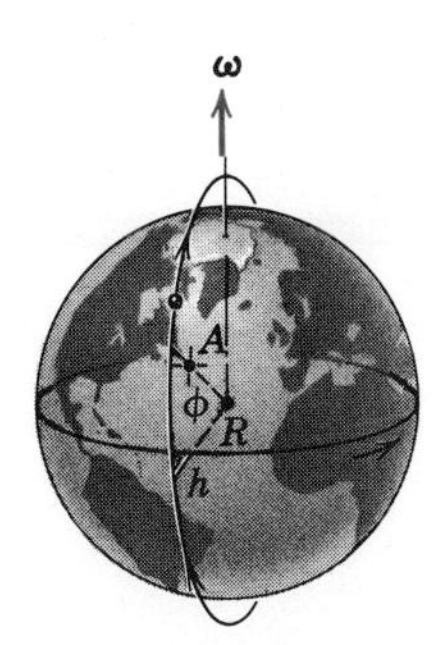

Problem 2/207

3 KINETICS OF PARTICLES

16 Introduction. When a particle is subjected to a force system which is unbalanced, the particle has accelerated motion. Kinetics is the study of the relations between unbalanced force systems and the changes in motion which they produce. The properties of force systems were covered in the study of statics, and the changes in the motions of particles were covered in Chapter 2. In Chapter 3 the laws governing the relations between force and motion are developed and illustrated. The basic concepts and definitions presented in Chapter 1 are fundamental to the development of kinetics, and a careful review of this introductory chapter is strongly recommended at this point.

The basic relation between force and acceleration is stated by Newton's second law of motion, Eq. 1. The proof for this law is entirely experimental, and the fundamental meaning of the law will be described by an ideal experiment in which force and acceleration are assumed to be measured without error. A mass particle is isolated in the primary inertial system* and is subjected to the action of a single force $\mathbf{F}_1$. The acceleration $\mathbf{a}_1$ of the particle is measured, and the ratio F_1/a_1 of the magnitudes of the force and the acceleration will be some number C_1 whose value depends on the units used for measurement of force and acceleration. The experiment is now repeated by subjecting the same particle to a different force $\mathbf{F}_2$ and measuring the corresponding acceleration $\mathbf{a}_2$. Again the ratio of the magnitudes F_2/a_2 will produce a number C_2. The experiment is repeated as many times as desired. Two important conclusions are drawn from the results. First, the ratios of applied force to corresponding acceleration all equal the *same* number, provided the units used for measurement are not changed in the experiments. Thus,

$$\frac{F_1}{a_1} = \frac{F_2}{a_2} = \cdots = \frac{F}{a} = C, \qquad \text{a constant}$$

The constant C is a measure of some property of the particle which does not change. This property is the *inertia* of the particle which is its *resistance to rate of change of velocity.* For a particle of high inertia (large C) the acceleration will be small for a given force F, and, conversely, if the inertia is small, the acceleration will be large. The mass m is used as a quantitative measure of inertia, and therefore the expression $C = km$ may be written, where k is

* The primary inertial system or astronomical frame of reference is an imaginary set of reference axes which are assumed to have no translation or rotation in space. See Art. 2, Chapter 1.

a constant to account for the units used. Thus the experimental relation becomes

$$F = kma \tag{42}$$

where F is the magnitude of the resultant force acting on the particle of mass m, and a is the magnitude of the resulting acceleration of the particle.

The second conclusion from the ideal experiment is that the acceleration is always in the direction of the applied force. Thus Eq. 42 is a *vector* relation and may be written

$$\mathbf{F} = km\mathbf{a} \tag{43}$$

Although an actual experiment cannot be performed in the ideal manner described, the conclusions are inferred from the measurements of countless accurately performed experiments where the results are correctly predicted from the hypothesis of the ideal experiment. One of the most accurate checks lies in the precise prediction of the motions of planets based on Eq. 43.

It should be understood that the results of the fundamental experiment may be obtained only if measurements are made relative to the "fixed" primary inertial system. Thus, if the experiment described were performed on the surface of the earth and all measurements were made relative to a reference system attached to the earth, the measured results would show a slight discrepancy upon substitution into Eq. 43. This discrepancy would be due to the fact that the measured acceleration would not be the correct absolute acceleration. The discrepancy would disappear when the corrections due to the acceleration components of the earth were accounted for. These corrections are negligible for most engineering problems which involve the motions of structures and machines on the surface of the earth. In such case the accelerations measured with respect to reference axes attached to the surface of the earth may be treated as "absolute," and Eq. 43 may be applied with negligible error to experimental measurements made on the surface of the earth.*

There are an increasing number of problems, particularly in the fields of rocket and spacecraft design, where the acceleration components of the earth are of primary concern. For this work it is essential that the fundamental basis of Newton's law be thoroughly understood and that the appropriate absolute acceleration components be employed.

* An example of the magnitude of the error introduced by neglect of the motion of the earth may be cited for the case of a particle which is allowed to fall from rest (relative to the earth) at a height h above the ground. It may be shown that the rotation of the earth gives rise to an eastward acceleration (Coriolis acceleration) relative to the earth and, neglecting air resistance, that the particle falls to the ground a distance

$$x = \frac{2}{3}\omega\sqrt{\frac{2h^3}{g}}\cos\gamma$$

east of the point on the ground directly under that from which it was dropped. The angular velocity of the earth is $\omega = 0.729(10^{-4})$ rad/s, and the latitude, north or south, is γ. At a latitude of 45° and from a height of 200 m, this eastward deflection would be $x = 43.9$ mm.

Before 1905 the laws of Newtonian mechanics had been verified by innumerable physical experiments and were considered the final description of the motion of bodies. The concept of *time,* considered an absolute quantity in the Newtonian theory, received a basically different interpretation in the theory of relativity announced by Einstein in 1905. The new concept called for a complete reformulation of the accepted laws of mechanics. The theory of relativity was subjected to early ridicule but has had experimental check and is now universally accepted by scientists the world over. Although the difference between the mechanics of Newton and that of Einstein is basic, there is a practical difference in the results given by the two theories only when velocities of the order of the speed of light ($300(10^6)$ m/s) are encountered.* Important problems dealing with atomic and nuclear particles, for example, involve calculations based on the theory of relativity and are of basic concern to both scientists and engineers.

It is customary to take k equal to unity in Eq. 43, thus putting the relation in the usual form of Newton's second law

$$\mathbf{F} = m\mathbf{a} \qquad [1]$$

A system of units for which k is unity is known as a *kinetic* system. Thus for a kinetic system the units of force, mass, and acceleration are not independent. In the SI system, as explained in Art. 4 of Chapter 1, the units of force (newtons N) are derived by Newton's second law from the base units of mass (kilograms kg) times acceleration (metres per second squared m/s^2). Thus $N = kg \cdot m/s^2$. This system is known as an *absolute* system since the unit for force is dependent on the absolute value of mass.

In the U.S.-British system of units, on the other hand, the units of mass (slugs) are derived from the units of force (pounds force lbf) divided by acceleration (feet per second squared ft/s^2). Thus the mass units are slugs = $lbf \cdot s^2/ft$. This system is known as a gravitational system since mass is derived from force as determined from gravitational attraction.

For measurements made relative to the rotating earth the relative value of g should be used. The internationally accepted value of g relative to the earth at sea level and at a latitude of 45° is 9.806 65 m/s^2. Except where greater precision is required the value of 9.81 m/s^2 will be used for g. For measurements relative to a nonrotating earth the absolute value of g should be used. At a latitude of 45° and at sea level the absolute value is 9.8236 m/s^2. The sea-level variation in both the absolute and relative values of g with latitude is shown in Fig. 1 of Art. 5.

In the U.S.-British system the standard value of g relative to the rotating earth at sea level and at a latitude of 45° is 32.1740 ft/s^2. The corresponding value relative to a nonrotating earth is 32.2295 ft/s^2.

*The theory of relativity demonstrates that there is no such thing as a preferred primary inertial system and that measurements which are made of time in two coordinate systems which have a velocity relative to one another are different. On this basis, for example, the principles of relativity show that a clock carried by the pilot of a spacecraft traveling around the earth in a circular polar orbit of 644 km altitude at a velocity of 27 080 km/h would be slow compared with a clock at the pole by 0.000 001 85 s for each orbit.

17 Equation of Motion. Consider a particle of mass m subjected to the action of the concurrent forces $\mathbf{F}_1$, $\mathbf{F}_2$, $\mathbf{F}_3, \ldots$ whose vector sum is $\Sigma\mathbf{F}$. Equation 1 becomes

$$\Sigma\mathbf{F} = m\mathbf{a} \tag{44}$$

In the solution of problems Eq. 44 is usually expressed in component form depending on the coordinate system used. The choice of the appropriate coordinate system is dictated by the type of motion involved and is a vital step in the formulation of the problem.

When the forces or motion are described by rectangular coordinates x, y, z, Eq. 44 will have the components

$$\begin{aligned} \Sigma F_x &= ma_x \\ \Sigma F_y &= ma_y \\ \Sigma F_z &= ma_z \end{aligned} \tag{45}$$

where

$$|\Sigma\mathbf{F}| = \sqrt{(\Sigma F_x)^2 + (\Sigma F_y)^2 + (\Sigma F_z)^2} \quad \text{and} \quad |\mathbf{a}| = \sqrt{a_x{}^2 + a_y{}^2 + a_z{}^2}$$

If the motion is rectilinear, the x-axis, for example, may be chosen to coincide with the direction of the acceleration $\mathbf{a}$, and the motion equations become $\Sigma F_x = ma$, $\Sigma F_y = 0$, $\Sigma F_z = 0$. The equation of motion may also be written as a differential equation $\Sigma F_x = m\ddot{x}$. This form of the motion equation may be used to describe problems where ΣF_x is a function of time, displacement, or velocity. Two successive integrations of the equation would be required to express x as a function of t. Once this functional relation is established, the motion is completely known.

For plane curvilinear motion of a particle where n- and t-components are used, the components of Eq. 44 may be written

$$\begin{aligned} \Sigma F_n &= ma_n \\ \Sigma F_t &= ma_t \end{aligned} \tag{46}$$

where $a_n = v\dot{\theta} = \rho\dot{\theta}^2 = v^2/\rho$ and $a_t = \dot{v} = \ddot{s}$ from Eqs. 12 and 13. For circular motion the radius of curvature ρ is constant, and the acceleration components may be written as in Eq. 16.

For plane curvilinear motion where r- and θ-components are used, the components of Eq. 44 may be written

$$\begin{aligned} \Sigma F_r &= ma_r \\ \Sigma F_\theta &= ma_\theta \end{aligned} \tag{47}$$

where $a_r = \ddot{r} - r\dot{\theta}^2$ and $a_\theta = r\ddot{\theta} + 2\dot{r}\dot{\theta}$ from Eq. 18.

Similarly, for space curvilinear motion the cylindrical coordinates of Eqs. 26 may be used with

$$\Sigma F_r = ma_r$$
$$\Sigma F_\theta = ma_\theta \tag{48}$$
$$\Sigma F_z = ma_z$$

or the spherical coordinates of Eqs. 28 may be used with

$$\Sigma F_R = ma_R$$
$$\Sigma F_\theta = ma_\theta \tag{49}$$
$$\Sigma F_\phi = ma_\phi$$

There are frequent occasions where the absolute acceleration **a** of a particle is best described in terms of the acceleration of a moving coordinate system and the acceleration relative to that moving system. This formulation will be discussed in Art. 21.

It should be clear that the choice of the coordinate system to use in a given problem is an important decision. Thus it is imperative that the student be familiar with the kinematics of particle motion in order that he may proceed with the kinetics of the problem. The choice of a coordinate system is frequently indicated by the number and geometry of the constraints. Thus, if a particle is free to move in space, as with the center of mass of a rocket in free flight, the particle is said to have *three degrees of freedom* which will require three independent coordinates to specify the position of the particle at any instant. All three of the scalar components, Eqs. 45, 48, or 49, of the equation of motion would have to be applied and integrated to obtain the space coordinates as a function of time. If a particle is constrained to move along a surface, as with a marble sliding on the curved surface of a bowl, only two coordinates are needed to specify its position, and in this case it is said to have *two degrees of freedom.* If a particle is constrained to move along a fixed linear path, as with a bead sliding along a fixed wire, its position may be specified by the coordinate measured along the wire. In this case the particle would have only *one degree of freedom.*

Equation 44, or any one of the component forms of the force-mass-acceleration equation, is usually referred to as the *equation of motion.* The equation of motion gives the instantaneous value of the acceleration corresponding to the instantaneous value of the forces which are acting. If the forces are variable, the acceleration will also be variable, and the changes in velocity and displacement of the particle during an interval of its motion may be computed by a direct integration of the corresponding equation of motion. Integration of the equation of motion by numerical or graphical means is often required when the functional relationship between the variable forces and the coordinates cannot be written. This is frequently the case when the forces are determined experimentally or from other approximate data.

In applying any of the force-mass-acceleration equations of motion it is absolutely necessary to account correctly for *all* forces acting on the particle. The only forces which may be neglected are those whose magnitudes are negligible compared with other forces acting, such as the forces of mutual attraction between two particles compared with their attraction to a celestial body such as the earth. The vector sum $\Sigma\mathbf{F}$ of Eq. 44 means the vector sum of *all* forces acting *on* the particle in question. Likewise, the corresponding scalar force summation in any one of the component directions means the sum of the components of *all* forces acting *on* the particle in that particular direction. The only reliable way to account accurately and consistently for every force is to *isolate* the particle under consideration from *all* contacting and influencing bodies and replace the bodies removed by the forces they exert on the particle isolated. The resulting *free-body diagram* is the means by which every force, known and unknown, which acts on the particle is represented and hence accounted for. Only after this vital step has been completed should the appropriate equation or equations of motion be written. The free-body diagram serves the same key purpose in dynamics as it does in statics. This purpose is simply to establish a *thoroughly reliable method* for the correct evaluation of the resultant of all real forces acting on the particle or body in question. In statics this resultant equals zero, whereas in dynamics it is equated to the product of mass and acceleration. If the student recognizes that the equations of motion must be interpreted literally and exactly, and if in so doing he respects the full scalar and vector meaning of the equals sign in the motion equation, then a minimum of difficulty will be experienced. Every experienced student of engineering mechanics recognizes that careful and consistent observance of the *free-body method* is the *most important single lesson* to be learned in the study of engineering mechanics. As a part of the drawing of a free-body diagram, the coordinate axes and their positive directions should be clearly indicated. When the equations of motion are written, all force summations should be consistent with the choice of these positive directions. Also, as an aid to the identification of external forces which act on the body in question, these forces are shown as heavy red vectors in the illustrations throughout the remainder of the book.

In solving problems the student often wonders how to get started and what sequence of steps to follow in arriving at the solution. This difficulty may be minimized if the student forms the habit of first recognizing some relationship between the quantity desired in the problem and other quantities, known and unknown. Additional relationships between these unknowns and other quantities, known and unknown, are then perceived. Finally the dependence upon the original data is established, and the procedure for the analysis and computation is indicated. A few minutes spent organizing the plan of attack through recognition of the dependence of one quantity upon another will be time well spent and will usually prevent groping for the answer with irrelevant calculations.

Sample Problems

3/1 A 75-kg man stands on a spring scale in an elevator. During the first 3 s of motion from rest, the tension T in the hoisting cable is 8300 N. Find the reading R of the scale during this interval and the upward velocity v of the elevator at the end of the 3 seconds. The total mass of the elevator, man, and scale is 750 kg.

Solution. The force registered by the scale and the velocity both depend on the acceleration of the elevator, which is constant during the interval for which the forces are constant. From the free-body diagram of the elevator, scale, and man taken together, the acceleration is found to be

$$[\Sigma F_y = ma_y] \qquad 8300 - 7360 = 750\, a_y, \qquad a_y = 1.257 \text{ m/s}^2$$

The scale reads the force exerted on the man's feet, and from the free-body diagram of the man alone the equation of motion gives

$$[\Sigma F_y = ma_y] \qquad R - 736 = 75(1.257), \qquad R = 830 \text{ N} \qquad \textit{Ans.}$$

The velocity reached at the end of the 3 seconds is

$$\left[\Delta v = \int a\, dt\right] \qquad v - 0 = \int_0^3 1.257\, dt, \qquad v = 3.77 \text{ m/s} \qquad \textit{Ans.}$$

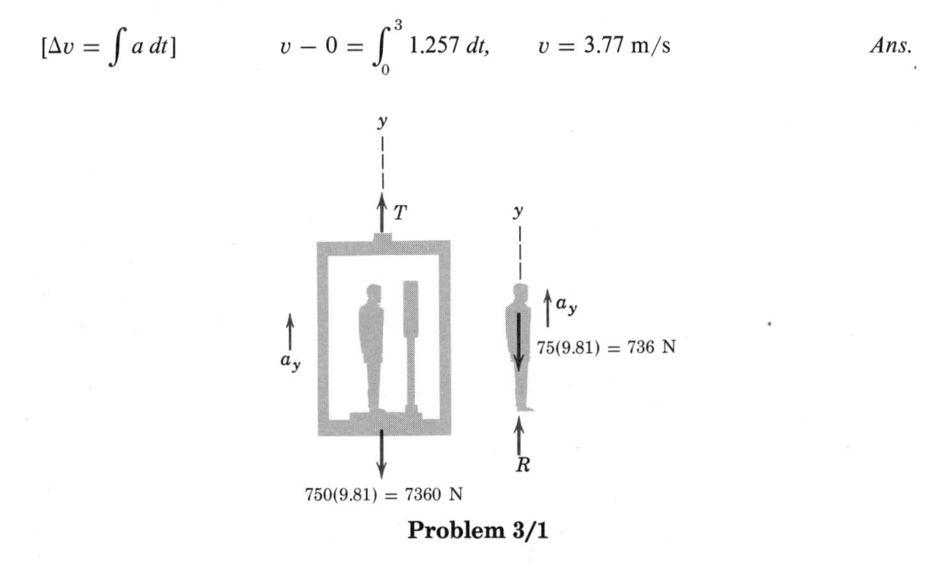

Problem 3/1

3/2 The 10-kg collar A is released from rest in the position for which $\theta = 30°$ and slides along the fixed horizontal shaft. The spring has a stiffness of 1750 N/m and is unstretched when $\theta = 0$. If the coefficient of friction between the shaft and the collar is 0.2, calculate the initial acceleration of the collar.

Solution. The collar will be considered as a particle with the resultant of all forces acting through its center. The weight of the spring and yoke will be neglected compared with the other forces. The free-body diagram of the isolated collar on the right of the figure shows the spring force of $1750(0.5 \sec 30° - 0.5) = 135.3$ N, the 98.1-N weight, a normal force N exerted by the shaft on the collar, and the corresponding friction force of $0.2N$ acting on the collar in the direction to oppose motion. Equilibrium in the vertical direction requires

$$[\Sigma F_y = 0] \qquad N - 98.1 - 135.3 \cos 30° = 0, \qquad N = 215 \text{ N}, \qquad 0.2N = 43.0 \text{ N}$$

The equation of motion in the x-direction gives

$$[\Sigma F_x = ma_x] \qquad 135.3 \sin 30° - 43.0 = 10a_x$$

$$a_x = 2.47 \text{ m/s}^2 \qquad \textit{Ans.}$$

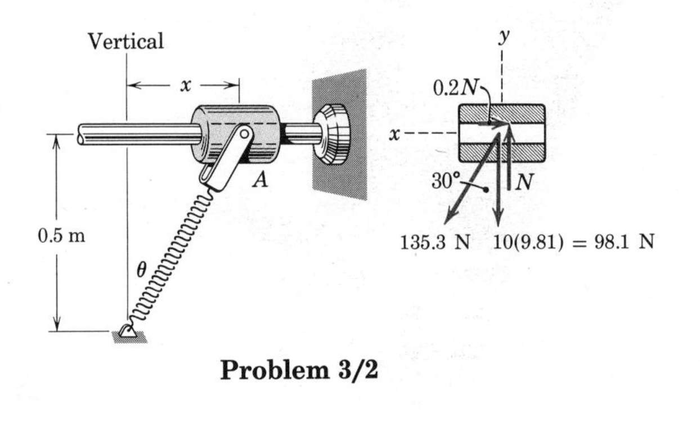

Problem 3/2

3/3 Small objects are released from rest at A and slide down the smooth circular surface of radius R to a conveyor B. Determine the expression for the normal contact force N between the guide and each object in terms of θ and specify the correct angular velocity ω of the conveyor pulley of radius r to prevent any sliding on the belt as the objects transfer to the conveyor.

Solution. The free-body diagram of the object is shown together with the coordinate directions n and t. The normal force N depends on the n-component of the acceleration which, in turn, depends on the velocity. The velocity will be cumulative according to the tangential acceleration a_t. Hence a_t will be found first for any general position.

$$[\Sigma F_t = ma_t] \qquad mg \cos \theta = ma_t \qquad a_t = g \cos \theta$$

Thus the velocity becomes

$$[v\,dv = a_t\,ds] \qquad \int_0^v v\,dv = \int_0^\theta g \cos \theta\, d(R\theta) \qquad v^2 = 2gR \sin \theta$$

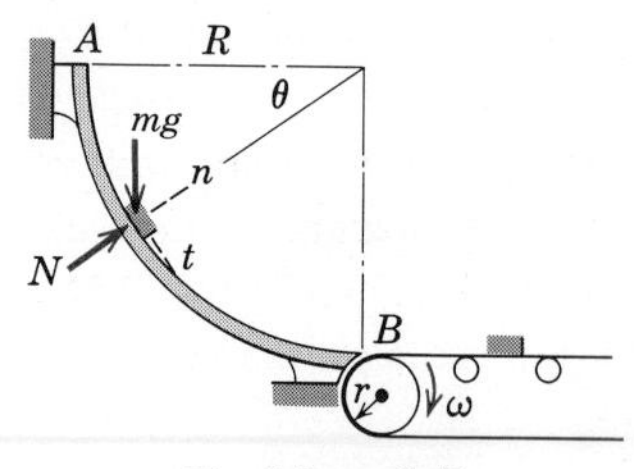

Problem 3/3

The normal force is obtained by summing forces in the positive n-direction which is the direction of the n-component of acceleration.

$$[\Sigma F_n = ma_n] \qquad N - mg\sin\theta = m\frac{v^2}{R} \qquad N = 3\,mg\sin\theta \qquad \textit{Ans.}$$

The conveyor pulley must turn at the rate $v = r\omega$ for $\theta = \pi/2$, so that

$$\omega = \sqrt{2gR}/r \qquad \textit{Ans.}$$

3/4 The slotted disk rotates in a vertical plane about O with a constant angular velocity ω, and a slider of mass m moves in the slot with negligible friction. If the slider starts from rest at $r = 0$ as the slot crosses the position $\theta = 0$, set up the equations of motion of the slider which would be necessary to determine both the normal force N and the radius r as functions of θ.

Solution. With r a variable, polar coordinates will be used to describe the motion. The free-body diagram of the slider is drawn, and Eqs. 47 yield

$$[\Sigma F_r = ma_r] \qquad mg\sin\theta = m(\ddot{r} - r\omega^2) \qquad \textit{Ans.}$$

$$[\Sigma F_\theta = ma_\theta] \qquad mg\cos\theta - N = m(0 + 2\dot{r}\omega) \qquad \textit{Ans.}$$

Solution of the first of these equations for r is prerequisite to the determination of N from the second equation.

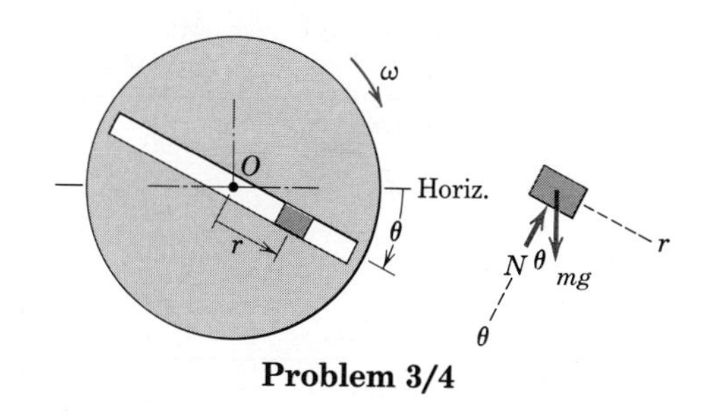

Problem 3/4

3/5 Wind-tunnel tests of the resistance of a sphere in a moving air stream for low velocities give the plotted curve shown in the full line. If the sphere has a mass of 100 g and is released from rest and falls freely in still air, use these data to predict the velocity v which it will acquire after dropping 2.5 m from rest.

Solution. The free-body diagram of the sphere is drawn as shown and includes the variable resistance R and the constant weight $W = mg = 0.981$ N. The positive x-direction is taken downward with origin at the point of release from rest. The equation of motion is

$$[\Sigma F_x = ma_x] \qquad mg - R = ma_x$$

Since the velocity at a particular value of displacement is required, an integration with respect to displacement is indicated. Multiplication of the equation by dx yields

$$(mg - R)\,dx = ma_x\,dx \qquad \text{or} \qquad (mg - R)\,dx = mv\,dv$$

which may be written as

$$(mg - R)\,dx = \frac{m}{2}d(v^2)$$

With the substitution of the numerical values of m and g the equation may be written

$$d(v^2) = (19.62 - 20R)\,dx$$

By using small but finite increments the expression may be approximated by

$$\Delta(v^2) = (19.62 - 20R)\,\Delta x$$

In this form integration of the equation of motion may be effected by a step-by-step process either on a digital computer or else by hand as shown in the following table. For an approximate hand solution the 2.5-m interval is divided into 0.25-m increments for the first 0.5 m where the velocity is changing most rapidly and into 0.5-m increments for the remaining interval. The value of v^2, and hence v, for any value of x equals v^2 at the previous displacement plus the increment $\Delta(v^2)$ determined from the defining relation expressed in terms of Δx. Smaller increments in Δx could be chosen for a more accurate solution. Values of R are read directly from the graph for the corresponding values of v.

The velocity at $x = 2.5$ m is found to be 4.03 m/s. The maximum velocity reached by the sphere occurs when the increment $\Delta(v^2) = 0$ or when $19.62 - 20R = 0$. This

x m	Δx m	R N	$(19.62 - 20R)$ m/s^2	$\Delta(v^2)$ (m/s)^2	v^2 (m/s)^2	v m/s
0		0	19.62		0	0
	0.25			4.90		
0.25		0.20	15.62		4.90	2.21
	0.25			3.90		
0.50		0.41	11.22		8.80	2.97
	0.50			5.61		
1.00		0.82	3.22		14.41	3.80
	0.50			1.61		
1.50		0.96	0.40		16.02	4.00
	0.50			0.20		
2.00		0.98	0.02		16.22	4.03
	0.50			0.01		
2.50					16.23	4.03

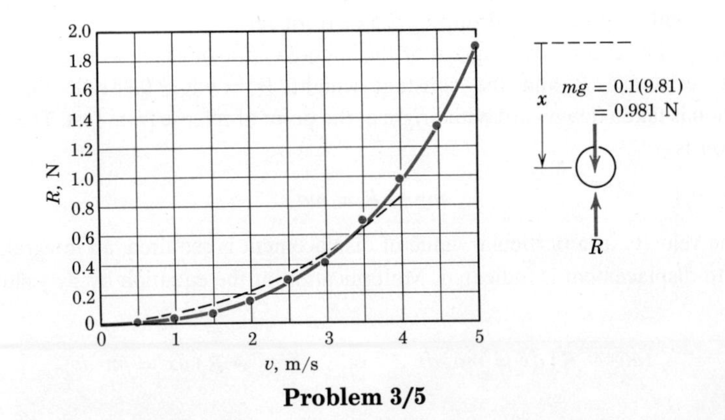

Problem 3/5

gives $R = 0.981$ N, which, from the graph, corresponds to a velocity of 4.05 m/s. Thus the calculation shows that the sphere has almost reached its maximum or terminal velocity.

An analytic solution may also be used to approximate the velocity by writing R as an appropriate function of v. By assuming the form $R = kv^2$ the equation of motion may be integrated, upon separation of the variables, to obtain

$$\int_0^x \frac{2}{m} dx = \int_0^{(v^2)} \frac{d(v^2)}{mg - kv^2} \quad \text{or} \quad \frac{2x}{m} = \frac{1}{k} \ln \frac{mg}{mg - kv^2}$$

from which

$$v = \sqrt{\frac{mg}{k}(1 - e^{-2kx/m})}$$

An appropriate value of k may be determined by assuming the function to agree with the given data at about $v = 3.25$ m/s which represents a fair average of the resistance over the interval involved. This point on the curve gives $k = 0.053\ \text{N}\cdot\text{s}^2/\text{m}^2$ and defines the dotted curve shown in the figure. Substitution of the numerical values into the expression for v gives

$$v = \sqrt{\frac{0.981}{0.053}\left(1 - e^{-\frac{2(0.053)(2.5)}{0.10}}\right)} = 4.15 \text{ m/s}$$

This approximation to v is slightly higher than the maximum (terminal) velocity. A closer approximation could be achieved by using a higher degree polynomial in v.

Problems

Rectilinear Motion

3/6 Determine the vertical acceleration a of the 150-kg mass for each of the two cases illustrated. Neglect friction and the mass of the pulleys.

Ans. (*a*) $a = 1.40$ m/s^2
(*b*) $a = 3.27$ m/s^2

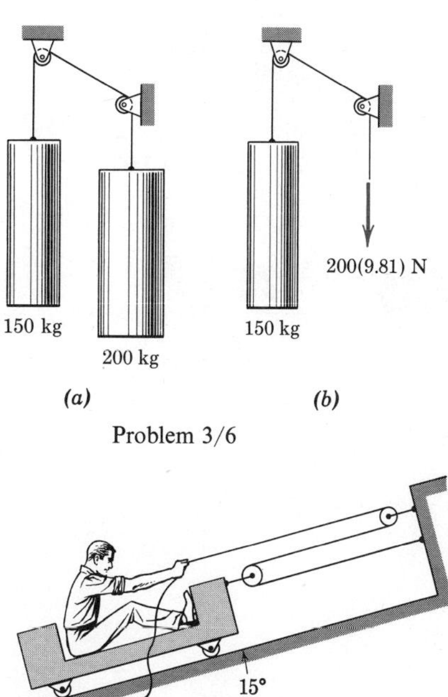

Problem 3/6

3/7 A man pulls himself up the 15° incline by the method shown. If the combined mass of the man and cart is 100 kg, determine the acceleration of the cart while the man exerts a pull of 250 N on the rope. Neglect all friction and the mass of the rope, pulleys, and wheels.

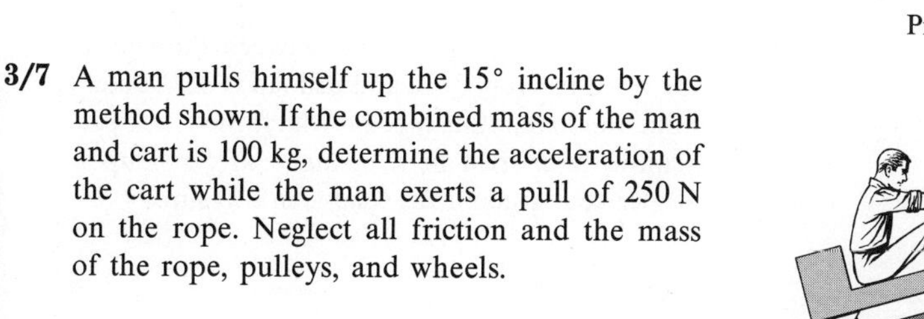

Problem 3/7

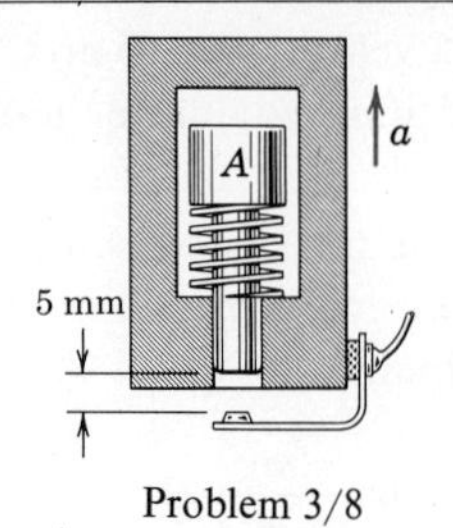

Problem 3/8

3/8 The device shown is used as an accelerometer and consists of a 0.1-kg plunger A which deflects the spring as the housing of the unit is given an upward acceleration a. Specify the necessary spring stiffness k which will permit the plunger to deflect 5 mm beyond the equilibrium position and touch the electrical contact when the steadily but slowly increasing upward acceleration reaches $5g$. Friction may be neglected.

Ans. $k = 981$ N/m

3/9 A cesium-ion engine for deep-space propulsion is designed to produce a constant thrust of 2.5 N for long periods of time. If the engine is to propel a 70-t spacecraft on an interplanetary mission, compute the time t for it to increase its speed from 40 000 km/h to 65 000 km/h. Also find the distance s traveled during this interval. Assume that the spacecraft is moving in a remote region of space where the thrust from its ion engine is the only force acting on the spacecraft in the direction of its motion.

3/10 Beginning at time $t = 0$ a force F_y is applied for 3 s in the positive y-direction to a 3.5-kg particle moving initially in the negative y-direction with a velocity of 3 m/s. If the force in newtons increases with the time in seconds according to $F_y = 16t$ and if F_y is the only force acting on the particle in the y-direction, determine the net displacement Δy of the particle during the 3 seconds. *Ans.* $\Delta y = 11.57$ m

Vertical

A

B

$a = 2g$

15°

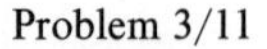

Problem 3/11

3/11 The smooth 25-kg metal cylinder is supported in a carriage which is given an acceleration of $2g$ up the 15° incline. Calculate the forces of contact at A and B.

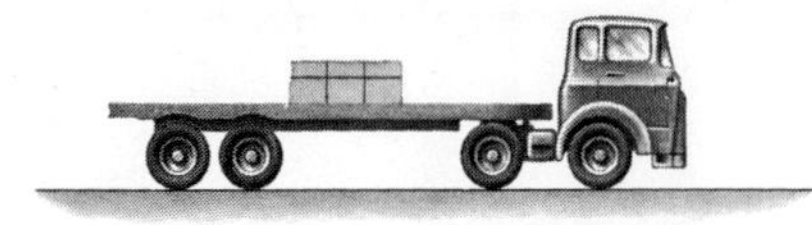

Problem 3/12

3/12 The coefficient of friction between the flat bed of the truck and the crate it carries is 0.30. Determine the minimum stopping distance s which the truck can have from a speed of 70 km/h with constant deceleration if the crate is not to slip forward. *Ans.* $s = 64.3$ m

3/13 If the coefficient of friction between the bed of the truck in Prob. 3/12 and its load is 0.30, determine the maximum speed v which the truck can acquire from rest in a distance of 50 m up a 10-per-cent grade if the load is not to slip.

3/14 Compute the distance s which the sliding block travels up the incline from its point of release, where its initial velocity is 9 m/s, to the point where its velocity is 6 m/s. The coefficient of friction between the incline and the block is 0.30.
Ans. $s = 4.18$ m

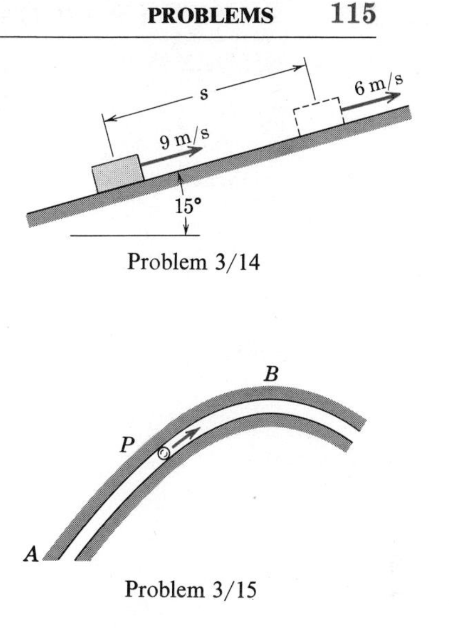

Problem 3/14

Problem 3/15

Normal and Tangential Coordinates (*n-t*)

3/15 A particle P having a mass of 2 kg is given an initial velocity at A in the smooth guide, which is curved in the vertical plane. If the particle has a horizontal velocity of 4.8 m/s at the top B of the guide, where its radius of curvature is 0.5 m, calculate the force R exerted on the particle by the guide at this point. *Ans.* $R = 72.5$ N down

3/16 An 80-kg stunt pilot executes a vertical inside loop. If his horizontal velocity is 300 km/h in the upside-down position and if the force between him and his seat has dropped to one-fourth of his weight at this point, ascertain the radius of curvature ρ of the top of the loop.

3/17 In the design of a space station to operate outside the earth's gravitational field, it is desired to give the structure a rotational speed N which will simulate the effect of the earth's gravity for members of the crew. If the centers of the crew's quarters are to be located 8 m from the axis of rotation, calculate the necessary rotational speed N of the space station in revolutions per minute.
Ans. $N = 10.57$ rev/min

3/18 The simple pendulum is released from the dotted position with an initial velocity. As it swings past the vertical, the tension T in the supporting wire is k times the weight of the pendulum. Determine the expression for the velocity v of the pendulum at the bottom position in terms of k.

3/19 Calculate the tension T in a metal hoop rotating in its plane about its geometric axis with a large rim speed v. Analyze a differential element of the hoop as a particle. The mass per unit length of rim is ρ.

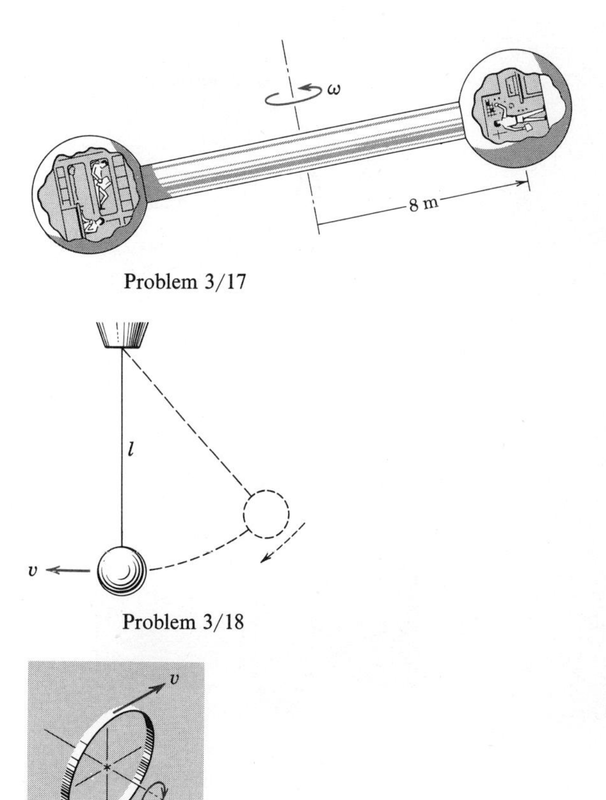

Problem 3/17

Problem 3/18

Problem 3/19

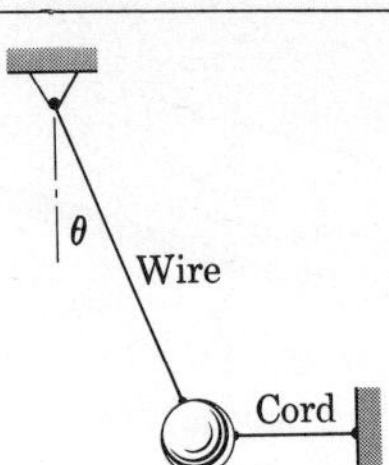

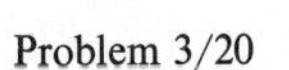

Problem 3/20

3/20 The small mass m and its supporting wire become a simple pendulum when the horizontal cord is severed. Determine the ratio k of the tension T in the supporting wire immediately after the cord is cut to that in the wire before the cord is cut.

Ans. $k = \cos^2 \theta$

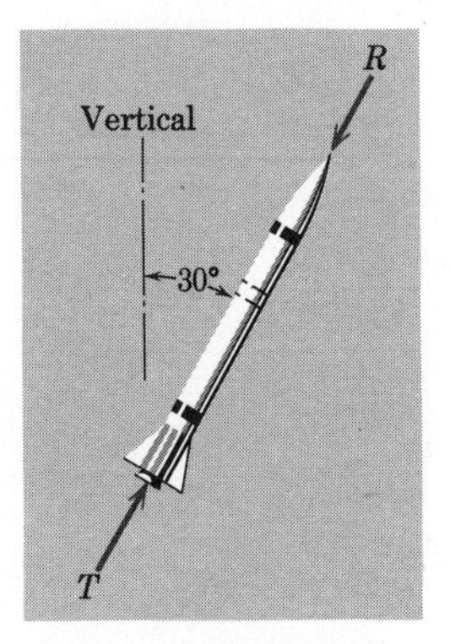

Problem 3/21

3/21 The rocket moves in a vertical plane and is being propelled by a thrust T of 32 kN. It is also subjected to an atmospheric resistance R of 9.6 kN. If the rocket has a velocity of 3 km/s and if the gravitational acceleration is 6 m/s^2 at the altitude of the rocket, calculate the radius of curvature ρ of its path for the position described.

Ans. $\rho = 3000$ km

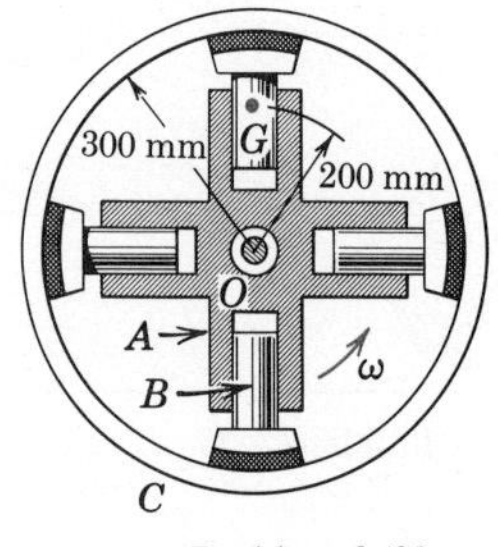

Problem 3/22

3/22 The figure shows a centrifugal clutch consisting in part of a rotating spider A which carries four plungers B. As the spider is made to rotate about its center with a speed ω, the plungers move outward and bear against the interior surface of the rim of wheel C, causing it to rotate. The wheel and spider are independent except for frictional contact. If each plunger has a mass of 2 kg with center of mass at G, and if the coefficient of friction between the plungers and the wheel is 0.40, calculate the maximum moment M which can be transmitted to wheel C for a spider speed of 3000 rev/min.

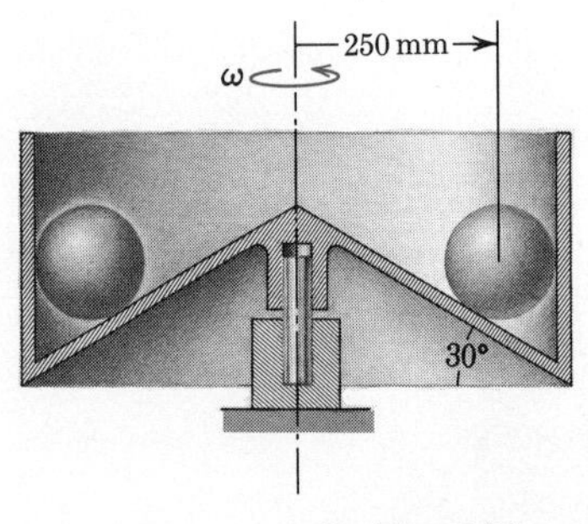

Problem 3/23

3/23 The conical dish rotates about its vertical axis at the constant speed of 60 rev/min and carries the two spheres with it. Calculate the force N between each of the 4-kg spheres and the vertical surface of the dish. Use only one force equation.

Ans. $N = 62.1$ N

3/24 A small slider is free to move along the circular rod AB with negligible friction. If the rod is rotating about the vertical axis OA with a constant angular velocity ω, determine the angle θ which locates the stable position assumed by the slider.

$$Ans.\ \theta = \cos^{-1}\frac{g}{r\omega^2},\quad \omega \geq \sqrt{g/r}$$

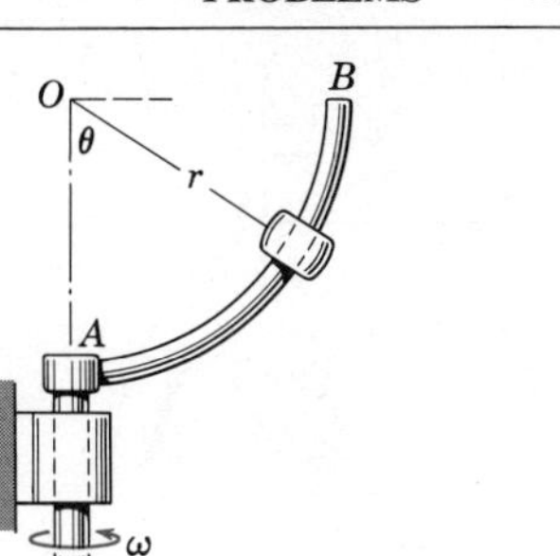

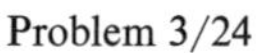

Problem 3/24

3/25 Calculate the necessary rotational speed N for the aerial ride in the amusement park in order that the arms of the gondolas will assume an angle $\theta = 60°$ with the vertical. Neglect the mass of the arms to which the gondolas are attached and treat each gondola as a particle.

Ans. $N = 10.7$ rev/min

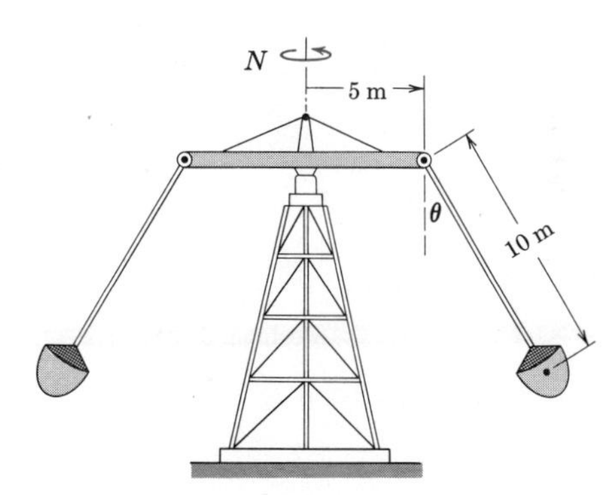

Problem 3/25

Polar Coordinates (r-θ)

3/26 A 2-kg particle moves in a horizontal plane with its polar coordinates given by $r = 50t^3 - 75t + 100$ and $\theta = 2 \sin \frac{1}{2}\pi t$ where r is in millimetres, θ is in radians, and t is in seconds. Compute the resultant horizontal force on the particle at the instant $t = 2$ s. *Ans.* $F = 8.72$ N

3/27 The 2-kg slider P moves with negligible friction along the radial slot of the arm, which rotates about a vertical axis through O. The radial motion of the slider is controlled by a light cord which is allowed to slip through a small hole in the shaft at O at a constant rate of $\dot{r} = 100$ mm/s. Determine the horizontal force F exerted by the slot on the slider when $r = 375$ mm if at this instant $\omega = 3$ rad/s and ω is decreasing at the rate of 2 rad/s per second. Does the slider bear against the A- or B-side of the slot? Also determine the tension T in the cord at the instant described.

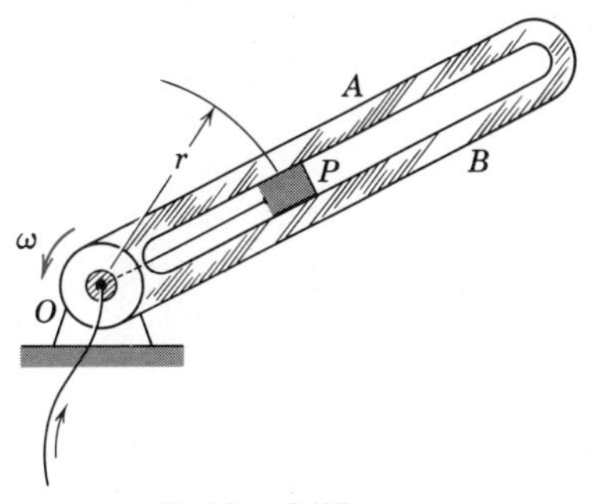

Problem 3/27

3/28 The slotted arm revolves with a constant angular velocity ω about a vertical axis through the center O of the fixed cam. The cam is cut so that the radius to the path of the center of the pin A varies according to $r = r_0 + b \sin N\omega t$ where N equals the number of lobes—six in this case. If $\omega = 12$ rad/s, $r_0 = 100$ mm, $b = 10$ mm, and the compression in the spring varies from 11.5 N to 19.1 N from valley to crest, calculate the force R between the cam and the 100-g pin A as it passes over the top of the lobe in the position shown. *Ans.* $R = 12.33$ N

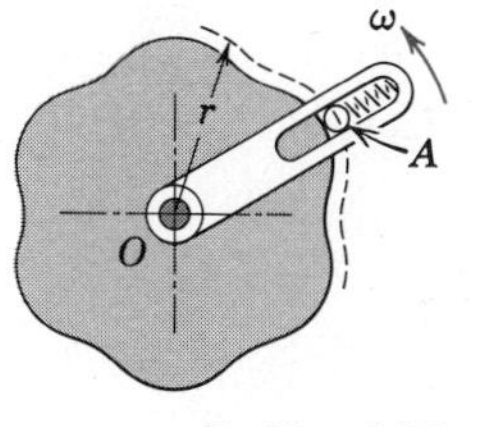

Problem 3/28

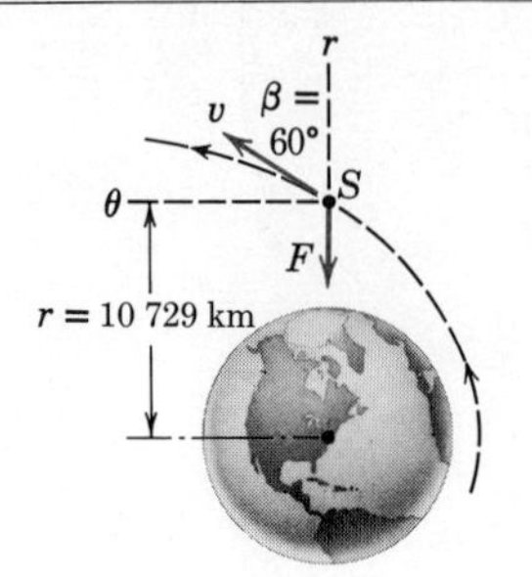

Problem 3/29

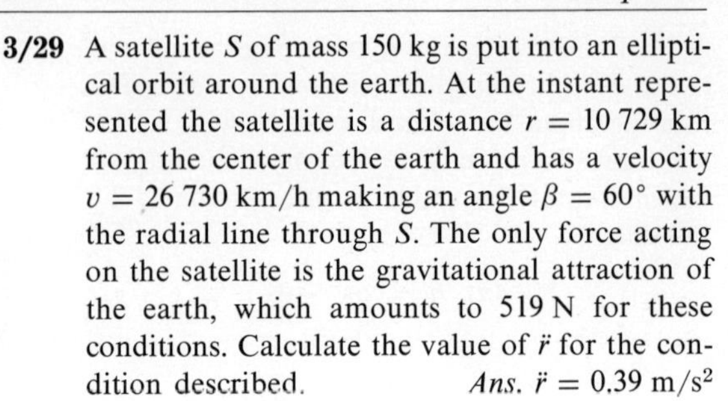

3/29 A satellite S of mass 150 kg is put into an elliptical orbit around the earth. At the instant represented the satellite is a distance $r = 10\ 729$ km from the center of the earth and has a velocity $v = 26\ 730$ km/h making an angle $\beta = 60°$ with the radial line through S. The only force acting on the satellite is the gravitational attraction of the earth, which amounts to 519 N for these conditions. Calculate the value of $\ddot{r}$ for the condition described. *Ans.* $\ddot{r} = 0.39\ \text{m/s}^2$

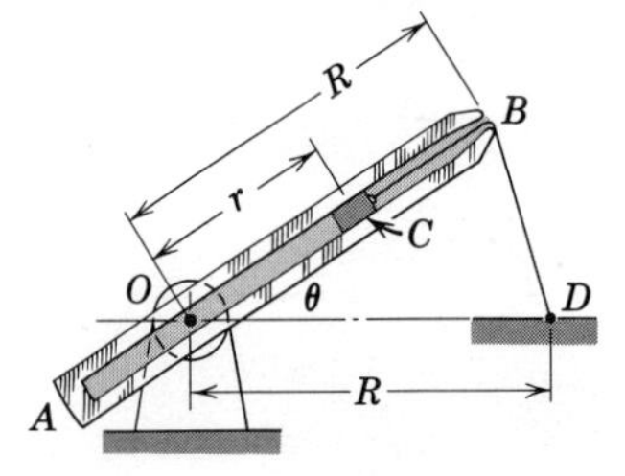

Problem 3/30

3/30 The mechanism of Prob. 2/115 is repeated here. If the arm rotates about a horizontal axis through O and if the slider C has a mass m, determine the expression in terms of θ for the tension T in the cord at its attachment to C. The angular velocity of OB is $\omega = \dot{\theta}$ and is constant during the motion interval concerned. Also $r = 0$ when $\theta = 0$. All surfaces are assumed to be smooth. What is the maximum speed $\dot{\theta}$, for a given θ, which the arm may have before T goes to zero?

General

3/31 A constant force $\mathbf{F} = 0.3\mathbf{j}$ N acts for 2 s on a 200-g particle which has an initial velocity $\mathbf{v} = 1.0\mathbf{i} + 4.5\mathbf{k}$ m/s at time $t = 0$. Compute the magnitude of the velocity of the particle for the instant when $t = 2$ s. *Ans.* $v = 5.5$ m/s

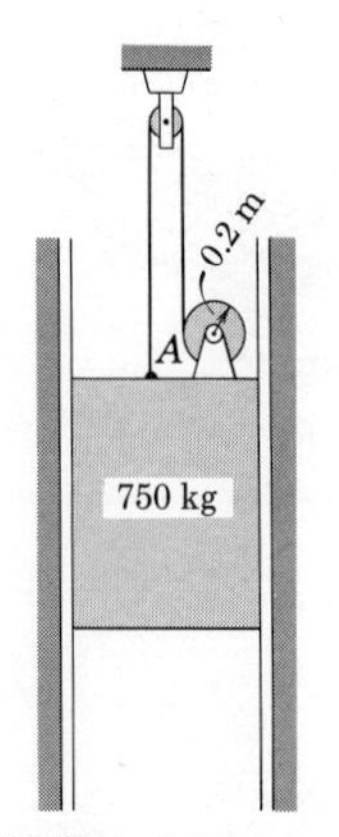

Problem 3/32

3/32 The 750-kg elevator is operated through the action of the rotating drum A around which is wrapped the hoisting cable. Determine the constant torque M which must be supplied to the shaft of the drum by its motor (not shown) in order to give the elevator an upward velocity of 3 m/s in a vertical rise of 4 m from rest. The mass of the drum is small, and it may be analyzed as though it were in rotational equilibrium. Neglect friction in the vertical guides.

3/33 When a V-belt drives a pulley at high speed, the centrifugal action of the belt tends to lessen its contact with the pulley and hence to reduce the capacity to transmit torque. A device to compensate for this effect is shown in the figure and consists of a cage and four balls which rotate with the pulley. Because of centrifugal action the balls press against the two 30° conical surfaces and force the inner side A to slide to the left toward the opposite side B, thus tightening the belt. Part A is splined to B along the hub, so that it must rotate with the remainder of the pulley but is free to slide on B in the axial direction. Compute the axial force F on A caused by the action of the balls for a speed of 600 rev/min, if the mass of each of the four balls is 2.5 kg.

Ans. $F = 5130$ N

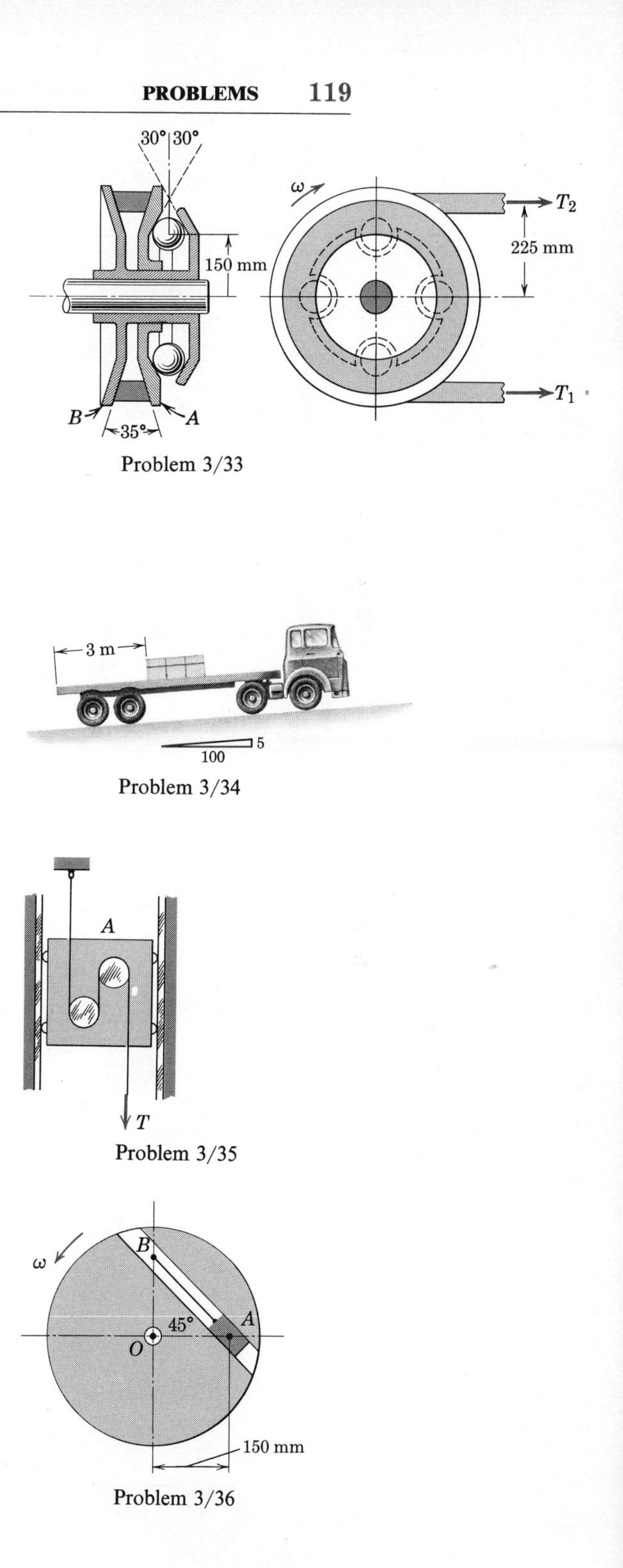

Problem 3/33

Problem 3/34

Problem 3/35

Problem 3/36

3/34 A 3.6-t flat-bed truck carries a 750-kg box. As the truck starts from rest with constant acceleration, the box slides 3 m to the edge of the bed in the time that it takes the truck to acquire a velocity of 40 km/h in a distance of 15 m up the incline. Determine the coefficient of friction f between the box and the truck bed.

3/35 The acceleration of the 50-kg carriage A in its smooth vertical guides is controlled by the tension T exerted on the control cable which passes around the two circular pegs fixed to the carriage. Determine the value of T required to limit the downward acceleration of the carriage to 1.2 m/s^2 if the coefficient of friction between the cable and the pegs is 0.20. *Ans.* $T = 171.3$ N

3/36 The 3-kg slider A fits loosely in the smooth 45° slot in the disk, which rotates in a horizontal plane about its center O. If A is held in position by a cord secured to point B, determine the tension T in the cord for a constant rotational velocity $\omega = 300$ rev/min. Would the direction of the velocity make any difference?

3/37 Determine the tension T in the cord in Prob. 3/36 as the disk starts from rest with $\dot{\omega} = 120$ rad/s^2. Also calculate the initial horizontal force N exerted by the slot on the slider.

Ans. $T = 38.2$ N, $N = 38.2$ N

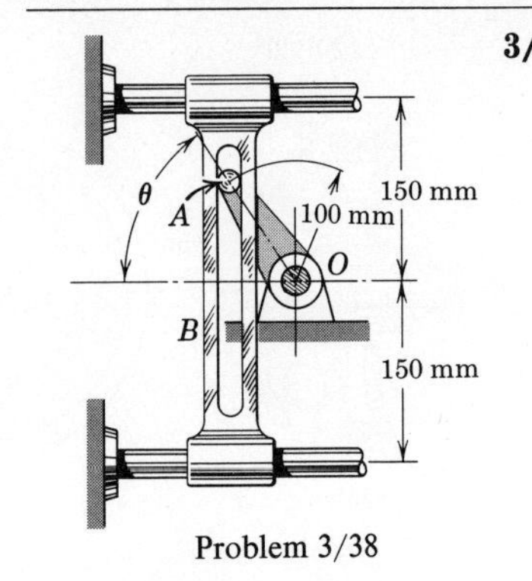

Problem 3/38

3/38 The 10-kg slotted guide B oscillates back and forth with negligible friction along the fixed horizontal shafts under the action of pin A on the rotating crank OA. If OA has a counterclockwise angular acceleration of 24 rad/s² and if it has a counterclockwise angular velocity of 9 rad/s when $\theta = 45°$, determine the force F of contact between the smooth slot and the pin A at this instant. Does the pin contact the right or the left side of the slot?

Ans. $F = 40.3$ N; right side

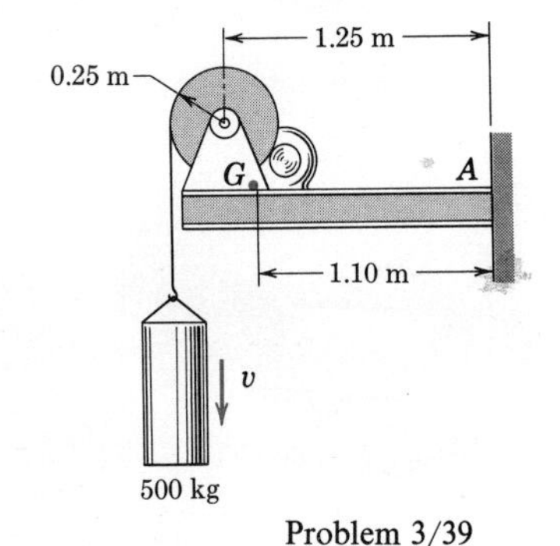

Problem 3/39

3/39 The mass of the hoisting drum, motor drive, and beam together is 240 kg with a combined center of mass 1.10 m to the left of the welded attachment at A. If the drum is lowering the 500-kg load with a velocity v which increases uniformly from 0.6 m/s to 6 m/s in 4 s, calculate the bending moment M supported by the beam at A during this interval.

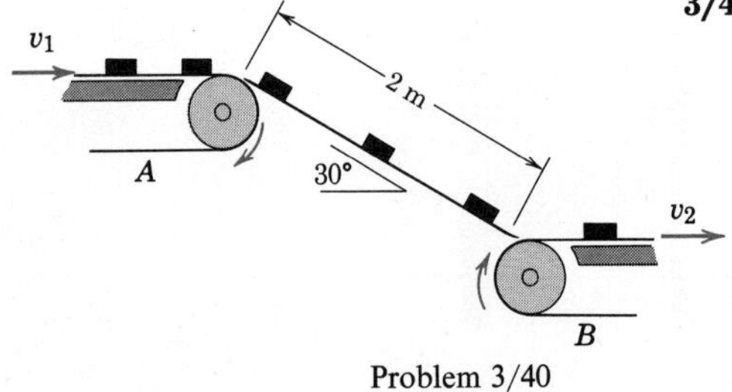

Problem 3/40

3/40 Small objects are delivered to the 2-m inclined chute by a conveyer belt A which moves at a speed $v_1 = 0.4$ m/s. If the conveyor belt B has a speed $v_2 = 1$ m/s and the objects are delivered to this belt with no slipping, calculate the coefficient of friction f between the objects and the chute.

Ans. $f = 0.553$

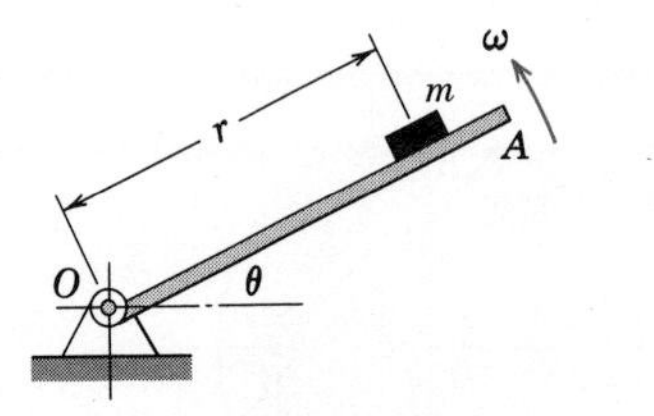

Problem 3/41

3/41 The member OA rotates about a horizontal axis through O with a constant counterclockwise velocity $\omega = 3$ rad/s. As it passes the position $\theta = 0$, a small mass m is placed upon it at a radial distance $r = 0.5$ m. If the mass is observed to slip at $\theta = 45°$, determine the coefficient of friction f between the mass and the member.

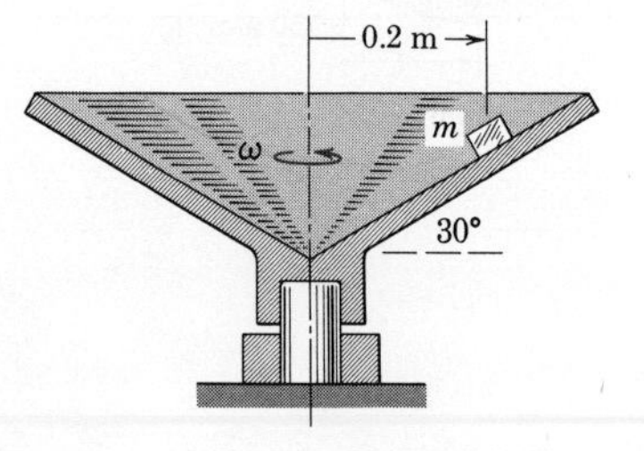

Problem 3/42

3/42 The small mass is placed on the surface of the conical dish at the radius shown. If the coefficient of friction between the mass and the conical surface is 0.30, for what range of rotational speeds about the vertical axis will the block remain on the dish without slipping? Assume that speed changes are made slowly so that any angular acceleration may be neglected.

Ans. $32.5 < N < 68.9$ rev/min

3/43 The 2-kg collar is released from rest against the light elastic spring, which has a stiffness of 1.75 kN/m and which has been compressed a distance of 0.15 m. Determine the acceleration a of the collar as a function of the vertical displacement x of the collar measured in metres from the point of release. Find the velocity v of the collar when $x = 0.15$ m. Friction is negligible.

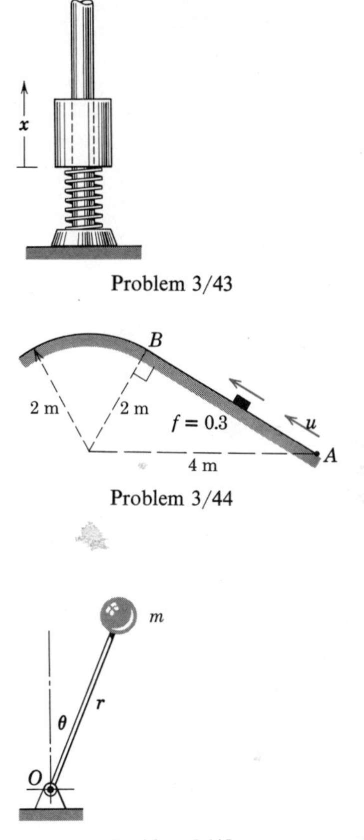

Problem 3/43

3/44 An object of mass m is given an initial velocity u up the incline at point A. An instant after it passes point B, the normal force of contact between it and the supporting surface drops to one-half of the value it had when the object was approaching B. The coefficient of friction between the object and the incline is 0.30. Calculate u. *Ans.* $u = 7.75$ m/s

Problem 3/44

3/45 The small sphere of mass m is fastened to the light rod hinged about a horizontal axis through O. The distance between O and the center of the sphere is r. If the sphere is released from rest with θ essentially zero, determine the value of θ at which the force in the bar changes from compression to tension. *Ans.* $\theta = 48.19°$

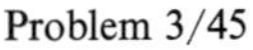

Problem 3/45

3/46 The position of a particle with respect to a convenient origin of coordinates is given in metres by the vector expression $\mathbf{r} = \mathbf{i}(\sin 2\pi t) - 16\mathbf{j}(1 - e^{-t/4}) + \frac{1}{4}\mathbf{k}t^2$ where t is the time of travel of the particle in seconds after passing the origin. If the mass of the particle is 2 kg, find as a function of time the x-component of the resultant $\mathbf{R}$ of all forces acting on the particle and determine the magnitude of $\mathbf{R}$ when $t = 4$ s.

3/47 The total resistance R to horizontal motion of a rocket test sled is shown by the solid line in the graph. Approximate R by the dotted line and determine the distance x which the sled must travel along the track from rest to reach a velocity of 400 m/s. The forward thrust T of the rocket motors is essentially constant at 300 kN, and the total mass of the sled is also nearly constant at 2 t. *Ans.* $x = 1037$ m

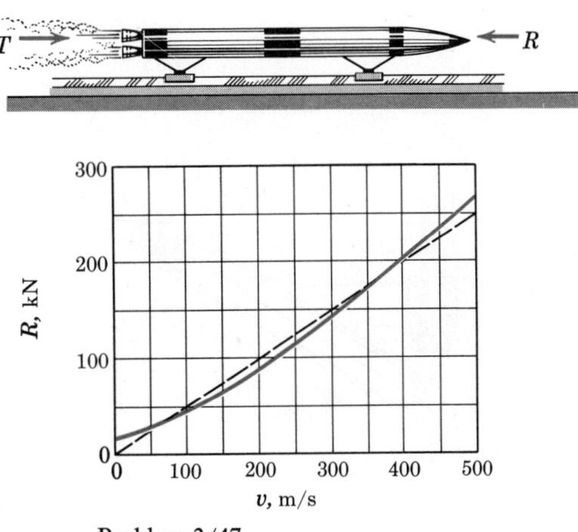

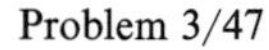

Problem 3/47

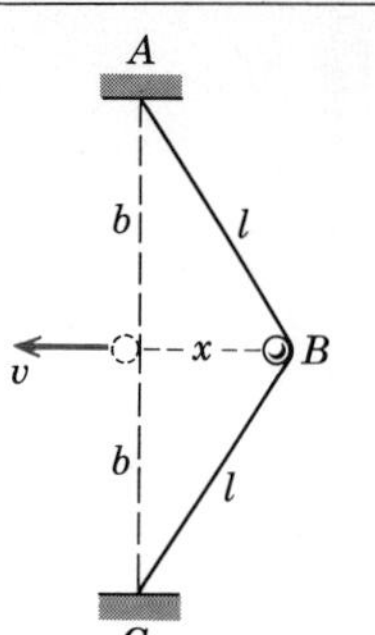

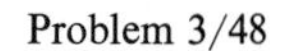
Problem 3/48

3/48 A particle of mass m is released from rest in the position shown and is accelerated to the left by the action of the elastic cord ABC which lies in the horizontal plane and which is unstretched at $x = 0$. The cord has a unit stiffness K, which is the ratio of its tension force to its unit elongation or strain (change in length divided by original length). Determine the expression for the velocity v reached by the particle at the instant x returns to zero. (Solve without using work-energy principles.)

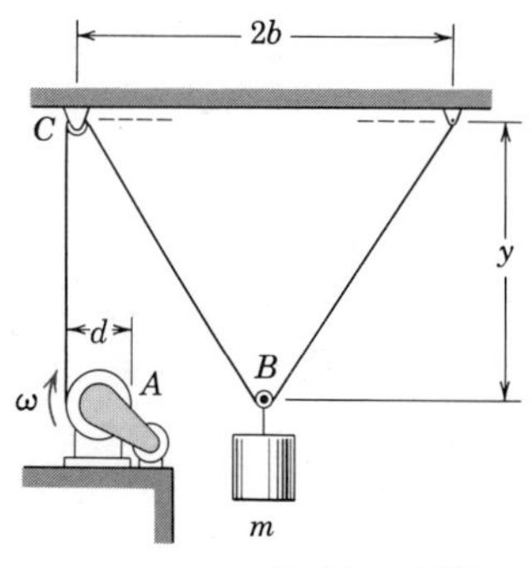

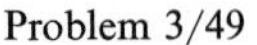
Problem 3/49

3/49 The hoisting drum A has a diameter d and turns clockwise with a constant angular velocity ω. Determine the tension T in the cable which attaches the mass m to the small pulley B. Express the result in terms of the variable y. The size, mass, and friction of the pulleys at C and B are negligible.

$$Ans.\ T = mg\left[1 + \frac{\omega^2 b^2 d^2}{16 g y^3}\right]$$

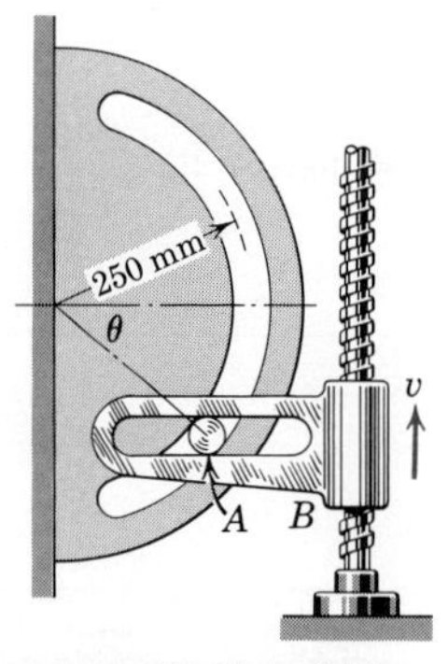

Problem 3/50

3/50 The motion of the 0.5-kg pin A in the circular slot is controlled by the guide B which is being elevated by its lead screw with a constant upward velocity v of 2 m/s for an interval of its motion. Calculate the force N exerted on the pin by the circular guide as the pin passes the position for which $\theta = 30°$. Friction is negligible.

Ans. $N = 14.22$ N

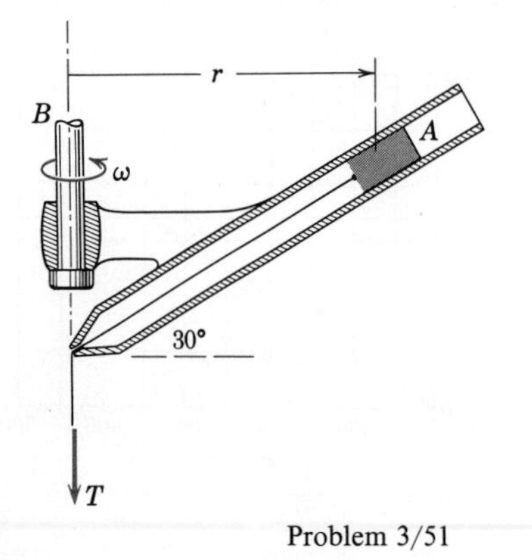

Problem 3/51

3/51 The slider A has a mass of 2 kg and moves with negligible friction in the smooth tube, shown in section. The tube is constrained to rotate about the fixed vertical shaft B with a constant rotational velocity ω of 90 rev/min. The cord, which holds the slider, is released through the lower end of the tube at the constant rate of 0.1 m/s. Compute the component N in the plane of the figure and the component H normal to the plane of the figure of the total force **F** exerted by the tube on A when the position $r = 0.25$ m is reached.

Ans. $N = 39.2$ N, $H = 3.26$ N

3/52 In theory an object projected vertically up from the surface of the earth with a sufficiently high velocity can escape from the earth's gravity field. Calculate the minimum value v of this escape velocity on the basis of the absence of an atmosphere to offer resistance due to air friction and considering the earth to be nonrotating. Take the radius of the earth to be 6370 km and the gravitational acceleration at the earth's surface to be 9.821 m/s^2.

3/53 The chain is released from rest in the position shown with barely enough overhanging links to initiate motion. The coefficient of friction between the links and the horizontal surface is f. Determine the velocity v of the chain when the last link leaves the edge. Neglect friction at the corner.

$$Ans.\ v = \sqrt{\frac{gL}{1+f}}$$

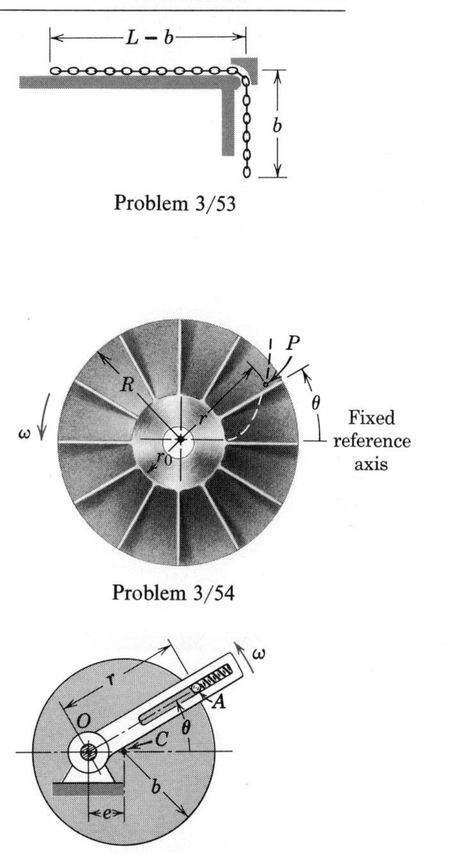

Problem 3/53

Problem 3/54

Problem 3/55

3/54 The centrifugal pump with smooth radial vanes rotates about its vertical axis with an angular velocity ω. Find the force N exerted by a vane on a particle of mass m as it moves out along the vane. The particle is introduced at $r = r_0$ without radial velocity. Assume that the particle contacts the side of the vane only.

3/55 The rotating slotted arm and the fixed circular cam of Prob. 2/121 are repeated here. If the arm revolves around a vertical axis through O with the constant speed ω and if the pin A has a mass m, write the expression for the contact force R exerted by the slotted arm on the pin when $\theta = \pi/2$ at which position the spring force is P. Neglect friction and pin diameter.

$$Ans.\ R = me\omega^2\frac{b^2}{b^2 - e^2} + \frac{Pe}{\sqrt{b^2 - e^2}}$$

3/56 A body at rest relative to the surface of the earth rotates with the earth and therefore moves in a circular path about the polar axis of the earth considered as fixed. Derive an expression for the ratio k of the apparent weight of such a body as measured by a spring scale at the equator (calibrated to read the actual force applied) to the true weight of the body which is the absolute gravitational attraction to the earth. The absolute acceleration due to gravity at the equator is $g = 9.815$ m/s^2. The radius of the earth at the equator is $R = 6378$ km, and the angular velocity of the earth is $\omega = 0.729(10^{-4})$ rad/s. If the true weight is 100 N, what is the apparent measured weight W'?

$$Ans.\ k = 1 - \frac{R\omega^2}{g},\quad W' = 99.655\text{ N}$$

3/57 The mean radius R of the earth is 6370 km, and the mean distance L between the centers of the earth and the moon is 384 398 km. Determine the period τ of the orbit of the moon about the earth considered as fixed. Assume a circular orbit.

$$\textit{Ans. } \tau = \frac{2\pi L}{R}\sqrt{\frac{L}{g}}, \quad \tau = 27.5 \text{ days}$$

3/58 Derive an expression for the magnitude of the velocity $\mathbf{v}$ of an artificial satellite which is to operate in a circular orbit at an altitude h above the surface of the earth. The altitude is sufficiently great to permit neglect of the effect of atmospheric resistance. The mean radius R of the earth is 6370 km. What is the period τ of the satellite's motion? Compute v and τ for $h = 640$ km.

$$\textit{Ans. } v = R\sqrt{\frac{g}{R+h}} = 27\ 143 \text{ km/h}$$

$$\tau = 2\pi\frac{(R+h)^{3/2}}{R\sqrt{g}} = 1 \text{ h } 37 \text{ min } 22 \text{ s}$$

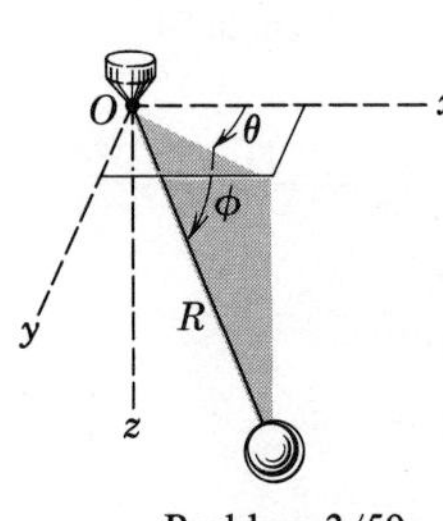

Problem 3/59

3/59 The spherical pendulum consists of a small ball of mass m which is secured to the fixed point O by a cord of length R and is given an initial velocity with both θ and ϕ varying. Set up the differential equations of motion in the three spherical coordinates shown. (Give consideration to the solution of these equations but limit the time spent.)

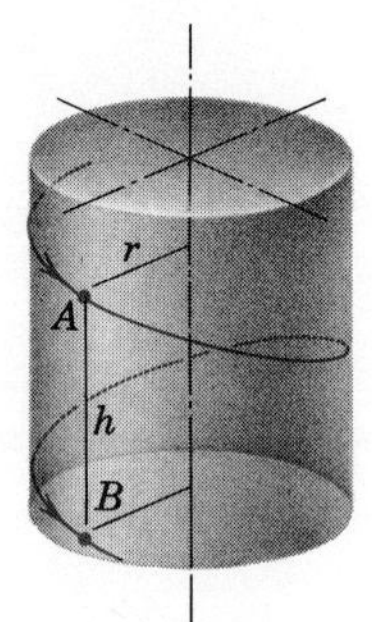

Problem 3/60

3/60 For the helical guide of Prob. 2/166, repeated here, prove that the acceleration, tangent to the path, of a particle of mass m which slides down the smooth path is $g \sin \gamma$. Also determine the expression for the contact force N between the guide and the particle as it passes B after one complete turn from its initial position at A where it was released from rest.

$$\textit{Ans. } N = mg\sqrt{\cos^2\gamma + 4\pi^2 \sin^2 2\gamma}$$

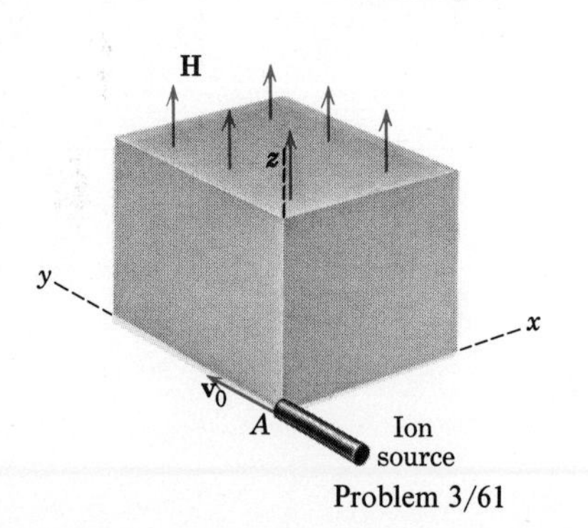

Problem 3/61

3/61 An ion of mass m and charge e is emitted from an ion source at A with a velocity $\mathbf{v}_0$ in the direction shown into a uniform magnetic field of strength $\mathbf{H} = H\mathbf{k}$. The force acting on the ion is given by $e\mathbf{v} \times \mathbf{H}$, where $\mathbf{v}$ is the velocity of the ion at any time. Describe the subsequent motion of the ion.

3/62 Solve the equations of motion of Sample Prob. 3/4 for r and N as functions of θ for the initial conditions as defined in the problem statement.

$$Ans.\ r = \frac{g}{2\omega^2}(\sinh\theta - \sin\theta)$$

$$N = mg(2\cos\theta - \cosh\theta)$$

3/63 A stretch of highway includes a succession of evenly spaced dips and humps, the contour of which may be represented by the relation $y = b\sin\dfrac{2\pi x}{L}$. What is the maximum speed at which the car A can go over a hump and still maintain contact with the road? If the car maintains this critical speed, what is the total reaction N under its wheels at the bottom of a dip? The mass of the car is m. $\quad Ans.\ v = \dfrac{L}{2\pi}\sqrt{g/b},\quad N = 2mg$

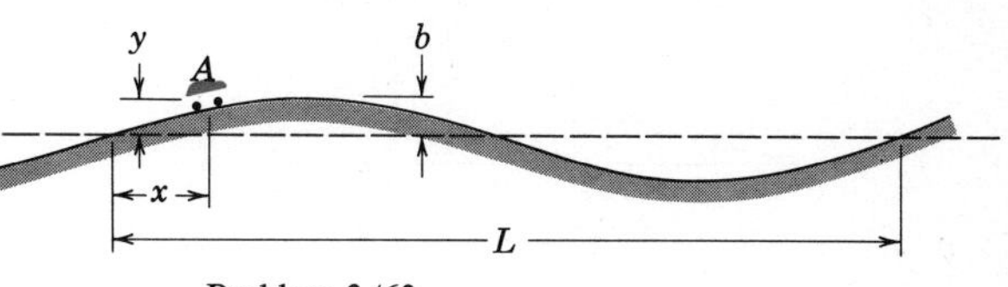

Problem 3/63

3/64 A flexible bicycle-type chain of length $\pi r/2$ has a mass ρ per unit length and is released after being held by its upper end in an initial rest condition in the smooth circular channel. Determine the acceleration a_t which all links experience just after release. Also find the expression for the tension T in the chain as a function of θ for the condition immediately following release. Isolate a differential element of the chain as a free body and apply the appropriate motion equation. $\quad Ans.\ a_t = \dfrac{2g}{\pi},\quad T = \rho gr\left[\dfrac{2\theta}{\pi} - \sin\theta\right]$

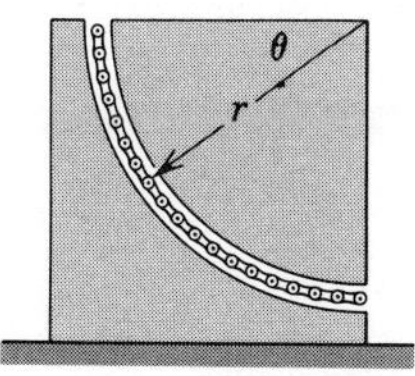

Problem 3/64

3/65 A small object is released from rest at A and slides with friction down the circular path. If the coefficient of friction is $\frac{1}{5}$, determine the velocity of the object as it passes B. [*Hint:* Write the equations of motion in the n- and t-directions, eliminate N, and substitute $v\,dv = a_t r\,d\theta$, first changing variables to $u = v^2$ so that $du = 2a_t r\,d\theta$. The resulting equation is a linear nonhomogeneous differential equation of the form $dy/dx + f(x)y = g(x)$, the solution of which is well known.] $\quad Ans.\ v = 5.52\ \text{m/s}$

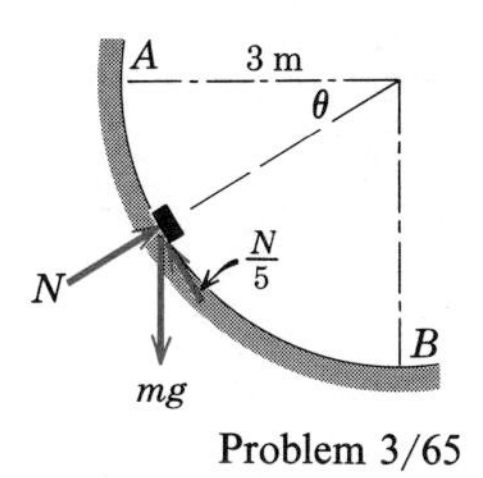

Problem 3/65

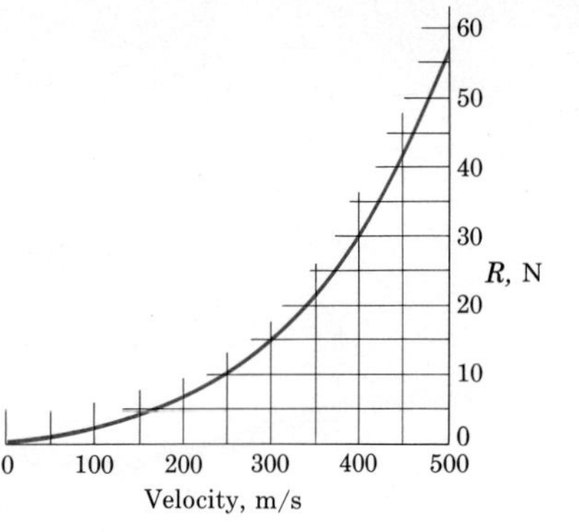

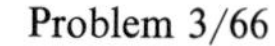
Problem 3/66

◀ **3/66** Tests on a certain 4-kg projectile show that the air resistance R varies with the velocity as shown in the graph. If the projectile is fired vertically up with an initial velocity of 500 m/s, determine the maximum height x to which it ascends and the total time t from firing until it returns to the ground. (Neglect the effect of altitude on both air resistance and gravitational attraction.)

Ans. $x = 8.0$ km, $t = 80$ s

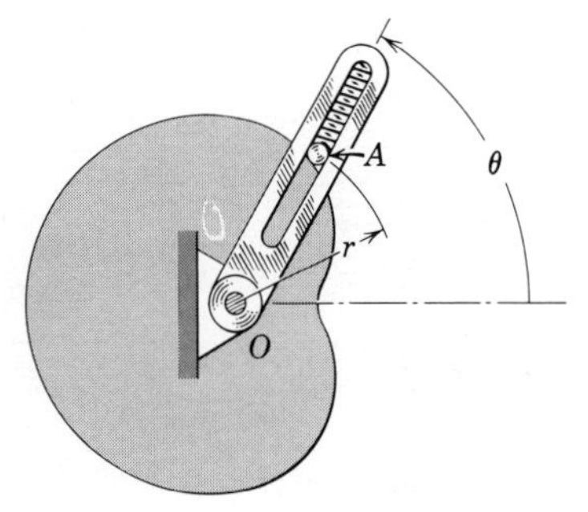

Problem 3/67

◀ **3/67** The slotted arm of Prob. 2/86, repeated here, rotates in a horizontal plane around the fixed cam with a constant counterclockwise velocity $\omega = 20$ rad/s. The spring has a stiffness of 5.4 kN/m and is uncompressed when $\theta = 0$. The cam has the shape $r = b - c\cos\theta$. If $b = 100$ mm, $c = 75$ mm, and the smooth roller A has a mass of 0.5 kg, find the force P exerted on A by the smooth sides of the slot for the position in which $\theta = 60°$. *Ans.* $P = 231$ N

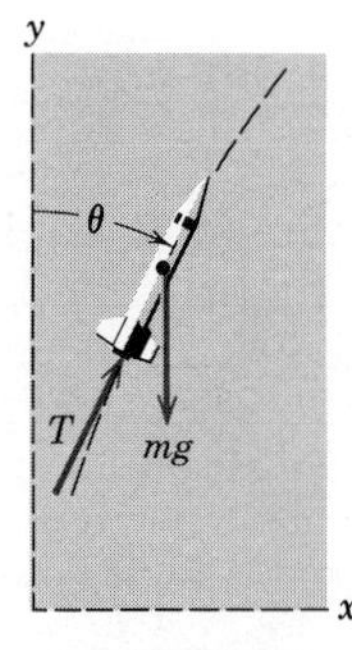

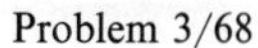
Problem 3/68

◀ **3/68** The motion of a certain rocket which operates above the earth's atmosphere is programmed for a constant pitching rate, $\theta = kt$, where k is a constant and t is the time. Take $x = y = 0$ at the position for which $t = 0$ and determine x and y as functions of the time. Assume that the gravitational attraction mg and thrust T remain constant in magnitude. The vertical y-velocity of the rocket was u when $t = 0$.

$$Ans.\ x = \frac{T}{mk}\left(t - \frac{1}{k}\sin kt\right)$$

$$y = \frac{T}{mk^2}(1 - \cos kt) - \tfrac{1}{2}gt^2 + tu$$

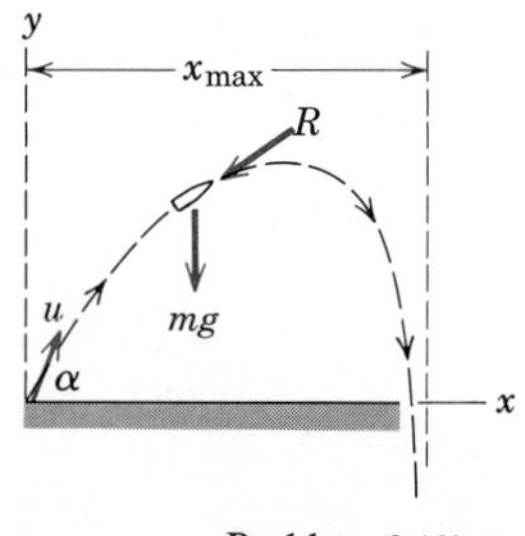

Problem 3/69

◀ **3/69** As an introduction to exterior ballistics, consider an unpowered projectile fired at an angle α with the horizontal in a fixed coordinate system. The constant gravitational attraction is mg, and the resistance R to motion is in the direction opposite to the velocity. For small velocities ($v < 30$ m/s), assume $R = kv$ and write the expressions for the coordinates of the projectile in terms of the time t after firing. What is the maximum possible horizontal displacement?

◀ 3/70 The 15-m boom OA is being elevated at a constant rate $\dot{\phi} = 0.3$ rad/s, and at the same time it is swiveled about the vertical z-axis at a constant rate $\omega = 0.2$ rad/s as indicated. Determine the magnitude of the total force F exerted by the end of the boom on a rigidly attached 30-kg ball at A at the instant that the position $\phi = 60°$ is passed. *Ans.* $F = 265$ N

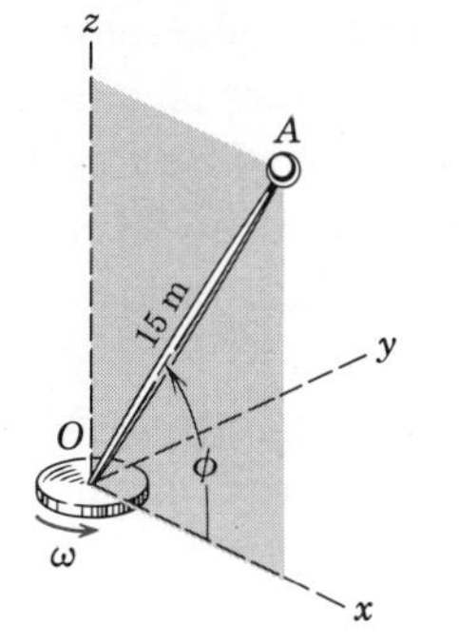

Problem 3/70

◀ 3/71 The telescoping spacecraft boom of length L carries a 4-kg ball on its end. During a ground test of the mechanism, it is revolving about the vertical z-axis with a speed of 60 rev/min while the boom is lowered at the rate $\dot{\beta} = 45°$/s and extended at the rate $\dot{L} = 0.6$ m/s, all speeds being constant over the range of motion considered. At the position $\beta = 60°$ and $L = 0.9$ m, determine the shear force Q induced in the boom at its attachment to the ball. What would Q be for these same conditions if the mechanism were operating on a spacecraft moving in space with a constant velocity at a remote distance from the earth?

Ans. $Q = 101.7$ N on earth
$Q = 72.5$ N in space

Problem 3/71

◀ 3/72 The lunar landing module has a mass of 17.5 t. As it descends onto the moon's surface, its retro engine is shut off when it reaches a hovering condition a few metres above the moon's surface. The module then falls under the action of lunar gravity, and the coiled springs in the landing pads cushion the impact. If the stiffness of each of the three springs is 15 kN/m, determine the maximum magnitude of the "jerk" J (time-rate-of-change of the acceleration) during compression of the springs. The acceleration due to lunar gravity is 1.62 m/s^2, and touchdown velocity is to be 1.5 m/s. *Ans.* $J = 4.65$ m/s^3

Problem 3/72

18 Work and Energy. In the previous article Newton's second law $\mathbf{F} = m\mathbf{a}$ was applied to various problems of particle motion to establish the instantaneous relationship between the net force acting on a particle and the resulting acceleration of the particle. When the change in the velocity or the corresponding displacement of the particle was desired, the computed acceleration was then integrated by using the appropriate kinematical equations.

There are two general classes of problems where the cumulative effects of unbalanced forces on a particle are of interest. These cases involve, respectively, integration of the forces with respect to the displacement of the particle and integration of the forces with respect to the time they are applied. The results of these integrations may be incorporated directly into the governing equations of motion, so that it becomes unnecessary to solve directly for the acceleration. Integration with respect to displacement leads to the equations of work and energy which are the subject of this article. Integration with respect to time leads to the equations of impulse and momentum which are discussed in Art. 19.

Part A: Work and Kinetic Energy

The concept of the work of a force was developed in the study of virtual work (Chapter 7 of *Statics*) but will be reviewed briefly here. Consider a particle of mass m moving along the dotted path shown in Fig. 25. Let $\mathbf{F}$ be the resultant $\Sigma\mathbf{F}$ of all forces acting on m. The position of m is established by the position vector $\mathbf{r}$, and its displacement along its path during time dt is represented by the change $d\mathbf{r}$ in its position vector. The work dU done by $\mathbf{F}$ during this displacement is defined by the dot product

$$dU = \mathbf{F} \cdot d\mathbf{r}$$

This expression represents the scalar magnitude $F_t|d\mathbf{r}| = F_t\,ds = (F\cos\alpha)\,ds$ where s is the scalar distance measured along the curve. Work is thus a scalar quantity. If F_t is in the direction of $d\mathbf{r}$, the work is positive. If F_t is in the direction opposite to $d\mathbf{r}$, the work is negative. The component F_n of the force normal to the direction of the path can do no work since its dot product with the displacement $d\mathbf{r}$ is zero.

The work done by $\mathbf{F}$ during a finite movement of the particle from points 1 to 2 is

$$U = \int \mathbf{F} \cdot d\mathbf{r} = \int_{s_1}^{s_2} F_t\,ds$$

where the limits specify the initial and final end points of the interval of motion involved.

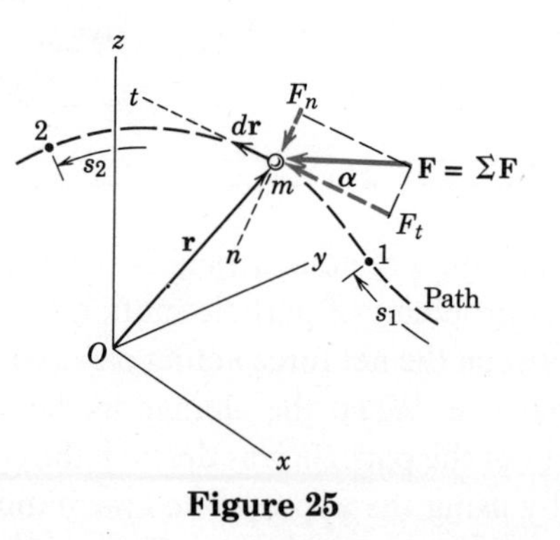

Figure 25

For a system of forces $\mathbf{F}_1$, $\mathbf{F}_2$, $\mathbf{F}_3$, ... acting on a particle which has a displacement $d\mathbf{r}$, the work done by the resultant $\Sigma\mathbf{F}$ of the system equals the sum of the scalar works done by each of its components. Thus

$$dU = \mathbf{F}_1 \cdot d\mathbf{r} + \mathbf{F}_2 \cdot d\mathbf{r} + \mathbf{F}_3 \cdot d\mathbf{r} + \dots$$
$$= (\mathbf{F}_1 + \mathbf{F}_2 + \mathbf{F}_3 + \dots) \cdot d\mathbf{r} = \Sigma\mathbf{F} \cdot d\mathbf{r}$$

from which it follows that $U = \int \Sigma\mathbf{F} \cdot d\mathbf{r}$.

Work has the units of force (N) times displacement (m) or N·m. This unit is given the special name *joule* (J) which is defined as the work done by a force of 1 N moving through a distance of 1 m in the direction of the force. Consistent use of the joule for work (and energy) rather than the units N·m will avoid possible ambiguity with the units of moment of a force or torque which are also written N·m.

If the functional relation between the tangential component F_t of a force and the distance s along the path is known, the integral for the work U may be evaluated mathematically. If the functional relationship is not known as a mathematical expression which can be integrated but is specified in the form of approximate or experimental data, then the work may be evaluated by carrying out a numerical or graphical integration which would be represented by the area under the curve of F_t versus s, as indicated in Fig. 26.

Although not needed in the discussion of the motion of a particle, the work done by a couple acting on a body which rotates will be covered here since it is analogous to that for the work of a force. Consider the couple $\mathbf{M}$ acting on the disk of Fig. 27 and consisting of the two equal and opposite forces $\mathbf{F}$ parallel to the x-axis and a distance $2r$ apart. Clearly no net work is done during a translation of the disk. Now consider an infinitesimal rotation $d\boldsymbol{\theta} = \mathbf{i}\, d\theta_x + \mathbf{j}\, d\theta_y + \mathbf{k}\, d\theta_z$ of the disk about *any* axis through its center, such as *a-a*. It is seen that no work is done by either force during components of rotation $d\theta_x$ and $d\theta_y$. During the rotation $d\theta_z$, however, the work done is $dU = 2Fr\, d\theta_z = M\, d\theta_z$ which may be written as

$$dU = \mathbf{M} \cdot d\boldsymbol{\theta}$$

The total work done during a finite rotation is, then,

$$U = \int \mathbf{M} \cdot d\boldsymbol{\theta} = \int (M_x\, d\theta_x + M_y\, d\theta_y + M_z\, d\theta_z)$$

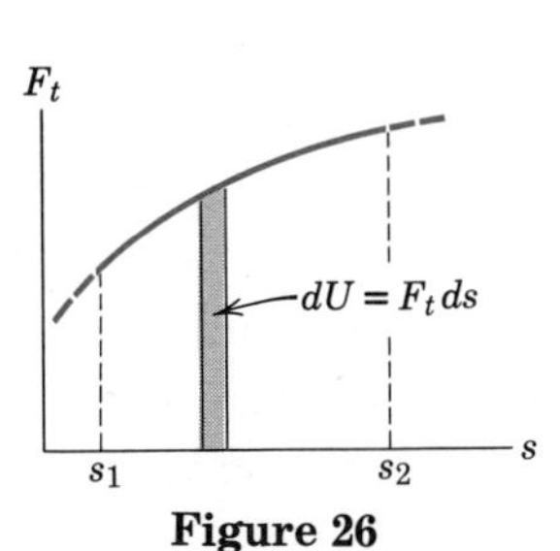

Figure 26

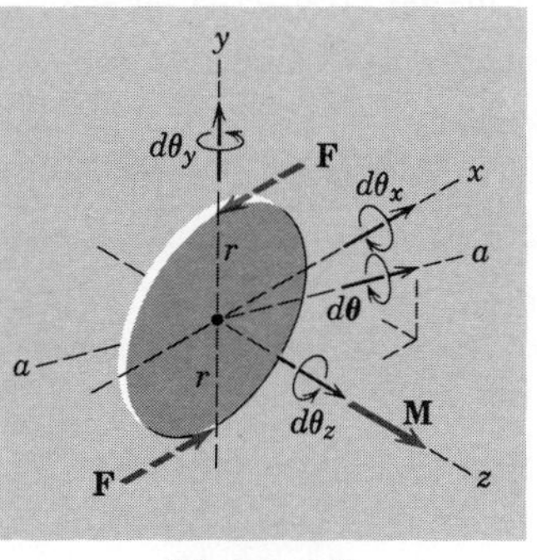

Figure 27

With the concept of the work of a force established, the work done by the resultant $\Sigma\mathbf{F}$ of all forces acting on the particle m of Fig. 25 during a finite interval of motion along its path may now be determined by substituting Newton's second law, Eq. 44, which gives

$$U = \int \Sigma\mathbf{F} \cdot d\mathbf{r} = \int m\mathbf{a} \cdot d\mathbf{r}$$

But $\mathbf{a} \cdot d\mathbf{r} = a_t\, ds$ where a_t is the tangential component of the acceleration of m and ds is the magnitude of $d\mathbf{r}$. In terms of the velocity v of the particle, Eq. 6 gives $a_t\, ds = v\, dv$. Thus the expression for the work of $\Sigma\mathbf{F}$ becomes

$$U = \int \Sigma\mathbf{F} \cdot d\mathbf{r} = \int mv\, dv$$

or

$$U = \int \Sigma\mathbf{F} \cdot d\mathbf{r} = \tfrac{1}{2}m(v_2^2 - v_1^2) \qquad \textbf{(50)}$$

where the integration is carried out between points 1 and 2 along the curve at which points the velocities have the magnitudes v_1 and v_2, respectively.

The *kinetic energy* T of the particle is defined as

$$T = \tfrac{1}{2}mv^2$$

and is the total work which must be done on the particle to bring it from a state of rest to a velocity v. Kinetic energy T is a scalar quantity with the units of N·m or *joules* (J) and is *always* positive irrespective of the direction of the velocity. Equation 50 may be restated simply as

$$U = \Delta T \qquad \textbf{(51)}$$

which is the *work-energy equation* for a particle.

The equation states that the *total work done* by all forces acting on a particle during an interval of its motion equals the corresponding *change in kinetic energy* of the particle.

It is now seen from Eq. 50 that a major advantage of the method of work and energy is that it avoids the necessity of computing the acceleration and leads directly to the velocity changes as functions of the forces which do work. Further, the work-energy equation involves only those forces which do work and give rise to changes in the magnitude of the velocities.

Consider two or more particles joined together by connections which are frictionless and incapable of elastic deformation. The forces in the connections occur in pairs of equal and opposite forces, and the points of application of these forces necessarily have identical displacement components in the direction of the forces. Hence the net work done by these internal forces is zero during any movement of the system of the two or more connected particles. Thus Eq. 51 is applicable to the entire system, where U is the total or net work done on the system by forces external to it and

ΔT is the change in the total kinetic energy of the system. The total kinetic energy is the sum of the kinetic energies of all elements of the system. It may now be observed that a further advantage of the work-energy method is that it permits the analysis of a system of particles joined in the manner described without dismembering the system.

Sample Problem

3/73 The 25-kg slider is released from the position shown with a velocity $v_0 = 0.6$ m/s on the inclined rail and slides under the influence of gravity and friction. The coefficient of friction between the slider and the rail is 0.5. Calculate the velocity of the slider as it passes the position for which the spring is compressed a distance $x = 100$ mm. The spring offers a compressive resistance C and is known as a "hardening" spring, since its stiffness increases with deflection as shown in the accompanying graph.

Solution. In addition to the weight of $25(9.81) = 245$ N, the normal reaction N, and the friction force $0.5N$, the free-body diagram of the slider shows the compressive force $C = 16x + 0.06x^2$ exerted by the spring on the slider after it contacts the spring. Inasmuch as the friction force $0.5N$ does work on the slider, the normal force must be determined. Since there is no acceleration in the y-direction,

$$[\Sigma F_y = 0] \quad N - 245 \cos 60° = 0, \qquad N = 122.6 \text{ N}, \qquad 0.5N = 61.3 \text{ N}$$

For the interval under consideration the displacement is $0.9 + 0.1 = 1.0$ m. The work-energy equation gives

$[U = \Delta T]$

$$(245 \sin 60° - 61.3)(1.00) - 10^{-3}\int_0^{100} (16x + 0.06x^2)\, dx = \frac{1}{2}25(v^2 - [0.6]^2)$$

Care must be exercised to preserve consistent units, and the factor 10^{-3} multiplies the integral to convert it from mJ to J. Solution of the above equation gives

$$v = 2.11 \text{ m/s} \qquad \textit{Ans.}$$

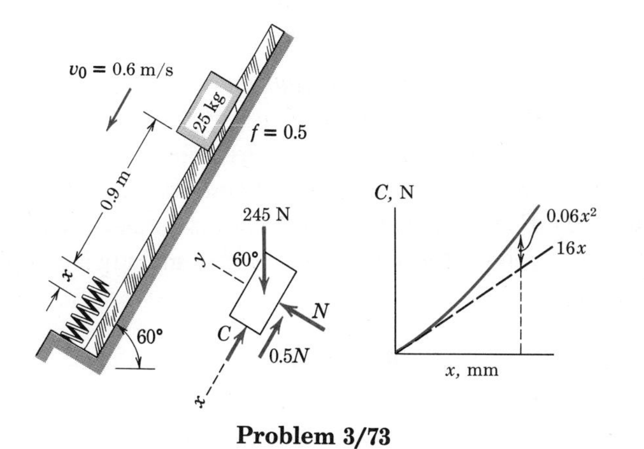

Problem 3/73

Part B: Potential Energy

The work done by a force $\mathbf{F}$ during a displacement $d\mathbf{r}$ of its point of application may be written as

$$U = \int \mathbf{F} \cdot d\mathbf{r} = \int (F_x \, dx + F_y \, dy + F_z \, dz)$$

The integral $\int \mathbf{F} \cdot d\mathbf{r}$ is a line integral dependent, in general, upon the particular path followed between any two points 1 and 2 in space. If, however, $\mathbf{F} \cdot d\mathbf{r}$ is an *exact differential** $-dV$ of some scalar function V of the coordinates, then

$$U = \int_{V_1}^{V_2} -dV = -(V_2 - V_1) \tag{52}$$

which depends only on the end points of the motion and which is thus *independent* of the path followed. The minus sign before dV is arbitrary but is chosen to agree with the customary designation of the sign of potential energy change in the earth's gravity field. If V exists, the differential change in V may be written as

$$dV = \frac{\partial V}{\partial x} dx + \frac{\partial V}{\partial y} dy + \frac{\partial V}{\partial z} dz$$

Comparison with $-dV = \mathbf{F} \cdot d\mathbf{r} = F_x \, dx + F_y \, dy + F_z \, dz$ yields

$$F_x = -\frac{\partial V}{\partial x} \qquad F_y = -\frac{\partial V}{\partial y} \qquad F_z = -\frac{\partial V}{\partial z}$$

The force may also be written as the vector

$$\mathbf{F} = -\nabla V \tag{53}$$

where the symbol ∇ stands for the vector operator "del" which is

$$\nabla = \mathbf{i}\frac{\partial}{\partial x} + \mathbf{j}\frac{\partial}{\partial y} + \mathbf{k}\frac{\partial}{\partial z}$$

The quantity V is known as the *potential function,* and the expression ∇V is known as the *gradient of the potential function.*

When the components of a force are derivable from a potential in the manner described, the force is said to be *conservative,* and it follows that the work done by $\mathbf{F}$ between any two points is independent of the path followed. For such a force, the work done by $\mathbf{F}$ in making a complete circuit of any closed path is

$$\oint \mathbf{F} \cdot d\mathbf{r} = 0$$

* Recall that a function $d\phi = P \, dx + Q \, dy + R \, dz$ is an exact differential in the coordinates x-y-z if

$$\frac{\partial P}{\partial y} = \frac{\partial Q}{\partial x}, \qquad \frac{\partial P}{\partial z} = \frac{\partial R}{\partial x}, \qquad \frac{\partial Q}{\partial z} = \frac{\partial R}{\partial y}.$$

When the force is a function of velocity or involves dissipative friction, it will be *nonconservative,* and the work done will be dependent upon the particular path followed. In this case a potential function does not exist.

The most common example of a conservative force field is that of the earth's gravity given by Eq. 2. For motion in close proximity to the earth's surface, the gravitational attraction is $F = mg$ and is essentially constant. The work done by this force acting on the attracted object, Fig. 28*a*, during an elevation of the object a vertical distance h as it travels along any path is $-mgh$. Equating the work to the negative of the potential change in accordance with Eq. 52 gives $-mgh = -(V_g - 0)$ or

$$V_g = mgh \tag{54}$$

where the potential energy is arbitrarily taken to be zero at the lower position and where the subscript g denotes gravitational potential. It should be clear that the datum plane for zero potential energy is arbitrary since it is only the *change* in potential energy that matters.

When large changes in altitude in the earth's field are encountered, Fig. 28*b*, the gravitational force $Kmm_0/r^2 = mgR^2/r^2$ is no longer constant. Equating the work done by this force acting on the object to the negative of the potential change, by Eq. 52, during any motion from a radial distance r from the center of the earth to a greater one r' gives

$$\int_r^{r'} -\frac{mgR^2}{r^2}\,dr = -(V_g' - V_g) \qquad \text{or} \qquad V_g - V_g' = mgR^2\left(\frac{1}{r'} - \frac{1}{r}\right)$$

It is customary to take $V_g' = 0$ when $r' = \infty$, so that with this datum, the gravitational potential becomes

$$V_g = -\frac{mgR^2}{r} \tag{55}$$

It may be proved that the gravitational law of Eq. 2 used in this proof holds with respect to the earth treated as a concentrated mass particle at its center.

A second common example of a conservative force is found with the deformation of an elastic body. For the one-dimensional elastic spring of stiffness k, Fig. 29, the force supported by the spring at any deformation x, compression or extension, from its undeformed position is $F = kx$. Thus the

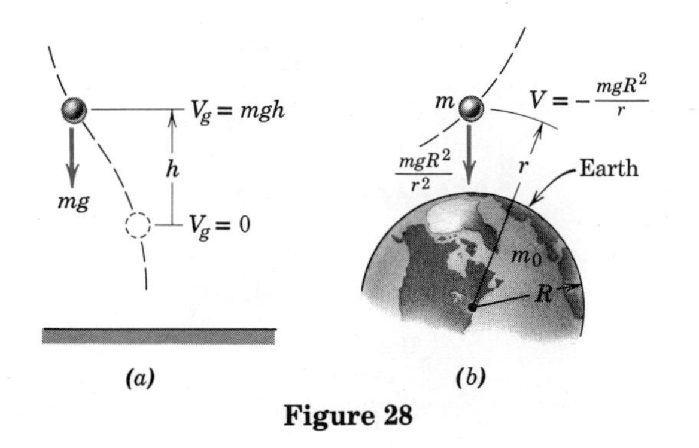

Figure 28

work done on an elastic spring during an interval of compression or extension is

$$\int F\,dx = \int_{x_1}^{x_2} kx\,dx = \tfrac{1}{2}kx_2^2 - \tfrac{1}{2}kx_1^2$$

which represents the trapezoidal area in the F-x diagram. It is noted that positive work is done on a spring by the force which deforms it whether it is extended or compressed. It follows that the work done by the equal and opposite force on the body to which the spring is attached is negative. Conversely, during a release from its tension or compression, negative work is done on the spring which means that the spring does positive work on the body to which it is attached. The work done on a spring to deform it an amount x from its undeformed state is stored in the spring and is known as its *elastic potential energy,* which is

$$V_e = \tfrac{1}{2}kx^2 \qquad \textbf{(56)}$$

The force F' exerted *by* the spring *on* the body which deforms it, from Eq. 53, is

$$F' = -\frac{\partial}{\partial x}(\tfrac{1}{2}kx^2) = -kx$$

which is the negative of the force F acting on the spring.

With the introduction of gravitational potential energy and elastic potential energy it is frequently convenient to replace the work done by the gravity force and the work done by the spring force by the negative of their respective changes in potential energy. If U stands for the work done on the particle by all forces *other* than gravity and spring forces, then Eq. 51 which relates work and kinetic energy change becomes $U + (-\Delta V_g) + (-\Delta V_e) = \Delta T$ or

$$U = \Delta T + \Delta V_g + \Delta V_e \qquad \textbf{(57)}$$

This alternative form of the work-energy equation is often far more convenient to use than Eq. 51, since the work of both gravity and spring forces is accounted for by focusing attention on the end-point positions of the

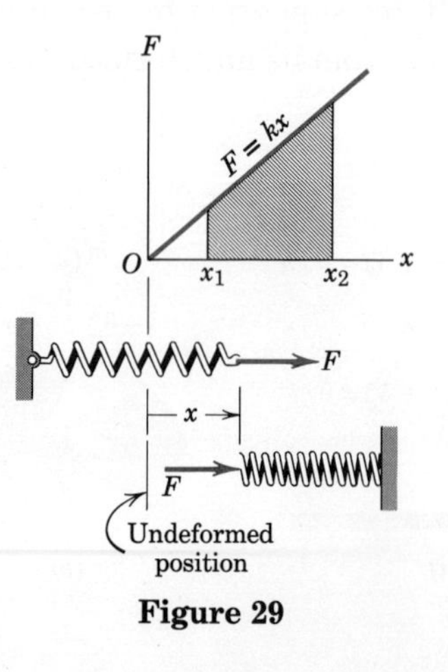

Figure 29

center of gravity and of the length of the elastic spring. The path followed between these end-point positions is of no consequence in the evaluation of ΔV_g and ΔV_e.

To help clarify the difference between the use of Eqs. 51 and 57, Fig. 30*a* shows schematically a particle of mass m constrained to move along a fixed path under the action of forces F_1 and F_2, the gravitational force $W = mg$, the spring force F, and the normal reaction N. In the *b*-part of the figure the particle is isolated with its free-body diagram, and the work done by each of the forces F_1, F_2, W, and the spring force $F = kx$ is evaluated for the interval of motion in question, say from A to B, and equated to the change ΔT in kinetic energy using Eq. 51. The constraint reaction N, if normal to the path, will do no work. In the *c*-part of the figure for the alternative approach the spring is included as a part of the isolated system. The work done during the interval by F_1 and F_2 constitutes the U-term of Eq. 57 with the changes in elastic and gravitational potential energies included on the energy side of the equation. It may be noted with the first approach that the work done by $F = kx$ could require a somewhat awkward integration to account for the changes in magnitude and direction of F as the particle moved from A to B. With the second approach, however, only the initial and final lengths of the spring would be required to evaluate ΔV_e.

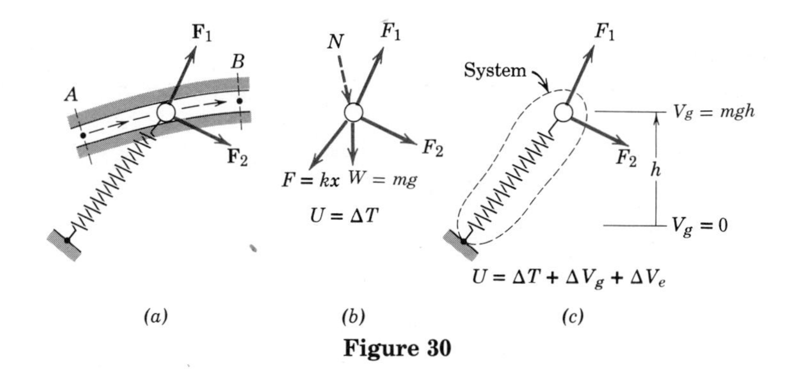

Figure 30

Sample Problem

3/74 The 10-kg slider A moves in the vertical plane with negligible friction along the inclined guide. The attached spring has a stiffness of 60 N/m and is stretched a distance of 0.6 m in position A where the slider is released from rest. A constant 250-N force is applied to the light connecting cord which passes over the small pulley at B. The pulley offers negligible resistance to motion of the cord. Calculate the velocity v of the slider as it passes point C.

Solution. The slider and inextensible cord together with the attached spring will be analyzed as a system, which permits the use of Eq. 57. The only force acting on this system which does work is the 250-N tension applied to the cord. While the slider moves from A to C the point of application of the 250-N force moves a distance of $\overline{AB} - \overline{BC}$ or $1.5 - 0.9 = 0.6$ m. Thus

$$U = 250(0.6) = 150 \text{ J}$$

The reactions of the guides on the slider are normal to the direction of motion and do no work.

The change in kinetic energy of the slider is

$$\Delta T = \tfrac{1}{2}m(v^2 - v_0^2) = \tfrac{1}{2}(10)v^2 \text{ J}$$

where the initial velocity v_0 is zero. The change in gravitational potential energy is positive since the center of mass of the slider has an upward component of displacement. Thus

$$\Delta V_g = mg(\Delta h) = 10(9.81)(1.2 \sin 30°) = 58.9 \text{ J}$$

The change in elastic potential energy is

$$\Delta V_e = \tfrac{1}{2}k(x_2^2 - x_1^2) = \tfrac{1}{2}(60)([1.2 + 0.6]^2 - [0.6]^2) = 86.4 \text{ J}$$

Substitution into the alternative work-energy equation gives

$[U = \Delta T + \Delta V_g + \Delta V_e]$ $\qquad 150 = \tfrac{1}{2}(10)v^2 + 58.9 + 86.4$

$$v = 0.974 \text{ m/s}$$ *Ans.*

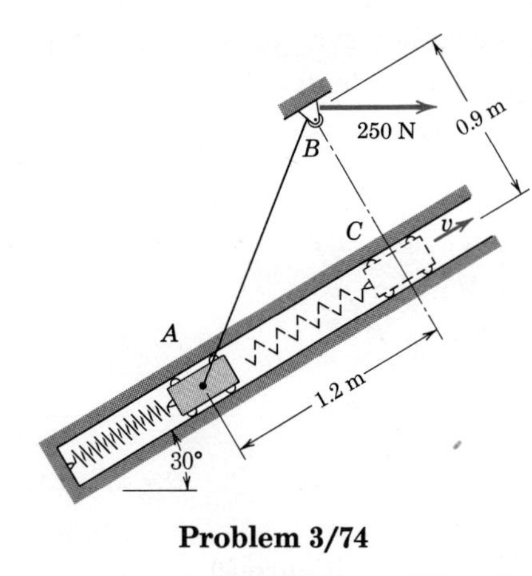

Problem 3/74

Part C: Conservation of Energy; Power

The alternative work-energy relation, Eq. 57, may be rewritten for a particle-and-spring system as

$$U = \Delta(T + V_g + V_e) = \Delta E \tag{57a}$$

where $E = T + V_g + V_e$ is the total mechanical energy of the particle and its attached linear spring. Equation 57*a* states that the net work done on the system by all forces other than gravitational forces and elastic forces equals the change in the total mechanical energy of the system. For problems where the only forces are gravitational, elastic, and nonworking constraint forces, the U-term will be zero, and the energy equation becomes merely

$$\Delta E = 0 \qquad \text{or} \qquad E = \text{const.} \tag{58}$$

When E is constant, it is seen that transfers of energy between kinetic and potential may take place as long as the total mechanical energy $T + V_g + V_e$ does not change. Equation 58 expresses the *law of conservation of dynamical energy.*

Power. The capacity of a machine is rated by its *power* which is its *time rate of doing work*. Thus if U stands for work done by a force $\mathbf{F}$, then the power P developed by this force is

$$P = \dot{U} = \mathbf{F} \cdot \dot{\mathbf{r}} = \mathbf{F} \cdot \mathbf{v} \tag{59}$$

where $\dot{\mathbf{r}} = \mathbf{v}$ is the velocity of the point of application of the force. Similarly, the power developed by a couple $\mathbf{M}$ acting on a body which has an angular velocity $\boldsymbol{\omega}$ is

$$P = \dot{U} = \mathbf{M} \cdot \boldsymbol{\omega} \tag{60}$$

Power is clearly a scalar quantity and has the units of N·m/s = J/s. The special unit for power is the *watt* (W) which equals one joule per second (J/s).

Sample Problem

3/75 The 3-kg slider is released from rest at point A and slides with negligible friction in a vertical plane along the circular rod. The attached spring has a stiffness of 350 N/m and has an unstretched length of 0.6 m. Determine the velocity of the slider as it passes position B.

Solution. The work done by the weight and the spring force on the slider will be treated as changes in the potential energies, and the reaction of the rod on the slider is normal to motion and does no work. Hence $U = 0$. The changes in the potential and kinetic energies for the system of slider and spring are

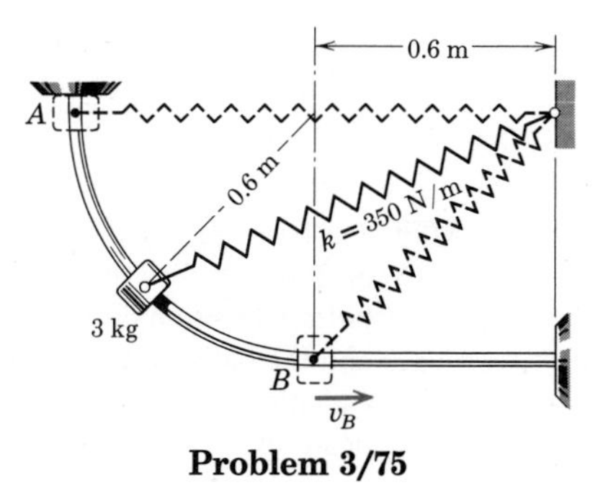

Problem 3/75

$$\Delta V_e = \tfrac{1}{2}k(x_B{}^2 - x_A{}^2) = \tfrac{1}{2}(350)\{(0.6[\sqrt{2} - 1])^2 - (0.6)^2\} = -52.2 \text{ J}$$

$$\Delta V_g = mg\,\Delta h = 3(9.81)(-0.6) = -17.66 \text{ J}$$

$$\Delta T = \tfrac{1}{2}m(v_B{}^2 - v_A{}^2) = \tfrac{1}{2}(3)(v_B{}^2 - 0) = 1.5\, v_B{}^2$$

$[\Delta T + \Delta V_g + \Delta V_e = 0]$ $\qquad 1.5v_B{}^2 - 17.66 - 52.2 = 0, \quad v_B = 6.82 \text{ m/s}$ *Ans.*

Note that the evaluation of the work done by the spring force acting on the slider by means of the integral $\int \mathbf{F} \cdot d\mathbf{r}$ would necessitate a lengthy computation to account for the change in the magnitude of the force along with the change in the angle between the force and the tangent to the path. Note further that v_B depends only on the end conditions of the motion and does not require knowledge of the shape of the path since the slider moves in a conservative field of force.

Problems

3/76 A car is traveling down a 10-per-cent grade at a speed of 50 km/h when the brakes on all four wheels lock causing the car to skid to a stop in a distance of 15 m. Calculate the coefficient of friction between the tires and the road. Treat the car as a particle. *Ans.* $f = 0.759$

3/77 The frame is moving horizontally with a constant velocity of 1 m/s with its small pendulum bob hanging in the vertical position. If the frame is brought to an abrupt stop, calculate the amplitude θ of the subsequent angular oscillations of the pendulum. Neglect frictional resistance to motion of the pendulum.

3/78 The 2-kg slider moves along the smooth fixed rod under the action of its weight and a constant externally applied force $\mathbf{F} = -15\mathbf{i} + 10\mathbf{j} + 15\mathbf{k}$ N. If the slider starts from rest at A, determine the magnitude of its velocity as it reaches B. *Ans.* $v_B = 4.39$ m/s

3/79 The ball is released from position A with a velocity of 2 m/s and swings in a vertical plane. At the bottom position the cord strikes the fixed bar at B, and the ball continues to swing in the dotted arc. Calculate the velocity v_C of the ball as it passes position C.

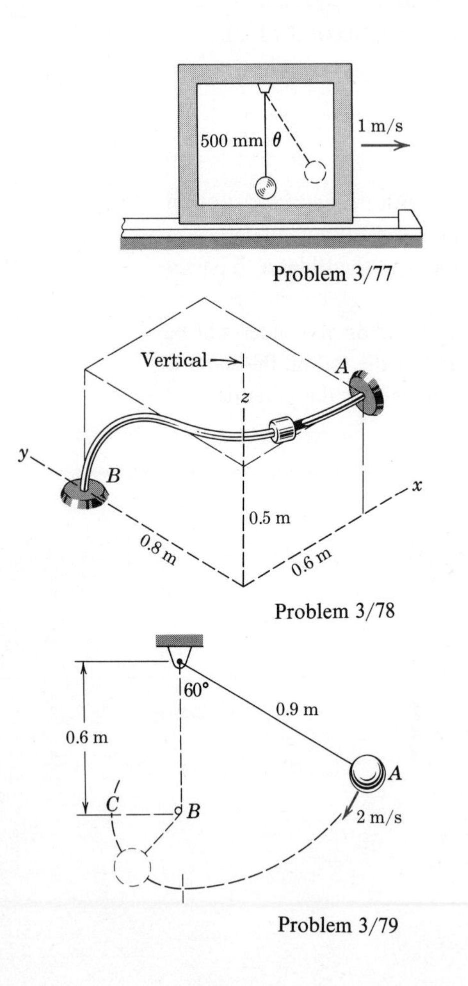

Problem 3/77

Problem 3/78

Problem 3/79

3/80 The 10-kg mass slides freely on the fixed circular guide. Determine the velocity v of the slider as it reaches B if it is elevated from rest at A by the action of a constant 250-N force in the cable.
Ans. $v = 8.54$ m/s

3/81 The 10-kg mass is released from rest in position (a) with the spring unstretched. Determine the distance h in position (b) where the mass has reached its lowest position. The stiffness of the spring is $k = 450$ N/m.

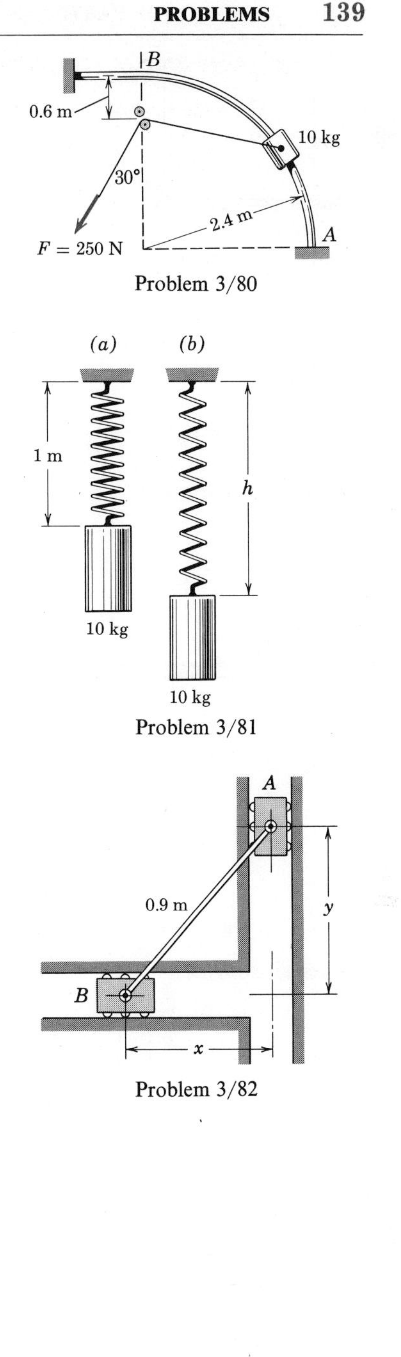

Problem 3/80

Problem 3/81

Problem 3/82

3/82 Sliders A and B are of equal mass and are confined to move, respectively, in the vertical and horizontal guides. Their connecting link has negligible mass. If they are released from rest in the position for which $x = y$ and if they slide with negligible friction, calculate the velocity v_A of A as it passes the horizontal line through B.
Ans. $v_A = 3.53$ m/s

3/83 The escalator in a department store is designed to move 100 people per minute from one floor to the next higher floor through a vertical distance of 6 m. If the mass of the average person is 70 kg and 30 per cent of the power is lost in friction, compute the required power output of the driving motor.

3/84 The position vector of a particle is given by $\mathbf{r} = 1.2t\mathbf{i} + 0.9t^2\mathbf{j} - 0.6(t^3 - 1)\mathbf{k}$ where t is the time in seconds from the start of the motion and where $\mathbf{r}$ is expressed in metres. For the condition when $t = 4$ s, determine the power P developed by the force $\mathbf{F} = 60\mathbf{i} - 25\mathbf{j} - 40\mathbf{k}$ N which acts on the particle.
Ans. $P = 1.044$ kW

3/85 Small metal blocks are discharged with a velocity of 0.45 m/s to a ramp by the upper conveyor shown. If the coefficient of friction between the blocks and the ramp is 0.30, calculate the angle θ which the ramp must make with the horizontal so that the blocks will transfer without slipping to the lower conveyor moving at the speed of 0.15 m/s.

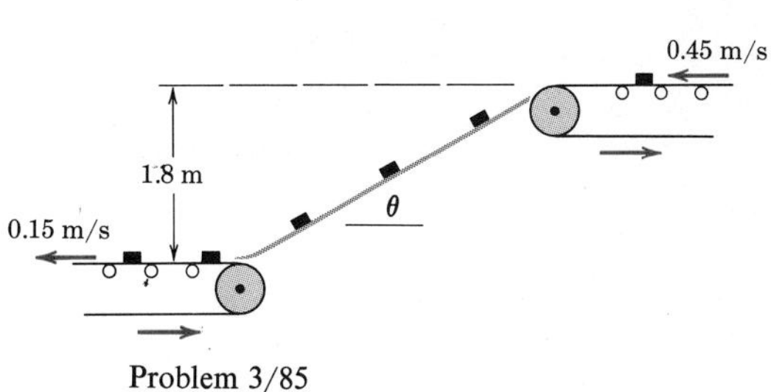

Problem 3/85

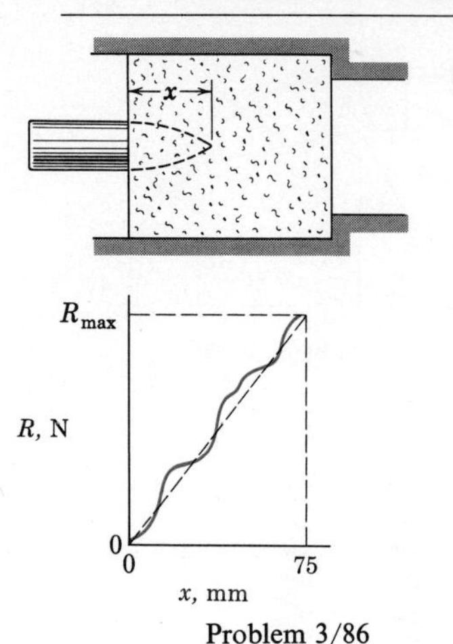

Problem 3/86

3/86 The resistance R to penetration x of a 0.25-kg projectile fired with a velocity of 600 m/s into a certain block of fibrous material is shown in the graph. Represent this resistance by the dotted line and compute the velocity v of the projectile for the instant when $x = 25$ mm if the projectile is brought to rest after a total penetration of 75 mm. *Ans.* $v = 566$ m/s

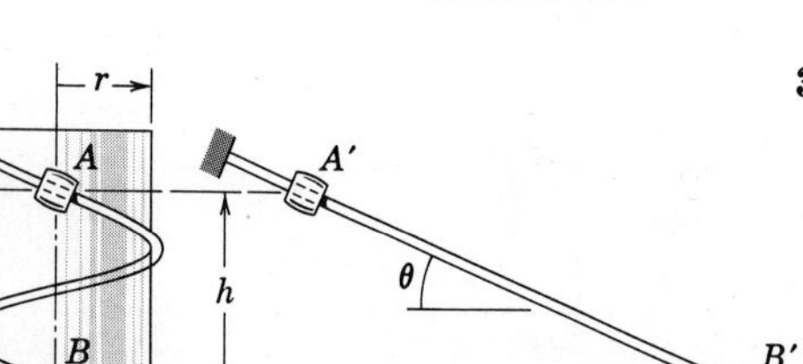

Problem 3/87

3/87 A bead is released from rest at point A and slides down the wire bent into the form of a helix, as shown in the a-part of the figure. The experiment is repeated in the b-part of the figure where the wire has been unwrapped to form a straight line with a slope equal to that of the helix. What is the velocity reached by the bead as it passes point B or B' in each case in the absence of friction? In the presence of some friction does the velocity depend on the path, and if so, for which case would the velocity at the lower position be greater?

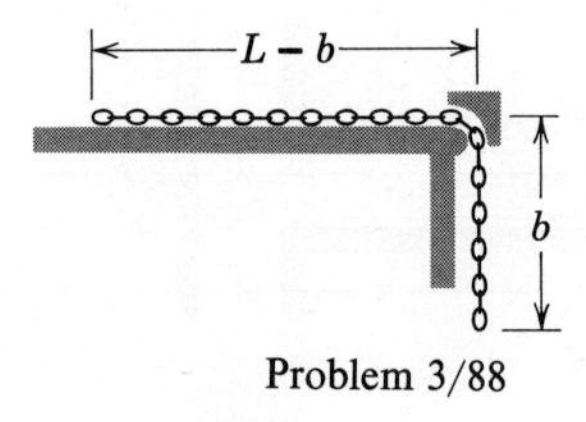

Problem 3/88

3/88 The figure for Prob. 3/53 is shown again here. Solve for the velocity v of the chain as the last link leaves the edge if the chain is released from rest in the position shown. Friction is negligible.

$$Ans.\ v = \sqrt{gL\left(1 - \frac{b^2}{L^2}\right)}$$

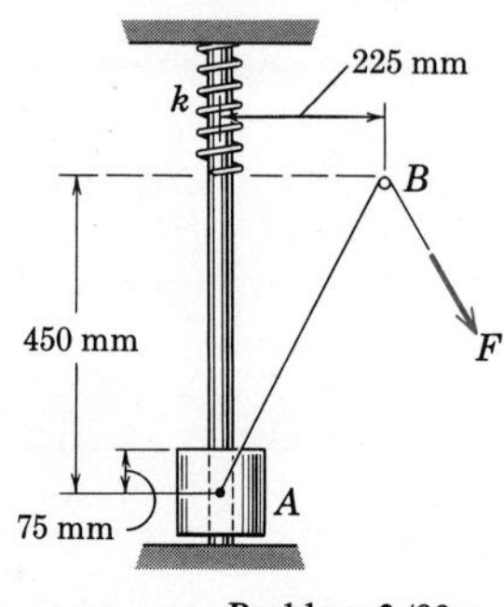

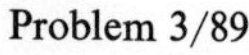

Problem 3/89

3/89 The 7-kg collar A slides with negligible friction on the fixed vertical shaft. When the collar is released from rest at the bottom position shown, it moves up the shaft under the action of the constant force $F = 200$ N applied to the cable. Calculate the stiffness k which the spring must have if its maximum compression is to be limited to 75 mm. The position of the small pulley at B is fixed. *Ans.* $k = 8.79$ kN/m

3/90 A small rocket-propelled test vehicle with a total mass of 100 kg starts from rest at A and moves with negligible friction along the track in the vertical plane as shown. If the propelling rocket exerts a constant thrust T of 2 kN from A to position B where it is shut off, determine the distance s which the vehicle rolls up the incline before stopping. The loss of mass due to the expulsion of gases by the rocket is small and may be neglected.

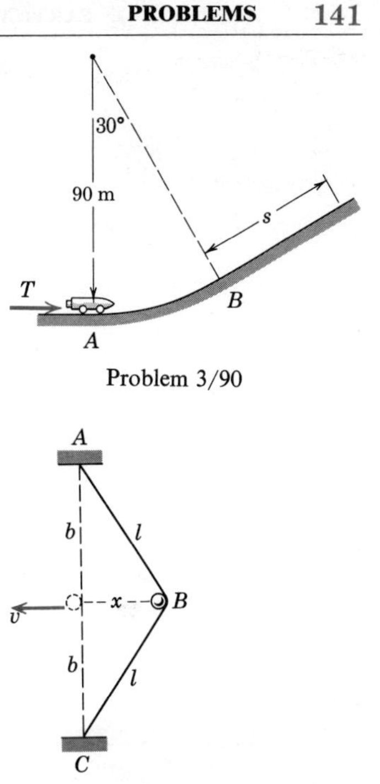

Problem 3/90

3/91 The figure for Prob. 3/48 is repeated here. Solve for the velocity v imparted to the particle of mass m under the conditions described with that problem.

Ans. $v = \sqrt{\frac{2K}{mb}}\,(\sqrt{b^2 + x^2} - b)$

Problem 3/91

3/92 A skier starts from rest at A and slides with a small amount of friction to a position for horizontal takeoff at B. He lands a distance s down the 45° slope from the base of the take-off ramp. Calculate the maximum possible value of s which would occur if ski friction and air resistance were both zero. *Ans.* $s = 92.8$ m

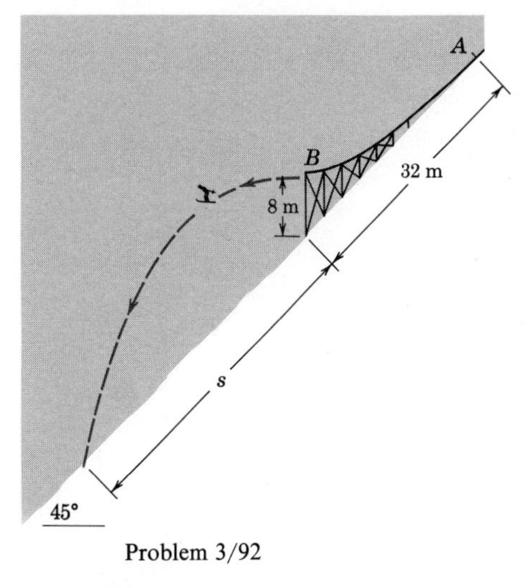

Problem 3/92

3/93 Calculate the maximum velocity reached by the 10-kg mass in Prob. 3/81 as it moves between the rest positions (a) and (b).

3/94 The 2-kg cylinder is released from rest against a coiled spring which has been compressed 0.5 m from its uncompressed position. If the stiffness of the spring is 120 N/m, determine (a) the maximum height h reached by the cylinder above its released position, and (b) the maximum velocity v reached by the cylinder. Neglect the mass of the spring. *Ans.* (a) $h = 0.764$ m
(b) $v = 2.61$ m/s

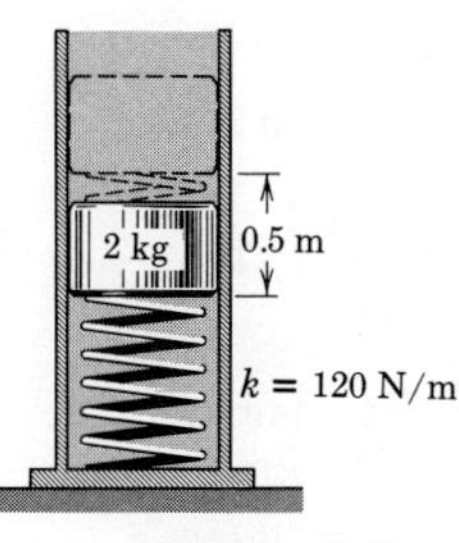

Problem 3/94

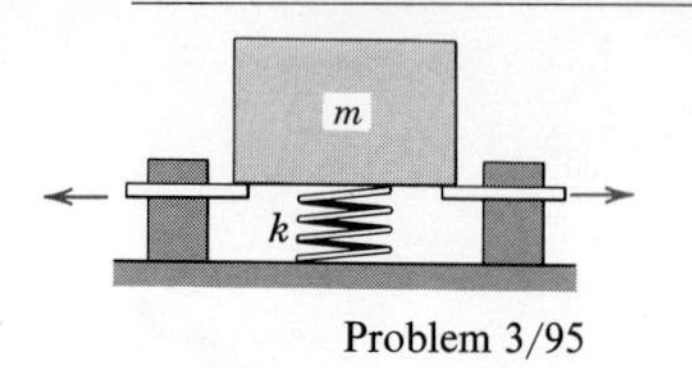

Problem 3/95

3/95 The block of mass m is supported by the two fingers and is barely in contact with the uncompressed elastic spring of stiffness k and negligible mass. If the fingers are suddenly removed, determine the maximum velocity v reached by the block, the maximum force R transmitted to the floor through the spring, and the maximum deformation δ of the spring.

3/96 Calculate the maximum velocity $v_{B\max}$ of slider B during the motion described in Prob. 3/82. Note that the velocities of A and B are related through the time derivatives $\dot{x}$ and $\dot{y}$ determined from the equation $x^2 + y^2 = (0.9)^2$.

Ans. $v_{B_{\max}} = 0.962$ m/s

3/97 If the power output P of a car of mass m remains constant during a certain interval, find the distance s required to increase its speed within this interval from v_1 to v_2 on a level road. Neglect frictional resistance and treat the car as a particle.

Problem 3/98

3/98 Determine the velocity v of the 50-kg sliding block after it has moved 1.2 m along the incline from rest under the action of the constant force $P = 480$ N. The light cables are wrapped securely around the integral pulleys of negligible mass. Pulley diameters are in the ratio of 2 : 1, friction in the bearing is negligible, and the coefficient of friction f between the block and the incline is 0.40. Note that the center of the pulley moves up the incline the same distance that the point of application of P on the cable moves down the incline. *Ans.* $v = 1.764$ m/s

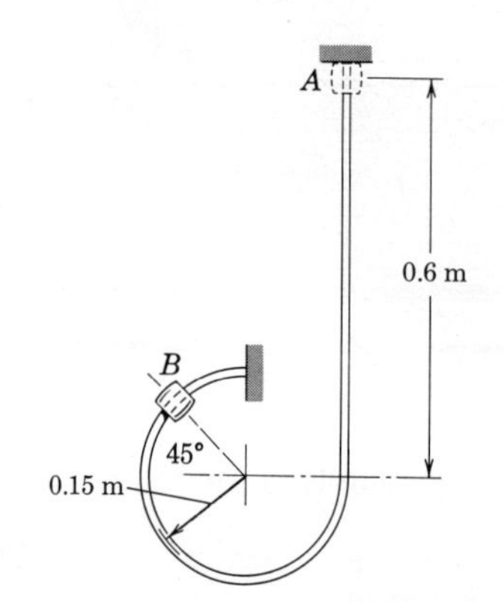

Problem 3/99

3/99 A bead with a mass of 0.25 kg is released from rest at A and slides down and around the fixed smooth wire. Determine the force N between the wire and the bead as it passes point B.

Ans. $N = 14.42$ N

$\overline{OA} = 16\,000$ km
$\overline{OB} = 7200$ km
$v_A = 7100$ km/h

Problem 3/100

3/100 Upon its return voyage to earth, a space capsule has an absolute velocity of 7100 km/h at point A which is 16 000 km from the center of the earth. Determine the absolute velocity of the capsule when it reaches point B which is 7200 km from the earth's center. The trajectory between these two points is outside the effect of the earth's atmosphere.

3/101 A rocket launches an unpowered space capsule at point A with an absolute velocity $v_A =$ 13 000 km/h at an altitude of 40 km. After the capsule has traveled a distance of 400 km measured along its absolute space trajectory, its velocity at B is 12 400 km/h and its altitude is 80 km. Determine the average resistance P to motion in the rarified atmosphere. The mass of the capsule is 22 kg, and the earth's mean radius is 6370 km. Consider the center of the earth as fixed in space. *Ans.* $P = 11.16$ N

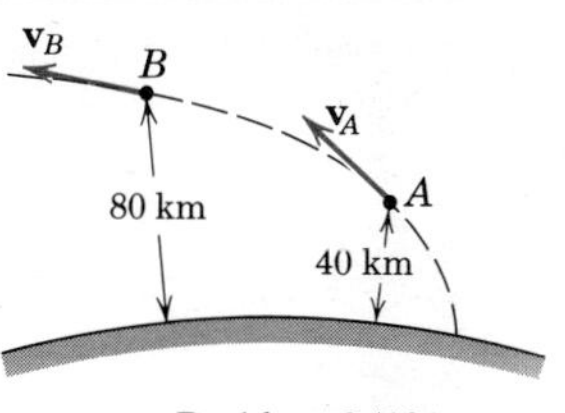

Problem 3/101

3/102 A spacecraft m is heading toward the center of the moon with a velocity of 3000 km/h at a distance from the moon's surface equal to the radius R of the moon. Compute the impact velocity v with the surface of the moon if the spacecraft is unable to fire its retro-rockets. Consider the moon fixed in space. The radius R of the moon is 1738 km, and the acceleration due to gravity at its surface is 1.62 m/s².

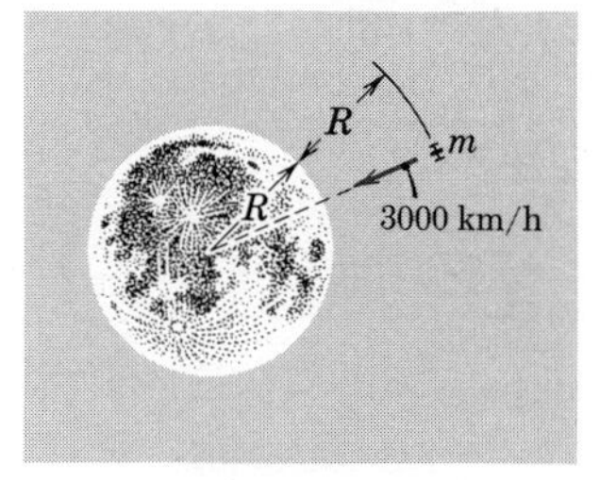

Problem 3/102

3/103 Derive an expression for the net amount of energy E which would be required to move an earth satellite of mass m from a rest condition on the earth's surface at the equator to a circular orbit of altitude h. Assume that the transfer is made without atmospheric friction loss. The radius of the earth is R and its angular velocity is ω. (See Prob. 3/58 for the satellite velocity.)

$$\textit{Ans.}\ E = \frac{1}{2}mR\left(g\frac{R + 2h}{R + h} - R\omega^2\right)$$

3/104 The 2.5-kg sliding collar C with attached spring moves with friction from A to B along the fixed rod. If C has a velocity of 1.8 m/s at A and a velocity of 2.4 m/s as it reaches B, determine the frictional energy loss. The spring has a stiffness of 30 N/m and an unstretched length of 0.9 m. Also calculate the average friction force F_{av} between the collar and the rod over the distance of travel.

3/105 The small bodies A and B each of mass m are connected and supported by the pivoted links of negligible mass. If A is released from rest in the position shown, calculate its velocity v_A as it crosses the vertical center line. Neglect any friction. *Ans.* $v_A = 2.30$ m/s

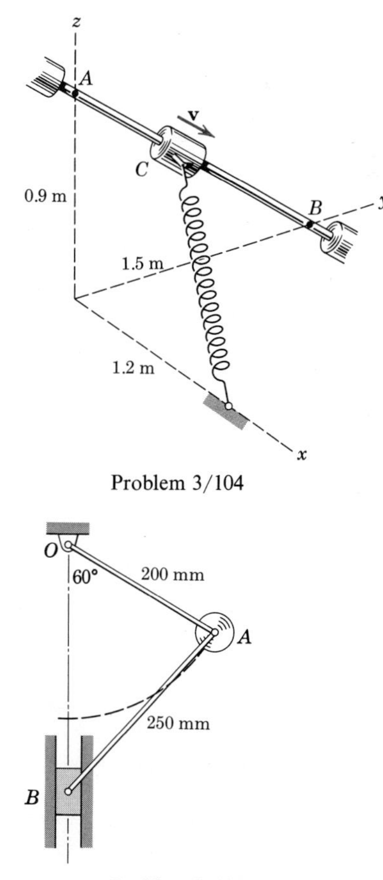

Problem 3/104

Problem 3/105

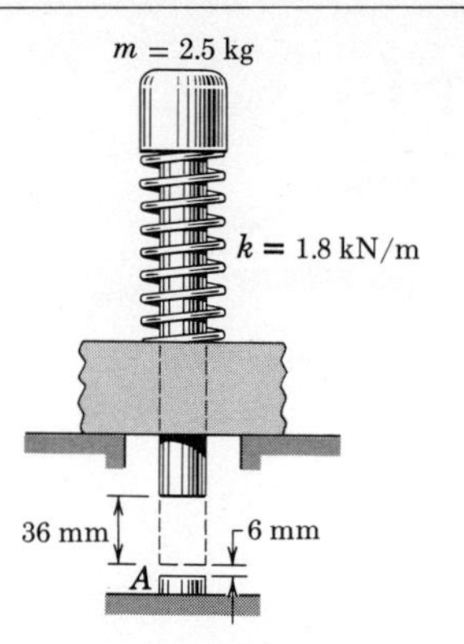

Problem 3/106

3/106 The shank of the 2.5-kg vertical plunger occupies the dotted position when resting in equilibrium against the spring of stiffness $k = 1.8$ kN/m. The upper end of the spring is welded to the plunger, and the lower end is welded to the base plate. If the plunger is lifted 36 mm above its equilibrium position and released from rest, calculate its velocity v as it strikes the button A. Friction is negligible. *Ans.* $v = 0.952$ m/s

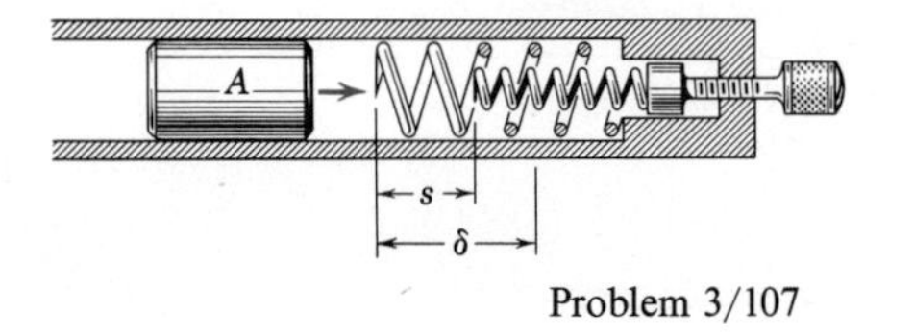

Problem 3/107

3/107 The nest of two springs is used to bring the 0.5-kg plunger A to a stop from a speed of 5 m/s and reverse its direction of motion. The inner spring increases the deceleration, and the adjustment of its position is used to control the exact point at which the reversal takes place. If this point is to correspond to a maximum deflection $\delta = 200$ mm for the outer spring, specify the adjustment of the inner spring by determining the distance s. The outer spring has a stiffness of 300 N/m and the inner one a stiffness of 150 N/m. *Ans.* $s = 142.3$ mm

3/108 A body rotates with an angular acceleration α about an axis fixed in space and is starting from rest with $\omega = 0$. At any point within the body a tangential force $F_t = r\alpha$ may be associated with a unit-mass particle in order to hold it in place relative to the body. Show that this force field is nonconservative and, hence, that a force potential does not exist.

3/109 A body rotates with a constant angular velocity ω about an axis fixed in space. At any point within the body a radial force $F_r = -r\omega^2$ may be associated with a unit-mass particle in order to hold it in place. Show that this force field is conservative and derive the potential function V from which the force components may be obtained by differentiation. Take $V = 0$ when $r = 0$.

◀ **3/110** Determine mathematically the maximum power developed by the spring in Prob. 3/94. Plot the power as a function of the recovery deformation of the spring. *Ans.* $P = 91.4$ W

3/111 Prove that the mass of a solid sphere whose density is a function only of the radial distance r from the center may be considered as though it were concentrated at the center insofar as gravitational attraction is concerned. (*Hint:* First consider the attraction of a spherical shell of the sphere on a unit mass at A. Start with the force dF' resulting from the attraction of an element of a differential ring of the shell. Note that only the component $dF = dF' \cos \alpha$ will remain.)

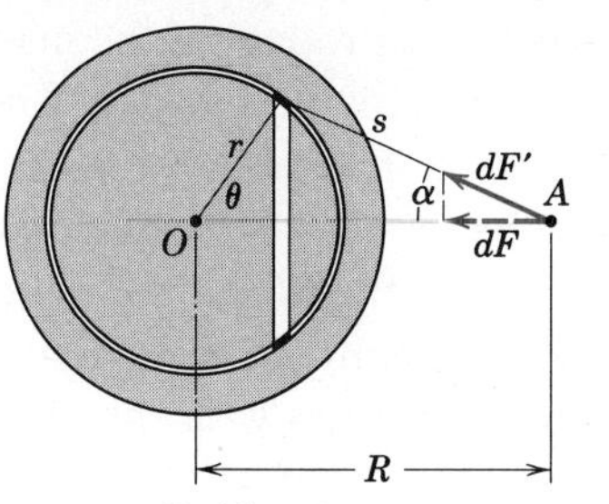

Problem 3/111

3/112 Show mathematically that the gravitational potential for a unit mass at *any* point within a hollow spherical shell of mass m_0 and radius a is $V = -Km_0/a$, where K is the universal gravitational constant. Take $V = 0$ at infinity.

3/113 Consider a diametral hole to be drilled along the polar axis of the earth and a particle to be dropped into the hole. From the results of Prob. 3/112 show that the force of attraction on the falling particle is due only to that part of the earth represented by the internal sphere whose radius r is the same as the distance from the center of the earth to the particle. If the variation of the earth's density with r were known, indicate how the velocity of the particle at any position could be determined, assuming no friction.

---

19 Impulse and Momentum. In the previous article attention was focused on the equations of work and energy which follow from integrating the equation of motion $\mathbf{F} = m\mathbf{a}$ with respect to the displacement of the particle. In the present article attention is devoted to the integral of the equation of motion with respect to time, which leads to the equations of impulse and momentum. It will be found that these equations greatly facilitate the solution of many problems where the applied forces act during specified intervals of time and for conditions of impact between particles where the time during which the forces act may be extremely small.

Part A: Equations of Impulse and Momentum

Consider again the general curvilinear motion in space of a particle of mass m, Fig. 31, where its position vector $\mathbf{r}$ is measured from a fixed origin O. The velocity of the particle is $\mathbf{v} = \dot{\mathbf{r}}$ and is tangent to its path (shown as a dashed line). The resultant $\Sigma\mathbf{F}$ of all forces on m is in the direction of its acceleration $\dot{\mathbf{v}}$. The basic equation of motion for the particle, Eq. 44, may be written

$$\Sigma\mathbf{F} = m\dot{\mathbf{v}} = \frac{d}{dt}(m\mathbf{v}) \quad \text{or} \quad \Sigma\mathbf{F} = \dot{\mathbf{G}} \tag{61}$$

where the product of the mass and velocity is known as the *linear momentum* $\mathbf{G} = m\mathbf{v}$ of the particle. The units of linear momentum $\mathbf{G}$ are seen to be kg·m/s which also equals N·s. Equation 61 states that the *resultant of all forces acting on a particle equals its time rate of change of linear momentum.*

Since Eq. 61 is a vector equation, it should be recognized that, in addition to the equality in the magnitudes of $\Sigma\mathbf{F}$ and $\dot{\mathbf{G}}$, the direction of the resultant force coincides with the direction of the *change* in linear momentum, which is the direction of the change in velocity. Equation 61 is one of the most useful and important relationships in dynamics, and it holds as long as the mass m of the particle is not changing with time. The case where m changes with time is discussed later in Chapter 10. The three scalar components of Eq. 61 may be written

$$\Sigma F_x = \dot{G}_x \qquad \Sigma F_y = \dot{G}_y \qquad \Sigma F_z = \dot{G}_z \tag{62}$$

and may be applied independently of one another.

The effect of the resultant force $\Sigma\mathbf{F}$ on the linear momentum of the particle over a finite period of time may be obtained by integrating Eq. 61 from time t_1 to time t_2. Multiplying the equation by dt gives $\Sigma\mathbf{F}\,dt = d\mathbf{G}$ which is integrated to obtain

$$\int_{t_1}^{t_2} \Sigma\mathbf{F}\,dt = \mathbf{G}_2 - \mathbf{G}_1 \tag{63}$$

where the velocity in $\mathbf{G}$ changes from $\mathbf{v}_1$ at time t_1 to $\mathbf{v}_2$ at time t_2. The product of force and time is defined as *linear impulse,* and Eq. 63 states that the *total linear impulse on m equals the corresponding change in linear momentum.*

The impulse integral is, in the general case, a vector which may change both in magnitude and direction during the time interval. Under these conditions it will be necessary to express $\Sigma\mathbf{F}$ and $\mathbf{G}$ in component form and then combine the integrated components. Thus the x-component of Eq. 63 becomes the scalar equation

$$\int_{t_1}^{t_2} \Sigma F_x\,dt = (mv_x)_2 - (mv_x)_1 \tag{64}$$

and similarly for the y- and z-components.

In evaluating the impulse it is necessary to include the effect of *all* forces

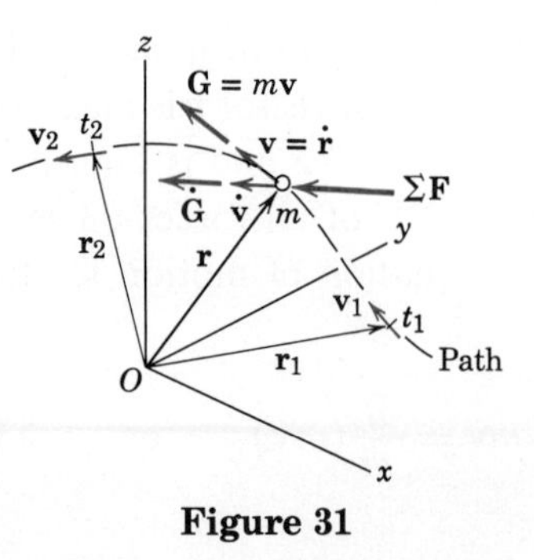

Figure 31

acting on m except those whose magnitudes are negligible. The student should be fully aware at this point that the only reliable method of accounting for the effects of all forces is to isolate the particle in question by drawing its *free-body diagram.*

In addition to the equations of linear impulse and linear momentum, there exists a parallel set of equations of angular impulse and angular momentum. The moment $\Sigma \mathbf{M}_O$ about the fixed point O of all forces on m in Fig. 31 is

$$\Sigma \mathbf{M}_O = \mathbf{r} \times \Sigma \mathbf{F} = \mathbf{r} \times m\dot{\mathbf{v}}$$

But $\frac{d}{dt}(\mathbf{r} \times m\mathbf{v}) = \mathbf{r} \times m\dot{\mathbf{v}}$ since $\dot{\mathbf{r}} \times m\dot{\mathbf{r}} = \mathbf{0}$. Thus the moment equation becomes

$$\Sigma \mathbf{M}_O = \frac{d}{dt}(\mathbf{r} \times m\mathbf{v}) \qquad \text{or} \qquad \Sigma \mathbf{M}_O = \dot{\mathbf{H}}_O \tag{65}$$

where the moment of the linear momentum of m is defined as its *angular momentum* $\mathbf{H}_O = \mathbf{r} \times m\mathbf{v} = \mathbf{r} \times \mathbf{G}$. Equation 65 states that the *moment about the fixed point O of all forces acting on m equals the time rate of change of angular momentum about O.* This relation, particularly when extended to a system of particles, rigid or nonrigid, constitutes one of the most powerful tools of analysis in all of dynamics. The units of angular momentum are clearly $\text{kg}\cdot\text{m}^2/\text{s}$ which also equals $\text{N}\cdot\text{m}\cdot\text{s}$.

Equation 65 is a vector equation with scalar components

$$\Sigma M_{O_x} = \dot{H}_{O_x} \qquad \Sigma M_{O_y} = \dot{H}_{O_y} \qquad \Sigma M_{O_z} = \dot{H}_{O_z} \tag{66}$$

The scalar components of angular momentum may be obtained from the expansion

$$\mathbf{H}_O = \mathbf{r} \times m\mathbf{v} = \mathbf{i}m(v_z y - v_y z) + \mathbf{j}m(v_x z - v_z x) + \mathbf{k}m(v_y x - v_x y)$$

or

$$\mathbf{H}_O = m \begin{vmatrix} \mathbf{i} & \mathbf{j} & \mathbf{k} \\ x & y & z \\ v_x & v_y & v_z \end{vmatrix} \tag{67}$$

so that

$$H_x = m(v_z y - v_y z), \qquad H_y = m(v_x z - v_z x), \qquad H_z = m(v_y x - v_x y)$$

Each of these expressions for angular momentum may be checked easily from Fig. 32, which shows the three linear momentum components, by taking the moments of these components about the respective axes.

To obtain the effect of the moment $\Sigma \mathbf{M}_O$ on the angular momentum of the particle over a finite period of time, Eq. 65 may be integrated from time t_1 to time t_2. Multiplying the equation by dt gives $\Sigma \mathbf{M}_O \, dt = d\mathbf{H}_O$ which may be integrated to obtain

$$\int_{t_1}^{t_2} \Sigma \mathbf{M}_O \, dt = \mathbf{H}_{O_2} - \mathbf{H}_{O_1} \tag{68}$$

where $\mathbf{H}_{O_2} = \mathbf{r}_2 \times m\mathbf{v}_2$ and $\mathbf{H}_{O_1} = \mathbf{r}_1 \times m\mathbf{v}_1$. The product of moment and time is defined as *angular impulse,* and Eq. 68 states that the *total angular impulse on m equals the corresponding change in angular momentum.*

As in the case of linear impulse and linear momentum, the equation of angular impulse and angular momentum is a vector equation where changes in direction as well as magnitude may occur during the interval of integration. Under these conditions it is necessary to express $\Sigma\mathbf{M}_O$ and $\mathbf{H}_O$ in component form and then combine the integrated components. Thus the x-component of Eq. 68 becomes

$$\int_{t_1}^{t_2} \Sigma M_{O_x}\, dt = (H_{O_x})_2 - (H_{O_x})_1$$
$$= m[(v_z y - v_y z)_2 - (v_z y - v_y z)_1]$$

where the subscripts 1 and 2 refer to the values of the respective quantities at times t_1 and t_2. Similar expressions exist for the y- and z-components of the angular momentum integral.

Equations 61 and 65 add no new basic information since they are merely alternative forms of Newton's second law. However, it will be discovered in subsequent chapters that the motion equations expressed in terms of time-rate-of-change-of-momentum are applicable to the motion of rigid and nonrigid bodies and constitute a very general and powerful approach to many problems. The full generality of Eq. 65 is usually not required to describe the motion of a single particle which frequently can be handled as a plane-motion problem where moments are taken about a single axis normal to the plane of motion.

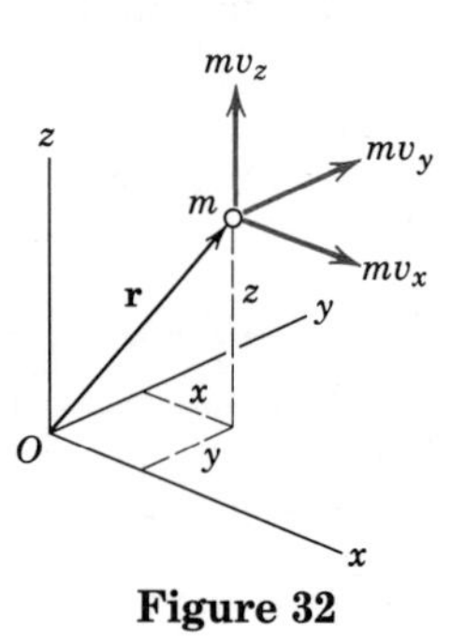

Figure 32

Sample Problem

3/114 A particle with a mass of 0.5 kg has a velocity $u = 10$ m/s in the x-direction at time $t = 0$. Forces $\mathbf{F}_1$ and $\mathbf{F}_2$ act on the particle, and their magnitudes change with time according to the graphical schedule shown. Determine the final velocity $\mathbf{v}$ of the particle at the end of the 3 seconds.

Solution. The impulse-momentum equation is applied in component form and gives for the x- and y-directions, respectively,

$\left[\int \Sigma F_x\, dt = m\, \Delta v_x\right]$ $\qquad -(4 + 4) = 0.5(v_x - 10), \qquad v_x = -6$ m/s

$\left[\int \Sigma F_y\, dt = m\, \Delta v_y\right]$ $\qquad (2 + 2) = 0.5(v_y - 0), \qquad v_y = 8$ m/s

Thus

$$\mathbf{v} = -6\mathbf{i} + 8\mathbf{j} \text{ m/s} \quad \text{and} \quad v = \sqrt{6^2 + 8^2} = 10 \text{ m/s}$$

$$\theta_x = \tan^{-1}\frac{8}{-6} = 126.9° \qquad \textit{Ans.}$$

It is important to note that the algebraic signs must be carefully respected in applying the momentum equations.

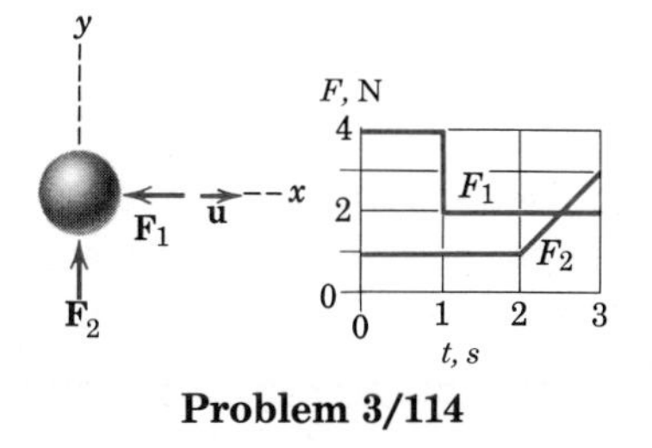

Problem 3/114

Part B: Conservation of Momentum

If the resultant force on a particle is zero during an interval of time, Eq. 61 requires that its linear momentum $\mathbf{G}$ remain constant. Similarly if the resultant moment about a fixed point O of all forces acting on a particle is zero during an interval of time, Eq. 65 requires that its angular momentum $\mathbf{H}_O$ about that point remain constant. In the first case the linear momentum of the particle is said to be conserved, and in the second case the angular momentum of the particle is said to be conserved. Linear momentum may be conserved in one coordinate direction, such as x, but not necessarily in the y- or z-directions. Also angular momentum may be conserved about one axis but not about another axis.

Consider now the motion of two particles of mass m_a and m_b moving with velocities $\mathbf{v}_a$ and $\mathbf{v}_b$. If the particles interact during an interval of time t and if the forces of interaction $\mathbf{F}$ and $-\mathbf{F}$ are the only forces acting on the particles during the interval, then Eq. 63 may be written for each particle as

$$\int_0^t \mathbf{F}\, dt = m_a\, \Delta\mathbf{v}_a \quad \text{and} \quad \int_0^t -\mathbf{F}\, dt = m_b\, \Delta\mathbf{v}_b$$

where $\Delta\mathbf{v}_a$ and $\Delta\mathbf{v}_b$ are the vector changes in the respective velocities of m_a and m_b during the period of interaction. The impulse integrals differ only in sign, so that $m_a\, \Delta\mathbf{v}_a = -m_b\, \Delta\mathbf{v}_b$. Thus $\Delta(m_a\mathbf{v}_a) + \Delta(m_b\mathbf{v}_b) = \mathbf{0}$ or

$$\blacktriangleright \qquad \Delta\mathbf{G} = \mathbf{0} \qquad \textbf{(69)}$$

where the total linear momentum of the system of two particles is $\mathbf{G} = \mathbf{G}_a + \mathbf{G}_b = m_a\mathbf{v}_a + m_b\mathbf{v}_b$. Equation 69 states that the linear momentum of the system remains unchanged during an interval if no forces external to the system act upon it. This statement constitutes the *principle of conservation of linear momentum.* The principle can apply to one direction only, such as the x-direction, if no external forces act on the system in that direction. The principle may be broadened to cover the motion of a system containing any number of interacting particles where all forces acting on the particles are actions and reactions internal to the system.

A result analogous to Eq. 69 also applies for the angular momentum of the interacting particles. The moments about the fixed point O of the forces of interaction $\mathbf{F}$ and $-\mathbf{F}$ necessarily cancel, and in the absence of other forces external to the system, Eq. 68 written for each of the particles would require $\Delta\mathbf{H}_{O_a} = -\Delta\mathbf{H}_{O_b}$ or

$$\Delta\mathbf{H}_O = \mathbf{0} \tag{70}$$

where the total angular momentum of the system of the two particles is $\mathbf{H}_O = \mathbf{H}_{O_a} + \mathbf{H}_{O_b}$. Equation 70 states that the angular momentum of the system about a fixed point O remains unchanged during an interval if no moments about O due to external forces are present. This statement constitutes the *principle of conservation of angular momentum.* The principle may be broadened to cover the motion of a system containing any number of interacting particles where all forces which exert moments about O are actions and reactions internal to the system.

Sample Problems

3/115 The 50-g bullet traveling at 600 m/s strikes the 4-kg block centrally and is embedded within it. If the block is sliding on a smooth horizontal plane with a velocity of 12 m/s in the direction shown just before the impact, determine the velocity $\mathbf{v}$ of the block and bullet combined and its direction θ immediately after impact.

Solution. Since the force of impact is internal to the system composed of the block and bullet and since there are no other external forces acting on the system in the plane

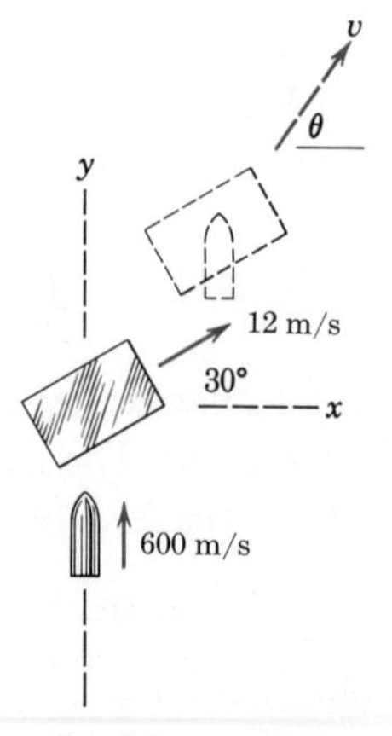

Problem 3/115

of motion, it follows that the linear momentum of the system is conserved in both the x- and y-directions. Thus

$$[\Delta G_x = 0] \qquad 4(12 \cos 30°) + 0 = (4 + 0.050)v_x$$

$$v_x = 10.26 \text{ m/s}$$

$$[\Delta G_y = 0] \qquad 4(12 \sin 30°) + 0.05(600) = (4 + 0.050)v_y$$

$$v_y = 13.33 \text{ m/s}$$

The final velocity is given by

$$[v = \sqrt{v_x^2 + v_y^2}] \qquad v = \sqrt{(10.26)^2 + (13.33)^2} = 16.83 \text{ m/s} \qquad \textit{Ans.}$$

The direction of the final velocity is given by

$$\left[\tan\theta = \frac{v_y}{v_x}\right] \qquad \tan\theta = \frac{13.33}{10.26} = 1.30, \qquad \theta = 52.4° \qquad \textit{Ans.}$$

3/116 A small mass particle is given an initial velocity $\mathbf{v}_0$ tangent to the horizontal rim of a smooth hemispherical bowl at a radius r_0 from the vertical center line, as shown at point A. As the particle slides past point B, a distance h below A and a distance r from the vertical center line, its velocity $\mathbf{v}$ makes an angle θ with the horizontal tangent to the bowl through B. Determine θ.

Solution. The forces on the particle are its weight and the normal reaction exerted by the smooth surface of the bowl. Neither force exerts a moment about the axis O-O, so that angular momentum is conserved about that axis. Thus

$$[\Delta H_O = 0] \qquad mv_0 r_0 = mvr \cos\theta$$

Also, energy is conserved so that

$$[\Delta T + \Delta V = 0] \qquad \tfrac{1}{2}m(v^2 - v_0^2) - mgh = 0, \qquad v = \sqrt{v_0^2 + 2gh}$$

Eliminating v and substituting $r^2 = r_0^2 - h^2$ give

$$v_0 r_0 = \sqrt{v_0^2 + 2gh}\,\sqrt{r_0^2 - h^2}\cos\theta$$

$$\theta = \cos^{-1}\frac{1}{\sqrt{1 + \dfrac{2gh}{v_0^2}}\sqrt{1 - \dfrac{h^2}{r_0^2}}} \qquad \textit{Ans.}$$

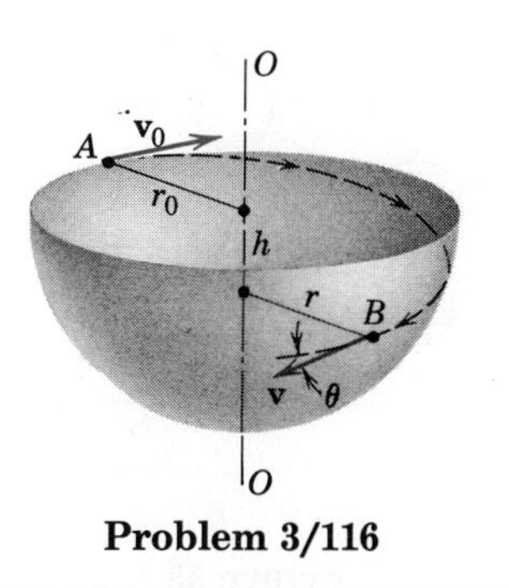

Problem 3/116

Part C: Impact

The principles of impulse and momentum find important use in describing the behavior of impacting bodies. *Impact* refers to the collision between two bodies and is characterized by the generation of relatively large contact forces which act over a very short interval of time. Impact theory has been exceedingly difficult to verify experimentally by virtue of the short time intervals available for measurements. However, with the advent of modern instrumentation reliable data on many impact problems have become available.

Consider the two particles 1 and 2 of masses m_1 and m_2, Fig. 33*a*, moving in the same plane and approaching each other on a collision course with velocities $\mathbf{v}_1$ and $\mathbf{v}_2$ as shown. The directions of the velocities are measured here from the direction tangent to the surfaces of contact during impact. The positions of the particles while they are in contact are shown in Fig. 33*b*. During this very short interval the contact area grows rapidly to a peak value as deformation increases and then reduces to zero during the restoration period. The final or rebound conditions are shown in Fig. 33*c*. If $\mathbf{F}$ is the contact force in the direction *n-n* on particle 1 at any instant during contact, Fig. 33*d*, then the force on particle 2 is $-\mathbf{F}$. From the principle of conservation of linear momentum, Eq. 69, it follows that the total linear momentum of the system of the two particles remains constant during the impact. It is assumed that any forces, other than the internal contact forces, which may act on the particles during impact are relatively small and produce negligible impulses compared with the impulses associated with the impact forces which are generally very large.

To carry out a solution for the final velocities after impact and their directions in this problem, the conservation-of-momentum requirement for the two particles taken together is used and yields for the *n*-direction

$$-m_1v_1 \sin\theta_1 + m_2v_2 \sin\theta_2 = m_1v_1' \sin\theta_1' - m_2v_2' \sin\theta_2'$$

For given masses and initial values of v_1, v_2, θ_1, and θ_2, this equation has four unknowns, v_1', v_2', θ_1', and θ_2'. Two equations may be written for the

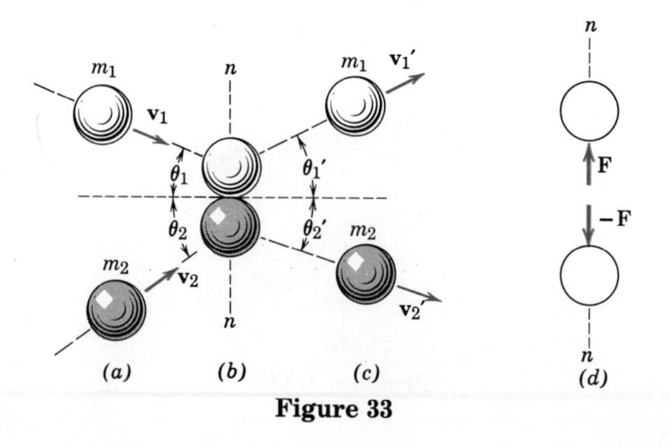

Figure 33

t-direction since, without impulse in the t-direction, each particle suffers no velocity change in this direction. Therefore

$$v_1 \cos \theta_1 = v_1' \cos \theta_1' \qquad \text{and} \qquad v_2 \cos \theta_2 = v_2' \cos \theta_2'$$

The fourth relation depends on the shape and material properties of the contacting bodies. An overall index of these combined effects has been defined as the *coefficient of restitution e* which is the ratio of the impulse during the period of restoration to the impulse during the period of deformation. If F_r and F_d stand for the contact forces during these two periods, respectively, and if v_0 represents the common velocity component of the two particles in the n-direction at the instant of transition from deformation to restoration, then the impulse-momentum equations for body 1 give

$$e = \frac{\int_{t_1}^{t} F_r \, dt}{\int_0^{t_1} F_d \, dt} = \frac{m_1(v_1' \sin \theta_1' - v_0)}{m_1(v_0 + v_1 \sin \theta_1)}$$

where t_1 is the time required for the deformation, t is the total time of contact, and the velocity changes are considered positive in the direction of the contact force on the particle.

A similar equation for particle 2 yields

$$e = \frac{m_2(v_2' \sin \theta_2' + v_0)}{m_2(-v_0 + v_2 \sin \theta_2)}$$

Elimination of v_0 between the two equations and solution for e give

$$e = \frac{v_1' \sin \theta_1' + v_2' \sin \theta_2'}{v_1 \sin \theta_1 + v_2 \sin \theta_2}$$

or, more simply, the coefficient of restitution may be expressed as

$$\blacktriangleright \qquad e = \frac{\text{relative velocity of separation}}{\text{relative velocity of approach}} \qquad \textbf{(71)}$$

where the velocity components are measured in the direction of the impact forces.

According to this classical theory of impact the value $e = 1$ means that the capacity of the two particles to recover equals their tendency to deform. This condition is one of *elastic impact* with no energy loss. The value $e = 0$, on the other hand, describes *inelastic* or *plastic impact* where the particles cling together after collision and the loss of energy is a maximum. All impact conditions lie somewhere in between these two extremes. Also it should be noted that a coefficient of restitution must be associated with a *pair* of contacting bodies.

The coefficient of restitution is often considered a constant for given geometries and a given combination of contacting materials. Actually it depends upon the impact velocity and approaches unity as the impact velocity approaches zero. A handbook value for e is generally unreliable.

Impact is a complex phenomenon involving a loss of initial energy through the generation of heat at the contact point, the generation of internal elastic waves within the bodies, and the generation of sound energy.

The conditions described in Fig. 33 are those for *oblique impact.* When the impact and rebound velocities are collinear, the impact is said to be *direct central,* and the angles shown in the figure are all 90°. The student should write the necessary equations for the simpler case.

Sample Problem

3/117 Spherical particle 1 has a velocity $v_1 = 6$ m/s in the direction shown in the *a*-part of the figure and collides with spherical particle 2 of equal mass and diameter and initially at rest. If the coefficient of restitution for these conditions is 0.6, determine the resulting motion of each particle following impact.

Solution. The geometry of the spheres indicates that the normal n to the contacting surfaces makes an angle of 30° with the direction of $\mathbf{v}_1$ as indicated in the *b*-part of the figure.

Conservation of momentum of the system of two particles in the n-direction gives

$[\Delta G_n = 0]$ $$(m_2 v_2' + m_1 v_1' \sin \theta_1') - m_1(6 \cos 30°) + 0 = 0$$

or

$$v_2' + v_1' \sin \theta_1' - 3\sqrt{3} = 0 \qquad (a)$$

Conservation of momentum for each particle in the t-direction occurs since the contact force on it has no t-component. Thus for particle 1

$[\Delta G_t = 0]$ $$m_1(6 \sin 30°) - m_1(v_1' \cos \theta_1') = 0 \qquad (b)$$

The given coefficient of restitution requires

$$0.6 = \frac{v_2' - v_1' \sin \theta_1'}{6 \cos 30° - 0} \qquad (c)$$

Solution of Eqs. *a*, *b*, and *c* for the unknown quantities gives

$$v_1' = 3.18 \text{ m/s} \qquad v_2' = 4.16 \text{ m/s} \qquad \theta_1' = 19.11° \qquad \textit{Ans.}$$

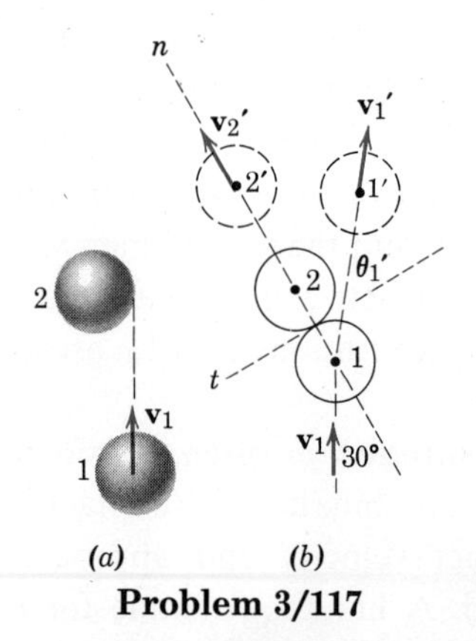

Problem 3/117

Problems

3/118 A jet-propelled airplane with a mass of 10 t is flying horizontally at a constant speed of 1000 km/h when the pilot ignites two rocket-assist units, each of which develops a forward thrust of 8 kN for 9 s. If the velocity of the airplane in its horizontal flight is 1050 km/h at the end of the 9 seconds, calculate the time-average increase ΔR in air resistance. The mass of the rocket fuel used is negligible compared with that of the airplane. *Ans.* $\Delta R = 568$ N

3/119 A particle of mass m starts from rest and moves in a horizontal straight line under the action of a constant force P. Resistance to motion is proportional to the square of the velocity and is $R = kv^2$. Determine the total impulse I on the particle from the time it starts until it reaches its maximum velocity.

3/120 The 180-g projectile is fired with a velocity of 3 km/s at the center of the 0.96-kg disk, which rests on a smooth support. If the projectile passes through the disk and emerges with a velocity of 1.5 km/s, determine the velocity v' of the disk immediately after the projectile clears the disk. *Ans.* $v' = 281$ m/s

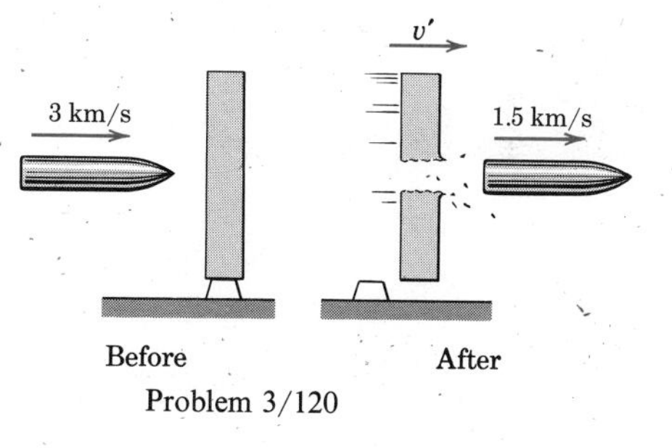

Problem 3/120

3/121 The linear momentum of a certain particle is given by the equation $\mathbf{G} = 6t^3\mathbf{i} + 15t^2\mathbf{j}$, where t is the time in seconds after the particle starts from rest and where the units of $\mathbf{G}$ are kg·m/s. Calculate the magnitude of the resultant force $\mathbf{F}$ that acts on the particle at time $t = 3$ s. *Ans.* $F = 185.3$ N

3/122 A 3-t rocket sled is propelled by six rocket motors each with an impulse rating of 100 kN·s. The rockets are fired at $\frac{1}{4}$-s intervals starting with the sled at rest, and the duration of each rocket firing is 1.5 s. If the velocity of the sled is 150 m/s in 3 s from the start, compute the time average R of the total resistance to motion. Neglect the loss of mass due to exhaust gases compared with the total mass of the sled.

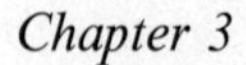

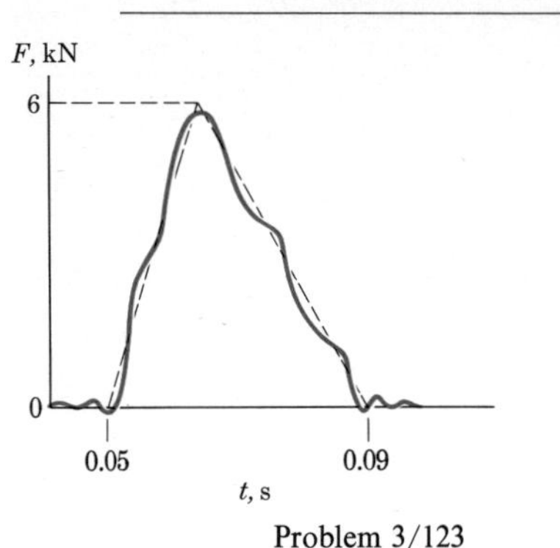

Problem 3/123

3/123 A 5-kg object, which is moving on a smooth horizontal plane with a velocity of 20 m/s to the right, is struck with an impulsive force F that acts to the left on the body. The magnitude of F is represented in the graph. Approximate the loading by the dotted line shown and determine the final velocity v of the object.

Ans. $v = 4.00$ m/s to the left

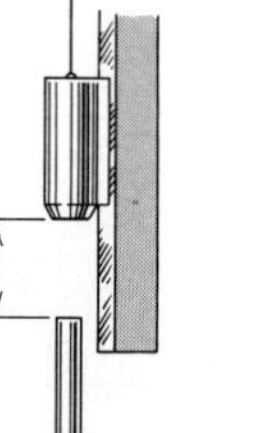

Problem 3/124

3/124 The 400-kg ram of a pile driver falls 1.2 m from rest and strikes the top of a 250-kg pile embedded 0.9 m in the ground. Upon impact the ram is seen to move with the pile with no noticeable rebound. Determine the velocity v of the pile and ram immediately after impact.

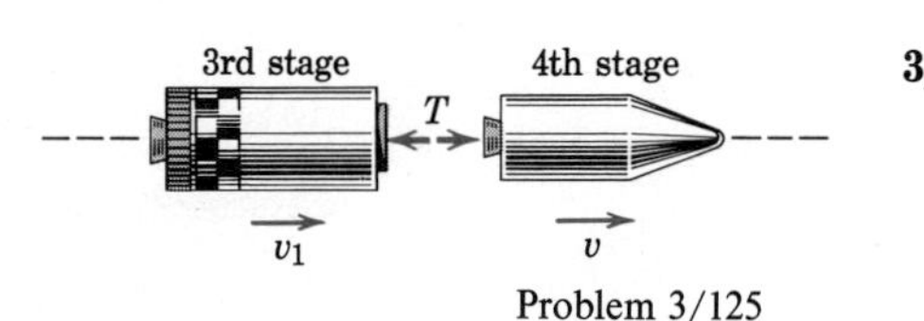

Problem 3/125

3/125 The third and fourth stages of a rocket are coasting in space with a velocity of 15 000 km/h when the fourth-stage engine ignites causing separation of the stages under a thrust T and its reaction. If the velocity v of the fourth stage is 10 m/s greater than the velocity v_1 of the third stage at the end of the $\frac{1}{2}$-s separation interval, calculate the average thrust T during this period. The masses of the third and fourth stages at separation are 30 kg and 50 kg, respectively. Assume that the entire blast of the fourth-stage engine impinges against the third stage with a force equal and opposite to T.

Ans. $T = 375$ N

3/126 A ball is released from rest and drops a distance h onto the horizontal surface of a heavy steel plate. If the ball rebounds to a height h', determine the coefficient of restitution e. What is the fraction n of the original energy which it lost?

3/127 Determine the coefficient of restitution e for a steel ball dropped from rest at a height h above a heavy horizontal steel plate if the height of the second rebound is h_2.

Ans. $e = (h_2/h)^{1/4}$

3/128 The lunar landing craft has a mass of 200 kg and is equipped with a retro-rocket that provides a constant upward thrust T capable of yielding a total impulse of 18 kN·s during a 10-s period. If the lunar craft was 150 m above the surface of the moon and was descending at the rate of 45 m/s, determine the time t required to reduce the descent rate to 1.5 m/s. The absolute acceleration due to gravity near the surface of the moon is 1.62 m/s^2.

3/129 A 0.5-kg particle has a velocity of 10 m/s in the direction shown in the horizontal x-y plane and encounters a steady flow of air in the y-direction at time $t = 0$. If the y-component of the force exerted on the particle by the air is essentially constant and equal to 0.4 N, determine the time t required for the particle to cross the fixed x-axis again. *Ans.* $t = 12.5$ s

3/130 A spacecraft with a mass of 260 kg is moving with a velocity $u = 30\,000$ km/h in the fixed x-direction remote from any attracting celestial body. The spacecraft is spin-stabilized and rotates about the z-axis at the constant rate $\dot{\theta} = \pi/10$ rad/s. During a quarter of a revolution from $\theta = 0$ to $\theta = \pi/2$, a jet is activated which produces a 600-N thrust of constant magnitude. Determine the y-component of the velocity of the spacecraft when $\theta = \pi/2$. Neglect the small change in mass due to the loss of exhaust gas through the control nozzle.

3/131 A 5-kg body is traveling in a horizontal straight line with a velocity of 8 m/s when a horizontal force P is applied to it at right angles to the initial direction of motion. If P varies according to the accompanying graph, remains constant in direction, and is the only force acting on the body in its plane of motion, find the magnitude of the velocity of the body when $t = 2$ s and the angle θ it makes with the direction of P.

Ans. $v = 12.81$ m/s, $\theta = 38.7°$

3/132 Three identical steel cylinders are free to slide horizontally on the fixed horizontal shaft. Cylinders 2 and 3 are at rest and are approached by cylinder 1 at a speed u. Express the final speed v of cylinder 3 in terms of u and the coefficient of restitution e.

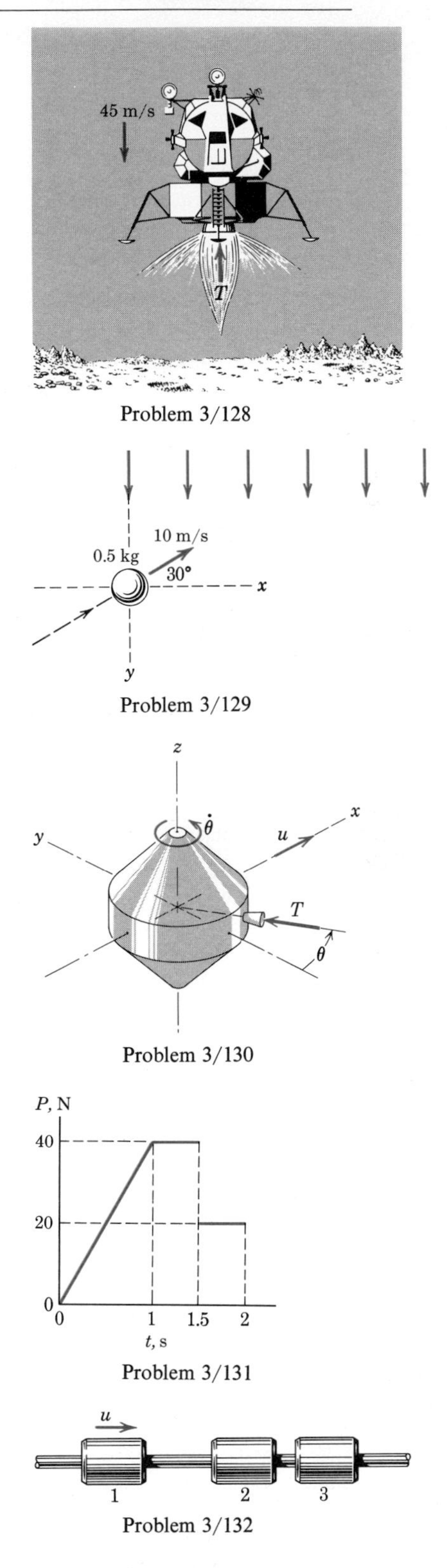

Problem 3/128

Problem 3/129

Problem 3/130

Problem 3/131

Problem 3/132

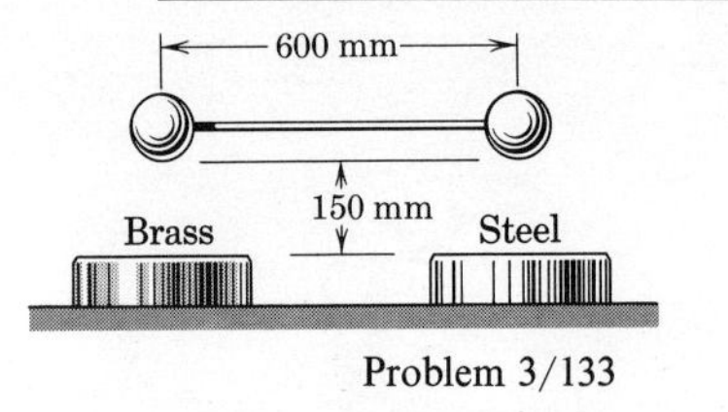

Problem 3/133

3/133 Two steel balls of the same diameter are connected by a rigid bar of negligible mass as shown and are dropped in the horizontal position from a height of 150 mm above the heavy steel and brass base plates. If the coefficient of restitution between the ball and the steel base is 0.6 and that between the other ball and the brass base is 0.4, determine the angular velocity ω of the bar immediately after impact. Assume the two impacts are simultaneous.

Ans. $\omega = 0.57$ rad/s counterclockwise

3/134 In the selection of the ram of a pile driver for a certain job it is desired that the ram lose all of its kinetic energy at each blow. Hence the velocity of the ram is zero immediately after impact. The mass of each pile to be driven is 300 kg, and experience has shown that a coefficient of restitution of 0.3 can be expected. What should be the mass m of the ram? Compute the velocity v of the pile immediately after impact if the ram is dropped from a height of 4 m onto the pile. Also compute the energy loss ΔE due to impact at each blow.

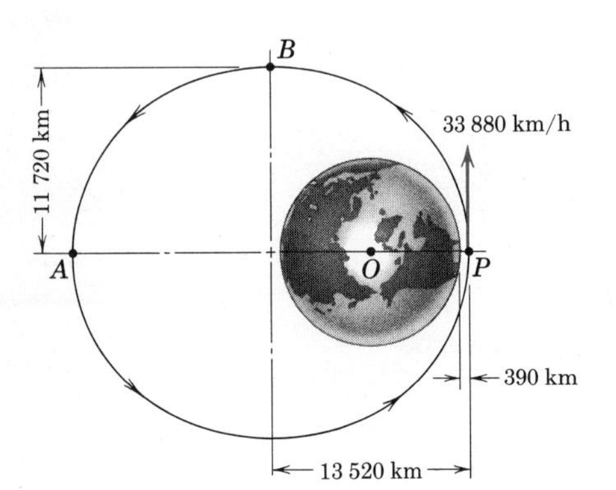

Problem 3/135

3/135 The central attractive force F on an earth satellite can exert no moment about the center O of the earth. For the particular elliptical orbit with major and minor axes as shown, a satellite will have a velocity of 33 880 km/h at the perigee altitude of 390 km. Determine the velocity of the satellite at point B and at apogee A. The radius of the earth is 6370 km.

Ans. $v_B = 19\,540$ km/h
$v_A = 11\,290$ km/h

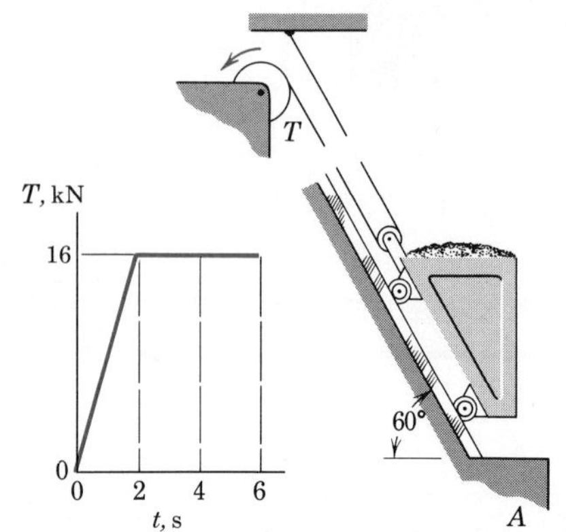

Problem 3/136

3/136 The loaded mine skip has a gross mass of 3 t. The hoisting drum produces a tension T in the cable according to the time schedule shown. If the skip is at rest against A when the drum is activated, determine the speed v of the skip when $t = 6$ s. Friction loss may be neglected.

Ans. $v = 9.13$ m/s

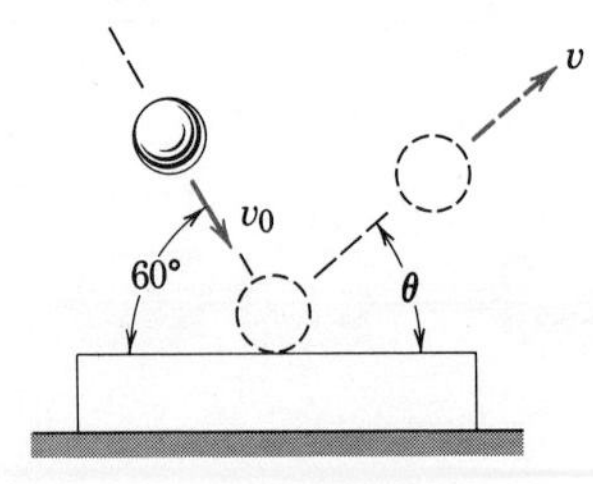

Problem 3/137

3/137 The steel ball strikes the heavy steel plate with a velocity $v_0 = 24$ m/s at an angle of 60° with the horizontal. If the coefficient of restitution is $e = 0.8$, compute the velocity v and its direction θ with which the ball rebounds from the plate.

3/138 The sphere of mass m travels with an initial velocity v_1 and collides centrally with a heavier sphere of mass $2m$ initially at rest. For a coefficient of restitution which makes the loss of kinetic energy of the system a maximum, determine the velocities v_1' and v_2' of m and $2m$ after impact. *Ans.* $v_1' = v_1/3$, $v_2' = v_1/3$

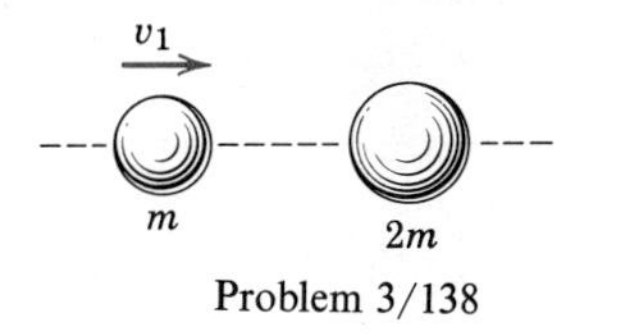

Problem 3/138

3/139 With a coefficient of restitution which would permit the maximum transfer of energy from the ram to the pile of Prob. 3/124, what would be the pile velocity v following impact? Would this value of e be reasonable?

3/140 Each of the 4-kg balls is mounted on the frame of negligible mass and is rotating freely at a speed of 90 rev/min about the vertical with $\theta = 60°$. If the force F on the vertical control rod is increased so that the frame rotates with $\theta = 30°$, determine the new rotational speed N and the work U done by F. Point O on the rotating collar remains fixed.
Ans. $N = 270$ rev/min, $U = 56.6$ J

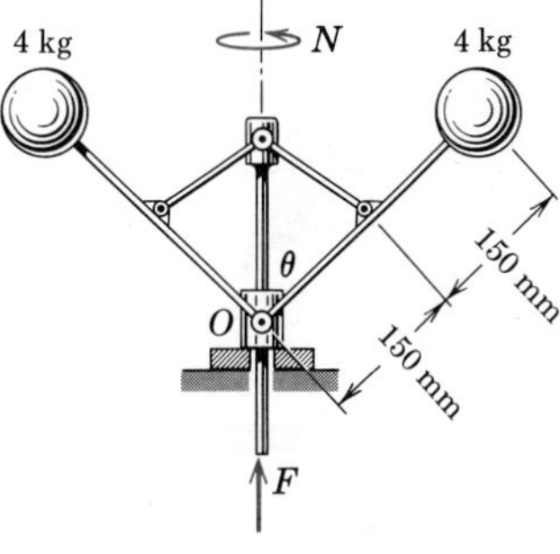

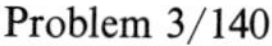
Problem 3/140

3/141 The small particle of mass m and its restraining cord are spinning with an angular velocity ω on the horizontal surface of a smooth disk, shown in section. As the force F is slightly relaxed, r increases and ω changes. Determine the rate of change of ω with respect to r and show that the work done by F during a movement dr equals the change in kinetic energy of the particle.

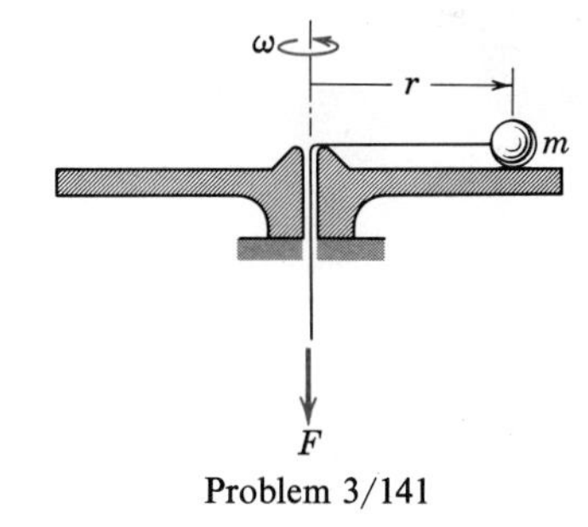

Problem 3/141

3/142 The small particle of mass m is given an initial high velocity in the horizontal plane and winds its cord around the fixed vertical shaft of radius a. All motion occurs essentially in the horizontal plane. If the angular velocity of the cord is ω_0 when the distance from the particle to the tangency point is r_0, determine the angular velocity ω of the cord and its tension T after it has turned through an angle θ. Does either of the principles of momentum conservation apply? Why?

$$Ans.\ \omega = \frac{\omega_0}{1 - \frac{a}{r_0}\theta}, \quad T = mr_0\omega_0\omega$$

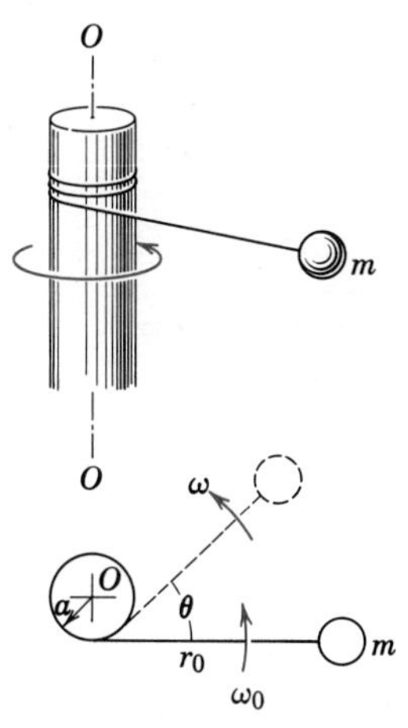

Problem 3/142

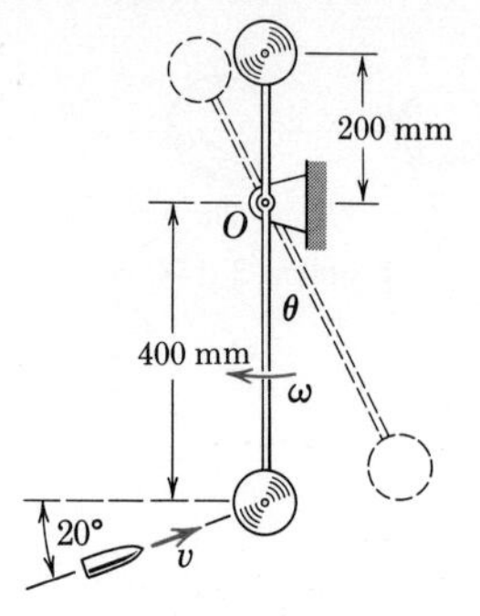

Problem 3/143

3/143 A pendulum consists of two 3.2-kg concentrated masses positioned as shown on a light but rigid bar. The pendulum is swinging through the vertical position with a clockwise angular velocity $\omega = 6$ rad/s when a 50-g bullet traveling with a velocity $v = 300$ m/s in the direction shown strikes the lower mass and is embedded in it. Calculate the angular velocity ω' which the pendulum has immediately after impact and find the maximum angular deflection θ of the pendulum.

Ans. $\omega' = 2.77$ rad/s counterclockwise
$\theta = 52.1°$

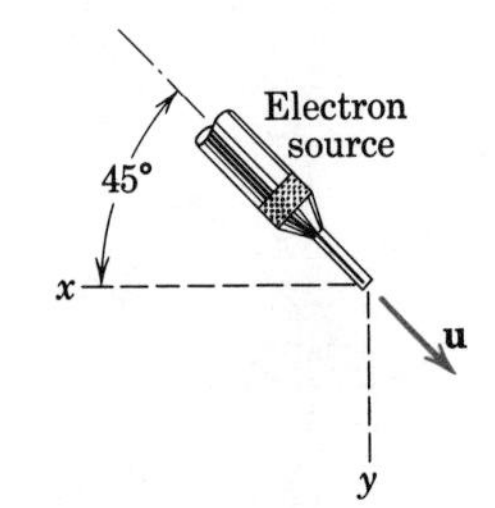

Problem 3/144

3/144 Electrons of mass m are introduced with a velocity $\mathbf{u} = (u/\sqrt{2})(-\mathbf{i} + \mathbf{j})$ into an electric field. The electric field or voltage gradient is the vector $\mathbf{E} = \mathbf{i}E_0 \sin pt + \mathbf{j}E_0 \cos pt$. If an electron is admitted into the field at time $t = 0$, determine its velocity $\mathbf{v}$ after $\frac{1}{4}$ cycle has elapsed $(t = \pi/2p)$. The force on the electron in the direction of the field equals $\mathbf{E}$ multiplied by the electron charge e.

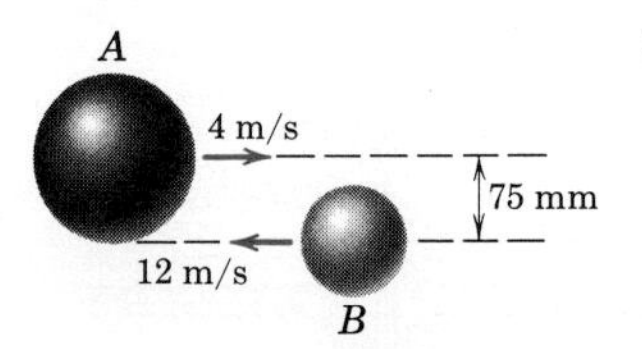

Problem 3/145

3/145 Sphere A has a mass of 23 kg and a radius of 75 mm, while sphere B has a mass of 4 kg and a radius of 50 mm. If the spheres are traveling initially along the parallel paths with the speeds shown, determine the velocities of the spheres immediately after impact. The coefficient of restitution is 0.4 for the conditions involved, and friction is neglected.

Ans. $v_A' = 2.46$ m/s
$v_B' = 9.16$ m/s

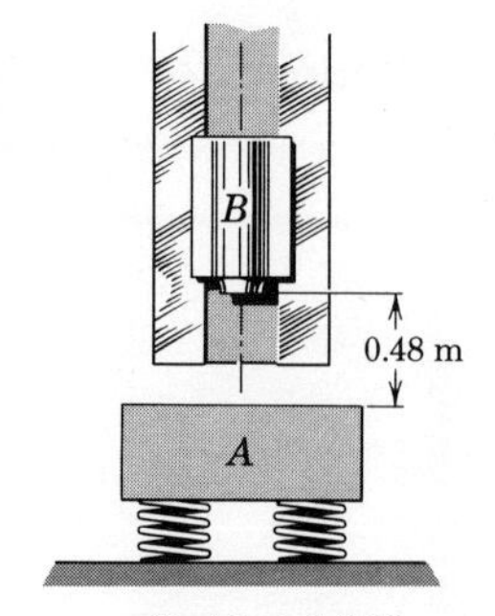

Problem 3/146

3/146 The 3-t anvil A for a drop forge is mounted on a nest of four springs with a combined stiffness of 2.88 MN/m. The 500-kg hammer B falls 0.48 m from rest and strikes the anvil which is observed to have a deflection of 20 mm from its equilibrium position. Determine the height h of rebound of the hammer and the coefficient of restitution e which applies.

Ans. $h = 21.5$ mm, $e = 0.414$

3/147 To pass inspection, steel balls for ball bearings must clear the fixed bar A at the top of their rebound when dropped from rest through the vertical distance $H = 900$ mm onto the heavy inclined steel plate. If balls which have a coefficient of restitution of less than 0.7 with the rebound plate are to be rejected, determine the position of the bar by specifying h and s. Neglect any friction during impact.

Ans. $h = 379$ mm, $s = 339$ mm

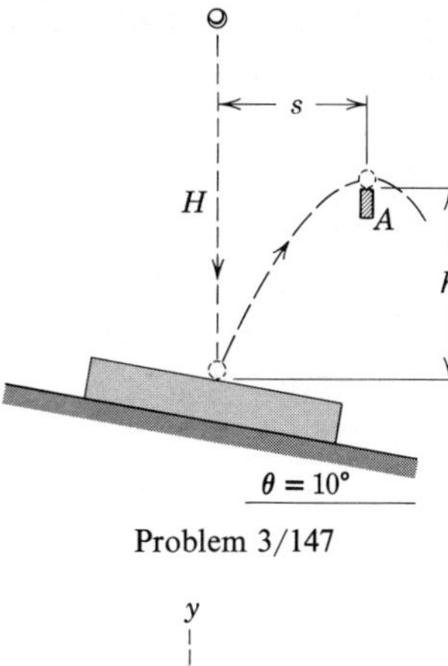

Problem 3/147

3/148 The two identical steel balls moving with initial velocities v_1 and v_2 as shown collide in such a way that the line joining their centers is in the direction of v_2. From previous experiment the coefficient of restitution is known to be 0.60. Determine the velocity of each ball immediately after impact and find the percentage loss of kinetic energy of the system as a result of the impact.

Ans. $v_1' = 3.70$ m/s, $v_2' = 3.36$ m/s
kinetic energy loss is 52%

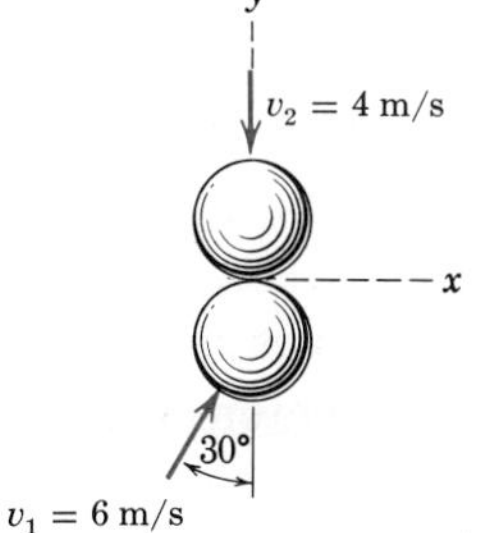

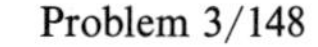

Problem 3/148

3/149 Show that the loss of energy due to direct central impact of two masses m_1 and m_2 having velocities v_1 and v_2 directed toward each other is given by

$$\Delta E = \frac{1 - e^2}{2} \frac{m_1 m_2}{m_1 + m_2} (v_1 + v_2)^2$$

where e is the coefficient of restitution for these particular impact conditions and the internal vibrational energy is neglected. (*Hint:* The energy loss depends on the relative impact velocity $v_1 + v_2$, Thus the center of gravity of the system may be taken at rest to simplify the algebra so that $m_1 v_1 = m_2 v_2$.)

---

20 Central-Force Motion. When a particle moves under the influence of a force directed toward a fixed center of attraction, the motion is called central-force motion. The most common example of central-force motion is found with the orbits of planets and satellites. The laws which govern this motion were deduced from observation of the motions of the planets by J. Kepler (1571–1630). The dynamics of central-force motion is basic to the design of high-altitude rockets, earth satellites, and space vehicles of all types.

Consider a particle of mass m, Fig. 34, moving under the action of the central gravitational attraction

$$F = K \frac{m m_0}{r^2}$$

where m_0 is the mass of the attracting body assumed to be fixed, K is the universal gravitational constant, and r is the distance between the centers of the masses. The particle of mass m could represent the earth moving about the sun, the moon moving about the earth, or a satellite in its orbital motion about the earth above the atmosphere. The most convenient coordinate system to use is polar coordinates in the plane of motion since **F** will always be in the negative r-direction and there is no force in the θ-direction.

Equations 47 may be applied directly for the r- and θ-directions to give

$$-K\frac{mm_0}{r^2} = m(\ddot{r} - r\dot{\theta}^2)$$
$$0 = m(r\ddot{\theta} + 2\dot{r}\dot{\theta}) \tag{72}$$

The second of the two equations when multiplied by r/m is seen to be the same as $d(r^2\dot{\theta})/dt = 0$ which is integrated to give

$$r^2\dot{\theta} = h, \quad \text{a constant} \tag{73}$$

The physical significance of Eq. 73 is made clear when it is noted that the angular momentum $\mathbf{r} \times m\mathbf{v}$ of m about m_0 has the magnitude $mr^2\dot{\theta}$. Thus Eq. 73 merely states that the angular momentum of m about m_0 remains constant, or is conserved. This statement is easily deduced from Eq. 65 where it is observed that the angular momentum $\mathbf{H}_O$ remains constant (is conserved) if there is no moment acting on the particle about a point O.

It is noted that during time dt the radius vector sweeps out an area, shaded in Fig. 34, equal to $dA = (\frac{1}{2}r)(r\,d\theta)$. Therefore the rate at which area is swept by the radius vector is $\dot{A} = \frac{1}{2}r^2\dot{\theta}$ which is constant according to Eq. 73. This conclusion is expressed in Kepler's *second law* of planetary motion which states that the areas swept through in equal times are equal.

The shape of the path followed by m may be obtained by solving the first of Eqs. 72 with the time t eliminated through combination with Eq. 73. To this end the mathematical substitution $r = 1/u$ is useful. Thus $\dot{r} = -(1/u^2)\dot{u}$ which from Eq. 73 becomes $\dot{r} = -h(\dot{u}/\dot{\theta})$ or $\dot{r} = -h(du/d\theta)$. The second time derivative is $\ddot{r} = -h(d^2u/d\theta^2)\dot{\theta}$, which, by combining with Eq. 73, becomes $\ddot{r} = -h^2u^2(d^2u/d\theta^2)$. Substitution into the first of Eqs. 72 now gives

$$-Km_0u^2 = -h^2u^2\frac{d^2u}{d\theta^2} - \frac{1}{u}h^2u^4$$

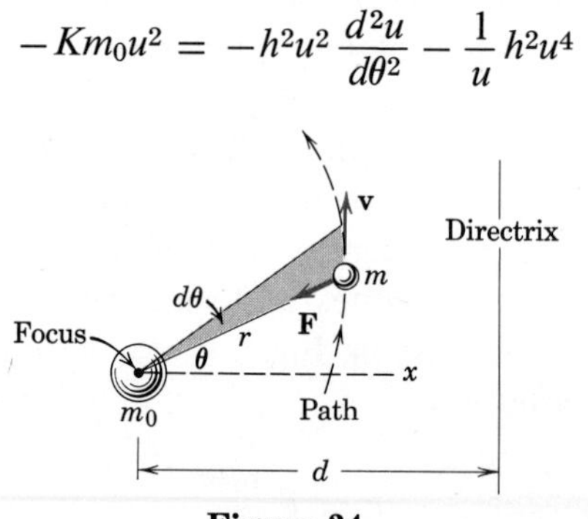

Figure 34

or

$$\frac{d^2u}{d\theta^2} + u = \frac{Km_0}{h^2} \tag{74}$$

which is a nonhomogeneous linear differential equation. The solution of this familiar second order equation is seen by inspection to be

$$u = \frac{1}{r} = C \cos(\theta + \delta) + \frac{Km_0}{h^2}$$

where C and δ are the two integration constants. The phase angle δ may be eliminated by choosing the x-axis so that r is a minimum when $\theta = 0$. Thus

$$\frac{1}{r} = C \cos\theta + \frac{Km_0}{h^2} \tag{75}$$

The interpretation of Eq. 75 requires a knowledge of the equations for conic sections. Recall that a conic section is formed by the locus of a point which moves so that the ratio e of its distance from a point (focus) to a line (directrix) is constant. Thus from Fig. 34, $e = r/(d - r\cos\theta)$, which may be rewritten as

$$\frac{1}{r} = \frac{1}{d}\cos\theta + \frac{1}{ed} \tag{76}$$

which is the same as Eq. 75. Hence it is seen that the motion of m is along a conic section with $d = 1/C$ and $ed = h^2/(Km_0)$ or

$$e = \frac{h^2C}{Km_0} \tag{77}$$

There are three cases to be investigated corresponding to $e < 1$ (ellipse), $e = 1$ (parabola), and $e > 1$ (hyperbola). The trajectory for each of these cases is shown in Fig. 35.

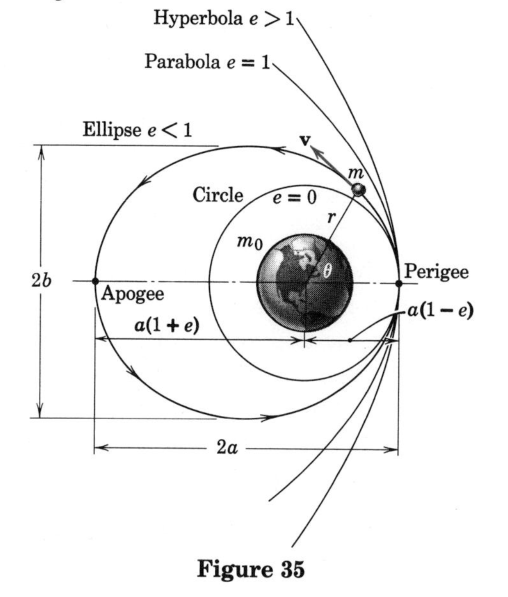

Figure 35

Case I: ellipse ($e < 1$). From Eq. 76 it is seen that r is a minimum when $\theta = 0$ and is a maximum when $\theta = \pi$. Thus,

$$2a = r_{\text{min}} + r_{\text{max}} = \frac{ed}{1+e} + \frac{ed}{1-e} \qquad \text{or} \qquad a = \frac{ed}{1-e^2}$$

With the distance d expressed in terms of a, Eq. 76 and the maximum and minimum values of r may be written as

$$\frac{1}{r} = \frac{1 + e\cos\theta}{a(1-e^2)}$$
$$r_{\text{min}} = a(1-e) \qquad r_{\text{max}} = a(1+e) \tag{78}$$

In addition, the relation $b = a\sqrt{1-e^2}$, which comes from the geometry of the ellipse, gives the expression for the semiminor axis. It is seen that the ellipse becomes a circle with $r = a$ when $e = 0$.

The period τ for the elliptical orbit is the total area A of the ellipse divided by the constant rate $\dot{A}$ at which area is swept through. Thus

$$\tau = A/\dot{A} = \frac{\pi ab}{\frac{1}{2}r^2\dot{\theta}} \qquad \text{or} \qquad \tau = \frac{2\pi ab}{h}$$

by Eq. 73. Substitution of Eq. 77, the identity $d = 1/C$, the geometric relationships $a = ed/(1-e^2)$ and $b = a\sqrt{1-e^2}$ for the ellipse, and the equivalence $Km_0 = gR^2$ yields upon simplification

$$\tau = 2\pi \frac{a^{3/2}}{R\sqrt{g}} \tag{79}$$

In this equation it is noted that R is the mean radius of the central attracting body and g is the absolute value of the acceleration due to gravity at the surface of the attracting body.

Equation 79 expresses Kepler's *third law* of planetary motion which states that the square of the period of motion is proportional to the cube of the semimajor axis of the orbit.

Case II: parabola ($e = 1$). Equations 76 and 77 become

$$\frac{1}{r} = \frac{1}{d}(1 + \cos\theta) \qquad \text{and} \qquad h^2C = Km_0$$

The radius vector and the dimension a become infinite as θ approaches π.

Case III: hyperbola ($e > 1$). From Eq. 76 it is seen that the radial distance r becomes infinite for the two values of the polar angle θ_1 and $-\theta_1$ defined by $\cos\theta_1 = -1/e$. Only branch I corresponding to $-\theta_1 < \theta < \theta_1$, Fig. 36, represents a physically possible motion. Branch II corresponds to angles in the remaining sector (with r negative). For this branch positive r's may be used if θ is replaced by $\theta - \pi$ and $-r$ by r. Thus Eq. 76 becomes

$$\frac{1}{-r} = \frac{1}{d}\cos(\theta - \pi) + \frac{1}{ed} \qquad \text{or} \qquad \frac{1}{r} = -\frac{1}{ed} + \frac{\cos\theta}{d}$$

But this expression contradicts the form of Eq. 75 where Km_0/h^2 is necessarily positive. Hence branch II does not exist (except for repulsive forces).

Now consider the energies of particle m. The system is conservative, and the constant energy E of m is the sum of its kinetic energy T and potential energy V. The kinetic energy is $T = \frac{1}{2}mv^2 = \frac{1}{2}m(\dot{r}^2 + r^2\dot{\theta}^2)$ and the potential energy from Eq. 55 is $V = -mgR^2/r$. Recall that g is the absolute acceleration due to gravity measured at the surface of the attracting body, R is the radius of the attracting body, and $Km_0 = gR^2$. Thus

$$E = \tfrac{1}{2}m(\dot{r}^2 + r^2\dot{\theta}^2) - \frac{mgR^2}{r}$$

This constant value of E can be determined from its value at $\theta = 0$ where $\dot{r} = 0$, $1/r = C + gR^2/h^2$ from Eq. 75, and $r\dot{\theta} = h/r$ from Eq. 73. Substitution into the expression for E and simplification yield

$$\frac{2E}{m} = h^2C^2 - \frac{g^2R^4}{h^2}$$

Now eliminate C by substitution of Eq. 77, which may be written as $h^2C = egR^2$, and obtain

$$e = +\sqrt{1 + \frac{2Eh^2}{mg^2R^4}} \tag{80}$$

The plus value of the radical is mandatory since by definition e is positive. It is now seen that for the

elliptical orbit	$e < 1$,	E is negative
parabolic orbit	$e = 1$,	E is zero
hyperbolic orbit	$e > 1$,	E is positive

These conclusions, of course, depend on the arbitrary selection of the datum condition for zero potential energy ($V = 0$ when $r = \infty$).

The expression for the velocity v of m may be found from the energy equation which is

$$\tfrac{1}{2}mv^2 - \frac{mgR^2}{r} = E$$

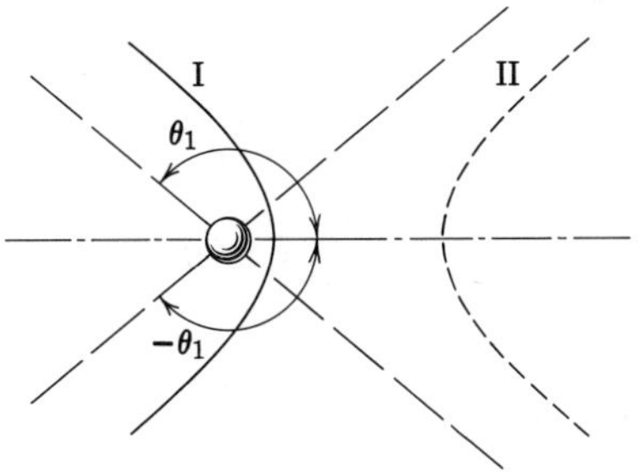

Figure 36

The total energy E is obtained from Eq. 80 by combining Eq. 77 and $1/C = d = a(1 - e^2)/e$ to give for the elliptical orbit

$$E = -\frac{gR^2m}{2a} \tag{81}$$

Substitution into the energy equation yields

$$v^2 = 2gR^2\left(\frac{1}{r} - \frac{1}{2a}\right) \tag{82}$$

from which the magnitude of the velocity may be computed for a particular orbit in terms of the radial distance r. Combination of the expressions for $r_{\min}$ and $r_{\max}$ corresponding to perigee and apogee, Eq. 78, with Eq. 82 gives for the respective velocities at these two positions for the elliptical orbit

$$\begin{aligned} v_p &= R\sqrt{\frac{g}{a}}\sqrt{\frac{1+e}{1-e}} = R\sqrt{\frac{g}{a}}\sqrt{\frac{r_{\max}}{r_{\min}}} \\ v_a &= R\sqrt{\frac{g}{a}}\sqrt{\frac{1-e}{1+e}} = R\sqrt{\frac{g}{a}}\sqrt{\frac{r_{\min}}{r_{\max}}} \end{aligned} \tag{83}$$

Selected numerical data pertaining to the solar system are included in Appendix C and will be found useful in applying the foregoing relationships to problems in planetary motion.

▼ ▼ ▼ ▼ ▼

In the foregoing analysis of planetary motion it is assumed that the central attracting mass is fixed. To account for the motion of m_0 due to the force exerted on it by m, consider the equations of motion for each of the two masses in a fixed system, Fig. 37.

The magnitude of the force of mutual attraction is F, and the vector expressions for this force acting on m_0 and m, respectively, may be written as $\mathbf{F} = (F/r)\mathbf{r}$ and $-\mathbf{F} = -(F/r)\mathbf{r}$ which preserve the magnitude and specify the correct direction of each force. The equations of motion for m_0 and m in that order are $(F/r)\mathbf{r} = m_0\ddot{\mathbf{r}}_2$ and $-(F/r)\mathbf{r} = m\ddot{\mathbf{r}}_1$, from which may be written

$$\ddot{\mathbf{r}} = \ddot{\mathbf{r}}_1 - \ddot{\mathbf{r}}_2 = -\left(\frac{1}{m} + \frac{1}{m_0}\right)\frac{F}{r}\mathbf{r}$$

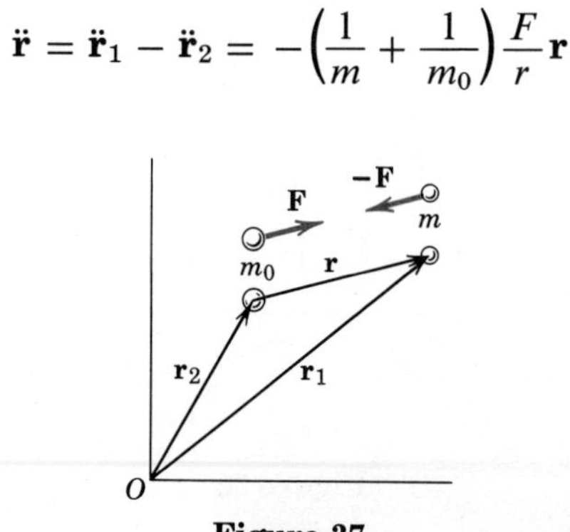

Figure 37

Now substitute $1/\mu = 1/m + 1/m_0$, and the expression may be written

$$\ddot{\mathbf{r}} = -\frac{1}{\mu}\frac{F}{r}\mathbf{r} \qquad \text{or} \qquad m\ddot{\mathbf{r}} = -\left(\frac{Km}{\mu}\right)\frac{mm_0}{r^2}\frac{\mathbf{r}}{r}$$

which is the equation of motion of m relative to m_0 since it contains the relative position vector $\mathbf{r}$. The equation of motion of m assuming m_0 to be fixed is

$$m\ddot{\mathbf{r}} = -\frac{F}{r}\mathbf{r} \qquad \text{or} \qquad m\ddot{\mathbf{r}} = -K\frac{mm_0}{r^2}\frac{\mathbf{r}}{r}$$

Therefore, if the K in this approximate equation is replaced by Km/μ, the correct equation results. Consequently the corrected expression for the period of m about m_0 from Eq. 79 before R^2g was substituted for Km_0 becomes

$$\tau = 2\pi\frac{a^{3/2}}{\sqrt{\dfrac{Km}{\mu}m_0}} = 2\pi\frac{a^{3/2}}{\sqrt{K(m+m_0)}} \tag{84}$$

Problems

(Unless otherwise indicated the velocities mentioned in the problems which follow are measured from a nonrotating reference frame moving with the center of the attracting body.)

3/150 Identify the location on the earth and the direction of flight trajectory which require the least expenditure of fuel for the launching of a satellite into an earth orbit.

3/151 Show that Eq. 82 reduces to the equation for the velocity of a satellite in a circular orbit of altitude H when $e = 0$ and $a = R + H$ (see Prob. 3/58).

3/152 For a certain satellite with a perigee altitude of 389 km the ratio of its maximum to its minimum orbital velocity is 1.5. Compute the apogee altitude H_a. *Ans.* $H_a = 3768$ km

3/153 Calculate the velocity of a spacecraft which orbits the moon in a circular path of 80-km altitude.

3/154 Calculate the time τ required for the spacecraft of Prob. 3/153 to make one complete circular orbit around the moon at the altitude of 80 km.
Ans. $\tau = 1$ h 56 min 2 s

3/155 A rocket is orbiting the earth in a circular path of altitude H. If the jet engine is activated to give the rocket a sudden burst of speed, write the expression for the velocity v which must be attained in order to escape from the influence of the earth.

Ans. $v = R\sqrt{\dfrac{2g}{R+H}}$

Problem 3/156

3/156 An earth satellite moves in an elliptical orbit with an eccentricity $e = 0.2$. If the satellite comes to within 145 km of the earth at its point of closest approach, calculate its velocity at the two points A in the orbit which are 320 km from the earth.

3/157 Compute the minimum velocity of an earth satellite in its orbit if its perigee and apogee altitudes above the earth are equal to the radius and three times the radius of the earth, respectively.

Ans. $v = 11\,630$ km/h

3/158 A space probe is to be put into an orbit around Venus with its closest approach to the surface of the planet being 640 km. If the orbit is to have an eccentricity of 0.25, calculate the maximum velocity of the probe in its orbit.

Ans. $v_{max} = 7.70$ km/s

3/159 For an earth satellite moving in an elliptical orbit, show that the distance r from the center of the earth to the satellite is given by the expression $r = r_{max}r_{min}/a$ for the position corresponding to an angular displacement $\theta = \pi/2$ of the r-vector beyond the perigee position. The distance between the perigee and apogee (major axis) is $2a$.

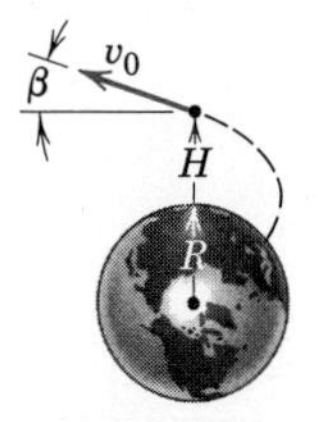

Problem 3/160

3/160 An artificial satellite is injected into orbit at an altitude H with an absolute velocity v_0 at a flight angle β as shown. Determine the expression for the velocity v of the satellite during its orbit in terms of its radial distance r from the center of the earth.

Ans. $v^2 = v_0{}^2 + 2gR^2\left(\dfrac{1}{r} - \dfrac{1}{R+H}\right)$

3/161 Write an expression for the period τ of an earth satellite for the given launching conditions of Prob. 3/160.

Ans. $\tau = \dfrac{2\pi g R^2}{\sqrt{[2gR^2/(R+H) - v_0{}^2]^3}}$

3/162 If the perigee altitude of an earth satellite is 240 km and the apogee altitude is 400 km, compute the eccentricity e of the orbit and the period τ of one complete orbit in space.

Ans. $e = 0.011\,96$, $\tau = 1$ h 30 min 46 s

3/163 If an earth satellite has a velocity v whose direction makes an angle α with the radius r from the center of the earth to the satellite, derive an expression for the radius of curvature ρ of the path at this particular position. The radius of the earth is R.

3/164 A space capsule which executes a polar orbit comes within 150 km of the north pole at its point of closest approach to the earth. If the capsule passes over the pole once every 90 min, calculate its velocity v over the north pole.

Ans. $v = 28\,420$ km/h

3/165 A satellite passes over the north pole at the perigee altitude of 500 km in an elliptical orbit of eccentricity $e = 0.7$. Calculate the absolute velocity v of the satellite as it crosses the equator.

3/166 An earth satellite moves in a polar orbit with an eccentricity of 0.2 and a minimum altitude of 300 km at the north pole. An observer A directly under the satellite as it passes over the equator at B notes a change $\Delta\beta$ in the longitude of the satellite when it returns to position B on its next trip around the earth. Calculate $\Delta\beta$.

Ans. $\Delta\beta = 31.7°$ west

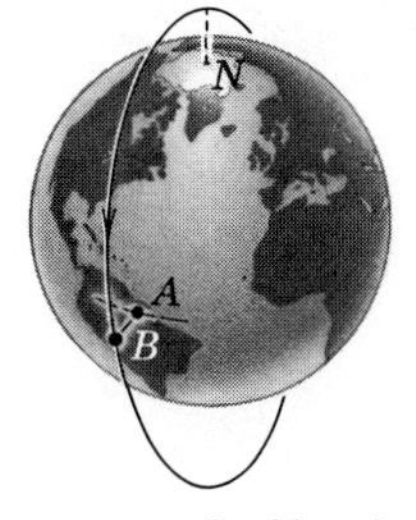

Problem 3/166

3/167 An artificial earth satellite with a mass of 450 kg is injected into a circular orbit at an altitude of 250 km in an easterly direction above the equator. Write an expression for the net energy ΔE from earth sources imparted to the satellite between launching and injection into orbit and compute its value.

$$\textit{Ans. } \Delta E = mgR\left[1 - \frac{R}{2(R+H)} - \frac{R\omega^2}{2g}\right] = 14.56 \text{ GJ}$$

3/168 It is proposed to place an observation satellite in a circular orbit around the earth so that it will remain above the same spot on the earth's surface at all times. Compute the required altitude H and absolute velocity v at which injection into orbit should occur. What limitations on latitude exist for the spot beneath the satellite?

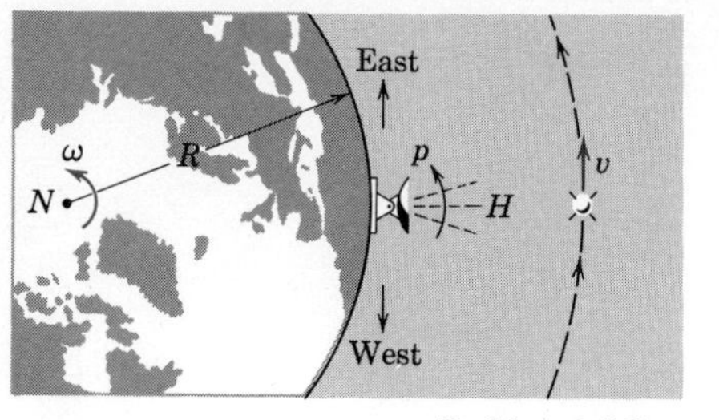

Problem 3/169

3/169 A satellite moving in a west-to-east equatorial orbit is observed by a tracking station located on the equator. If the satellite has a perigee altitude $H = 150$ km with velocity v directly over the station and an apogee altitude of 1500 km, determine an expression for the angular rate p (relative to the earth) at which the antenna dish must be rotated when the satellite is directly overhead. Compute p. The angular velocity of the earth is ω.

$$Ans.\ p = \frac{v - R\omega}{H} - \omega = 0.0513 \text{ rad/s}$$

3/170 Determine the expression for the absolute velocity v with which an object must be projected from the earth's surface in order to escape from the earth's gravity. Assume absence of an atmosphere. What is the effect of latitude and angle of firing? Compute v.

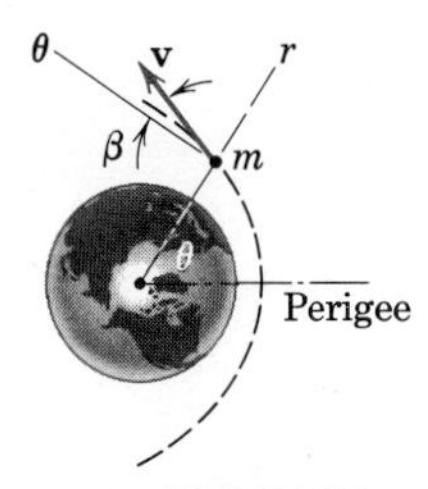

Problem 3/171

3/171 Determine the angle β made by the velocity vector $\mathbf{v}$ with respect to the θ-direction for an earth satellite traveling in an elliptical orbit of eccentricity e. Express β in terms of the angle θ measured from perigee.

$$Ans.\ \tan \beta = \frac{e \sin \theta}{1 + e \cos \theta}$$

3/172 Assume that a tower could be erected to a height of 800 km at the equator. Compute the muzzle velocity u (relative to the gun) of a bullet to be fired horizontally from the tower in order that it orbit the earth in a circular path if the bullet is aimed (a) south and (b) west.

Ans. (a) $u = 26\,770$ km/h
(b) $u = 28\,680$ km/h

3/173 Derive an expression for the radius of curvature ρ of the path of an earth satellite at its nearest point of approach to the earth. The perigee altitude above the earth is H, the radius of the earth is R, and the eccentricity of the orbit is e.

$$Ans.\ \rho = (R + H)(1 + e)$$

3/174 If the earth were suddenly deprived of its orbital velocity around the sun, how long would it take for the earth to "fall" into the sun? (*Hint:* The time would be one half the period of a degenerate elliptical orbit around the sun with the semiminor diameter approaching zero.) Assume a fixed sun.

3/175 An earth satellite of mass m is injected into orbit at an altitude H above the earth with an absolute velocity v_p normal to the radial vector and somewhat greater than is needed for a circular orbit. Determine expressions for the distance r_{max} of the apogee from the center of the earth and the velocity v_a at apogee.

3/176 An earth satellite moves in a circular orbit of altitude H over the two poles. When the satellite is passing over the equator heading for the south pole, an observer measures an apparent angle β west of south for the direction of the flight. Write an expression for β using R for the radius of the earth and ω for the angular velocity of the earth. Compute β for $R = 6370$ km and $H = 300$ km.

Ans. $\beta = \tan^{-1}\dfrac{\omega\sqrt{R + H}}{\sqrt{g}}, \quad \beta = 3.44°$

3/177 A satellite is moving in a circular orbit at an altitude of 320 km above the surface of the earth. If a rocket motor on the satellite is activated to produce a velocity increase of 300 m/s in the direction of its motion during a very short interval of time, calculate the altitude H of the satellite at its new apogee position.

Ans. $H = 1470$ km

3/178 The 300-kg spacecraft A approaches the planet Mars along the trajectory a-a with an absolute velocity in space of 27.0 km/s in the orbital plane of Mars. The planet has a velocity of 24.1 km/s in space along the trajectory b-b. The closest point of approach along the two trajectories occurs at A' when the relative velocity v_r of the spacecraft with respect to Mars is at right angles to the line joining them. At this condition the relative velocity is 5.0 km/s, and the spacecraft is 8000 km from the center of Mars. Also at this point retrorockets are fired in the direction opposite to v_r in order to lower the relative speed of the spacecraft so that it will orbit Mars with a trajectory which will bring it some 5000 km from the center of Mars on the opposite side. Determine the time t during which the 500-N-thrust retroengines should be fired in order to produce the new Mars orbit.

Ans. $t = 29$ min 43 s

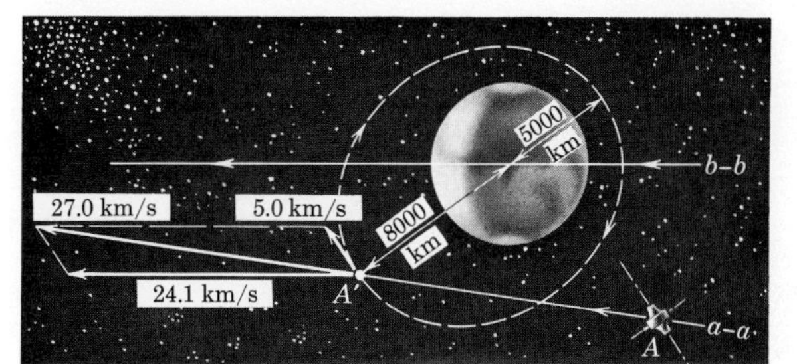

Problem 3/178

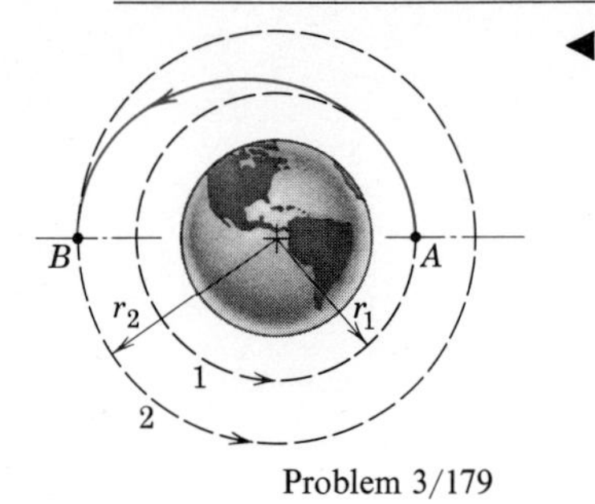

Problem 3/179

◀ **3/179** A space vehicle moving in a circular orbit of radius r_1 transfers to a larger circular orbit of radius r_2 by means of an elliptical path between A and B. (This transfer path is known as the Hohmann transfer ellipse.) The transfer is accomplished by a burst of speed Δv_A at A and a second burst of speed Δv_B at B. Write expressions for Δv_A and Δv_B in terms of the radii shown and the value g of the acceleration due to gravity at the earth's surface. If each Δv is positive, how can the velocity for path 2 be less than that for path 1? Compute each Δv if $r_1 = (6370 + 500)$ km and $r_2 = (6370 + 1000)$ km.

$$\textit{Ans. } \Delta v_A = R\sqrt{\frac{g}{r_1}}\left(\sqrt{\frac{2r_2}{r_1 + r_2}} - 1\right)$$
$$= 132.6 \text{ m/s}$$
$$\Delta v_B = R\sqrt{\frac{g}{r_2}}\left(1 - \sqrt{\frac{2r_1}{r_1 + r_2}}\right)$$
$$= 130.2 \text{ m/s}$$

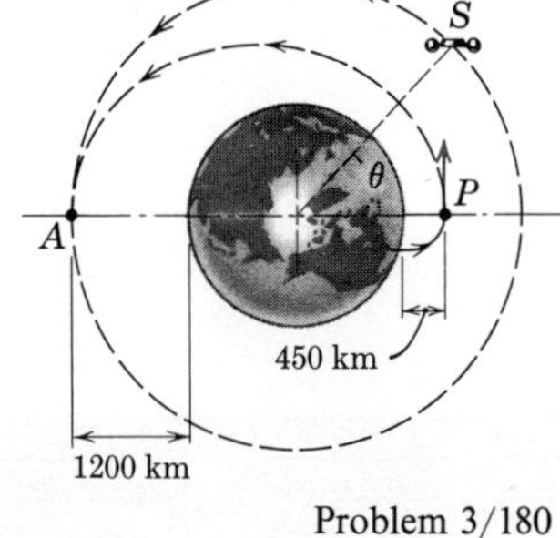

Problem 3/180

◀ **3/180** A space station S is being assembled in its circular orbit 1200 km above the earth. The next payload P of parts along with its carrier rocket has a mass of 800 kg and is injected into a coasting orbit at P, which is 450 km above the earth. Determine the angle θ which specifies the relative position between S and P at injection in order that rendezvous will occur at A with parallel paths. If the carrier rocket can develop a thrust of 900 N, determine the time t during which the engines should be fired as P approaches S at A in order that the relative velocity between P and S will be zero at rendezvous.

Ans. $\theta = 13.2°$, $t = 170$ s

◀ **3/181** For the given launching conditions of Prob. 3/160, derive an expression for the eccentricity e of the resulting orbit.

$$\textit{Ans. } e = +\sqrt{1 - \mu(2 - \mu)\cos^2\beta}$$
$$\text{where } \mu = \frac{(R + H)v_0^2}{gR^2}$$

◀ **3/182** The attempt is made to launch an earth satellite in a circular orbit of altitude H. If the orbit injection angle β (see figure for Prob. 3/160) is not zero and the injection velocity v_0 is different from the velocity v_c for a circular orbit, an elliptical orbit will result. Determine the expressions for r_{max} and r_{min} resulting when (*a*) $\beta = 0$, $\eta \neq 1$, and (*b*) $\beta \neq 0$, $\eta = 1$, where $\eta = v_0/v_c$.

Ans. (*a*) $r_{\substack{max \\ min}} = R + H \quad \begin{pmatrix} \eta < 1 \\ \eta > 1 \end{pmatrix}$

$$r_{\substack{max \\ min}} = \frac{R + H}{2 - \eta^2}\eta^2 \quad \begin{pmatrix} \eta > 1 \\ \eta < 1 \end{pmatrix}$$

(*b*) $r_{\substack{max \\ min}} = (R + H)(1 \pm \sin \beta)$

3/183 Determine the error introduced in the calculation of the period of the earth about the sun with the assumption of a fixed sun. Neglect the influence of all other planets in the solar system.
Ans. Uncorrected period 47 s/year longer than corrected period

3/184 Determine the error introduced in the calculation of the period of the moon about the earth with the assumption of a fixed earth.
Ans. 4.0090 h too great

---

21 Motion Relative to Moving Axes. The determination of the relationships between the forces acting on a particle and its motion relative to a moving reference frame is a problem often encountered. The solution of this problem requires the isolation of the particle with its free-body diagram as previously treated and the application of the equations for relative acceleration derived in the previous chapter.

Part A: Equations of Motion

Consider a particle A of mass m whose motion is observed from a set of axes x-y-z translating with an acceleration $\mathbf{a}_B$ where B is the origin of the moving system. From the relative acceleration relation, Eq. 37 (or Eq. 20), the equation of motion of m, which is $\Sigma \mathbf{F} = m\mathbf{a}_A$, may be written as

$$\Sigma \mathbf{F} = m(\mathbf{a}_B + \mathbf{a}_{rel}) \tag{85}$$

where $\mathbf{a}_{rel} = \mathbf{a}_{A/B}$ is the acceleration of the particle A measured by the observer relative to the moving axes.

For measurements made from a reference frame x-y-z rotating with an angular velocity ω and having an acceleration $\mathbf{a}_B$ of its origin, the principle of relative acceleration of Eq. 41 (or Eq. 23) is used to write

$$\Sigma \mathbf{F} = m[\mathbf{a}_B + \dot{\omega} \times \mathbf{r} + \omega \times (\omega \times \mathbf{r}) + 2\omega \times \mathbf{v}_{rel} + \mathbf{a}_{rel}] \tag{86}$$

The vector $\mathbf{r} = \mathbf{r}_{A/B}$ locates the particle A in x-y-z, and $\mathbf{v}_{rel}$ is the velocity of the particle measured by the observer relative to the moving axes.

In the case of motion relative to the earth where axes x-y-z are attached to the earth's surface, the term $\mathbf{a}_B$ in Eq. 86 has a magnitude of $R\omega^2 \cos \gamma$, where γ is the latitude, R is the earth's radius, and ω is the angular rate of rotation of the earth as well as of x-y-z. The term $\mathbf{a}_B$ varies from $3.387(10^{-2})$ m/s^2 at the equator to zero at the poles and is usually neglected in practical calculations. Also, $\dot{\boldsymbol{\omega}}$ is essentially zero for all practical purposes. Also, if the particle in question is close to the earth's surface, then $\mathbf{r}$ is essentially zero. Hence, with the assumptions noted, Eq. 86 becomes

$$\Sigma\mathbf{F} = m(2\boldsymbol{\omega} \times \mathbf{v}_{\text{rel}} + \mathbf{a}_{\text{rel}}) \tag{87}$$

If the particle is falling in a vacuum, $\Sigma\mathbf{F} = m\mathbf{g}$ so that Eq. 87 gives for the measured or relative acceleration due to gravity $\mathbf{a}_{\text{rel}} = \mathbf{g}'$

$$\mathbf{g}' = \mathbf{g} - 2\boldsymbol{\omega} \times \mathbf{v}_{\text{rel}} \tag{88}$$

where $\mathbf{g}$ is the absolute acceleration due to gravity as measured from a nonrotating earth or from nonrotating axes attached to the center of the earth. Equation 88 gives the effect of the Coriolis acceleration on the measured or relative acceleration $\mathbf{g}'$ due to gravity.

If $\mathbf{a}_B$ is not neglected but $\mathbf{v}_{\text{rel}} = \mathbf{0}$, then the equation reduces to

$$\mathbf{g}' = \mathbf{g} - \mathbf{a}_B \tag{89}$$

where $|\mathbf{a}_B| = R\omega^2 \cos\gamma$. The magnitudes of the gravitational acceleration terms, g' relative to the rotating earth and g relative to a nonrotating earth ("absolute"), are plotted in Fig. 1 of Chapter 1 as a function of the latitude γ. The vector relationship expressed by Eq. 89 is shown in Fig. 38. The maximum value of θ is found to be 6 min of arc and occurs for $\gamma = 45$ deg.

As previously indicated, the force sum $\Sigma\mathbf{F}$ is obtained from the essential step of constructing the free-body diagram of the particle. The choice of moving coordinate system will depend on the particular problem, and the coordinate axes may or may not be attached to the particle.

Part B: D'Alembert's Principle

When the particle is observed from a fixed set of axes X-Y-Z, Fig. 39*a*, its absolute acceleration $\mathbf{a}$ is measured, and the familiar relation $\Sigma\mathbf{F} = m\mathbf{a}$ is applied. When the particle is observed from a moving system x-y-z

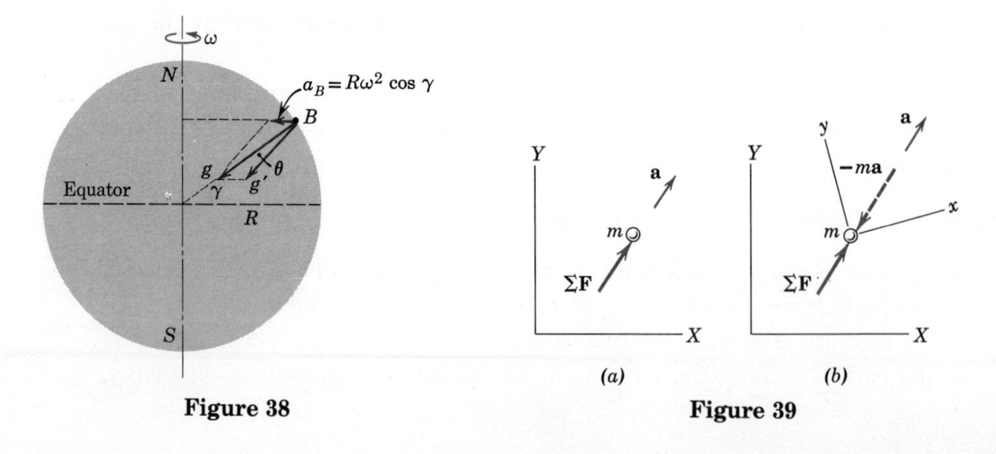

Figure 38

Figure 39

(rotating or nonrotating) attached to the particle at the origin, Fig. 39*b*, the relative motion terms as well as $\mathbf{r}$ are zero, and the observer attached to x-y-z notes that the particle is at rest relative to his reference frame. If the observer claims that the particle is in "equilibrium" in x-y-z, then he concludes that a force $-m\mathbf{a}$ acts to balance $\Sigma\mathbf{F}$. This point of view which permits the treatment of a dynamics problem by the methods of statics was an outgrowth of the work of D'Alembert contained in his *Traité de Dynamique* published in 1743. This approach merely amounts to rewriting the equation of motion as $\Sigma\mathbf{F} - m\mathbf{a} = \mathbf{0}$, which assumes the form of a zero force summation if $-m\mathbf{a}$ is treated as a force. This fictitious force is known as the *inertia force,* and the artificial state of equilibrium created is known as *dynamic equilibrium.* The apparent transformation of a problem in dynamics to one in statics has become known as *D'Alembert's principle.*

Opinion differs concerning the original interpretation of D'Alembert's principle, but the principle in the form in which it is generally known is regarded in this book as being mainly of historical interest. It was evolved during a time when understanding and experience with dynamics were extremely limited and was a means of explaining dynamics in terms of the principles of statics which during earlier times were more fully understood. This excuse for using an artificial situation to describe a real one is open to question, as there is today a wealth of knowledge and experience with the phenomena of dynamics to support strongly the direct approach of thinking in terms of dynamics rather than in terms of statics. It is somewhat difficult to justify the long persistence in the acceptance of statics as a way of understanding dynamics particularly in view of the continued search for the understanding and description of physical phenomena in their most direct form.

Only one simple example of the method known as D'Alembert's principle will be cited. The conical pendulum of mass m, Fig. 40*a*, is swinging in a horizontal circle with its radial line r having an angular velocity ω. In the straightforward application of the equation of motion $\Sigma\mathbf{F} = m\mathbf{a}_n$ in the direction n of the acceleration, the free-body diagram in the *b*-part of the figure shows that $T \sin\theta = mr\omega^2$. When combined with the equilibrium requirement in the y-direction, $T\cos\theta - mg = 0$, the unknowns T and θ can be found. If the reference axes are attached to the particle, then the particle would appear

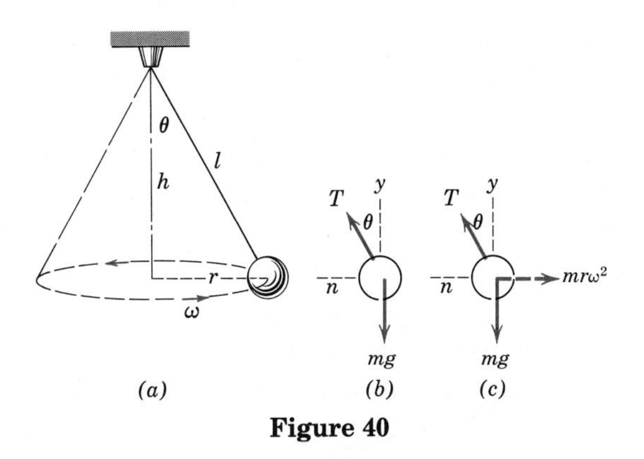

Figure 40

to be in equilibrium relative to these axes. Accordingly, the inertia force $-m\mathbf{a}$ would be added which amounts to visualizing the application of $mr\omega^2$ in the direction opposite to the acceleration, as shown in the c-part of the figure. With this pseudo free-body diagram a zero force summation in the n-direction gives $T \sin \theta - mr\omega^2 = 0$ which, of course, gives the same result as before. It may be concluded that no advantage results from this alternative formulation. The author recommends against using it since it introduces no simplification and adds a nonexistent force to the diagram. In the case of a particle moving in a circular path this hypothetical inertia force is known as the *centrifugal force* since it is directed away from the center and is opposite to the direction of the acceleration. The student is urged to recognize that there is no actual centrifugal force acting on the particle. The only actual force which may properly be called centrifugal is the horizontal component of the tension T exerted on the cord by the particle.

Part C: Motion Relative to Constant-Velocity Translating Systems

In discussing particle motion relative to moving reference systems, it is of importance to note the special case where the motion is observed from a translating reference frame which moves with a constant velocity. If the motion of a particle A of mass m is observed relative to the constant-velocity frame x-y with origin B, the acceleration of the particle relative to this frame is identical with the absolute acceleration of the particle $\mathbf{a}_A = \mathbf{a}_{rel}$ since $\mathbf{a}_B = \mathbf{0}$. Hence Newton's second law of motion holds for measurements made in a constant-velocity system and may be written

$$\Sigma \mathbf{F} = m\mathbf{a}_{rel}$$

Observers in the moving system and in the fixed system will also agree on the designation of the resultant force acting on the particle from their identical free-body diagrams, provided they avoid the use of any so-called "inertia forces".

The parallel question concerning the validity of the work-energy equation and the impulse-momentum equation relative to a constant-velocity reference system will now be examined. For this purpose it is entirely sufficient to consider a one-dimensional motion of the particle A in, say, the X-direction of a fixed reference frame X-Y. The coordinate of A is x in the parallel system x-y, which moves with a constant velocity v_0 in the x-direction. It is clear that the time derivatives of the two coordinates of A are related by the equations $\dot{X} = v_0 + \dot{x}$ and $\ddot{X} = \ddot{x}$. With the equivalence of the acceleration measurements in mind, the work-energy relation and the impulse-momentum equation will now be written and compared for measurements made in both the moving x-y system and the fixed X-Y system.

Relative to x-y the work done by the resultant force ΣF acting on the particle during an infinitesimal movement dx relative to the moving system is

$dU_r = \Sigma F\,dx$. But $\Sigma F = m\ddot{X} = m\ddot{x}$, so $dU_r = m\ddot{x}\,dx = m\dot{x}\,d\dot{x}$ which is merely $dU_r = d(\frac{1}{2}m\dot{x}^2)$. Thus, if the kinetic energy relative to x-y is defined as $T_r = \frac{1}{2}m\dot{x}^2$, it is seen that

$$dU_r = dT_r \qquad \text{and} \qquad U_r = \Delta T_r \tag{90}$$

Therefore the work-energy equation holds for measurements of particle motion relative to a translating system moving with a constant velocity.

Relative to x-y the impulse on the particle during time dt is $\Sigma F\,dt = m\ddot{x}\,dt = m\,d(\dot{x}) = d(m\dot{x})$. Thus, if the momentum relative to x-y is defined as $G_r = m\dot{x}$, the impulse-momentum equation becomes

$$\Sigma F\,dt = dG_r \qquad \text{and} \qquad \int_0^t \Sigma F\,dt = \Delta G_r \tag{91}$$

Consequently the impulse-momentum equation also holds in the constant-velocity system as long as the relative momentum is used.

Although the work-energy and impulse-momentum equations hold relative to a system translating with a constant velocity, the individual expressions for work, kinetic energy, and momentum differ between the fixed and the moving systems. Thus

$$(dU = F\,dX) \neq (dU_r = F\,dx)$$
$$(T = \tfrac{1}{2}m\dot{X}^2) \neq (T_r = \tfrac{1}{2}m\dot{x}^2)$$
$$(G = m\dot{X}) \neq (G_r = m\dot{x})$$

The foregoing demonstrations for one-dimensional motion may be extended to cover two- and three-dimensional motion of a particle relative to a constant-velocity translating system. The work-energy and impulse-momentum equations may also be written for accelerating and rotating reference axes, but the results are somewhat awkward and not too useful so will not be developed here.

Sample Problem

3/185 The arm OB and rigidly attached rod of lengh l rotate with a constant angular velocity ω about a vertical axis through the fixed bearing at O. A small slider of mass m is released from rest (relative to the rod) at the dotted position x_0 and slides with negligible friction along the rotating rod. Determine the distance x as a function of the time t following release. Also determine the expression for the horizontal component N of the force exerted by the rod on the slider as a function of x.

Solution. The rotating reference frame x-y attached to the rod is seen to be a convenient set of axes in which to measure the motion of the slider. The free-body diagram of the slider shows only the horizontal component N of the force exerted by the rod on the slider. (The vertical component equals mg and is perpendicular to the plane of the diagram so appears as a point.) The equation of relative motion for rotating axes, Eq. 86, governs the solution and is

$$\Sigma \mathbf{F} = m(\mathbf{a}_B + \dot{\omega} \times \mathbf{r} + \omega \times (\omega \times \mathbf{r}) + 2\omega \times \mathbf{v}_{\text{rel}} + \mathbf{a}_{\text{rel}})$$

In the application of this equation the position vector $\mathbf{r}$ measured in the moving system is $\mathbf{r} = \mathbf{i}x$ and the angular velocity of the axes is $\boldsymbol{\omega} = \mathbf{k}\omega$. The separate terms become

$$\mathbf{a}_B = -b\omega^2\mathbf{j}$$

$$\dot{\boldsymbol{\omega}} \times \mathbf{r} = 0$$

$$\boldsymbol{\omega} \times (\boldsymbol{\omega} \times \mathbf{r}) = \omega\mathbf{k} \times (\omega\mathbf{k} \times \mathbf{i}x) = -x\omega^2\mathbf{i}$$

$$2\boldsymbol{\omega} \times \mathbf{v}_{\text{rel}} = 2\omega\mathbf{k} \times \dot{x}\mathbf{i} = 2\omega\dot{x}\mathbf{j}$$

$$\mathbf{a}_{\text{rel}} = \ddot{x}\mathbf{i}$$

Substitution into the equation of motion yields

$$-N\mathbf{j} = m(-b\omega^2\mathbf{j} - x\omega^2\mathbf{i} + 2\omega\dot{x}\mathbf{j} + \ddot{x}\mathbf{i})$$

Separation of the **i**- and **j**-terms yields the two scalar equations

$$\ddot{x} - x\omega^2 = 0 \qquad \text{and} \qquad N = m(b\omega^2 - 2\omega\dot{x})$$

The general solution of the first equation is seen by direct substitution to be $x = A \sinh \omega t + B \cosh \omega t$. The integration constants A and B are found from the conditions $x = x_0$ and $\dot{x} = 0$ when $t = 0$ which give $A = 0$ and $B = x_0$. Thus the x-coordinate is

$$x = x_0 \cosh \omega t \qquad \textit{Ans.}$$

The expression for the normal force N contains the relative velocity $\dot{x} = x_0\omega \sinh \omega t$ which may be expressed in terms of x by substituting the identity $\cosh^2 \omega t - \sinh^2 \omega t = 1$. Thus $\sinh \omega t = \sqrt{\cosh^2 \omega t - 1} = \sqrt{(x/x_0)^2 - 1}$, and $\dot{x} = x_0\omega\sqrt{(x/x_0)^2 - 1}$. The expression for the normal force N becomes

$$N = m\omega^2(b - 2\sqrt{x^2 - x_0^2}) \qquad \textit{Ans.}$$

which is limited to values of $x > x_0$.

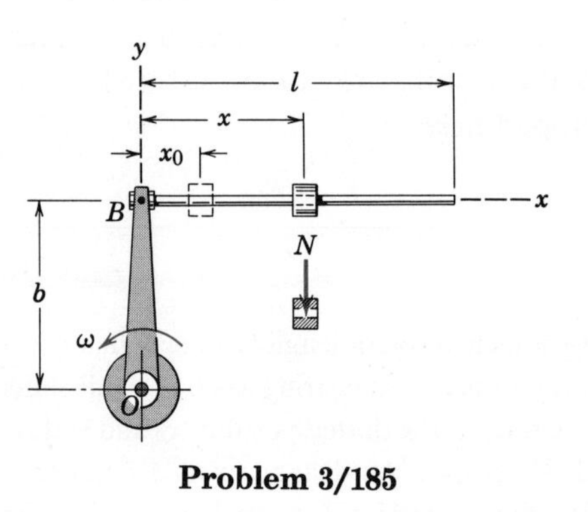

Problem 3/185

Problems

3/186 A particle is released from rest from a position fixed to the earth's surface at the equator. The absolute acceleration due to gravity at the equator is 9.815 m/s^2. Calculate the measured or apparent gravitational acceleration g' of the freely falling particle. (Compare the values with Fig. 1.)

3/187 The aircraft carrier is moving at a constant speed and launches a 3-t jet plane in a distance of 75 m along the deck by means of a steam-driven catapult. If the plane leaves the deck with a velocity of 240 km/h relative to the carrier and if the jet thrust is constant at 22 kN during take-off, compute the constant force P exerted by the catapult on the airplane during the 75-m travel of the launch carriage.

Ans. $P = 66.9$ kN

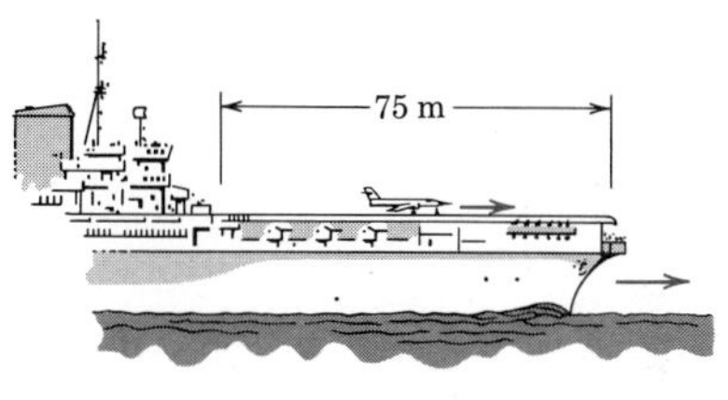

Problem 3/187

3/188 Two jet planes A and B are flying alongside one another at a constant speed of 1000 km/h when the pilot of A ignites a constant-thrust jet-assist unit which burns for 5 s. The navigator of B observes that A is traveling 16 km/h faster than B at the end of the 5 s. Calculate the constant forward thrust T of the jet-assist unit if plane A has an average total mass of 5 t during the acceleration period. Neglect the small increment of air resistance due to the higher speed. What is the momentum G_r of A in B's system at the end of the 5-s period?

3/189 The slider A with a mass of 2 kg moves with negligible friction in the slot of the vertical plate. If the plate has a constant acceleration of 8 m/s^2 to the left and the slider is released from rest relative to the plate, calculate the force N exerted by the side of the slot on the slider during the motion. Does contact occur on side B or side C? Also find the acceleration a_{rel} of the slider relative to the slot.

Ans. $N = 8.99$ N (side C)
$a_{\text{rel}} = 11.83$ m/s^2

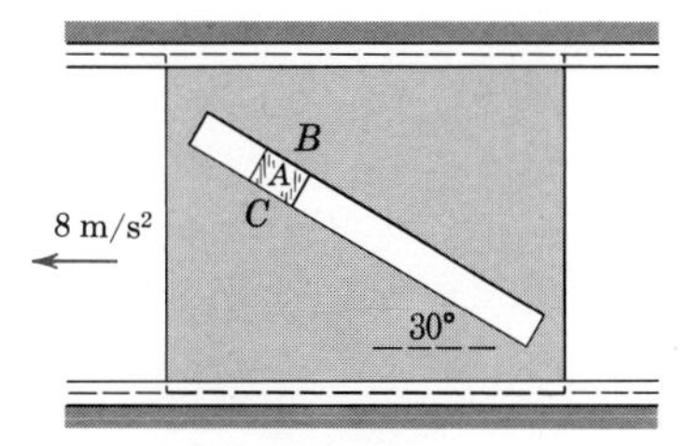

Problem 3/189

3/190 The slotted disk rotates about a vertical axis through its center O with a constant angular velocity $\omega = 20$ rad/s. The 0.5-kg slider A oscillates in the slot under the action of springs (not shown). If the slider has a velocity $\dot{x} = 0.9$ m/s as it passes the center of the disk at $x = 0$, determine the horizontal component N of the force exerted by one side of the slot on the slider as it passes the center position.

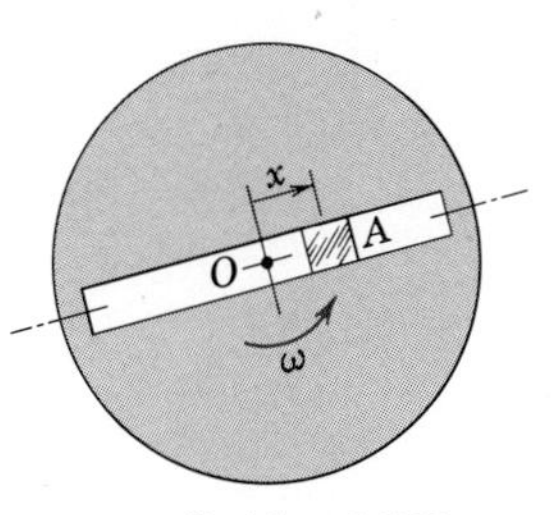

Problem 3/190

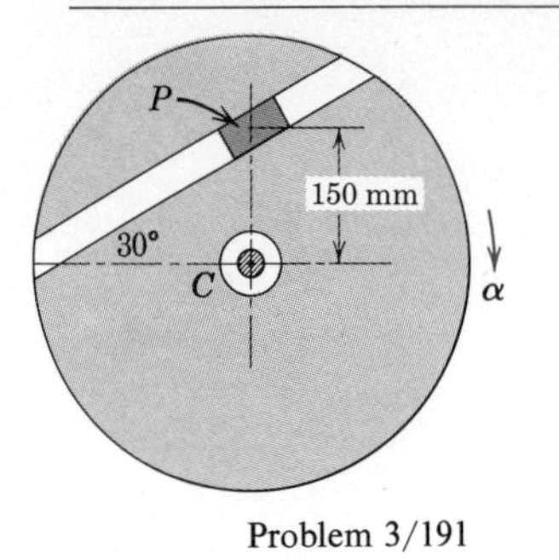

Problem 3/191

3/191 The slotted disk rotates in a horizontal plane about its shaft at C, and the 0.5-kg slider P moves in the slot with negligible friction. The disk is given a clockwise angular acceleration of 40 rad/s^2 starting from rest with the slider initially at rest in the position shown. Determine the horizontal force F exerted on P by the slot at the start of the motion and the initial acceleration of P relative to the slot.

Ans. $F = 1.5$ N
$a_{rel} = 5.20$ m/s^2

3/192 A car with a mass of 1.5 t is driven east at 150 km/h on a straight road situated on the equator. If the road is a true tangent to the earth's surface at a point on the equator and, hence, has no curvature, calculate the value of the total normal force exerted by the road on the tires. The absolute gravitational acceleration at the equator is 9.815 m/s^2, and the equatorial diameter of the earth is 12 756 km.

Ans. $N = 14.66$ kN

3/193 For the car of Prob. 3/192, if the road follows the curvature of the earth rather than a straight path tangent to the equatorial circle, determine the change ΔN in the total normal force exerted by the road on the tires between the conditions of travel on the straight road and travel on the curved road.

3/194 A 1.5-t rocket sled is traveling due north on a level track at latitude 30° north. If the speed reaches 2400 km/h, compute the side thrust T exerted on the guide rails by the sled as a result of the rotation of the earth. *Ans.* $T = 72.9$ N

3/195 If the aircraft carrier of Prob. 3/187 is moving with a constant speed u and its catapult launches a plane of mass m with a velocity v relative to the carrier and in the direction of u, the kinetic energy of the plane with respect to the land is $\frac{1}{2}m(v+u)^2 = \frac{1}{2}mv^2 + \frac{1}{2}mu^2 + muv$ when launching is complete. If the plane is launched by the same catapult with the same accelerating force when the carrier is not moving, the kinetic energy of the plane is $\frac{1}{2}mv^2$. Explain the difference between the two kinetic-energy expressions.

3/196 A plumb bob is suspended from a fixed point on the surface of the earth at north latitude γ. Determine the expression for the small angle θ made by the cord which supports the bob with the true vertical due to the angular velocity ω of the earth. The radius of the earth is R. Calculate the value of θ for $\gamma = 30°$. Let g stand for the absolute acceleration due to gravity. (See Fig. 1.)

$$\textit{Ans.}\ \tan\theta = \frac{R\omega^2}{2g}\frac{\sin 2\gamma}{1 - (R\omega^2/g)\cos^2\gamma}$$

$$\theta = 5'$$

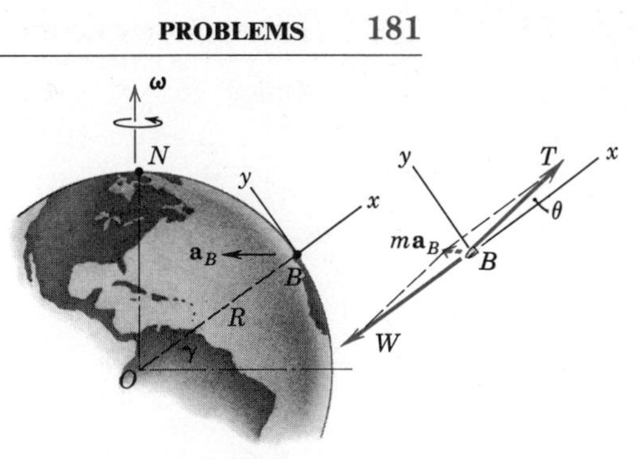

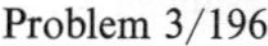

Problem 3/196

3/197 A ball of mass m is dropped down a fixed vertical tube of height h situated at the equator. If frictional resistance is negligible, determine the contact force N between the ball and the tube as a function of the time t of fall from rest relative to the top of the tube. Also specify the side of the tube against which contact takes place. Neglect h compared with the radius R of the earth but retain the square of the angular velocity ω of the earth. *Ans.* $N = 2m\omega(g - R\omega^2)t$ against east side of tube

3/198 The point of support B for a simple pendulum of mass m and length l has a constant horizontal acceleration a as shown. If the pendulum is released from rest relative to the moving system with $\theta = 0$, determine the expression for the tension T in the cord of the pendulum as a function of θ.

$$\textit{Ans.}\ T = mg\left[3\sin\theta + \frac{a}{g}(3\cos\theta - 2)\right]$$

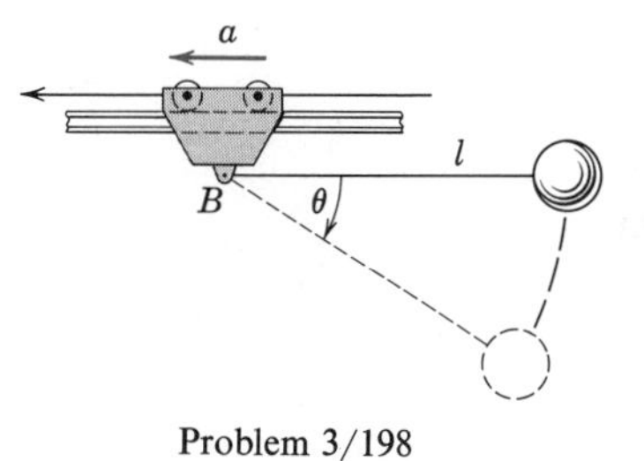

Problem 3/198

3/199 A particle is dropped from a tower of height h located at north latitude γ. Derive an expression for the distance s from the point on the ground directly under the point of release to the spot at which the particle lands. Neglect air resistance and the square of the earth's angular velocity ω.

$$\textit{Ans.}\ s = \frac{2}{3}\omega h\sqrt{\frac{2h}{g}}\cos\gamma,\ \text{east}$$

3/200 At north latitude γ a particle is projected vertically upward with an initial velocity u from a point A on the ground. Write an expression for the distance s from the point A to the point of impact B upon return to the ground. Neglect air resistance and the square of the earth's angular velocity ω.

$$\textit{Ans.}\ s = \frac{4}{3}\frac{\omega u^3}{g^2}\cos\gamma,\ \text{west}$$

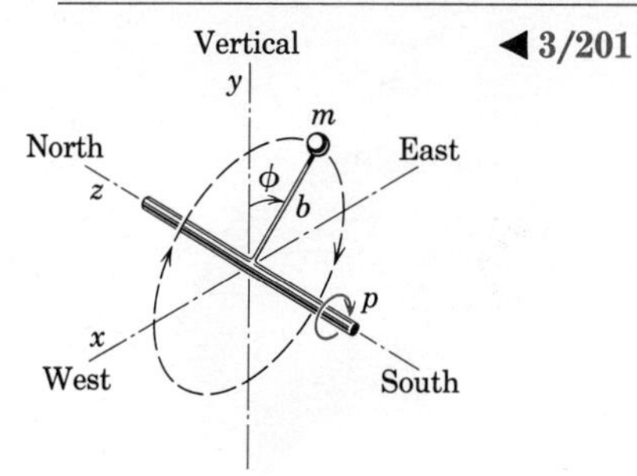

Problem 3/201

◀ **3/201** A particle of mass m is attached to an arm of length b which rotates with a constant angular velocity p about a horizontal axis which points north and south as shown. The latitude is γ north, and the radius and angular velocity of the earth are, respectively, R and ω. Write expressions for the x-, y-, and z-components of the force exerted by the arm on m at the instant that $\phi = 90°$.

Ans. $F_x = mb(\omega^2 + p^2 + 2p\omega \cos \gamma)$
$F_y = m(g - R\omega^2 \cos^2 \gamma)$
$F_z = \frac{1}{2}mR\omega^2 \sin 2\gamma$

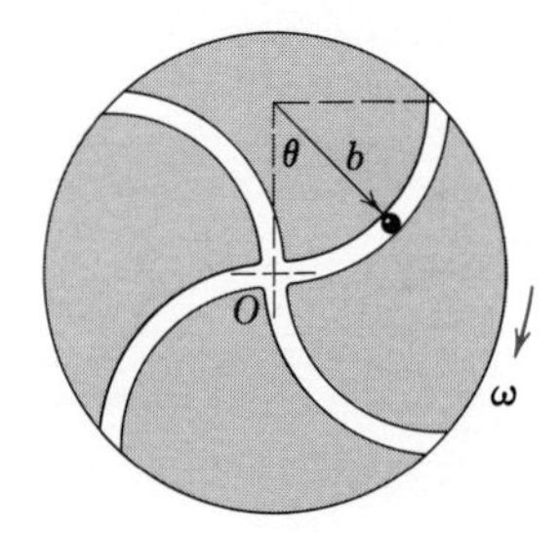

Problem 3/202

◀ **3/202** Particles each of mass m move with negligible friction out along the circular slots in the disk, which rotates in the horizontal plane with a constant angular velocity ω. Derive an expression for the force F exerted by the slot on a particle in terms of θ. Also determine the velocity u of the particle relative to the slot just prior to leaving the slot. Assume each particle starts essentially from rest at $\theta = 0$.

Ans. $F = 2mb\omega^2 \sin \frac{\theta}{2}(3 \sin \frac{\theta}{2} - 2)$

$u = b\omega\sqrt{2}$ at exit

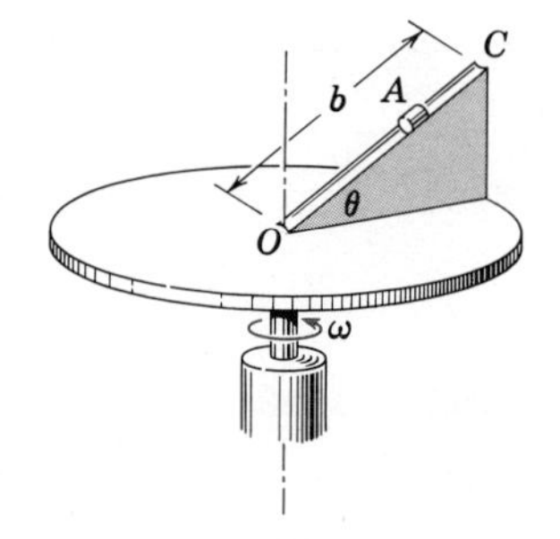

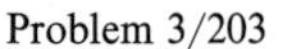
Problem 3/203

◀ **3/203** The particle A is released from rest (relative to the disk) at position C and slides in the inclined groove toward the center O of the disk while the disk rotates with a constant angular velocity ω. If friction is negligible, determine the time t required for the particle to reach O and specify the limiting value of ω beyond which the particle will not move inward.

Ans. $t = \dfrac{1}{\omega \cos \theta} \cosh^{-1} \dfrac{1}{1 - K}$ where

$K = \dfrac{b\omega^2 \cos^2 \theta}{g \sin \theta}$; $\omega < \sec \theta \sqrt{\dfrac{g}{b} \sin \theta}$

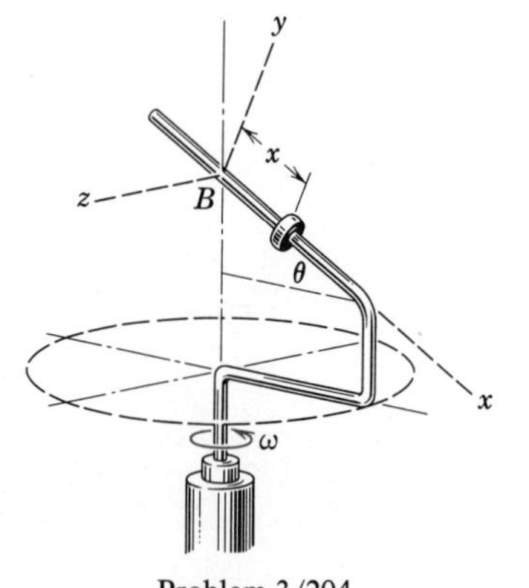

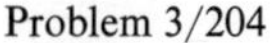
Problem 3/204

◀ **3/204** The bent shaft is made to revolve about the vertical axis with a constant angular speed ω. The slider of mass m is released from rest relative to the rod at $x = 0$. Determine the distance x as a function of the time t after release. Also specify the components of the contact force between the shaft and the slider in terms of t. Neglect friction.

Ans. $\mathbf{N}_1 = \dfrac{mg[1 - \sin^2 \theta \cosh (\omega t \cos \theta)]}{\cos \theta}\mathbf{j}$

$\mathbf{N}_2 = -2mg \sin \theta \sinh (\omega t \cos \theta)\mathbf{k}$

3/205 The centrifugal distributor rotates at a constant speed ω about a fixed vertical axis. Small particles each of mass m are released from rest at A at essentially zero radius and move with negligible friction out along the rotating circular path between the radial vanes. Determine the expression for the magnitude v of the absolute velocity of each particle as it leaves the vane horizontally at B. Also find the horizontal force N exerted by the vertical vane on each particle as it approaches B.

$$Ans.\ v = \sqrt{2r(g + r\omega^2)}$$
$$N = 2m\omega\sqrt{r(2g + r\omega^2)}$$

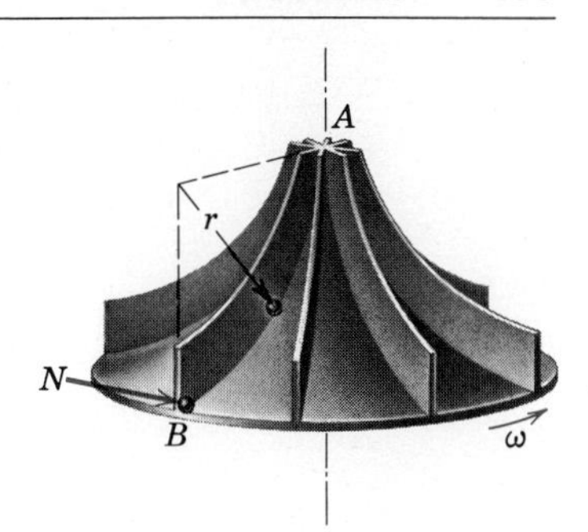

Problem 3/205

4 KINETICS OF SYSTEMS OF PARTICLES

22 **Introduction.** The previous two chapters have dealt with the principles of kinematics and kinetics for the motion of a particle. Although attention was focused primarily on the kinetics of a single particle in Chapter 3, limited reference to the motion of two particles, considered together as a system, was made in the discussions of work-energy and impulse-momentum. The next major step in the development of the subject of dynamics is to extend these principles for the motion of a single particle to describe the motion of a general system of particles. This extension provides a unity to the remaining sections of *Dynamics,* which include the motion of rigid bodies in Part II and the motion of nonrigid systems in Part III. It is recalled that a rigid body is a solid system of particles wherein the distances between particles remain essentially unchanged. The overall motions found with machines, land and air vehicles, rockets and spacecraft, and many moving structures provide examples of rigid-body problems. A nonrigid body, on the other hand, may be a solid body where the object of investigation is the time dependence of the changes in shape due to elastic or nonelastic deformations. Or a nonrigid body may be a defined mass of liquid or gaseous particles which have a time-dependent rate of flow. Examples would be the air and fuel flowing through the turbine of an aircraft engine, the burned gases issuing from the nozzle of a rocket motor, or the water passing through a rotary pump.

Although the extension of the equations for single-particle motion to a general system of particles is accomplished without undue labor, it cannot be expected that the full generality and significance of these extended principles can be adequately understood at the outset, particularly without considerable problem experience. For this reason it is strongly recommended that the appropriate articles of this chapter be read or reread when each of the remaining chapters of the book is studied. In this way the unity contained in Chapter 4 will be brought into greater focus and a broader view of dynamics will be acquired.

23 **Equations of Motion.** Newton's second law of motion for a single particle, Eq. 1 or Eq. 44, will now be extended to cover a general system of n mass particles bounded by a closed surface in space, Fig. 41. This bounding envelope, for example, might be the exterior surface of a given rigid body, the bounding surface of an arbitrary portion of the body, the exterior surface of a rocket containing both rigid and moving particles, or a particular volume of fluid particles. In each case the system considered is the mass within the envelope, and the mass must be clearly defined and isolated.

Figure 41 shows a representative particle of mass m_i of the system isolated with forces $\mathbf{F}_1, \mathbf{F}_2, \mathbf{F}_3, \cdots$ acting on m_i from sources *external* to the envelope, and forces $\mathbf{f}_1, \mathbf{f}_2, \mathbf{f}_3, \cdots$ acting on m_i from sources *internal* to the envelope. The external forces are due to contact with external bodies or to external gravitational, electric, or magnetic forces. The internal forces are forces of reaction with other mass particles within the envelope. The particle m_i is located by its position vector $\mathbf{r}_i$ measured from a fixed Newtonian set of reference axes. The center of mass G of the system of particles isolated is located by the position vector $\bar{\mathbf{r}}$ which, from Varignon's theorem in statics, is given by

$$m\bar{\mathbf{r}} = \Sigma m_i \mathbf{r}_i$$

where the total system mass is $m = \Sigma m_i$. The summation sign Σ represents the summation over all n particles.

Equation 44 when applied to m_i gives

$$\mathbf{F}_1 + \mathbf{F}_2 + \mathbf{F}_3 + \cdots + \mathbf{f}_1 + \mathbf{f}_2 + \mathbf{f}_3 + \cdots = m_i \ddot{\mathbf{r}}_i$$

where $\ddot{\mathbf{r}}_i$ is the acceleration of m_i. A similar equation may be written for each of the particles of the system. If these equations written for *all* particles of the system are added together, there results

$$\Sigma \mathbf{F} + \Sigma \mathbf{f} = \Sigma m_i \ddot{\mathbf{r}}_i$$

The term $\Sigma\mathbf{F}$ then becomes the vector sum of *all* forces acting on all particles of the isolated system from sources external to the system, and $\Sigma\mathbf{f}$ becomes the vector sum of all forces on all particles produced by the internal actions and reactions between particles. This last sum is identically zero since all internal forces occur in pairs of equal and opposite actions and reactions. By differentiating the equation defining $\bar{\mathbf{r}}$ twice with time, there results $m\ddot{\bar{\mathbf{r}}} = \Sigma m_i \ddot{\mathbf{r}}_i$ where m has no time derivative as long as mass is not entering or leaving the system.* Substitution into the summation of the

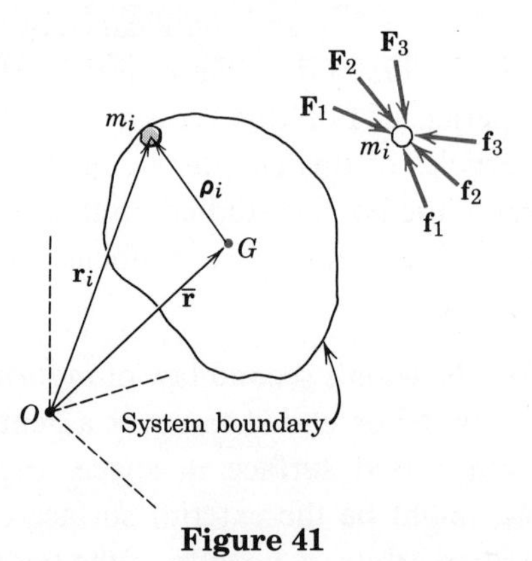

Figure 41

* If m is a function of time, a more complex situation develops; this situation is discussed in Chapter 10 under the heading of variable mass.

equations of motion gives

$$\Sigma \mathbf{F} = m\ddot{\bar{\mathbf{r}}} \qquad \text{or} \qquad \Sigma \mathbf{F} = m\bar{\mathbf{a}} \tag{92}$$

where $\bar{\mathbf{a}}$ is the acceleration $\ddot{\bar{\mathbf{r}}}$ of the center of mass of the system.

Equation 92 is the generalized Newton's second law of motion for a mass system and is referred to as the *equation of motion* of m. The equation states that the resultant of the external forces on any system of mass equals the total mass of the system times the acceleration of the center of mass. This law expresses the so-called *principle of motion of the mass center.* It should be observed that $\bar{\mathbf{a}}$ is the acceleration of the mathematical point which represents instantaneously the position of the mass center for the given n particles. For a rigid body this acceleration will also be the acceleration of a particular mass particle located at the mass center. For a nonrigid body this acceleration need not represent the acceleration of any particular particle. It should be noted further that Eq. 92 holds for each instant of time and is therefore an instantaneous relationship. Equation 92 for the mass system may not be inferred directly from Eq. 44 for the single particle but must be proved. Equation 92 may be expressed in component form, as in Eqs. 92*a*, using whatever coordinate system is most convenient for the problem at hand:

$$\Sigma F_x = m\bar{a}_x \qquad \Sigma F_y = m\bar{a}_y \qquad \Sigma F_z = m\bar{a}_z \tag{92a}$$

Although Eq. 92, as a vector equation, requires that the acceleration vector $\bar{\mathbf{a}}$ must have the same direction as the resultant external force $\Sigma\mathbf{F}$, it does not follow that $\Sigma\mathbf{F}$ necessarily passes through G. In general, in fact, $\Sigma\mathbf{F}$ does not pass through G.

24 **Work and Energy.** In Art. 18 the work-energy relation for a single particle was developed. Applicability to a system of two or more joined particles was also noted in Part A of that article. Attention is now turned to the general system of Fig. 41, where the work-energy relation for the representative particle of mass m_i is $U_i = \Delta T_i$. Here U_i is the work done on m_i by all forces $\mathbf{F}_i = \mathbf{F}_1 + \mathbf{F}_2 + \mathbf{F}_3 + \cdots$ applied from sources external to the system and by all forces $\mathbf{f}_i = \mathbf{f}_1 + \mathbf{f}_2 + \mathbf{f}_3 + \cdots$ internal to the system. The change in kinetic energy of m_i is $\Delta T_i = \Delta(\frac{1}{2}m_i v_i^2)$ where v_i is the magnitude of the particle velocity $\mathbf{v}_i = \dot{\mathbf{r}}_i$.

For the entire system the sum of the work-energy equations written for all particles is $\Sigma U_i = \Sigma\, \Delta T_i$ which may be represented by Eq. 51 of Art. 18, namely,

$$U = \Delta T \qquad [51]$$

where $U = \Sigma U_i$, the work done by all forces on all particles, and ΔT is the change in the total kinetic energy $T = \Sigma T_i$ of the system.

For a rigid body or a system of rigid bodies joined by ideal connections in which no net work is done by the internal interacting forces or moments in the connections, the work done by all pairs of internal forces $\mathbf{f}_i$ and $-\mathbf{f}_i$ of the system is zero since their points of application have identical displacement components in the direction of the forces. For this situation U becomes the work done on the system by the external forces.

For a nonrigid mechanical system which has elastic members capable of storing energy, a part of the work done by the external forces goes into changing the internal elastic potential energy V_e. Also, if the works done by the gravity forces are excluded from the U-term and are accounted for by the changes in gravitational potential energy V_g, the work-energy equation for the nonrigid mechanical system may be written as Eq. 57 which is

$$U = \Delta T + \Delta V_g + \Delta V_e \qquad [57]$$

If no work is done on the system, then its mechanical energy $E = T + V_e + V_g$ remains constant.

▼ ▼ ▼ ▼ ▼

A more detailed analysis of the work-energy relationships sheds further light on the motion of a general mass system. The total work done on m_i may be written as $U_i = \int (\mathbf{F}_i + \mathbf{f}_i) \cdot d\mathbf{r}_i$ where $d\mathbf{r}_i$ is the infinitesimal displacement of m_i. The total work done on the system by both external and internal forces is, accordingly,

$$\begin{aligned} U = \Sigma U_i &= \Sigma \int (\mathbf{F}_i + \mathbf{f}_i) \cdot d\mathbf{r}_i = \Sigma \int (\mathbf{F}_i + \mathbf{f}_i) \cdot (d\bar{\mathbf{r}} + d\boldsymbol{\rho}_i) \\ &= \Sigma \int (\mathbf{F}_i + \mathbf{f}_i) \cdot d\bar{\mathbf{r}} + \Sigma \int (\mathbf{F}_i + \mathbf{f}_i) \cdot d\boldsymbol{\rho}_i \end{aligned}$$

where the position vector $\boldsymbol{\rho}_i$ of m_i with respect to G, from Fig. 41, has been introduced by the substitution $\mathbf{r}_i = \bar{\mathbf{r}} + \boldsymbol{\rho}_i$. In the first summation $d\bar{\mathbf{r}}$ is a common factor in all of the terms, and this expression may be written $\Sigma \int (\mathbf{F}_i + \mathbf{f}_i) \cdot d\bar{\mathbf{r}} = \int \{(\Sigma \mathbf{F}_i) + (\Sigma \mathbf{f}_i)\} \cdot d\bar{\mathbf{r}}$. The vector sum $\Sigma \mathbf{F}_i$ of all external forces acting on the system will be written merely as $\Sigma \mathbf{F}$, and the vector sum $\Sigma \mathbf{f}_i$ of all internal forces acting on all particles is necessarily zero since the internal forces exist in pairs of equal and opposite forces. Thus the expression for the total work on the system becomes

$$U = \int \Sigma \mathbf{F} \cdot d\bar{\mathbf{r}} + \int \Sigma \{(\mathbf{F}_i + \mathbf{f}_i) \cdot d\boldsymbol{\rho}_i\}$$

The kinetic energy for the system will now be written with the aid of the identity $v_i^2 = \dot{\mathbf{r}}_i \cdot \dot{\mathbf{r}}_i$ for the square of the velocity of the representative particle. Thus

$$\begin{aligned} T = \Sigma T_i &= \Sigma \tfrac{1}{2} m_i \dot{\mathbf{r}}_i \cdot \dot{\mathbf{r}}_i = \Sigma \tfrac{1}{2} m_i (\dot{\bar{\mathbf{r}}} + \dot{\boldsymbol{\rho}}_i) \cdot (\dot{\bar{\mathbf{r}}} + \dot{\boldsymbol{\rho}}_i) \\ &= \Sigma \tfrac{1}{2} m_i \dot{\bar{\mathbf{r}}} \cdot \dot{\bar{\mathbf{r}}} + \dot{\bar{\mathbf{r}}} \cdot \Sigma m_i \dot{\boldsymbol{\rho}}_i + \Sigma \tfrac{1}{2} m_i \dot{\boldsymbol{\rho}}_i \cdot \dot{\boldsymbol{\rho}}_i \end{aligned}$$

But $\dot{\bar{\mathbf{r}}} \cdot \dot{\bar{\mathbf{r}}} = \bar{v}^2$, the square of the magnitude of the velocity of the mass center. Also $\Sigma m_i \dot{\boldsymbol{\rho}}_i = d(\Sigma m_i \boldsymbol{\rho}_i)/dt = \mathbf{0}$ since $\Sigma m_i \boldsymbol{\rho}_i$ is necessarily zero with $\boldsymbol{\rho}_i$ measured from the mass center G. Also $\dot{\boldsymbol{\rho}}_i \cdot \dot{\boldsymbol{\rho}}_i = |\dot{\boldsymbol{\rho}}_i|^2$ where $\dot{\boldsymbol{\rho}}_i$ is the velocity of

m_i with respect to the mass center G. Therefore, the total kinetic energy becomes

$$T = \tfrac{1}{2}m\bar{v}^2 + \Sigma\tfrac{1}{2}m_i|\dot{\boldsymbol{\rho}}_i|^2 \tag{93}$$

This equation expresses the fact that the total kinetic energy of a mass system equals the energy of mass-center translation of the system as a whole plus the energy due to motion of all particles relative to the mass center.

The work-energy equation for a general mass system may now be written by equating the expressions for U and ΔT which gives

$$\int \Sigma\mathbf{F} \cdot d\bar{\mathbf{r}} + \int \Sigma\{(\mathbf{F}_i + \mathbf{f}_i) \cdot d\boldsymbol{\rho}_i\} = \Delta(\tfrac{1}{2}m\bar{v}^2) + \Delta(\Sigma\tfrac{1}{2}m_i|\dot{\boldsymbol{\rho}}_i|^2)$$

The respective first terms on the two sides of this equation equal one another as may be seen by taking the dot product of Eq. 92 and $d\bar{\mathbf{r}}$ and integrating This step becomes clear when it is recognized that $m\ddot{\bar{\mathbf{r}}} \cdot d\bar{\mathbf{r}} = d(\tfrac{1}{2}m\dot{\bar{\mathbf{r}}} \cdot \dot{\bar{\mathbf{r}}}) = d(\tfrac{1}{2}m\bar{v}^2)$. Thus the complete work-energy relation may be written as the two independent equations

$$\begin{aligned} \int \Sigma\mathbf{F} \cdot d\bar{\mathbf{r}} &= \Delta(\tfrac{1}{2}m\bar{v}^2) \\ \int \Sigma\{(\mathbf{F}_i + \mathbf{f}_i) \cdot d\boldsymbol{\rho}_i\} &= \Delta(\Sigma\tfrac{1}{2}m_i|\dot{\boldsymbol{\rho}}_i|^2) \end{aligned} \tag{94}$$

The first of Eqs. 94 describes the motion of the mass center which responds as though the external force resultant acts through G, and the second equation describes the motion relative to the mass center as a function of the work of both external and internal forces. For a rigid body or for a system of rigid bodies joined by ideal connections in which no net work is done by the interacting forces or moments in the connections, the work done by all pairs of internal forces acting within the system is zero since their points of application have identical displacement components in the direction of the forces. For this case, the second of Eqs. 94 reduces to

$$\int \Sigma\mathbf{F}_i \cdot d\boldsymbol{\rho}_i = \Delta(\Sigma\tfrac{1}{2}m_i|\dot{\boldsymbol{\rho}}_i|^2)$$

It will be seen in Chapters 6 and 8 that this last term represents that part of the kinetic energy of a rigid body which is due to the rotation of the body about its mass center.

25 Linear and Angular Momentum. Equations 61 and 65, which represent the linear and angular impulse-momentum relations for a single particle, will now be extended to the general mass system represented in Fig. 41. The linear momentum $\mathbf{G}$ of the system is defined as the vector sum of the linear momenta of all of its particles. Thus

$$\mathbf{G} = \Sigma m_i\mathbf{v}_i \tag{95}$$

where $\mathbf{v}_i$ is the velocity $\dot{\mathbf{r}}_i$ of the representative particle. The time derivative of $\mathbf{G}$ is $\dot{\mathbf{G}} = \Sigma m_i \mathbf{a}_i$, where the acceleration of the representative particle is $\mathbf{a}_i = \dot{\mathbf{v}}_i = \ddot{\mathbf{r}}_i$. But from Art. 23 the resultant $\Sigma \mathbf{F}$ of the external forces on the system also equals $\Sigma m_i \mathbf{a}_i$. Hence,

$$\Sigma \mathbf{F} = \dot{\mathbf{G}} \tag{96}$$

which has the same form as Eq. 61 for a single particle. Equation 96 states that the resultant of the external forces on any mass system equals the time rate of change of the linear momentum of the system and is an alternative form of the generalized second law of motion, Eq. 92. In Eq. 96 the total mass was held constant during differentiation with time, so that the equation does not apply to systems whose mass changes with time.

The angular momentum of the mass system of Fig. 41 about a fixed point O is defined as the vector sum of the moments of the linear momenta about O of all particles of the system and is

$$\mathbf{H}_O = \Sigma(\mathbf{r}_i \times m_i \mathbf{v}_i)$$

The time derivative is $\dot{\mathbf{H}}_O = \Sigma(\dot{\mathbf{r}}_i \times m_i \mathbf{v}_i) + \Sigma(\mathbf{r}_i \times m_i \dot{\mathbf{v}}_i)$. The first summation vanishes since the cross product of two identical vectors $\dot{\mathbf{r}}_i = \mathbf{v}_i$ is zero. The second summation is $\Sigma(\mathbf{r}_i \times m_i \mathbf{a}_i) = \Sigma(\mathbf{r}_i \times \mathbf{F}_i)$ which is the vector sum of the moments about O of all forces acting on all particles of the system. This moment sum $\Sigma \mathbf{M}_O$ represents only the moments of forces external to the system, since the internal forces cancel one another. Thus the moment sum becomes

$$\Sigma \mathbf{M}_O = \dot{\mathbf{H}}_O \tag{97}$$

which has the same form as Eq. 65 for a single particle. Equation 97 states that the resultant vector moment about any fixed point of all external forces on any system of mass equals the time rate of change of angular momentum of the system about the fixed point. As in the linear momentum case, Eq. 97 does not apply if the total mass of the system is changing with time.

The principle expressed by Eq. 97 also holds when the mass center G is chosen as the reference point. The angular momentum about G is

$$\overline{\mathbf{H}} = \Sigma(\boldsymbol{\rho}_i \times m_i \mathbf{v}_i)$$

and its derivative is $\dot{\overline{\mathbf{H}}} = \Sigma(\dot{\boldsymbol{\rho}}_i \times m_i \mathbf{v}_i) + \Sigma(\boldsymbol{\rho}_i \times m_i \dot{\mathbf{v}}_i)$. But from Fig. 41 $\boldsymbol{\rho}_i = \mathbf{r}_i - \bar{\mathbf{r}}$, so that the time derivative gives $\dot{\boldsymbol{\rho}}_i = \dot{\mathbf{r}}_i - \dot{\bar{\mathbf{r}}}$. Hence the first summation becomes $\Sigma(\dot{\mathbf{r}}_i \times m_i \mathbf{v}_i) - \Sigma(\dot{\bar{\mathbf{r}}} \times m_i \mathbf{v}_i)$, the first term of which vanishes since $\dot{\mathbf{r}}_i = \mathbf{v}_i$. The second term also vanishes when it is rearranged as

$$-\dot{\bar{\mathbf{r}}} \times \frac{d}{dt} \Sigma m_i \mathbf{r}_i = -\dot{\bar{\mathbf{r}}} \times \frac{d}{dt}(m\bar{\mathbf{r}}) = -\dot{\bar{\mathbf{r}}} \times m\dot{\bar{\mathbf{r}}} = \mathbf{0}$$

The second summation in the expression for $\dot{\overline{\mathbf{H}}}$ becomes $\Sigma(\boldsymbol{\rho}_i \times m_i \mathbf{a}_i)$ which is the sum of the moments $\Sigma \overline{\mathbf{M}}$ of all external forces acting on the system about the mass center G. Thus

$$\Sigma \overline{\mathbf{M}} = \dot{\overline{\mathbf{H}}} \tag{98}$$

which proves that the moment sum may always be taken about the mass center regardless of the motion.

▼ ▼ ▼ ▼ ▼

A more general formulation of the angular momentum relations is obtained by considering the reference axes of Fig. 41 to be translating with a velocity $\mathbf{v}_O$ and an acceleration $\mathbf{a}_O$ of the origin O. The absolute velocity of the representative particle is, then, $\mathbf{v}_i = \mathbf{v}_O + \dot{\mathbf{r}}_i$, so that the time derivative of the absolute angular momentum $\mathbf{H}_O = \Sigma(\mathbf{r}_i \times m_i\mathbf{v}_i)$ about O becomes

$$\dot{\mathbf{H}}_O = \Sigma(\dot{\mathbf{r}}_i \times m_i\mathbf{v}_O) + \Sigma(\dot{\mathbf{r}}_i \times m_i\dot{\mathbf{r}}_i) + \Sigma(\mathbf{r}_i \times m_i\mathbf{a}_i)$$

The first summation becomes

$$-\mathbf{v}_O \times \Sigma m_i\dot{\mathbf{r}}_i = -\mathbf{v}_O \times \frac{d}{dt}(m\bar{\mathbf{r}}) = -\mathbf{v}_O \times m(\bar{\mathbf{v}} - \mathbf{v}_O) = -\mathbf{v}_O \times m\bar{\mathbf{v}}$$

The second summation is identically zero, and the third term is the summation $\Sigma\mathbf{M}_O$ of the moments about O of all external forces acting on the system. Thus

$$\Sigma\mathbf{M}_O = \dot{\mathbf{H}}_O + \mathbf{v}_O \times m\bar{\mathbf{v}} \tag{99}$$

It may now be observed that this more general moment equation reduces to the form of Eq. 97 under any one of the following three conditions.

$$\blacktriangleright \quad \Sigma\mathbf{M}_O = \dot{\mathbf{H}}_O \text{ if } \begin{cases} 1.\ \mathbf{v}_O = \mathbf{0} \\ 2.\ \bar{\mathbf{v}} = \mathbf{0} \\ 3.\ \mathbf{v}_O \text{ and } \bar{\mathbf{v}} \text{ are parallel} \end{cases} \tag{99a}$$

It is also useful to express the momentum relations in terms of the angular momentum *relative* to the translating axes $\mathbf{H}_{O_r} = \Sigma(\mathbf{r}_i \times \mathbf{m}_i\dot{\mathbf{r}}_i)$ where the relative velocity $\dot{\mathbf{r}}_i$ of m_i with respect to the moving point O is used in place of the absolute velocity. The absolute acceleration of m_i is $\mathbf{a}_i = \mathbf{a}_O + \ddot{\mathbf{r}}_i$, so that the time derivative of the relative angular momentum becomes

$$\dot{\mathbf{H}}_{O_r} = \Sigma(\dot{\mathbf{r}}_i \times m_i\dot{\mathbf{r}}_i) + \Sigma(\mathbf{r}_i \times m_i\mathbf{a}_i) - \Sigma(\mathbf{r}_i \times m_i\mathbf{a}_O)$$

The first summation is identically zero, and the second summation is the moment sum $\Sigma\mathbf{M}_O$ about O of all external forces on the system. The third summation may be reduced to $\Sigma(\mathbf{r}_i \times m_i\mathbf{a}_O) = -\mathbf{a}_O \times \Sigma m_i\mathbf{r}_i = -\mathbf{a}_O \times m\bar{\mathbf{r}}$. Thus

$$\Sigma\mathbf{M}_O = \dot{\mathbf{H}}_{O_r} + \bar{\mathbf{r}} \times m\mathbf{a}_O \tag{100}$$

From this relation it is observed that the form of Eq. 97 holds under any one of the following three conditions

$$\blacktriangleright \quad \Sigma\mathbf{M}_O = \dot{\mathbf{H}}_{O_r} \text{ if } \begin{cases} 1.\ \mathbf{a}_O = \mathbf{0} \\ 2.\ \bar{\mathbf{r}} = \mathbf{0} \text{ so that } \Sigma\overline{\mathbf{M}} = \dot{\overline{\mathbf{H}}}_r \\ 3.\ \mathbf{a}_O \text{ and } \bar{\mathbf{r}} \text{ are parallel } (\mathbf{a}_O \text{ directed} \\ \quad \text{toward or away from } G) \end{cases} \tag{100a}$$

Comparison of the second condition with Eq. 98 yields the conclusion that $\overline{\mathbf{H}}_r = \overline{\mathbf{H}}$ which is easily proved directly if desired.

It is frequently convenient to write the expression for angular momentum about some particular point in terms of the angular momentum about the mass center. The absolute angular momentum about O may be written $\mathbf{H}_O = \Sigma([\bar{\mathbf{r}} + \boldsymbol{\rho}_i] \times m_i\mathbf{v}_i)$. The first term is $\bar{\mathbf{r}} \times \Sigma m_i\mathbf{v}_i = \bar{\mathbf{r}} \times m\bar{\mathbf{v}}$, and the second term is $\Sigma(\boldsymbol{\rho}_i \times m_i\mathbf{v}_i) = \overline{\mathbf{H}}$. Thus

$$\mathbf{H}_O = \overline{\mathbf{H}} + \bar{\mathbf{r}} \times m\bar{\mathbf{v}}$$

Similarly, the relative angular momentum is $\mathbf{H}_{O_r} = \Sigma([\bar{\mathbf{r}} + \boldsymbol{\rho}_i] \times m_i\dot{\mathbf{r}}_i)$. The first term is $\bar{\mathbf{r}} \times \Sigma m_i\dot{\mathbf{r}}_i = \bar{\mathbf{r}} \times m\dot{\bar{\mathbf{r}}} = \bar{\mathbf{r}} \times m\bar{\mathbf{v}}_r$, and the second term is $\overline{\mathbf{H}}_r$, so that

$$\mathbf{H}_{O_r} = \overline{\mathbf{H}}_r + \bar{\mathbf{r}} \times m\bar{\mathbf{v}}_r$$

which has the same form as that for the absolute angular momentum. It is noted, therefore, that the angular momentum (absolute or relative) about any point O is the corresponding angular momentum about the mass center plus the moment about O of the linear momentum (absolute or relative) treated as a vector passing through the mass center.

26 Conservation of Energy and Momentum. A mass system is said to be *conservative* if it does not lose energy by virtue of internal friction forces which do negative work or by virtue of nonelastic members which dissipate energy upon cycling. If no work is done on a conservative system during an interval of motion by external forces (other than gravity or other potential forces), then no part of the energy of the system is lost. In such a case Eq. 57 gives

$$\blacktriangleright \qquad \Delta T + \Delta V_e + \Delta V_g = 0 \qquad \textbf{(101)}$$

which expresses the *law of conservation of dynamical energy*. This law holds only in the ideal case where internal kinetic friction is sufficiently small to be neglected.

If, for a certain interval of time, the resultant external force $\Sigma\mathbf{F}$ acting on a conservative or nonconservative mass system is zero, Eq. 96 requires that $\dot{\mathbf{G}} = \mathbf{0}$, so that during this interval

$$\blacktriangleright \qquad \Delta\mathbf{G} = \mathbf{0} \qquad \textbf{(102)}$$

which expresses the *principle of conservation of linear momentum.* Thus in the absence of an external impulse the linear momentum of a system remains unchanged.

Similarly, if the resultant moment about a fixed point O or about the mass center G of all external forces on any mass system is zero, Eq. 97 or 98 requires, respectively, that

$$\blacktriangleright \qquad \Delta\mathbf{H}_O = \mathbf{0} \quad \text{or} \quad \Delta\overline{\mathbf{H}} = \mathbf{0} \qquad \textbf{(103)}$$

These relations express the *principle of conservation of angular momentum* for a general mass system in the absence of an angular impulse. Thus, if there is no angular impulse about a fixed point (or about the mass center), the angular momentum of the system about the fixed point (or about the mass center) remains unchanged. Either equation may hold without the other.

Equations 92 through 103 are among the most important of the basic derived laws of mechanics. In this chapter these laws have been derived for the most general system of constant mass in order that the generality of the laws will be established. These laws find repeated use when applied to specific mass systems such as rigid and nonrigid solids and certain fluid systems which are discussed in the chapters that follow. The reader is urged to study these laws carefully and to compare them with their more restricted forms encountered earlier in Chapter 3 on the kinetics of particles.

II DYNAMICS OF RIGID BODIES

5 PLANE KINEMATICS OF RIGID BODIES

27 Introduction. The description of the motion of rigid bodies is useful in two important ways. First, it is frequently necessary to generate, transmit, or control certain desired motions by the use of cams, gears, and linkages of various types. Here the description of the motion is necessary to determine the design geometry of the mechanical linkage. As a result of the motion generated, forces are frequently developed which must be accounted for in the design of the linkage. Second, it is often necessary to determine the motion of a rigid body resulting from the forces applied to it. Calculation of the trajectory of a rocket under the influence of its jet thrust and gravitational attraction is an example of such a problem. In both situations it is necessary to have command of the principles of rigid-body kinematics before a determination may be made of the accompanying forces. In this chapter the kinematics of motion which may be analyzed as taking place in a single plane is developed. In Chapter 7 the kinematics of motion which requires three spatial coordinates for its description is covered.

A rigid body was defined in the previous chapter as a system of particles for which the distances between the particles remain unchanged. Thus if each particle of such a body is located from reference axes attached to and rotating with the body, there would be no change in any position vector as measured from these axes. This formulation is, of course, an ideal one since all solid materials change shape to some extent when forces are applied to them. Nevertheless, if the movements associated with the changes in shape are very small compared with the overall movements of the body as a whole, then the ideal concept of rigidity is quite acceptable. As mentioned in Chapter 1 the displacements due to the flutter of an aircraft wing, for instance, are of no consequence in the description of the total flight path of the aircraft for which the assumption of a rigid body is clearly in order. On the other hand, if the problem is one of describing, as a function of time, the internal wing stress due to wing flutter, then the relative motions of portions of the wing become of prime importance and cannot be neglected. In this case the wing may not be considered a rigid body. In the present chapter and in the three that follow essentially all of the material is based upon the assumption of rigidity. In Chapter 10 an introduction to the dynamics of nonrigid systems is given.

The plane motion of a rigid body may be divided into several categories as represented in Fig. 42. *Translation* is defined as any motion in which every line fixed in the body remains parallel to its original position at all times.

Rectilinear translation, part *a*, is translation in which all points in the body move in straight lines. *Curvilinear translation,* part *b*, is translation in which all points move on congruent curves. In curvilinear translation there is *no rotation of any line in the body.* It should be noted that in each case of translation the motion of the body is completely specified by the motion of any point in the body, since all points have the same motion.

Rotation about a fixed axis, part *c*, is the angular motion about the axis. It follows that all particles move in circular paths about the axis of rotation, and all lines in the body (including those that do not pass through the axis) rotate through the same angle in the same time.

General plane motion of a rigid body, part *d*, is a combination of translation and rotation.

In each of the examples cited all particles in the body move in parallel planes. The motion, however, is represented by its projection onto a single plane parallel to the motion called the *plane of motion.* This plane is usually considered as passing through the center of mass of the body.

The angular displacement of a rigid body in plane motion will now be considered. Figure 43 shows a rigid body which undergoes plane motion in the plane of the figure. The angular positions of any two lines 1 and 2 attached to the body are specified by θ_1 and θ_2 measured from any fixed reference direction which is convenient. Since the angle β is invariant, the relation $\theta_2 = \theta_1 + \beta$ gives $\dot{\theta}_2 = \dot{\theta}_1$ and $\ddot{\theta}_2 = \ddot{\theta}_1$, or, during a finite interval $\Delta\theta_2 = \Delta\theta_1$. Thus all lines in a rigid body in its plane of motion have the same angular

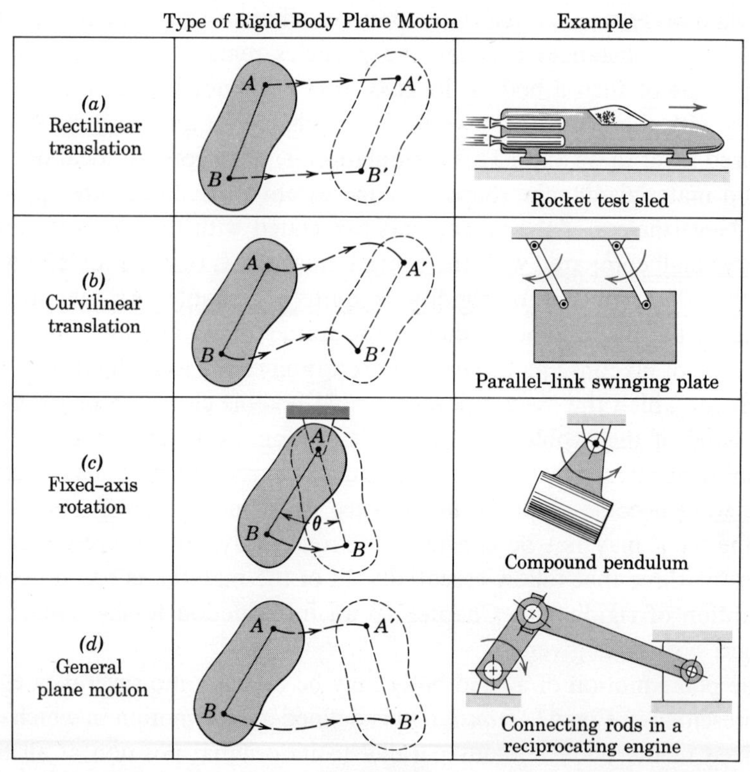

Figure 42

displacement, the same angular velocity, and the same angular acceleration. It should be noted that the angular motion of a line depends only on its angular displacement with respect to any arbitrary fixed reference and on the time derivatives of the displacement. Angular motion does not require the presence of a fixed axis normal to the plane of motion about which the line and the body rotate.

Determination of the plane motion of rigid bodies is accomplished by either a direct calculation of the absolute displacements and their time derivatives from the absolute geometry involved or else by utilizing the principles of relative motion with translating or rotating reference axes. Each method is important and useful and will be taken up in turn.

28 Absolute Motion. The determination of velocities and accelerations, both linear and angular, in rigid-body plane motion by direct differentiation of the equations for displacements is a straightforward and direct approach. The selection of this method in preference to solution by relative-motion methods will depend mainly on the relative simplicity of the geometry involved. Choice is best indicated after experience has been acquired with both approaches.

In the absolute-motion analysis of plane motion use is made of the differential relations, Eqs. 4, 5, and 6, developed in Art. 10 for the rectilinear motion of particles, and Eqs. 7, developed in Art. 11 for the rotation of lines.

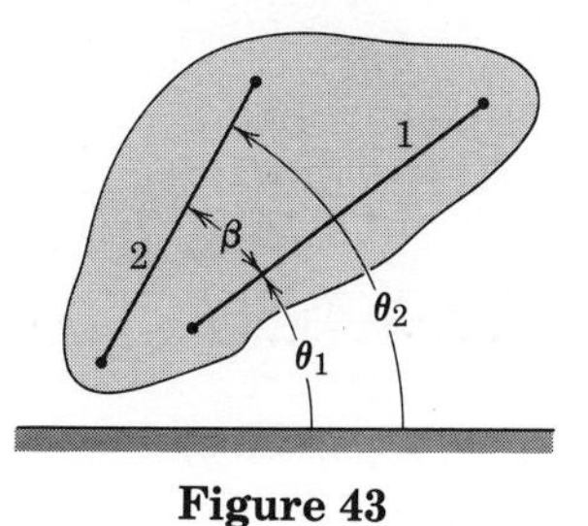

Figure 43

Sample Problems

5/1 A wheel of radius r rolls on a flat surface without slipping. Determine the angular motion of the wheel in terms of the linear motion of its center O. Also determine the acceleration of a point on the rim of the wheel as the point comes into contact with the surface upon which the wheel rolls.

Solution. The figure shows the wheel rolling to the right from the dotted to the full position without slipping. The linear displacement of the center O is s, which is also the arc length $C'A$ along the rim upon which the wheel rolls. The radial line CO rotates through the angle θ, where θ is measured from the vertical direction. The displacement relationship and its two time derivatives give

$$s = r\theta$$

$$v_O = r\omega \qquad \textit{Ans.}$$

$$a_O = r\alpha$$

where $v_O = \dot{s}$, $a_O = \dot{v}_O = \ddot{s}$, $\omega = \dot{\theta}$, and $\alpha = \dot{\omega} = \ddot{\theta}$. (The relations between the linear motion of the center O of a rolling wheel and the angular motion of the wheel find repeated use and should be learned thoroughly.) The acceleration a_O will be directed in the sense opposite to that of v_O if the wheel is slowing down. In this event the angular acceleration α will be directed oppositely to ω.

The origin of fixed coordinates is taken arbitrarily but conveniently at the point of contact between C on the rim of the wheel and the ground. When point C has moved along its cycloidal path to C', its coordinates and their time derivatives become

$$x = s - r\sin\theta = r(\theta - \sin\theta) \qquad y = r - r\cos\theta = r(1 - \cos\theta)$$

$$\dot{x} = r\dot{\theta}(1 - \cos\theta) = v_O(1 - \cos\theta) \qquad \dot{y} = r\dot{\theta}\sin\theta = v_O\sin\theta$$

$$\ddot{x} = \dot{v}_O(1 - \cos\theta) + v_O\dot{\theta}\sin\theta \qquad \ddot{y} = \dot{v}_O\sin\theta + v_O\dot{\theta}\cos\theta$$

$$= a_O(1 - \cos\theta) + r\omega^2\sin\theta \qquad = a_O\sin\theta + r\omega^2\cos\theta$$

For the desired instant of contact, $\theta = 0$ and

$$\ddot{x} = 0 \quad \text{and} \quad \ddot{y} = r\omega^2 \qquad \textit{Ans.}$$

Thus the acceleration of the point C on the rim at the instant of contact with the ground depends only on r and ω and is directed toward the center of the wheel. If desired, the velocity and acceleration of C at any position θ may be obtained by writing the expressions $\mathbf{v} = \mathbf{i}\dot{x} + \mathbf{j}\dot{y}$ and $\mathbf{a} = \mathbf{i}\ddot{x} + \mathbf{j}\ddot{y}$.

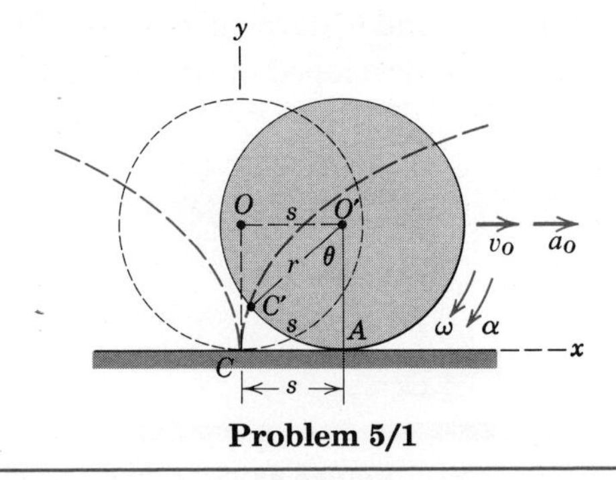

Problem 5/1

5/2 Motion of the equilateral triangular plate ABC in its plane is controlled by the hydraulic cylinder D. If the piston rod in the cylinder is moving upward at the constant rate of 0.3 m/s during an interval of its motion, calculate for the instant when $\theta = 30°$ the velocity and acceleration of the center of the roller B in the horizontal guide and the angular velocity and angular acceleration of edge CB.

Solution. With the x-y coordinates chosen as shown the given motion of A is $v_A = \dot{y} = 0.3$ m/s and $a_A = \ddot{y} = 0$. The accompanying motion of B is given by x and its time derivatives which may be obtained from $x^2 + y^2 = b^2$. Differentiating gives

$$x\dot{x} + y\dot{y} = 0 \qquad \dot{x} = -\frac{y}{x}\dot{y}$$

$$x\ddot{x} + \dot{x}^2 + y\ddot{y} + \dot{y}^2 = 0 \qquad \ddot{x} = -\frac{\dot{x}^2 + \dot{y}^2}{x} - \frac{y}{x}\ddot{y}$$

With $y = b\sin\theta$, $x = b\cos\theta$, and $\ddot{y} = 0$ the expressions become

$$v_B = \dot{x} = -v_A\tan\theta$$

$$a_B = \ddot{x} = -\frac{v_A{}^2}{b}\sec^3\theta$$

Substituting the numerical values $v_A = 0.3$ m/s and $\theta = 30°$ gives

$$v_B = -0.3\left(\frac{1}{\sqrt{3}}\right) = -0.1732 \text{ m/s} \qquad \textit{Ans.}$$

$$a_B = -\frac{(0.3)^2(2/\sqrt{3})^3}{0.2} = -0.693 \text{ m/s}^2 \qquad \textit{Ans.}$$

The negative signs indicate that the velocity and the acceleration of B are both to the right since x and its derivatives are positive to the left.

The angular motion of CB is the same as that of every line on the plate including AB. Differentiating $y = b \sin\theta$ gives

$$\dot{y} = b\dot{\theta}\cos\theta, \qquad \omega = \dot{\theta} = \frac{v_A}{b}\sec\theta$$

The angular acceleration is

$$\alpha = \dot{\omega} = \frac{v_A}{b}\dot{\theta}\sec\theta\tan\theta = \frac{v_A{}^2}{b^2}\sec^2\theta\tan\theta$$

Substitution of the numerical values gives

$$\omega = \frac{0.3}{0.2}\frac{2}{\sqrt{3}} = 1.732 \text{ rad/s} \qquad \textit{Ans.}$$

$$\alpha = \frac{(0.3)^2}{(0.2)^2}\left(\frac{2}{\sqrt{3}}\right)^2\frac{1}{\sqrt{3}} = 1.732 \text{ rad/s}^2 \qquad \textit{Ans.}$$

Both ω and α are counterclockwise since their signs are positive in the sense of positive measurement of θ.

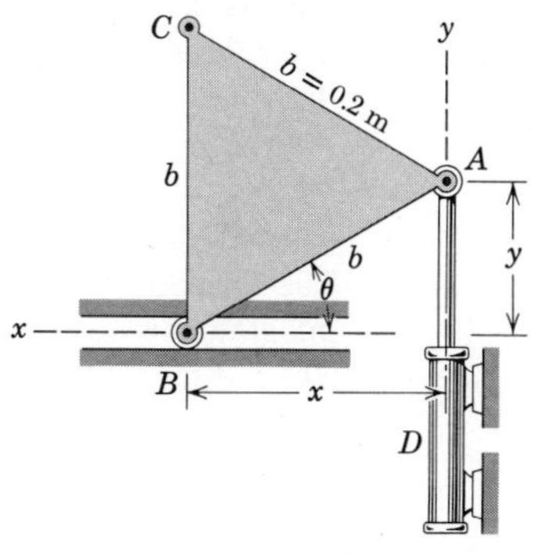

Problem 5/2

Problems

5/3 The load L is hoisted by the pulley-and-cable combination shown. If the system starts from rest and the upper cable acquires a velocity $v =$ 4 m/s with constant acceleration when the load is 6 m above its starting position, calculate the acceleration of the load and find its velocity at this instant.

Ans. $a_L = 0.0208 \text{ m/s}^2$, $v_L = 0.5$ m/s

Problem 5/3

Problem 5/4

5/4 The cables at A and B are wrapped securely around the rims and the hub of the integral pulley as shown. If the cables at A and B are given upward velocities of 4 m/s and 3 m/s, respectively, calculate the velocity of the center O and the angular velocity of the pulley.

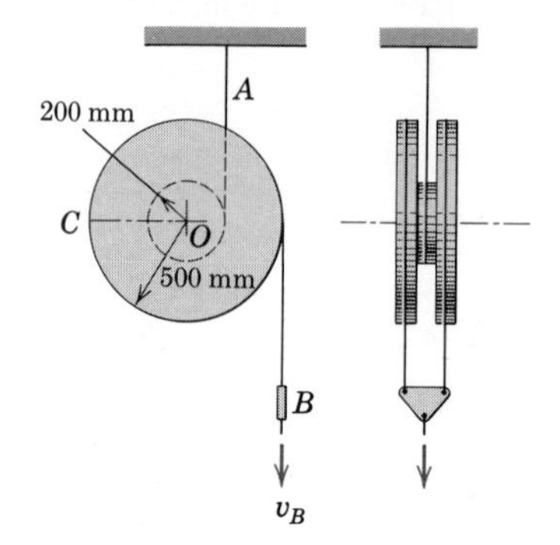

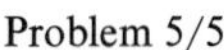
Problem 5/5

5/5 The spool rolls on its hub up the inner cable A as the equalizer plate B pulls the outer cables down. The three cables are wrapped securely around their respective peripheries and do not slip. If, at the instant represented, B has moved down a distance of 800 mm from rest with a constant acceleration of 100 mm/s^2, determine the velocity of point C and the acceleration of the center O for this particular instant.

Ans. $a_O = 66.7$ mm/s^2
$v_C = 933$ mm/s

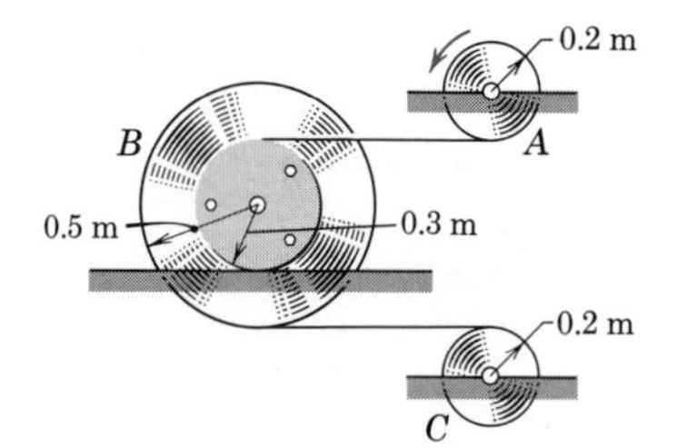

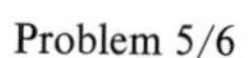
Problem 5/6

5/6 The cable from drum A turns the double wheel B which rolls on its hubs without slipping. Determine the angular velocity ω and the angular acceleration α of the drum C at the instant when the angular velocity and angular acceleration of A are 4 rad/s and 3 rad/s^2, respectively, both in the counterclockwise direction.

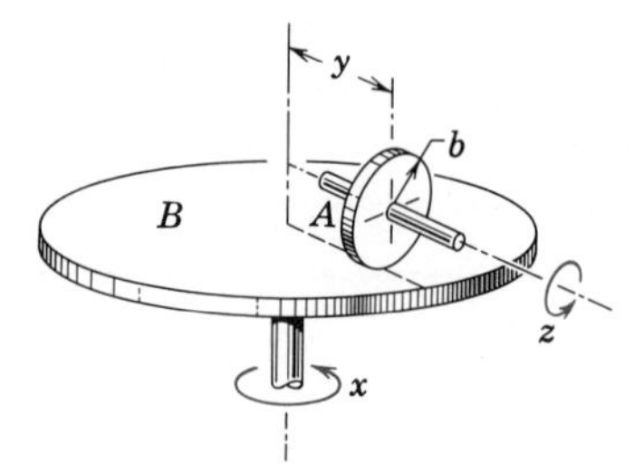

Problem 5/7

5/7 The elements of a wheel-and-disk mechanical integrator are shown in the figure. The integrator wheel A turns about its fixed shaft and is driven by friction from disk B with no slipping occurring tangent to its rim. The distance y is a variable and can be controlled at will. Show that the angular displacement of the integrator wheel is given by $z = (1/b) \int y\, dx$, where x is the angular displacement of the disk B.

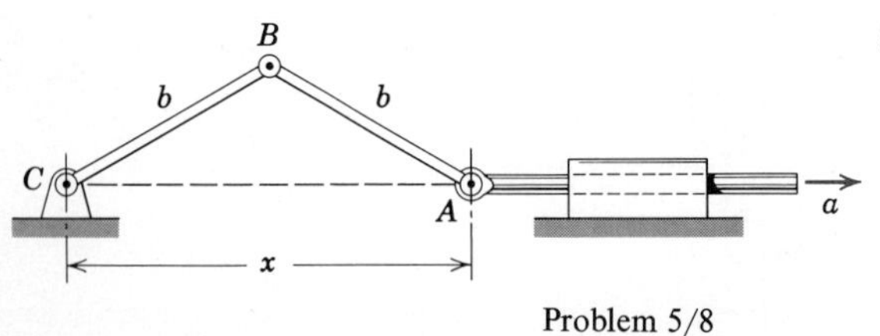

Problem 5/8

5/8 Point A is given a constant acceleration a to the right starting from rest with x essentially zero. Determine the angular velocity ω of link AB in terms of x and a.

Ans. $\omega = \dfrac{\sqrt{2ax}}{\sqrt{4b^2 - x^2}}$ counterclockwise

5/9 Calculate the angular velocity ω of the slender bar AB as a function of the distance x and the constant angular velocity ω_0 of the drum.

5/10 A reel of film is being unwound by pulling the film through the guide rollers with a constant velocity v. The reel speeds up as its radius diminishes. Determine the angular acceleration α of the reel if the thickness of the film is t and the diminishing radius of the roll of film is r. The distance s is very large compared with r.

$$\textit{Ans. } \alpha = \frac{tv^2}{2\pi r^3}$$

5/11 The circular cam is mounted eccentrically about its fixed bearing at O and turns counterclockwise at the constant angular velocity ω. The cam causes the fork A and attached control rod to oscillate in the horizontal x-direction. Write the expressions for the velocity v_x and acceleration a_x of the control rod in terms of the angle θ measured from the vertical. The contact surfaces of the fork are vertical.

5/12 The angular motion of the hinged block OA is controlled by the horizontal motion of the rod BC in its guide. If BC has a velocity v to the right, determine the expression for the corresponding angular velocity ω of OA in terms of the distance x.

$$\textit{Ans. } \omega = \frac{rv}{x\sqrt{x^2 - r^2}}$$

5/13 The slotted arm OA rotates with a constant angular velocity $\omega = \dot{\theta}$ during a limited interval of its motion and moves the pivoted slider block along the horizontal slot. Write the expressions for the velocity v_B and acceleration a_B of the pin B in the slider block in terms of θ.

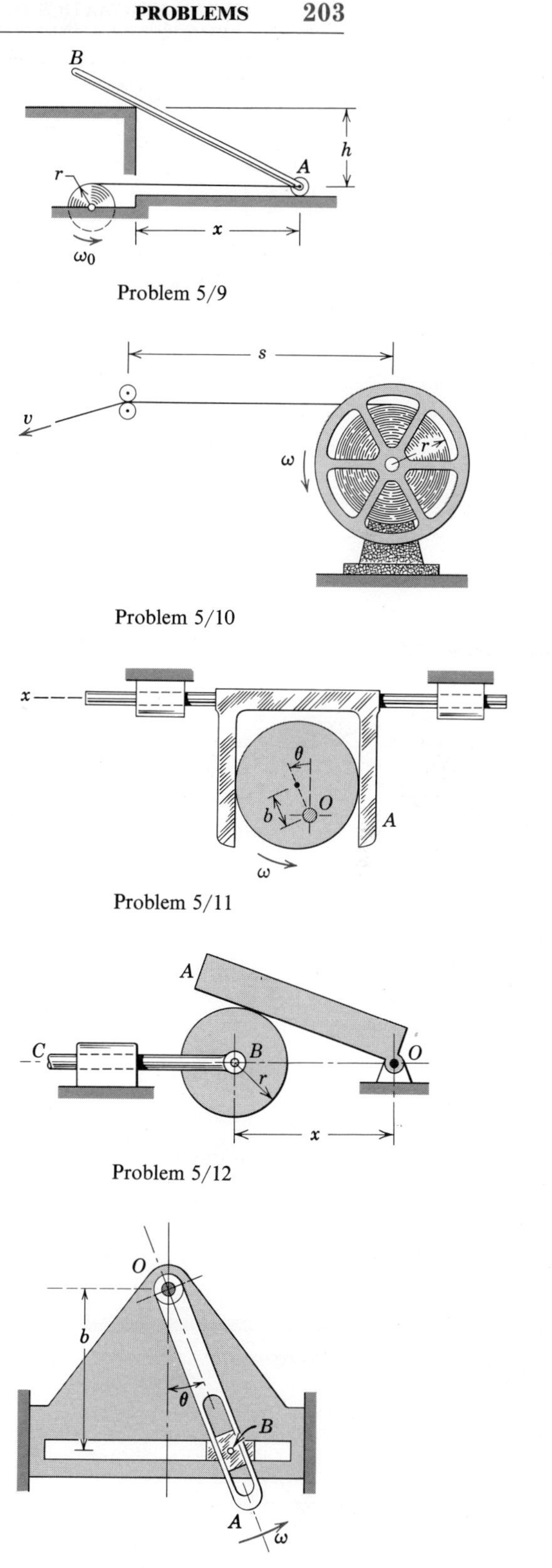

Problem 5/9

Problem 5/10

Problem 5/11

Problem 5/12

Problem 5/13

Problem 5/14

5/14 The triangular plate rotates about O under the action of the pivoted hydraulic cylinder and its piston rod. If the piston moves so that the distance between A and C is increasing at the constant rate of 0.15 m/s, calculate the normal component of the acceleration of point B in its circular motion about O as the plate passes the position shown. *Ans.* $(a_B)_n = 0.30\ \text{m/s}^2$

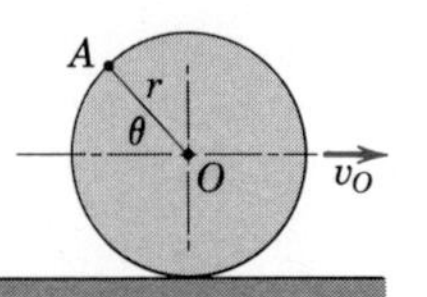

Problem 5/15

5/15 The wheel rolls to the right without slipping, and its center O has a constant velocity v_O. Determine the velocity v and acceleration a of a point A on the rim of the wheel in terms of the angle θ measured clockwise from the horizontal.

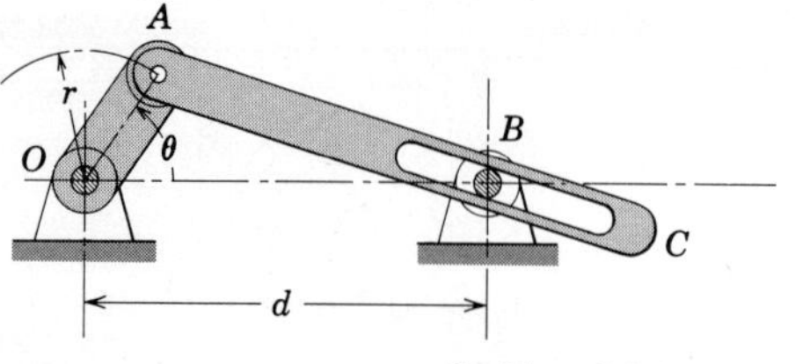

Problem 5/16

5/16 The crank OA revolves counterclockwise at the constant angular rate $\omega_0 = \dot{\theta}$. Derive the expression for the angular velocity ω of the link AC in terms of θ.

$$Ans.\ \omega = r\omega_0 \frac{d \cos \theta - r}{d^2 - 2rd \cos \theta + r^2}$$

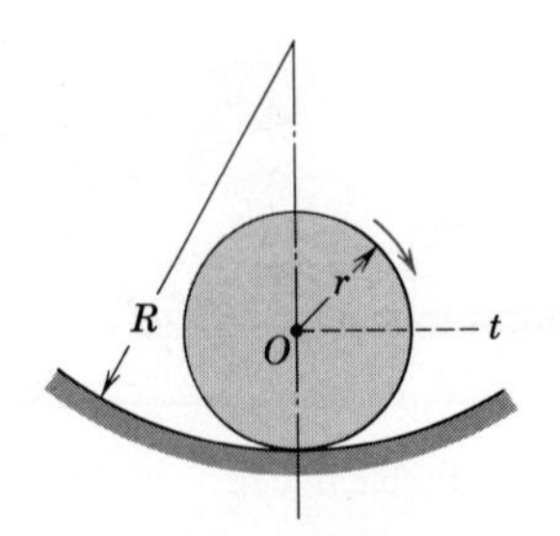

Problem 5/17

5/17 Show that the expressions $v = r\omega$ and $a_t = r\alpha$ hold for the motion of the center O of the wheel which rolls on the circular arc, where ω and α are the absolute angular velocity and acceleration, respectively, of the wheel.

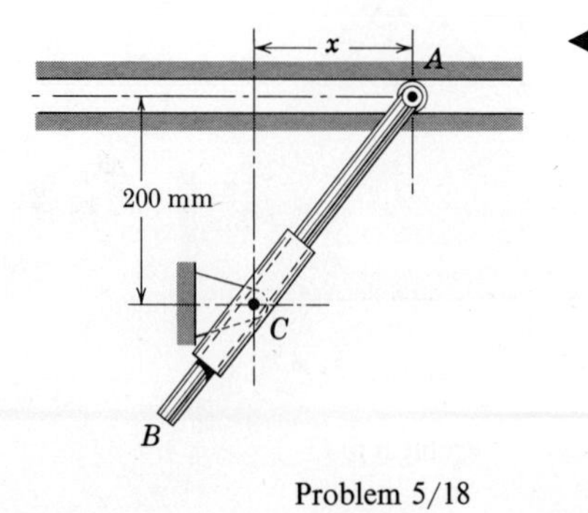

Problem 5/18

◀ **5/18** The rod AB slides through the pivoted collar as end A is moved along the slot. If A starts from rest at $x = 0$ and moves to the right with a constant acceleration of 100 mm/s², calculate the angular acceleration α of AB at the instant when $x = 150$ mm.

Ans. $\alpha = 0.1408\ \text{rad/s}^2$ counterclockwise

◀**5/19** One of the most common mechanisms is the slider-crank. Express the angular velocity ω and angular acceleration α of the connecting rod AB in terms of the crank angle θ for a given constant crank speed ω_0. Take ω and α to be positive counterclockwise.

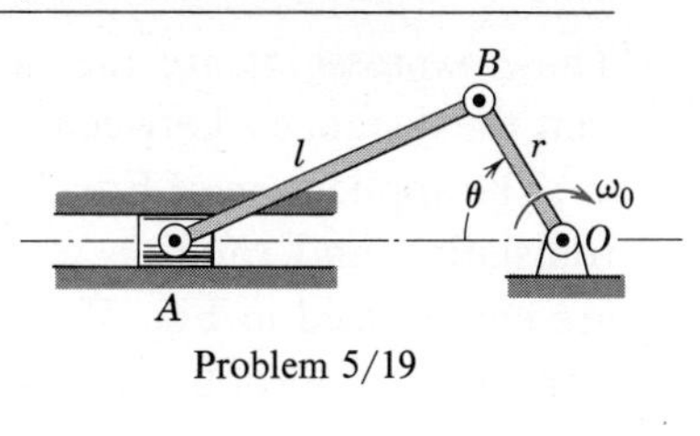

Problem 5/19

Ans. $\omega = \dfrac{r\omega_0}{l}\dfrac{\cos\theta}{\sqrt{1-\dfrac{r^2}{l^2}\sin^2\theta}}$

$$\alpha = \frac{r\omega_0^2}{l}\sin\theta\frac{\dfrac{r^2}{l^2}-1}{\left(1-\dfrac{r^2}{l^2}\sin^2\theta\right)^{3/2}}$$

29 Relative Motion; Translating Axes.

29 Relative Motion; Translating Axes. In Part A of Art. 13 in Chapter 2 the principles of relative motion were developed and applied to the motions of two points A and B. If these two points are fixed in a given rigid body in the plane of motion, then the distance between them remains constant. Consequently the motion of one point observed from a translating coordinate system attached to the other appears to be circular. This fact is seen from Fig. 44, where the rigid body AB moves to $A'B'$. This movement may be considered as occurring in two parts. First, the body translates to the parallel position $A''B'$ with the displacement $\Delta\mathbf{r}_B$. Second, the body rotates about B' through the angle $\Delta\theta$. From the nonrotating reference axes x'-y' attached to the reference point B', it is seen that this remaining motion of the body is one of simple rotation about B' giving rise to the displacement $\Delta\mathbf{r}_{A/B}$ of A with respect to B.

The total displacement of A is seen to be

$$\Delta\mathbf{r}_A = \Delta\mathbf{r}_B + \Delta\mathbf{r}_{A/B}$$

where $|\Delta\mathbf{r}_{A/B}|$ may be replaced by $r\,\Delta\theta$ in the limit as $\Delta\theta$ approaches zero. It is noted that the *relative linear motion* $\Delta\mathbf{r}_{A/B}$ is accompanied by the *absolute angular motion* $\Delta\theta$, as seen from the translating axes x'-y'. Dividing the expression for $\Delta\mathbf{r}_A$ by the corresponding time interval Δt, passing to the limit, and then differentiating again, give the relative velocity and relative acceleration equations

$$\mathbf{v}_A = \mathbf{v}_B + \mathbf{v}_{A/B} \quad \text{and} \quad \mathbf{a}_A = \mathbf{a}_B + \mathbf{a}_{A/B} \tag{104}$$

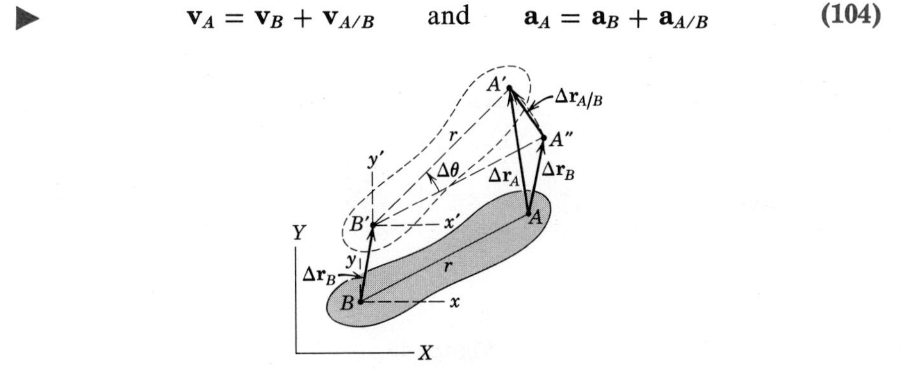

Figure 44

These expressions are the same as Eqs. 19 and 20 with the one restriction that the distance r between A and B remains constant.

The application of Eqs. 104 is clarified by the visualization of the separate translation and rotation components of the equations. These components are emphasized in Fig. 45*a* where the velocity of A is the vector sum of the translational portion $\mathbf{v}_B$ plus the rotational portion $\mathbf{v}_{A/B} = \boldsymbol{\omega} \times \mathbf{r}$ which has the magnitude $v_{A/B} = r\omega$, where $|\boldsymbol{\omega}| = \dot{\theta}$, the *absolute* angular velocity of AB. The fact that the *relative linear velocity* is *always perpendicular* to the line joining the two points in question is an important key to the solution of many problems.

In the case of acceleration, Fig. 45*b*, the relative acceleration term is that arising from the apparent circular motion of point A from nonrotating axes attached to B. The relative acceleration term is, then,

$$\mathbf{a}_{A/B} = (\mathbf{a}_{A/B})_n + (\mathbf{a}_{A/B})_t$$

where the component vectors are

$$(\mathbf{a}_{A/B})_n = \boldsymbol{\omega} \times (\boldsymbol{\omega} \times \mathbf{r})$$

and

$$(\mathbf{a}_{A/B})_t = \boldsymbol{\alpha} \times \mathbf{r}$$

with magnitudes $(a_{A/B})_n = r\omega^2 = v_{A/B}{}^2/r$ toward B and $(a_{A/B})_t = r\alpha$ directed along the tangent to the arc of the relative circle generated by A about B. It should be noted that the angular acceleration $\boldsymbol{\alpha} = \dot{\boldsymbol{\omega}}$ of AB is the *absolute* angular acceleration of the body. Inasmuch as the relative normal

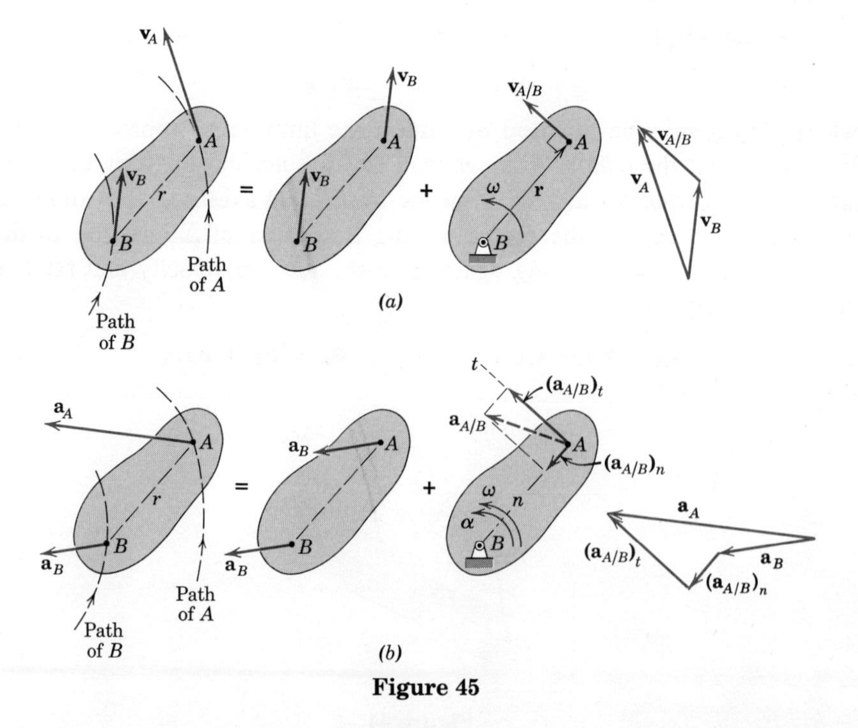

Figure 45

acceleration $(\mathbf{a}_{A/B})_n$ in the acceleration equation involves the angular velocity ω of the body, it follows that it is generally necessary to determine the angular velocity first by solving the relative velocity equation.

In terms of the cross-product equivalences of the relative motion terms, Eqs. 104 may be written alternatively as

$$\mathbf{v}_A = \mathbf{v}_B + \boldsymbol{\omega} \times \mathbf{r} \quad \text{and} \quad \mathbf{a}_A = \mathbf{a}_B + \boldsymbol{\omega} \times (\boldsymbol{\omega} \times \mathbf{r}) + \boldsymbol{\alpha} \times \mathbf{r} \qquad (105)$$

Equations 104 and 105 may be solved graphically or by scalar or vector algebra. In many problems of mechanism analysis the graphical solution is found to be useful. For either the graphical or the algebraic solution, it is helpful to note that the single vector equation in two dimensions is equivalent to two scalar equations, so that two scalar unknowns can be solved for in a two-dimensional problem. The unknowns, for instance, might be the magnitude of one vector and the direction of another. If the equations are solved graphically, the known vectors should be constructed first, and the unknown vectors will be the closing legs to the polygon.

The selection of *B* as the reference point in Figs. 44 and 45 is quite arbitrary, as *A* could have been chosen just as well. The student should draw his own diagram corresponding to Figs. 44 and 45 for *A* as the reference point. It will be seen that the angular velocity ω and angular acceleration α are necessarily unchanged regardless of reference point. The reference point is generally chosen as some point whose acceleration is either known or can be easily found.

The first of Eqs. 104 for relative velocity may often be simplified if the reference point has zero velocity at the instant considered. The relative velocity then becomes the absolute velocity. Figure 46 shows the location of a point *C* which has zero velocity, since it fulfills the condition that an instantaneous point about which a body would momentarily be rotating would be the intersection of two radial lines normal to the velocities of the two points. Point *C* is known as the *instantaneous center of zero velocity*. The triangle CAB may be momentarily considered a rigid extension of the body *AB* and, hence, lines *CA*, *CB*, *AB*, and any other line on the body in the plane of motion have the identical angular velocity $\omega = v_A/r_A = v_B/r_B$. The student should construct the relations which exist when *C* is between *A* and *B* and when *C* is on the line *AB* extended.

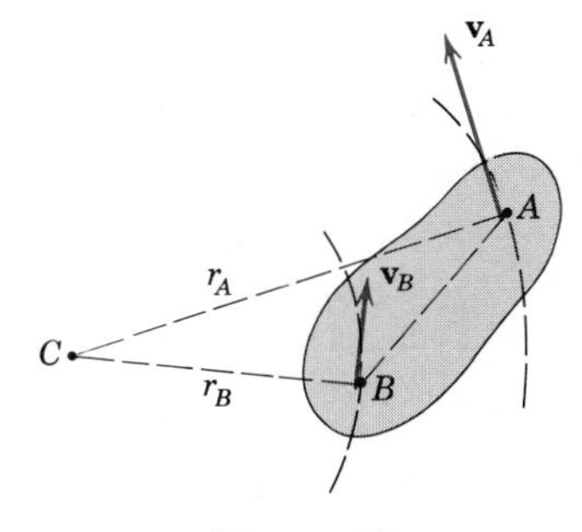

Figure 46

In general, a new instantaneous center C will exist for each new position of the body during its motion. The locus of these centers in space is known as the *space centrode,* and the locus on the body (or body extended) is known as the *body centrode.* During the motion the body centrode curve is seen to roll on the space centrode curve. The absolute velocity of the point, attached to the body, which becomes the instantaneous center at a certain instant is zero at this instant, but its acceleration is *not* zero. Thus the instantaneous center of zero velocity, considered a point attached to the body, *cannot* be used as an instantaneous center of zero acceleration in a manner analogous to its use for finding velocity.

An instantaneous center of zero acceleration exists for bodies in general plane motion. The reader who wishes to pursue the study of this zero-acceleration point may consult references which specialize in mechanism kinematics.

It is evident from the diagram of Fig. 46 that once the instantaneous center C and the angular velocity ω of the body are determined, the velocity of any point on the body is easily constructed in direction since it must be normal to the line joining the point with C. The magnitude of the velocity of the point is simply the radial distance from the point to C times the angular velocity.

Sample Problems

5/20 The wheel of radius r rolls to the left without slipping, and at the instant considered the center O has a velocity $\mathbf{v}_O$ and an acceleration $\mathbf{a}_O$ to the left. Determine the velocity and acceleration of points A and C on the wheel at the instant shown.

Solution. Since the wheel does not slip, point C has zero velocity at the instant of contact with the ground and is therefore the instantaneous center of zero velocity. Thus the velocity of A is

$$[v = r\omega] \qquad v_A = \overline{AC}\omega_{AC} = \frac{\overline{AC}}{r}v_O \qquad \textit{Ans.}$$

The velocity of A may also be found using O as the reference point from the equation

$$\mathbf{v}_A = \mathbf{v}_O + \mathbf{v}_{A/O}$$

where $v_{A/O} = r_O\omega = r_O v_O/r$. The vector addition shown gives the same $\mathbf{v}_A$ as before.

The acceleration of A is given by

$$\mathbf{a}_A = \mathbf{a}_O + \mathbf{a}_{A/O}$$

where the relative acceleration term has the components $(a_{A/O})_n = r_O\omega^2$ directed from A to O and $(a_{A/O})_t = r_O\alpha = r_O(a_O/r)$ directed along t. The addition of the vectors gives $\mathbf{a}_A$, as shown in the upper right-hand diagram.

The acceleration of the instantaneous center C of zero velocity considered a point on the wheel is obtained by the expression

$$\mathbf{a}_C = \mathbf{a}_O + \mathbf{a}_{C/O}$$

where the components of the relative acceleration term are $(a_{C/O})_n = r\omega^2$ directed from C to O and $(a_{C/O})_t = r\alpha$ directed to the right on account of the counterclockwise angular acceleration of line CO about O. The terms are added together in the lower right-hand diagram, and it is seen that

$$a_C = r\omega^2 \qquad \textit{Ans.}$$

Thus the acceleration of C is independent of α and is directed toward the center of the circle. The conclusion is a useful result to remember.

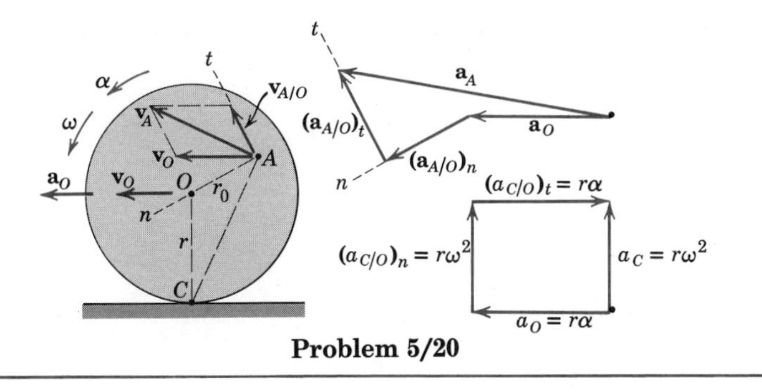

Problem 5/20

5/21 Crank OB of the linkage oscillates about O through a limited arc, causing crank AC to oscillate about C. When the linkage passes the position shown with OB normal to the x-axis and CA normal to the y-axis, the angular velocity of OB is 2 rad/s clockwise and constant. For this instant calculate the angular accelerations of CA and AB.

Solution. The motions of the three links may be described by equating the motion of A in its absolute circular path about C to the motion of A determined from its motion relative to B. The corresponding equations are

$$\mathbf{v}_A = \mathbf{v}_B + \mathbf{v}_{A/B} \qquad \mathbf{a}_A = \mathbf{a}_B + \mathbf{a}_{A/B}$$

The velocity equation may be written as

$$\boldsymbol{\omega}_{CA} \times \mathbf{r}_A = \boldsymbol{\omega}_{OB} \times \mathbf{r}_B + \boldsymbol{\omega}_{AB} \times \mathbf{r}_{A/B}$$

where $\boldsymbol{\omega}_{CA} = \omega_{CA}\mathbf{k}$, $\boldsymbol{\omega}_{OB} = 2\mathbf{k}$ rad/s, $\boldsymbol{\omega}_{AB} = \omega_{AB}\mathbf{k}$, $\mathbf{r}_A = 75\mathbf{i}$ mm, $\mathbf{r}_B = 100\mathbf{j}$ mm, and $\mathbf{r}_{A/B} = -175\mathbf{i} + 100\mathbf{j}$ mm. Substitution gives

$$(\omega_{CA}\mathbf{k} \times 75\mathbf{i}) = (2\mathbf{k} \times 100\mathbf{j}) + \omega_{AB}\mathbf{k} \times (-175\mathbf{i} + 100\mathbf{j})$$

$$75\omega_{CA}\mathbf{j} = -200\mathbf{i} - 175\omega_{AB}\mathbf{j} - 100\omega_{AB}\mathbf{i}$$

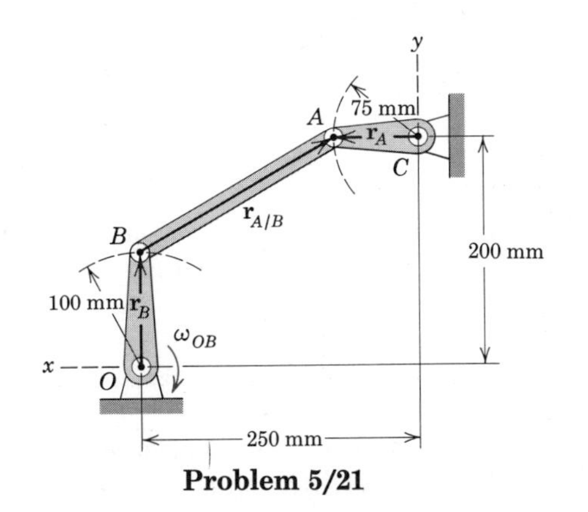

Problem 5/21

Equating the respective coefficients of the **i**- and **j**-terms gives

$$0 = -200 - 100\omega_{AB} \qquad \text{and} \qquad 75\omega_{CA} = -175\omega_{AB}$$

the solutions of which are

$$\omega_{AB} = -2 \text{ rad/s} \qquad \text{and} \qquad \omega_{CA} = 4.67 \text{ rad/s}$$

Since the unit vector **k** points into the paper in the positive z-direction, it is seen that the angular velocity of AB is counterclockwise and that of CA is clockwise.

The acceleration equation is now solved in a similar manner. Its terms are

$$\begin{aligned}
\mathbf{a}_A &= \boldsymbol{\alpha}_{CA} \times \mathbf{r}_A + \boldsymbol{\omega}_{CA} \times (\boldsymbol{\omega}_{CA} \times \mathbf{r}_A) \\
&= \alpha_{CA}\mathbf{k} \times 75\mathbf{i} + 4.67\mathbf{k} \times (4.67\mathbf{k} \times 75\mathbf{i}) \\
&= 75\alpha_{CA}\mathbf{j} - 1633\mathbf{i} \text{ mm/s}^2 \\
\mathbf{a}_B &= \boldsymbol{\alpha}_{OB} \times \mathbf{r}_B + \boldsymbol{\omega}_{OB} \times (\boldsymbol{\omega}_{OB} \times \mathbf{r}_B) \\
&= \mathbf{0} \times 100\mathbf{j} + 2\mathbf{k} \times (2\mathbf{k} \times 100\mathbf{j}) \\
&= -400\mathbf{j} \text{ mm/s}^2 \\
\mathbf{a}_{A/B} &= \boldsymbol{\alpha}_{AB} \times \mathbf{r}_{A/B} + \boldsymbol{\omega}_{AB} \times (\boldsymbol{\omega}_{AB} \times \mathbf{r}_{A/B}) \\
&= \alpha_{AB}\mathbf{k} \times (-175\mathbf{i} + 100\mathbf{j}) + (-2\mathbf{k}) \times [(-2\mathbf{k}) \times (-175\mathbf{i} + 100\mathbf{j})] \\
&= -175\alpha_{AB}\mathbf{j} - 100\alpha_{AB}\mathbf{i} + 700\mathbf{i} - 400\mathbf{j} \text{ mm/s}^2
\end{aligned}$$

Substitution into the acceleration equation and equating the respective coefficients of the **i**- and **j**-terms give

$$-1633 = 700 - 100\alpha_{AB}$$

$$75\alpha_{CA} = -800 - 175\alpha_{AB}$$

the solutions of which are

$$\alpha_{AB} \doteq 23.3 \text{ rad/s}^2 \qquad \text{and} \qquad \alpha_{CA} = -65.1 \text{ rad/s}^2 \qquad \textit{Ans.}$$

Since the unit vector **k** points into the paper in the positive z-direction, it is seen that the angular acceleration of AB is clockwise and that of CA is counterclockwise.

The student may use the instantaneous center of zero velocity of AB as an alternative approach to finding the velocities. He should also visualize the relative acceleration term $\mathbf{a}_{A/B}$ as descriptive of the acceleration which A would appear to have in its circular motion relative to B considered fixed.

5/22 For the slider-crank mechanism shown, the crank OB has a constant clockwise rotational velocity of 1200 rev/min. For the instant when the crank angle is $\theta = 30°$, determine the velocity and acceleration of both the piston A and the center of mass G of the connecting rod. Also find the angular acceleration of the rod AB. Solve graphically.

Solution. The instantaneous center of zero velocity C of AB is located by the intersection of the normals to the known directions of the velocities of the two points A and B on the rod. The radial distances from A, G, and B to C are scaled or computed. The velocity of B in its circular motion about O is

$$[v = r\omega] \qquad v_B = 0.2\frac{(1200)(2\pi)}{60} = 25.1 \text{ m/s}$$

The angular velocity of AB is the same as the angular velocity of triangle CAB considered an extension of the rigid body AB and is

$\left[\omega = \frac{v}{r}\right]$ $\qquad \omega_{AB} = \omega_{CB} = \frac{v_B}{\overline{CB}} = \frac{25.1}{0.566} = 44.4 \text{ rad/s counterclockwise}$

The linear velocities of A and G are, therefore,

$[v = r\omega]$ $\qquad v_A = (0.383)(44.4) = 17.0 \text{ m/s}$ *Ans.*

$\qquad v_G = (0.395)(44.4) = 17.5 \text{ m/s}$ *Ans.*

where $\mathbf{v}_G$ is normal to GC.

The velocities may be found alternatively with the aid of the relative velocity equation

$$\mathbf{v}_A = \mathbf{v}_B + \mathbf{v}_{A/B}$$

The term $\mathbf{v}_B$ is calculated in magnitude as before and laid off to scale as shown in the upper right-hand diagram. Next, the known direction of $\mathbf{v}_{A/B}$, normal to AB, is constructed through the head of $\mathbf{v}_B$. Finally the known direction of $\mathbf{v}_A$ is constructed through the starting point of the vector polygon, and the intersection P of the two lines determines the solution to the equation. The values scaled from the figure are

$$v_A = 17.0 \text{ m/s} \qquad \text{and} \qquad v_{A/B} = 22.2 \text{ m/s}$$

With A considered rotating about B (upper middle diagram), the absolute angular velocity of AB is

$\left[\omega = \frac{v}{r}\right]$ $\qquad \omega_{AB} = \frac{v_{A/B}}{\overline{AB}} = \frac{22.2}{0.5} = 44.4 \text{ rad/s counterclockwise}$

The velocity of G is given by

$$\mathbf{v}_G = \mathbf{v}_B + \mathbf{v}_{G/B}$$

where $v_{G/B} = \overline{GB}\omega_{GB} = \overline{GB}\omega_{AB} = (0.3)(44.4) = 13.32$ m/s. The vector sum is indicated by the dotted vectors on the previous velocity polygon, and $v_G = 17.5$ m/s is scaled from the figure. The student should become thoroughly familiar with both methods of solving for the velocities.

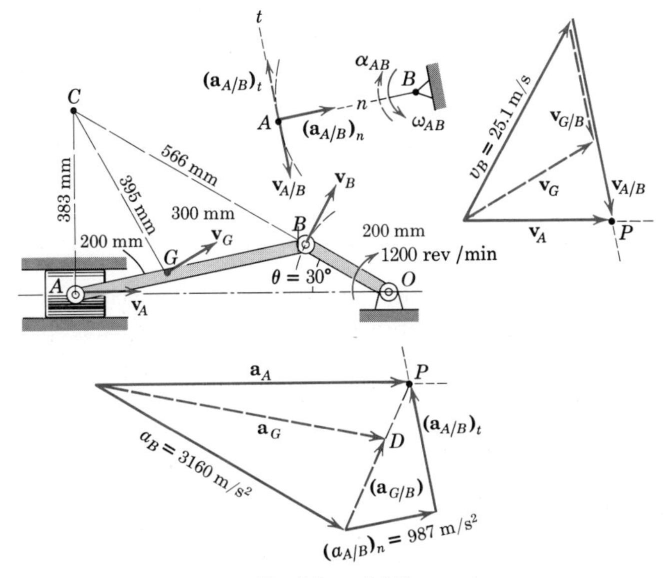

Problem 5/22

The acceleration of A is given by

$$\mathbf{a}_A = \mathbf{a}_B + \mathbf{a}_{A/B}$$

The acceleration of B has only the normal component directed toward O since the angular acceleration of OB is zero. This normal component is

$$[a_n = r\omega^2] \qquad a_B = 0.2\left[\frac{(1200)(2\pi)}{60}\right]^2 = 3160 \text{ m/s}^2$$

and is laid off to scale in the lower diagram. The term $\mathbf{a}_{A/B}$ is determined by considering A to have circular motion about B, as shown in the upper middle diagram. The normal component is directed from A to B and is

$$[a_n = r\omega^2] \qquad (a_{A/B})_n = (\overline{AB})\omega_{AB}{}^2 = (0.5)(44.4)^2 = 987 \text{ m/s}^2$$

This vector is added to $\mathbf{a}_B$. Next the known direction of $(\mathbf{a}_{A/B})_t$ is constructed through the head of $(\mathbf{a}_{A/B})_n$ even though the sense of the vector is unknown. Finally the known direction of $\mathbf{a}_A$, determined by the constraint of the horizontal guide, is constructed through the starting point of the polygon. The point P at the intersection of the last two lines drawn determines the solution to the polygon. By scaling the figure the acceleration of the piston A is found to be

$$a_A = 3420 \text{ m/s}^2 \qquad \textit{Ans.}$$

and the relative tangential acceleration is

$$(a_{A/B})_t = 1415 \text{ m/s}^2$$

The sense of this last vector requires that the angular acceleration of AB, which is

$$\left[\alpha = \frac{a_t}{r}\right] \qquad \alpha_{AB} = \frac{(a_{A/B})_t}{\overline{AB}} = \frac{1415}{0.5} = 2830 \text{ rad/s}^2 \qquad \textit{Ans.}$$

be clockwise, as shown in the upper middle figure by the dotted lines.

The acceleration of G is determined from

$$\mathbf{a}_G = \mathbf{a}_B + \mathbf{a}_{G/B}$$

The two components of the relative acceleration term are each proportional to the distance $\overline{GB}$. Hence $a_{G/B} = (\overline{GB}/\overline{AB})a_{A/B} = 0.6a_{A/B}$. This ratio locates point D on the first acceleration polygon which is also used for the present equation. The magnitude of $\mathbf{a}_G$ is scaled from the polygon and is

$$a_G = 3200 \text{ m/s}^2 \qquad \textit{Ans.}$$

with the direction shown.

The solution to this sample problem embodies most of the major elements of plane relative motion, and thorough mastery will be a great aid in working the problems which follow. Before proceeding the student might consider what modifications to the problem would be required if the crank OB had an angular acceleration in addition to its angular velocity and if the point A were confined to move in an arc of known radius rather than in a straight line.

A graphical solution has been employed in this problem to stress the physical and geometrical relationships that exist. It is equally acceptable to solve the vector equation by resolving it into its scalar components in any two mutually perpendicular directions which are convenient. The two scalar equations which result will permit the solution for two scalar unknowns. If the two equations contain more than two unknowns, then it must be concluded that information necessary to the solution is missing. A solution using the complete notation of vector algebra would require the expression of all vectors in the plane of the motion in terms of their **i**- and **j**-components and the angular velocity and angular acceleration vectors in terms of their **k**-components. This procedure offers no particular advantage here, but for three-dimensional kinematics it is a very useful way in which to account for the more complex geometry.

Problems

5/23 At a particular instant end B of the bar AB has a velocity of 10 m/s in the direction shown. What is the minimum possible velocity which end A can have? What would be the corresponding angular velocity of the bar?

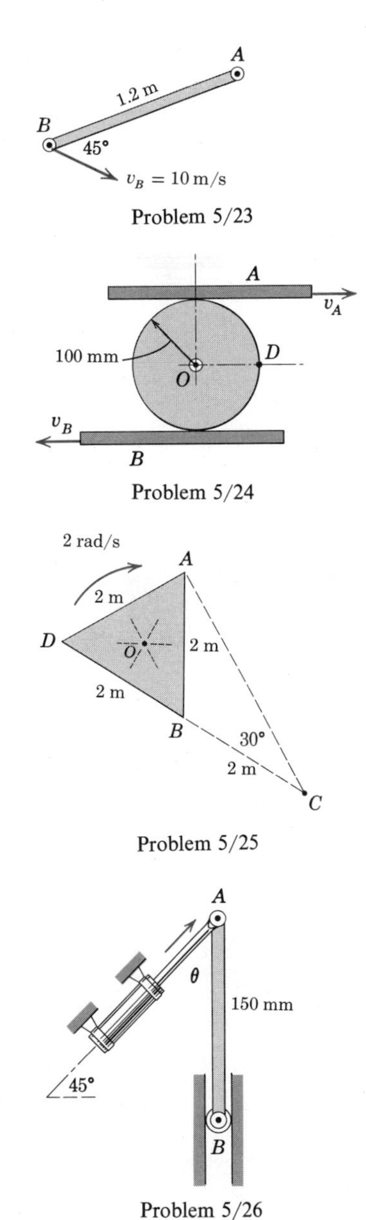

Problem 5/23

Problem 5/24

Problem 5/25

Problem 5/26

5/24 The circular disk rolls without slipping on the two plates A and B, which move parallel to each other but in opposite directions. If $v_A = 2$ m/s and $v_B = 4$ m/s, locate the instantaneous center of zero velocity for the disk and determine the velocity of point D at the instant represented.

Ans. $v_D = 3.16$ m/s

5/25 At the instant represented, the instantaneous center of zero velocity for the plane motion of the equilateral triangular plate is located at C. If the plate has a clockwise angular velocity of 2 rad/s at this instant, determine the corresponding velocity of the center O of the plate.

5/26 The piston rod of the hydraulic cylinder is moving with a constant velocity of 80 mm/s in the direction indicated. For the instant when the rod AB reaches the vertical position with $\theta = 45°$, calculate the acceleration of B.

Ans. $a_B = 21.3$ mm/s^2

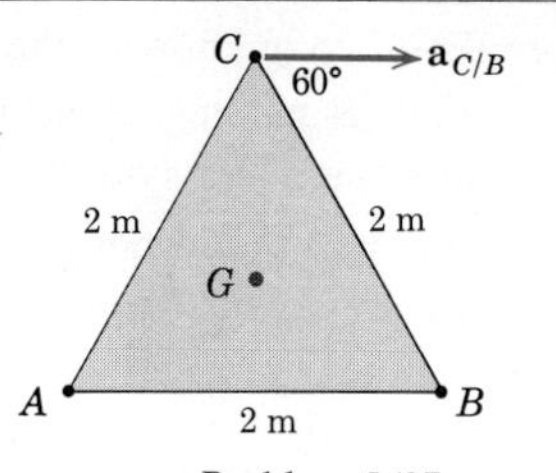

Problem 5/27

5/27 At the instant represented with the plane motion of the triangular plate, the acceleration of C with respect to B is $a_{C/B} = 16\ \text{m/s}^2$ in the direction shown. For this instant determine the angular velocity and angular acceleration of line GA where G is the center of mass of the plate.

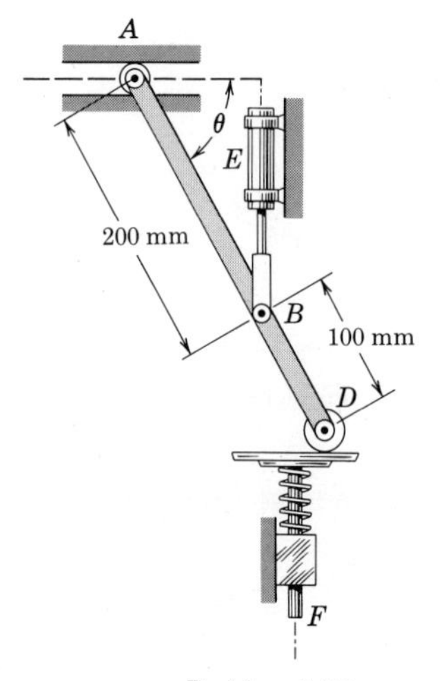

Problem 5/28

5/28 Vertical oscillation of the spring-loaded plunger F is controlled by a periodic change in pressure in the vertical hydraulic cylinder E. For the position $\theta = 60°$ determine the angular velocity of AD and the linear velocity of the roller A in its horizontal guide for a downward velocity of 2 m/s of the plunger F.

Ans. $\omega_{AD} = 13.33$ rad/s
$v_A = 2.31$ m/s

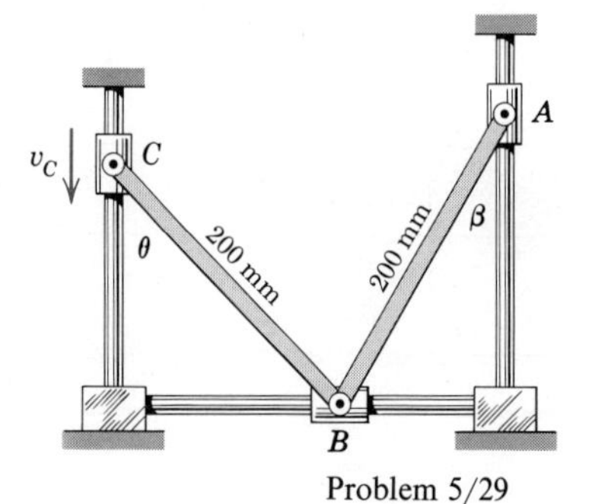

Problem 5/29

5/29 Collars A and C are confined to slide along the vertical rods while collar B slides along the horizontal rod. If C has a downward velocity of 0.2 m/s as it reaches the position for which $\theta = 45°$ and $\beta = 30°$, determine the corresponding angular velocity of AB.

Ans. $\omega_{AB} = 1.155$ rad/s counterclockwise

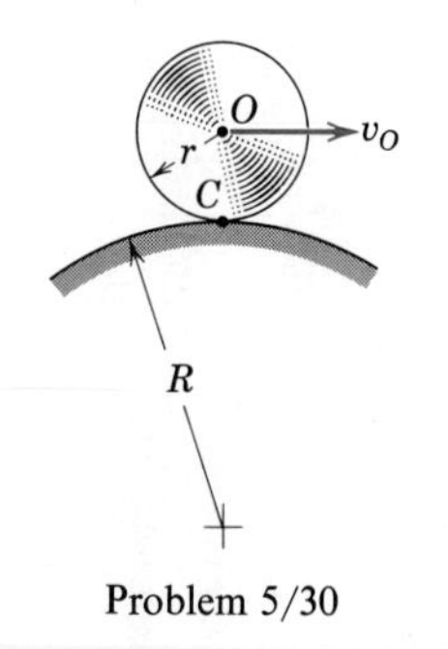

Problem 5/30

5/30 Determine the acceleration of the instantaneous center C of zero velocity for the wheel which rolls over the circular surface with a velocity of constant magnitude v_O of its center.

5/31 The rotation of the gear is controlled by the horizontal motion of end A of the rack AB. If the piston rod has a constant velocity $\dot{x} = 300$ mm/s during a short interval of motion, determine the angular velocity ω_0 of the gear and the angular velocity ω of AB at the instant when $x = 800$ mm.

Ans. $\omega = 0.0968$ rad/s counterclockwise
$\omega_0 = 1.45$ rad/s clockwise

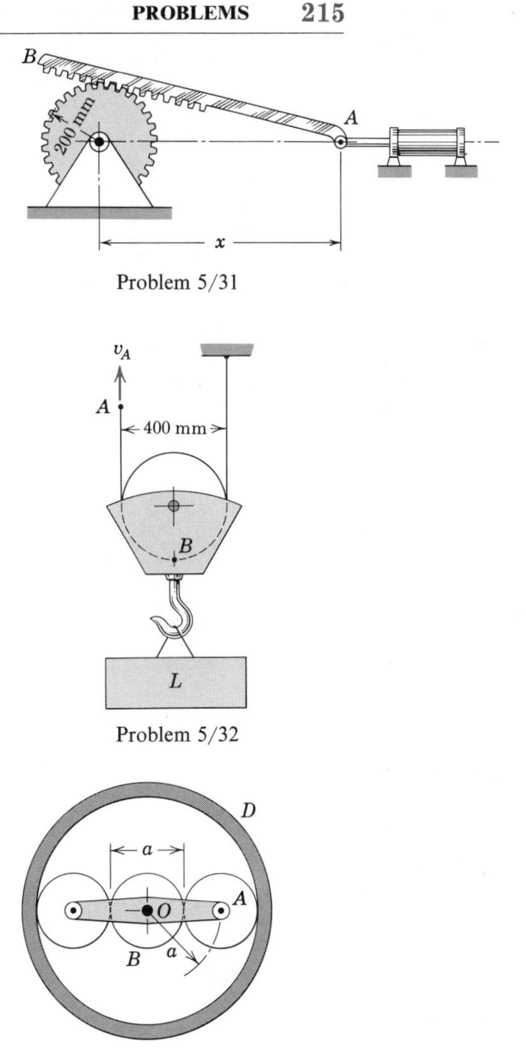

Problem 5/31

Problem 5/32

5/32 At the instant represented the load L has an upward acceleration of 3 m/s^2, and point A of the hoisting cable has a velocity of 1.8 m/s in the same direction. For this instant compute the acceleration of point B on the bottom of the rim of the sheave.

5/33 In Prob. 5/24 if plate B is slowing down at the rate of 8 m/s^2 while plate A maintains its velocity of 2 m/s, calculate the acceleration of the point D at the instant represented.

Ans. $a_D = 86.1$ m/s^2

5/34 The shaft at O drives the arm OA at a clockwise speed of 90 rev/min about the fixed bearing at O. Use the method of the instantaneous center of zero velocity to determine the rotational speed of gear B (gear teeth not shown) if (*a*) ring gear D is fixed and (*b*) ring gear D rotates counterclockwise about O with a speed of 80 rev/min.

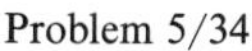

Problem 5/34

5/35 The large roller bearing rolls to the left on its outer race with a velocity of its center O of 0.9 m/s. At the same time the central shaft and inner race rotate counterclockwise with an angular speed of 240 rev/min. Determine the angular velocity ω of each of the rollers.

Ans. $\omega = 10.73$ rad/s clockwise

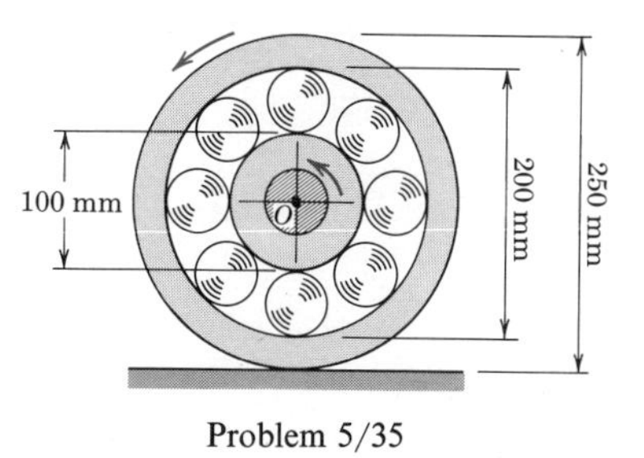

Problem 5/35

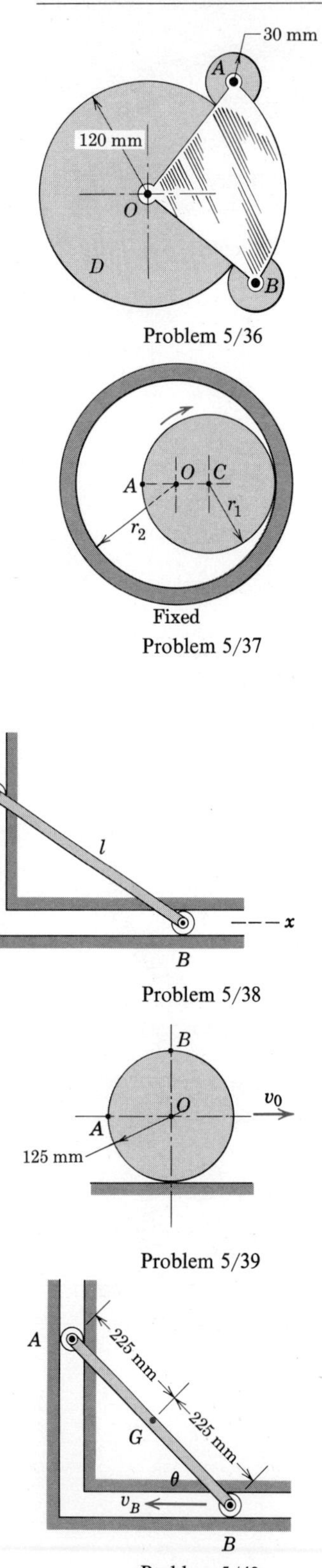

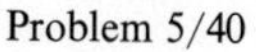

Problem 5/36

Problem 5/37

Problem 5/38

Problem 5/39

Problem 5/40

5/36 The gear D (teeth not shown) rotates counterclockwise about O with a constant angular velocity of 3 rad/s. The 90° sector AOB is mounted on an independent shaft at O, and each of the small gears at A and B meshes with gear D. If the sector has a clockwise angular velocity of 4 rad/s and a counterclockwise angular acceleration of 12 rad/s^2 at the instant represented, determine the corresponding angular velocity ω and the angular acceleration α of each of the small gears.

5/37 The inner gear rolls around the inner periphery of the outer fixed ring gear with constant speed. If the time required for C to make one complete circle about O is τ, write the expression for the acceleration of point A in the position shown.

Ans. $a_A = (r_2 - r_1)\left(2 - \dfrac{r_2}{r_1}\right)\left(\dfrac{2\pi}{\tau}\right)^2$

5/38 Construct the space centrode and the body centrode for the link AB within the limits of its constrained motion. Show that the motion is described by rolling the body centrode on the space centrode.

5/39 At the instant represented the center O of the rolling wheel has a velocity of 0.5 m/s to the right, and the velocity is diminishing at the rate of 1.5 m/s^2. Determine the magnitudes of the accelerations of points A and B for this instant. The wheel rolls without slipping.

Ans. $a_A = 1.581$ m/s^2, $a_B = 3.61$ m/s^2

5/40 End B of the 450-mm bar is given a constant velocity $v_B = 1.8$ m/s to the left. Calculate the acceleration of the center of mass G of the bar for the position $\theta = 45°$.

5/41 If the velocity of end B of the bar in Prob. 5/40 is zero and its acceleration is $a_B = 1.2$ m/s^2 to the right when the angle θ is 45°, calculate the corresponding acceleration of the center of mass G.

Ans. $a_G = 0.849$ m/s^2

5/42 The slider at B has a constant horizontal velocity v_B to the left as link AB passes the vertical position and link AO momentarily becomes horizontal. For this instant determine the angular acceleration of AO.

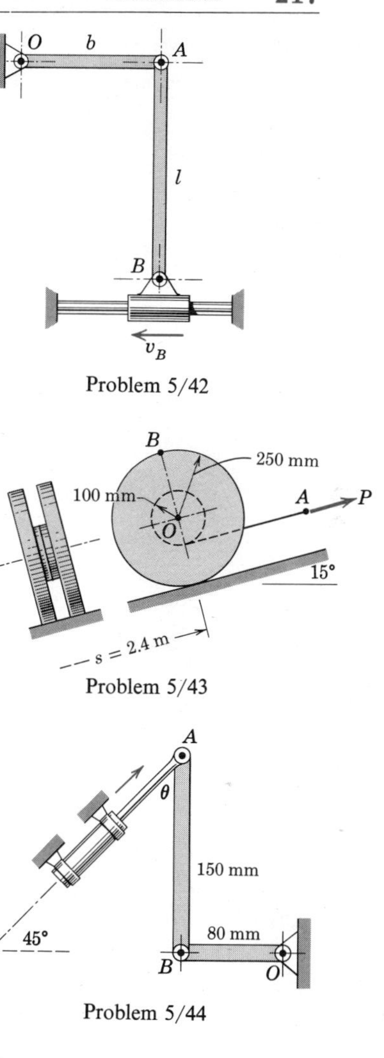

Problem 5/42

Problem 5/43

Problem 5/44

5/43 The center O of the spool starts from rest and acquires a velocity of 1.2 m/s up the incline with constant acceleration in a distance s of 2.4 m under the action of a steady pull P applied to point A of the cable. The cable is wrapped securely around the hub, and the wheel rolls without slipping. For the 2.4-m position shown, calculate the acceleration of point A on the cable and point B on the spool.

Ans. $a_A = 0.18 \text{ m/s}^2, \quad a_B = 5.79 \text{ m/s}^2$

5/44 The linkage shown is the same as that of Prob. 5/26 except that end B is guided by the pivoted link BO rather than the straight vertical slot. The piston rod has a constant velocity of 80 mm/s in the direction indicated, and at the instant represented $\theta = 45°$ and BO is horizontal. Compute the acceleration of B for this position.

5/45 At the instant represented $\theta = 45°$, and the link ABC has a counterclockwise angular velocity of 20 rad/s and a clockwise angular acceleration of 100 rad/s^2. Determine the corresponding velocity v and acceleration a of control rod CE in this position.

Ans. $v = 4.24 \text{ m/s}, \quad a = 106.1 \text{ m/s}^2$

5/46 The small vehicle is driven by the 300-mm friction wheel, which acquires a rotational speed $N = 150$ rev/min while the vehicle moves through a distance of 6 m from rest. Calculate the constant acceleration of the vehicle during the motion and find the acceleration, expressed in vector form, of point A on the top of the front wheel as the vehicle reaches the 6-m position. No slipping occurs at any of the rolling surfaces.

Ans. $a_O = 0.463 \text{ m/s}^2$

$\mathbf{a}_A = 0.925\mathbf{i} - 14.80\mathbf{j} \text{ m/s}^2$

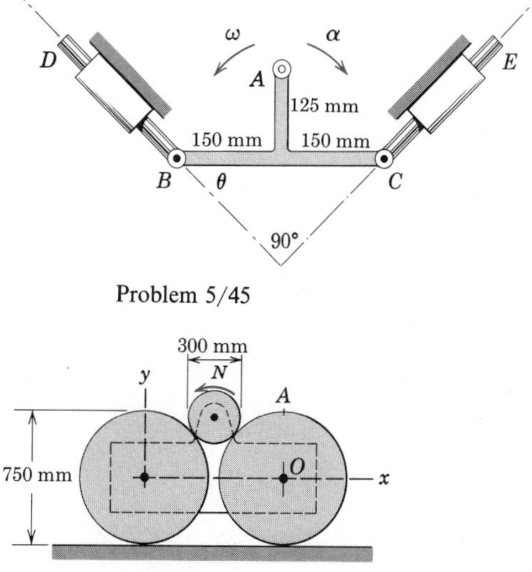

Problem 5/45

Problem 5/46

5/47 At the instant represented the triangular plate ABD has a clockwise angular velocity of 3 rad/s, and OA has a constant angular velocity. For this instant determine the angular acceleration of link BC.

Problem 5/47

5/48 Calculate the angular acceleration of link AB for the conditions described in Prob. 5/29 if, in addition, collar C has no acceleration as it passes the specified position.

Ans. $\alpha_{AB} = 4.04\text{ rad/s}^2$ clockwise

5/49 The elements of a power hacksaw are shown in the figure. The saw blade is mounted in a frame which slides along the horizontal guide. If the motor turns the flywheel at a constant counterclockwise speed of 60 rev/min, determine the acceleration of the blade for the position where $\theta = 90°$, and find the corresponding angular acceleration of the link AB.

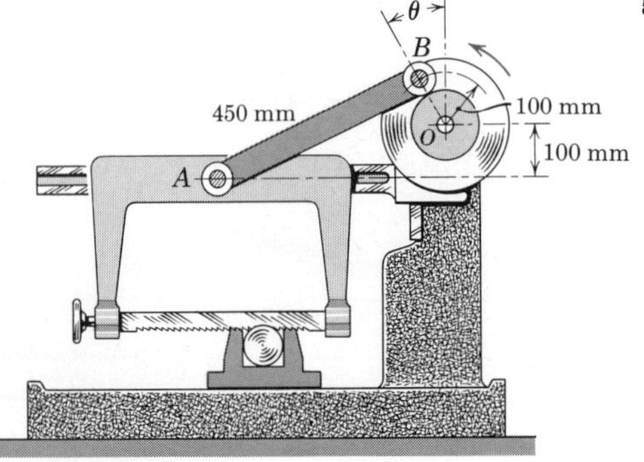

Problem 5/49

5/50 The plunger P is given a constant velocity $v = 1.2$ m/s during a short interval of its motion. Determine the angular velocity ω and angular acceleration α of link AB at the instant when $x = 100$ mm.

Ans. $\omega = 3.23$ rad/s counterclockwise
$\alpha = 101.5\text{ rad/s}^2$ clockwise

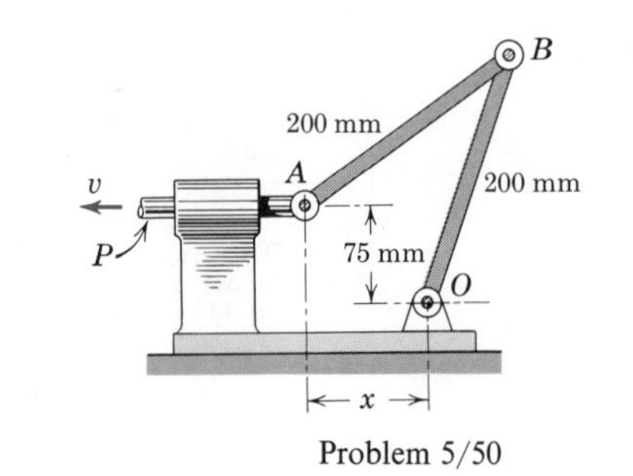

Problem 5/50

◀**5/51** The control rods R and S are given oscillatory motions in their respective parallel guides. Determine the acceleration of B at the instant when $x = 75$ mm if R has a constant velocity of 1.2 m/s to the right and S has a constant velocity of 1.2 m/s to the left during a short interval of the motion. Solve graphically.

Ans. $v_B = 3.09$ m/s, $a_B = 120.6\text{ m/s}^2$

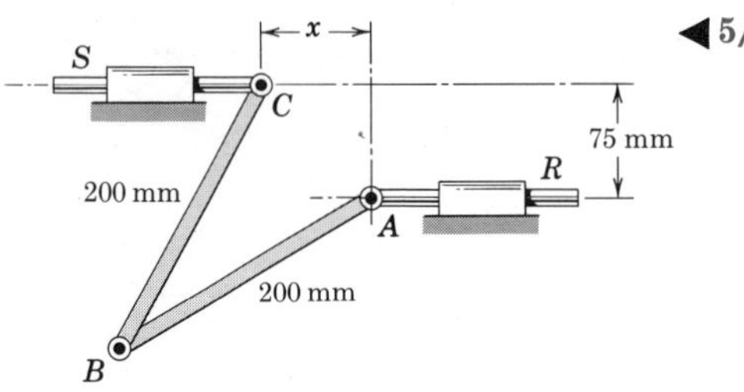

Problem 5/51

◀**5/52** The steel band D is attached to the pivoted sector and leads tangentially away from it. The band has a constant velocity of 3 m/s for an interval of motion in the neighborhood of the position shown. Determine the angular velocity and angular acceleration of link AB for the instant represented when OA is horizontal and BC is vertical.

Ans. $\omega_{AB} = 5$ rad/s counterclockwise
$\alpha_{AB} = 4.69$ rad/s^2 clockwise

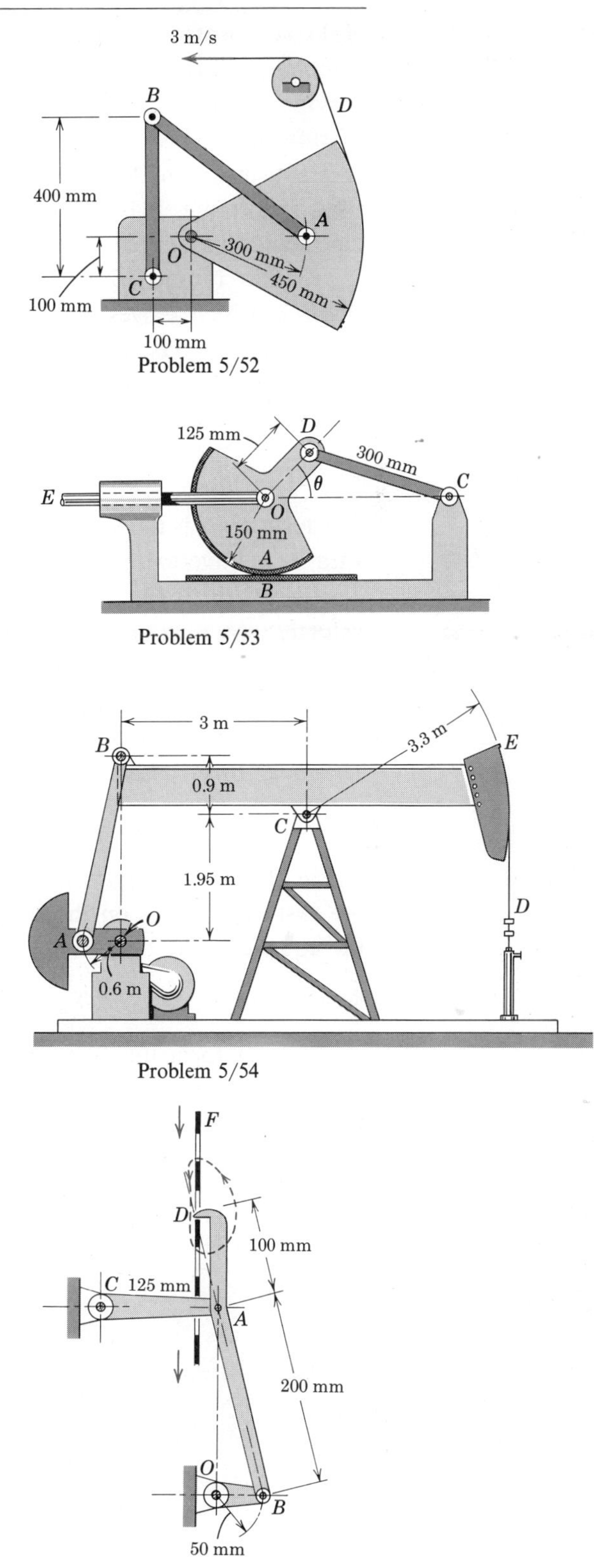

Problem 5/52

Problem 5/53

Problem 5/54

Problem 5/55

◀**5/53** A device which tests the resistance to wear of two materials A and B is shown. If the link EO has a velocity of 1.2 m/s and an acceleration of 1.8 m/s^2 both to the right when $\theta = 45°$, determine the rubbing velocity v_A and the time rate of change of the rubbing velocity.

Ans. $v_A = 2.76$ m/s
$\dot{v}_A = -8.57$ m/s^2

◀**5/54** An oil pumping rig is shown in the figure. The flexible pump rod D is fastened to the sector at E and is always vertical as it enters the fitting below D. The link AB causes the beam BCE to oscillate as the weighted crank OA revolves. If OA has a constant clockwise speed of 1 rev every 3 s, determine the acceleration of the pump rod D when the beam and the crank OA are both in the horizontal position shown.

Ans. $a_D = 0.568$ m/s^2 down

◀**5/55** An intermittent-drive mechanism for perforated tape F consists of the link DAB driven by the crank OB. The trace of the motion of the finger at D is shown by the dotted line. Determine the acceleration of D at the instant shown when both OB and CA are horizontal if OB has a constant clockwise rotational velocity of 120 rev/min.

Ans. $a_D = 1997$ mm/s^2

30 Relative Motion; Rotating Axes. In Part B of Art. 13, Chapter 2, the principles of relative motion were developed and applied to two points moving in a plane, with the observations being made from a rotating frame of reference. This same analysis is quite useful for a class of problems in mechanism kinematics where sliding contact occurs along a guide which itself is rotating. It is suggested, therefore, that the reader review the principles and procedures in Part B of Art. 13 before proceeding.

Figures 15 and 16 are reproduced again here for convenient reference. Point A, which moves along the rotating path, will represent a point on a rigid body which is moving with plane motion. If the velocity of A is referred to some point B, taken as the origin of the rotating coordinates, Fig. 15, then the absolute velocity of A, as shown in Art. 13, becomes

$$\blacktriangleright \quad \mathbf{v}_A = \mathbf{v}_B + \omega \times \mathbf{r} + \mathbf{v}_{\text{rel}} \qquad [22]$$

The term $\omega \times \mathbf{r}$ is the velocity of the point P, fixed to the path and coincident with A, relative to B. The term $\mathbf{v}_{\text{rel}}$ has the magnitude $\dot{s}$ and is the velocity of A relative to the path, measured from the rotating system. The total velocity of A may also be written as

$$\blacktriangleright \quad \mathbf{v}_A = \mathbf{v}_P + \mathbf{v}_{A/P} \qquad [22a]$$

where $\mathbf{v}_{A/P}$ is the same as $\mathbf{v}_{\text{rel}}$.

The absolute acceleration of A was shown in Art. 13 to be

$$\blacktriangleright \quad \mathbf{a}_A = \mathbf{a}_B + \dot{\omega} \times \mathbf{r} + \omega \times (\omega \times \mathbf{r}) + 2\omega \times \mathbf{v}_{\text{rel}} + \mathbf{a}_{\text{rel}} \qquad [23]$$

Again, if the coincident point P is used as the reference point rather than B, the first three terms on the right-hand side of the equation become $\mathbf{a}_P$, and the equation may be written

$$\blacktriangleright \quad \mathbf{a}_A = \mathbf{a}_P + 2\omega \times \mathbf{v}_{\text{rel}} + \mathbf{a}_{\text{rel}} \qquad [23b]$$

It is in this form that the relative acceleration equation is most frequently used for the kinematic analysis of sliding motion along rotating paths. If the path is curved, the relative acceleration term will have the two components $(a_{\text{rel}})_n = v_{\text{rel}}^2/\rho$ and $(a_{\text{rel}})_t = \ddot{s}$, where ρ is the radius of curvature of the

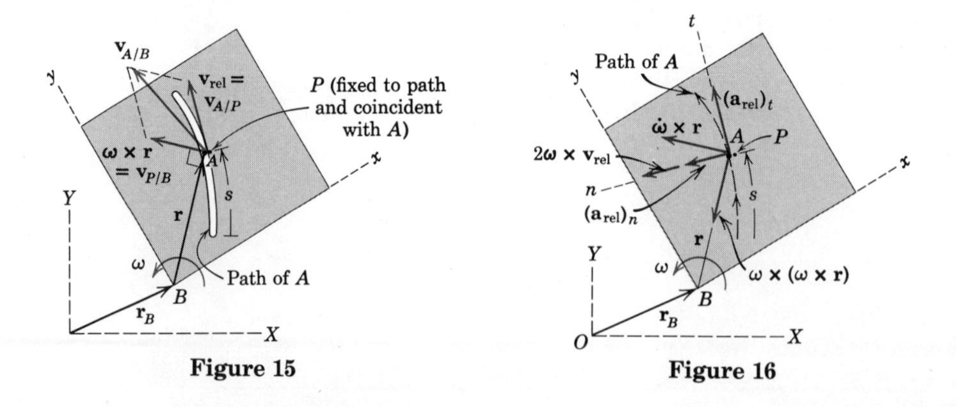

Figure 15

Figure 16

rotating path. For plane motion in the x-y plane the angular velocity of the rotating path will be expressed by $\boldsymbol{\omega} = \omega\mathbf{k}$.

It should be noted again that as long as the relative velocity is measured with respect to the coincident point P, the relative velocity term is the same in both rotating and nonrotating axes. Such is not the case with acceleration, however, as the Coriolis acceleration $2\boldsymbol{\omega} \times \mathbf{v}_{rel}$ represents the difference between the relative acceleration as measured from rotating and nonrotating axes.

The use of the foregoing relative motion relations for sliding contact with rotation is best illustrated by example.

Sample Problem

5/56 The pin A of the hinged link AC is confined to move in the rotating slot of link BO. The angular velocity of BO is $\omega = 2$ rad/s clockwise and is constant for the interval of motion concerned. For the position where $\theta = 45°$ with AC horizontal, determine the angular velocity of AC, the velocity of A relative to the rotating slot in BO, the angular acceleration of AC, and the acceleration of A relative to the rotating slot in BO.

Solution. A point P, which is attached to the rotating slot and is coincident with A for the position in question, is designated. The velocity of A from Eq. 22*a* is written

$$\mathbf{v}_A = \mathbf{v}_P + \mathbf{v}_{A/P}$$

The velocity of the point P on member BO is

$[v = r\omega]$ $\qquad v_P = \overline{OP}\omega = 225\sqrt{2}(2) = 450\sqrt{2}$ mm/s

The relative velocity $\mathbf{v}_{A/P}$, which is the same as $\mathbf{v}_{rel}$, is seen from the b-part of the figure to be along the slot toward O. This conclusion becomes clear when it is observed that A is approaching P along the slot from below before coincidence and is receding from

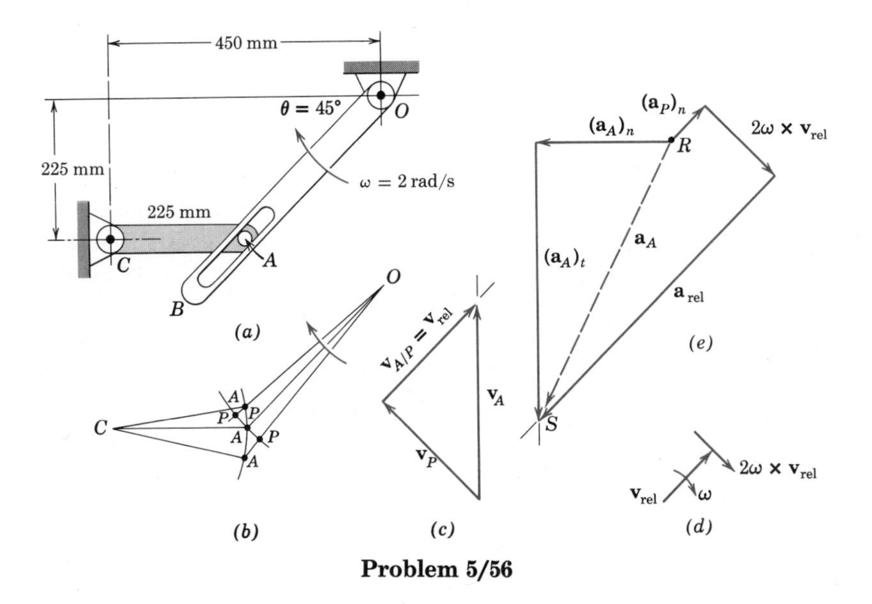

Problem 5/56

P upward along the slot following coincidence. The velocity of A is tangent to its circular arc about C. The vector equation may now be solved since there are only two remaining scalar unknowns, namely, the magnitude of $\mathbf{v}_{A/P}$ and the magnitude of $\mathbf{v}_A$. For the 45° position, the c-part of the figure requires

$$v_{A/P} = v_{\text{rel}} = 450\sqrt{2}\tan 45° = 450\sqrt{2} \text{ mm/s} \qquad \textit{Ans.}$$

$$v_A = 450\sqrt{2}(\sqrt{2}) = 900 \text{ mm/s}$$

each in its direction shown. The angular velocity of AC is now determined as

$$\left[\omega = \frac{v}{r}\right] \qquad \omega_{AC} = \frac{v_A}{\overline{AC}} = \frac{900}{225} = 4 \text{ rad/s counterclockwise} \qquad \textit{Ans.}$$

The accelerations are determined from Eq. 23b which is written

$$\mathbf{a}_A = \mathbf{a}_P + 2\boldsymbol{\omega} \times \mathbf{v}_{\text{rel}} + \mathbf{a}_{\text{rel}}$$

The terms in the equation are

$$(a_A)_n = \overline{AC}\omega_{AC}^2 = 225(4)^2 = 3600 \text{ mm/s}^2 \text{ directed toward } C$$

$$(a_A)_t = \overline{AC}\alpha_{AC} = 225\alpha_{AC} \text{ normal to } AC\text{, sense unknown}$$

$$(a_P)_n = \overline{OP}\omega^2 = 225\sqrt{2}(2)^2 = 900\sqrt{2} \text{ mm/s}^2 \text{ directed toward } O$$

$$(a_P)_t = \overline{OP}\alpha = 0 \text{ since } \alpha = \dot{\omega} = 0$$

$$|2\boldsymbol{\omega} \times \mathbf{v}_{\text{rel}}| = 2\omega v_{\text{rel}} = 2(2)(450\sqrt{2}) = 1800\sqrt{2} \text{ mm/s}^2 \text{ directed as in the } d\text{-part of figure}$$

$$\mathbf{a}_{\text{rel}} = \text{vector measured along slot}$$

The vector equation may now be solved since there are only two remaining scalar unknowns, namely, the magnitudes of $(\mathbf{a}_A)_t$ and $\mathbf{a}_{\text{rel}}$. The solution of Eq. 23b, shown in the e-part of the figure, is begun at point R and ends at point S where the lines with the known directions of $(\mathbf{a}_A)_t$ and $\mathbf{a}_{\text{rel}}$ intersect. From the diagram the two magnitudes are found to be

$$(a_A)_t = 7200 \text{ mm/s}^2 \quad \text{and} \quad a_{\text{rel}} = 8910 \text{ mm/s}^2 \qquad \textit{Ans.}$$

from which

$$\alpha_{AC} = \frac{(a_A)_t}{\overline{AC}} = \frac{7200}{225} = 32 \text{ rad/s}^2 \qquad \textit{Ans.}$$

With $(\mathbf{a}_A)_t$ pointing down in the figure, α_{AC} is seen to be clockwise.

The foregoing results may be obtained by expressing each of the terms in vector notation, using **i**- and **j**-components corresponding to any convenient choice of reference axes x-y. Equating coefficients of the **i**-terms and then of the **j**-terms would produce two algebraic equations which could be solved for the two unknowns in the usual way. In this and similar problems the vector diagram can help to identify any possible shortcut to the solution. By equating components normal to $\mathbf{a}_{\text{rel}}$, for example, the term a_{rel} is eliminated automatically from the equation, and $(a_A)_t$ is obtained immediately without resorting to the labor of a simultaneous solution. Thus

$$\frac{(a_A)_t}{\sqrt{2}} = \frac{(a_A)_n}{\sqrt{2}} + |2\boldsymbol{\omega} \times \mathbf{v}_{\text{rel}}|$$

$$(a_A)_t = 3600 + 1800\sqrt{2}(\sqrt{2}) = 7200 \text{ mm/s}^2$$

Similarly, equating components normal to $(\mathbf{a}_A)_t$ gives

$$\frac{a_{rel}}{\sqrt{2}} = (a_A)_n + \frac{(a_P)_n}{\sqrt{2}} + \frac{|2\boldsymbol{\omega} \times \mathbf{v}_{rel}|}{\sqrt{2}}$$

$$a_{rel} = (3600 + 900 + 1800)\sqrt{2} = 8910 \text{ mm/s}^2$$

It should be noted in this problem that, if OB had had an angular acceleration in addition to an angular velocity, the acceleration of the point P on OB would have had a tangential as well as a normal component. Further, if the slot had been curved with a radius of curvature ρ, the term $\mathbf{a}_{rel}$ would have had a component v_{rel}^2/ρ normal to the slot and toward the center of curvature in addition to its component along the slot.

Problems

5/57 The slider blocks are pinned together at point A with one block confined to move in the horizontal slot of the fixed plate while the other block slides along the rotating rod OC. If the rod has an angular velocity $\dot{\theta} = 3$ rad/s when the position $\theta = 30°$ is passed, determine the corresponding velocity of point A as measured by an observer attached to the rotating axes x-y. Let P be a point attached to OC and coincident with point A at the instant considered. Visualize the proper direction and sense of $\mathbf{v}_{A/P}$ by constructing the positions of P and A just prior to, at, and just after the instant considered.

Ans. $v_{A/P} = 400$ mm/s

5/58 Solve for the acceleration of pin A in Prob. 5/57 by using Eq. 23*b* and the results of Prob. 5/57. Assume $\dot{\theta}$ is constant.

Ans. $a_A = 2770$ mm/s^2

5/59 The shaft slides through the guide bearing which is pivoted at O, and pin B at the end of the shaft has a constant upward velocity $v = 450$ mm/s in the fixed slot for an interval of motion. For the instant when $\theta = 30°$, determine the acceleration a_A of a point A which is on the center line of the shaft and is coincident with the point O at this instant. (*Hint:* Express the acceleration of A in terms of the known motion of B and also in terms of the motion of A relative to O.)

Ans. $a_A = 893$ mm/s^2

5/60 In the position shown the bar DC is rotating counterclockwise at the constant rate $N = 2$ rad/s. Determine the angular velocity ω and the angular acceleration α of EBO at this instant.

Ans. $\omega = 2$ rad/s counterclockwise
$\alpha = 8$ rad/s^2 clockwise

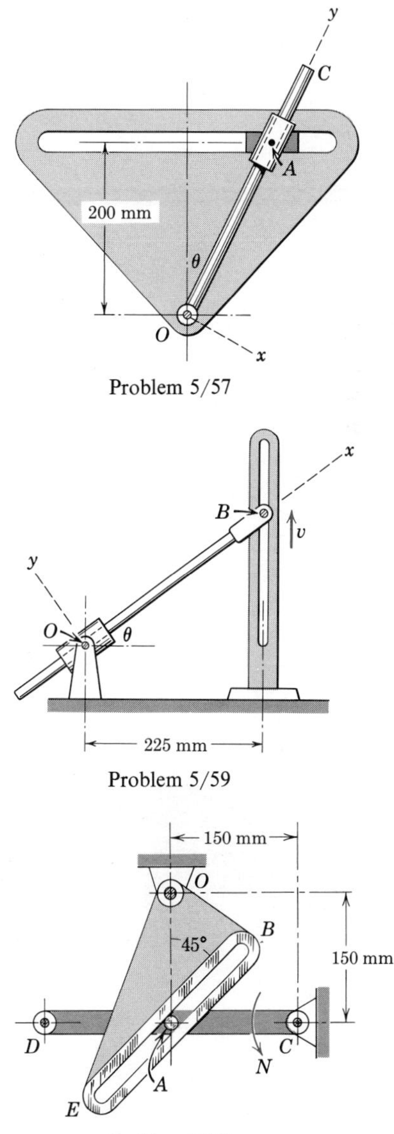

Problem 5/57

Problem 5/59

Problem 5/60

Problem 5/61

5/61 The crank OA revolves clockwise with a constant angular velocity of 10 rad/s within a limited arc of its motion. For the position $\theta = 30°$ determine the angular velocity of the slotted link CB and the acceleration of A as measured relative to the slot in CB.

Ans. $\omega_{BC} = 5$ rad/s clockwise
$a_{\text{rel}} = 8.66$ m/s^2 toward C

5/62 If the slotted link BC of Prob. 5/61 has a constant clockwise angular velocity of 10 rad/s in the position for which $\theta = 30°$, determine the angular velocity of OA and the acceleration of A as measured relative to the slot in CB.

Ans. $\omega_{AO} = 20$ rad/s clockwise
$a_{\text{rel}} = 34.6$ m/s^2 toward C

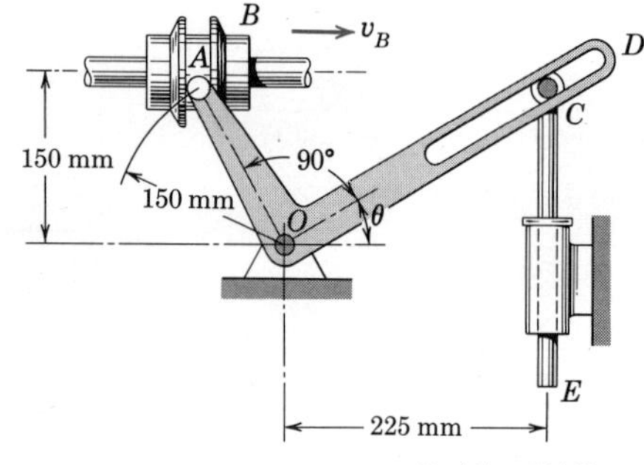

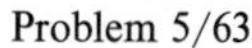
Problem 5/63

5/63 The pin A in the bell crank AOD is guided by the flanges of the collar B, which slides with a constant velocity v_B of 0.9 m/s along the fixed shaft. For the position $\theta = 30°$ determine the acceleration of the plunger CE, whose upper end is positioned by the radial slot in the bell crank.

Ans. $a_C = 24.9$ m/s^2 up

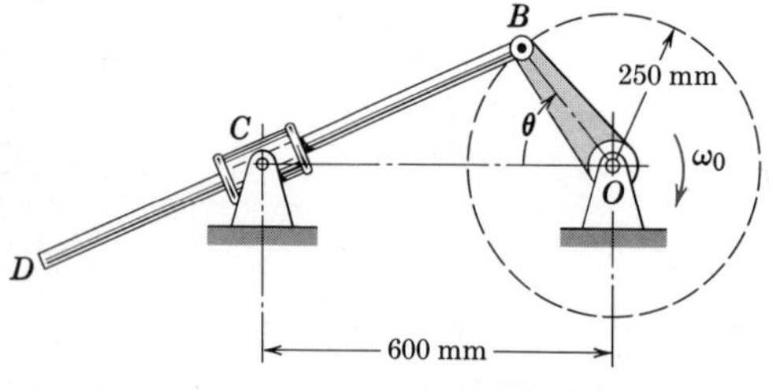

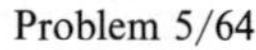
Problem 5/64

5/64 The crank OB revolves clockwise at the constant rate ω_0 of 5 rad/s. For the instant when $\theta = 90°$ determine the angular acceleration α of the rod BD, which slides through the pivoted collar at C.

Ans. $\alpha = 6.25$ rad/s^2 clockwise

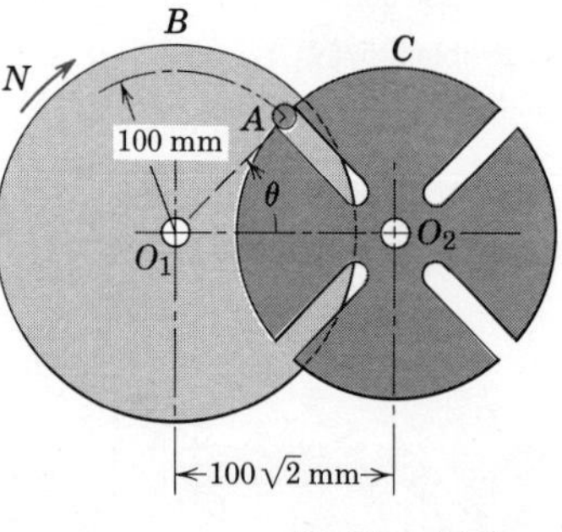

Problem 5/65

5/65 Intermittent rotary motion can be obtained with the Geneva mechanism shown, where the driving pin of disk B engages the radial slots of disk C in the 45° position and turns the disk through 90° before disengaging itself from the slot. If driver B has a constant clockwise rotational velocity $N = 60$ rev/min, determine the angular velocity ω and angular acceleration α of C for the instant when $\theta = 30°$.

Ans. $\omega = 2.57$ rad/s counterclockwise
$\alpha = 92.1$ rad/s^2 counterclockwise

5/66 The figure illustrates a commonly used quick-return mechanism which produces a slow cutting stroke of the tool (attached to D) and a rapid return stroke. If the driving crank OA is turning at the constant rate of $\dot{\theta} = 3$ rad/s, determine the velocity of point B and the angular acceleration of BC for the instant when $\theta = 30°$.

Ans. $v_B = 288$ mm/s
$\alpha_{BC} = 1.11$ rad/s^2 counterclockwise

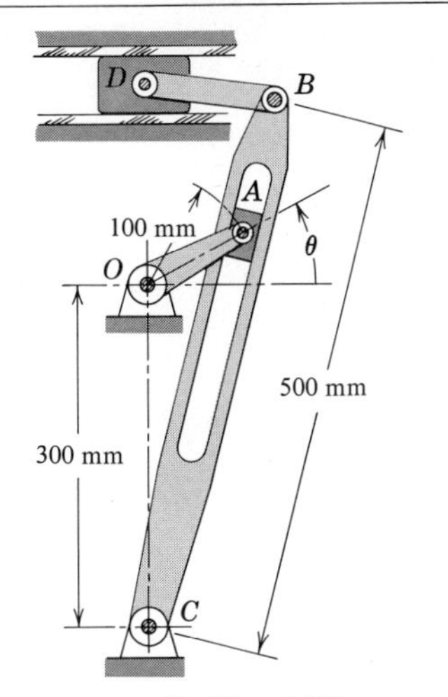

Problem 5/66

5/67 Determine the angular acceleration of link OB in the position shown if $\omega = 2$ rad/s, $\dot{\omega} = 6$ rad/s^2, and if $\theta = \beta = 60°$. Pin A is a part of link OB, and the curved slot has a radius of curvature of 125 mm.

Ans. $\alpha_{OB} = 9.96$ rad/s^2 counterclockwise

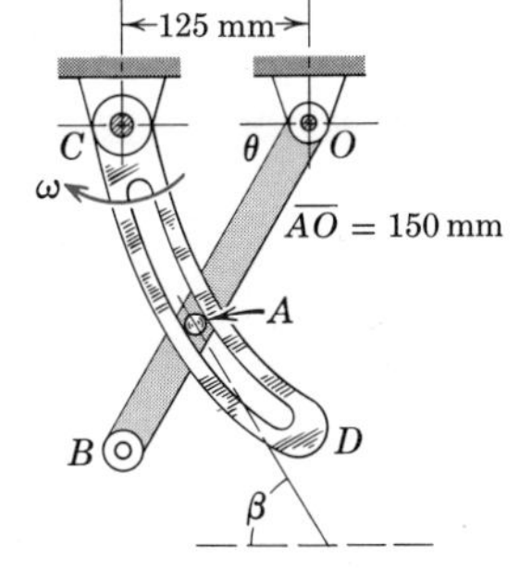

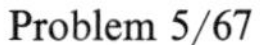

Problem 5/67

5/68 The mechanism shown is a device to produce high torque on the shaft at O. The gear unit, pivoted at C, turns the right-handed lead screw at a constant speed $N = 100$ rev/min in the direction shown which advances the threaded collar at A along the screw toward C. Determine the time rate of change $\dot{\omega}$ of the angular velocity of AO as it passes the vertical position shown. The screw has a single thread with a pitch of 2.5 mm. *Ans.* $\dot{\omega} = -3.88(10^{-4})$ rad/s^2

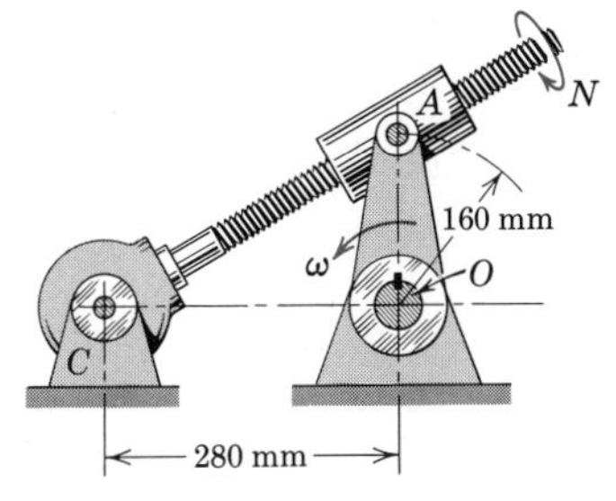

Problem 5/68

5/69 The link AC is given an oscillatory rotation about C which causes the slotted member to oscillate about O. When $\theta = 30°$, the angular motion of link AC is given by $\dot{\theta} = 6$ rad/s and $\ddot{\theta} = -30$ rad/s^2. For this condition determine the angular velocity ω and angular acceleration α of the slotted member.

Ans. $\omega = 3.53$ rad/s counterclockwise
$\alpha = 17.4$ rad/s^2 clockwise

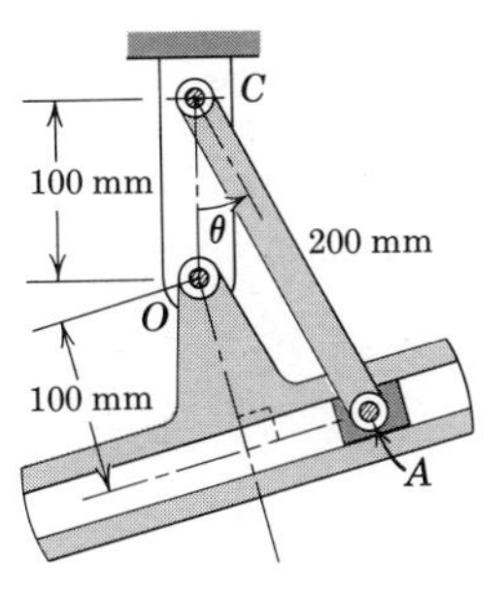

Problem 5/69

6 PLANE KINETICS OF RIGID BODIES

31 Introduction. The kinetics of rigid bodies deals with the relationships between the forces which act upon them from external sources and the corresponding translational and rotational motions of the bodies. In Chapter 5 the kinematical relationships for the plane motion of rigid bodies were developed, and these relationships will be used extensively in this present chapter where the effects of forces on the two-dimensional motion of rigid bodies are examined.

In Chapter 3 it was found that two force equations of motion were required to define the plane motion of a particle whose motion has two linear components. For the plane motion of a rigid body an additional equation is needed to specify the state of rotation of the body. Thus two force equations and one moment equation or their equivalent are required to determine the state of rigid-body plane motion.

The relationships which form the basis for most of the analysis of rigid-body motion were developed in Chapter 4 for a general system of mass particles. Frequent reference will be made to these equations as they are further developed in Chapter 6 and applied specifically to the plane motion of rigid bodies. The student is advised to refer to the developments of Chapter 4 frequently as he studies Chapter 6. The student will also be well advised not to proceed further unless he has a firm grasp of the calculation of velocities and accelerations as developed in Chapter 5 for rigid-body plane motion. Without the ability to determine accelerations correctly from the principles of kinematics, it is frequently useless to attempt to apply the force principles of motion. Consequently, the student is urged to master the necessary kinematics, including the calculation of relative accelerations, before proceeding.

Basic to the approach to kinetics is the isolation of the body or system to be analyzed. This isolation was illustrated and used in Chapter 3 for particle kinetics and will be employed consistently in the present chapter. For problems involving the instantaneous relationships among force, mass, and acceleration or momentum, the body or system should be explicitly defined by isolating it with its *free-body diagram.* When the principles of work and energy are employed, an *active-force diagram* which shows only those external forces that do work on the system may be used in lieu of the free-body diagram. No solution of a problem should be attempted without first defining the complete external boundary of the body or system and identifying all external forces that act on it.

32 Mass Moments of Inertia about an Axis. The equation of moments about an axis normal to the plane of motion for a rigid body in plane motion contains an integral which depends on the distribution of mass with respect to the moment axis. This integral occurs whenever there is an angular acceleration of a rigid body. This important integral will be described in this article prior to the derivation in the next article of the moment equation of motion in which the integral appears.

Consider a body of mass m, Fig. 47, rotating about an axis O-O with an angular acceleration α. All particles of the body move in parallel planes which are normal to the rotation axis O-O. Any one of the planes may be considered the plane of motion, although the one containing the center of mass is usually the one so designated. An element of mass dm has a component of acceleration tangent to its circular path equal to $r\alpha$, and by Newton's second law of motion the resultant tangential force on this element equals $r\alpha\, dm$. The moment of this force about the axis O-O is $r^2\alpha\, dm$, and the sum of the moments of these forces for all elements is $\int r^2\alpha\, dm$. For a rigid body α is the same for all radial lines in the body and may be taken outside the integral sign. The remaining integral is known as the moment of inertia I of the mass m about the axis O-O and is

$$I = \int r^2\, dm \tag{106}$$

This integral represents an important property of a body and is involved in the force analysis of any body that has rotational acceleration about a given axis. Just as the mass m of a body is a measure of the resistance to translational acceleration, the moment of inertia is a measure of resistance to rotational acceleration of the body.

The moment-of-inertia integral may be expressed alternatively as

$$I = \Sigma r_i^2 m_i \tag{106a}$$

where r_i is the radial distance from the inertia axis to the representative particle of mass m_i and where the summation is taken over all particles of the body.

If the density ρ is constant throughout the body, the moment of inertia becomes

$$I = \rho \int r^2\, dV$$

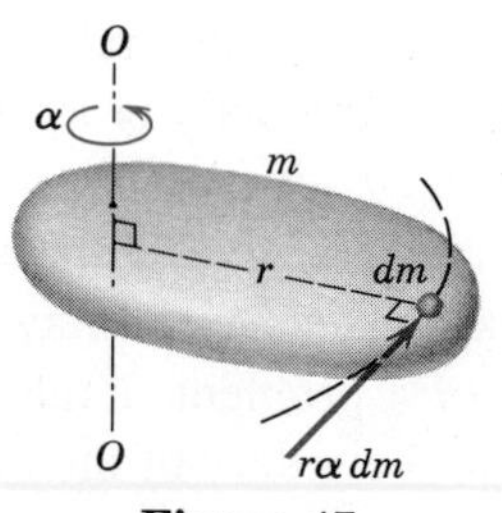

Figure 47

where dV is the element of volume. In this case the integral by itself defines a purely geometrical property of the body. When the density is not constant but is expressed as a function of the coordinates of the body, it must be left within the integral sign and its effect accounted for in the integration process.

In general the coordinates which best fit the boundaries of the body should be used in the integration. It is particularly important to make a good choice of the element of volume dV. An element of lowest possible order should be chosen, and the correct expression for the moment of inertia of the element about the axis involved should be used. For example, in finding the moment of inertia of a right circular cone about its central axis, an element in the form of a circular slice of infinitesimal thickness may be used, Fig. 48*a*. The differential moment of inertia for this element is the correct expression for the moment of inertia of a circular cylinder of infinitesimal altitude about its central axis. Alternatively an element in the form of a cylindrical shell of infinitesimal thickness may be chosen as shown in Fig. 48*b*. Since all of the mass of the element is at the same distance r from the inertia axis, the differential moment of inertia for this element is merely $r^2\,dm$ where dm is the differential mass of the elemental shell.

The dimensions of mass moments of inertia are (mass)(distance)^2 and are expressed in the units $kg \cdot m^2$.

Radius of Gyration. The radius of gyration k of a mass m about an axis for which the moment of inertia is I is

$$k = \sqrt{\frac{I}{m}} \quad \text{or} \quad I = k^2 m \tag{107}$$

Thus k is a measure of the distribution of mass of a given body about the axis in question, and its definition is analogous to the definition of the radius of gyration for second moments of area. If all the mass m could be concentrated at a distance k from the axis, the correct moment of inertia would be $k^2 m$. The moment of inertia of a body about a particular axis is frequently indicated by specifying the mass of the body and the radius of gyration of the body about the axis. The moment of inertia is then calculated from Eq. 107.

Transfer of Axes. If the moment of inertia of a body is known about a centroidal axis, it may be determined easily about any parallel axis. To prove

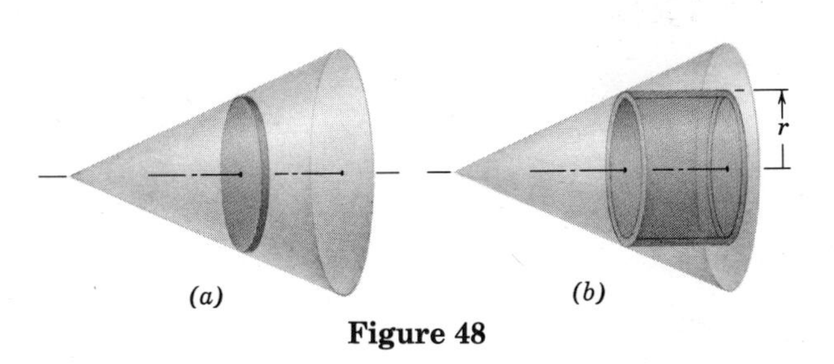

Figure 48

this statement consider the two parallel axes in Fig. 49, one of which is a centroidal axis through the mass center G. The radial distances from the two axes to any element of mass dm are r_o and r, and the separation of the axes is d. Substituting the law of cosines $r^2 = r_o^2 + d^2 + 2r_o d \cos\theta$ into the definition for the moment of inertia about the noncentroidal axis gives

$$I = \int r^2\, dm = \int (r_o^2 + d^2 + 2r_o d \cos\theta)\, dm$$
$$= \int r_o^2\, dm + d^2 \int dm + 2d \int u\, dm$$

The first integral is the moment of inertia $\bar{I}$ about the mass-center axis, the second term is md^2, and the third integral equals zero, since the u-coordinate of the mass center with respect to the axis through G is zero. Thus the parallel-axis theorem is

$$I = \bar{I} + md^2 \tag{108}$$

It must be remembered that the transfer cannot be made unless one axis passes through the center of mass and unless the axes are parallel. When the expressions for the radii of gyration are substituted in Eq. 108, there results

$$k^2 = \bar{k}^2 + d^2 \tag{108a}$$

which is the parallel-axis theorem for obtaining the radius of gyration k about an axis a distance d from a parallel centroidal axis for which the radius of gyration is $\bar{k}$.

The similarity between the defining expressions for mass moments of inertia and area moments of inertia is easily observed. An exact relationship between the two moment of inertia expressions exists in the case of flat plates. Consider the flat plate of uniform thickness in Fig. 50. If the constant thickness is t and the density is ρ, the mass moment of inertia I_{zz} of the plate about the z-axis normal to the plate is

$$I_{zz} = \int r^2\, dm = \rho t \int r^2\, dA = \rho t J_z \tag{109}$$

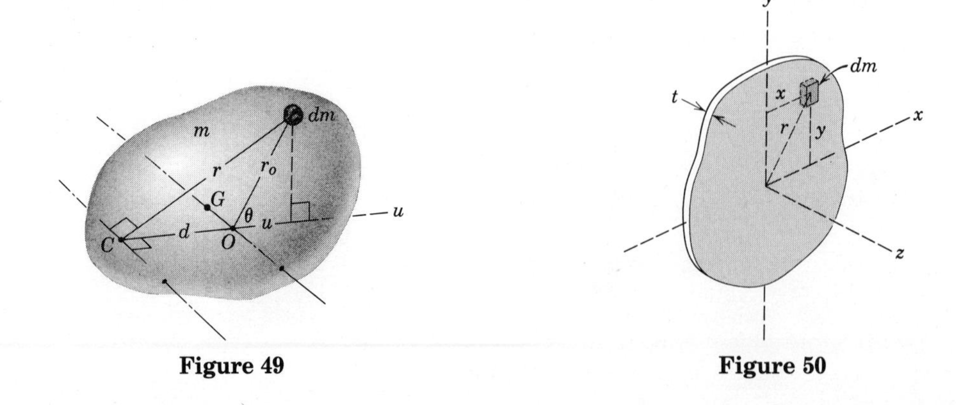

Figure 49

Figure 50

Thus the mass moment of inertia about the z-axis equals the mass per unit area, ρt, times the polar moment of inertia J_z of the plate area about the z-axis. If t is small compared with the dimensions of the plate in its plane, the mass moments of inertia I_{xx} and I_{yy} of the plate about the x- and y-axes are closely approximated by

$$\begin{aligned} I_{xx} &= \int y^2\,dm = \rho t \int y^2\,dA = \rho t I_x \\ I_{yy} &= \int x^2\,dm = \rho t \int x^2\,dA = \rho t I_y \end{aligned} \tag{110}$$

Hence the mass moments of inertia equal the mass per unit area ρt times the corresponding area moments of inertia. The double subscript for mass moments of inertia distinguish these quantities from area moments of inertia.

Inasmuch as $J_z = I_x + I_y$ for area moments of inertia, it follows that

$$I_{zz} = I_{xx} + I_{yy} \tag{111}$$

which holds *only* for a thin flat plate. This restriction is observed from Eqs. 110, which do not hold true unless the thickness t or z-coordinate of the element is negligible compared with the distance of the element from the corresponding x- or y-axis. Equation 111 is very useful when dealing with a differential mass element taken as a flat slice of differential thickness, say dz. In this case Eq. 111 holds exactly and becomes

$$dI_{zz} = dI_{xx} + dI_{yy} \tag{111a}$$

Composite Bodies. The defining integral, Eq. 106, involves the square of the distance from the axis to the element and so is always positive. Thus, as in the case of area moments of inertia, the mass moment of inertia of a composite body is the sum of the moments of inertia of the individual parts about the same axis. It is often convenient to consider a composite body as defined by positive volumes and negative volumes. The moment of inertia of a negative element, such as a hole, must be considered a negative quantity.

Thorough familiarity with the calculation of mass moments of inertia for rigid bodies about axes normal to the plane of motion is an absolute necessity for the study of the dynamics of this motion.

A summary of some of the more useful formulas for mass moments of inertia of various masses of common shape is given in Table C5, Appendix C.

Sample Problems

6/1 Determine the moment of inertia and radius of gyration of a homogeneous right circular cylinder of mass m and radius r about its central axis O–O.

Solution. An element of mass in cylindrical coordinates is $dm = \rho\,dV = \rho t r_o\,dr_o\,d\theta$. The moment of inertia about the axis of the cylinder is

$$I = \int r_o^2\,dm = \rho t \int_0^{2\pi}\int_0^{r} r_o^3\,dr_o\,d\theta = \rho t\frac{\pi r^4}{2} = \tfrac{1}{2}mr^2 \qquad \textit{Ans.}$$

The radius of gyration is

$$k = \sqrt{\frac{I}{m}} = \frac{r}{\sqrt{2}} \qquad \textit{Ans.}$$

The result $I = \frac{1}{2}mr^2$ applies *only* to a solid homogeneous circular cylinder and cannot be used for any other wheel of circular periphery.

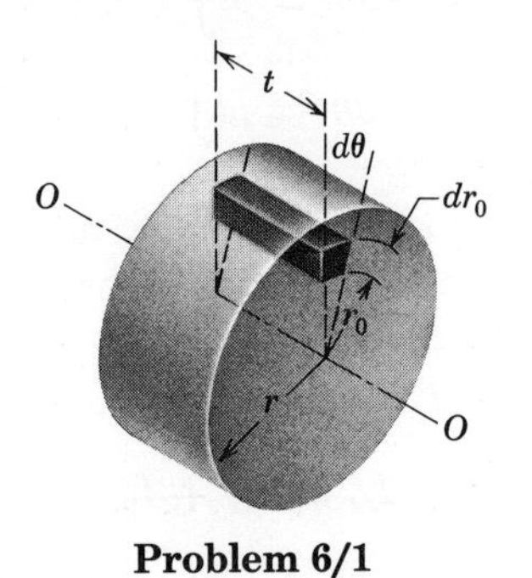

Problem 6/1

6/2 Determine the moment of inertia and radius of gyration of a homogeneous solid sphere of mass m and radius r about a diameter.

Solution. A circular slice of radius y and thickness dx is chosen as the volume element. From the results of Sample Prob. 6/1 the moment of inertia about the x-axis of the elemental cylinder is

$$dI_{xx} = \tfrac{1}{2}(dm)y^2 = \tfrac{1}{2}(\pi\rho y^2\,dx)y^2 = \frac{\pi\rho}{2}(r^2 - x^2)^2\,dx$$

where ρ is the constant density of the sphere. The total moment of inertia about the x-axis is

$$I_{xx} = \frac{\pi\rho}{2}\int_{-r}^{r}(r^2 - x^2)^2\,dx = \tfrac{8}{15}\pi\rho r^5 = \tfrac{2}{5}mr^2 \qquad \textit{Ans.}$$

The radius of gyration is

$$k = \sqrt{\frac{I}{m}} = \sqrt{\frac{2}{5}}\,r \qquad \textit{Ans.}$$

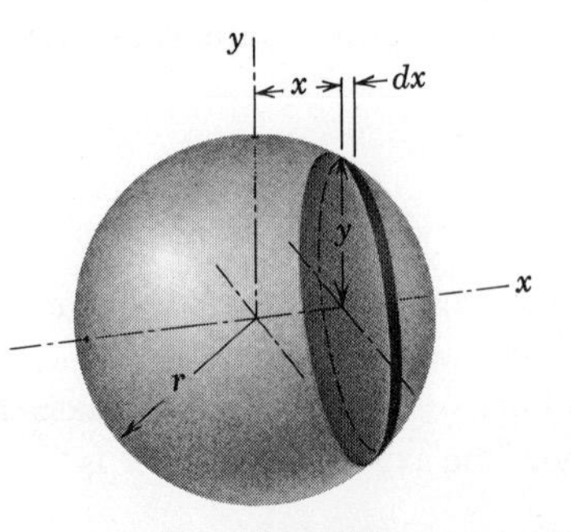

Problem 6/2

6/3 Determine the moments of inertia of the homogeneous rectangular parallelepiped of mass m about the centroidal x_o- and z-axes and about the x-axis through one end.

Solution. A transverse slice of thickness dz is selected as the element of volume. The moment of inertia of this slice of infinitesimal thickness equals the moment of inertia of the area of the section times the mass per unit area $\rho\, dz$. Thus the moment of inertia of the transverse slice about the y'-axis is

$$dI_{y'y'} = (\rho\, dz)(\tfrac{1}{12}ab^3)$$

and that about the x'-axis is

$$dI_{x'x'} = (\rho\, dz)(\tfrac{1}{12}a^3b)$$

As long as the element is a plate of differential thickness, the principle given by Eq. 111*a* may be applied to give

$$dI_{zz} = dI_{x'x'} + dI_{y'y'} = (\rho\, dz)\frac{ab}{12}(a^2 + b^2)$$

These expressions may now be integrated to obtain the desired results.

The moment of inertia about the z-axis is

$$I_{zz} = \int dI_{zz} = \frac{\rho ab}{12}(a^2 + b^2)\int_0^l dz = \tfrac{1}{12}m(a^2 + b^2) \qquad \textit{Ans.}$$

where m is the mass of the block. By interchanging symbols the moment of inertia about the x_o-axis is

$$I_{x_ox_o} = \tfrac{1}{12}m(a^2 + l^2) \qquad \textit{Ans.}$$

The moment of inertia about the x-axis may be found by the parallel-axis theorem, Eq. 108. Thus

$$I_{xx} = I_{x_ox_o} + m\left(\frac{l}{2}\right)^2 = \tfrac{1}{12}m(a^2 + 4l^2) \qquad \textit{Ans.}$$

This last result may be obtained by expressing the moment of inertia of the elemental slice about the x-axis and integrating the expression over the length of the bar. Again by the parallel-axis theorem

$$dI_{xx} = dI_{x'x'} + z^2\, dm = (\rho\, dz)(\tfrac{1}{12}a^3b) + z^2\rho ab\, dz = \rho ab\left(\frac{a^2}{12} + z^2\right)dz$$

Integrating gives the result obtained previously,

$$I_{xx} = \rho ab\int_0^l\left(\frac{a^2}{12} + z^2\right)dz = \frac{\rho abl}{3}\left(l^2 + \frac{a^2}{4}\right) = \tfrac{1}{12}m(a^2 + 4l^2)$$

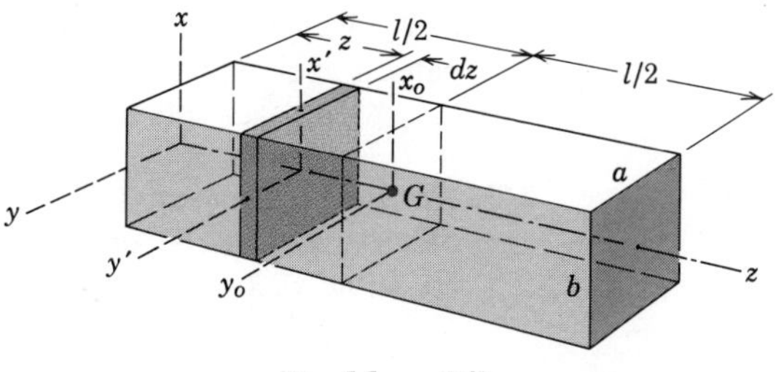

Problem 6/3

The expression for I_{xx} may be simplified for a long prismatical bar or slender rod whose transverse dimensions are small compared with the length. In this case a^2 may be neglected compared with $4l^2$, and the moment of inertia of such a slender bar about an axis through one end normal to the bar becomes $I = \frac{1}{3}ml^2$. By the same approximation the moment of inertia about a centroidal axis normal to the bar is $I = \frac{1}{12}ml^2$.

Problems

6/4 A bar 250 mm long has a square cross section 25 mm on a side. Determine the percentage error e in using the approximate formula $I = \frac{1}{3}ml^2$ for the moment of inertia about an axis which is normal to the bar and passes through the center of one end parallel to an edge. (See Sample Prob. 6/3.)

6/5 The moment of inertia of a solid homogeneous cylinder of radius r about an axis parallel to the central axis of the cylinder may be obtained approximately by multiplying the mass of the cylinder by the square of the distance d between the two axes. What per cent error e results if (*a*) $d = 10r$, (*b*) $d = 2r$?

Ans. (*a*) $e = 0.498$ per cent
(*b*) $e = 11.1$ per cent

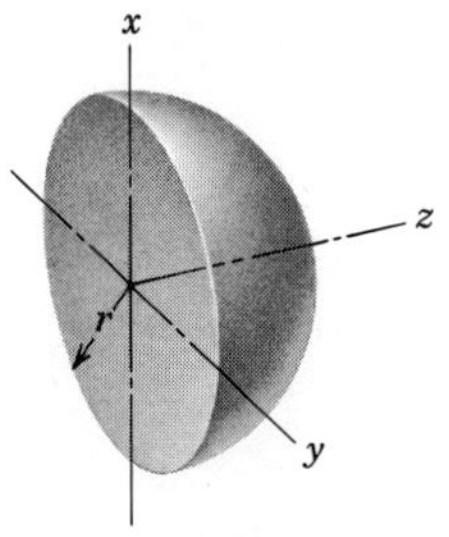

Problem 6/6

6/6 From the results of Sample Prob. 6/2 state without computation the moments of inertia of the solid homogeneous hemisphere of mass m about the x- and z-axes.

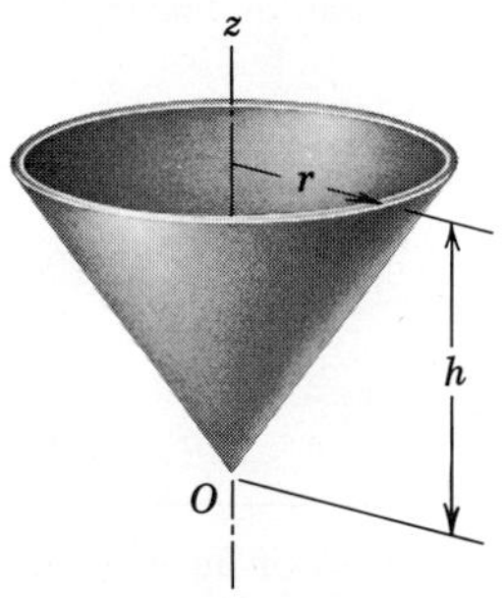

Problem 6/7

6/7 State without calculation the moment of inertia about the z-axis of the thin conical shell of mass m and radius r from the results of Sample Prob. 6/1 applied to a circular disk. Observe the radial distribution of mass by viewing the cone along the z-axis.

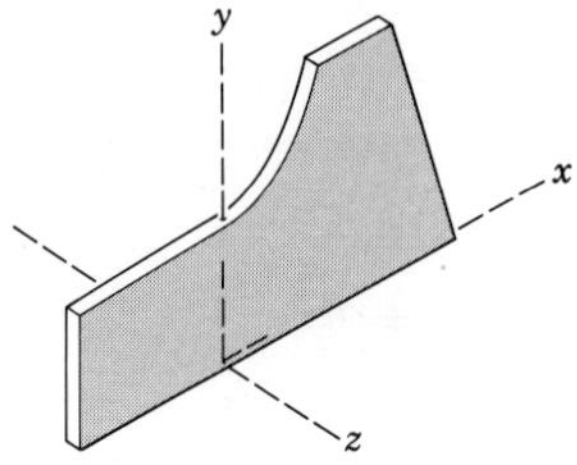

Problem 6/8

6/8 The pattern is cut from sheet metal which has a mass of 20.5 kg/m^2. If the area moments of inertia of its face about the x- and y-axes are $1.40(10^8)$ mm^4 and $3.05(10^8)$ mm^4, respectively, determine the mass moment of inertia of the pattern about the z-axis normal to its surface.

Ans. $I_{zz} = 9.12(10^{-3})$ kg·m^2

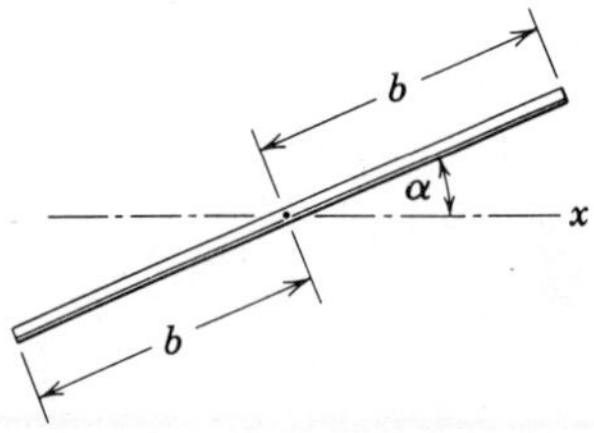

6/9 Determine the moment of inertia of the inclined uniform slender rod of length $2b$ and mass m about the x-axis through its center.

Problem 6/9

6/10 Calculate the mass moment of inertia about the axis O–O for the uniform 250-mm block of steel with cross-section dimensions of 150 mm and 200 mm. *Ans.* $I_{OO} = 2.21\ \text{kg}\cdot\text{m}^2$

Problem 6/10

6/11 The semi-circular disk is made of cast iron and has a mass of 45 kg. Calculate its moment of inertia about the axis A–A.

Problem 6/11

6/12 A uniform brass rod having a mass of 0.6 kg is bent into the shape shown. Calculate the moment of inertia of the rod about an axis through O normal to the plane of the figure.
Ans. $I_{OO} = 11.54(10^{-3})\ \text{kg}\cdot\text{m}^2$

Problem 6/12

6/13 Calculate the radius of gyration about axis O–O for the steel disk with the hole.

Problem 6/13

6/14 Determine the moment of inertia of the half-ring of mass m about its diametral axis a-a and about axis b-b through the midpoint of the arc normal to the plane of the ring. The radius of the circular cross section is small compared with r.

Ans. $I_{aa} = \frac{1}{2}mr^2, \quad I_{bb} = 2mr^2\left(1 - \frac{2}{\pi}\right)$

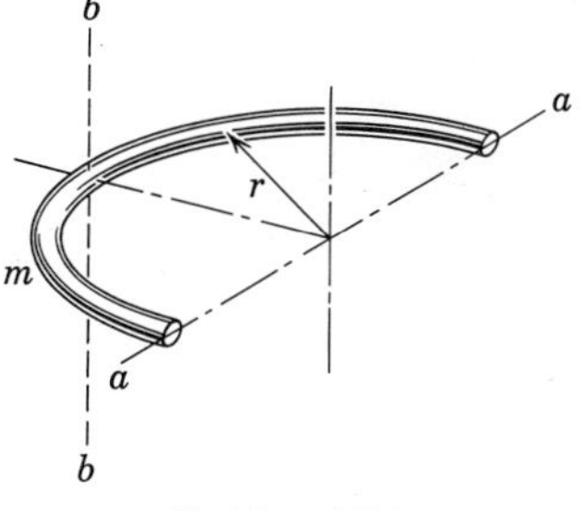

Problem 6/14

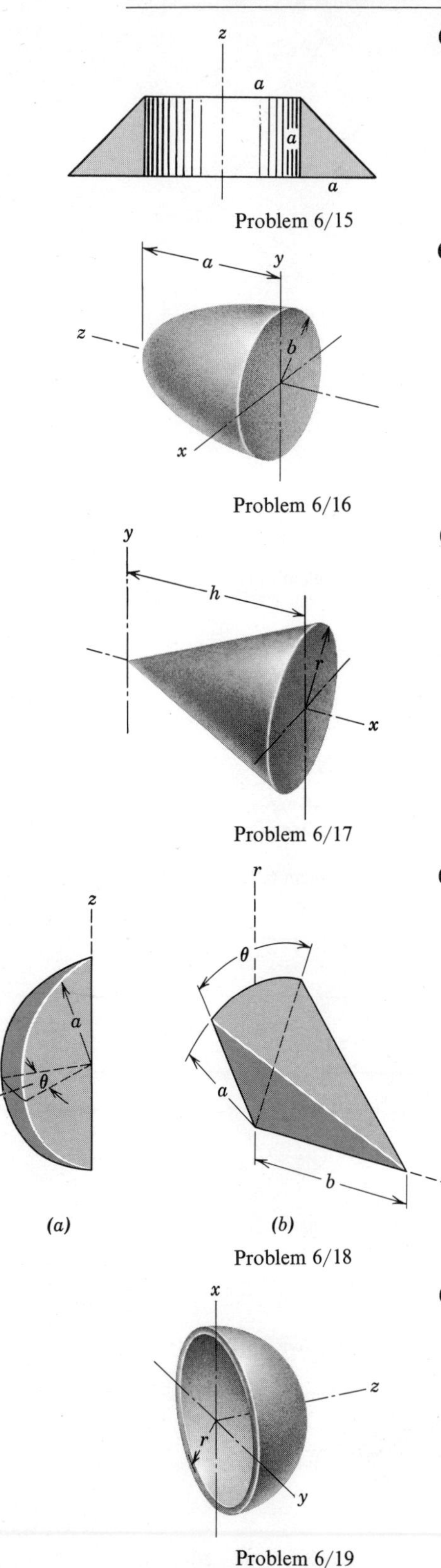

Problem 6/15

Problem 6/16

Problem 6/17

Problem 6/18

Problem 6/19

6/15 A homogeneous solid of mass m, shown in section, is formed by revolving the 45° right triangle about the z-axis. Determine the moment of inertia of the solid about the z-axis.

6/16 Determine the moment of inertia about the z-axis of the homogeneous solid paraboloid of revolution of mass m. *Ans.* $I_{zz} = \frac{1}{3}mb^2$

6/17 Calculate the moment of inertia of the homogeneous right circular cone of mass m, base radius r, and altitude h about the cone axis x and about the y-axis through its vertex.

Ans. $I_{xx} = \frac{3}{10}mr^2, \quad I_{yy} = \frac{3}{5}m\left(\frac{r^2}{4} + h^2\right)$

6/18 Without integrating determine from the results of Sample Prob. 6/2 and Prob. 6/17 the moments of inertia about the z-axis for (a) the spherical wedge and (b) the conical wedge. Each wedge has a mass m.

Ans. (a) $I_{zz} = \frac{2}{5}ma^2$, (b) $I_{zz} = \frac{3}{10}mr^2$

6/19 Determine the moments of inertia of the half spherical shell with respect to the x- and z-axes. The mass of the shell is m, and its thickness is negligible compared with the radius r.

6/20 Determine by integration the moment of inertia of the thin conical shell of Prob. 6/7 about an axis through the vertex O normal to the z-axis. The mass of the shell is m and its altitude is h.

Ans. $I_{OO} = \frac{m}{4}(r^2 + 2h^2)$

6/21 Determine the moment of inertia of the circular sector of mass m and radius r about the tangent line O–O. The thickness of the sector is small compared with r.

6/22 In the study of high-speed re-entry into the earth's atmosphere small solid cones are fired at high velocities into low-density gas. A condition of critical stability occurs when the moment of inertia of the cone about its axis of generation a-a equals that about a transverse axis b-b through the mass center. Determine the critical value of the cone angle α for this condition.

Ans. $\alpha = 26°34'$

6/23 Find the moment of inertia of the tetrahedron of mass m about the z-axis.

6/24 Determine the moments of inertia of the homogeneous right circular cylinder of mass m about the x_o-, x-, and z'-axes shown.

Ans. $I_{x_ox_o} = \frac{1}{12}m(3r^2 + l^2)$
$I_{xx} = \frac{1}{12}m(3r^2 + 4l^2)$
$I_{z'z'} = \frac{3}{2}mr^2$

6/25 Determine by integration the moment of inertia of the half-cylindrical shell of mass m about the axis a-a. The thickness of the shell is small compared with r.

Ans. $I_{aa} = \frac{m}{2}\left(r^2 + \frac{l^2}{6}\right)$

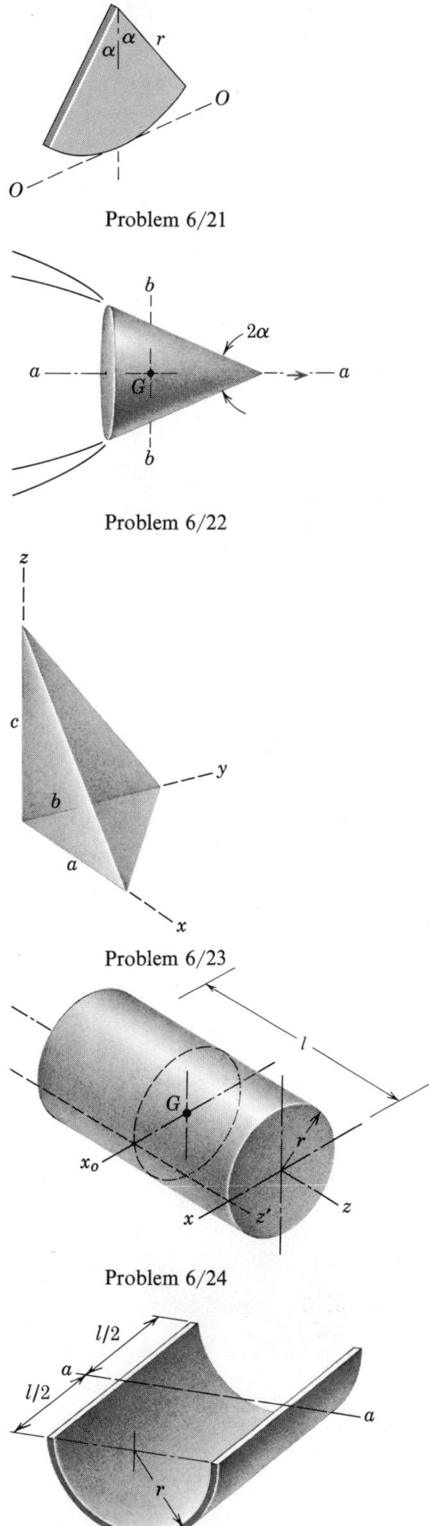

Problem 6/21

Problem 6/22

Problem 6/23

Problem 6/24

Problem 6/25

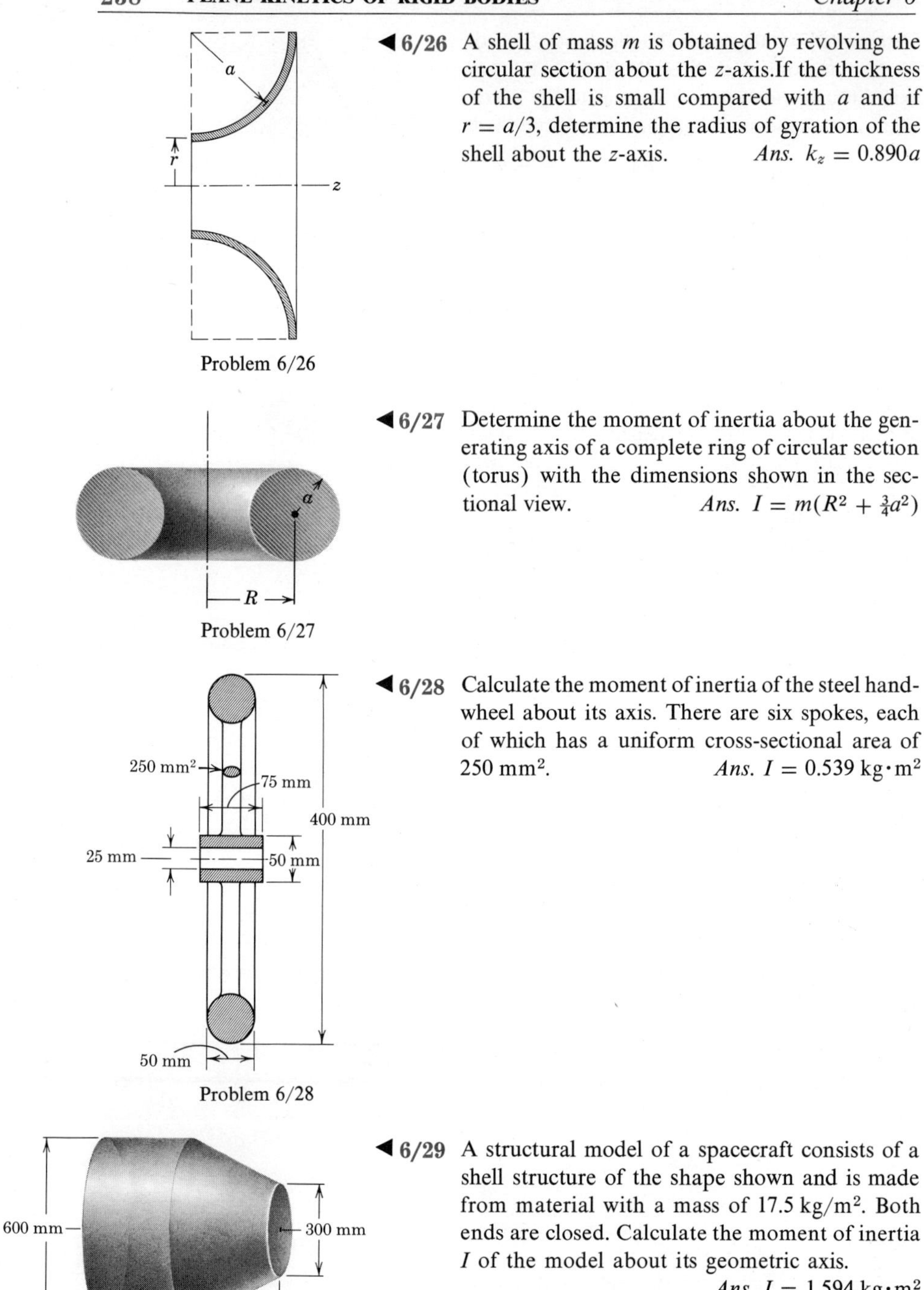

Problem 6/26

Problem 6/27

Problem 6/28

Problem 6/29

◀**6/26** A shell of mass m is obtained by revolving the circular section about the z-axis.If the thickness of the shell is small compared with a and if $r = a/3$, determine the radius of gyration of the shell about the z-axis. *Ans.* $k_z = 0.890a$

◀**6/27** Determine the moment of inertia about the generating axis of a complete ring of circular section (torus) with the dimensions shown in the sectional view. *Ans.* $I = m(R^2 + \frac{3}{4}a^2)$

◀**6/28** Calculate the moment of inertia of the steel handwheel about its axis. There are six spokes, each of which has a uniform cross-sectional area of 250 mm². *Ans.* $I = 0.539\ \text{kg}\cdot\text{m}^2$

◀**6/29** A structural model of a spacecraft consists of a shell structure of the shape shown and is made from material with a mass of 17.5 kg/m². Both ends are closed. Calculate the moment of inertia I of the model about its geometric axis.

Ans. $I = 1.594\ \text{kg}\cdot\text{m}^2$

◀6/30 The part shown is made of mild steel. Calculate its moment of inertia about the axis *a-a*.

Ans. $I_{aa} = 0.461 \text{ kg}\cdot\text{m}^2$

◀6/31 The cube with semicircular grooves in two opposite faces is cast of lead. Calculate the moment of inertia of the solid about the *a-a* axis.

Ans. $I_{aa} = 0.367 \text{ kg}\cdot\text{m}^2$

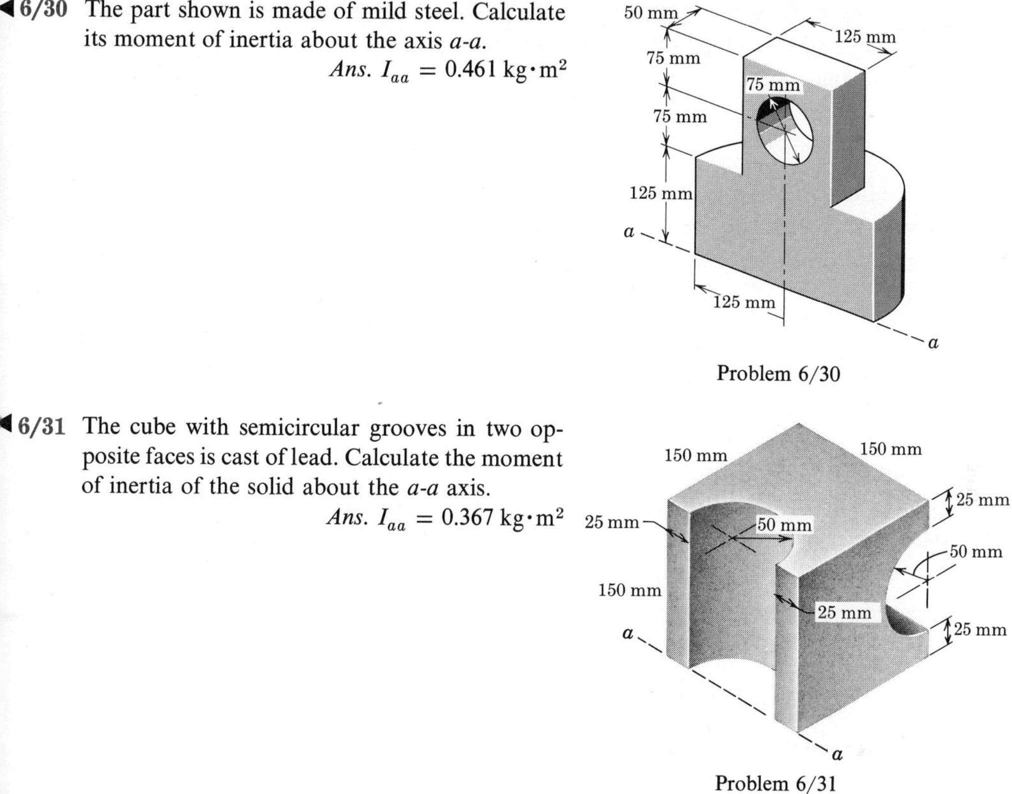

Problem 6/30

Problem 6/31

33 Force, Mass, and Acceleration. The position of a rigid body in its plane of motion requires the specification of three scalar coordinates. Thus the two coordinates of the mass center, or some other convenient reference point, and the angular position of the body about the reference point selected will determine uniquely the position of each point in the body. Therefore, plane motion requires three independent scalar equations for its description. Two of these equations, the *x*- and *y*-components of Eq. 92*a* for motion in the *x-y* plane, have already been described for a general mass system, which includes a rigid body. The third equation may be obtained by simplifying Eq. 98 applied about the *z*-axis through the mass center *G* or by simplifying Eq. 97 applied about a *z*-axis through a point *O* which has no acceleration. It is somewhat more instructive in this case, however, to derive the moment equation directly from basic principles.

To obtain the moment equation, consider the rigid body shown in Fig. 51 which illustrates plane motion in the *X-Y* plane under the action of forces $\mathbf{F}_1, \mathbf{F}_2, \mathbf{F}_3, \ldots$ applied from sources external to the body. The acceleration of the mass center *G* is represented by the vector $\bar{\mathbf{a}}$, and the angular velocity and angular acceleration of the body have the magnitudes ω and α, respectively. A representative particle of the body with mass m_i has acceleration components equal to $\bar{a}$ plus the relative terms $\rho_i\omega^2$ and $\rho_i\alpha$, where the mass center *G* is used as the reference point. It follows that the resultant of all forces on m_i has the components $m_i\bar{a}$, $m_i\rho_i\omega^2$, and $m_i\rho_i\alpha$ in the directions shown in Fig. 51. The sum of the moments of these force components about G in the sense of α is

$$\overline{M}_i = m_i\bar{a}\rho_i \cos \beta + m_i\rho_i^2\alpha$$

where the force $m_i\rho_i\omega^2$ passes through G and has no moment about G. Similar expressions may be written for all particles in the body. The sum of the moments about G for the resultant forces acting on all particles may be written as

$$\Sigma\overline{M} = \Sigma m_i\bar{a}\rho_i \cos \beta + \Sigma m_i\rho_i^2\alpha$$

The first term on the right-hand side of the equation may be written as $\bar{a}\Sigma m_i y_i$ where the y-coordinate of m_i is $y_i = \rho_i \cos \beta$. But from the definition of the mass center, $\Sigma m_i y_i = \bar{y}m = 0$, so that the first moment summation is zero. The resultant sum of moments about G due to all forces acting on all particles is then merely

$$\Sigma\overline{M} = \Sigma m_i\rho_i^2\alpha = \alpha\Sigma m_i\rho_i^2 = \bar{I}\alpha$$

where

$$\bar{I} = \Sigma m_i\rho_i^2 \qquad \text{or} \qquad \bar{I} = \int \rho^2 \, dm$$

This integral was developed in the previous article and is the mass moment of inertia of the body about an axis through G normal to the plane of motion.

The moment sum $\Sigma\overline{M}$ includes the moments of forces external and internal to the body. The contribution to the moment sum of the internal forces is zero since for every internal force there is an equal and opposite reaction. Thus the moment sum represents the sum of the moments about G of *all external forces* acting on the body. The moment equation and the two scalar components of the generalized Newton's second law of motion, Eq. 92*a*, give

$$\blacktriangleright \qquad \Sigma\overline{M} = \bar{I}\alpha \qquad \Sigma F_x = m\bar{a}_x \qquad \Sigma F_y = m\bar{a}_y \qquad \textbf{(112)}$$

Equations 112 are the general equations of motion for a rigid body in plane motion. Figure 52*a* illustrates schematically the free-body diagram of a rigid body in plane motion with angular acceleration α and mass-center accelera-

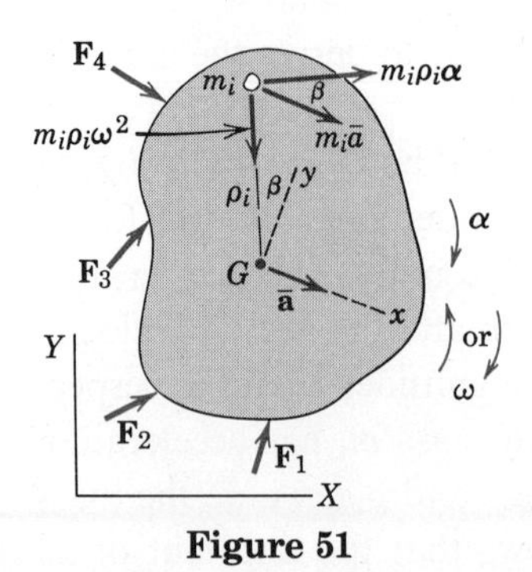

Figure 51

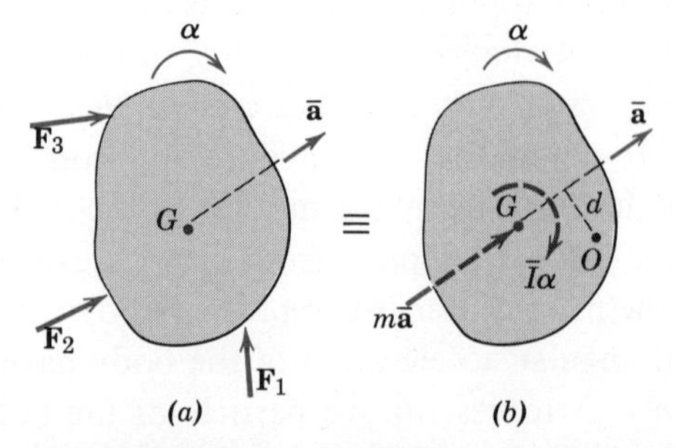

Figure 52

tion $\bar{\mathbf{a}}$ at the instant considered. Figure 52*b* illustrates the equivalent resultants of the applied force system which are a resultant force $m\bar{\mathbf{a}}$ through G in the direction of $\bar{\mathbf{a}}$ and a resultant couple $\bar{I}\alpha$ in the sense of α. This figure illustrates one of the most important and useful of the conclusions in dynamics. This equivalence between the external force system which produces plane motion of a rigid body, Fig. 52*a*, and the resultants of this system, Fig. 52*b*, makes it possible to write the necessary equations that establish the instantaneous relations between the forces and the accelerations for any given problem. Representation of the resultants $m\bar{\mathbf{a}}$ and $\bar{I}\alpha$ for every problem in plane motion will ensure that the force and moment sums as disclosed from the free-body diagram are equated to their proper resultants. This representation permits complete freedom of choice of a convenient moment center. If point O in Fig. 52*b* is such a convenient point, the summation of moments of the external forces about O would give

$$\Sigma M_O = \bar{I}\alpha + m\bar{a}d \tag{113}$$

If the moment center is chosen on the opposite side of $m\bar{a}$, it is clear that the sign of the $m\bar{a}d$ term would be negative for a clockwise summation in the sense of α. Equation 113 is merely an expression of the principle of moments. It states that the sum of the moments about some point O of all external forces acting on the body in the plane of motion equals the moment of their resultant. The resultant is expressed as the couple $\bar{I}\alpha$ and the force $m\bar{\mathbf{a}}$ through G.

The foregoing discussion will now be applied to the three cases of motion in a plane, namely, translation, fixed-axis rotation, and general plane motion.

Part A: Translation

Rigid-body translation in plane motion was described in Art. 27 and illustrated in Figs. 42*a* and 42*b* where it is seen that every line in a translating body remains parallel to its original position at all times. In rectilinear translation all points move in straight lines, whereas in curvilinear translation all points move on congruent curved paths. In either case there can be no angular motion of the translating body. With zero angular acceleration Eqs. 112 become

$$\Sigma F_x = m\bar{a}_x \qquad \Sigma F_y = m\bar{a}_y \qquad \Sigma \bar{M} = 0 \tag{114}$$

which eliminates all reference to the moment of inertia. Figure 53 represents the free-body diagram and its equivalent resultant-force diagram for a translating body. With the equivalence established between the two diagrams, it is readily seen that the moment equation may be written in any one of the three following ways:

$$\Sigma \bar{M} = 0 \qquad \Sigma M_A = 0 \qquad \Sigma M_O = m\bar{a}d \tag{115}$$

where O is any point not on a line through G in the direction of $\bar{\mathbf{a}}$ and where

A is any point on this line. For curvilinear translation the resultant $m\bar{\mathbf{a}}$ may be represented by its normal and tangential components $m\bar{a}_n$ and $m\bar{a}_t$.

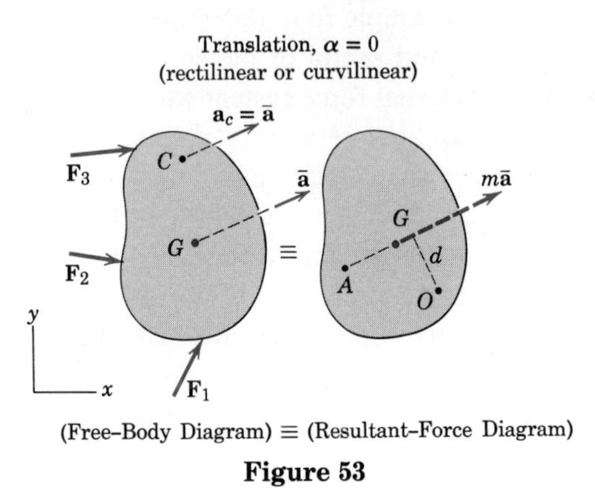

(Free-Body Diagram) $\equiv$ (Resultant-Force Diagram)

Figure 53

Sample Problem

6/32 The vertical bar AB has a mass of 150 kg with center of mass G half way between the ends. The bar is elevated from rest at $\theta = 0$ by means of the parallel links of negligible mass with a constant couple $M = 5$ kN·m applied to the lower link at C. Determine the angular acceleration α of the links as a function of θ, and find the force B in the link DB at the instant when $\theta = 30°$.

Solution. The motion of the bar is seen to be curvilinear translation since the bar itself does not rotate during the motion. With negligible mass of the links the tangential component A_t of the force at A is $A_t = M/\overline{AC} = 5/1.5 = 3.33$ kN, and the force at B is along the link. All applied forces are shown on the free-body diagram, and the resultant-force diagram is also indicated where the resultant force is shown in terms of its two components.

A summation of forces in the t-direction will eliminate all unknowns other than α, so that for the general position shown

$$[\Sigma F_t = m\bar{a}_t] \qquad 3.33 - 0.15(9.81)\cos\theta = 0.15(1.5\alpha)$$

$$\alpha = 14.81 - 6.54\cos\theta \text{ rad/s}^2 \qquad \textit{Ans.}$$

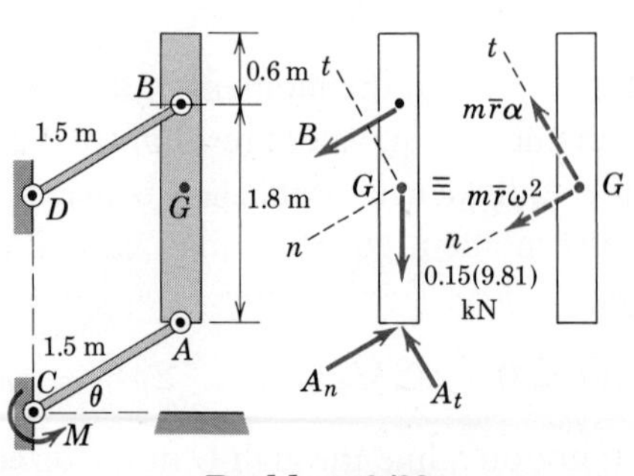

Problem 6/32

With α a known function of θ, the angular velocity ω of the links is obtained from

$$[\omega\,d\omega = \alpha\,d\theta] \qquad \int_0^\omega \omega\,d\omega = \int_0^\theta (14.81 - 6.54\cos\theta)\,d\theta \qquad \omega^2 = 29.6\theta - 13.08\sin\theta$$

Substitution for $\theta = 30°$ gives

$$(\omega^2)_{30°} = 8.97\ (\text{rad/s})^2 \qquad \alpha_{30°} = 9.15\ \text{rad/s}^2$$

and

$$m\bar{r}\omega^2 = 0.15(1.5)(8.97) = 2.02\ \text{kN}$$

$$m\bar{r}\alpha = 0.15(1.5)(9.15) = 2.06\ \text{kN}$$

The force B may be obtained by a moment summation about A, which eliminates A_n and A_t and the weight. Or a moment summation may be taken about the intersection of A_n and the line of action of $m\bar{r}\alpha$, which eliminates A_n and $m\bar{r}\alpha$. Using A as a moment center gives

$$[\Sigma M_A = m\bar{a}d] \qquad 1.8\frac{\sqrt{3}}{2}B = 2.02(1.2)\frac{\sqrt{3}}{2} + 2.06(0.6)$$

$$B = 2.14\ \text{kN} \qquad \textit{Ans.}$$

The component A_n could be obtained from a force summation in the n-direction or from a moment summation about G or about the intersection of B and the line of action of $m\bar{r}\alpha$.

Part B: Fixed-axis Rotation

The resultant-force system for the general case of plane motion, shown in Fig. 52, is easily adapted to the plane rotation of a rigid body about a fixed axis, as shown in Fig. 54*a*. If the angular acceleration and angular velocity at the instant considered are α and ω, respectively, then the acceleration $\bar{\mathbf{a}}$ of the mass center G has the normal and tangential components $\bar{a}_n = \bar{r}\omega^2$ and $\bar{a}_t = \bar{r}\alpha$ as shown. The *b*-part of the figure represents the free-body diagram upon which are shown all forces acting on the isolated body from external sources. Included are the force exerted on the body by the bearing at O and the weight, if the body is not rotating in the horizontal plane. This external force system is equivalent to the resultant-force system $m\bar{a}$ and $\bar{I}\alpha$ which is

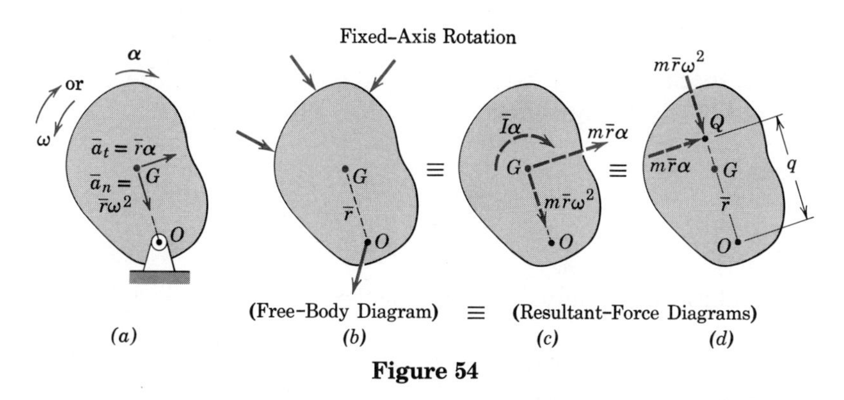

Figure 54

represented in the c-part of the figure. Here the resultant force is replaced by its components $m\bar{r}\omega^2$ and $m\bar{r}\alpha$ which act through G in the direction of its normal and tangential accelerations. The equivalence between the free-body diagram and the resultant-force system of Fig. 54c is sufficient to solve any problem of fixed-axis rotation.

It is usually more convenient to combine the couple $\bar{I}\alpha$ and the force $m\bar{r}\alpha$ by moving the force to a parallel position through a point Q a distance q from the point of rotation O and in line with G and O, Fig. 54d. The distance q is calculated to preserve the same resultant moment about O. Thus $m\bar{r}^2\alpha + \bar{I}\alpha = m\bar{r}\alpha q$. Substituting $\bar{I} = m\bar{k}^2$ and cancelling $m\alpha$ give $\bar{r}^2 + \bar{k}^2 = \bar{r}q$ so that $q = (\bar{r}^2 + \bar{k}^2)/\bar{r} = k_O{}^2/\bar{r}$ where k_O is the radius of gyration of the body about O. Point Q is a unique point for a given body with a given fixed point O and is called the *center of percussion.* The sum of moments about point O may now be written $\Sigma M_O = m\bar{r}\alpha(k_O{}^2/\bar{r}) = I_O\alpha$ where $I_O = mk_O{}^2$. The three equations of motion for fixed-axis rotation, written in their most useful form, are, therefore

$$\begin{aligned} \Sigma F_n &= m\bar{r}\omega^2 \\ \Sigma F_t &= m\bar{r}\alpha \\ \Sigma M_O &= I_O\alpha \end{aligned} \qquad \textbf{(116)}$$

It is important to note that, whereas the moment equation about O does not involve the bearing force exerted on the body at O, each of the force equations must include any component of this bearing force in its respective n- or t-direction. Thus the bearing force must not be omitted from the free-body diagram.

With the equivalence established between the free-body diagram and either of the two resultant-force diagrams there is complete freedom in choosing a moment center which is most convenient in any particular problem. It is readily apparent from Fig. 54d that the center of percussion Q provides a unique moment center since a moment sum about Q must be zero. The student should now observe the alternative moment equations about the three special points O, G, and Q. They are

$$\Sigma M_O = I_O\alpha \qquad \Sigma\bar{M} = \bar{I}\alpha \qquad \Sigma M_Q = 0$$

These equations are not independent, but any two of them may be used to replace the second and third of Eqs. 116.

For the special case of rotation of a rigid body about a fixed axis through the mass center G, the quantity $\bar{r}$ is zero, and Eqs. 116 become

$$\begin{aligned} \Sigma F_x &= 0 \\ \Sigma F_y &= 0 \\ \Sigma\bar{M} &= \bar{I}\alpha \end{aligned} \qquad (117)$$

where x and y are any mutually perpendicular directions and where O and G are the same point. Since the resultant force is zero, the resultant for this case of centroidal rotation is a couple. It follows that $\Sigma\bar{M}$ is the same as a moment summation about any axis parallel to the centroidal axis.

Sample Problem

6/33 The pendulum has a mass of 7.5 kg with center of mass at G and has a radius of gyration about the pivot O of 295 mm. If the pendulum is released from rest at $\theta = 0$, determine the total force supported by the bearing at the instant when $\theta = 30°$. Friction in the bearing is negligible.

Solution. The free-body diagram of the pendulum is constructed as shown in the b-part of the figure, and the resultants of the external forces are shown in the c-part of the figure. Note that these resultants are shown acting through the center of percussion each in the sense of its acceleration component of G. If they had been shown acting through G, it would have been necessary to add the couple $\bar{I}\alpha$, as shown in Fig. 54c.

The normal component O_n is found from a force equation in the n-direction which involves the normal acceleration $\bar{r}\omega^2$. Since the angular velocity ω of the pendulum is found from the integral of the angular acceleration and since O_t depends on the tangential acceleration $\bar{r}\alpha$, it follows that α should be obtained first. To this end, the moment equation about O gives

$[\Sigma M_O = I_O\alpha]$ $\qquad 7.5(9.81)(0.25)\cos\theta = 7.5(0.295)^2\alpha \qquad \alpha = 28.2\cos\theta \text{ rad/s}^2$

and for $\theta = 30°$

$[\omega\, d\omega = \alpha\, d\theta]$ $\qquad \int_0^{\omega} \omega\, d\omega = \int_0^{\pi/6} 28.2\cos\theta\, d\theta \qquad \omega^2 = 28.2 \text{ (rad/s)}^2$

The remaining two equations of motion applied to the 30° position yield

$[\Sigma F_n = m\bar{r}\omega^2]$ $\qquad O_n - 7.5(9.81)(0.5) = 7.5(0.25)(28.2) \qquad O_n = 89.6 \text{ N}$

$[\Sigma F_t = m\bar{r}\alpha]$ $\qquad 7.5(9.81)(0.866) - O_t = 7.5(0.25)(28.2)(0.866) \qquad O_t = 17.96 \text{ N}$

The total force on the bearing at O for this position is, then,

$$O = \sqrt{(89.6)^2 + (17.96)^2} = 91.4 \text{ N} \qquad \textit{Ans.}$$

The proper sense for O_t may be observed at the outset by applying the alternative moment equation $\Sigma\bar{M} = \bar{I}\alpha$ where the moment about G due to O_t must be clockwise to agree with α. The proper sense of O_t may also be seen initially by the relation $\Sigma M_Q = 0$ which equation may be used to obtain O_t directly without obtaining α.

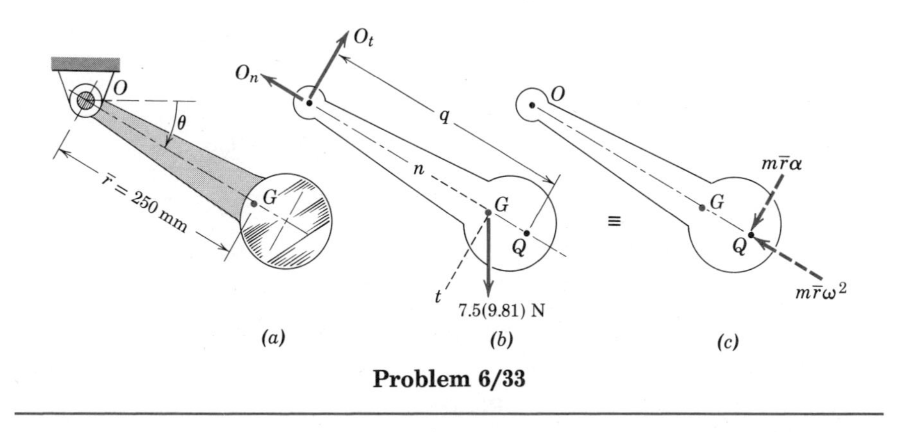

Problem 6/33

Part C: General Plane Motion

The general case of plane motion was introduced in the first part of this article. It was illustrated in Figs. 51 and 52 and described by Eqs. 112. Equation 113 represented the form of an alternative moment equation about any convenient point O other than the mass center G. As in the previous examples it is only necessary to establish the equivalence between the external force system, as disclosed by the free-body diagram, and the force resultants to solve the plane-motion problem.

Mention should be made of two special cases in plane motion which occur with sufficient frequency to warrant special attention. The first case occurs when the moment center O, as a point on the body or body extended, has no acceleration. The moment equation about O then becomes

$$\Sigma M_O = I_O\alpha \tag{118}$$

which meets the same conditions as for a body rotating about a fixed axis through O. Point O need not be fixed but could have a constant velocity.

The second case of frequent occurrence exists when a moment center O is chosen which has an acceleration directed toward G, Fig. 55*a*. The acceleration of G, written in terms of the acceleration of O, has the components a_O, $\bar{r}\omega^2$, and $\bar{r}\alpha$ so that the resultant force $m\bar{\mathbf{a}}$ has the components ma_O, $m\bar{r}\omega^2$, and $m\bar{r}\alpha$ as shown in the *b*-part of the figure. The moment sum about O becomes $\Sigma M_O = \bar{I}\alpha + m\bar{r}^2\alpha$. Substitution of $I_O = \bar{I} + m\bar{r}^2$ gives

$$\Sigma M_O = I_O\alpha \tag{118a}$$

Figure 55*c* shows the frequently-encountered example of the foregoing situation which occurs for a rolling wheel with mass center G at the geometric center. Here the instantaneous center C of zero velocity has an acceleration directed toward the mass center, and therefore the equation

$$\Sigma M_C = I_C\alpha \tag{119}$$

may be used for the moment equation as long as the wheel is not slipping. If the wheel slips or if the mass center is not the geometric center, then the acceleration of the contact point C would not in general pass through G and Eq. 119 would not hold.

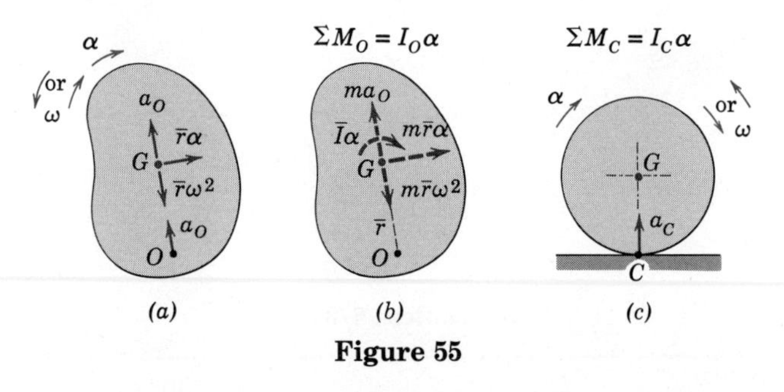

Figure 55

Strong emphasis should be placed on the importance of a clear choice of the body to be isolated and the representation of this isolation by a correct free-body diagram. Only after this vital step has been completed can the equivalence between the external forces and their resultants be properly evaluated. Of equal importance in the analysis of plane motion is a clear understanding of the kinematics involved. Very often the difficulties experienced by students at this point have to do with kinematics. The student will find that a thorough review of the relative acceleration relations for plane motion will be most helpful. In formulating the solution to a problem it should be recognized that the directions of certain forces or accelerations may not be known at the outset, so that it may be necessary to make initial assumptions whose validity will be proved or disproved when the solution is carried out. It is essential, however, that all assumptions made are consistent with the principle of action and reaction and with any kinematical requirements, which are also called conditions of constraint. Thus if a wheel is rolling on a horizontal surface, its center is constrained to move in a horizontal line. Furthermore, if the unknown linear acceleration a of the center of the wheel is assumed positive to the right, the unknown angular acceleration α must be positive in a clockwise sense in order that $a = +r\alpha$, assuming the wheel does not slip. Also, it should be noted that for a wheel which rolls without slipping $a = r\alpha$, but the friction force F between the wheel and its supporting surface is generally less than its maximum value, so that $F \neq fN$. But if the wheel slips as it rolls, $a \neq r\alpha$ although the friction force has reached its limiting value so that $F = fN$. It may be necessary to test the validity of either assumption in a given problem.

Dynamic equilibrium. In the preceding discussion of this article the basic relationships among force, mass, and acceleration have been developed and applied to various cases of rigid-body motion. In this analysis the resultants of the applied force system were determined for each type of motion encountered, and the equivalence between the external forces and their resultants was used in the solution of each problem. This approach is straightforward and treats dynamics in a direct manner by always relating the external forces to the accelerations, described from fixed reference axes.

An alternative viewpoint for the solution of problems involving the relationship between forces and accelerations is contained in an approach known as *D'Alembert's principle* which was described briefly in Part B of Art. 21 on the kinetics of particles as viewed from reference axes moving with the particle. This principle may easily be extended to describe the plane motion of rigid bodies. By attaching reference axes to the rigid body of Fig. 52 which executes general plane motion, an observer who is fixed to these axes can measure no motion of the body with respect to his moving reference frame. Relative to these moving axes, then, the observer can conclude that the body is in "equilibrium." Clearly, the only way in which the applied forces can be made to have no unbalanced resultant and be consistent with this "equilibrium" is to impose a fictitious force $-m\bar{\mathbf{a}}$ on the body to counterbalance the resultant or *effective force* which equals $m\bar{\mathbf{a}}$. This fictitious force is referred to as the *inertia force* or *reversed effective force,* and it must be applied through the mass center G in the direction opposite to the acceleration $\bar{\mathbf{a}}$ of G. Similarly, the only way in which the applied forces can be made to have no unbalanced moment is to impose a fictitious *inertia couple* $-\bar{I}\alpha$ on the body in the rotational sense opposite to the angular acceleration α.

Thus the applied system of actual forces $\mathbf{F}_1$, $\mathbf{F}_2$, $\mathbf{F}_3, \ldots$ along with the fictitious inertia force and inertia couple would constitute an equilibrium system and would create an artificial state known as *dynamic equilibrium.* The familiar equilibrium equations $\Sigma\mathbf{F} = 0$ and $\Sigma M = 0$ may then be applied to the free-body diagram using any convenient reference directions for the zero force summations and any convenient moment center for the zero moment summation.

If the method known as D'Alembert's principle is used to create dynamic equilibrium, the free-body diagram should include the inertia force and the inertia couple, and the principles of equilibrium should be stated and applied. If the dynamic equilibrium method is not used, the free-body diagram should disclose only the actual applied forces which should then be equated to their resultant force and their resultant couple with the aid of a resultant-force diagram and by direct application of the equations of motion appropriately chosen from Eqs. 112 through 119. This latter and more direct description is employed in this book in preference to the method of dynamic equilibrium.

Sample Problems

6/34 The drum A is given a constant angular acceleration α_0 of 3 rad/s^2 and causes the 70-kg spool B to roll on the horizontal surface by means of the connecting cable which wraps around the inner hub of the spool. The radius of gyration $\bar{k}$ of the spool about its mass center G is 250 mm, and the coefficient of friction between the spool and the horizontal surface is 0.25. Determine the tension T in the cable and the friction force F exerted by the horizontal surface on the spool.

Solution: The free-body diagram and the resultant-force diagram of the spool are drawn as shown. The correct direction of the friction force may be assigned in this problem by observing from both diagrams that with counterclockwise angular acceleration a moment sum about point G (and also about point D) must be counterclockwise. A point on the connecting cable has an acceleration $a_t = r\alpha = 0.25(3) = 0.75$ m/s^2, which is also the horizontal component of the acceleration of point D on the spool. It will be assumed initially that the spool rolls without slipping in which case it would have a counterclockwise angular acceleration $\alpha = (a_D)_x/\overline{DC} = 0.75/0.30 = 2.5$ rad/s^2. The acceleration of the mass center G is, therefore, $\bar{a} = r\alpha = 0.45(2.5) = 1.125$ m/s^2.

With the kinematics determined, the forces may be found from the equivalence between the free-body diagram and the resultant-force diagram. Thus

$$[\Sigma M_C = \bar{I}\alpha + m\bar{a}r] \qquad 0.3T = 70(0.25)^2(2.5) + 70(1.125)(0.45)$$

$$T = 154.6 \text{ N} \qquad \textit{Ans.}$$

The friction force may be obtained by the force summation

$[\Sigma F_x = m\bar{a}_x]$ $\quad 154.6 - F = 70(1.125) \quad F = 75.8$ N *Ans.*

Establishing the validity of the answer requires a check of the assumption that no slipping occurs. With

$[\Sigma F_y = 0]$ $\quad N = 687$ N

the surfaces are capable of supporting a maximum friction force equal to $F = fN = 0.25(687) = 171.7$ N. Thus the assumption that the spool rolls without slipping is valid since a friction force of only 75.8 N is required to maintain pure rolling.

If the coefficient of friction had been 0.1, for example, then the friction force would have been limited to $0.1(687) = 68.7$ N which is less than 75.8 N, and the spool would slip. In this event the kinematical relation $\bar{a} = r\alpha$ would no longer hold. With $(a_D)_x$ known, the angular acceleration would be $\alpha = [\bar{a} - (a_D)_x]/\overline{GD}$. The correct acceleration $\bar{a}$ of G would then be obtained from a moment equation about point D, and a force equation in the x-direction would determine the new value of T.

Alternatively, the given problem with the assumption of no slipping could be solved by using Eq. 119 directly. This equation is identical with the moment equation about C which was obtained from the resultant-force diagram. Also, an equation for moments about point D could be written initially to give F after which T could be obtained from a force equation.

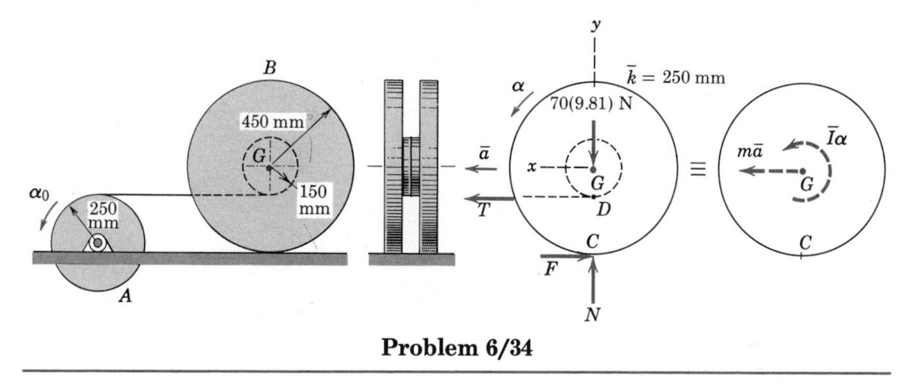

Problem 6/34

6/35 Determine the forces on the piston pin A and crank pin B of the slider-crank mechanism for the crank position $\theta = 30°$ and for a constant clockwise crank speed of 1200 rev/min. The connecting rod AB has a mass of 12 kg with center of mass at G and has a centroidal radius of gyration $\bar{k} = 160$ mm. Friction and pressure on the 5-kg piston are considered negligible. The plane of motion is horizontal.

Solution. The forces acting on the rod will depend on its acceleration which was determined in Sample Prob. 5/22 for this same configuration. The values obtained for this position are $a_A = 3420$ m/s² to the right, $\alpha_{AB} = 2830$ rad/s² clockwise, and $\omega_{AB} = 44.4$ rad/s counterclockwise. Additionally the angle ϕ is

$$\phi = \sin^{-1}\left(\tfrac{2}{5}\sin 30°\right) = \sin^{-1} 0.2 = 11.54°$$

The horizontal component of the piston-pin force at A is determined by the mass and acceleration of the piston. Thus

$[\Sigma F = ma]$ $\quad P = 5(3420) = 17\,100$ N

The free-body diagram of the rod is drawn, and the forces at A and B are represented by their components. The weights of the members are omitted since they are negligible compared with the other forces which act. The resultant-force diagram is shown to the right of the free-body diagram. The resultant $m\bar{a}$ is represented here by its components where the acceleration terms are written from the relative acceleration equation with respect to A. These components of $m\bar{a}$ and the resultant couple are

$$ma_A = 12(3420) = 41\,000 \text{ N}$$

$$m\bar{r}\omega^2 = 12(0.2)(44.4)^2 = 4730 \text{ N}$$

$$m\bar{r}\alpha = 12(0.2)(2830) = 6790 \text{ N}$$

$$\bar{I}\alpha = 12(0.16)^2(2830) = 869 \text{ N}\cdot\text{m}$$

The three remaining unknown forces on the rod may now be found from the three equations of motion. In evaluating the first equation it is noted that the $m\bar{a}d$ term is expressed in terms of its components.

$[\Sigma M_B = \bar{I}\alpha + m\bar{a}d]$

$$-R(0.5 \cos 11.54°) + 17.10(0.2 \sin 30°) = 0.869 - 41.0(0.3 \sin 11.54°) - 6.79(0.3)$$

$$R = 10.90 \text{ kN}$$

$[\Sigma F_n = m\bar{a}_n]$ $\quad N + 17.10 \cos 11.54° + 10.90 \sin 11.54° = 4.73 - 41.0 \cos 11.54°$

$$N = -54.4 \text{ kN}$$

$[\Sigma F_t = m\bar{a}_t]$ $\quad T - 17.10 \sin 11.54° + 10.90 \cos 11.54° = 6.79 + 41.0 \sin 11.54°$

$$T = 7.74 \text{ kN}$$

Thus the respective pin forces are

$$A = \sqrt{(10.90)^2 + (17.10)^2} = 20.3 \text{ kN} \qquad \textit{Ans.}$$

$$B = \sqrt{(-54.4)^2 + (7.74)^2} = 55.0 \text{ kN} \qquad \textit{Ans.}$$

It is not always possible to assign the unknown force components initially in the correct direction on the free-body diagram, and a negative sign from the computation, as in the present case with N, will indicate that the direction of the force is opposite to that assumed.

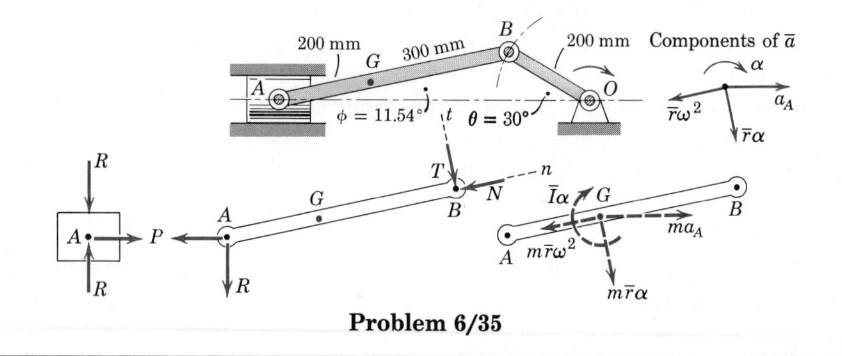

Problem 6/35

Problems

Translation

6/36 The arm OA of the classifying accelerometer has a mass of 0.2 kg with center of mass at G. The adjusting screw and spring are preset to a force of 8 N at B. At what acceleration a would the electrical contacts at A be on the verge of opening? Motion is in the vertical plane of the figure. *Ans.* $a = 7.12g$

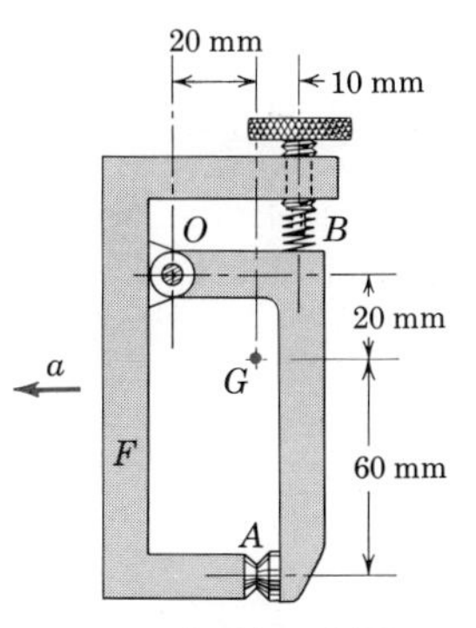

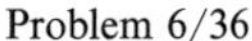
Problem 6/36

6/37 Solid homogeneous cylinders 400 mm high and 250 mm in diameter are supported by a flat conveyor belt which moves horizontally. If the speed of the belt increases according to $v = 1.2 + 0.9t^2$ m/s, where t is the time in seconds measured from the instant the increase begins, calculate the value of t for which the cylinders begin to tip over. Cleats on the belt prevent the cylinders from slipping.

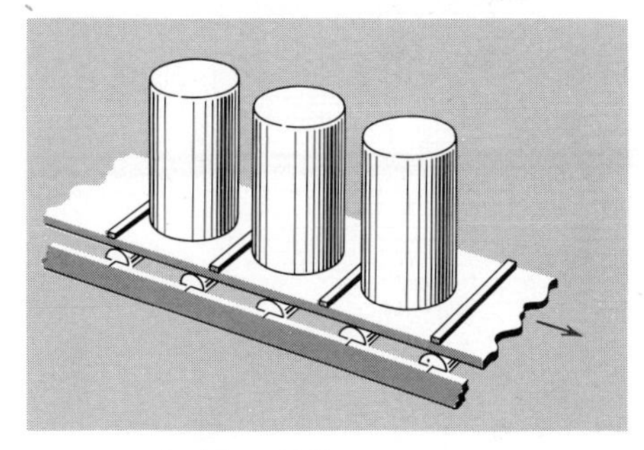

Problem 6/37

6/38 The vertical 800-kg plate with mass center at G is held in the position shown by the parallel cables A and B and the horizontal cable C. If the cable C is suddenly released, calculate the tension in cable B immediately after release. *Ans.* $T_B = 1792$ N

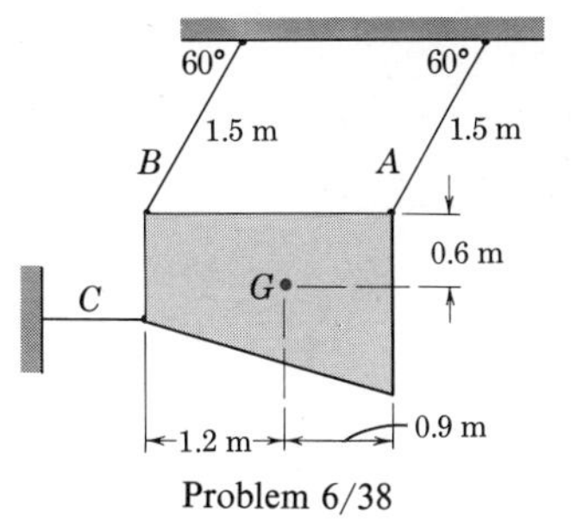

Problem 6/38

6/39 The block A and attached rod have a combined mass of 50 kg and are confined to move along the 60° guide under the action of the 600-N applied force. The uniform horizontal rod has a mass of 15 kg and is welded to the block at B. Friction in the guide is negligible. Compute the bending moment M exerted by the weld on the rod at B.

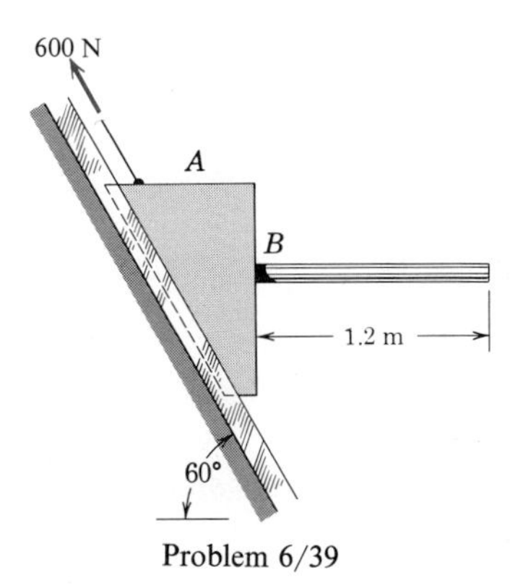

Problem 6/39

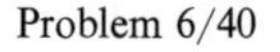

Problem 6/40

6/40 The loaded trailer has a mass of 900 kg with center of mass at G and is attached at A to a rear-bumper hitch. If the car and trailer reach a velocity of 60 km/h on a level road in a distance of 30 m from rest with constant acceleration, compute the vertical component of the force supported by the hitch at A. Neglect the small friction force exerted on the relatively light wheels. *Ans.* $A_y = 1389$ N

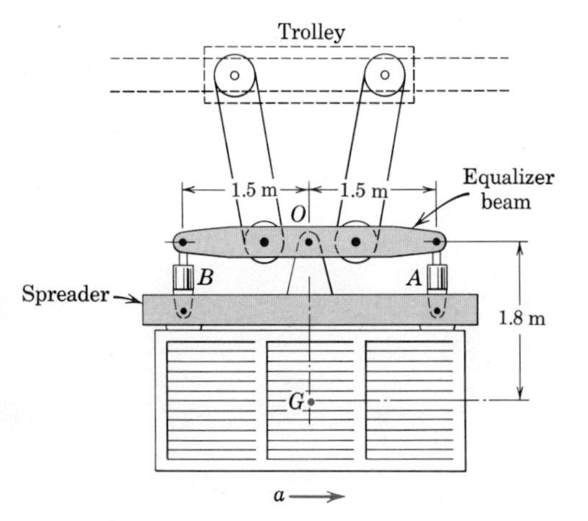

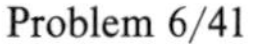

Problem 6/41

6/41 An antiswing system for unloading cargo containers from ships consists of the moving trolley and its suspended equalizer beam and spreader, to which the loaded container is secured. Rotation of the spreader relative to the equalizer beam is controlled by the hydraulic cylinders A and B which have a piston diameter of 100 mm. During the horizontal acceleration of the system with a load of 4 t the oil pressure in cylinder A is 400 kPa and that in cylinder B is 500 kPa. Compute the acceleration a of the system. The center of mass of the load is at G.

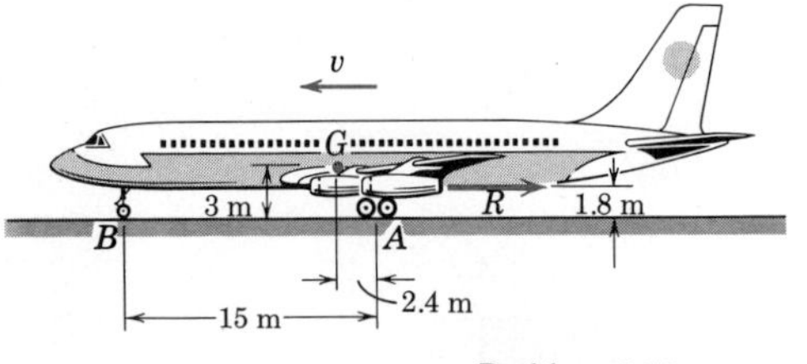

Problem 6/42

6/42 A jet transport with a landing speed of 200 km/h reduces its speed to 50 km/h with a negative thrust R from its jet thrust reversers in a distance of 450 m along the runway with constant deceleration. The aircraft has a total mass 125 t with center of mass at G. Compute the reaction N under the nose wheel B toward the end of the braking interval and prior to the application of mechanical braking. At the lower speed aerodynamic forces on the aircraft are small and may be neglected. *Ans.* $N = 228$ kN

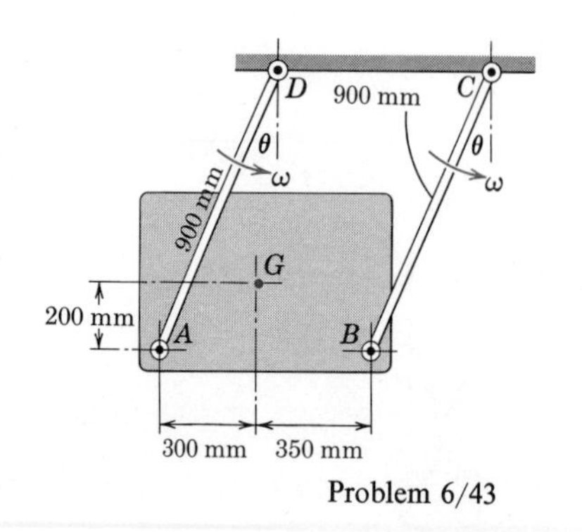

Problem 6/43

6/43 The 50-kg plate with center of mass at G is suspended by the light parallel links AD and BC and is free to swing in the vertical plane. At what angle θ of the links can the plate be released from rest and have zero force in link BC immediately after release? At this instant what is the force in link AD?

6/44 The 50-kg plate of Prob. 6/43 is given an initial velocity so that the two links have an angular velocity of 4 rad/s as they cross the vertical position with $\theta = 0$. For this instant compute the force in AD. *Ans.* $AD = 652$ N

6/45 Determine the expression for the maximum velocity v which the micro-bus can reach in a distance s from rest without slipping its rear driving wheels if the coefficient of friction between the tires and the pavement is f. Neglect the mass of the wheels.

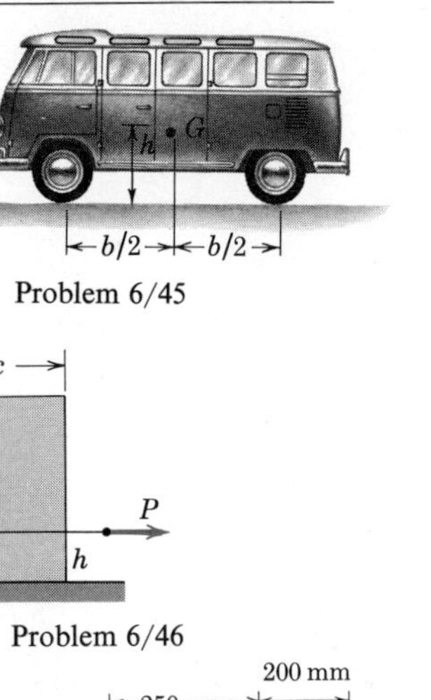

Problem 6/45

Problem 6/46

6/46 A force P is applied to the homogeneous rectangular box of mass m and causes it to move. If the coefficient of friction is f, determine the limiting values of h so that the box will slide without tipping about either the front edge or the rear edge.

$$Ans.\ h_{\binom{\max}{\min}} = \frac{1}{2}\left[b - \frac{mg}{P}(fb \mp c)\right]$$

6/47 The two identical wheels are mounted on the central frame and roll to the left without slipping with a constant velocity $v = 2.4$ m/s. The 7.5-kg connecting rod AB is pinned to the forward wheel at A, and the pin B of the second wheel fits into a smooth horizontal slot in AB. Determine the total forces exerted on the rod by pins A and B for the position $\theta = 60°$.

Ans. $A = 116.5$ N, $B = 60.9$ N

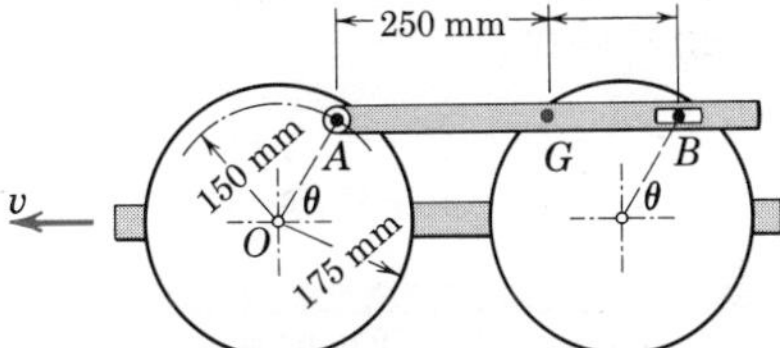

Problem 6/47

Fixed-Axis Rotation

6/48 Derive the moment equation of motion $\Sigma M_O = I_O\alpha$ for the plane rotation of a general rigid body about the fixed axis O. Start with the equation of motion for a representative particle of mass m_i of the body.

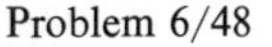

Problem 6/48

6/49 Each of the two drums and connected hubs of 250-mm radius has a mass of 100 kg and a radius of gyration about its center of 375 mm. Calculate the angular acceleration of each drum. Friction in each bearing is negligible.

Ans. $\alpha_a = 3.20$ rad/s^2, $\alpha_b = 3.49$ rad/s^2

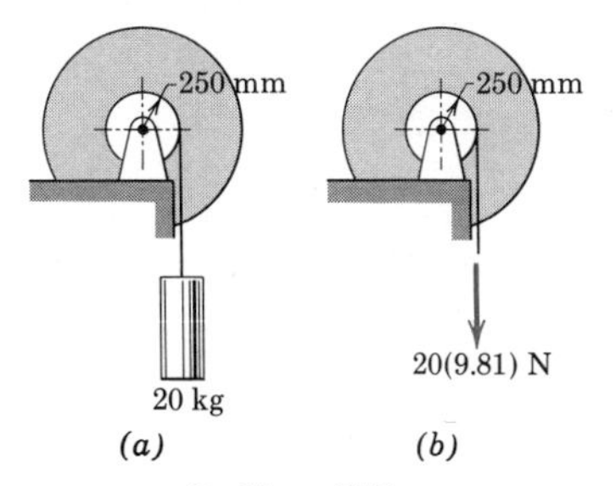

Problem 6/49

6/50 A total of 120 m of cable having a mass of 0.863 kg/m is wrapped around the drum A, which turns with negligible friction about its horizontal axis O–O. If the drum is released from rest with $x = 15$ m of cable unwound from the drum, calculate the initial angular acceleration α of the drum. The drum alone has a mass of 110 kg with radius of gyration about O–O of 546 mm.

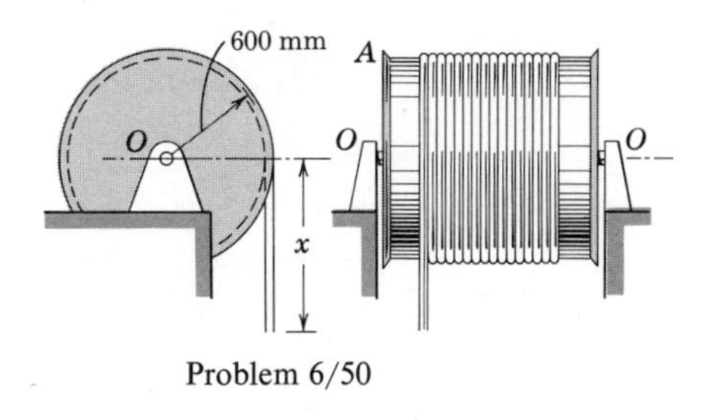

Problem 6/50

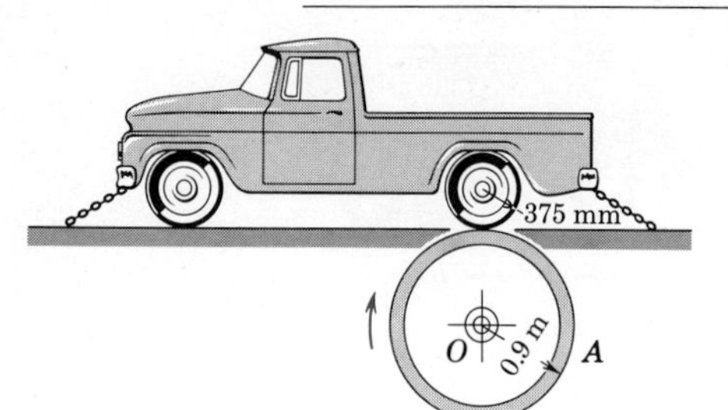

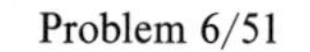
Problem 6/51

6/51 The automotive dynamometer is able to simulate road conditions for an acceleration of $0.6g$ for the loaded pick-up truck with a gross mass of 2.8 t. Calculate the required moment of inertia of the dynamometer drum about its center O assuming that the drum turns freely during the acceleration phase of the test.

Ans. $I_O = 2.27\ \text{t}\cdot\text{m}^2$

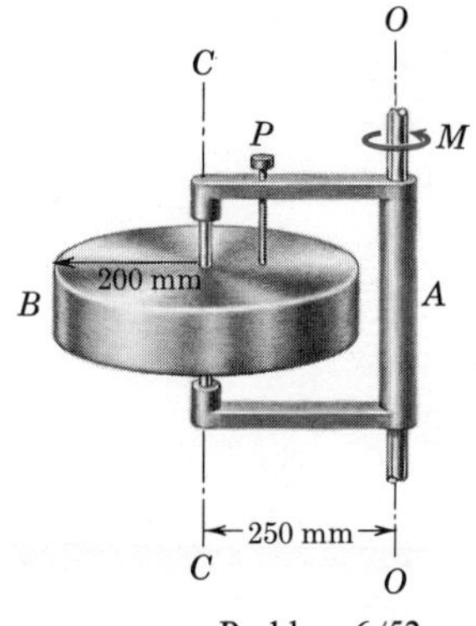

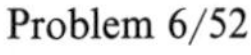
Problem 6/52

6/52 The solid cylindrical rotor B has a mass of 43 kg and is mounted on its central axis C-C. The frame A rotates about the fixed vertical axis O-O under the applied torque $M = 30$ N·m. The rotor may be unlocked from the frame by withdrawing the locking pin P. Calculate the angular acceleration α of the frame A if the locking pin is (a) in place and (b) withdrawn. Neglect all friction and the mass of the frame.

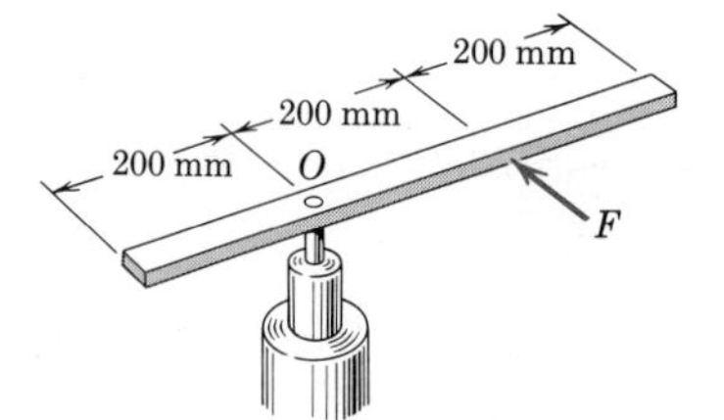

Problem 6/53

6/53 The 600-mm slender bar has a mass of 10 kg and is pivoted freely about a vertical axis at O. A horizontal force F, which has an initial value of 150 N, is applied to the bar when it is at rest. Calculate the initial value of the horizontal component of the reaction at O by using only one force or moment equation of motion.

Ans. $O = 75$ N

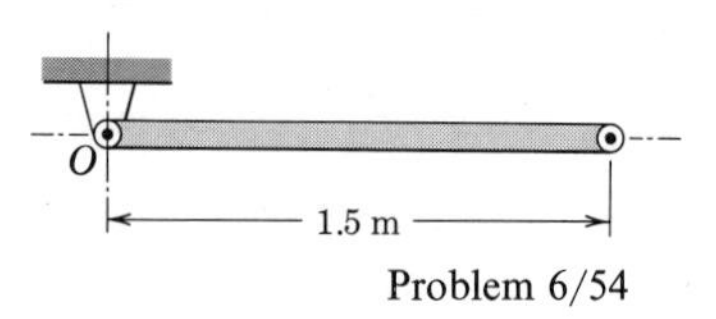

Problem 6/54

6/54 The uniform slender bar has a mass of 20 kg and is pivoted about a horizontal axis through O. The bearing at O fits closely on its shaft and exerts a constant frictional moment of 6 N·m on the bar opposing rotation. If the bar is released from rest in the horizontal position, compute the initial reaction on the bar at O.

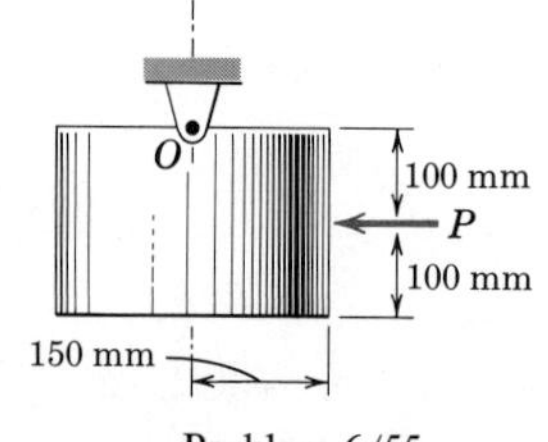

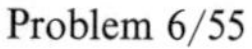
Problem 6/55

6/55 The solid 50-kg cylinder of 150-mm radius is pivoted about a horizontal diameter at O so that it can swing freely about this axis. An impact force P is applied as shown to the cylinder initially at rest, and a horizontal reaction results at the bearing. If the peak value of this reaction is measured with a sensitive transducer and found to be 100 N, determine the corresponding peak value of P.

Ans. $P = 212$ N

6/56 The uniform slender bar is suspended by a light cord from its end and swings as a pendulum. With the aid of a free-body diagram show that the bar cannot remain in line with the cord during the motion.

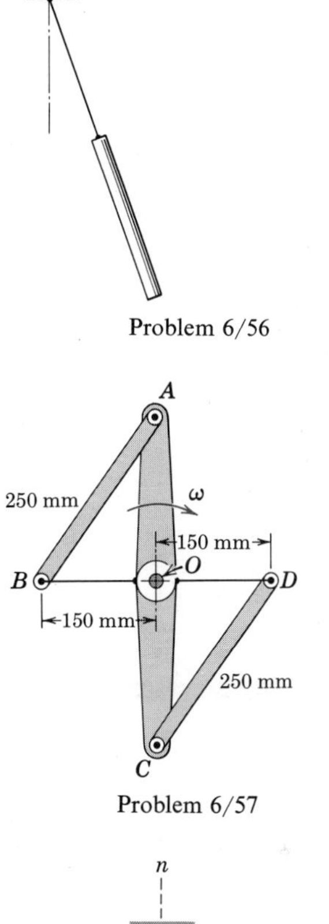

Problem 6/56

6/57 Link AC is made to revolve with a constant angular velocity $\omega = 10$ rad/s about a fixed vertical axis through its center O. Each of the uniform links AB and CD has a mass of 4 kg and is held in the configuration shown by a cord leading perpendicularly to the rotating link AC. Calculate the tension T in BO and DO.

Ans. $T = 30$ N

Problem 6/57

6/58 The slender bar of mass m and length l is acted upon by a torsion spring at O which exerts a counterclockwise couple M on the bar as it swings past the vertical position. If the angular velocity of the bar is ω in the vertical position, write the expressions for the n- and t-components of the bearing reaction on the bar at this instant.

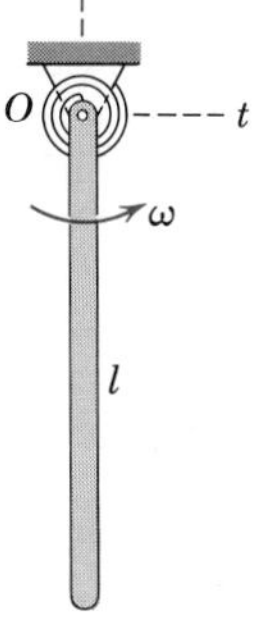

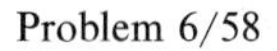

Problem 6/58

6/59 The 0.4-kg link B with center of mass 55 mm from O–O has a radius of gyration about O–O of 69 mm. The link is welded to the steel tube and is free to rotate about the fixed horizontal shaft at O–O. The mass of the tube is 0.92 kg. If the tube is released from rest with the link in the horizontal position, calculate the resulting initial angular acceleration α and the reaction O exerted by the shaft on the link.

Ans. $\alpha = 58.3\ \text{rad/s}^2$, $O = 3.28$ N

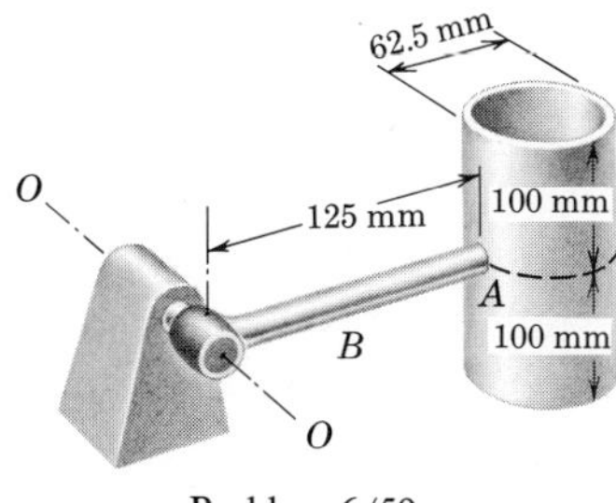

Problem 6/59

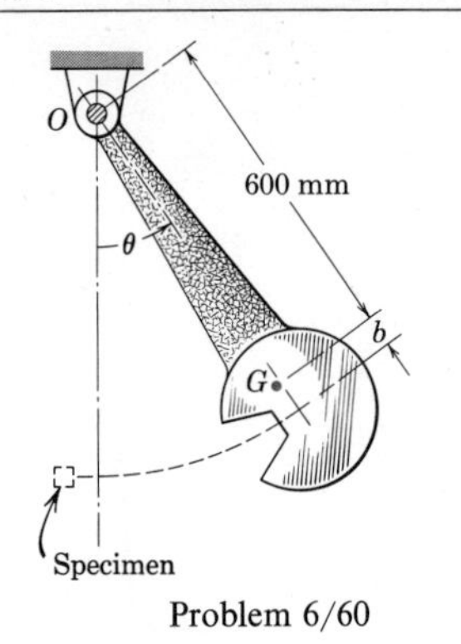

Problem 6/60

6/60 The pendulum for the impact-testing machine has a mass of 35 kg with center of mass at G and has a radius of gyration about O of 650 mm. The pendulum is designed so that the force on the bearing at O has no horizontal component during impact with the specimen at the bottom of the swing. Determine the distance b. Also calculate the total force on the bearing at O for the instant after the pendulum is released from rest at $\theta = 60°$.

Ans. $b = 104.2$ mm, $O = 177.2$ N

General Plane Motion

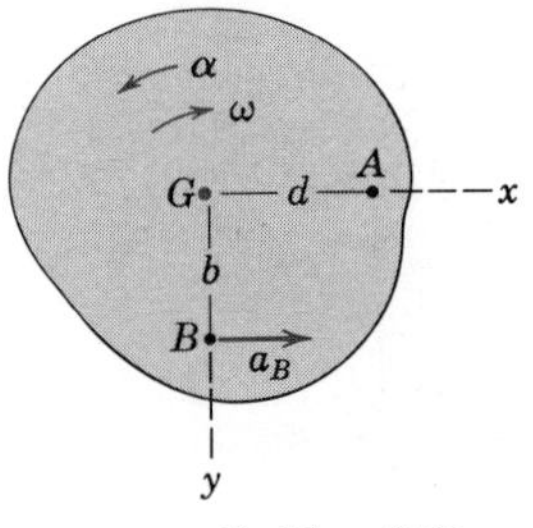

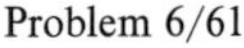
Problem 6/61

6/61 A rigid body of mass m with center of mass at G and centroidal moment of inertia $\bar{I}$ is moving in the plane of the figure under a system of external forces not shown. At the instant represented B has an absolute acceleration a_B, and the angular velocity and angular acceleration of the body are as indicated. From the resultant-force diagram indicate the expressions for the moment sums in a counterclockwise sense of external forces about points A and B.

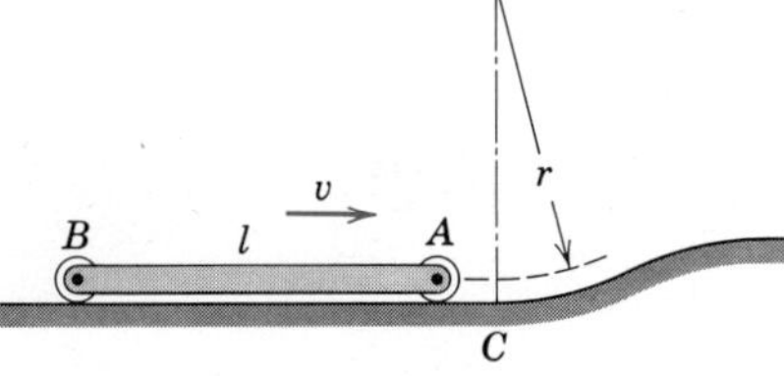

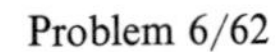
Problem 6/62

6/62 The uniform heavy bar AB of mass m is moving on its light end rollers along the horizontal with a velocity v when end A passes point C and begins to move on the curved portion of the path with radius r. Determine the force exerted by the path on the roller A immediately after it passes C.

Ans. $A = mg\left(\frac{1}{2} + \frac{v^2}{3gr}\right)$

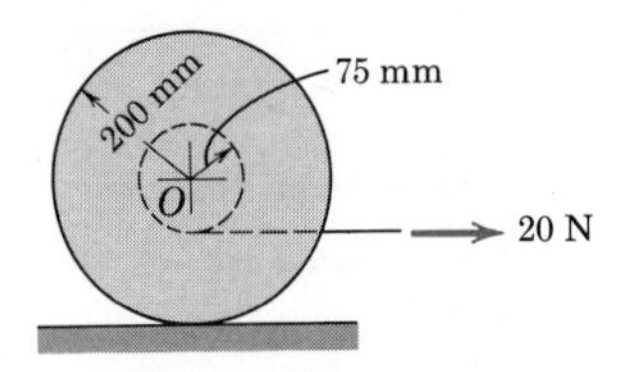

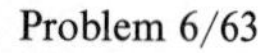
Problem 6/63

6/63 The circular disk of 200-mm radius has a mass of 25 kg with centroidal radius of gyration of 175 mm and has a concentric circular groove of 75 mm radius cut into it. If a steady horizontal force of 20 N is applied to a cord wrapped around the groove as shown, calculate the angular acceleration α of the disk as it starts from rest. The coefficient of friction between the disk and the horizontal surface is 0.10. (Be sure to recognize that the wheel rolls clockwise and *not* counterclockwise. Assume first that the wheel does not slip and then verify your assumption with the results.)

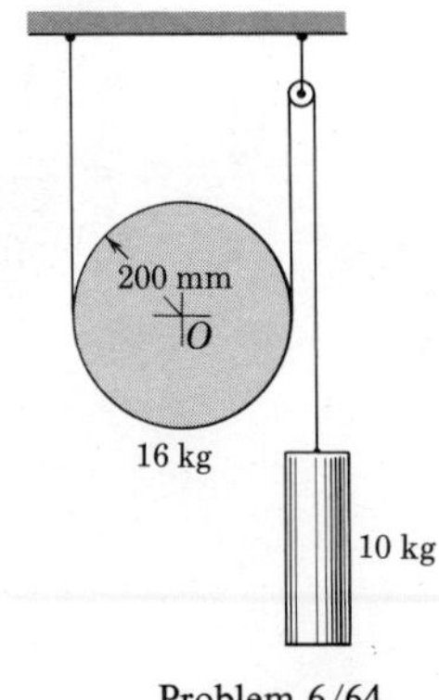

Problem 6/64

6/64 Calculate the upward acceleration a of the center O of the 16-kg solid circular disk of 200-mm radius when the 10-kg counterweight is allowed to fall.

Ans. $a = 0.613$ m/s^2

6/65 The uniform slender bar of length l and mass m is resting on a smooth horizontal surface. If a horizontal force F is suddenly applied normal to the bar at end A, determine the expression for the acceleration a_B of end B during application of F.

Problem 6/65

6/66 The spacecraft is spinning with a constant angular velocity ω about the z-axis at the same time that its mass center O is traveling with a velocity v_O in the y-direction. If a tangential hydrogen-peroxide jet is fired when the craft is in the position shown, determine the expression for the absolute acceleration of point A on the spacecraft rim at the instant the jet force is F. The radius of gyration of the craft about the z-axis is k, and its mass is m.

$$\textit{Ans. } \mathbf{a}_A = -\frac{Fr^2}{mk^2}\mathbf{i} - \left(\frac{F}{m} - r\omega^2\right)\mathbf{j}$$

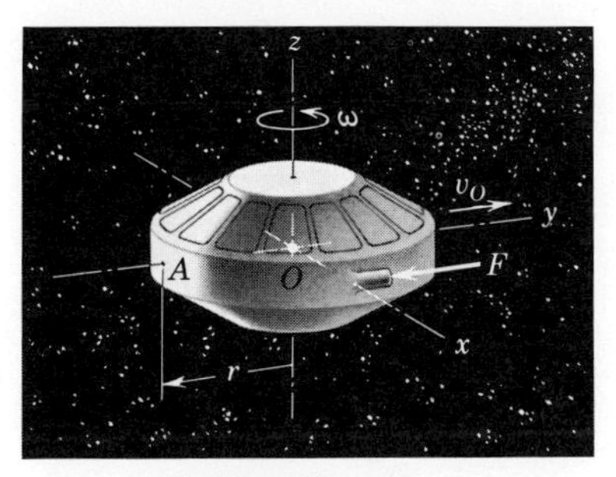

Problem 6/66

6/67 A 15-kg roll of wrapping paper in the form of a solid circular cylinder with a diameter of 300 mm is resting on a horizontal surface. If a force of 40 N is applied to the paper at the angle $\theta = 30°$, determine the initial acceleration a of the center of the roll and the corresponding angular acceleration α. The coefficient of friction between the paper and the horizontal surface is 0.2.

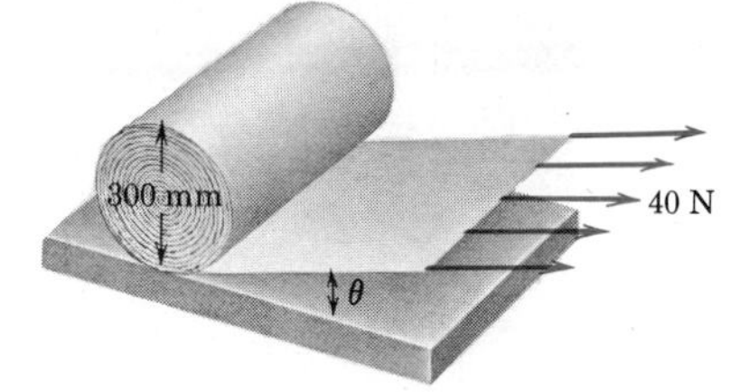

Problem 6/67

6/68 The slender rod of mass m and length l is released from rest in the vertical position with the small roller at end A resting on the incline. Determine the initial acceleration of A.

$$\textit{Ans. } a_A = \frac{g \sin\theta}{1 - \frac{3}{4}\cos^2\theta}$$

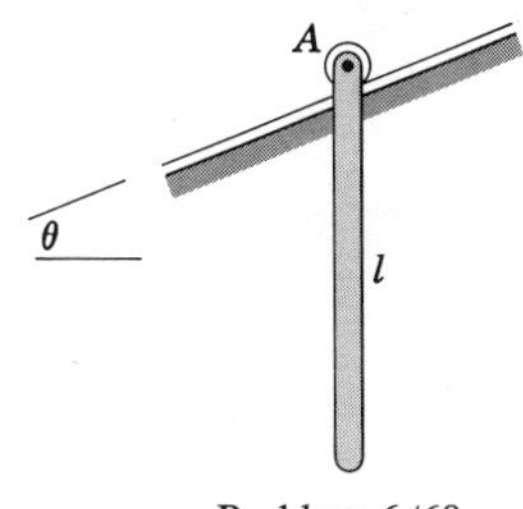

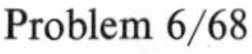

Problem 6/68

6/69 A Mars probe consists of a capsule of instruments and a rocket motor which gives it an acceleration a in the direction of its axis. During an acceleration period prior to approach to Mars, the exposure doors are opened automatically. Write an expression for the torque M which must be applied to the door at its hinge axis A–A for a given angle θ in order to limit the angular acceleration to a specified value of $\ddot{\theta}$. The door may be treated as a uniform plate with a mass ρ per unit area of plate.

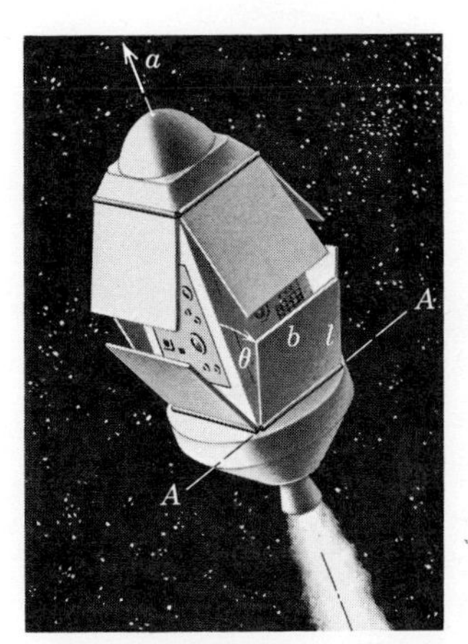

Problem 6/69

Problem 6/70

6/70 The uniform steel beam is 5 m long and has a mass of 500 kg. If the supporting cable CB breaks, determine the tension T in the remaining cable AC an instant after the break occurs. The beam may be treated as a slender bar. *Ans.* $T = 1387$ N

Problem 6/71

6/71 The hemispherical shell of mass m and radius r is released from rest in the position shown and rocks on the horizontal supporting surface. If friction is sufficient to prevent slipping, determine the value of the angular acceleration α an instant after release.

$$Ans.\ \alpha = \frac{3g}{2r}\frac{\sin\theta}{5 - 3\cos\theta}$$

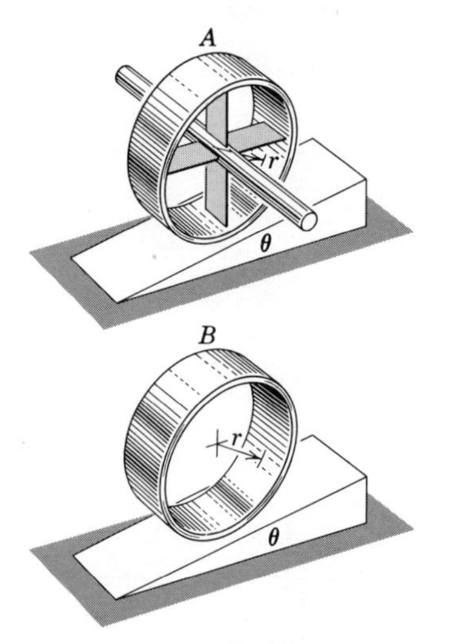

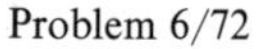
Problem 6/72

6/72 Determine the angular acceleration of each of the two wheels as they roll without slipping down the inclines. For wheel A investigate the case where the mass of the rim and spokes is negligible and the mass of the bar is concentrated along its center line. For wheel B assume that the thickness of the rim is negligible compared with its radius so that all of the mass is concentrated in the rim. Also specify the minimum coefficient of friction f required to prevent each wheel from slipping.

$$Ans.\ \alpha_A = \frac{g\sin\theta}{r},\quad f_A = 0$$

$$\alpha_B = \frac{g\sin\theta}{2r},\quad f_B = \tfrac{1}{2}\tan\theta$$

General

6/73 The collar A and attached rods B are given a simple harmonic oscillation $x = x_0 \sin pt$ along the horizontal shaft. The amplitude of the motion measured from the middle position of the oscillation is $x_0 = 10$ mm, and the frequency of the vibration is 40 Hz (frequency in cycles per second is represented by the unit *hertz,* Hz). Each rod has a mass of 4 kg per metre of length. Determine an expression for the bending moment M in each rod in newton-metres in terms of the distance r in metres from the shaft axis for an instant of time when bending is greatest.

Ans. $M = 1263(0.4 - r)^2$ N·m, r in metres

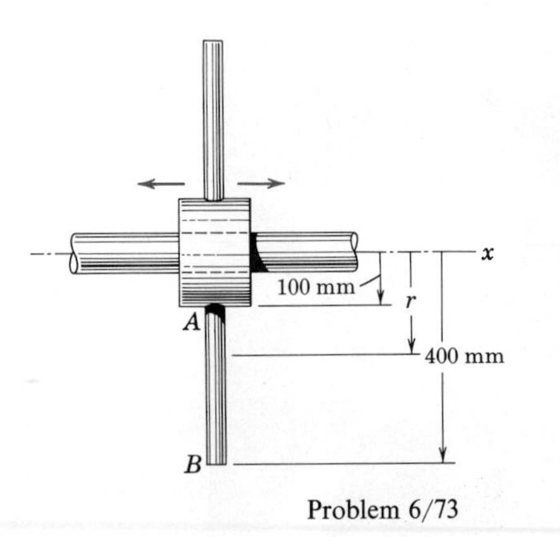

Problem 6/73

6/74 In the design of a space station, remote from gravitational attraction, a long tubular antenna is to be deployed in the following way. Thin spring-steel tape of length L and mass ρ per unit of length is preformed into a single split tube, after which it is flattened and wound around a spool of radius r. The bending moment in the tape as it leaves the spool tangentially is M. This moment acts to unroll the tape, and the spool is pushed away from the clamped end A of the tube. The bracket and guide rolls contain the tape and keep it tangent to the spool as it unwinds. Neglect the mass of the spool and bracket compared with the mass of the tape and determine the acceleration a of the center of the spool and the compressive force P in the tube as the roll accelerates away from A. Neglect any variation in r and assume that the clamp A is fixed in space. The roll may be analyzed with the free-body diagram shown.

Ans. $a = \dfrac{M}{2\rho r(L - x)}, \quad P = \dfrac{M}{2r}$

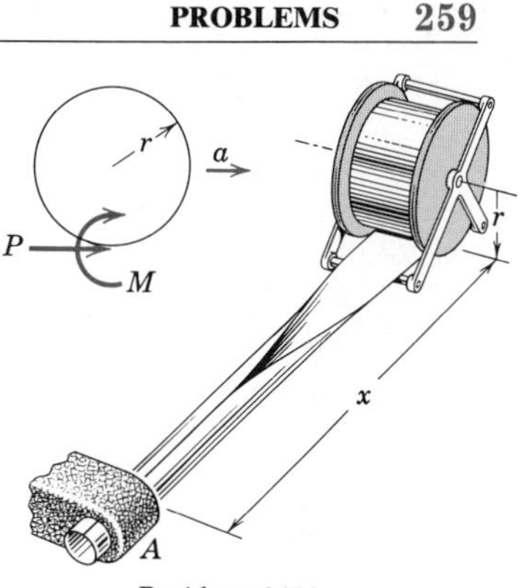

Problem 6/74

6/75 The access end of a large vacuum tank consists of a uniform hemispherical shell whose mass is 1.7 t. The shell is hinged about a horizontal axis through O and is controlled by the hydraulic cylinder AB. If the link AB exerts a pull of 25 kN on the bracket at B, find the initial angular acceleration α of the shell and the x-component of the hinge reaction at O at the instant immediately following release of the closing latch at C when the seal around the flange has been broken. The mass of the bracket, flange, and cylinder may be neglected compared with the mass of the shell.

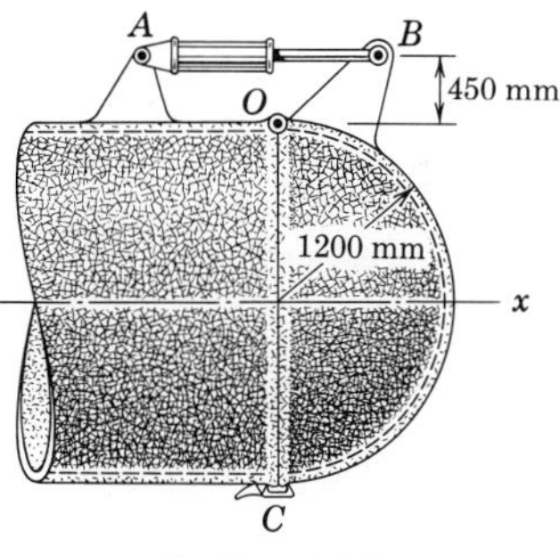

Problem 6/75

6/76 The riding power mower has a mass of 140 kg with center of mass at G_1. The mass of the operator is 90 kg with center of mass at G_2. Calculate the minimum coefficient of friction f which will permit the front wheels of the mower to lift off the ground as the mower starts to move forward. *Ans.* $f = 0.598$

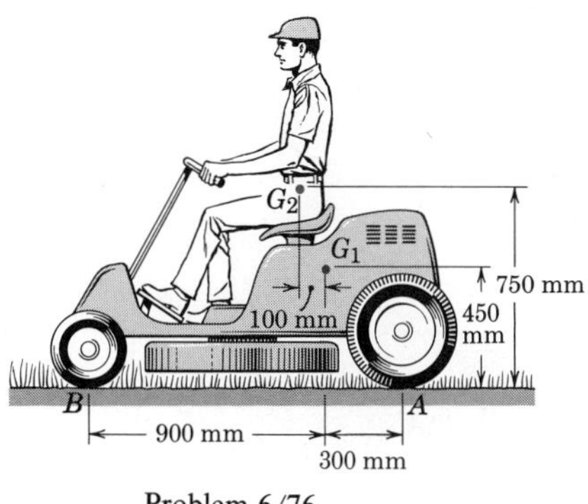

Problem 6/76

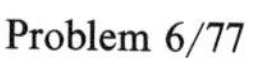

Problem 6/77

6/77 The fairing which covers the spacecraft package in the nose of the booster rocket is jettisoned when the rocket is in space where gravitational attraction is negligible. A mechanical actuator moves the two halves slowly from the closed position I to position II at which point the fairings are released to rotate freely about their hinges at O under the influence of a constant acceleration a of the rocket. When position III is reached, the hinge at O is released and the fairings drift away from the rocket. Determine the angular velocity ω of the fairings at the 90-deg position. The mass of each fairing is m with center of mass at G and radius of gyration k_O about O.

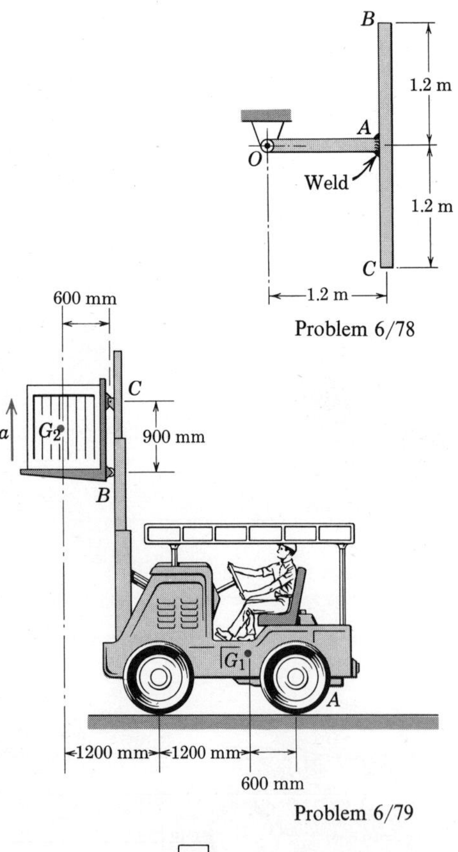

Problem 6/78

6/78 The uniform slender bars OA and BC of mass 25 kg and 50 kg, respectively, are welded at right angles to each other at A. If the configuration is released from rest in the position shown with OA horizontal and BC vertical, calculate the initial force exerted on OA by the pin at O and the bending moment M_A applied to bar BC by OA through the weld at A. Neglect friction in the bearing.

Ans. $O = 225$ N, $M_A = 163.5$ N·m

Problem 6/79

6/79 The fork-lift truck with center of mass at G_1 has a mass of 1.6 t including the vertical mast. The fork and load have a combined mass of 900 kg with center of mass at G_2. The roller guide at B is capable of supporting horizontal force only, whereas the connection at C, in addition to supporting horizontal force, also transmits the vertical elevating force. If the fork is given an upward acceleration which is sufficient to reduce the force under the rear wheels at A to zero, calculate the corresponding reaction at B.

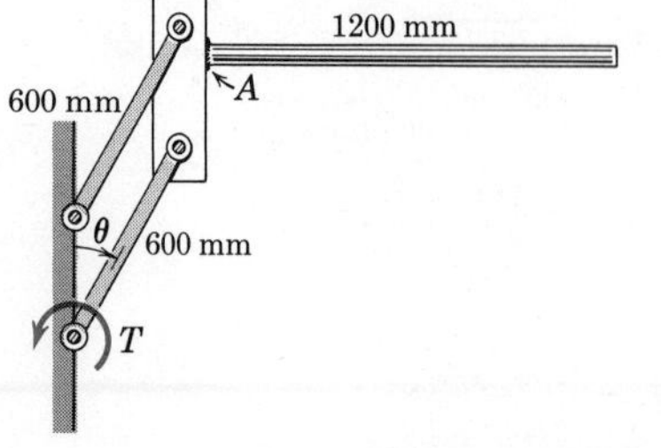

Problem 6/80

6/80 The torque T produces an angular acceleration $\ddot{\theta} = -10$ rad/s^2 along with an angular velocity $\dot{\theta} = -6$ rad/s for the rotating links in their vertical plane of motion at the position $\theta = 30°$. For this instant compute the bending moment M supported by the weld at A which attaches the uniform 1200-mm horizontal rod of mass 30 kg to the support. *Ans.* $M = 106.1$ N·m

6/81 The uniform semicircular bar of mass m and radius r is hinged freely about a horizontal axis through A. If the bar is released from rest in the position shown where AB is horizontal, determine the initial angular acceleration α of the bar and the expression for the force exerted on the bar by the pin at A. (Note carefully that the initial tangential acceleration of the mass center is not vertical.)

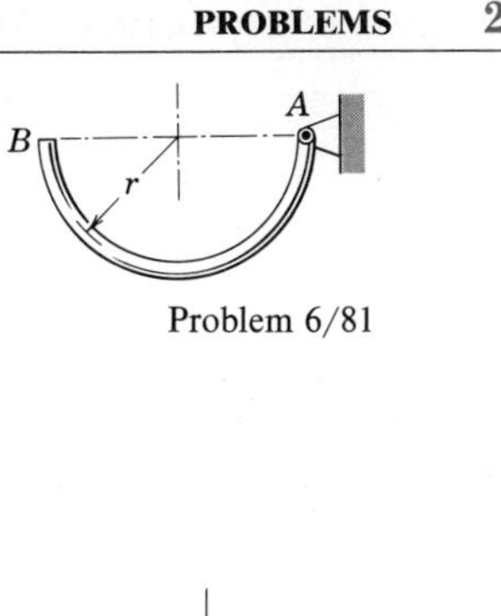

Problem 6/81

6/82 A uniform rod of length l and mass m is secured to a circular hoop of radius l as shown. The mass of the hoop is negligible. If the rod and hoop are released from rest on a horizontal surface in the position illustrated, determine the initial values of the friction force F and normal force N under the hoop if friction is sufficient to prevent slipping. *Ans.* $F = \frac{3}{8}mg$, $N = \frac{13}{16}mg$

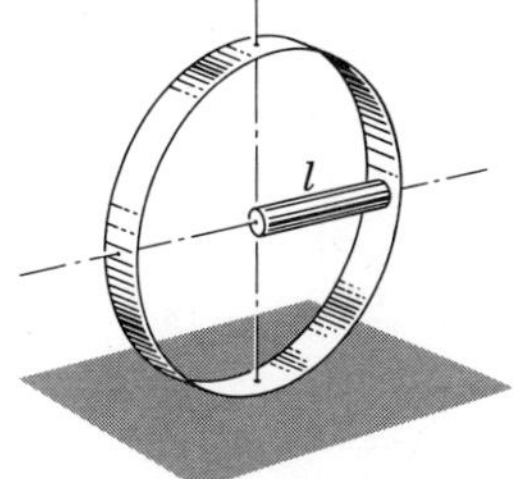

Problem 6/82

6/83 The car seen from the rear is traveling at a speed v around a turn of mean radius r banked inward at an angle θ. The coefficient of friction between the tires and the road is f. Determine (a) the proper bank angle for a given v to eliminate any tendency to slip or tip, and (b) the maximum speed v before the car tips or slips for a given θ. Note that the forces and the acceleration lie in the plane of the figure so that the problem may be treated as one of plane motion even though the velocity is normal to this plane.

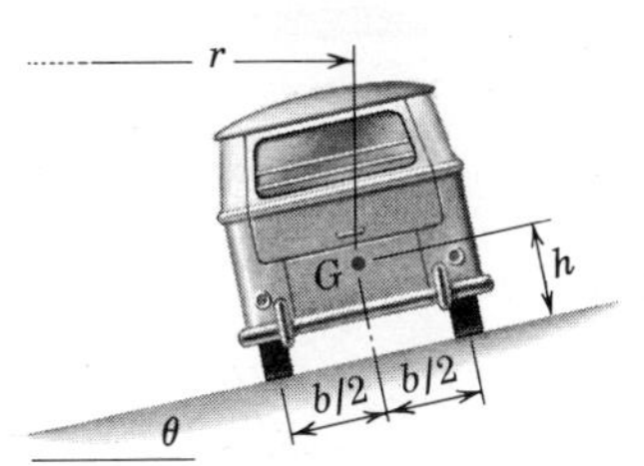

Problem 6/83

6/84 The solid circular cylinder is released from rest on the 60° incline. Calculate the angular velocity ω of the cylinder and the linear velocity v of its center G after it has moved 3 m down the incline. The coefficient of friction is $f = 0.30$.
Ans. $v = 6.49$ m/s, $\omega = 18.13$ rad/s

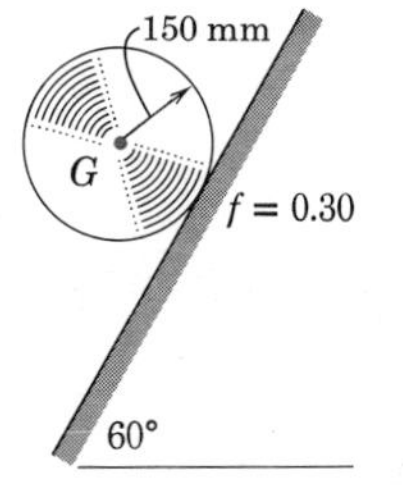

Problem 6/84

6/85 The uniform I-beam has a mass of 1200 kg, and the power drum unit mounted on the beam has a mass of 240 kg with center of mass at the center of the drum. If the clockwise starting torque of the drum is 4000 N·m, determine the momentary bending moment M in the beam at section C–C. Treat the beam as a slender bar.
Ans. $M = 27.4$ kN·m

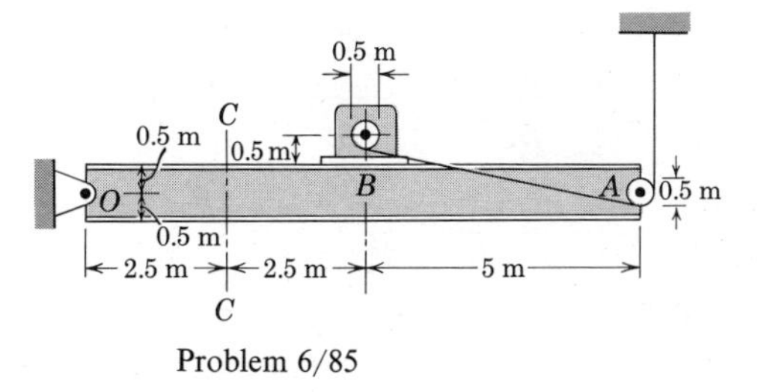

Problem 6/85

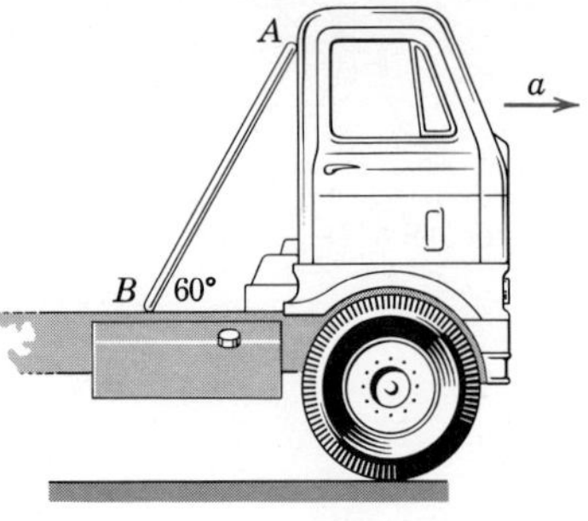

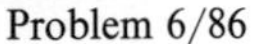
Problem 6/86

6/86 The coefficient of friction at both ends of the uniform bar is 0.40. Determine the maximum horizontal acceleration a which the truck may have without causing the bar to slip. Solve by using only one equation, a moment equation. The location of the moment center may be determined graphically.

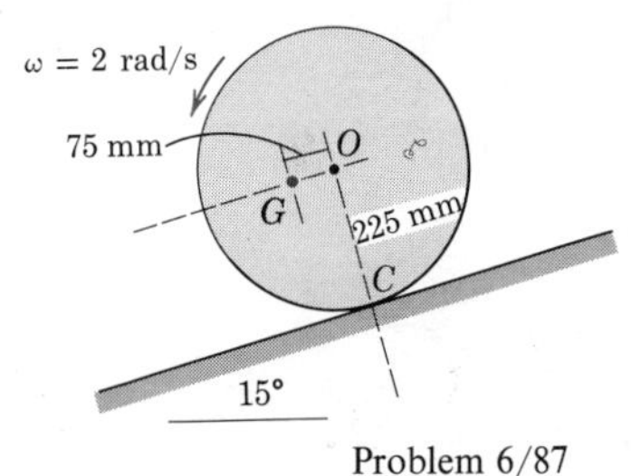

Problem 6/87

6/87 The unbalanced wheel has a mass of 30 kg with center of mass located 75 mm from the center O. The radius of gyration about G is 200 mm. Compute the normal component of the force of contact at C for the position shown as the wheel rolls without slipping down the 15° incline. The wheel has an angular velocity $\omega = 2$ rad/s at this particular instant. *Ans.* $N = 253$ N

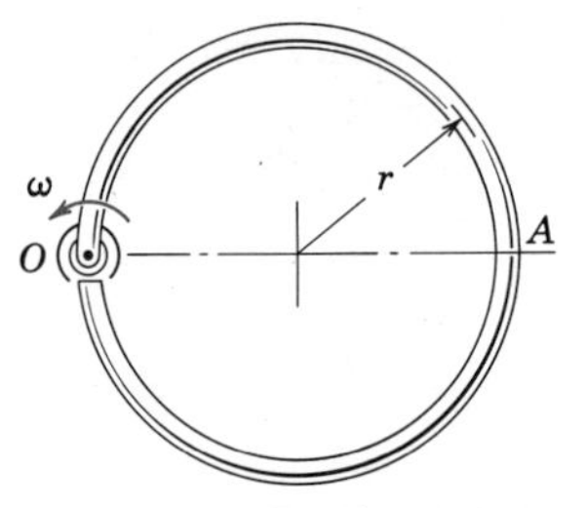

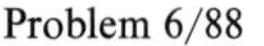
Problem 6/88

6/88 The split ring of mass m and radius r is rotating in its plane with an angular velocity ω about a fixed vertical axis through one end at O. Determine the bending moment M and shear force V in the ring at A if ω is constant.

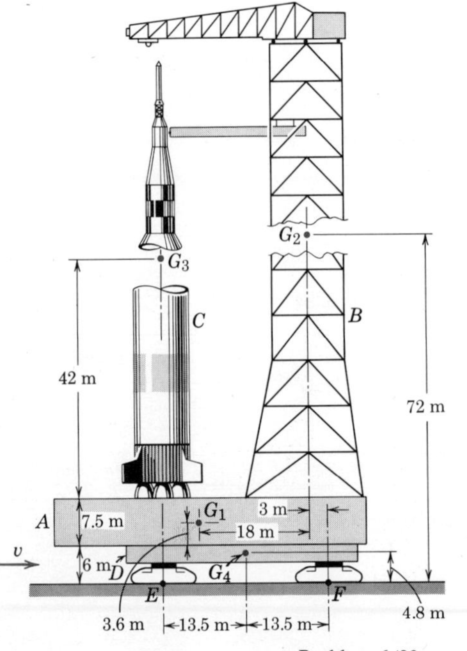

Problem 6/89

6/89 The figure shows the Saturn V mobile launch platform A together with the umbilical tower B, unfueled rocket C, and crawler-transporter D which carries the system to the launch site. The approximate dimensions of the structure and locations of the mass centers G are given. The approximate masses are $m_A = 3$ kt, $m_B = 3.3$ kt, $m_C = 0.23$ kt, $m_D = 3$ kt. The minimum stopping distance from the top speed of 1.5 km/h is 0.1 m. Compute the vertical component of the reaction under the front crawler unit F during the period of maximum deceleration.

Ans. $F = 59.5$ MN

6/90 The horizontal force F gives the collar a velocity of 9 m/s along the fixed rod in a distance of 1.2 m from rest. The collar has a mass of 4 kg with center of mass at G and fits loosely on the shaft. If the coefficient of friction between the collar and shaft is 0.5, calculate the force N_A normal to the shaft exerted on the collar at A by the shaft. *Ans.* $N_A = 81.6$ N

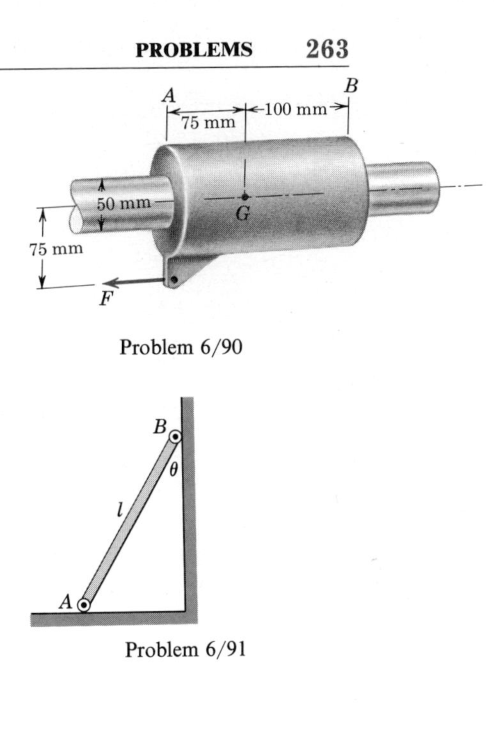

Problem 6/90

Problem 6/91

6/91 The uniform slender bar of mass m and length l is released from rest in the position shown. If friction against the vertical and horizontal surfaces is negligible, determine the expression for the initial angular acceleration α of the bar.

6/92 If the split ring of Prob. 6/88 starts from rest under the action of a counterclockwise torque T applied to it through its vertical shaft at O, determine the bending moment M and shear V acting in the ring at A at the start of the motion.

$$Ans.\ M = 0, \quad V = \frac{T}{2\pi r}$$

6/93 The rectangular block, which is solid and homogeneous, is supported at its corners by small rollers resting on horizontal surfaces. If the supporting surface at B is suddenly removed, determine the expression for the initial acceleration of corner A.

$$Ans.\ a_A = \frac{3g}{4\frac{b}{h} + \frac{h}{b}}$$

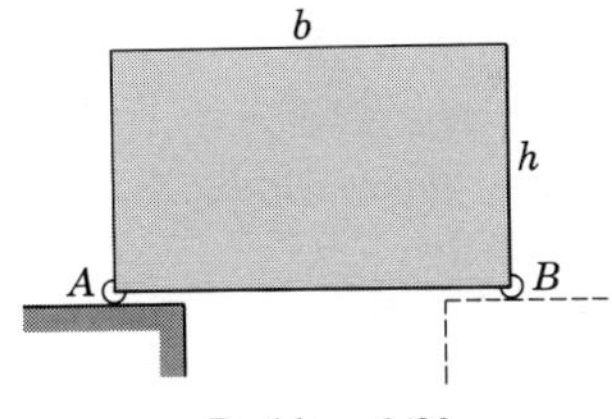

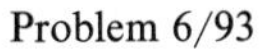

Problem 6/93

6/94 The circular disk of mass m and radius r is released from rest with θ essentially zero and rolls without slipping on the circular guide of radius R. Determine the expression for the normal force N between the disk and the guide in terms of θ.

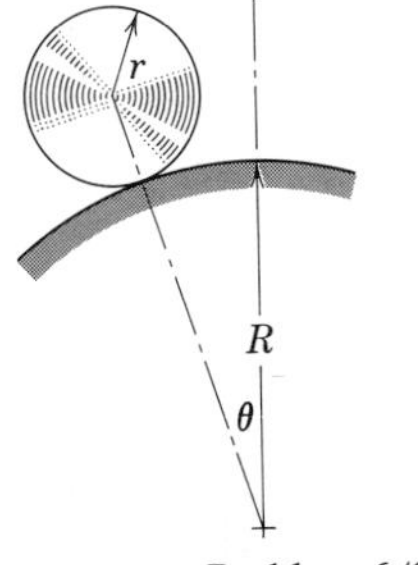

Problem 6/94

Problem 6/95

6/95 Model tests of the landing of the lunar excursion module (LEM) are conducted using the pendulum which consists of the model suspended by the two parallel wires A and B. If the model has a mass of 10 kg with center of mass at G, and if $\dot{\theta} = 2$ rad/s when $\theta = 60°$, calculate the tension in each of the wires at this instant.

Ans. $T_A = 147.9$ N, $T_B = 21.1$ N

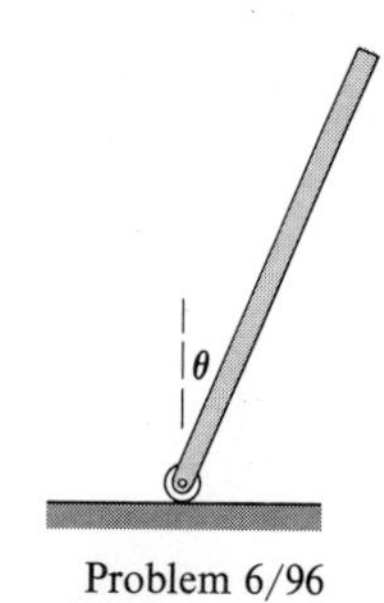

Problem 6/96

6/96 The slender bar of mass m is released from rest in the position shown. Determine the normal force N exerted on the roller by the supporting horizontal surface an instant after release.

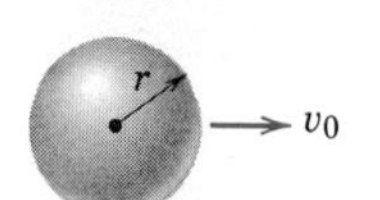

Problem 6/97

6/97 A bowling ball is released with a velocity v_0 but with no angular velocity as it contacts the alley surface. If the radius of the ball is r and its radius of gyration about a central axis is $\bar{k}$, determine the expression for the distance s traveled by the ball while it is still slipping on the alley floor. The coefficient of kinetic friction is f.

$$Ans.\ s = \frac{v_0{}^2}{2fg}\frac{\bar{k}^2(\bar{k}^2 + 2r^2)}{(\bar{k}^2 + r^2)^2}$$

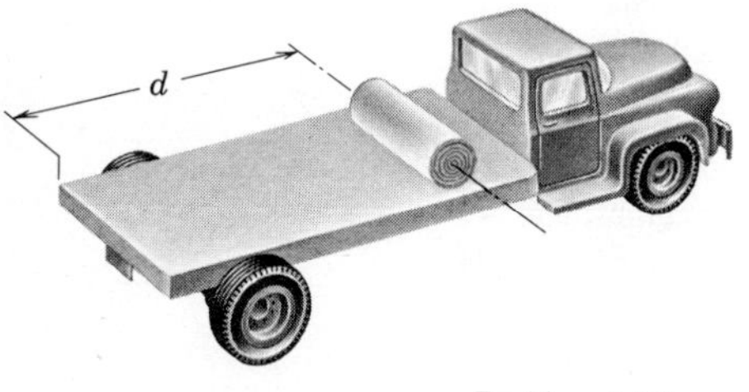

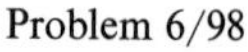

Problem 6/98

◀**6/98** The truck, initially at rest with a solid cylindrical roll of paper in the position shown, moves forward with a constant acceleration a. Find the distance s which the truck goes before the paper rolls off the edge of its horizontal bed. Friction is sufficient to prevent slipping.

Ans. $s = 3d/2$

◀**6/99** If the roll of paper in Prob. 6/98 is placed on the truck bed with its axis turned at an angle θ from the position shown, find the angular acceleration α of the roll if its radius is r and if the acceleration of the truck is a. Assume no slipping of the roll on the truck bed.

$$Ans.\ \alpha = \frac{2a\cos\theta}{3r}$$

6/100 In a study of head injury against the dashboard of a car during sudden or crash stops where lap belts without shoulder straps are used, the segmented human model shown in the figure is analyzed. The hip joint O is assumed to remain fixed relative to the car, and the torso above the hip is treated as a rigid body of mass m freely pivoted at O. The center of mass of the torso is at G with the initial position of OG taken as vertical. The radius of gyration of the torso about O is k_O. If the car is brought to a sudden stop with a constant deceleration a, determine the velocity v relative to the car with which the model's head strikes the instrument panel. Substitute the values $m = 50$ kg, $\bar{r} = 450$ mm, $r = 800$ mm, $k_O = 550$ mm, $\theta = 45°$, and $a = 10g$ and compute v. *Ans.* $v = 11.73$ m/s

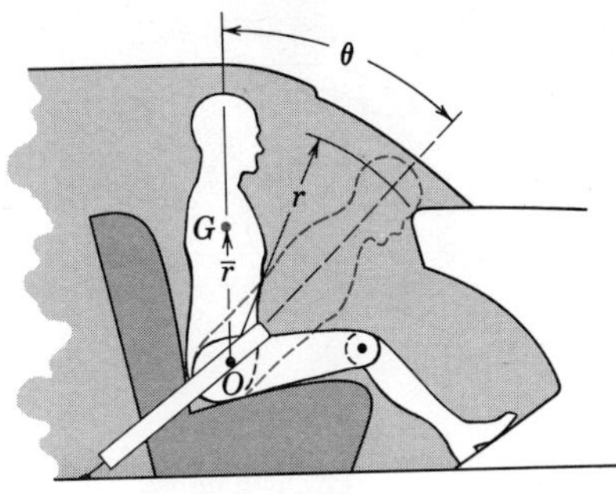

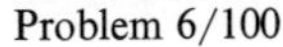
Problem 6/100

6/101 A load L with a mass of 480 kg and center of mass at G is supported by two links of negligible mass from the platform P which slides on the fixed horizontal shaft. Initially both platform and load are moving with a constant velocity $v = 1.5$ m/s to the left. If the motion of P is suddenly arrested so that it comes to an abrupt stop, (*a*) calculate the force on pin D an instant after P has stopped. Also (*b*) calculate the force on pin D at the instant L reaches its maximum amplitude of swing.

Ans. (*a*) $D = 3250$ N, (*b*) $D = 520$ N

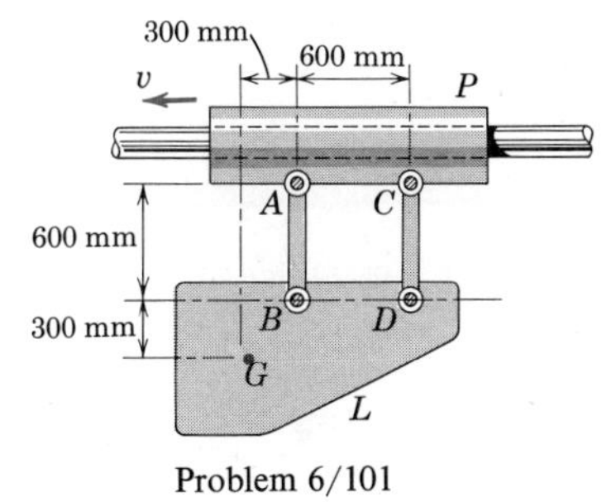

Problem 6/101

6/102 The uniform slender rod of length l and mass m is welded at its end tangent to the rim of the circular disk of radius r which rotates about a vertical axis through O. Determine the bending moment M, the shear force N, and the axial force T which the weld exerts on the rod (*a*) for a constant angular velocity ω of the disk and (*b*) as the disk starts from rest with a counterclockwise angular acceleration α.

Ans.

$$(a)\ M = \frac{mrl\omega^2}{2},\ N = mr\omega^2,\ T = \frac{ml\omega^2}{2}$$

$$(b)\ M = -\frac{ml^2\alpha}{3},\ N = -\frac{ml\alpha}{2},\ T = mr\alpha$$

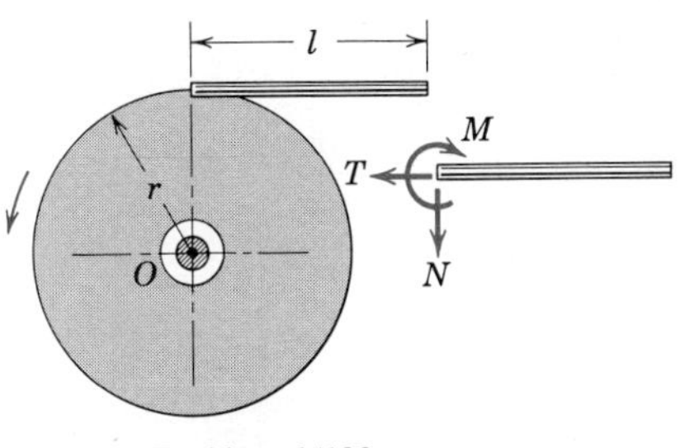

Problem 6/102

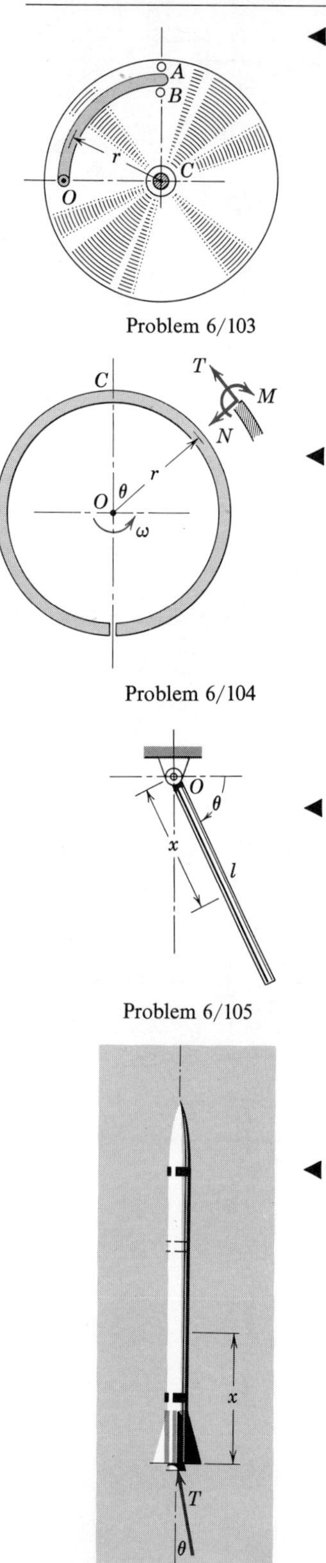

Problem 6/103

Problem 6/104

Problem 6/105

Problem 6/106

◀**6/103** The curved bar of mass m is hinged to the rotating disk at O and bears against one of the smooth pins A and B which are fastened to the disk. If the disk rotates about its vertical axis C, determine the force exerted on the bar by the hinge at O and the reaction A or B on the bar (a) if the disk has a constant angular velocity ω and (b) as the disk starts from rest with a counterclockwise angular acceleration α.

Ans. (a) $O = \dfrac{2mr\omega^2}{\pi}$, $A = \dfrac{2mr\omega^2}{\pi}$

(b) $O = mr\alpha\sqrt{1 + \dfrac{4}{\pi^2}}$, $B = mr\alpha\left(1 - \dfrac{2}{\pi}\right)$

◀**6/104** The split ring of radius r is rotating about a vertical axis through its center O with a constant angular velocity ω. Use a differential element of the ring and derive expressions for the shear force N and rim tension T in the ring in terms of the angle θ. Determine the bending moment M_C at point C by using one half of the ring as a free body. The mass of the ring per unit length of rim is ρ.

Ans. $N = \rho r^2\omega^2 \sin\theta$

$T = \rho r^2\omega^2(1 + \cos\theta)$

$M_C = 2\rho r^3\omega^2$

◀**6/105** The uniform slender bar of mass m and length l is pivoted freely at its end about a horizontal axis through O and released from rest at $\theta = 0$. Write the expression for the bending moment M in the bar in terms of x and θ. For a given θ, find the maximum value of M and the value of x at which it occurs.

Ans. $M = \dfrac{mgx}{4}\left(1 - \dfrac{x}{l}\right)^2 \cos\theta$

$M_{\text{max}} = \dfrac{mgl}{27}\cos\theta$ at $x = \dfrac{l}{3}$

◀**6/106** A rocket is given a thrust T at an angle θ with its axis in order to change the direction of its motion. Treat the rocket as a uniform slender bar of length l and mass m, and write the expression for the bending moment M in the rocket as a function of x when the rocket is in the vertical position shown.

Ans. $M = \left(\dfrac{l - x}{l}\right)^2 xT\sin\theta$

34 Work and Energy. It was seen in Art. 18 of Chapter 3 on particle motion that the equations of work and energy were useful in describing motion resulting from the cumulative effect of forces acting through distances. The work-energy equation, the concepts of potential energy and force fields, and power were all discussed in this article. A review of this material before proceeding is recommended. In Art. 24 of Chapter 4 the work-energy equation for a single particle was expanded to encompass a general system of particles. In the optional material of Art. 24 the expression for the kinetic energy of any mass system was represented by Eq. 93. This relation will be seen subsequently to cover the expressions for kinetic energy for the three classes of rigid-body plane motion which are derived as follows.

Translation. The translating rigid body of Fig. 56*a* has a mass m and all of its particles have a common velocity v. The kinetic energy of any particle of mass m_i of the body is $T_i = \frac{1}{2}m_i v^2$, so for the entire body $T = \Sigma\frac{1}{2}m_i v^2 = \frac{1}{2}v^2\Sigma m_i$ or

▶ $$T = \tfrac{1}{2}mv^2 \tag{120}$$

This expression holds, of course, for both rectilinear and curvilinear translation.

Fixed-Axis Rotation. The rigid body of Fig. 56*b* rotates with an angular velocity ω about the fixed axis through O. The kinetic energy of a representative particle of mass m_i is $T_i = \frac{1}{2}m_i(r_i\omega)^2$ so that $T = \frac{1}{2}\omega^2\Sigma m_i r_i^2$. But the moment of inertia of the body about O is $I_O = \Sigma m_i r_i^2$, so

▶ $$T = \tfrac{1}{2}I_O\omega^2 \tag{121}$$

The similarity in form of the kinetic energy expressions for translation and rotation should be noted. The reader should verify that the dimensions of the two expressions are identical.

General Plane Motion. The rigid body in Fig. 56*c* executes plane motion where, at the instant considered, the velocity of its mass center G is $\bar{v}$ and the angular velocity is ω. The velocity v_i of a representative particle of mass m_i may be expressed in terms of the mass-center velocity $\bar{v}$ and the velocity $\rho_i\omega$ relative to the mass center as shown. With the aid of the law of cosines the kinetic energy of the body may be written as the sum ΣT_i of the kinetic

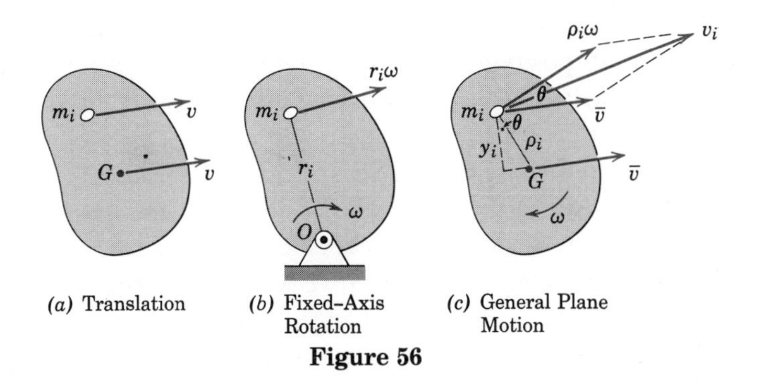

(*a*) Translation (*b*) Fixed-Axis Rotation (*c*) General Plane Motion

Figure 56

energies of all its particles. Thus

$$T = \Sigma\tfrac{1}{2}m_i v_i^2 = \Sigma\tfrac{1}{2}m_i(\bar{v}^2 + \rho_i^2\omega^2 + 2\bar{v}\rho_i\omega\cos\theta)$$

The third term in the expression becomes

$$\omega\bar{v}\Sigma m_i\rho_i\cos\theta = \omega\bar{v}\Sigma m_i y_i = 0$$

since $\Sigma m_i y_i = m\bar{y} = 0$. The kinetic energy of the body is, then, $T = \frac{1}{2}\bar{v}^2\Sigma m_i + \frac{1}{2}\omega^2\Sigma m_i\rho_i^2$ or

$$T = \tfrac{1}{2}m\bar{v}^2 + \tfrac{1}{2}\bar{I}\omega^2 \tag{122}$$

where $\bar{I}$ is the moment of inertia of the body about its mass center. This expression for kinetic energy clearly shows the separate contributions to the total kinetic energy resulting from the translational velocity $\bar{v}$ of the mass center and the rotational velocity ω about the mass center.

The kinetic energy of plane motion may also be expressed in terms of the rotational velocity about the instantaneous center C of zero velocity. Since C momentarily has zero velocity, the proof leading to Eq. 121 holds equally well for point C as for the fixed point O, so that, alternatively, the kinetic energy of a rigid body in plane motion may be written

$$T = \tfrac{1}{2}I_C\omega^2 \tag{123}$$

The expression for the kinetic energy of any mass system, rigid or nonrigid, Eq. 93 of Art. 24, is now easily seen to be equivalent to Eq. 122 when the mass system is rigid. For a rigid body the quantity $|\dot{\rho}_i|$ in Eq. 93 is the relative velocity $\rho_i\omega$ of the representative particle with respect to the mass center G. The second term in Eq. 93 now becomes $\Sigma\frac{1}{2}m_i(\rho_i\omega)^2 = \frac{1}{2}\omega^2\Sigma m_i\rho_i^2 = \frac{1}{2}\bar{I}\omega^2$ which brings Eq. 93 into agreement with Eq. 122.

The work-energy relation, Eq. 57, was introduced in Art. 18 of Chapter 3 for particle motion and was generalized in Art. 24 of Chapter 4 to include the motion of a general system of particles. This equation,

$$U = \Delta T + \Delta V_g + \Delta V_e \tag{[57]}$$

applies to any mechanical system. For application to the motion of a single rigid body the term ΔV_e disappears, and the total work U done on the body by the external forces (other than gravity forces) equals the corresponding change ΔT in the kinetic energy of the body plus the change ΔV_g in its potential energy of position in the gravitational field. Alternatively the equation may be written $U = \Delta T$ provided that the work of gravitational forces is included in the expression for U.

When applied to an interconnected system of rigid bodies, Eq. 57 will include the change ΔV_e in the stored elastic energy in the connections. The term U will include the work of all forces external to the system (other than gravitational forces) including the negative work of internal friction forces if any. The term ΔT is the sum of the changes in kinetic energy of all moving parts during the interval of motion in question, and ΔV_g is the sum of the changes in gravitational potential energy for the various members. Alternatively, if the work of gravitational forces is included in the U-term, then the ΔV_g-term must be omitted.

When applying the work-energy principles to a single rigid body, either a *free-body diagram* or an *active-force diagram* should be used. In the case of an interconnected system of rigid bodies, an active-force diagram of the entire system should be drawn in order to isolate the system and disclose all forces which do work on the system.

▼ ▼ ▼ ▼ ▼

In addition to the determination of velocities resulting from the action of forces acting over finite intervals of motion, the work-energy equation has two additional uses which embody the same advantage of analysis of a system as a whole without dismembering it. The equation may be used to establish the instantaneous accelerations of the members of a system of interconnected bodies as a result of the active forces applied. Or the equation may be modified to establish the configuration of such a system when it is subjected to prescribed accelerations.

Equation 57 when written for an infinitesimal interval of motion becomes

$$dU = dT + dV$$

The term dU represents the total work done by all active forces acting on the system under consideration during the infinitesimal change in the displacements. If the subscript i is used to denote a representative body of the interconnected system, the differential change in kinetic energy T for the entire system becomes

$$dT = d(\Sigma\tfrac{1}{2}m_i\bar{v}_i^2 + \Sigma\tfrac{1}{2}\bar{I}_i\omega_i^2) = \Sigma m_i\bar{v}_i\,d\bar{v}_i + \Sigma\bar{I}_i\omega_i\,d\omega_i$$

where $d\bar{v}_i$ and $d\omega_i$ are the respective changes in the magnitudes of the velocities and where the summation is taken over all bodies of the system. But for each body $m_i\bar{v}_i\,d\bar{v}_i = m_i\bar{\mathbf{a}}_i \cdot d\bar{\mathbf{s}}_i$ and $\bar{I}_i\omega_i\,d\omega_i = \bar{I}_i\alpha_i\,d\theta_i$, where $d\bar{\mathbf{s}}_i$ represents the infinitesimal linear displacement of the center of mass and where $d\theta_i$ represents the infinitesimal angular displacement of the body in the plane of motion. It should be noted that $\bar{\mathbf{a}}_i \cdot d\bar{\mathbf{s}}_i$ is identical to $(\bar{a}_i)_t\,d\bar{s}_i$ where $(\bar{a}_i)_t$ is the component of $\bar{\mathbf{a}}_i$ along the tangent of the curve described by the mass center of the body in question. Also α_i represents $\ddot{\theta}_i$, the angular acceleration of the representative body. Consequently for the entire system

$$dT = \Sigma m_i\bar{\mathbf{a}}_i \cdot d\bar{\mathbf{s}}_i + \Sigma\bar{I}_i\alpha_i\,d\theta_i$$

This change may also be written as

$$dT = \Sigma\mathbf{R}_i \cdot d\bar{\mathbf{s}}_i + \Sigma\overline{\mathbf{M}}_i \cdot d\boldsymbol{\theta}_i$$

where $\mathbf{R}_i$ and $\overline{\mathbf{M}}_i$ are the resultant force and resultant couple acting on each body and where $d\boldsymbol{\theta}_i = d\theta_i\mathbf{k}$. These last two equations merely show that the differential change in kinetic energy equals the differential work done on the body by the resultant force and resultant couple.

The term dV represents the differential change in the total gravitational potential energy V_g and the total elastic potential energy V_e and has the form

$$dV = d(\Sigma m_i g h_i + \Sigma \tfrac{1}{2} k_i x_i^2) = \Sigma m_i g \, dh_i + \Sigma k_j x_j \, dx_j$$

where h_i represents the vertical distance of the center of mass of the representative body of mass m_i above any convenient datum plane and where x_j stands for the deformation, tensile or compressive, of a representative elastic member of the system (spring) whose stiffness is k_j.

The complete expression for dU may now be written as

$$dU = \Sigma m_i \bar{\mathbf{a}}_i \cdot d\mathbf{s}_i + \Sigma \bar{I}_i \alpha_i \, d\theta_i + \Sigma m_i g \, dh_i + \Sigma k_j x_j \, dx_j \qquad (124)$$

In applying Eq. 124 to a system of one degree of freedom, which is any system whose configuration or position is uniquely determined by the value of a single coordinate, the terms $m_i \bar{\mathbf{a}}_i \cdot d\mathbf{s}_i$ and $\bar{I}_i \alpha_i \, d\theta_i$ will be positive if the accelerations are in the same direction as the respective displacements and negative if in the opposite direction. If it is more convenient to specify the position of the center of mass of a body in terms of its distance h_i below rather than above some arbitrary datum plane, then the sign of the $m_i g \, dh_i$ term must be reversed. Equation 124 has the advantage of relating the accelerations to the active forces directly which eliminates the need for dismembering the system and then eliminating the internal forces and reactive forces by simultaneous solution of the force-mass-acceleration equations written for each separate member of the system. Equation 124 is a further example of the fundamental principle used frequently in mechanics of stating the equivalence between the external forces on a mass system and the resultants of these forces.

In Eq. 124 the differential motions are differential charges in the real or actual displacements which occur. For mechanical systems whose configuration is an unknown function of a prescribed motion and for systems where there are more than one degree of freedom, it is often convenient to introduce the concept of *virtual work* when applying the work-energy relationship. The concepts of virtual work and virtual displacement have been introduced and used for the solution of equilibrium configurations for static systems of interconnected bodies (see Chapter 7 of *Statics*). A virtual displacement is any assumed and arbitrary displacement, linear or angular, away from the natural or actual position. For a system of connected bodies the virtual displacements must be consistent with the constraints of the system. For example, when one end of a link is hinged about a fixed pivot, the virtual displacement of the other end must be normal to the line joining the two ends. Or, if two links are freely pinned together, any virtual displacement of the joint considered as a point on one link must be identical with the virtual displacement of the joint considered as a point on the other link. Such requirements for displacements consistent with the constraints are purely kinematical, i.e., depend solely on the geometry of possible motions, and provide what are known as the *equations of constraint*. If a set of virtual displacements satisfying the equations of constraint and therefore consistent with the constraints is given to a mechanical system, the proper relationship between the coordinates which specify the configuration of the system will be established by applying the work-energy equation expressed by Eq. 124. This equation must be applied as many times as there are degrees of freedom of the system. For each application only one of the coordinate variables is allowed to change at a time, and the others are held constant. The virtual displacements will be the deviations from the steady-state configuration as seen from coordinate axes moving with the system. It is customary to use the differential symbol d to refer to differential changes in the *real* displacements, whereas the symbol δ is used to signify differential changes which are assumed or *virtual* changes.

The use of the concept of virtual work with Eq. 124 provides a very general and powerful method for the solution of complex problems involving conservative mechanical systems and finds special use for determining steady-state configurations of mechanisms under constant acceleration.

Sample Problems

6/107 The wheel rolls up the incline on its hubs without slipping and is pulled by the 100-N force applied to the cord wrapped around its outer rim. If the wheel starts from rest, compute its angular velocity ω after its center has moved a distance of 3 m up the incline. The wheel has a mass of 40 kg with center of mass at O and has a centroidal radius of gyration of 150 mm.

Solution. Of the four forces shown on the free-body diagram of the wheel only the 100-N pull and the weight of 40(9.81) = 392 N do work. The friction force does no work as long as the wheel does not slip. By use of the concept of the instantaneous center C of zero velocity it is seen that a point on the cord to which the 100-N force is applied moves a distance of (200 + 100)/100 times as far as the center O. Thus, with the effect of the weight included in the U-term, the work done on the wheel and its change in kinetic energy become

$$U = 100\frac{200 + 100}{100}(3) - 392 \sin 15^\circ\ (3) = 595 \text{ J}$$

$[T = \frac{1}{2}m\bar{v}^2 + \frac{1}{2}\bar{I}\omega^2]$

$$\Delta T = \frac{1}{2}40(0.10\omega)^2 + \frac{1}{2}40(0.15)^2\omega^2 - 0 = 0.650\omega^2$$

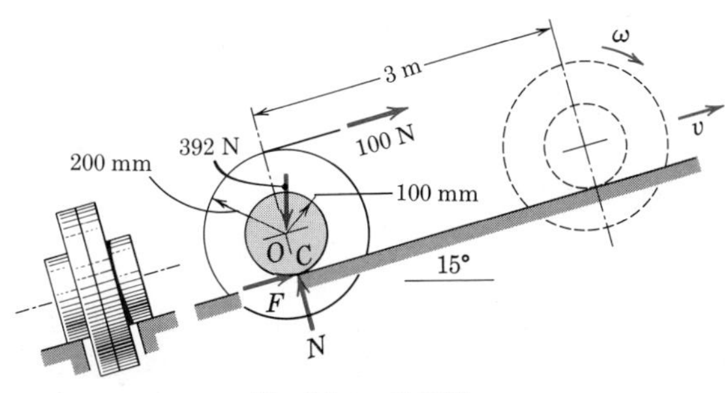

Problem 6/107

The work-energy equation gives

$[U = \Delta T]$ $\qquad 595 = 0.650\omega^2 \qquad \omega = 30.3$ rad/s *Ans.*

Alternatively the kinetic energy of the wheel may be written

$$[T = \tfrac{1}{2}I_C\omega^2] \qquad T = \frac{1}{2}40[(0.15)^2 + (0.10)^2]\omega^2 = 0.650\omega^2$$

6/108 In the mechanism shown each of the two wheels has a mass of 30 kg and a centroidal radius of gyration of 100 mm. Each link OB has a mass of 10 kg and may be treated as a slender bar. The 7-kg collar at B slides on the fixed vertical shaft with negligible friction. The spring has a stiffness $k = 30$ kN/m and is contacted by the bottom of the collar when the links reach the horizontal position. If the collar is released from rest at the position $\theta = 45°$ and if friction is sufficient to prevent the wheels from slipping, determine (*a*) the velocity v of the center O of the left-hand wheel at the instant that the position $\theta = 30°$ is passed, (*b*) the velocity v_B of the collar as it first strikes the spring, and (*c*) the maximum deformation x of the spring.

Solution. The mechanism executes plane motion and is conservative with the neglect of kinetic friction losses.

(*a*) The velocities of the collar and link for $\theta = 30°$ are first expressed in terms of the unknown velocity v of point O. From the relative velocity diagram it is seen that

$$\omega_{\text{link}} = \frac{v_{B/O}}{\overline{BO}} = \frac{v/\sin\theta}{\overline{BO}} = \frac{v}{0.5(0.375)} = 5.33v$$

$$v_B = v \operatorname{ctn}\theta = v\sqrt{3}$$

$$\bar{v}_{\text{link}} = \tfrac{1}{2}v_{B/O} = \frac{v}{2(0.5)} = v$$

The kinetic energies at the 30° position are

$$T_{\text{wheels}} = 2(\tfrac{1}{2}m\bar{v}^2 + \tfrac{1}{2}\bar{I}\omega^2) = 2\left(\frac{1}{2}30v^2 + \frac{1}{2}30[0.10]^2\left[\frac{v}{0.15}\right]^2\right) = 43.3v^2$$

$$T_{\text{links}} = 2(\tfrac{1}{2}m\bar{v}^2 + \tfrac{1}{2}\bar{I}\omega^2) = 2\left(\frac{1}{2}10v^2 + \frac{1}{2}\frac{1}{12}10[0.375]^2[5.33v]^2\right) = 13.33v^2$$

$$T_{\text{collar}} = \tfrac{1}{2}mv_B{}^2 = \tfrac{1}{2}(7)(3v^2) = 10.5v^2$$

Thus $\Delta T = (43.3v^2 + 13.33v^2 + 10.5v^2) - 0 = 67.2v^2$

The change in potential energy is computed using the arbitrary zero datum through O. The drop of end B from the 45° to the 30° position is $(0.375/\sqrt{2} - 0.375/2) =$ 0.0778 m and the potential energy change of the links and collar is

$$\Delta V_g = -2(10)(9.81)\frac{0.0778}{2} - 7(9.81)(0.0778) = -12.95 \text{ J}$$

Since there are no active applied forces $U = 0$, and the work-energy equation gives

$[U = \Delta T + \Delta V] \qquad 0 = 67.2v^2 - 12.95 \qquad v = 0.439$ m/s *Ans.*

(*b*) For the interval from $\theta = 45°$ to $\theta = 0$, it is noted that ΔT_{wheels} is zero since each wheel starts from rest and momentarily comes to rest at $\theta = 0$. Also, at the lower position each link is merely rotating about its point O so that

$$\Delta T = [2(\tfrac{1}{2}I_O\omega^2) - 0]_{\text{links}} + [\tfrac{1}{2}mv^2 - 0]_{\text{collar}}$$

$$= \frac{1}{3}10\,(0.375)^2\left(\frac{v_B}{0.375}\right)^2 + \frac{1}{2}7v_B{}^2 = 6.83v_B{}^2$$

The collar at B drops a distance $0.375/\sqrt{2} = 0.265$ m so that

$$\Delta V = \Delta V_g = 0 - 2(10)(9.81)\frac{0.265}{2} - 7(9.81)(0.265) = -44.2 \text{ J}$$

Also, $U = 0$. Hence,

$[U = \Delta T + \Delta V]$ $\quad 0 = 6.83v_B{}^2 - 44.2 \qquad v_B = 2.54$ m/s *Ans.*

(*c*) At the condition of maximum deformation x of the spring, all parts are momentarily at rest which again makes $\Delta T = 0$. Thus

$[U = \Delta T + \Delta V_g + \Delta V_e]$

$$0 = 0 - 2(10)(9.81)\left(\frac{0.265}{2} + \frac{x}{2}\right) - 7(9.81)(0.265 + x) + \tfrac{1}{2}(30)(10^3)x^2$$

Solution for the positive value of x gives

$$x = 60.1 \text{ mm}$$ *Ans.*

It should be noted that the results of parts (*b*) and (*c*) involve a very simple net energy change despite the fact that the mechanism has undergone a fairly complex sequence of motions. Solution of this and similar problems by other than a work-energy approach is not an inviting prospect.

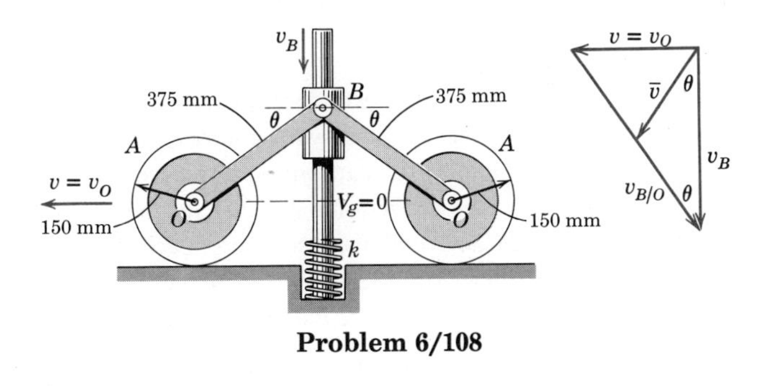

Problem 6/108

6/109 The bar A of mass $m = 30$ kg is elevated by the action of the couple $M = 270$ N·m applied to one end of the two parallel links B and C as shown. Each link has a mass $m_0 = 10$ kg and a length $l = 600$ mm. Determine the angular acceleration α of the links for any particular angle θ.

Solution. The figure serves as the active-force diagram for the interconnected mechanism as a whole. The differential terms of Eq. 124 are evaluated separately.

$$dU = M\,d\theta = 270\,d\theta$$

$$dT_{\text{bar}} = m\bar{\mathbf{a}}\cdot d\bar{\mathbf{s}} = m(l\alpha)\,d(l\theta) = ml^2\alpha\,d\theta$$

$$= 30(0.60)^2\alpha\,d\theta = 10.8\alpha\,d\theta$$

$$dT_{\text{links}} = \Sigma I_O \alpha \, d\theta = 2\left(\frac{1}{3} m_0 l^2\right) \alpha \, d\theta = \frac{2}{3} m_0 l^2 \alpha \, d\theta$$

$$= \frac{2}{3} 10(0.60)^2 \alpha \, d\theta = 2.40\alpha \, d\theta$$

$$(dV_g)_{\text{bar}} = mg \, dh = mg \, d(l \sin \theta) = mgl \cos \theta \, d\theta$$

$$= 30(9.81)(0.60) \cos \theta \, d\theta$$

$$(dV_g)_{\text{links}} = \Sigma m_0 g \, dh = 2m_0 g \, d\left(\frac{l}{2} \sin \theta\right) = m_0 gl \cos \theta \, d\theta$$

$$= 10(9.81)(0.60) \cos \theta \, d\theta$$

The work-energy equation for a differential movement becomes

$[dU = dT + dV_g]$ $\qquad 270 \, d\theta = 13.20\alpha \, d\theta + 235 \cos \theta \, d\theta$

Solution for α gives

$$\alpha = 20.5 - 17.8 \cos \theta \text{ rad/s}^2$$ *Ans.*

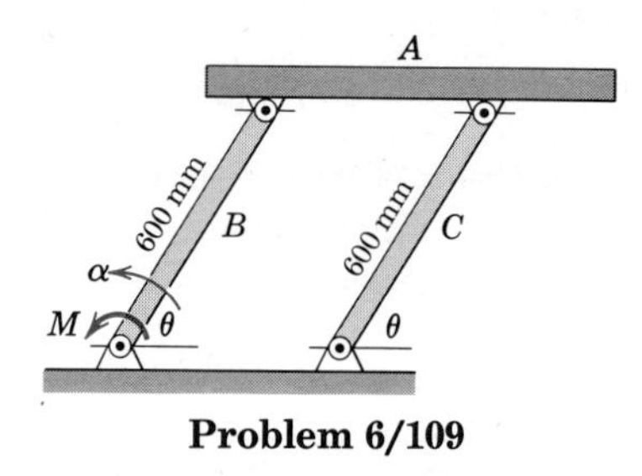

Problem 6/109

6/110 A constant force P is applied to end A of the two identical and uniform links and causes them to move to the right in their vertical plane with a horizontal acceleration a. Determine the steady-state angle θ made by the bars with one another.

Solution. The figure constitutes the active-force diagram for the system. To find the steady-state configuration, consider a virtual displacement of each bar from the natural position assumed during the acceleration. Measurement of the displacement with respect to end A eliminates any work done by force P during the virtual displacement, and

$$\delta U = 0$$

The virtual change in kinetic energy is

$$\delta T = \Sigma m\bar{\mathbf{a}} \cdot \delta \mathbf{s} = ma(-\delta s_1) + ma(-\delta s_2)$$

$$= -ma\left[\delta\left(\frac{l}{2} \sin \frac{\theta}{2}\right) + \delta\left(\frac{3l}{2} \sin \frac{\theta}{2}\right)\right]$$

$$= -ma\left(l \cos \frac{\theta}{2} \delta\theta\right)$$

The virtual change in potential energy is

$$\delta V_g = \delta\left(-2mg\frac{l}{2} \cos \frac{\theta}{2}\right) = \frac{mgl}{2} \sin \frac{\theta}{2} \delta\theta$$

Substitution into the work-energy equation for virtual changes gives

$[\delta U = \delta T + \delta V_g]$ $$0 = -mal\cos\frac{\theta}{2}\delta\theta + \frac{mgl}{2}\sin\frac{\theta}{2}\delta\theta$$

from which $$\theta = 2\tan^{-1}\frac{2a}{g}$$ *Ans.*

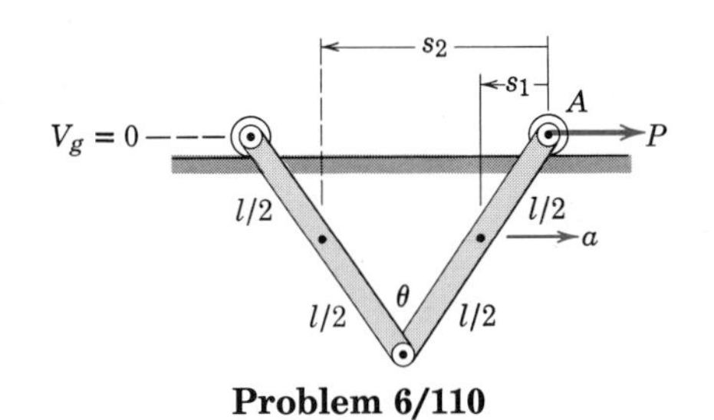

Problem 6/110

Problems

6/111 A landing on the moon is being made by a spacecraft which approaches the lunar surface along a vertical line (normal to the moon's surface). The spacecraft is slowed down by its retro rocket to a velocity of 1.8 m/s at a height of 6 m above the lunar surface. At this point the rocket thrust is cut off, and the craft falls the remaining 6 m. Determine the impact velocity v of the spacecraft with the lunar surface. (Consult Table C2, Appendix C, for relevant moon data.)

Ans. $v = 4.76$ m/s

6/112 The uniform slender bars are hinged at B and are confined to move in the vertical plane. If they are released from rest in the positions shown, determine the expression for the velocity v with which B strikes the horizontal plane. Neglect all friction.

6/113 The steel I-beam AB is being positioned by the horizontal force F in the cable at B. If this cable breaks in the position shown where the cable AC is horizontal, determine the velocity of end A as it reaches A'. Neglect friction.

Ans. $v = 11.29$ m/s

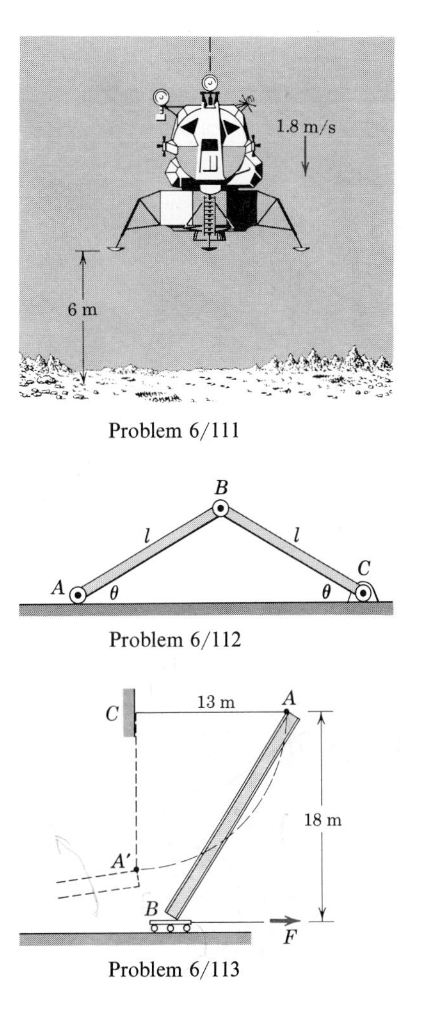

Problem 6/111

Problem 6/112

Problem 6/113

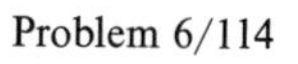
Problem 6/114

6/114 The two hinged links are released from rest with OA in the horizontal position shown. Calculate the velocity of end B along the horizontal surface for the instant when OA reaches the vertical position. Link OA has twice the mass of link AB, and both may be treated as slender bars. Neglect all friction and the mass of the small roller at B.

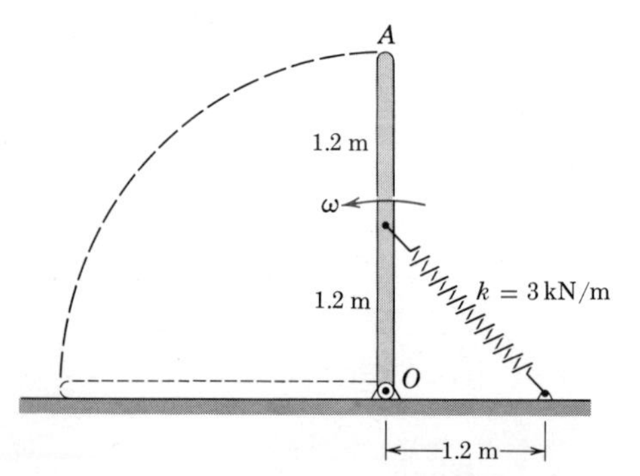

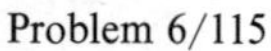
Problem 6/115

6/115 Calculate the initial angular velocity of the 30-kg slender bar OA in the vertical position so that the bar will reach the dotted horizontal position with zero velocity under the action of the spring. In the initial position the spring is unstretched.

Ans. $\omega = 3.67$ rad/s

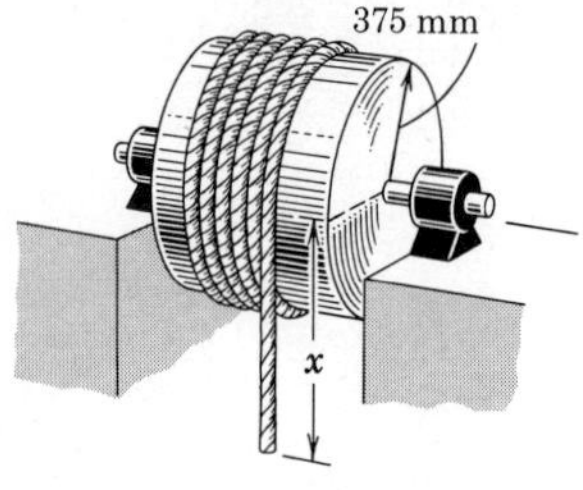

Problem 6/116

6/116 The drum of 375-mm radius and its shaft have a mass of 41 kg and a radius of gyration of 300 mm about the axis of rotation. A total of 18 m of flexible steel cable with a mass of 3.08 kg per metre of length is wrapped around the drum with one end secured to the surface of the drum. The free end of the cable has an initial overhang $x = 0.6$ m as the drum is released from rest. Determine the angular velocity ω of the drum for the instant when $x = 6$ m. Assume that the center of mass of the portion of cable remaining on the drum at any time lies on the shaft axis.

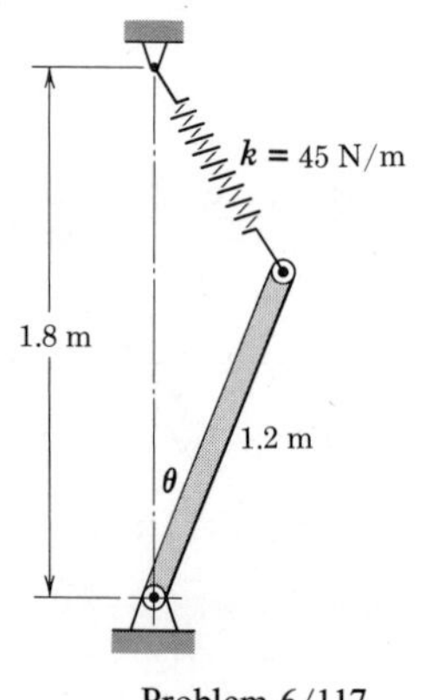

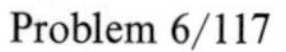
Problem 6/117

6/117 What initial clockwise angular velocity ω must the uniform and slender 7.5-kg bar have as it crosses the vertical position ($\theta = 0$) in order that it just reach the horizontal position ($\theta = 90°$)? The spring has a stiffness of 45 N/m and is unstretched when $\theta = 0$.

Ans. $\omega = 2.45$ rad/s

6/118 A slender metal rod of length l and mass m is welded to the rim of a hoop of radius l. If the hoop is released from rest in the position shown where the rod is normal to the supporting surface, determine the speed v of the center of the hoop after it has made one complete revolution. Assume no slipping and continuous contact between the hoop and its supporting surface. Also, neglect the mass of the hoop.

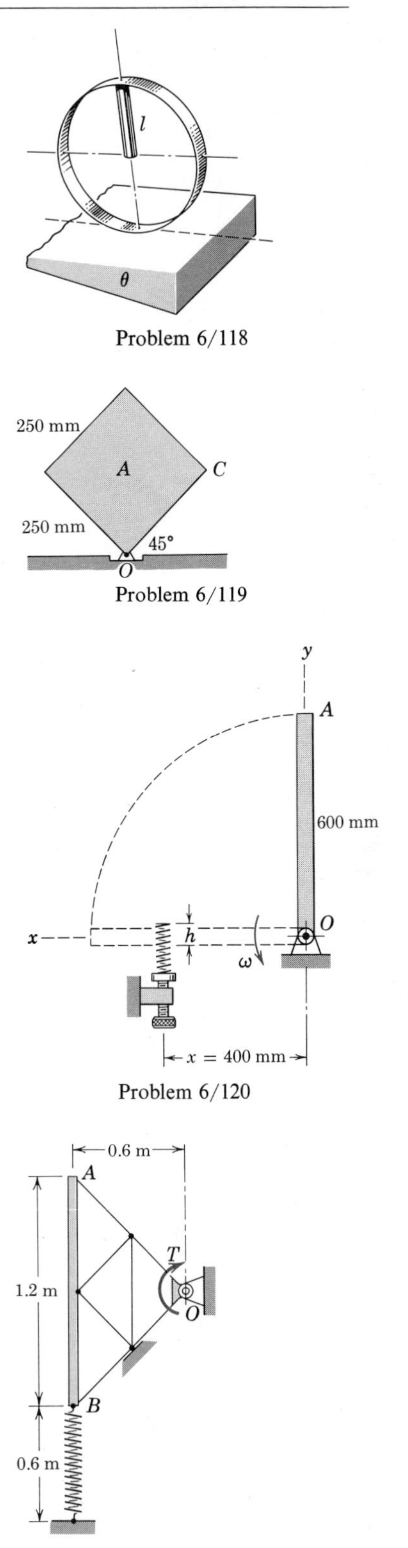

Problem 6/118

Problem 6/119

Problem 6/120

Problem 6/121

6/119 The solid square block is hinged about a horizontal axis at O and is released from rest in the position shown. Calculate the velocity v_C with which corner C hits the horizontal surface.

Ans. $v_C = 1.234$ m/s

6/120 The 15-kg slender bar OA is released from rest in the vertical position and compresses the spring of stiffness $k = 20$ kN/m as the horizontal position is passed. Determine the proper setting of the spring by specifying the distance h which will result in the bar having an angular velocity $\omega = 4$ rad/s as it crosses the horizontal position. What is the effect of x on the dynamics of the problem?

6/121 The figure shows the cross section AB of a 100-kg door which is a 1.2 m by 1.8 m panel of uniform thickness. The door is supported by a framework of negligible mass hinged about a horizontal shaft at O. In the position shown, the spring, which has a stiffness $k = 450$ N/m, is unstretched. If a constant torque $T = 880$ N·m is applied to the frame through its shaft at O starting from the rest position shown, determine the angular velocity of the door when it reaches the horizontal position. *Ans.* $\omega = 4.42$ rad/s

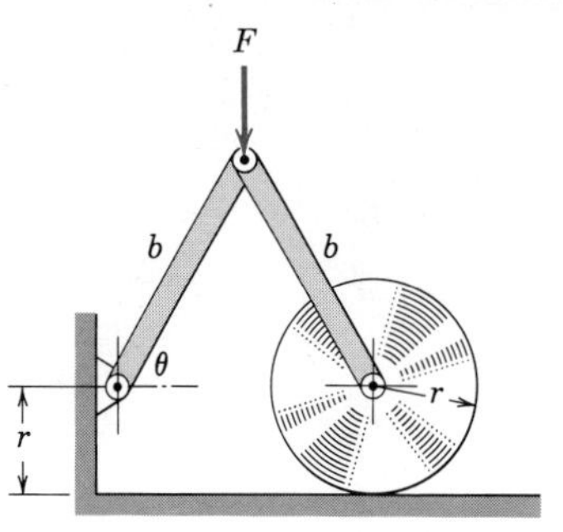

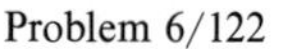
Problem 6/122

6/122 A force of constant magnitude F is applied in the vertical direction to the symmetrical linkage in the rest position shown. Determine the angular velocity ω which the links acquire as they reach the position $\theta = 0$. Each link has a mass m_0. The wheel is a solid circular disk of mass m and rolls on the horizontal surface without slipping.

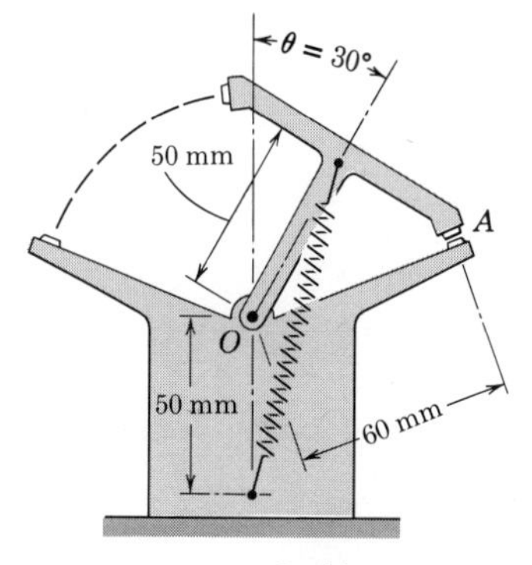

Problem 6/123

6/123 Specify the unstretched length l_0 of the spring of stiffness $k = 1400$ N/m which will result in a velocity of 0.25 m/s for the contact at A if the toggle is given a slight nudge from its null position at $\theta = 0$. The toggle has a mass of 1.5 kg and a radius of gyration about O of 55 mm. The toggle rotates in the horizontal plane. *Ans.* $l_0 = 90.0$ mm

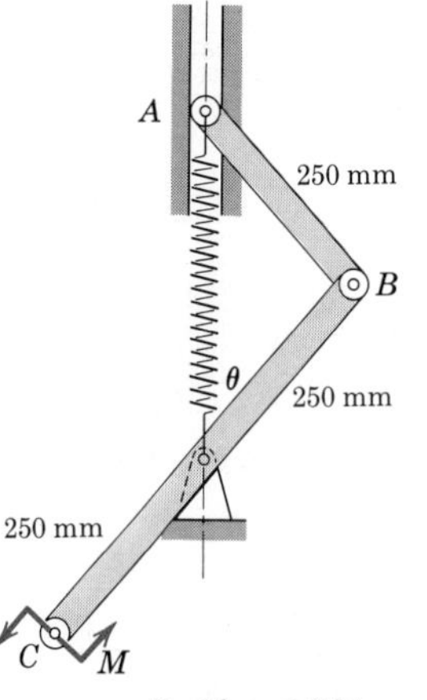

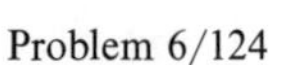
Problem 6/124

6/124 A couple $M = 12$ N·m is applied at C to the spring toggle mechanism which is released from rest in the position $\theta = 45°$. In this position the spring, which has a stiffness of 140 N/m, is stretched 150 mm. Bar AB has a mass of 3 kg and BC a mass of 6 kg. Calculate the angular velocity ω of BC as it crosses the position $\theta = 0$. Motion is in the vertical plane, and friction is negligible.

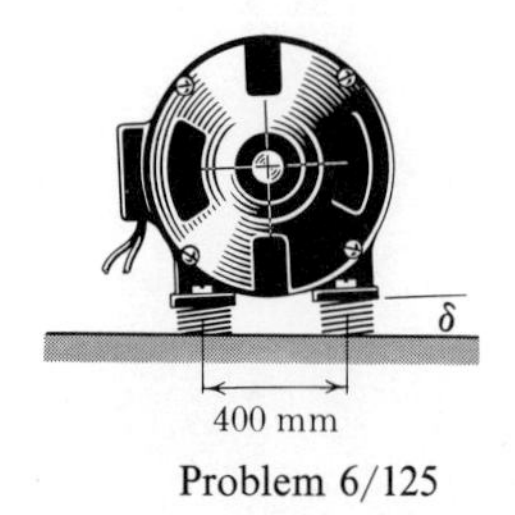

Problem 6/125

6/125 The electric motor shown is delivering 4.5 kW at 1725 rev/min to a pump which it drives. Calculate the angle δ through which the motor deflects under load if the stiffness of each of its four spring mounts is 10 kN/m. In what direction does the motor turn?

Ans. $\delta = 0.892°$, motor turns clockwise

6/126 The 700-mm horizontal bar has a mass of 24 kg and is freely pinned to the two identical wheels. Each wheel is a solid circular disk with a mass of 16 kg. If the system is released from rest with the bar in essentially the top position, compute the maximum angular velocity ω of the wheels during the subsequent motion. The wheels roll without slipping.

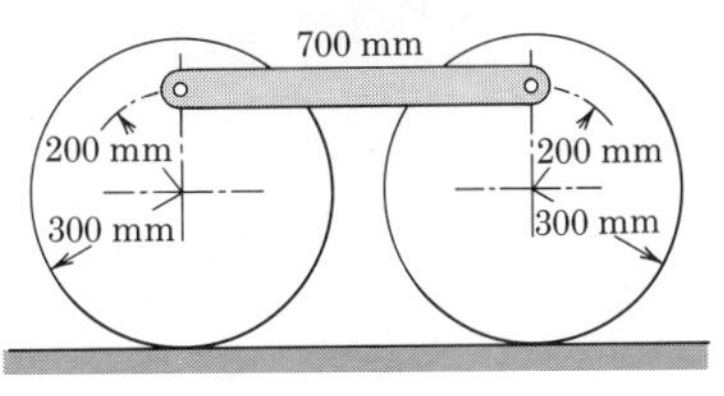

Problem 6/126

6/127 The center of the 100-kg wheel with radius of gyration of 100 mm has a velocity of 0.6 m/s down the incline in the position shown. Calculate the normal reaction N under the wheel as it rolls past position A. Assume no slipping occurs.

Ans. $N = 1696$ N

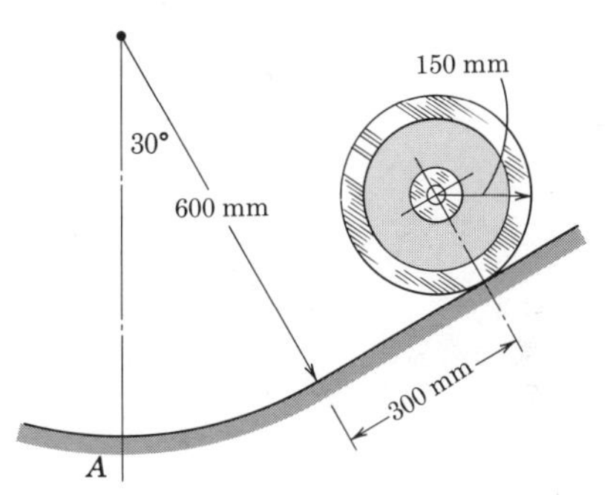

Problem 6/127

6/128 The solid semicircular disk of radius r is released from rest in the position shown. If no slipping occurs between the disk and the horizontal surface, determine the expression for the angular velocity ω reached by the disk when its kinetic energy is a maximum.

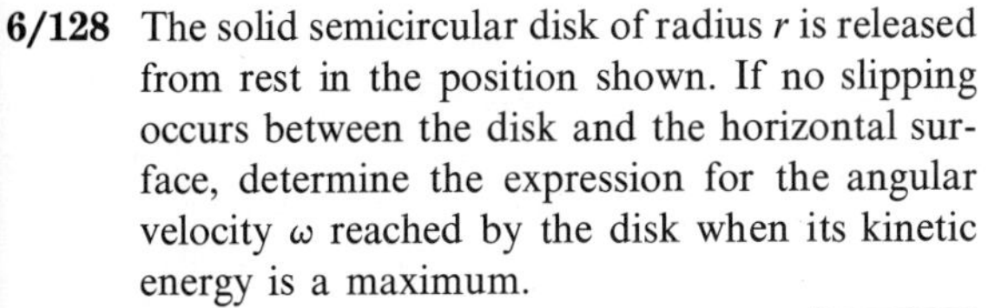

Ans. $\omega = 4\sqrt{\dfrac{g/r}{9\pi - 16}}$

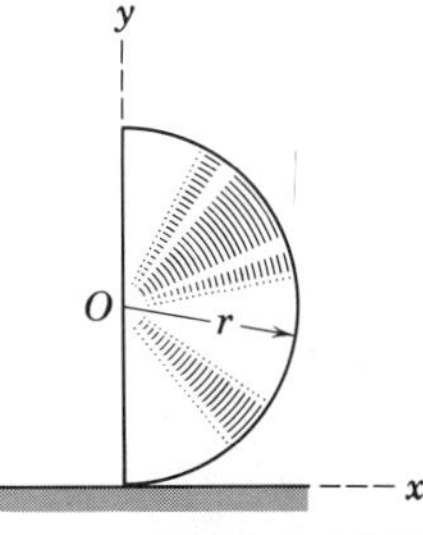

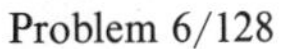
Problem 6/128

6/129 The light circular hoop carries a uniform heavy band of mass m around half of its circumference. If the hoop is released from rest in the position shown, write an expression for the normal force N under the hoop for the position where the kinetic energy of the hoop is greatest. Friction is sufficient to prevent slipping.

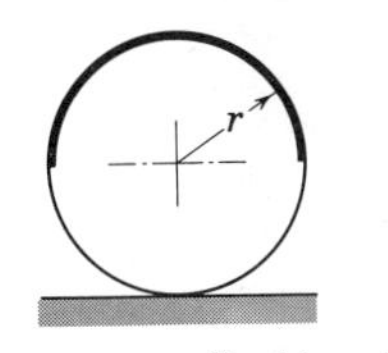

Problem 6/129

6/130 A hemispherical shell of radius r is released from rest with its axis of symmetry inclined an angle θ from the vertical. Determine the angular velocity ω of the shell as it rocks past the equilibrium position $\theta = 0$. Assume no slipping.

Ans. $\omega = \sqrt{\dfrac{3g}{2r}(1 - \cos\theta)}$

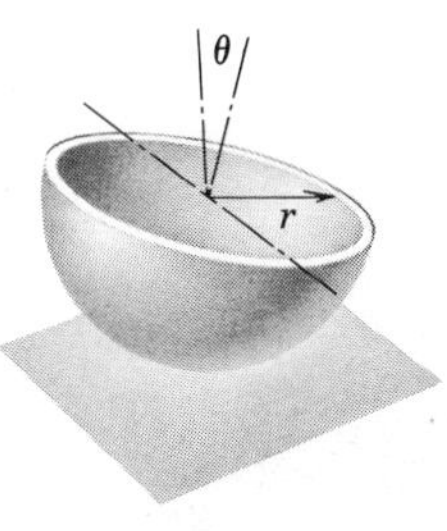

Problem 6/130

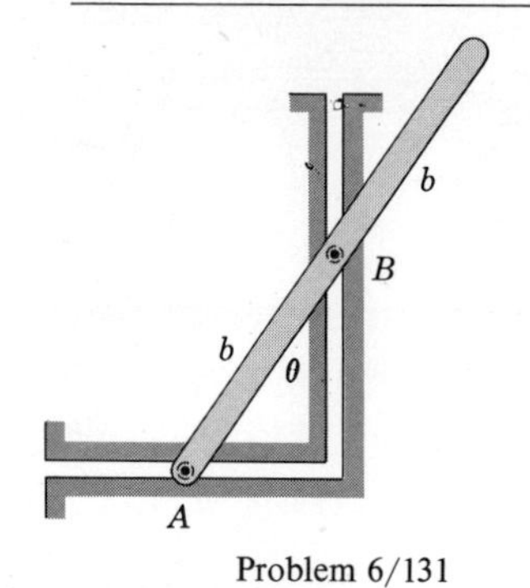

Problem 6/131

6/131 Motion of the slender bar of mass m is constrained by its two pins A and B which move in the horizontal and vertical slots, respectively. If the bar is released from rest with θ essentially zero, determine the velocity of end A of the bar as it reaches an angle θ. Friction in the guides is negligible.

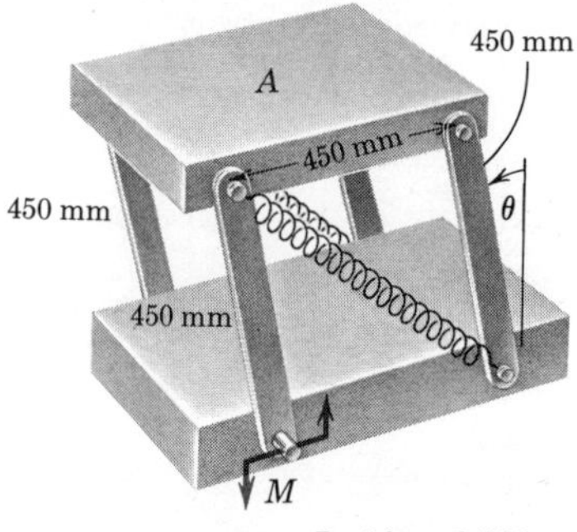

Problem 6/132

6/132 The horizontal platform A has a mass of 15 kg, and each of its uniform slender legs has a mass of 3 kg. When the legs are vertical at $\theta = 0$, each of the two springs of stiffness $k = 700$ N/m is unstretched. If a constant torque $M = 18$ N·m is applied to the one leg as shown, starting from rest in the position $\theta = 0$, determine the angular velocity ω of the legs when the position $\theta = 45°$ is passed. *Ans.* $\omega = 2.77$ rad/s

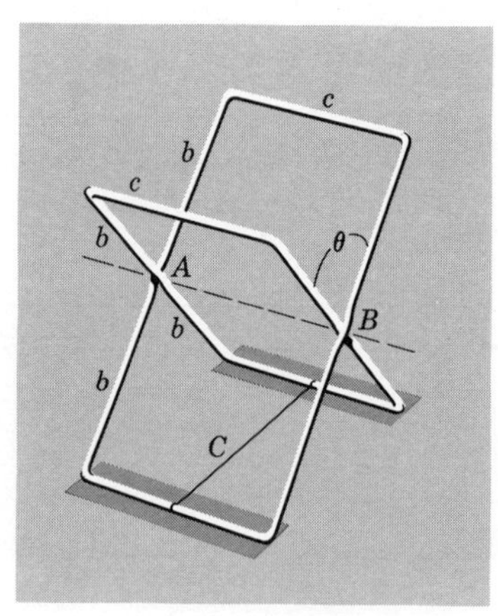

Problem 6/133

6/133 The two identical steel frames with the dimensions shown are fabricated from the same bar stock and are hinged at the midpoints A and B of their sides. If the frame is resting in the position shown on a horizontal surface with negligible friction, determine the velocity v with which each of the upper ends of the frame hits the horizontal surface if the cord at C is cut.

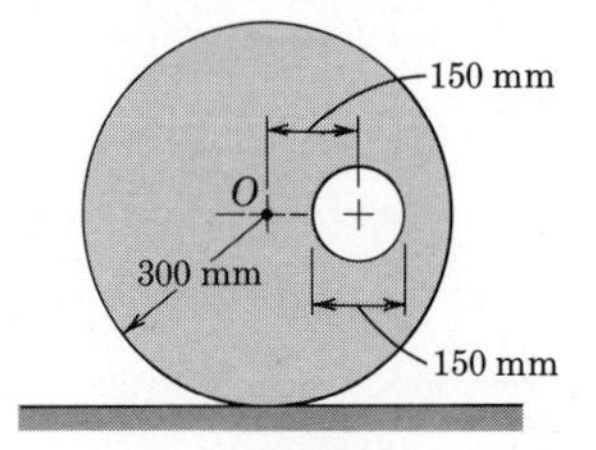

Problem 6/134

6/134 The circular disk of 300-mm radius with a 150-mm-diameter hole cut from it has a net mass of 75 kg. The disk is released from rest on a horizontal surface from the position shown. Calculate the maximum angular velocity ω reached by the wheel as it rolls without slipping under the influence of gravity. *Ans.* $\omega = 1.227$ rad/s

6/135 The small vehicle is designed for high-speed travel over the snow. The endless tread for each side of the vehicle has a mass ρ per unit length and is driven by the front wheels. Determine that portion M of the constant front-axle torque required to give both vehicle treads their motion corresponding to a vehicle velocity v achieved with constant acceleration in a distance s from rest on level terrain.

$$Ans.\ M = \frac{4\rho r v^2}{s}(\pi r + b)$$

6/136 The solid circular cylinder of mass m and radius r is resting on a horizontal surface which is given a constant acceleration a to the right from rest. Determine the work done on the cylinder during the interval in which it has rotated through 360°. The cylinder rolls without slipping. *Ans.* $U = mra\pi$

6/137 Determine the constant force P required to give the center of the pulley a velocity of 1.2 m/s in an upward movement of the center of 0.9 m from the rest position shown. The pulley has a mass of 15 kg with a radius of gyration of 250 mm, and the cable has a total length of 4.5 m with a mass of 3 kg/m. *Ans.* $P = 178.3$ N

6/138 The solid square block is supported on the horizontal plane by a small roller with negligible friction. The block is released from rest in the position shown. Calculate the angular velocity ω of the block and the linear velocity of corner O as corner C hits the horizontal surface.

Ans. $\omega = 6.25$ rad/s, $v_O = 0.781$ m/s

6/139 The uniform solid panel of mass m is hinged on the two links AB and CD so as to reduce its swinging space. Pins E and F in the panel at the midpoints of the upper and lower edges are confined to move in smooth horizontal guides across the opening. Determine the angular velocity ω of the panel as E approaches A and F approaches C if the panel is opened from rest in the closed position (A, B, and E collinear) by a force P of constant magnitude.

$$Ans.\ \omega = \frac{3}{4}\sqrt{\frac{P\pi}{mb}}$$

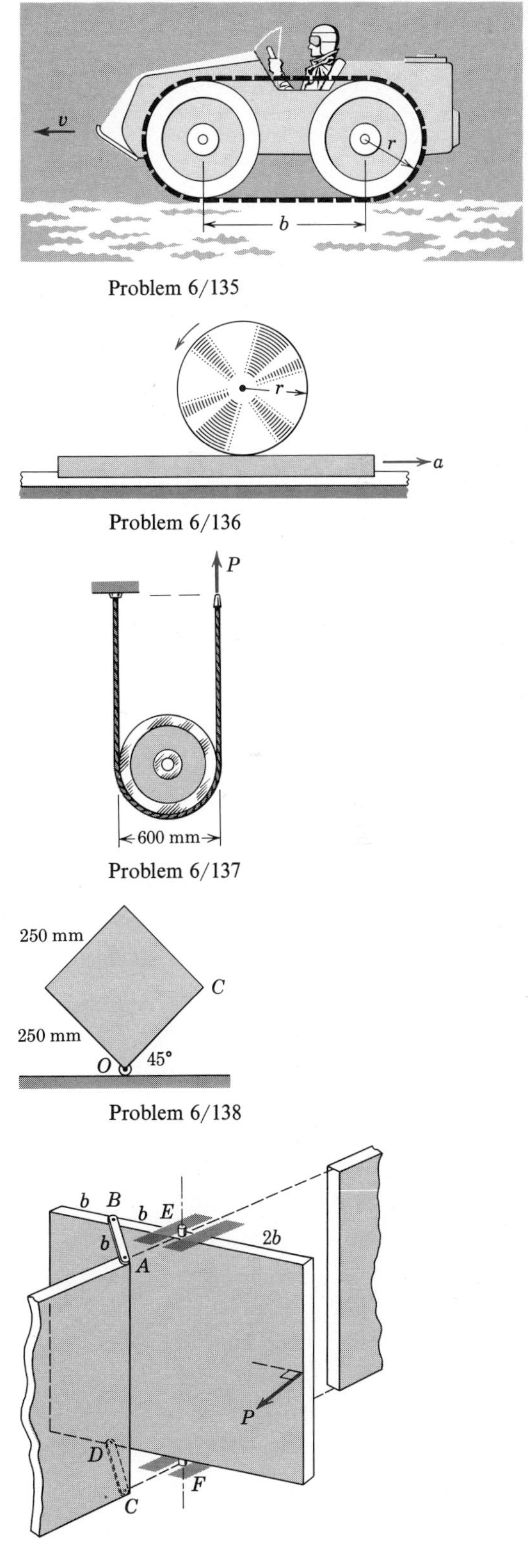

Problem 6/139

◀**6/140** Show that a nonrotating observer moving with a constant velocity $\mathbf{v}_0$ relative to a fixed reference system concludes from his observation of the accelerated motion of a rigid body that the work-energy equation $U = \Delta T$ holds even though he disagrees with a fixed observer on their respective measurements of both the work term and the kinetic-energy term.

6/141 Determine the initial angular acceleration of the links in Prob. 6/132 as the mechanism starts from rest at $\theta = 0$.

6/142 A car of mass m has an acceleration a up an incline θ. Each of the four wheels has a moment of inertia I and a radius r. Write the expression for the power P delivered by the engine to the rear driving wheels when the car reaches a speed v. Friction is sufficient to prevent slipping of the wheels. The horizontal force required to tow the car on a level road at the speed v with engine disengaged is R.

Ans. $P = \left(m + \frac{4I}{r^2}\right)av + mgv \sin\theta + Rv$

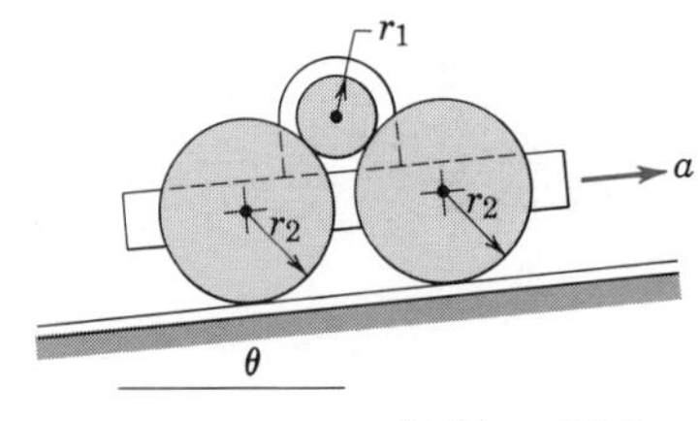

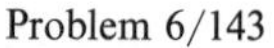
Problem 6/143

6/143 The small vehicle has a total mass m, and each of its four wheels has a moment of inertia I about its center. The vehicle is propelled by an electric motor which supplies a torque M to the frictional-drive pinion of radius r_1. The rotor of the electric motor and attached pinion have a moment of inertia I_O about the shaft axis. Determine the torque M required to give the vehicle an acceleration a up the incline. No slipping occurs between any of the frictional surfaces at the wheels or pinion.

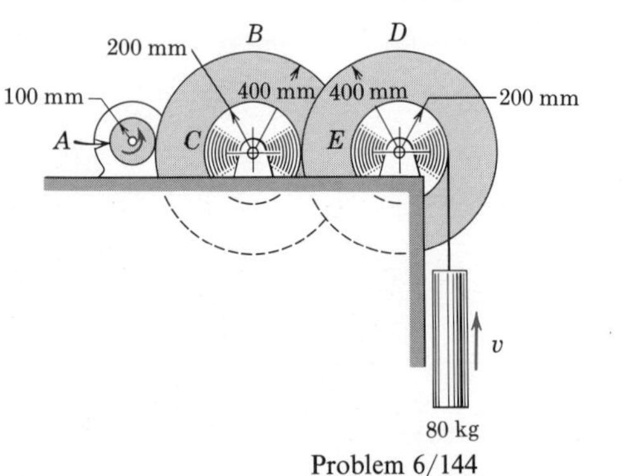

Problem 6/144

6/144 Pinion A of the electric motor turns gear B and its attached pinion C. This unit meshes with gear D and its attached hoisting drum E. The motor rotor and pinion A have a combined mass of 30 kg and a radius of gyration of 150 mm. The unit B and C, as well as the unit D and E, has a mass of 95 kg with a radius of gyration of 300 mm. The motor receives 2 kW of electrical power, 94 per cent of which is converted into mechanical power. For an instant when the upward velocity of the 80-kg mass is 0.9 m/s, calculate the upward acceleration a of the load.

Ans. $a = 0.585$ m/s^2

6/145 The aerial tower shown is designed to elevate a workman in a vertical direction. An internal mechanism at B maintains the angle between AB and BC at twice the angle θ between BC and the ground. If the combined mass of the man and the cab is 200 kg and if all other masses are neglected, determine the torque M applied to BC at C and the torque M_B in the joint at B required to give the cab an initial vertical acceleration of 1.2 m/s² when started from rest in the position $\theta = 30°$.

Ans. $M_B = 11.44$ kN·m
$M = 0$

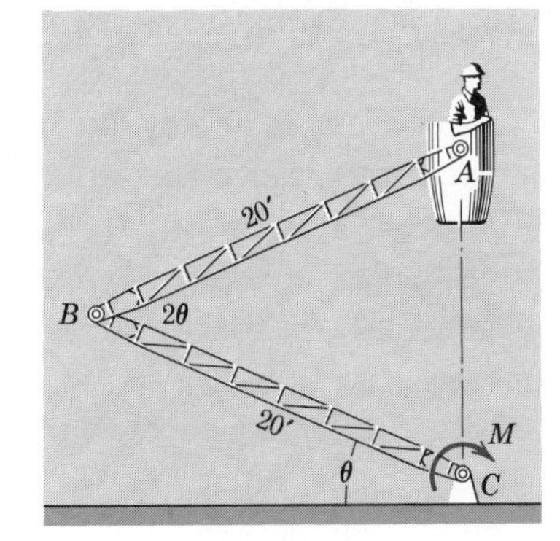

Problem 6/145

6/146 The vertical rod AB is subjected to a steady upward force F which is n times the combined weight of the assembly. Each of the four uniform links has a mass m and is freely pivoted as shown. The collar C of negligible mass is free to slide on the rod. The spring has a stiffness k and is uncompressed when $\theta = 0$. Determine the expression for the angle θ assumed by the links under the steady vertical acceleration.

Ans. $\theta = 2 \sin^{-1} \dfrac{nmg}{kl}$

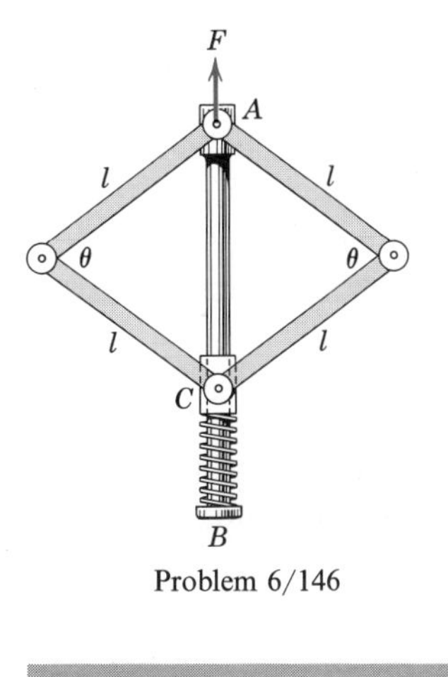

Problem 6/146

6/147 Each of the two uniform links has a mass of 3 kg. The rod AC has a mass of 2.5 kg and is free to slide through the pivoted collar at B. The spring has negligible mass, so that the total mass of the assembly is 8.5 kg. Determine the steady-state angle θ assumed by the links under a constant accelerating force $F = 50$ N. The links are suspended in the vertical plane by the rollers, and the spring has a stiffness $k = 350$ N/m. Also, the spring is uncompressed when θ is essentially zero for $F = 0$.

Ans. $\theta = 27.0°$

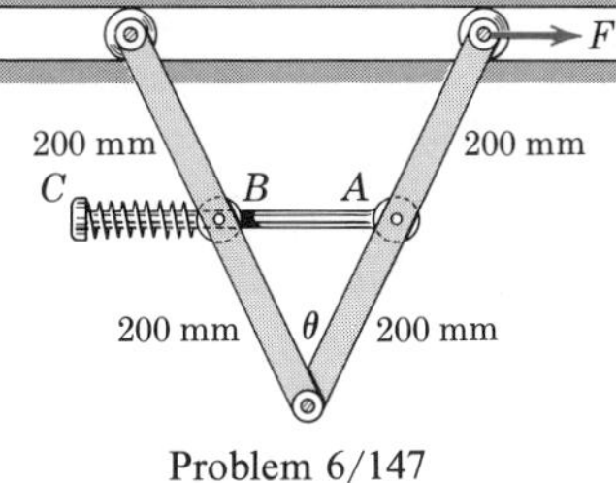

Problem 6/147

6/148 The 36-kg pulley has a radius of gyration about its center of 250 mm. The chain has a mass of 7.5 kg/m and has a length of 2.4 m plus the half-circumferential portion over the pulley. For a very slight unbalance starting from rest, the pulley turns clockwise and the chain piles up on the platform. As each link strikes the platform and comes to rest, it is unable to transmit any force to the links above it. Compute the angular velocity ω of the pulley at the instant that $x =$ 1.2 m. (*Hint:* Analyze the energy change during a differential interval of motion noting the energy loss of each increment of chain which comes to rest on the platform.)

Ans. $\omega = 4.53$ rad/s

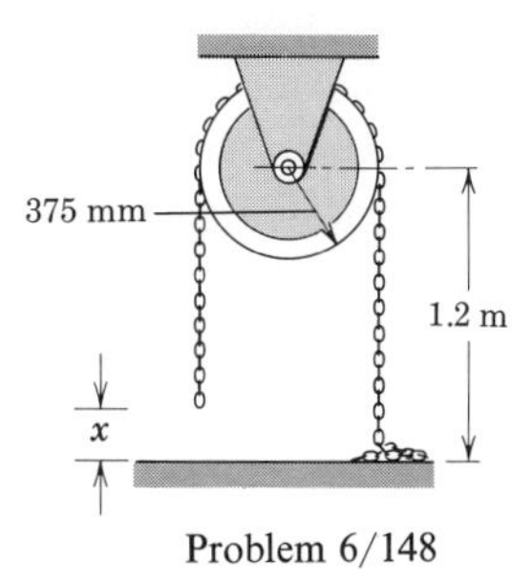

Problem 6/148

35 Impulse and Momentum. The principles of impulse and momentum were developed and used in Art. 19 of Chapter 3 in the description of particle motion. In this treatment it was observed that these principles were of particular importance when the applied forces were expressible as functions of the time and when interactions between particles were present, particularly during short periods of time. Similar advantages result when the impulse-momentum principles are developed and applied for the motion of rigid bodies.

In Art. 25 of Chapter 4 the impulse-momentum principles were extended to cover any defined system of mass particles without restriction as to the connections between the particles of the system. The linear momentum of the general mass system was defined as the vector sum of the linear momenta of all particles of the system and was written as $\mathbf{G} = \Sigma m_i \mathbf{v}_i$ where the velocity of the representative particle is $\mathbf{v}_i = \dot{\mathbf{r}}_i$. For a system of constant total mass the linear momentum may be written $\mathbf{G} = \dfrac{d}{dt} \Sigma m_i \mathbf{r}_i$. Substituting the moment principle $m\bar{\mathbf{r}} = \Sigma m_i \mathbf{r}_i$ gives $\mathbf{G} = \dfrac{d}{dt}(m\bar{\mathbf{r}})$ or

$$\blacktriangleright \qquad \mathbf{G} = m\bar{\mathbf{v}} \tag{125}$$

which holds for a rigid or nonrigid mass system.

The force-momentum relation Eq. 96 for the general mass system and, hence, for a rigid body and the integrated form of the equation are

$$\blacktriangleright \qquad \Sigma \mathbf{F} = \dot{\mathbf{G}} \qquad \text{and} \qquad \int_{t_1}^{t_2} \Sigma \mathbf{F}\, dt = \mathbf{G}_2 - \mathbf{G}_1 \tag{126}$$

The integral is the total linear impulse acting on the rigid body or system during the interval $t_2 - t_1$.

Equation 126 may be written in its scalar component form which, for plane motion in the x-y plane, gives

$$\begin{aligned} \Sigma F_x &= \dot{G}_x \qquad & \int_{t_1}^{t_2} \Sigma F_x\, dt &= G_{x_2} - G_{x_1} \\ & \qquad \text{and} & & \\ \Sigma F_y &= \dot{G}_y \qquad & \int_{t_1}^{t_2} \Sigma F_y\, dt &= G_{y_2} - G_{y_1} \end{aligned} \tag{126a}$$

As in the force-mass-acceleration formulation, the force summations in Eqs. 126 must include all forces acting externally on the body considered.

The angular momentum of the general mass system about its mass center was expressed in Art. 25 as $\overline{\mathbf{H}} = \Sigma(\boldsymbol{\rho}_i \times m_i \mathbf{v}_i)$ which is merely the vector sum of the moments of the linear momenta of all particles about the mass center. This moment sum will now be restricted to the case of a rigid body moving with plane motion, Fig. 57. The mass center G has a velocity $\bar{\mathbf{v}}$, and the body has an angular velocity $\boldsymbol{\omega} = \omega\mathbf{k}$ where $\mathbf{k}$ is a unit vector directed into the paper for the sense of $\boldsymbol{\omega}$ shown. The velocity $\mathbf{v}_i$ of the representative particle of mass m_i is that of the mass center $\bar{\mathbf{v}}$ plus the velocity $\dot{\boldsymbol{\rho}}_i = \boldsymbol{\omega} \times \boldsymbol{\rho}_i$ relative to the mass center as shown in the figure. The magnitude of this relative velocity is $|\dot{\boldsymbol{\rho}}_i| = \rho_i\omega$.

The angular momentum about the mass center may now be written $\overline{\mathbf{H}} = \Sigma(\boldsymbol{\rho}_i \times m_i[\dot{\boldsymbol{\rho}}_i + \bar{\mathbf{v}}])$. The first term is $\Sigma\rho_i{}^2m_i\omega\mathbf{k} = \bar{I}\omega\mathbf{k}$ where $\bar{I} = \Sigma m_i\rho_i{}^2$ is the moment of inertia of the body about its mass center. The second term is $\Sigma(\boldsymbol{\rho}_i \times m_i\bar{\mathbf{v}}) = -\bar{\mathbf{v}} \times \Sigma m_i\boldsymbol{\rho}_i = \mathbf{0}$ since $\Sigma m_i\boldsymbol{\rho}_i = m\bar{\boldsymbol{\rho}} = \mathbf{0}$. Because the angular momentum vector is always normal to the plane of motion, the vector notation is generally unnecessary, and the resulting angular momentum about the mass center may be written

$$\blacktriangleright \qquad \overline{H} = \bar{I}\omega \tag{127}$$

This angular momentum appears in the moment-angular-momentum relation, Eq. 98, which, along with its integrated form, is

$$\blacktriangleright \qquad \Sigma\overline{M} = \dot{\overline{H}} \qquad \text{and} \qquad \int_{t_1}^{t_2} \Sigma\overline{M}\,dt = \overline{H}_2 - \overline{H}_1 \tag{128}$$

The integral is the total angular impulse acting on the body about its mass center during the interval $t_2 - t_1$.

With the moments about G of the linear momenta of all particles accounted for by $\overline{H} = \bar{I}\omega$, it follows that the linear momentum vector $\mathbf{G} = m\bar{\mathbf{v}}$ is taken through the mass center G, as shown in Fig. 58*a*. Thus $\mathbf{G}$ and $\overline{\mathbf{H}}$ have vector properties analogous to those of the resultant force and couple.

With the establishment of the linear and angular momentum resultants in Fig. 58*a*, the angular momentum H_O about any point O is easily written as

$$\blacktriangleright \qquad H_O = \bar{I}\omega + m\bar{v}d \tag{129}$$

This expression holds at any particular instant of time about O, which may be a fixed or moving point on or off the body.

The angular momentum about O is related to the moment sum about O by Eq. 99 for the general case. The restricted form of this expression, given by Eq. 99*a*, is more useful. This moment equation and its integrated form, together with the three conditions of which one must hold for the equations

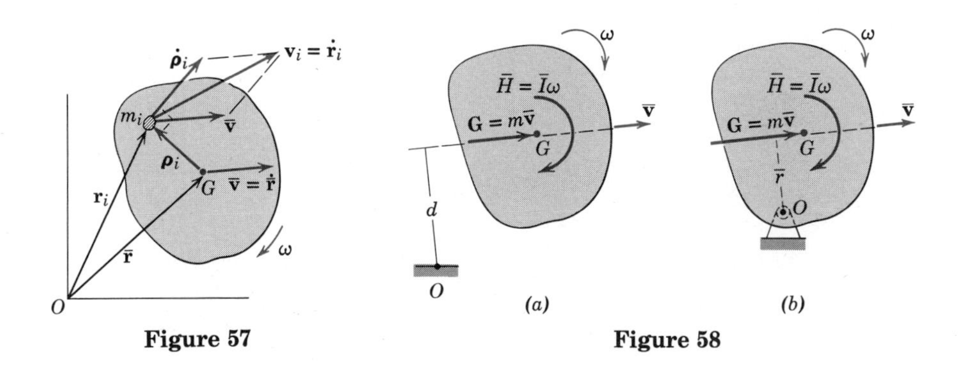

Figure 57 **Figure 58**

to be valid, are

$$\blacktriangleright \qquad \Sigma M_O = \dot{H}_O \qquad \text{and} \qquad \int_{t_1}^{t_2} \Sigma M_O \, dt = H_{O_2} - H_{O_1} \qquad \textbf{(130)}$$

$$\text{if} \begin{cases} 1. \; v_O = 0 \\ 2. \; \bar{v} = 0 \\ 3. \; v_O \text{ and } \bar{v} \text{ are parallel} \end{cases}$$

Use of the first relation is limited by the convenience of obtaining the expression $\dot{H}_O = \frac{d}{dt}(\bar{I}\omega + m\bar{v}d)$, and use of the integrated form is limited to the requirement that one of the three conditions expressed with Eq. 130 must hold throughout the entire motion interval.

When a body rotates about a fixed point O on the body or body extended, as shown in Fig. 58*b*, the relations $\bar{v} = \bar{r}\omega$ and $d = \bar{r}$ may be substituted into the expression for H_O giving $H_O = (\bar{I}\omega + m\bar{r}^2\omega)$. But $\bar{I} + m\bar{r}^2 = I_O$ so that

$$\blacktriangleright \qquad H_O = I_O\omega \qquad \textbf{(131)}$$

Thus, for rotation about a fixed point, the first of the conditions of Eq. 130 holds, and the moment equation and its integrated form are

$$\blacktriangleright \qquad \Sigma M_O = I_O\dot{\omega} \qquad \text{and} \qquad \int_{t_1}^{t_2} \Sigma M_O \, dt = I_O(\omega_2 - \omega_1) \qquad \textbf{(132)}$$

In the application of the linear and angular impulse-momentum equations for rigid-body motion, it is essential that a complete free-body diagram be drawn in order that the force and moment summations may be correctly evaluated. It is also useful to indicate the resultant linear-momentum vector and the angular-momentum couple. The student is cautioned again not to add linear momentum and angular momentum for the same reason that force and moment cannot be added directly.

The equations of impulse and momentum may also be used for a system of interconnected rigid bodies since the momentum principles are applicable to any general system of constant mass. In Fig. 59 are shown two of any given number of interconnected bodies. Equations 96 and 97, which are $\Sigma\mathbf{F} = \dot{\mathbf{G}}$ and $\Sigma\mathbf{M}_O = \dot{\mathbf{H}}_O$ where O is a fixed reference point, may be written for each member of the system and added. The sums would be

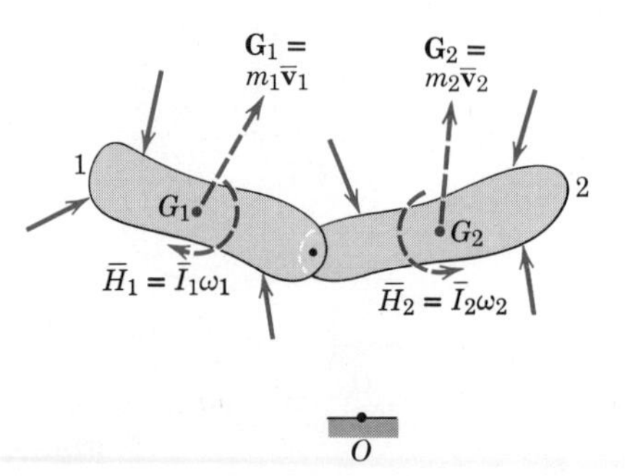

Figure 59

$$\Sigma\mathbf{F} = \dot{\mathbf{G}}_1 + \dot{\mathbf{G}}_2 + \cdots$$
$$\Sigma\mathbf{M}_O = \dot{\mathbf{H}}_{O_1} + \dot{\mathbf{H}}_{O_2} + \cdots \qquad (133)$$

In integrated form for a finite time interval, these expressions are

$$\int_{t_1}^{t_2} \Sigma\mathbf{F}\, dt = (\Delta\mathbf{G})_{\text{system}} \qquad \int_{t_1}^{t_2} \Sigma\mathbf{M}_O\, dt = (\Delta\mathbf{H}_O)_{\text{system}} \qquad (134)$$

It is noted that the equal and opposite actions and reactions in the connections are internal to the system and cancel one another so will not be involved in the force and moment summations. Also point O is one fixed reference point for the entire system.

In Art. 26 of Chapter 4 the principles of conservation of momentum for a general mass system were expressed by Eqs. 102 and 103. These principles are applicable to either a single rigid body or to a system of interconnected rigid bodies. Thus, if $\Sigma\mathbf{F} = \mathbf{0}$ for a given interval of time, then

▶ $$\Delta\mathbf{G} = \mathbf{0} \qquad [102]$$

which says that the linear momentum vector undergoes no change in the absence of a resultant linear impulse. For the system of interconnected rigid bodies there may be linear momentum changes of individual parts of the system during the interval, but there will be no resultant momentum change for the system if there is no resultant linear impulse.

Similarly, if the resultant moment about a given fixed point O or about the mass center is zero during a particular interval of time for a single rigid body or for a system of interconnected rigid bodies, then

▶ $$\Delta\mathbf{H}_O = \mathbf{0} \qquad \text{or} \qquad \Delta\overline{\mathbf{H}} = \mathbf{0} \qquad [103]$$

which says that the angular momentum either about the fixed point or about the mass center undergoes no change in the absence of a corresponding resultant angular impulse. Again, in the case of the interconnected system, there may be angular momentum changes of individual components during the interval, but there will be no resultant angular momentum change for the system if there is no resultant angular impulse about the fixed point or the mass center. Either of Eqs. 103 may hold without the other. In the case of the interconnected system the use of the center of mass for the system is generally inconvenient. As was illustrated previously in Art. 19 in the chapter on particle motion, the use of momentum principles greatly facilitates the analysis of situations where forces and couples act for very short periods of time.

Sample Problems

6/149 The center of the circular cylinder of mass m and radius r is given a velocity v_0 up the incline at time $t = 0$. If the cylinder does not slip, determine the velocity v of its center after time t.

Solution. The free-body diagram is constructed as shown, and a fixed point O is selected as a moment center which will eliminate F. The expression for H_O is $H_O = \bar{I}\omega + m\bar{v}r$. With the substitution of $\bar{v} = r\omega$ and $I_C = \bar{I} + mr^2$, the angular momentum about O becomes $H_O = I_C\omega$. The impulse-momentum principle is

$$\left[\int \Sigma M_O \, dt = \Delta H_O\right] \qquad \int_0^t mgr \sin\theta \, dt = I_C(\omega - [-\omega_0])$$

where the moments of N and $mg \cos\theta$ cancel.

Thus $$mgrt \sin\theta = \frac{3}{2}mr^2 \frac{v + v_0}{r}$$

Solution for v gives

$$v = \tfrac{2}{3}gt \sin\theta - v_0 \qquad \textit{Ans.}$$

If the angular momentum is referred to the mass center instead of O, there results

$$\left[\int \Sigma \bar{M} \, dt = \Delta \bar{H}\right] \qquad \int_0^t Fr \, dt = \bar{I}(\omega - [-\omega_0])$$

$$\left[\int \Sigma F \, dt = \Delta G\right] \qquad \int_0^t (mg \sin\theta - F) \, dt = m(v - [-v_0])$$

Elimination of F between the integrated equations gives the same result as previously but with somewhat more labor.

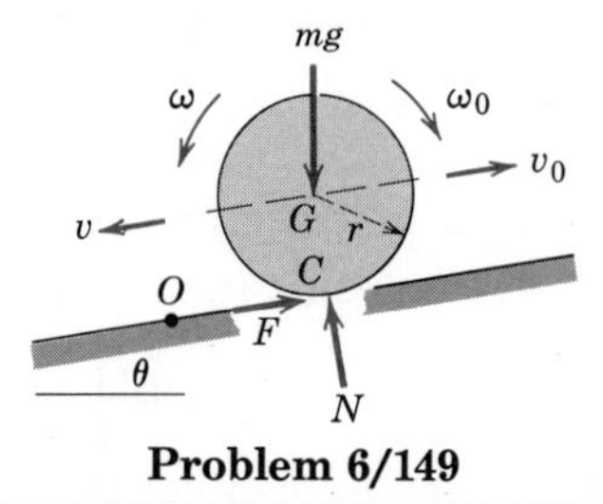

Problem 6/149

6/150 The sheave of the hoisting rig shown has a mass of 30 kg and a centroidal radius of gyration of 250 mm. The 40-kg load D which is carried by the sheave has an initial downward velocity $v_0 = 1.2$ m/s at the instant when a clockwise torque is applied to the hoisting drum to maintain essentially a constant force $F = 375$ N in the cable at B. Compute the angular velocity ω of the sheave 5 s after the torque is applied to the drum. Neglect all friction.

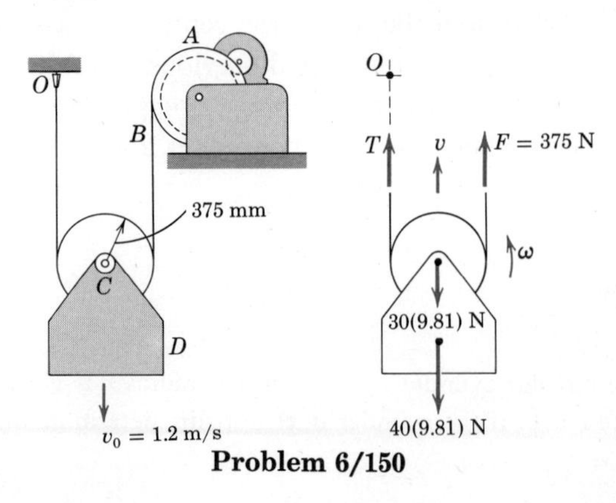

Problem 6/150

Solution. The load and the sheave taken together constitute the system, and its free-body diagram is shown in the right-hand part of the figure. A moment equation in the counterclockwise sense about the fixed point O will eliminate the unknown force T which was not asked for. Thus

$$\left[\int \Sigma M_O \, dt = \Delta H_O\right]$$

$$\big(375(0.75) - 70(9.81)(0.375)\big)5 = 70\big(0.375\omega - [-1.2]\big)0.375 + 30(0.25)^2\left(\omega - \left[-\frac{1.2}{0.375}\right]\right)$$

$$\omega = 6.93 \text{ rad/s counterclockwise} \qquad \textit{Ans.}$$

The unknown tension T could now be found if desired by applying the linear impulse-momentum relation to the system.

6/151 The uniform rectangular block of dimensions shown is sliding to the left on the horizontal surface with a velocity v when it strikes the small step in the surface. Assume negligible rebound at the step, and compute the minimum value of v which will permit the block to pivot about the edge of the step and just reach the standing position A with no velocity. Compute the energy loss ΔE for $b = c$.

Solution. It will be assumed that the edge of the step O acts as a latch on the corner of the block, so that the block pivots about O. Furthermore, the height of the step is assumed negligible compared with the dimensions of the block. During impact the only force which exerts a moment about O is the weight mg, but the angular impulse due to the weight is extremely small since the time of impact is negligible. Thus angular momentum about O may be said to be conserved.

The initial angular momentum of the block about O just before impact is the moment of its linear momentum and is $H_O = mv(b/2)$. The velocity of the center of mass G immediately after impact is $\bar{v}$, and the angular velocity is $\omega = \bar{v}/\bar{r}$. The angular momentum about O just after impact when the block is starting its rotation about O is

$$[H_O = I_O\omega] \qquad H_O = \left[\frac{1}{12}m(b^2 + c^2) + m\left(\left[\frac{c}{2}\right]^2 + \left[\frac{b}{2}\right]^2\right)\right]\omega = \frac{m}{3}(b^2 + c^2)\omega$$

Conservation of angular momentum gives

$$[\Delta H_O = 0] \qquad \frac{m}{3}(b^2 + c^2)\omega = mv\frac{b}{2} \qquad \omega = \frac{3vb}{2(b^2 + c^2)}$$

This angular velocity will be sufficient to raise the block just past position A if the kinetic

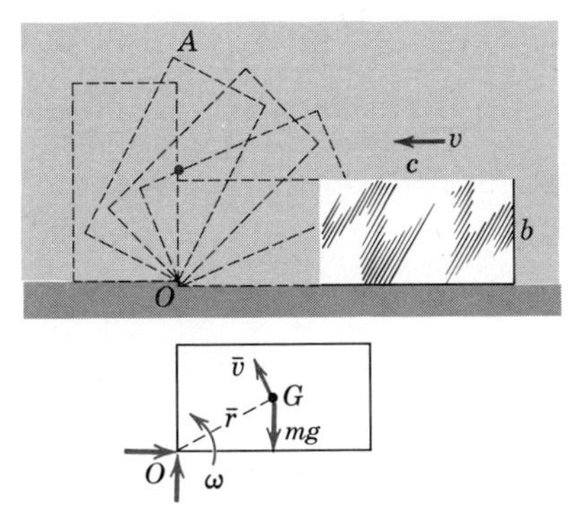

Problem 6/151

energy of rotation equals the increase in potential energy. Thus

$$[\Delta T + \Delta V_g = 0] \qquad \frac{1}{2}I_O\omega^2 - mg\left(\sqrt{\left(\frac{b}{2}\right)^2 + \left(\frac{c}{2}\right)^2} - \frac{b}{2}\right) = 0$$

$$\frac{1}{2}\frac{m}{3}(b^2 + c^2)\left[\frac{3vb}{2(b^2 + c^2)}\right]^2 - \frac{mg}{2}(\sqrt{b^2 + c^2} - b) = 0$$

$$v = 2\sqrt{\frac{g}{3}\left(1 + \frac{c^2}{b^2}\right)(\sqrt{b^2 + c^2} - b)} \qquad \textit{Ans.}$$

The percentage loss of energy is

$$\frac{\Delta E}{E} = \frac{\frac{1}{2}mv^2 - \frac{1}{2}I_O\omega^2}{\frac{1}{2}mv^2} = 1 - \frac{k_O{}^2\omega^2}{v^2} = 1 - \left(\frac{b^2 + c^2}{3}\right)\left[\frac{3b}{2(b^2 + c^2)}\right]^2$$

$$= 1 - \frac{3}{4\left(1 + \frac{c^2}{b^2}\right)}, \qquad \Delta E/E = 62.5 \text{ per cent for } b = c \qquad \textit{Ans.}$$

Problems

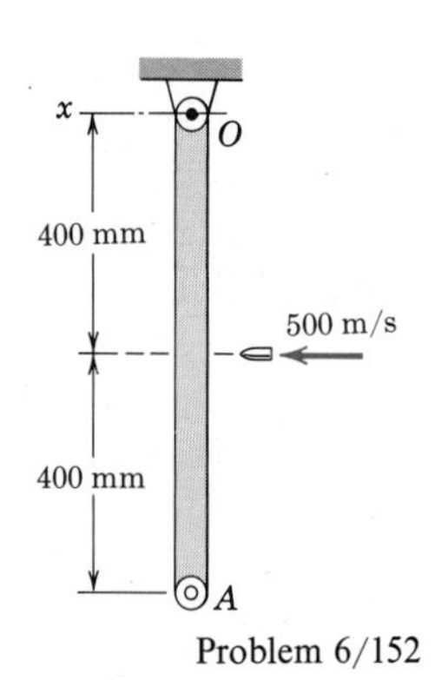

Problem 6/152

6/152 The 30-g bullet has a horizontal velocity of 500 m/s as it strikes the 10-kg slender bar OA, which is suspended from point O and is initially at rest. Calculate the angular velocity ω which the bar with its embedded bullet has acquired immediately after impact.

Ans. $\omega = 2.81$ rad/s

6/153 If the bullet of Prob. 6/152 takes 0.010 s to embed itself in the bar, calculate the time average of the horizontal force O_x exerted by the pin on the bar at O during the interaction between the bullet and the bar. Use the results cited for Prob. 6/152.

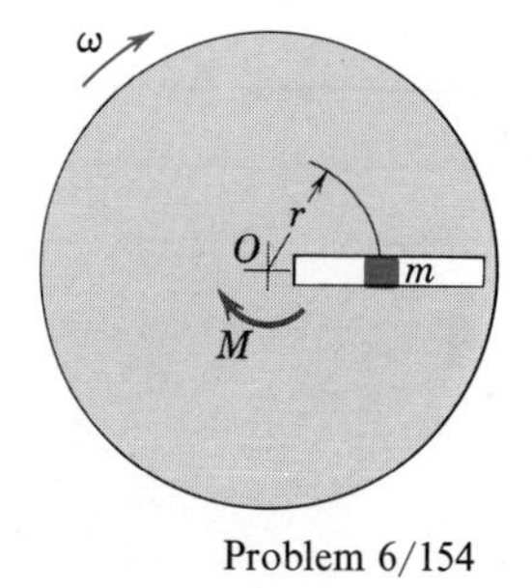

Problem 6/154

6/154 The small block of mass m slides along the radial slot of the disk while the disk rotates in the horizontal plane about its center O. The block is released from rest relative to the disk and moves outward with an increasing velocity $\dot{r}$ along the slot as the disk turns. Determine the expression in terms of r and $\dot{r}$ for the torque M which must be applied to the disk to maintain a constant angular velocity ω of the disk.

6/155 The uniform circular disk of 150-mm radius has a mass of 30 kg and is mounted on the rotating bar OA in three different ways. In each case the bar rotates about its vertical shaft at O with a clockwise angular velocity $\omega = 4$ rad/s. In case (a) the disk is welded to the bar. In case (b) the disk, which is pinned freely at A, moves with curvilinear translation and therefore has no rigid-body rotation. In case (c) the relative angle between the disk and the bar is increasing at the rate $\dot{\theta} = 8$ rad/s. Calculate the angular momentum of the disk about point O for each case.

Ans. (a) $H_O = 12.15\ \text{kg}\cdot\text{m}^2/\text{s}$
(b) $H_O = 10.80\ \text{kg}\cdot\text{m}^2/\text{s}$
(c) $H_O = 9.45\ \text{kg}\cdot\text{m}^2/\text{s}$

6/156 The unbalanced wheel is made to roll to the right without slipping with a constant velocity of 0.9 m/s of its center O. The wheel has a mass of 8 kg with center of mass at G and has a radius of gyration about O of 150 mm. Determine the angular momentum H_O of the wheel about O at the instant (a) when G passes directly over O with $\theta = 0$ and (b) when G passes the horizontal line through O where $\theta = 90°$.

6/157 The 24-kg slender bar is released from rest in the horizontal position shown. If point A of the bar becomes attached to the pivot at B upon impact after dropping through the 0.9-m distance, calculate the angular velocity ω of the bar immediately after impact. *Ans.* $\omega = 3.50$ rad/s

6/158 The wheel shown rolls on its 150-mm-diameter hubs on inclined rails without slipping. If the wheel has a clockwise angular velocity of 4 rad/s at time $t = 0$, calculate its angular velocity ω at $t = 16$ s. The radius of gyration of the wheel about its center is 200 mm.

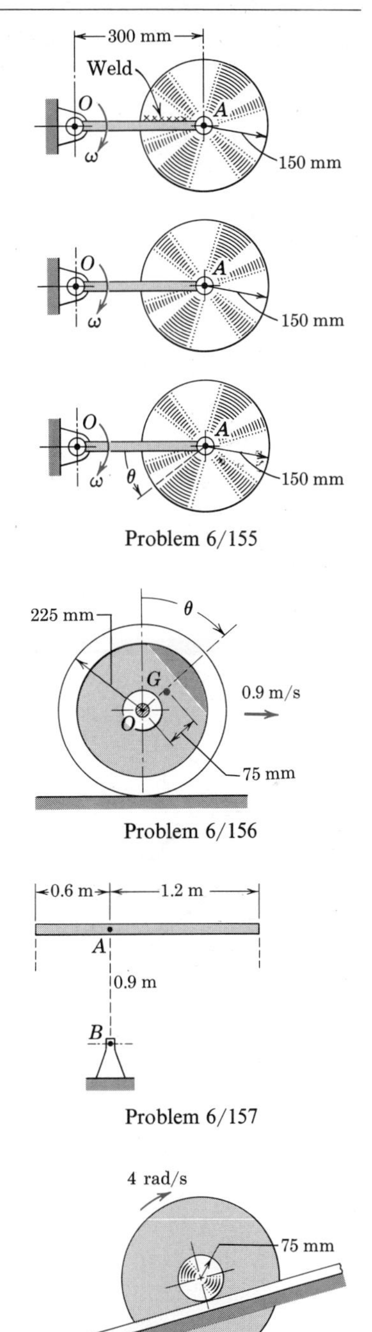

Problem 6/155

Problem 6/156

Problem 6/157

Problem 6/158

6/159 The 225-mm-radius wheel with rigidly attached 150-mm-radius hub has a mass of 60 kg and is released from rest on the 60° incline. The cord is securely wrapped around the hub and fastened to the fixed point A. Calculate the velocity of the center O for the position reached 3 seconds after release. *Ans.* $v_O = 3.88$ m/s

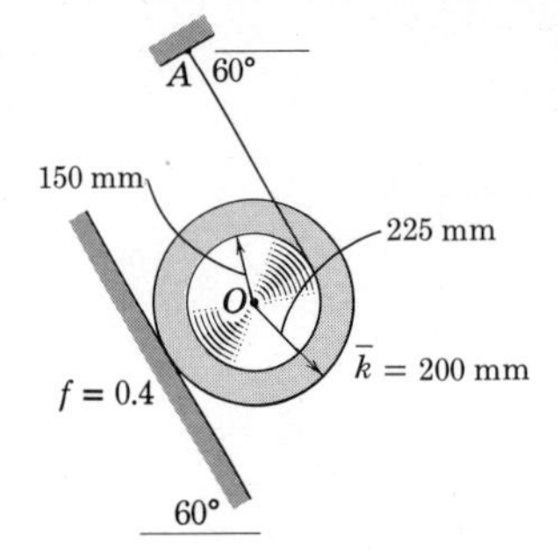

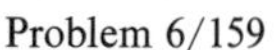
Problem 6/159

6/160 The slotted circular disk whose mass is 6 kg has a radius of gyration about O of 175 mm. The disk carries the four steel balls, each of mass 0.15 kg and located as shown, and rotates freely about a vertical axis through O with an angular speed of 120 rev/min. Each of the small balls is held in place by a latching device not shown. If the balls are released while the disk is rotating and come to rest in the dotted positions relative to the slots, compute the new angular velocity ω of the disk. Also find the magnitude $|\Delta E|$ of the energy loss due to the impact of the balls with the ends of the slots. Neglect the diameter of the balls and discuss this approximation.

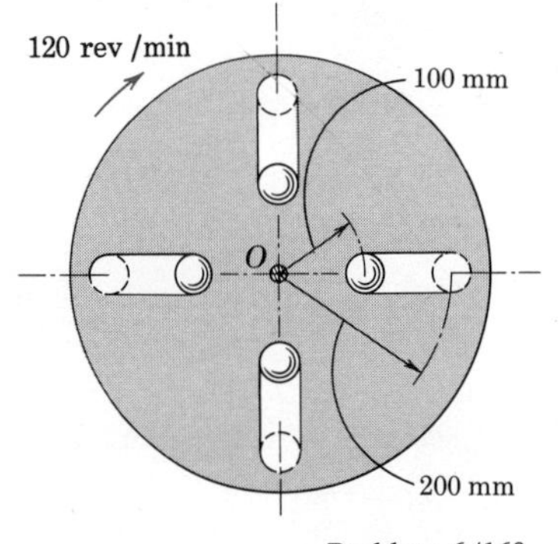

Problem 6/160

6/161 The 17.5-t lunar landing module with center of mass at G has a radius of gyration of 1.8 m about G. The module is designed to contact the lunar surface with a vertical free-fall velocity of 8 km/h. If one of the four legs hits the lunar surface on a small incline and suffers no rebound, compute the angular velocity ω of the module immediately after impact as it pivots about the contact point. The 9-m dimension is the distance across the diagonal of the square formed by the four feet as corners. *Ans.* $\omega = 0.308$ rad/s

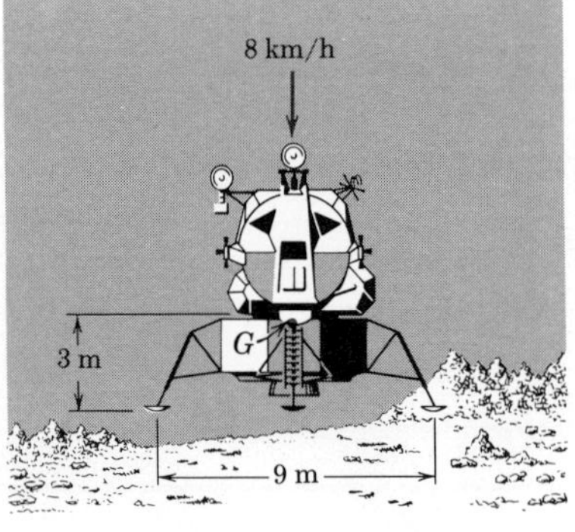

Problem 6/161

6/162 The body of the spacecraft has a mass of 160 kg and a radius of gyration about its z-axis of 0.45 m. Each of the two solar panels may be treated as a uniform flat plate of 8-kg mass. If the spacecraft is rotating about its z-axis at the angular rate of 1.0 rad/s with $\theta = 0$, determine the angular rate ω after the panels are rotated to the position $\theta = \pi/2$ by an internal mechanism. Neglect the small momentum change of the body about the y-axis. *Ans.* $\omega = 0.974$ rad/s

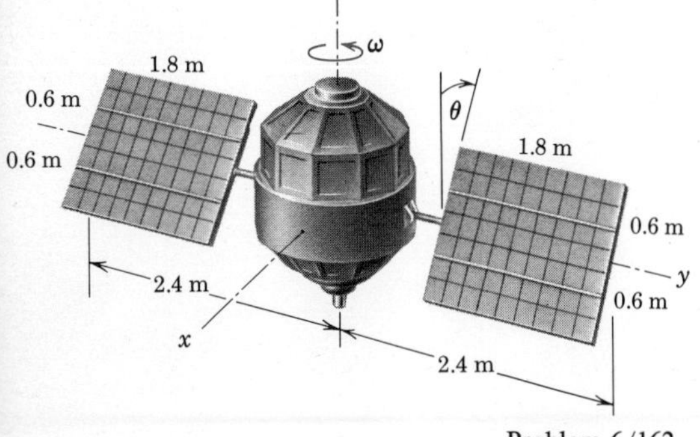

Problem 6/162

6/163 The uniform circular disk D with a mass of 4 kg is free to turn about the bearing axis C–C. The 2.5-kg arm B is fastened to the vertical shaft O–O. The arm may be approximated as a slender bar 300 mm in length, and the moment of inertia of the vertical shaft O–O may be neglected. The rod A has a mass of 3 kg and is fastened securely to the arm. If the initial angular velocity of D is $\omega_0 = 7$ rad/s in the direction shown and the arm B is at rest, determine the angular velocity ω_B of the arm after a torque $M = 1.5$ N·m has been applied to the shaft for 4 s.

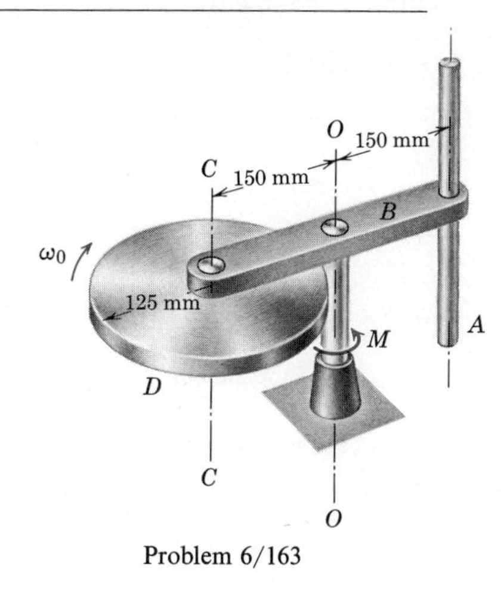

Problem 6/163

6/164 The solid circular cylinder of radius r is at rest on the flat belt when a force P is applied to the belt. If P is sufficient to cause slipping between the belt and the cylinder at all times, determine the time t required for the cylinder to reach the dotted position. Also determine the angular velocity ω of the cylinder in this same position. The coefficient of friction between the cylinder and the belt is f.

Ans. $t = \sqrt{\dfrac{2s}{fg}}, \quad \omega = \dfrac{2\sqrt{2fgs}}{r}$

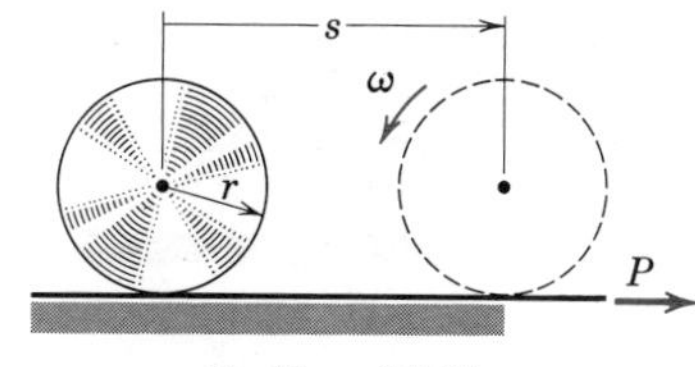

Problem 6/164

6/165 The small gear is made to rotate in a horizontal plane about the large stationary gear by means of the torque M applied to the arm OA. The small gear has a mass of 3 kg and may be treated as a circular disk. The 2-kg arm OA has a radius of gyration about the fixed bearing at O of 150 mm. Determine the constant torque M required to give the arm OA an absolute angular velocity of 20 rad/s in 3 s, starting from rest. Neglect friction. *Ans.* $M = 1.819$ N·m

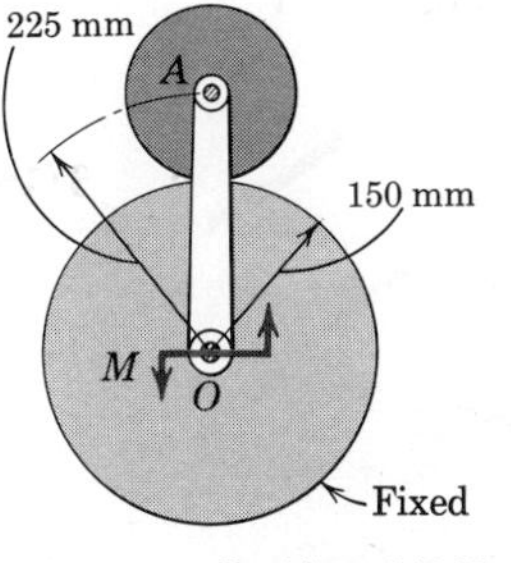

Problem 6/165

6/166 A uniform circular disk which rolls without slipping with a velocity v encounters an abrupt change in the direction of its motion as it rolls onto the incline θ. Determine the new velocity v' of the center of the disk as it starts up the incline, and find the fraction n of the initial energy which is lost due to contact with the incline if $\theta = 10°$.

Ans. $v' = \dfrac{v}{3}(1 + 2\cos\theta), \quad n = 0.020$

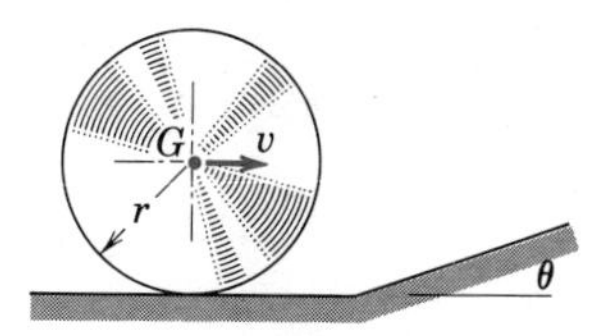

Problem 6/166

6/167 Determine the minimum velocity v which the wheel may have and just roll over the obstruction. The centroidal radius of gyration of the wheel is k, and it is assumed that the wheel does not slip.

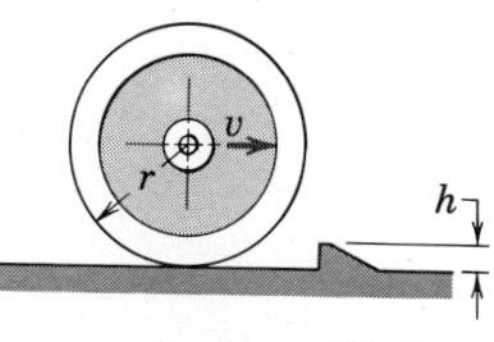

Problem 6/167

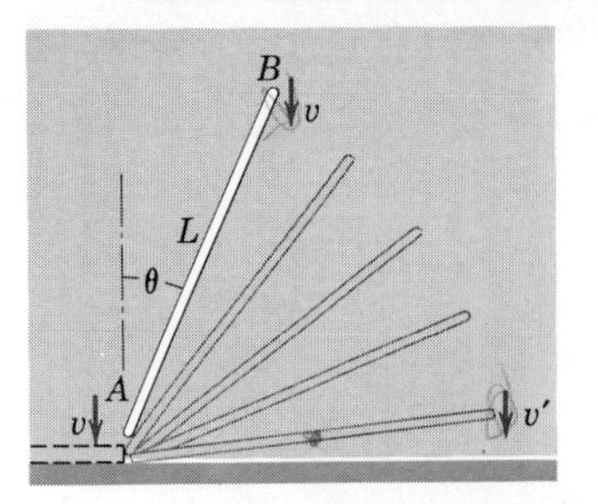

Problem 6/168

6/168 A uniform pole of length L is dropped at an angle θ with the vertical, and both ends have a velocity v as end A hits the ground. If end A pivots about its contact point during the remainder of the motion, determine the velocity v' with which end B hits the ground.

$$Ans.\ v' = \sqrt{\frac{9v^2}{4}\sin^2\theta + 3gL\cos\theta}$$

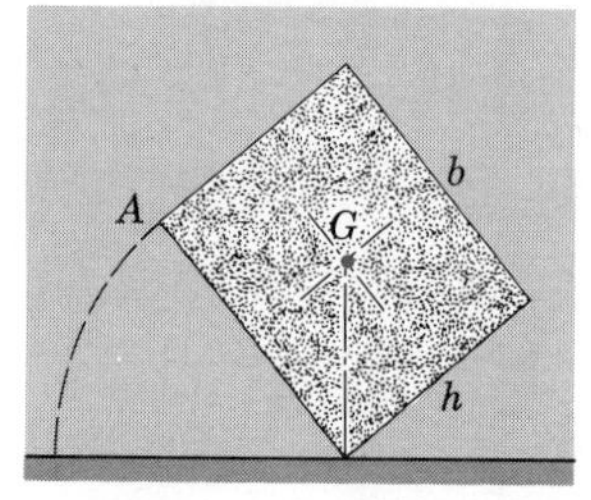

Problem 6/169

6/169 The uniform stone block with $b = 1.2$ m and $h = 0.9$ m is released from rest with its center of mass G directly above the supporting corner. Determine the angular velocity ω' of the block about corner A immediately after impact assuming that A remains in contact with the ground. Also assume that contact occurs at the corners only and that no slipping takes place. What fraction $\Delta E/E$ of the energy is lost due to impact?

Ans. $\omega' = 0.112$ rad/s
$\Delta E/E = 0.998$

6/170 Determine the maximum value of b/h for the stone block of Prob. 6/169 for which any rotation about corner A is possible following impact. Assume corner contact and no slipping.

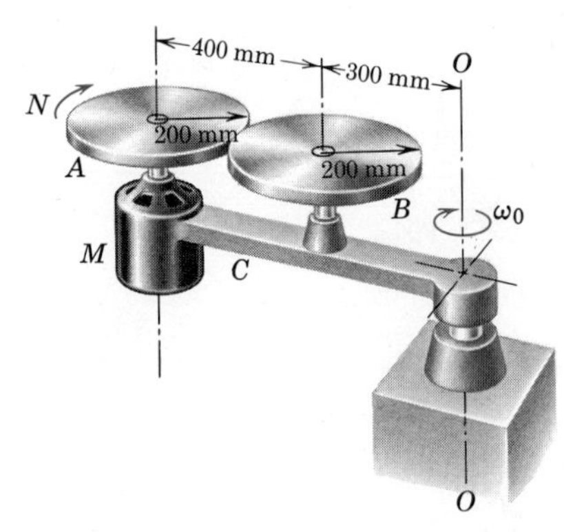

Problem 6/171

◀ **6/171** The motor M drives disk A which turns disk B with no slipping. Disk A and its attached shaft and motor armature have a mass of 18 kg and a combined radius of gyration of 85 mm. Disk B has a mass of 5 kg and a radius of gyration of 140 mm. The motor housing and attached arm C have a combined mass of 24 kg and a radius of gyration about the axis O–O of 450 mm. Before the motor is turned on, the entire assembly is rotating as a unit about O–O with a rotational velocity $\omega_0 = 30$ rev/min in the direction shown. The motor M has an operating speed $N = 1720$ rev/min in the direction shown as measured with C fixed. Determine the new rotational velocity ω of arm C if the motor is turned on. *Ans.* $\omega = 26.2$ rev/min

6/172 A mechanical device for reducing the spin of a satellite consists of two masses m attached to wires or thin metal tapes which unwind when released and acquire additional angular momentum at the expense of the angular momentum of the satellite. The masses are then released when unwound. Each tape has a length $R\phi$ and is fastened to its point A. If the satellite is spinning at an initial speed ω_0 when the masses are released with $\beta = 0$, determine the rate $\dot{\beta}$ of unwinding and find the new angular velocity ω_ϕ of the satellite when $\beta = \phi$. The moment of inertia of the satellite exclusive of the m's is I. (*Hint:* Equate the angular momentum and the kinetic energy of the system to their respective initial values.) *Ans.* $\dot{\beta} = \omega_0$ constant

$$\omega_\phi = \frac{I + 2mR^2(1 - \phi^2)}{I + 2mR^2(1 + \phi^2)}\omega_0$$

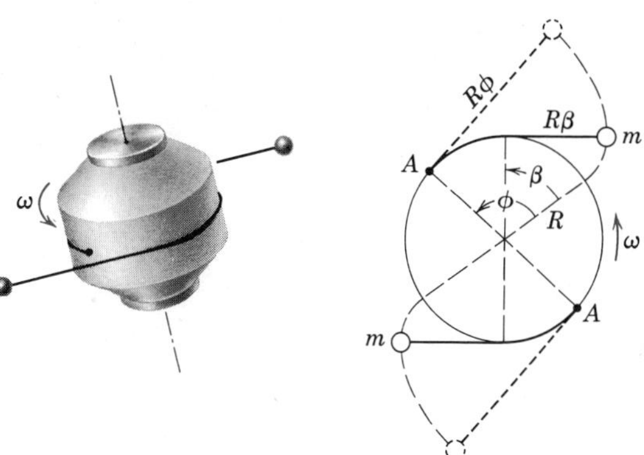

Problem 6/172

6/173 The two slender bars each having a mass of 4 kg are hinged at B and pivoted at C. If a horizontal impulse of $\int F\,dt = 14$ N·s is applied to the end A of the lower bar during an interval of 0.1 s during which the bars are still essentially in their vertical rest positions, compute the angular velocity ω_2 of the upper bar immediately after the impulse. *Ans.* $\omega_2 = 2.50$ rad/s

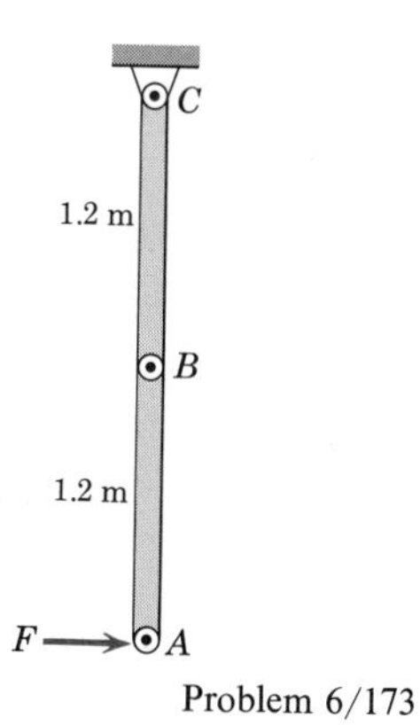

Problem 6/173

7 SPACE KINEMATICS OF RIGID BODIES

36 Introduction. Although a large percentage of dynamics problems in engineering lend themselves to solution by means of the principles of plane motion, modern developments have focused increasing attention upon problems which call for the analysis of motion in three dimensions. Inclusion of the third dimension adds considerable complexity to the kinematic and kinetic relationships. Not only does the added dimension introduce a third component to vectors which represent force, linear velocity, linear acceleration, and linear momentum, but the introduction of the third dimension adds the possibility for two additional components for vectors representing angular quantities including moments of forces, angular velocity, angular acceleration, and angular momentum. To a large extent this added mathematical complexity is handled by making full use of vector analysis and by introducing where needed matrix and tensor analysis.

A good background in the dynamics of plane motion is extremely useful in the study of spatial dynamics, as the approach to problems and many of the terms in the three-dimensional relationships are the same as or analogous to those in two dimensions. If the study of spatial dynamics is undertaken without the benefit of prior study of plane-motion dynamics, more time will be required to master the principles and to become familiar with the approach to problems.

The study of the space dynamics of rigid bodies follows the same plan as that for plane motion, namely, kinematics followed by kinetics.

37 Absolute Motion. The case of the *translation* of a rigid body in three-dimensional space, Fig. 60, will be mentioned first. Any two points in the body, such as A and B, will move along parallel straight lines if the motion is one of *rectilinear translation* or will move along congruent curves if the motion is one of *curvilinear translation.* In either case every line in the body,

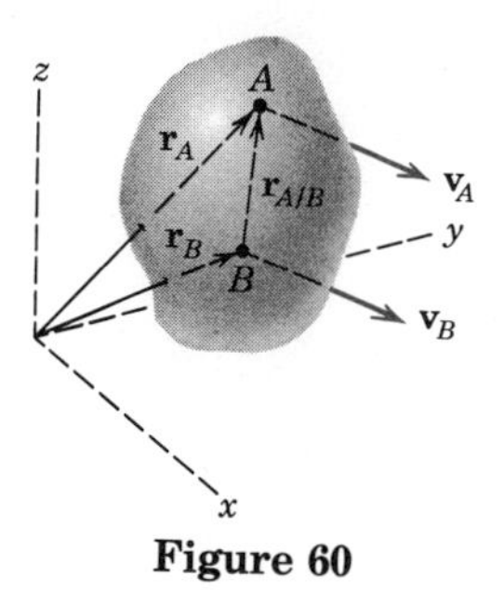

Figure 60

such as AB, remains parallel to its original position. The position vectors and their first and second time derivatives are

$$\mathbf{r}_A = \mathbf{r}_B + \mathbf{r}_{A/B} \qquad \mathbf{v}_A = \mathbf{v}_B \qquad \mathbf{a}_A = \mathbf{a}_B$$

where $\mathbf{r}_{A/B}$ remains constant and therefore has no time derivative. Thus all points in the body have the same velocity and the same acceleration.

Consider now the *rotation* of a rigid body about a fixed axis n-n in space with an angular velocity ω, as shown in Fig. 61. The origin O of the fixed coordinate system is arbitrarily chosen on the rotation axis. Any point such as A that is not on the axis moves in a circular arc in a plane normal to the axis and has a velocity

$$\blacktriangleright \qquad \mathbf{v} = \omega \times \mathbf{r} \tag{135}$$

which may be seen by replacing $\mathbf{r}$ by $\mathbf{a} + \mathbf{b}$ and noting that $\omega \times \mathbf{a} = \mathbf{0}$. The acceleration of A is given by the time derivative of Eq. 135. Thus,

$$\blacktriangleright \qquad \mathbf{a} = \dot{\omega} \times \mathbf{r} + \omega \times (\omega \times \mathbf{r}) \tag{136}$$

where $\dot{\mathbf{r}}$ has been replaced by its equal $\mathbf{v} = \omega \times \mathbf{r}$. The normal and tangential components of $\mathbf{a}$ for the circular motion have the familiar magnitudes $a_n = |\omega \times (\omega \times \mathbf{r})| = b\omega^2$ and $a_t = |\dot{\omega} \times \mathbf{r}| = b\alpha$, where $\alpha = \dot{\omega}$.

Next consider the motion of a rigid body which is confined to rotate about a fixed point rather than a fixed axis. It may be shown that any finite motion of the body about the fixed point is equivalent to a rotation of the body about some particular axis through this point. This statement is known as *Euler's theorem*, and its validity may be demonstrated with the aid of Fig. 62 where O is the fixed point and points 1 and 2 are any other two points in the body. After an arbitrary finite motion about O, the triangle O-1-2 has moved to position O-1′-2′. Line O-1 may revolve to its new position O-1′ by rotating about any axis which passes through O and also lies in a plane A which bisects the angle 1-O-1′ and which is normal to the plane 1-O-1′. Similarly line O-2 may revolve about any axis which passes through O and which also lies in a plane B which bisects the angle 2-O-2′

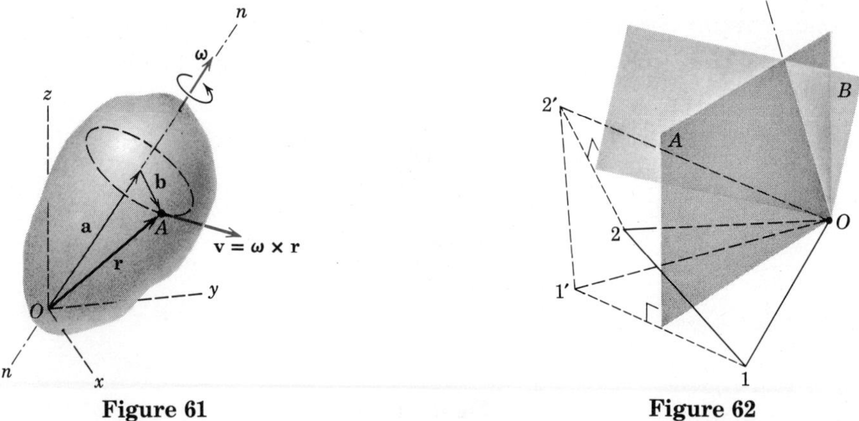

Figure 61 **Figure 62**

and which is normal to the plane 2-O-2′. The intersection O-n of these two planes is, then, the unique rotation axis which would permit the total movement to take place by a single rotation of the body through an angle $\Delta\theta$. It is interesting to note that during a finite motion of a rigid body about a fixed pivot, there is one line all points of which suffer zero net displacement. In general the points on this line may have motion during the displacement but will return to their original positions on the rotation axis O-n.

During time Δt the body will rotate through an angle $\Delta\theta$ about the unique rotation axis. As Δt approaches zero, the ratio $\Delta\theta/\Delta t$ becomes in the limit the magnitude of the angular velocity $\boldsymbol{\omega}$ of the body. The direction of $\boldsymbol{\omega}$ is along the rotation axis in the sense established by the right-hand rule.

Finite rotations of a rigid body are not commutative as may be seen in Fig. 63 which represents a sphere cut from the rigid body with the center O as the fixed point. The x-y-z axes here are fixed in space so do not rotate with the body. In the a-part of the figure two successive 90-deg rotations of the sphere about, first, the x-axis and, second, the y-axis result in the motion of a point which is initially on the y-axis in position 1, to positions 2 and 3, respectively. On the other hand, if the order of the rotations is reversed, the point suffers no motion during the y-rotation but moves to point 3 during the 90-deg rotation about the x-axis. Thus the two cases do not yield the same final position, and it is evident from this one special example that finite rotations do not generally obey the parallelogram law of vector addition and are not commutative. Thus finite rotations may not be treated as vectors.

Infinitesimal rotations, however, do obey the parallelogram law of vector addition as was shown in Art. 15 of Chapter 2 for the rotation of the rigid framework x-y-z of Fig. 21a. This fact is again shown in Fig. 64 which represents the combined effect of two infinitesimal rotations $d\boldsymbol{\theta}_1$ and $d\boldsymbol{\theta}_2$ of a rigid body about the respective axes through the fixed point O. Because of $d\boldsymbol{\theta}_1$ point A has a displacement $d\boldsymbol{\theta}_1 \times \mathbf{r}$, and likewise because of $d\boldsymbol{\theta}_2$ point A has a displacement $d\boldsymbol{\theta}_2 \times \mathbf{r}$. Either order of addition of these infinitesimal displacements clearly produces the same resultant displacement which is $d\boldsymbol{\theta}_1 \times \mathbf{r} + d\boldsymbol{\theta}_2 \times \mathbf{r} = (d\boldsymbol{\theta}_1 + d\boldsymbol{\theta}_2) \times \mathbf{r}$. Hence the two rotations are equivalent

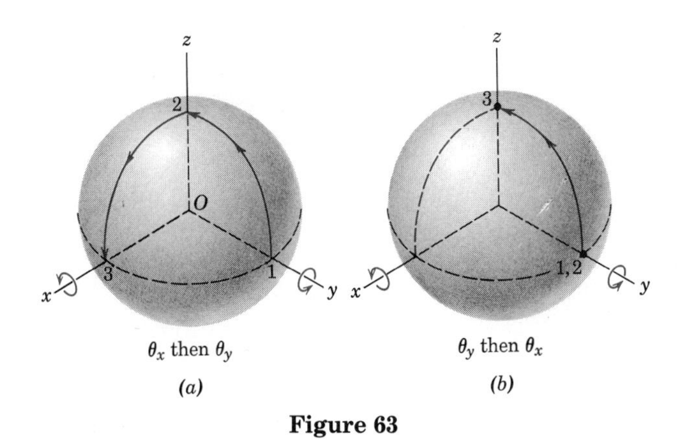

Figure 63

to the single rotation $d\boldsymbol{\theta} = d\boldsymbol{\theta}_1 + d\boldsymbol{\theta}_2$. It follows that the angular velocities $\omega_1 = \dot{\boldsymbol{\theta}}_1$ and $\omega_2 = \dot{\boldsymbol{\theta}}_2$ may be added vectorially to give $\boldsymbol{\omega} = \dot{\boldsymbol{\theta}} = \omega_1 + \omega_2$. It may be concluded that at any instant of time a body with one fixed point is rotating instantaneously about a particular axis passing through the fixed point.

To aid in visualizing the concept of the instantaneous axis of rotation a specific example will be cited. Figure 65 represents a solid cylindrical rotor made of clear plastic containing many black particles entrapped in the plastic. The rotor is spinning about its shaft axis, and its shaft, in turn, is rotating about the fixed vertical axis, with rotations in the directions indicated. If the rotor is photographed at a certain instant during its motion, the resulting picture would show one line of black dots in sharp focus indicating that, momentarily, their velocity was zero. This line of points with no velocity establishes the instantaneous position of the axis of rotation *O-n*. All other dots, such as the one at *P*, would appear out of focus, and their movements would show as short blurred streaks in the form of small circular arcs in planes normal to the axis *O-n*. Thus all particles of the body, except those on line *O-n*, are momentarily rotating in circular arcs about the instantaneous axis of rotation. If a succession of photographs were taken, it would be observed that the rotation axis would be defined by a new series of dots in focus and that the axis would change position both in space and relative to the body. For the general rotation of a rigid body about a fixed point, then, it is seen that the rotation axis is in general not a line fixed in the body.

Figure 61, which illustrated a body rotating about the axis *O-n* fixed both in the body and in space, will now be used to represent the general rotation of the rigid body about the fixed point *O* where the axis of rotation is no longer attached to the body or fixed in space. The instantaneous velocity **v** and acceleration **a** of any point *A* in the body are given by the expressions

$$\mathbf{v} = \boldsymbol{\omega} \times \mathbf{r} \tag{135}$$

$$\mathbf{a} = \dot{\boldsymbol{\omega}} \times \mathbf{r} + \boldsymbol{\omega} \times (\boldsymbol{\omega} \times \mathbf{r}) \tag{136}$$

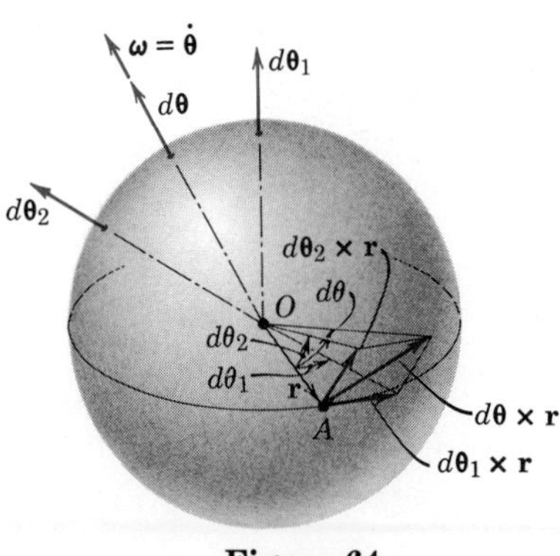

Figure 64

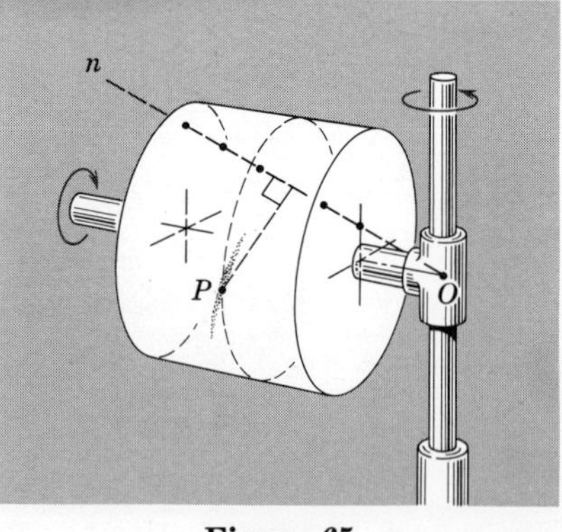

Figure 65

as in the case where the rotation axis was fixed. The only difference between the two cases lies in the $\dot{\omega}$-term, which is defined as the *angular acceleration* $\boldsymbol{\alpha}$ of the body. When the axis is fixed in the body and in space, the vector $\boldsymbol{\alpha} = \dot{\boldsymbol{\omega}}$ is directed along the fixed rotation axis and represents the rate of change of the magnitude of $\boldsymbol{\omega}$. When the rotation axis is not fixed in the body or in space, the vector $\boldsymbol{\alpha} = \dot{\boldsymbol{\omega}}$ reflects the change in direction of $\boldsymbol{\omega}$ as well as its change in magnitude and is not directed along the rotation axis.

In general, for a rigid body which rotates about a fixed point, the instantaneous axis of rotation will change position both in space and within or relative to the body. As the axis moves in space it generates a *space cone* (or *herpolhode cone*), and as the axis moves with respect to the body, it also generates a cone within or relative to the body known as the *body cone* (or *polhode cone*), as indicated in Fig. 66 where $\boldsymbol{\omega}$ is the instantaneous absolute angular velocity of the body. These cones are tangent along the instantaneous axis of rotation, and the body cone is seen to roll on the fixed space cone during the motion. If the body cone is external to the space cone, as illustrated in Fig. 66, the motion is known as *direct precession.* If the space cone lies inside the body cone or if the body cone lies inside the space cone, the motion is known as *retrograde precession.* These precessions will be studied again under the heading of gyroscopic motion in a later article.

The tip of the angular velocity vector $\boldsymbol{\omega}$ follows an absolute path p on the space cone as shown in Fig. 66. The angular acceleration $\boldsymbol{\alpha}$, which represents the time-rate-of-change of $\boldsymbol{\omega}$, will, then, be a vector in the direction of the change in $\boldsymbol{\omega}$ which is tangent to p as shown.

If a body moves parallel to a plane, it may be considered to rotate about a point at infinity. The space and body cones then become cylindrical surfaces, and the intersection of these surfaces with the plane of motion becomes the space and body centrodes described in Art. 29 for plane motion. Also the point of intersection of the instantaneous axis with the plane of motion becomes the instantaneous center of zero velocity discussed in that same article.

Consider now a rigid body whose rotation about a fixed point O is constrained by the specified motion of a point A on the body a distance r from O. Point A necessarily must move on the surface of a sphere of radius r and with center at O. This situation may be illustrated with the rigid link OA

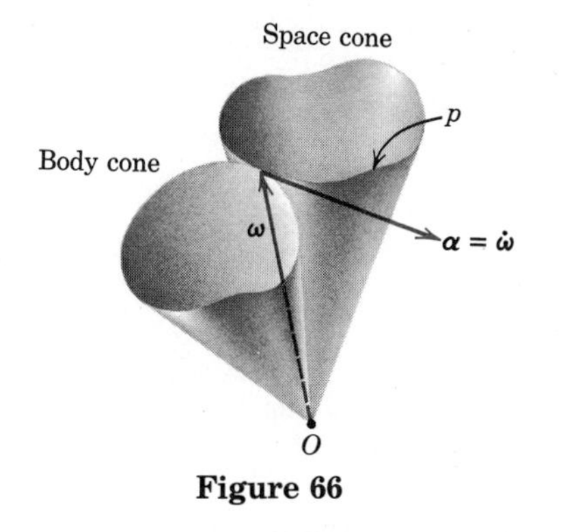

Figure 66

shown in Fig. 67*a* where the connection at each end is a ball-and-socket joint. In general the body will have an angular velocity $\boldsymbol{\omega} = \boldsymbol{\omega}_t + \boldsymbol{\omega}_n$. The component $\boldsymbol{\omega}_t$ represents the rotation of the body about the axis OA, and $\boldsymbol{\omega}_n$ represents the rate at which the axis OA revolves about the y-axis, which is normal to OA. As point A moves along its path to A', the axis OA revolves through the angle $d\theta$, and the rate of this rotation becomes $\dot{\theta} = \omega_n$. Also, the velocity of A is given by $\mathbf{v}_A = \boldsymbol{\omega} \times \mathbf{r} = \boldsymbol{\omega}_n \times \mathbf{r}$.

The angular acceleration $\boldsymbol{\alpha}$ of the body is the time derivative of $\boldsymbol{\omega}$. To obtain a clear picture of $\boldsymbol{\alpha}$ each term in the differentiation will be considered separately. With the subscripts m and d representing magnitude and direction, respectively, $\boldsymbol{\alpha}$ may be written

$$\boldsymbol{\alpha} = \dot{\boldsymbol{\omega}} = \dot{\boldsymbol{\omega}}_t + \dot{\boldsymbol{\omega}}_n = \dot{\boldsymbol{\omega}}_t]_m + \dot{\boldsymbol{\omega}}_t]_d + \dot{\boldsymbol{\omega}}_n]_m + \dot{\boldsymbol{\omega}}_n]_d$$

The differential increments in $\boldsymbol{\omega}_t$ and $\boldsymbol{\omega}_n$ which produce these acceleration components are shown in Fig. 67*b*. The first of the four terms is due to the change in magnitude of $\boldsymbol{\omega}_t$, so that $|\dot{\boldsymbol{\omega}}_t]_m| = \dot{\omega}_t$. The second term comes from the increment in $\boldsymbol{\omega}_t$ due to its change in direction which is $|d\boldsymbol{\omega}_t]_d| = \omega_t\, d\theta$. Division by dt and substitution of $\omega_n = \dot{\theta}$ gives $|\dot{\boldsymbol{\omega}}_t]_d| = \omega_t\omega_n$. The third term is due to the increment $d\omega_n$ in the magnitude $\dot{\theta}$ of $\boldsymbol{\omega}_n$, and division by dt gives $|\dot{\boldsymbol{\omega}}_n]_m| = \ddot{\theta}$. The fourth term is due to the change in direction of $\boldsymbol{\omega}_n$ which depends on the curvature of the path of A. In Fig. 67*a* the radius of curvature of the path of A is the radius r' of a circle which has a center at B and which "kisses" the path at A. The angle made by OA with the plane of this circle is denoted by γ. As A moves to A' the vector $\boldsymbol{\omega}_n$ generates a cone of semivertex angle γ as it rotates around OB. The radius $\omega_n \sin\gamma$ of the cone revolves about OB at the same angular rate v_A/r' as does BA. Thus the acceleration due to the change in direction of $\boldsymbol{\omega}_n$ becomes $|\dot{\boldsymbol{\omega}}_n]_d| = (\omega_n \sin\gamma)(v_A/r')$. Substitution of $r' = r\cos\gamma$ and $v_A = r\omega_n$ gives $|\dot{\boldsymbol{\omega}}_n]_d| = \omega_n{}^2 \tan\gamma$.

All four components of the angular acceleration $\boldsymbol{\alpha}$ of the body are shown in Fig. 67*c*. It is seen that the component $\boldsymbol{\alpha}_n$ normal to OA includes the effects

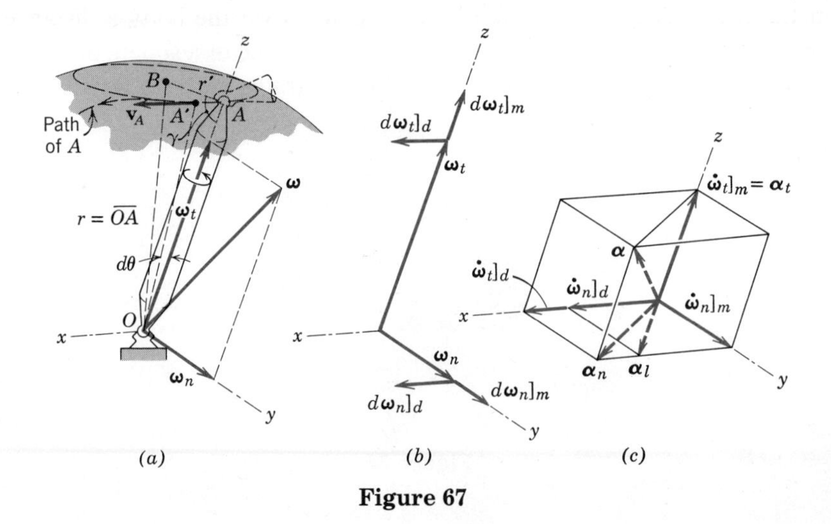

Figure 67

of changes in both $\boldsymbol{\omega}_n$ and $\boldsymbol{\omega}_t$. The vector $\boldsymbol{\alpha}_l$ is defined here as the angular acceleration of the line OA and depends only on the changes in $\boldsymbol{\omega}_n$. Thus $\boldsymbol{\alpha}_l = \dot{\boldsymbol{\omega}}_n$.

It is observed that the kinematical function of the link OA depends only on its length and on the motion of A and is in no way affected by its rotation about OA. This observation leads to a definition of the angular motion of a line which eliminates consideration of nonessential rotational components. It may be stated, therefore, that $\boldsymbol{\omega}_n$ and $\boldsymbol{\alpha}_l$ are the angular velocity and angular acceleration, respectively, of line OA.

From the foregoing analysis with $\boldsymbol{\alpha}_l$ perpendicular to $\mathbf{r}$ and $\boldsymbol{\omega}_n$ perpendicular to $\mathbf{v}_A$ it is apparent that

$$\boldsymbol{\alpha}_l \cdot \mathbf{r} = 0 \quad \text{and} \quad \boldsymbol{\omega}_n \cdot \mathbf{v}_A = 0$$

for a line segment OA which rotates about its fixed end O.

The discussion will now return to the motion of a rigid body and take up the important case of rigid-body motion about a fixed point. This situation occurs when a symmetrical rotor spins about its axis which itself rotates about the fixed point O, Fig. 68. For this type of motion it is convenient to introduce a new set of coordinates to specify the motion. In Fig. 68 the coordinates X-Y-Z are fixed in space, and plane A contains the X-Y axes and the fixed point O on the rotor axis. Plane B contains point O and is always normal to the rotor axis. Angle θ measures the inclination of the rotor axis from the vertical Z-axis and is also a measure of the angle between planes A and B. The intersection of the two planes is the x-axis which is located by the angle ψ from the X-axis. The y-axis lies in plane B, and the z-axis coincides with the rotor axis. The angles θ and ψ completely specify the position of the rotor axis. Rotation around the rotor axis is described by the axes x'-y'-z' which are rigidly fixed to the rotor and turn with it about the common z-axis. The angular displacement of the rotor in plane B is specified by the angle ϕ between the x'-axis attached to the rotor and the x-axis.

The three angles θ, ψ, ϕ completely specify the position of the rotor and are known as *Euler's angles*. The time rates of change of these angles, $\dot{\theta}$, $\dot{\psi}$, $\dot{\phi}$, specify, respectively, the *nutation, precession,* and *spin* velocities of the rotor.

In writing the expressions for the angular velocity of the rotating reference

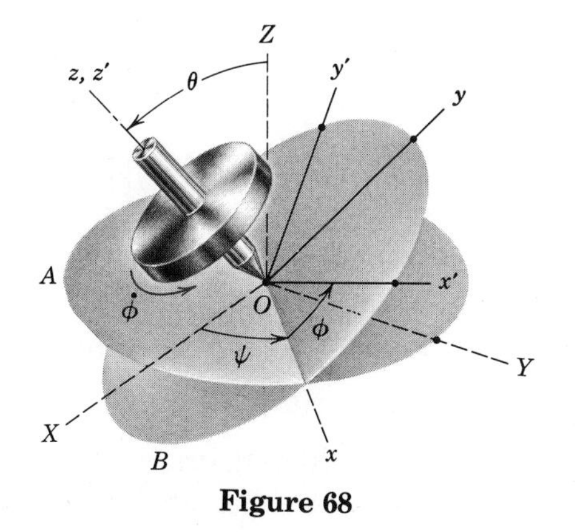

Figure 68

axes x-y-z and the angular velocity of the rotor, a distinction in the notation should be observed to avoid confusion between these different quantities. The symbol $\mathbf{\Omega}$ will be used henceforth to denote the angular velocity of the rotating reference axes, whereas the symbol $\boldsymbol{\omega}$ will continue to be used to denote the angular velocity of the rigid body, which is the rotor in this instance.

Thus the absolute angular velocity $\mathbf{\Omega}$ of the x-y-z axes is seen to be

$$\mathbf{\Omega} = \mathbf{i}\Omega_x + \mathbf{j}\Omega_y + \mathbf{k}\Omega_z = \mathbf{i}\dot{\theta} + \mathbf{j}\dot{\psi}\sin\theta + \mathbf{k}\dot{\psi}\cos\theta \tag{137}$$

where $\dot{\psi}$ itself produces a velocity whose vector is in the Z-direction with y- and z-components of $\dot{\psi}\sin\theta$ and $\dot{\psi}\cos\theta$ respectively. If the rotor were not spinning about its own axis, that is, if $\dot{\phi}$ were zero, then it would have velocity components identical to those of x-y-z. The spin velocity produces an added z-component of angular velocity. Thus the rotor has an absolute angular velocity $\boldsymbol{\omega}$ which becomes

$$\boldsymbol{\omega} = \mathbf{i}\omega_x + \mathbf{j}\omega_y + \mathbf{k}\omega_z = \mathbf{i}\dot{\theta} + \mathbf{j}\dot{\psi}\sin\theta + \mathbf{k}(\dot{\phi} + \dot{\psi}\cos\theta) \tag{138}$$

Euler's angles and the foregoing kinematic relationships find important application to gyroscopic motion which will be discussed in a later article. It is difficult to gain an adequate understanding of gyroscopic motion without a thorough grasp of the geometry of the kinematic relations developed in this article, and, consequently, thorough preparation at this point will be extremely helpful for later work.

The relationships between the x-y-z and the X-Y-Z coordinate systems of Fig. 68 can be expressed conveniently with the aid of a transformation matrix in a manner similar to that used for the coordinate transformations in Art. 14 of Chapter 2. The position of the x-y-z axes is established by two successive rotations from the X-Y-Z axes. The first is a rotation ψ with $\theta = 0$ to a position x_1-y_1-z_1 (not shown). Then the x_1-y_1-z_1 axes are rotated about the x_1-axis by an amount θ to the final x-y-z position. A vector $\mathbf{V}$ expressed in these coordinate systems would have components expressed by the matrix transformations

$$\{V_{x_1y_1z_1}\} = [T_\psi]\{V_{XYZ}\} \qquad \{V_{xyz}\} = [T_\theta]\{V_{x_1y_1z_1}\}$$

or

$$\{V_{xyz}\} = [T_\theta][T_\psi]\{V_{XYZ}\} \tag{139}$$

where

$$[T_\psi] = \begin{bmatrix} \cos\psi & \sin\psi & 0 \\ -\sin\psi & \cos\psi & 0 \\ 0 & 0 & 1 \end{bmatrix} \qquad [T_\theta] = \begin{bmatrix} 1 & 0 & 0 \\ 0 & \cos\theta & \sin\theta \\ 0 & -\sin\theta & \cos\theta \end{bmatrix}$$

and

$$[T_\theta][T_\psi] = \begin{bmatrix} \cos\psi & \sin\psi & 0 \\ -\cos\theta\sin\psi & \cos\theta\cos\psi & \sin\theta \\ \sin\theta\sin\psi & -\sin\theta\cos\psi & \cos\theta \end{bmatrix}$$

The inverse of the foregoing transfer matrices may be written by interchanging corresponding elements on either side of the main diagonal. Thus, for example, $\sin\psi$ and $-\cos\theta\sin\psi$ would be interchanged when writing the matrix for $\{V_{XYZ}\} = [T_\psi]^{-1}[T_\theta]^{-1}\{V_{xyz}\}$.

Transformation matrices between the x-y-z and the x'-y'-z' coordinate systems may be written, and this work is left to the student to complete.

The general case of space motion where no one point on the rigid body is fixed will be discussed in the next article on relative motion.

Sample Problem

7/1 The electric motor with attached disk is running at a constant low speed of 240 rev/min in the direction shown. Its housing and mounting base are initially at rest. The entire assembly is next set in rotation about the vertical Z-axis at the constant rate of $N = 60$ rev/min with a fixed angle γ of 30°. Determine the angular velocity ω_z of the rotor and disk about the z-axis and find the total angular velocity $\boldsymbol{\omega}$ and the angular acceleration $\boldsymbol{\alpha}$ of the rotor and disk. Construct the space and body cones for the motion.

Solution. Point O is the fixed point around which the rotor and disk are pivoted. The x-y-z axes are attached to the motor housing and revolve with it around the vertical Z-axis. The x-axis lies in the horizontal plane. This choice of reference axes agrees with that adopted in Fig. 68, with $\gamma = 90° - \theta$.

The angular velocity of the rotor and disk about the z-axis, from Eq. 138, is its spin $\dot{\phi} = 240(2\pi)/60 = 8\pi$ rad/s plus the z-component of the precession $\dot{\psi} = 60(2\pi)/60 = 2\pi$ rad/s. Thus,

$$\omega_z = \dot{\phi} + \dot{\psi}\cos(90° - \gamma) = 8\pi + 2\pi(\tfrac{1}{2}) = 9\pi \text{ rad/s} \qquad \textit{Ans.}$$

The total angular velocity of the rotor and disk from Eq. 138 is

$$\begin{aligned}\boldsymbol{\omega} &= \mathbf{i}\dot{\theta} + \mathbf{j}\dot{\psi}\cos\gamma + \mathbf{k}(\dot{\phi} + \dot{\psi}\sin\gamma)\\ &= 0 + (2\pi)(\sqrt{3}/2)\mathbf{j} + (8\pi + 2\pi\tfrac{1}{2})\mathbf{k} = \pi(\sqrt{3}\mathbf{j} + 9\mathbf{k}) \text{ rad/s} \qquad \textit{Ans.}\end{aligned}$$

The vector $\boldsymbol{\omega}$ is shown in the figure.

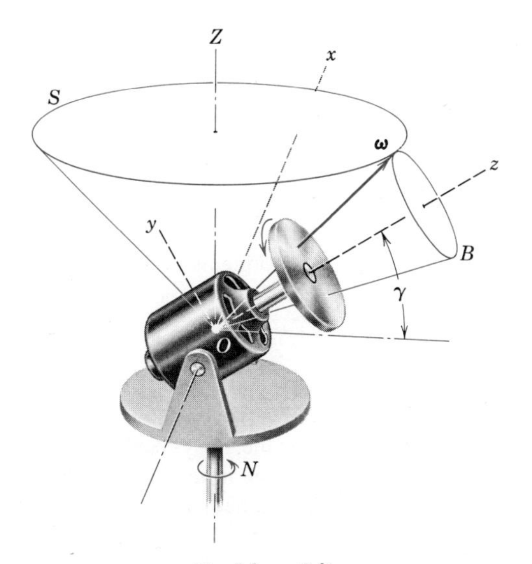

Problem 7/1

The angular acceleration by definition is $\boldsymbol{\alpha} = \dot{\boldsymbol{\omega}}$. Thus,

$$\boldsymbol{\alpha} = 0 + \dot{\mathbf{j}}\dot{\psi}\cos\gamma + \dot{\mathbf{k}}(\dot{\phi} + \dot{\psi}\sin\gamma)$$

where $\dot{\psi}$, γ, $\dot{\phi}$ are constant. But $\dot{\mathbf{j}} = \boldsymbol{\Omega} \times \mathbf{j}$ and $\dot{\mathbf{k}} = \boldsymbol{\Omega} \times \mathbf{k}$, where $\boldsymbol{\Omega}$ is the angular velocity of the x-y-z axes. This quantity becomes $\boldsymbol{\Omega} = \dot{\psi}(\mathbf{j}\cos\gamma + \mathbf{k}\sin\gamma)$ since, in this case, the axes are revolving only about the vertical Z-axis. Thus,

$$\begin{aligned}\boldsymbol{\alpha} &= \dot{\psi}(\mathbf{j}\cos\gamma + \mathbf{k}\sin\gamma) \times [\mathbf{j}\dot{\psi}\cos\gamma + \mathbf{k}(\dot{\phi} + \dot{\psi}\sin\gamma)] \\ &= \mathbf{i}\dot{\psi}\dot{\phi}\cos\gamma\end{aligned}$$

so that

$$\alpha = (2\pi)(8\pi)\frac{\sqrt{3}}{2} = 136.8 \text{ rad/s}^2 \qquad \textit{Ans.}$$

The vector $\boldsymbol{\alpha}$ is tangent to the two cones and is in the positive x-direction. It represents the vector change in $\boldsymbol{\omega}$ which, in this problem, involves a change in the direction only of $\boldsymbol{\omega}$. The total angular velocity $\boldsymbol{\omega}$ of the rotor and disk is the element common to both the body cone B and the space cone S as shown.

Problems

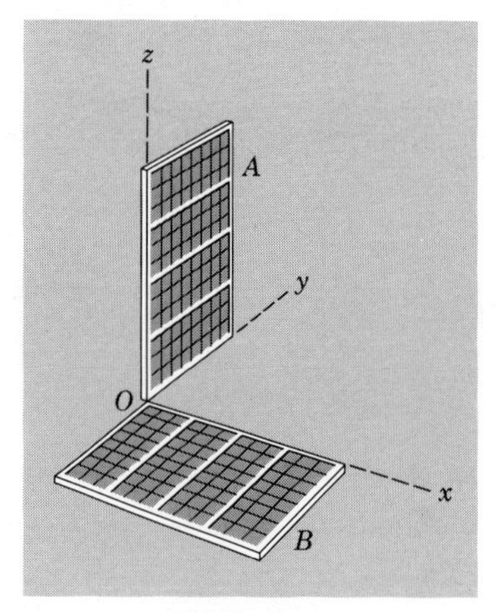

Problem 7/2

7/2 Determine the single fixed axis through O about which the solar panel of a certain spacecraft can be rotated in order to change its position from A to B in a single rotation. The side of the panel facing the plus x-direction in position A must face the plus z-direction in position B. Designate the axis of rotation by a unit vector $\mathbf{n}$ along the axis and specify the magnitude $\Delta\theta$ of the angle through which rotation about the axis takes place.

Ans. $\mathbf{n} = \frac{1}{\sqrt{2}}(\mathbf{i} + \mathbf{k}), \quad \Delta\theta = \pi$

7/3 In Fig. 68 a point C (not labeled) on the rotor axis a distance of 75 mm from O has at a certain instant a velocity $\mathbf{v}_C = 75\mathbf{i} + 100\mathbf{j}$ mm/s, where $\mathbf{i}$ and $\mathbf{j}$ are unit vectors along the x- and y-axes respectively. At this same instant $\psi = 60°$ and $\theta = 30°$. For this condition write the X-Y-Z components of $\mathbf{v}_C$. *Ans.* $v_X = -75/2$ mm/s
$v_Y = 125\sqrt{3}/2$ mm/s
$v_Z = 50$ mm/s

7/4 Write the transfer matrix to express a vector $\mathbf{V}$ for the body-fixed coordinate system x'-y'-z' of Fig. 68 in terms of the same vector for the fixed coordinate system X-Y-Z.

7/5 The rod A is hinged about the axis O-O of the clevis, which is attached to the end of the rotating vertical shaft. The shaft rotates with an angular velocity ω_0 as shown, and the x-y axes are attached to the vertical shaft. If θ is changing at the rate $\dot{\theta} = -p$, write the expressions for the angular velocity $\boldsymbol{\omega}_n$ of the center line of the rod and the angular velocity $\boldsymbol{\omega}$ of the rod as a three-dimensional body.

Ans. $\boldsymbol{\omega}_n = \omega_0 \sin\theta\,(\mathbf{i}\cos\theta + \mathbf{k}\sin\theta) + p\mathbf{j}$
$\boldsymbol{\omega} = p\mathbf{j} + \omega_0\mathbf{k}$

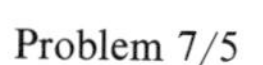
Problem 7/5

7/6 The shaft OA of the bevel gear B rotates about the fixed axis O-x with a constant speed $N =$ 60 rev/min in the direction shown. Gear B meshes with the bevel gear C along its pitch cone of semivertex angle $\gamma = \tan^{-1}\frac{3}{2}$ as shown. Determine the angular velocity $\boldsymbol{\omega}$ and angular acceleration $\boldsymbol{\alpha}$ of gear B if gear C is fixed and does not rotate. The y-axis revolves with the shaft OA.

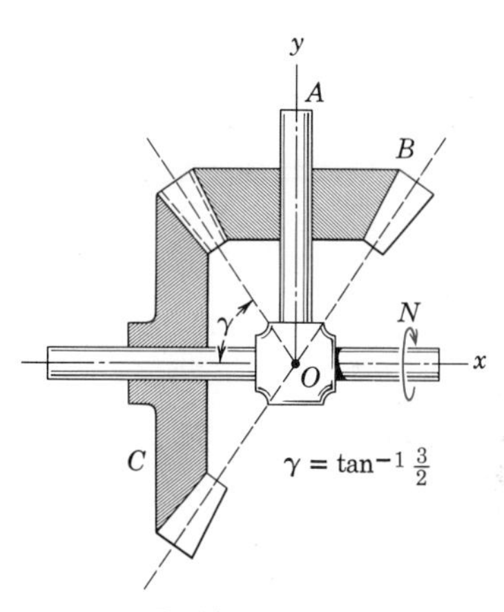

Problem 7/6

7/7 If gear C of Prob. 7/6 has a constant rotational velocity of 20 rev/min about the axis O-x in the same sense as N while OA maintains its constant rotational speed $N = 60$ rev/min, calculate the angular velocity $\boldsymbol{\omega}$ and angular acceleration $\boldsymbol{\alpha}$ of gear B.

Ans. $\boldsymbol{\omega} = 2\pi(-\mathbf{i} + \mathbf{j})$ rad/s
$\boldsymbol{\alpha} = -4\pi^2\mathbf{k}$ rad/s^2

7/8 The disk rotates with a spin velocity of 15 rad/s about its horizontal z-axis first in the direction (a) and second in the direction (b). The assembly rotates with the velocity $N = 10$ rad/s about the vertical axis. Construct the space and body cones for each case.

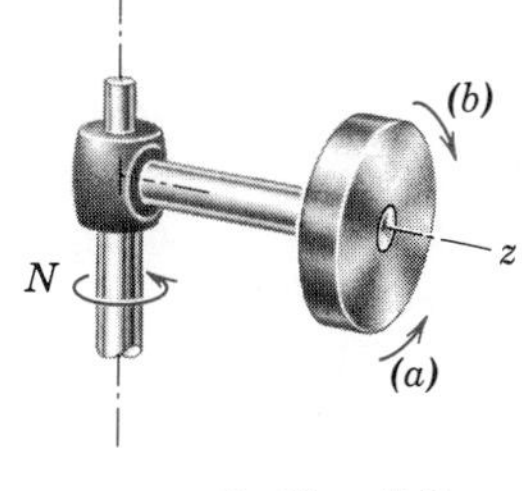

Problem 7/8

7/9 The space station is stabilized by giving it a rotation about its geometric axis A-A in the direction indicated. A small and slow precession of this axis about a fixed Z-axis in space is observed. Sketch the relative positions of the space and body cones. The station is remote from any attracting planet. Thus its mass center O has essentially zero acceleration, and the station may be considered a body which rotates about O as a point fixed in a reference frame that translates with constant velocity.

Problem 7/9

7/10 If the direction of rotation of the electric motor of Sample Prob. 7/1 is reversed but all other conditions remain unchanged, determine the expression for the total angular velocity ω of the rotor and disk. Also sketch the space and body cones and identify the type of precession.

Ans. $\boldsymbol{\omega} = \pi(\sqrt{3}\mathbf{j} - 7\mathbf{k})$ rad/s
retrograde precession

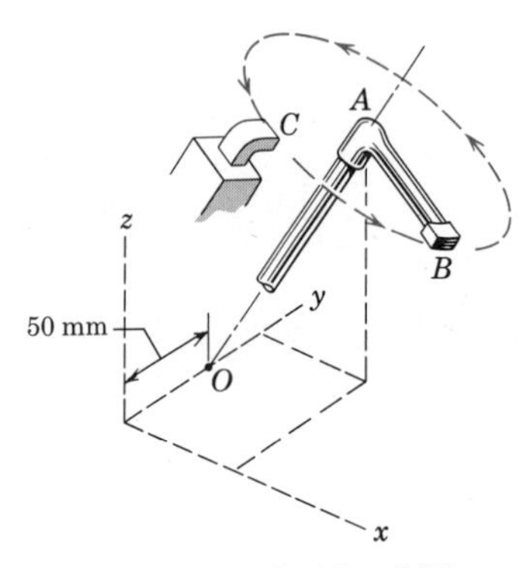

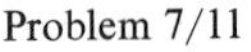
Problem 7/11

7/11 A timing mechanism consists of the rotating distributor arm AB and the fixed contact C. If the arm rotates about the fixed axis OA with a constant angular velocity $\boldsymbol{\omega} = 3(3\mathbf{i} + 2\mathbf{j} + 6\mathbf{k})$ rad/s, and if the coordinates of the contact C expressed in millimetres are (75, 75, 150), determine the magnitude of the acceleration of the tip B of the distributor arm as it passes point C.

Ans. $|\mathbf{a}_B| = 10.57\ \text{m/s}^2$

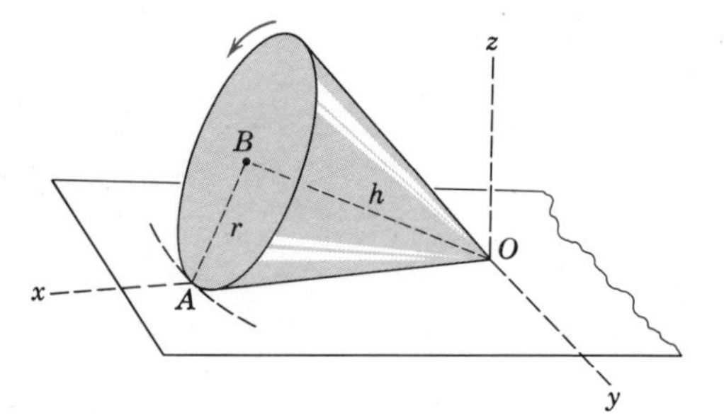

Problem 7/12

7/12 The solid right circular cone of base radius r and height h rolls on a flat surface without slipping. The center B of the circular base moves in a circular path around the z-axis with a constant speed v. Determine (*a*) the angular velocity $\boldsymbol{\omega}$ of the solid cone, (*b*) the angular velocity $\boldsymbol{\omega}_{OB}$ of the center line of the cone, (*c*) the angular velocity $\boldsymbol{\omega}_{OA}$ of the cone element momentarily in contact with the plane, and (*d*) the angular acceleration $\boldsymbol{\alpha}$ of the solid cone.

Ans. (*a*) $\boldsymbol{\omega} = v\sqrt{\dfrac{1}{r^2} + \dfrac{1}{h^2}}\,\mathbf{i}$

(*b*) $\boldsymbol{\omega}_{OB} = \dfrac{v}{h\sqrt{r^2 + h^2}}(\mathbf{i}r - \mathbf{k}h)$

(*c*) $\boldsymbol{\omega}_{OA} = \mathbf{0}$

(*d*) $\boldsymbol{\alpha} = -\dfrac{v^2}{h^2}\left(\dfrac{r}{h} + \dfrac{h}{r}\right)\mathbf{j}$

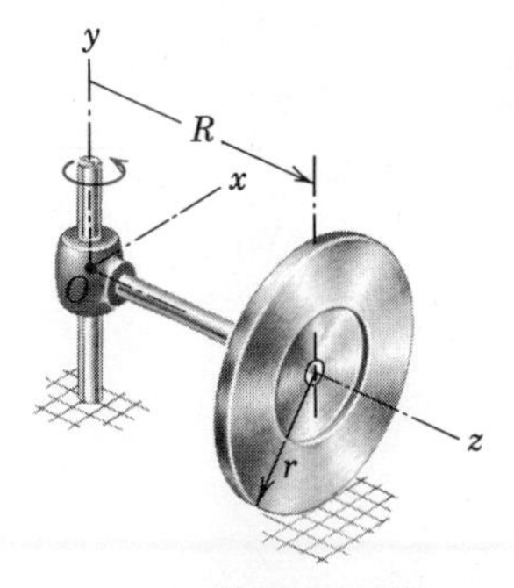

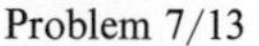
Problem 7/13

7/13 The wheel rolls without slipping in a circular arc of radius R and makes one complete turn about the vertical y-axis with constant speed in time τ. Determine the vector expression for the angular acceleration $\boldsymbol{\alpha}$ of the wheel and construct the space and body cones.

7/14 For the figure for Prob. 7/1 assume that the motor shaft and attached disk are rotating about the z-axis with a spin $\dot{\phi}$ of 30 rev/min which is increasing at the rate of 10 rad/s² at the position $\gamma = 30°$. At the same time $\dot{\gamma}$ is 12 rad/s and is increasing at the rate of 15 rad/s². Determine the vector expression for the angular acceleration $\boldsymbol{\alpha}$ of the rotor at this instant. There is no rotation about the Z-axis.

Ans. $\boldsymbol{\alpha} = -15\mathbf{i} + 12\pi\mathbf{j} + 10\mathbf{k}$ rad/s²

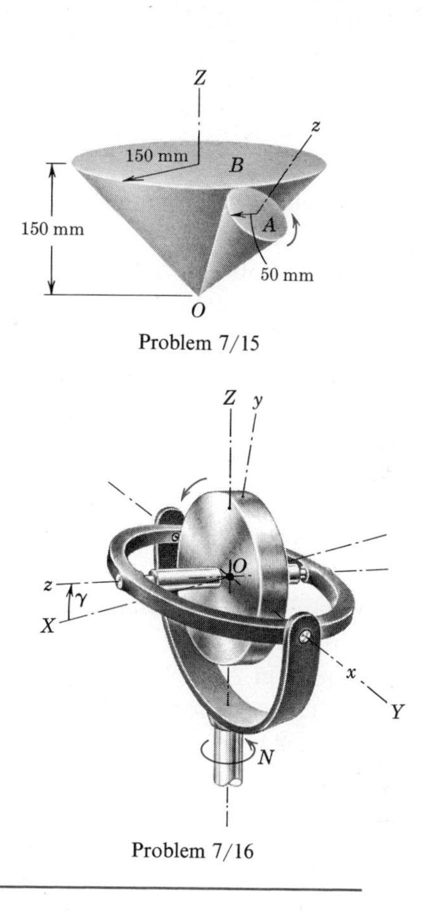

Problem 7/15

Problem 7/16

◀ **7/15** The right circular cone A rolls on the fixed right circular cone B at a constant rate and makes one complete trip around B every 4 s. Compute the magnitude of the angular acceleration $\boldsymbol{\alpha}$ of cone A.

Ans. $\alpha = 6.32$ rad/s²

◀ **7/16** The gyro rotor shown is spinning at the constant rate of 100 rev/min relative to the x-y-z axes in the direction indicated. If the angle γ between the gimbal ring and the horizontal X-Y plane is made to increase at the constant rate of 4 rad/s and if the unit is forced to precess about the vertical at the constant rate $N = 20$ rev/min, calculate the magnitude of the angular acceleration $\boldsymbol{\alpha}$ of the rotor when $\gamma = 30°$.

Ans. $\alpha = 42.8$ rad/s²

---

38 Relative Motion. In Art. 15 of Chapter 2 the kinematical equations for the space motion of a particle relative to translating axes were developed. These relations may be applied directly to the case of a rigid body, Fig. 69, where the origin of the translating system x-y-z is any convenient point B in the body and where point A is some other point fixed to the body. The body has an absolute angular velocity $\boldsymbol{\omega}$ at the instant considered. The ex-

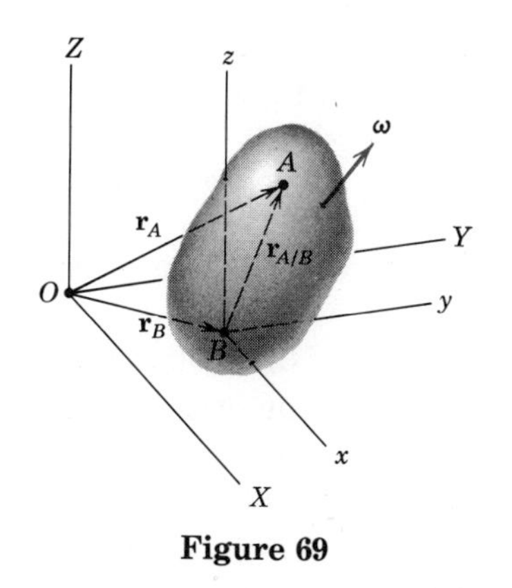

Figure 69

pressions for relative velocity and acceleration, Eqs. 36 and 37, developed in Art. 15 for the relative motion of two points A and B, are

$$\mathbf{v}_A = \mathbf{v}_B + \mathbf{v}_{A/B} \qquad \text{and} \qquad \mathbf{a}_A = \mathbf{a}_B + \mathbf{a}_{A/B}$$

In applying these relations to rigid-body motion it is noted that the distance $\overline{AB}$ remains constant. Thus, from an observer's position on x-y-z, A appears to rotate about the point B and to lie on a spherical surface with B as the center. Consequently, the general motion may be viewed as a translation of the body with the motion of B plus a rotation of the body about B.

The relative motion terms represent the effect of the rotation about B and are identical to the velocity and acceleration expressions discussed in the previous article for rotation of a rigid body and a line about a fixed point. Therefore the relative velocity and acceleration equations may be written

$$\begin{aligned} \mathbf{v}_A &= \mathbf{v}_B + \boldsymbol{\omega} \times \mathbf{r}_{A/B} \\ \mathbf{a}_A &= \mathbf{a}_B + \dot{\boldsymbol{\omega}} \times \mathbf{r}_{A/B} + \boldsymbol{\omega} \times (\boldsymbol{\omega} \times \mathbf{r}_{A/B}) \end{aligned} \tag{140}$$

where $\boldsymbol{\omega}$ is the instantaneous angular velocity of the body. These same expressions were written alternatively as Eqs. 105 for the case of plane motion in Art. 29.

The selection of the reference point B is quite arbitrary in theory. In practice point B is chosen for convenience as some point in the body whose motion is known in whole or in part. If point A is chosen as the reference point, the relative motion equations become

$$\begin{aligned} \mathbf{v}_B &= \mathbf{v}_A + \boldsymbol{\omega} \times \mathbf{r}_{B/A} \\ \mathbf{a}_B &= \mathbf{a}_A + \dot{\boldsymbol{\omega}} \times \mathbf{r}_{B/A} + \boldsymbol{\omega} \times (\boldsymbol{\omega} \times \mathbf{r}_{B/A}) \end{aligned}$$

where $\mathbf{r}_{B/A} = -\mathbf{r}_{A/B}$. It should be clear that $\boldsymbol{\omega}$ and, hence, $\dot{\boldsymbol{\omega}}$ are the same vectors for either formulation since the absolute angular motion of the body is independent of the choice of reference point.

When the foregoing relationships are applied to the space motion of a line segment, the conclusions deduced in the previous article for the angular motion of a line about a fixed point on the line will now apply to the relative motion terms in the relative velocity and relative acceleration relations. Thus the angular velocity $\boldsymbol{\omega}_n$ and angular acceleration $\boldsymbol{\alpha}_l$ of the line segment, as defined in the previous article, must obey the relations

$$\boldsymbol{\omega}_n \cdot \mathbf{v}_{A/B} = 0 \qquad \boldsymbol{\alpha}_l \cdot \mathbf{r}_{A/B} = 0$$

When the line segment is a part of a rigid body, then these same conclusions apply.

A more general formulation of the motion of a rigid body in space calls for the use of reference axes which rotate as well as translate. The description of Fig. 69 is modified in Fig. 70 to show reference axes attached to the reference point B as before but which rotate with an absolute angular velocity $\boldsymbol{\Omega}$. The absolute angular velocity of the body is $\boldsymbol{\omega}$.

The relations that describe the relative motion of a point A from a rotating position moving with a point B were developed in Art. 15 of Chapter 2 as Eqs. 40 and 41 and are repeated here with a substitution of $\mathbf{\Omega}$ for the angular velocity of the rotating reference axes x-y-z in place of ω.

$$\begin{aligned} \mathbf{v}_A &= \mathbf{v}_B + \mathbf{\Omega} \times \mathbf{r}_{A/B} + \mathbf{v}_{\text{rel}} \\ \mathbf{a}_A &= \mathbf{a}_B + \dot{\mathbf{\Omega}} \times \mathbf{r}_{A/B} + \mathbf{\Omega} \times (\mathbf{\Omega} \times \mathbf{r}_{A/B}) + 2\mathbf{\Omega} \times \mathbf{v}_{\text{rel}} + \mathbf{a}_{\text{rel}} \end{aligned} \qquad \textbf{(141)}$$

Again observe that $\mathbf{\Omega}$ is the angular velocity of the axes x-y-z. Equations 141 may be applied directly to the case of rigid-body motion by using the fact that $\mathbf{r}_{A/B}$ is constant in magnitude. It is noted that, if x-y-z are rigidly attached to the body, $\mathbf{\Omega} = \omega$ and $\mathbf{v}_{\text{rel}}$ and $\mathbf{a}_{\text{rel}}$ are both zero, which makes the equations identical to Eqs. 140.

A review of Art. 15 is suggested at this point so that the meaning of each term in the velocity and acceleration equations is understood. It should be noted that the relationship between the time derivative of a vector [ ] as measured in X-Y-Z and the time derivative as measured in x-y-z may, from Eq. 39, be expressed by the vector-operator equation

$$\left(\frac{d[\ \ ]}{dt}\right)_{XYZ} = \left(\frac{d[\ \ ]}{dt}\right)_{xyz} + \mathbf{\Omega} \times [\ \ ] \qquad \textbf{(142)}$$

where $\mathbf{\Omega}$ is the angular velocity of x-y-z. Application of this operator on $\mathbf{r}_{A/B}$ gives $\mathbf{v}_{\text{rel}} + \mathbf{\Omega} \times \mathbf{r}_{A/B}$ which is the relative velocity as measured from X-Y-Z. Thus $\mathbf{\Omega} \times \mathbf{r}_{A/B}$ is the vector difference between the relative velocity measured from nonrotating axes X-Y-Z and that measured from rotating axes x-y-z.

The student should verify that the application of the vector operator upon itself yields the second-order operator

$$\begin{aligned} \left(\frac{d^2[\ \ ]}{dt^2}\right)_{XYZ} &= \dot{\mathbf{\Omega}} \times [\ \ ] + \mathbf{\Omega} \times (\mathbf{\Omega} \times [\ \ ]) \\ &\quad + 2\mathbf{\Omega} \times \left(\frac{d[\ \ ]}{dt}\right)_{xyz} + \left(\frac{d^2[\ \ ]}{dt^2}\right)_{xyz} \end{aligned} \qquad (143)$$

If this operator is applied to the vector $\mathbf{r}_{A/B}$, the second of Eqs. 141 results. It

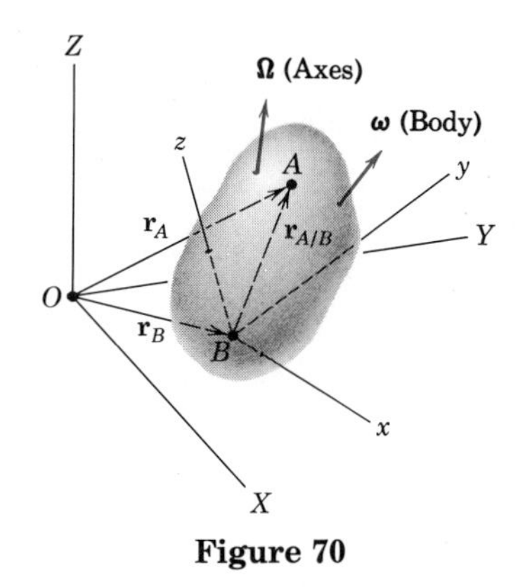

Figure 70

is seen, therefore, that the terms

$$\dot{\boldsymbol{\Omega}} \times \mathbf{r}_{A/B} + \boldsymbol{\Omega} \times (\boldsymbol{\Omega} \times \mathbf{r}_{A/B}) + 2\boldsymbol{\Omega} \times \left(\frac{d\mathbf{r}_{A/B}}{dt}\right)_{xyz}$$

constitute the difference between the relative acceleration measured from nonrotating axes, $(d^2\mathbf{r}_{A/B}/dt^2)_{XYZ}$, and that measured from rotating axes, $(d^2\mathbf{r}_{A/B}/dt^2)_{xyz}$.

Equations 141 are particularly useful when the reference axes are attached to a moving body or vehicle within which a relative motion occurs, the absolute value of which is sought. In Art. 15, Sample Prob. 2/180 and a number of examples in the problem set of that article illustrated the use of Eqs. 141 where attention was focused on the motion of a particle. Several similar problems are also included in the problem set of this present article with attention focused on rigid-body motion.

Sample Problems

7/17 Crank CB rotates about the horizontal axis with an angular velocity $\omega_1 = 6$ rad/s which is constant for a short interval of motion which includes the position shown. The link AB has a ball-and-socket fitting on each end and connects crank DA with CB. For the instant shown, determine the angular velocity ω_2 and angular acceleration $\dot{\omega}_2$ of crank DA. Also calculate the angular velocity $\boldsymbol{\omega}_n$ and angular acceleration $\boldsymbol{\alpha}_l = \dot{\boldsymbol{\omega}}_n$ of the center line of link AB.

Solution. The relative velocity relation, Eq. 140, will be solved first using translating reference axes attached to B. The equation is

$$\mathbf{v}_A = \mathbf{v}_B + \boldsymbol{\omega}_n \times \mathbf{r}_{A/B}$$

where $\boldsymbol{\omega}_n$ is the angular velocity of the line AB and is normal to AB. The velocities of A and B are

$[v = r\omega]$ $\qquad \mathbf{v}_A = 50\omega_2\mathbf{j} \qquad \mathbf{v}_B = 100(6)\mathbf{i} = 600\mathbf{i}$ mm/s

Also $\mathbf{r}_{A/B} = 50\mathbf{i} + 100\mathbf{j} + 100\mathbf{k}$ mm. Substitution into the velocity relation gives

$$50\omega_2\mathbf{j} = 600\mathbf{i} + \begin{vmatrix} \mathbf{i} & \mathbf{j} & \mathbf{k} \\ \omega_{n_x} & \omega_{n_y} & \omega_{n_z} \\ 50 & 100 & 100 \end{vmatrix}$$

Expanding the determinant and equating the coefficients of the $\mathbf{i}$, $\mathbf{j}$, $\mathbf{k}$ terms give

$$\begin{aligned} -6 &= \qquad\quad + \omega_{n_y} - \omega_{n_z} \\ \omega_2 &= -2\omega_{n_x} \qquad\quad + \omega_{n_z} \\ 0 &= 2\omega_{n_x} - \omega_{n_y} \end{aligned}$$

These equations may be solved for ω_2, which becomes

$$\omega_2 = 6 \text{ rad/s} \qquad \textit{Ans.}$$

As they stand the three equations incorporate the fact that $\boldsymbol{\omega}_n$ is normal to $\mathbf{v}_{A/B}$, but they cannot be solved until the requirement that ω_n be normal to $\mathbf{r}_{A/B}$ is included. Thus,

$$[\boldsymbol{\omega}_n \cdot \mathbf{r}_{A/B} = 0] \qquad 50\omega_{n_x} + 100\omega_{n_y} + 100\omega_{n_z} = 0$$

Combination with two of the three previous equations yields the solutions

$$\omega_{n_x} = -\tfrac{4}{3}\text{ rad/s} \qquad \omega_{n_y} = -\tfrac{8}{3}\text{ rad/s} \qquad \omega_{n_z} = \tfrac{10}{3}\text{ rad/s}$$

Thus

$$\boldsymbol{\omega}_n = \tfrac{2}{3}(-2\mathbf{i} - 4\mathbf{j} + 5\mathbf{k})\text{ rad/s} \qquad \textit{Ans.}$$

with

$$\omega_n = \tfrac{2}{3}\sqrt{2^2 + 4^2 + 5^2} = 2\sqrt{5}\text{ rad/s} \qquad \textit{Ans.}$$

The accelerations of the links may be found from the second of Eqs. 140, which may be written

$$\mathbf{a}_A = \mathbf{a}_B + \dot{\boldsymbol{\omega}}_n \times \mathbf{r}_{A/B} + \boldsymbol{\omega}_n \times (\boldsymbol{\omega}_n \times \mathbf{r}_{A/B})$$

In terms of their normal and tangential components, the accelerations of A and B are

$$\mathbf{a}_A = 50\omega_2^2\mathbf{i} + 50\dot{\omega}_2\mathbf{j} = 1800\mathbf{i} + 50\dot{\omega}_2\mathbf{j}\text{ mm/s}^2$$

$$\mathbf{a}_B = 100\omega_1^2\mathbf{k} + (0)\mathbf{i} = 3600\mathbf{k}\text{ mm/s}^2$$

Also

$$\boldsymbol{\omega}_n \times (\boldsymbol{\omega}_n \times \mathbf{r}_{A/B}) = -\omega_n^2\mathbf{r}_{A/B} = -20(50\mathbf{i} + 100\mathbf{j} + 100\mathbf{k})\text{ mm/s}^2$$

$$\dot{\boldsymbol{\omega}}_n \times \mathbf{r}_{A/B} = \mathbf{i}(100\dot{\omega}_{n_y} - 100\dot{\omega}_{n_z}) + \mathbf{j}(50\dot{\omega}_{n_z} - 100\dot{\omega}_{n_x}) + \mathbf{k}(100\dot{\omega}_{n_x} - 50\dot{\omega}_{n_y})$$

Substitution into the relative acceleration equation and equating respective coefficients of **i**, **j**, **k** give

$$\begin{aligned} 28 &= \dot{\omega}_{n_y} - \dot{\omega}_{n_z} \\ \dot{\omega}_2 + 40 &= -2\dot{\omega}_{n_x} + \dot{\omega}_{n_z} \\ -32 &= 2\dot{\omega}_{n_x} - \dot{\omega}_{n_y} \end{aligned}$$

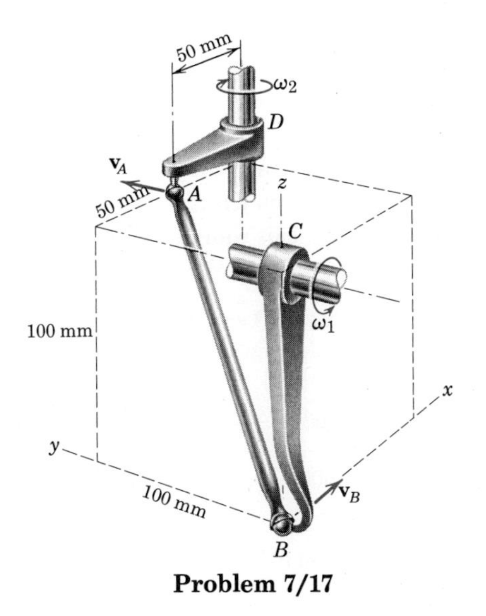

Problem 7/17

Solution of these equations for $\dot{\omega}_2$ gives

$$\dot{\omega}_2 = -36 \text{ rad/s}^2 \qquad \textit{Ans.}$$

The vector $\dot{\omega}_n$ is normal to $\mathbf{r}_{A/B}$ but is not normal to $\mathbf{v}_{A/B}$ as was the case with ω_n. (The component of $\dot{\omega}_n$ which is not normal to $\mathbf{v}_{A/B}$ gives rise to the change in direction of $\mathbf{v}_{A/B}$.) Thus,

$$[\dot{\boldsymbol{\omega}}_n \cdot \mathbf{r}_{A/B} = 0] \qquad 2\dot{\omega}_{n_x} + 4\dot{\omega}_{n_y} + 4\dot{\omega}_{n_z} = 0$$

which, when combined with the preceding relations for these same quantities, gives

$$\dot{\omega}_{n_x} = -8 \text{ rad/s}^2 \qquad \dot{\omega}_{n_y} = 16 \text{ rad/s}^2 \qquad \dot{\omega}_{n_z} = -12 \text{ rad/s}^2$$

Thus

$$\boldsymbol{\alpha}_l = \dot{\boldsymbol{\omega}}_n = 4(-2\mathbf{i} + 4\mathbf{j} - 3\mathbf{k}) \text{ rad/s}^2 \qquad \textit{Ans.}$$

and

$$|\boldsymbol{\alpha}_l| = 4\sqrt{2^2 + 4^2 + 3^2} = 4\sqrt{29} \text{ rad/s}^2 \qquad \textit{Ans.}$$

Attention is called to the fact that, if AB is considered a rigid body as distinguished from a line, then its angular velocity vector would have an additional component along AB due to any rotation of the link about its axis AB.

7/18 The housing of the electric motor swivels about a horizontal shaft (x-axis) which is mounted on the rotating bracket. The motor shaft and attached disk have a constant angular velocity of spin $p = 8$ rad/s with respect to the motor frame, and the bracket revolves around the vertical axis with the constant angular rate $\omega_0 = 3$ rad/s. If γ is constant at 60°, determine the angular acceleration $\boldsymbol{\alpha}$ of the disk and the linear acceleration of a point A at the top of the disk at the instant shown.

Solution. The rotating reference axes x-y-z are selected as shown to provide a simple description of the motion of the rotor and disk relative to these axes. This x-y-z frame rotates about the vertical, and has the angular-velocity components

$$\Omega_x = 0$$

$$\Omega_y = \omega_0 \cos\gamma = 3(\tfrac{1}{2}) = \tfrac{3}{2} \text{ rad/s}$$

$$\Omega_z = \omega_0 \sin\gamma = 3(\sqrt{3}/2) \text{ rad/s}$$

The fixed axes X-Y-Z are chosen to be momentarily in the position shown. The velocity and acceleration terms may be expressed in terms of X-Y-Z components with unit vectors $\mathbf{I}$, $\mathbf{J}$, $\mathbf{K}$, or in terms of x-y-z components with unit vectors $\mathbf{i}$, $\mathbf{j}$, $\mathbf{k}$.

The relative velocity equation is

$$\mathbf{v}_A = \mathbf{v}_B + \boldsymbol{\Omega} \times \mathbf{r}_{A/B} + \mathbf{v}_{\text{rel}}$$

and its terms are

$$\mathbf{v}_B = \boldsymbol{\Omega} \times \mathbf{R}_B = 3\mathbf{K} \times (200\mathbf{J} + 200\mathbf{I}) = -600(\mathbf{I} - \mathbf{J})$$

$$= -600\mathbf{i} + (600 \sin\gamma)\mathbf{j} - (600 \cos\gamma)\mathbf{k} = 600\left(-\mathbf{i} + \frac{\sqrt{3}}{2}\mathbf{j} - \frac{1}{2}\mathbf{k}\right) \text{ mm/s}$$

$$\boldsymbol{\Omega} \times \mathbf{r}_{A/B} = \left(\frac{3}{2}\mathbf{j} + \frac{3\sqrt{3}}{2}\mathbf{k}\right) \times (100\mathbf{j} + 100\mathbf{k}) = 150(1 - \sqrt{3})\mathbf{i} \text{ mm/s}$$

$$\mathbf{v}_{\text{rel}} = \mathbf{p} \times \mathbf{r}_A = 8\mathbf{j} \times 100\mathbf{k} = 800\mathbf{i} \text{ mm/s}$$

Combination of the terms produces

$$\mathbf{v}_A = 90.2\mathbf{i} + 520\mathbf{j} - 300\mathbf{k} \text{ mm/s}$$

The acceleration of A is given by

$$\mathbf{a}_A = \mathbf{a}_B + \dot{\mathbf{\Omega}} \times \mathbf{r}_{A/B} + \mathbf{\Omega} \times (\mathbf{\Omega} \times \mathbf{r}_{A/B}) + 2\mathbf{\Omega} \times \mathbf{v}_{\text{rel}} + \mathbf{a}_{\text{rel}}$$

and its terms are

$$\mathbf{a}_B = \mathbf{\Omega} \times (\mathbf{\Omega} \times \mathbf{R}_B) = 3\mathbf{K} \times 600(\mathbf{J} - \mathbf{I}) = -1800(\mathbf{I} + \mathbf{J}) \text{ mm/s}^2$$

$$= 1800\left(-\mathbf{i} - \frac{\sqrt{3}}{2}\mathbf{j} + \frac{1}{2}\mathbf{k}\right) \text{mm/s}^2$$

$$\dot{\mathbf{\Omega}} = \mathbf{0}$$

$$\mathbf{\Omega} \times (\mathbf{\Omega} \times \mathbf{r}_{A/B}) = \left(\frac{3}{2}\mathbf{j} + \frac{3\sqrt{3}}{2}\mathbf{k}\right) \times 150(1 - \sqrt{3})\mathbf{i} = 225(1 - \sqrt{3})(\sqrt{3}\mathbf{j} - \mathbf{k}) \text{ mm/s}^2$$

$$2\mathbf{\Omega} \times \mathbf{v}_{\text{rel}} = 3(\mathbf{j} + \sqrt{3}\mathbf{k}) \times 800\mathbf{i} = 2400(\sqrt{3}\mathbf{j} - \mathbf{k}) \text{ mm/s}^2$$

$$\mathbf{a}_{\text{rel}} = \mathbf{p} \times (\mathbf{p} \times \mathbf{r}_A) = 8\mathbf{j} \times 800\mathbf{i} = -6400\mathbf{k} \text{ mm/s}^2$$

Combination of these terms gives

$$\mathbf{a}_A = -1800\mathbf{i} + 2310\mathbf{j} - 7740\mathbf{k} \text{ mm/s}^2 \qquad \text{and} \qquad a_A = 8270 \text{ mm/s}^2 \qquad \textit{Ans.}$$

The total angular velocity of the rotor is

$$\mathbf{\omega} = \mathbf{\Omega} + \mathbf{p} = (p + \omega_0 \cos \gamma)\mathbf{j} + (\omega_0 \sin \gamma)\mathbf{k}$$

and the angular acceleration is

$$\mathbf{\alpha} = \dot{\mathbf{\omega}} = (p + \omega_0 \cos \gamma)\dot{\mathbf{j}} + (\omega_0 \sin \gamma)\dot{\mathbf{k}}$$

$$= (p + \omega_0 \cos \gamma)(\mathbf{\Omega} \times \mathbf{j}) + (\omega_0 \sin \gamma)(\mathbf{\Omega} \times \mathbf{k})$$

$$= \mathbf{\Omega} \times [(p + \omega_0 \cos \gamma)\mathbf{j} + (\omega_0 \sin \gamma)\mathbf{k}] = \mathbf{\Omega} \times \mathbf{\omega}$$

Substitution of the values for $\mathbf{\Omega}$ and $\mathbf{\omega}$ gives

$$\mathbf{\alpha} = -12\sqrt{3}\mathbf{i} \text{ rad/s}^2 \qquad \textit{Ans.}$$

The result represents the effect of change of direction only of $\mathbf{\omega}$ since ω_0, p, and γ remain constant. It may be noted that the result $\mathbf{\alpha} = \mathbf{\Omega} \times \mathbf{\omega}$ can be obtained directly from Eq. 142, where $(d\mathbf{\omega}/dt)_{xyz}$ is zero since $\mathbf{\Omega}$ and $\mathbf{p}$ have no derivatives in x-y-z.

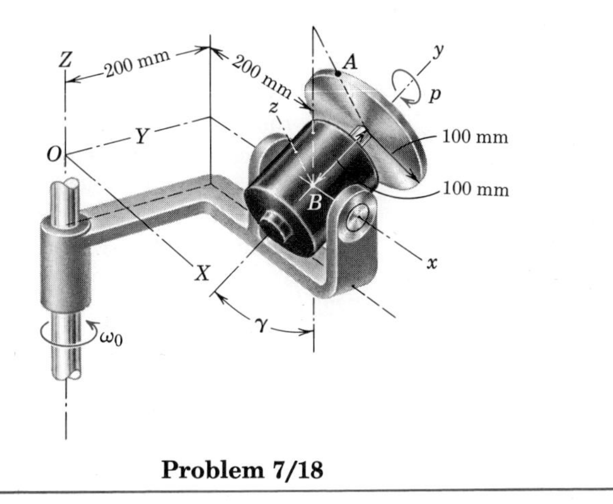

Problem 7/18

Problems

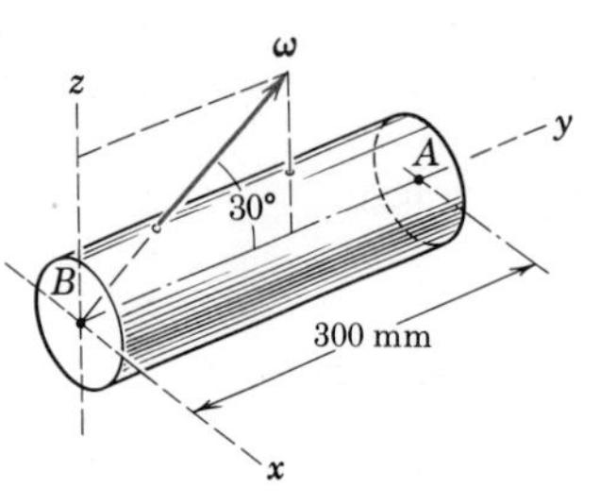

Problem 7/19

7/19 The solid cylinder shown has an angular velocity $\boldsymbol{\omega}$ whose magnitude is 40 rad/s. Calculate the velocity of A with respect to B. Represent $\mathbf{v}_{A/B}$ on a sketch. Also determine the rate p at which the cylinder is spinning about its central axis.

Ans. $\mathbf{v}_{A/B} = -6000\mathbf{i}$ mm/s, $p = 34.6$ rad/s

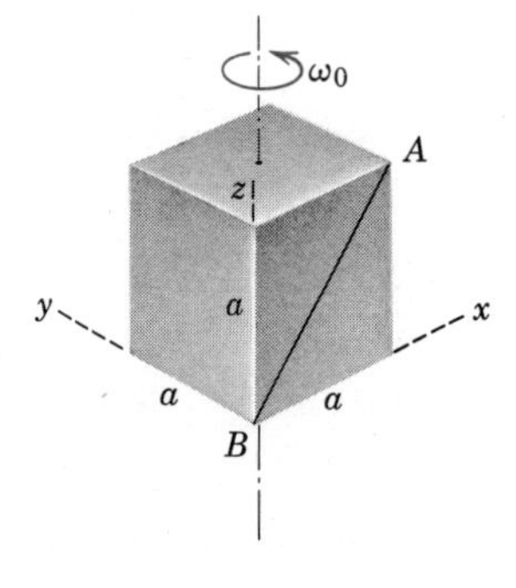

Problem 7/20

7/20 The rigid cube of side a rotates about its central axis with a constant angular velocity ω_0. Determine the expression for the angular velocity $\boldsymbol{\omega}_{AB}$ and angular acceleration $\boldsymbol{\alpha}_{AB}$ of the side diagonal AB treated as a line.

Problem 7/21

7/21 The helicopter is nozing over at the constant rate q rad/s. If the rotor blades revolve at the constant speed p rad/s, write the expression for the angular acceleration $\boldsymbol{\alpha}$ of the rotor. Take the y-axis to be attached to the fuselage and pointing forward perpendicular to the rotor axis.

Ans. $\boldsymbol{\alpha} = pq\mathbf{j}$

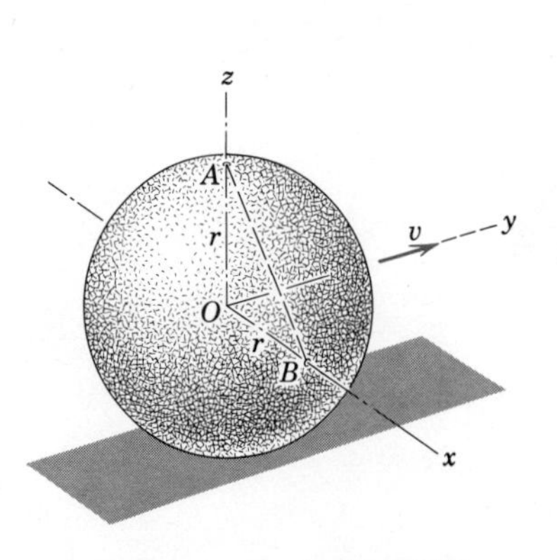

Problem 7/22

7/22 The center O and all points on the diametral x-axis of a solid sphere of radius r have a constant velocity v in the y-direction as the sphere rolls on a horizontal surface parallel to the x-y plane. Determine the expression for the instantaneous angular velocity $\boldsymbol{\omega}_n$ of the line AB which connects point A on the top of the sphere to point B on the x-axis at the surface of the sphere.

7/23 Determine the expression for the angular acceleration $\boldsymbol{\alpha}_l = \dot{\boldsymbol{\omega}}_n$ of line AB in Prob. 7/22 for the instant represented. (*Hint:* Apply Eq. 142 to the expression for $\boldsymbol{\omega}_n$ obtained in Prob. 7/22.)

Ans. $\boldsymbol{\alpha}_l = -\dfrac{v^2}{2r^2}\mathbf{j}$

7/24 The collars at the ends of the telescoping link AB slide along the parallel fixed shafts shown. During an interval of motion $v_A = 125$ mm/s and $v_B = 50$ mm/s. Determine the vector expression for the angular velocity $\boldsymbol{\omega}_n$ of the center line of the link for the position where $y_A = 100$ mm and $y_B = 50$ mm.

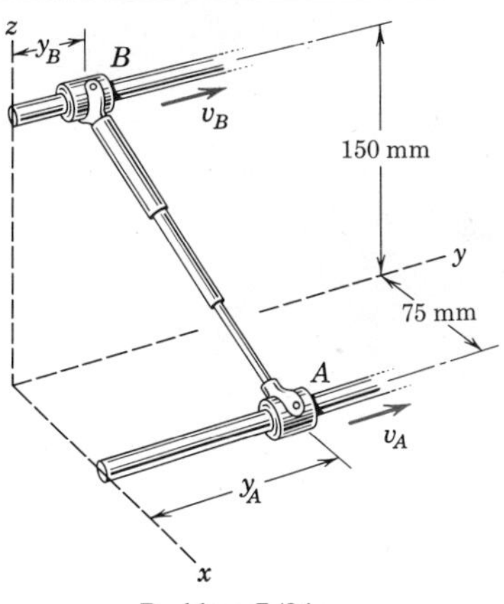

Problem 7/24

7/25 The spacecraft is revolving about its z-axis, which has a fixed space orientation, at the constant rate $p = \frac{1}{10}$ rad/s. Simultaneously its solar panels are unfolding at the rate $\dot{\beta}$ which is programmed to produce the variation shown in the graph. Determine the angular acceleration $\boldsymbol{\alpha}$ of panel A an instant (a) before and an instant (b) after it reaches the position $\beta = 18°$.

Ans. (a) $\boldsymbol{\alpha} = -0.00388\mathbf{i} - 0.00349\mathbf{j}$ rad/s^2
(b) $\boldsymbol{\alpha} = -0.00349\mathbf{j}$ rad/s^2

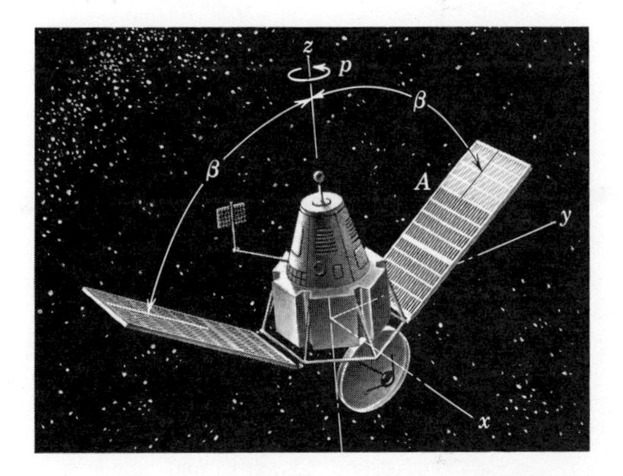

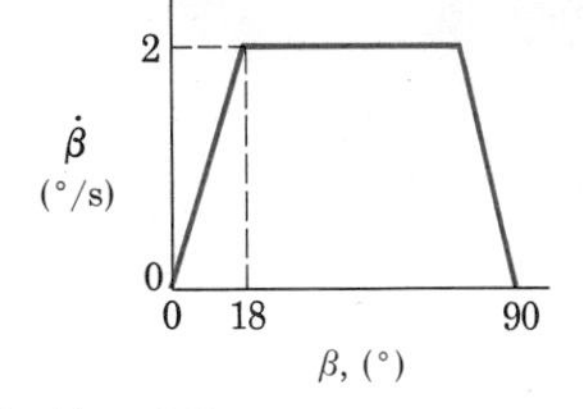

Problem 7/25

7/26 The end of the link A is attached to the collar C by the yoke which permits rotation of A about the z-axis. The collar C in turn may rotate about the fixed shaft S as well as slide along it. Irrespective of the motion of the other end of link A, show that the angular velocity $\boldsymbol{\omega}$ of the link must obey the relation $\boldsymbol{\omega} \cdot \mathbf{h} \times (\mathbf{r} \times \mathbf{h}) = 0$. Vectors $\mathbf{r}$ and $\mathbf{h}$ are any vectors, respectively, along the link and along the fixed shaft. The axis of the link A is normal to the yoke axis and, hence, lies in the x-y plane.

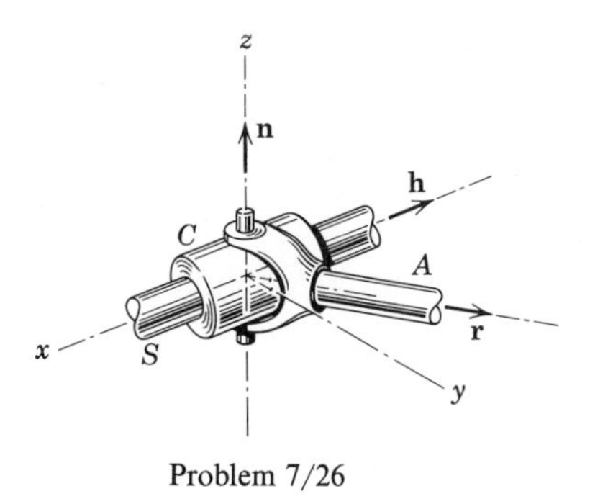

Problem 7/26

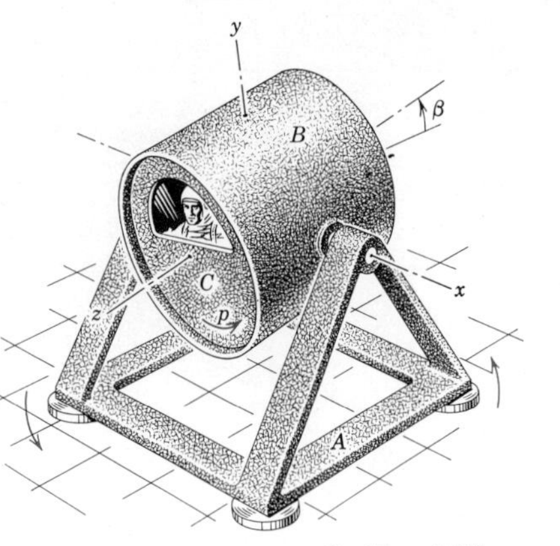

Problem 7/27

7/27 A simulator for perfecting the docking procedure for spacecraft consists of the frame A which is mounted on four air-bearing pads so that it can translate and rotate freely on the horizontal surface. Mounted in the frame is a drum B which can rotate about the horizontal axis of the frame A. The coordinate axes x-y-z are attached to the drum B, and the z-axis of the drum makes an angle β with the horizontal. Inside the drum is the simulated command module C which can rotate within the drum about its z-axis at a rate p. For a certain test run frame A is rotating on the horizontal surface in the direction shown with a constant angular velocity of 0.2 rad/s. Simultaneously the drum B is rotating about the x-axis at the constant rate $\dot{\beta} = 0.15$ rad/s, and the module C is turning inside the drum at the constant rate $p = 0.9$ rad/s in the direction indicated. For these conditions determine the angular velocity $\boldsymbol{\omega}$ and the angular acceleration $\boldsymbol{\alpha}$ of the simulator C as it passes the position $\beta = 0$.

Ans. $\boldsymbol{\omega} = 0.15\mathbf{i} + 0.2\mathbf{j} + 0.9\mathbf{k}$ rad/s
$\boldsymbol{\alpha} = 0.18\mathbf{i} - 0.135\mathbf{j} - 0.03\mathbf{k}$ rad/s^2

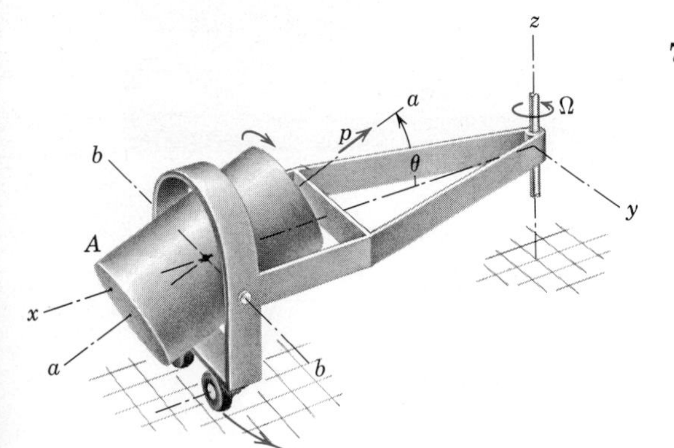

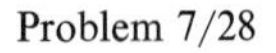

Problem 7/28

7/28 The test chamber of a flight simulator consists of a drum which spins about the a-a axis at a constant angular rate p relative to the cylindrical housing A. The housing in turn is mounted on transverse horizontal bearings and rotates about the axis b-b at the constant rate $\dot{\theta}$. The entire assembly is made to rotate about the fixed vertical z-axis at the constant rate Ω. Write the expression for the angular acceleration $\boldsymbol{\alpha}$ of the spinning drum during the compounded motion for a given value of θ.

Ans. $\boldsymbol{\alpha} = \dot{\theta}(p \sin\theta - \Omega)\mathbf{i} - (p\,\Omega \cos\theta)\mathbf{j} + (p\,\dot{\theta} \cos\theta)\mathbf{k}$

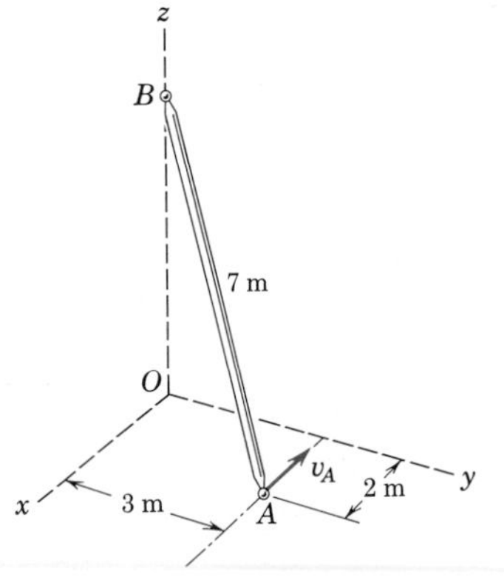

Problem 7/29

7/29 End A of the rigid link is confined to move in the $-x$-direction while end B is confined to move along the z-axis. Determine the angular velocity ω_n of the center line of the link as it passes the position shown with $v_A = 3$ m/s.

7/30 The center O of the spacecraft is moving through space with a constant velocity. During the period of motion prior to stabilization, the spacecraft has a constant rotational rate $\Omega = \frac{1}{2}$ rad/s about its z-axis. The x-y-z axes are attached to the body of the craft, and the solar panels rotate about the y-axis at the constant rate $\dot{\theta} = \frac{1}{4}$ rad/s with respect to the spacecraft. If $\boldsymbol{\omega}$ is the absolute angular velocity of the solar panels, determine $\dot{\boldsymbol{\omega}}$. Also find the acceleration of point A when $\theta = 30°$.

Ans. $\dot{\boldsymbol{\omega}} = \frac{1}{8}\mathbf{i}$ rad/s^2
$\mathbf{a}_A = 0.0938\mathbf{i} - 0.730\mathbf{j} - 0.0325\mathbf{k}$ m/s^2

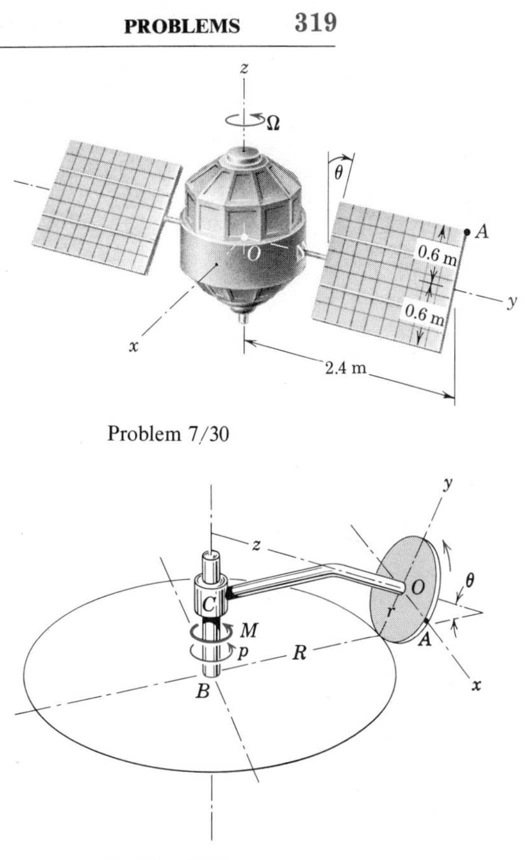

Problem 7/30

◀ **7/31** The wheel of radius r is free to rotate about the bent axle CO which turns about the vertical axis at the constant rate p rad/s. If the wheel rolls without slipping on the horizontal circle of radius R, determine the expressions for the angular velocity $\boldsymbol{\omega}$ and angular acceleration $\boldsymbol{\alpha}$ of the wheel. The x-axis is always horizontal.

$$Ans.\ \boldsymbol{\omega} = p\left[\mathbf{j}\cos\theta + \mathbf{k}\left(\sin\theta + \frac{R}{r}\right)\right]$$

$$\boldsymbol{\alpha} = \mathbf{i}\frac{Rp^2}{r}\cos\theta$$

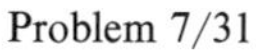

Problem 7/31

◀ **7/32** The motor turns the disk at the constant rate $p = 30$ rad/s. The motor is also swiveling about the horizontal axis BO (y-axis) at the constant rate $\dot{\theta} = 10$ rad/s. Simultaneously the entire assembly is rotating about the vertical axis C-C at the constant rate $q = 8$ rad/s. For the instant that $\theta = 30°$, determine the angular acceleration $\boldsymbol{\alpha}$ of the disk and the acceleration of point A at the bottom of the disk. Axes x-y-z are attached to the motor housing, and plane O-x_0-y is horizontal.

Ans. $\mathbf{a}_A = -66.5\mathbf{i} - 7.74\mathbf{j} + 129.8\mathbf{k}$ m/s^2
$\boldsymbol{\alpha} = 20(2\sqrt{3}\mathbf{i} + 6\sqrt{3}\mathbf{j} + 13\mathbf{k})$ rad/s^2

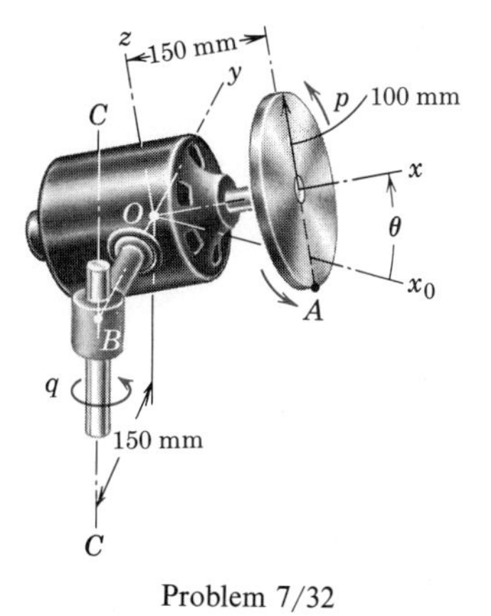

Problem 7/32

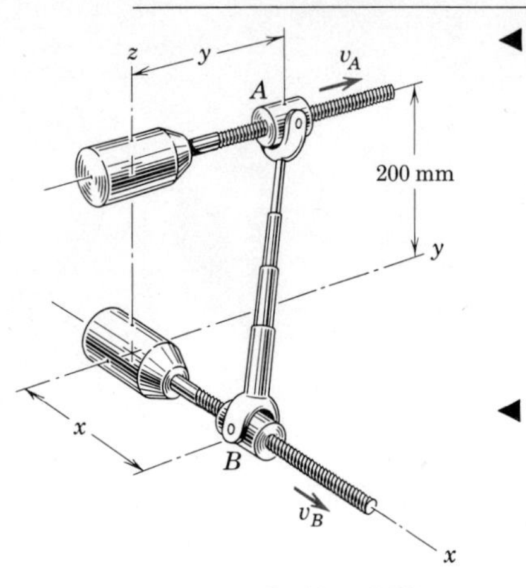

Problem 7/33

◀**7/33** The ends A and B of the telescoping link are secured to their respective collars whose positions on the shafts are controlled by the two lead screws. If the screws are turning so that both A and B are moving at the rate of 0.15 m/s in the directions shown, determine the angular velocity $\boldsymbol{\omega}_n$ of the center line AB of the link at an instant when $x = y = 200$ mm.

Ans. $\boldsymbol{\omega}_n = -\frac{1}{4}(\mathbf{i} + \mathbf{j})$ rad/s

◀**7/34** Determine the angular velocity $\boldsymbol{\omega}_A$ of the yoke at A for the telescoping link of Prob. 7/33 under the conditions specified. (See Prob. 7/26.)

Ans. $\boldsymbol{\omega}_A = -\frac{1}{8}(\mathbf{i} + 3\mathbf{j} + \mathbf{k})$ rad/s

8 SPACE KINETICS OF RIGID BODIES

39 Introduction. In Chapter 6 the general motion equations which were developed in Chapter 4 for any system of constant mass were reduced and applied to the case of rigid-body plane motion. These same general motion equations will be applied in this present chapter to the description of rigid-body motion in three dimensions, and a review of Chapter 4 along with the study of Chapter 8 is recommended. It will be found that the formulation of theory for three-dimensional motion is greatly facilitated by making full use of vector notation. Most readers will also find that the physical insight into dynamics gained from a thorough study of plane motion will be of considerable value in developing and understanding the description of three-dimensional motion.

The dependency of three-dimensional kinetics on the relationships of space kinematics is no less important than the parallel dependency in the case of plane motion. Thus the student will find it greatly to his advantage to have a firm grasp of the material in Chapter 7 before proceeding. Also the step in defining and isolating the body in question without ambiguity by means of the free-body diagram is as important in space kinetics as it was in plane kinetics.

40 Angular Momentum. The force equation for any mass system, rigid or nonrigid, Eq. 92 or 96, is the generalization of Newton's second law for the motion of a particle and should require no further explanation. The moment equation, however, is not nearly as simple as the first of Eqs. 112 for plane motion since the change of angular momentum has a number of additional components in three-dimensional motion which are absent in plane motion.

Consider now a rigid body moving with any general motion in space, Fig. 71*a*. Axes *x-y-z* are *attached* to the body with origin at the mass center *G*. The angular velocity $\boldsymbol{\omega}$ of the body becomes the angular velocity of the *x-y-z* frame as observed from the fixed reference axes *X-Y-Z*. The absolute angular momentum $\overline{\mathbf{H}}$ of the body about its mass center *G* is the sum of the moments about *G* of the linear momenta and was expressed in Art. 25 as $\overline{\mathbf{H}} = \Sigma(\boldsymbol{\rho}_i \times m_i\mathbf{v}_i)$, where $\mathbf{v}_i$ is the absolute velocity of m_i. But for the rigid body $\mathbf{v}_i = \overline{\mathbf{v}} + \boldsymbol{\omega} \times \boldsymbol{\rho}_i$ where $\boldsymbol{\omega} \times \boldsymbol{\rho}_i$ is the relative velocity of m_i with respect to *G*. Thus $\overline{\mathbf{H}}$ may be written as

$$\overline{\mathbf{H}} = -\overline{\mathbf{v}} \times \Sigma m_i\boldsymbol{\rho}_i + \Sigma(\boldsymbol{\rho}_i \times m_i[\boldsymbol{\omega} \times \boldsymbol{\rho}_i])$$

The first term is zero since $\Sigma m_i\boldsymbol{\rho}_i = m\overline{\boldsymbol{\rho}} = \mathbf{0}$, and the second term, with the substitution of dm for m_i and $\boldsymbol{\rho}$ for $\boldsymbol{\rho}_i$, gives

$$\overline{\mathbf{H}} = \int (\boldsymbol{\rho} \times [\boldsymbol{\omega} \times \boldsymbol{\rho}])\, dm \tag{144}$$

Before expanding the integrand of Eq. 144, consider also the case of a rigid body rotating about a fixed point O, Fig. 71b. The x-y-z axes are attached to the body, and both body and axes have an angular velocity $\boldsymbol{\omega}$. The angular momentum about O was expressed in Art. 25 and is $\mathbf{H}_O = \Sigma(\mathbf{r}_i \times m_i\mathbf{v}_i)$ where, for the rigid body, $\mathbf{v}_i = \boldsymbol{\omega} \times \mathbf{r}_i$. Thus, with the substitution of dm for m_i and $\mathbf{r}$ for $\mathbf{r}_i$, the angular momentum is

$$\mathbf{H}_O = \int (\mathbf{r} \times [\boldsymbol{\omega} \times \mathbf{r}])\, dm \tag{145}$$

It is observed now that for the two cases of Figs. 71a and 71b, the position vector $\boldsymbol{\rho}_i$ or $\mathbf{r}_i$ is given by the same expression $\mathbf{i}x + \mathbf{j}y + \mathbf{k}z$. Thus Eqs. 144 and 145 are identical in form, and the symbol $\mathbf{H}$ will be used here for either case. The expansion of the integrand in the two expressions for angular momentum is now carried out with recognition of the fact that the components of $\boldsymbol{\omega}$ are invariant with respect to the integrals over the body and, hence, become constant multipliers of the integrals. The cross-product expansion applied to the triple vector product upon collection of terms gives

$$\begin{aligned} d\mathbf{H} = \mathbf{i}[&(y^2 + z^2)\omega_x - xy\omega_y - xz\omega_z]\, dm \\ +\mathbf{j}[&-yx\omega_x + (z^2 + x^2)\omega_y - yz\omega_z]\, dm \\ +\mathbf{k}[&-zx\omega_x - zy\omega_y + (x^2 + y^2)\omega_z]\, dm \end{aligned}$$

Now let

$$\begin{aligned} I_{xx} &= \int (y^2 + z^2)\, dm & I_{xy} &= \int xy\, dm \\ I_{yy} &= \int (z^2 + x^2)\, dm & I_{xz} &= \int xz\, dm \\ I_{zz} &= \int (x^2 + y^2)\, dm & I_{yz} &= \int yz\, dm \end{aligned} \tag{146}$$

The quantities I_{xx}, I_{yy}, I_{zz} are known as the *moments of inertia* of the body

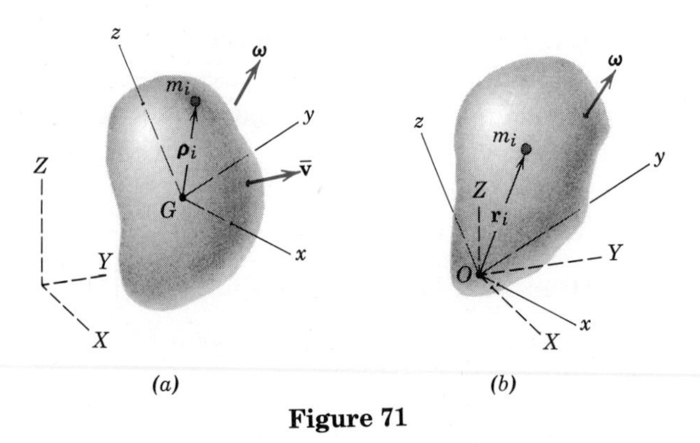

Figure 71

about the respective axes, and I_{xy}, I_{xz}, I_{yz} are known as the *products of inertia* with respect to the coordinate axes. These quantities describe the manner in which the mass of a rigid body is distributed with respect to the chosen axes. The double subscripts for the moments of inertia preserve a symmetry of notation which has special meaning in the tensor formulation discussed in the next article. It is observed that $I_{xy} = I_{yx}$, $I_{xz} = I_{zx}$, $I_{yz} = I_{zy}$. With the substitutions of Eqs. 146, the expression for $\mathbf{H}$ becomes

$$\begin{aligned} \mathbf{H} = \quad & \mathbf{i}(\ \ I_{xx}\omega_x - I_{xy}\omega_y - I_{xz}\omega_z) \\ & +\mathbf{j}(-I_{yx}\omega_x + I_{yy}\omega_y - I_{yz}\omega_z) \\ & +\mathbf{k}(-I_{zx}\omega_x - I_{zy}\omega_y + I_{zz}\omega_z) \end{aligned} \tag{147}$$

and the components of $\mathbf{H}$ are clearly

$$\begin{aligned} H_x &= \ \ I_{xx}\omega_x - I_{xy}\omega_y - I_{xz}\omega_z \\ H_y &= -I_{yx}\omega_x + I_{yy}\omega_y - I_{yz}\omega_z \\ H_z &= -I_{zx}\omega_x - I_{zy}\omega_y + I_{zz}\omega_z \end{aligned} \tag{148}$$

Equation 147 is the general expression for the angular momentum about either the mass center G or about a fixed point O for a rigid body rotating with an instantaneous angular velocity $\boldsymbol{\omega}$.

Attention is called to the fact that in each of the two cases represented, the reference axes x-y-z are *attached* to the rigid body. This attachment makes the moment-of-inertia integrals and the product-of-inertia integrals of Eq. 146 invariant with time. If the x-y-z axes were to rotate with respect to an irregular body, then these inertia integrals would be functions of the time which would introduce an undesirable complexity into the angular momentum relations. An important exception occurs when a rigid body is spinning about an axis of symmetry in which case the inertia integrals are not affected by the angular position of the body about its spin axis. Thus it is frequently convenient to permit a body with axial symmetry to rotate relative to the reference system about one of the coordinate axes. In addition to the momentum components due to the angular velocity $\boldsymbol{\Omega}$ of the reference axes, then, an added angular momentum component along the spin axis due to the relative spin about the axis would have to be accounted for.

41 Inertial Properties. Before the angular momentum relations of the previous article are related to the applied moments, it is necessary to examine further the inertial properties of a rigid body as defined by Eqs. 146.

The transformation from the vector $\boldsymbol{\omega}$ to the vector $\mathbf{H}$ in Eq. 147 may be written in matrix form as

$$\begin{Bmatrix} H_x \\ H_y \\ H_z \end{Bmatrix} = \begin{bmatrix} I_{xx} & -I_{xy} & -I_{xz} \\ -I_{yx} & I_{yy} & -I_{yz} \\ -I_{zx} & -I_{zy} & I_{zz} \end{bmatrix} \begin{Bmatrix} \omega_x \\ \omega_y \\ \omega_z \end{Bmatrix}$$

or merely

$$\{H_{xyz}\} = [I]\{\omega_{xyz}\} \tag{149}$$

The array $[I]$ is called the *inertia tensor* or *inertia matrix.* The values of the elements in the matrix depend on the particular origin and orientation of axes chosen. If the subscripts x, y, z are replaced by the notation 1, 2, 3, the matrix or tensor transformation may be written as

$$H_i = \sum_{j=1}^{3} \pm I_{ij}\omega_j \qquad i = 1, 2, 3 \tag{150}$$

where the plus sign for I_{ij} is used where $i = j$ and the minus sign is used where $i \neq j$. The inertia tensor is a symmetric tensor since $I_{ij} = I_{ji}$.

To examine the effect on the inertial properties of orientation of axes for a given origin O of coordinates, consider the moment of inertia I_M of a rigid body, Fig. 72, about any line M through the origin O. The direction cosines of M are l, m, n, and a unit vector $\boldsymbol{\lambda}$ along M may be written $\boldsymbol{\lambda}=\mathbf{i}l+\mathbf{j}m+\mathbf{k}n$. The moment of inertia about M is

$$I_M = \int h^2\, dm = \int (\mathbf{r} \times \boldsymbol{\lambda}) \cdot (\mathbf{r} \times \boldsymbol{\lambda})\, dm$$

where $|\mathbf{r} \times \boldsymbol{\lambda}| = h$. The cross product is

$$(\mathbf{r} \times \boldsymbol{\lambda}) = \mathbf{i}(yn - zm) + \mathbf{j}(zl - xn) + \mathbf{k}(xm - yl)$$

and the sum of the squares of the terms gives

$$(\mathbf{r} \times \boldsymbol{\lambda}) \cdot (\mathbf{r} \times \boldsymbol{\lambda}) = h^2 = (y^2 + z^2)l^2 + (x^2 + z^2)m^2 + (x^2 + y^2)n^2 - 2xylm - 2xzln - 2yzmn$$

Thus, with the substitution of the expressions of Eqs. 146, there results

$$I_M = I_{xx}l^2 + I_{yy}m^2 + I_{zz}n^2 - 2I_{xy}lm - 2I_{xz}ln - 2I_{yz}mn \tag{151}$$

This expression gives the moment of inertia about any axis OM in terms of the moments and products of inertia about the coordinate directions. If a point Q is located on OM a distance $q = 1/\sqrt{I_M}$ from O, then the expression for I_M may be written

$$1 = I_{xx}x_1^2 + I_{yy}y_1^2 + I_{zz}z_1^2 - 2I_{xy}x_1y_1 - 2I_{xz}x_1z_1 - 2I_{yz}y_1z_1$$

where x_1, y_1, z_1 are the coordinates of Q. This expression is the equation of

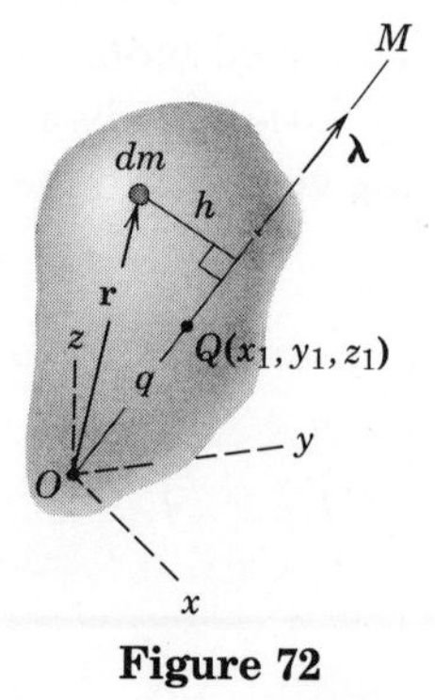

Figure 72

a quadric surface about O. Furthermore, the surface must define an ellipsoid, Fig. 73*a*, since, for a finite rigid body there can be no orientation of OM for which I_M is zero and q infinite. This surface is known as the *ellipsoid of inertia* for the particular point O of the given rigid body. The geometry of the ellipsoid completely defines the inertial properties of the body about O. In general a different ellipsoid of inertia will be associated with each different point in the body.

Inasmuch as an ellipsoid has three axes of symmetry, it is always possible to orient the coordinate directions to coincide with these axes, as indicated in Fig. 73*b*. The moments of inertia about these axes are known as the *principal moments of inertia* I_1, I_2, I_3, and the axes are referred to as the *principal axes of inertia*. For this orientation of axes, the products of inertia vanish, and the equation for the quadric surface becomes

$$1 = I_1x^2 + I_2y^2 + I_3z^2 \qquad \text{or} \qquad \frac{x^2}{q_1^2} + \frac{y^2}{q_2^2} + \frac{z^2}{q_3^2} = 1$$

The quantities $q_1 = 1/\sqrt{I_1}$, $q_2 = 1/\sqrt{I_2}$, $q_3 = 1/\sqrt{I_3}$, are the three stationary values of q which give the lengths of the semi-minor axis, the intermediate axis, and the semi-major axis of the ellipsoid. With the product of inertia terms zero for this orientation of axes, the inertia tensor takes the form

$$[I] = \begin{bmatrix} I_1 & 0 & 0 \\ 0 & I_2 & 0 \\ 0 & 0 & I_3 \end{bmatrix}$$

and is said to be diagonalized. From the definition of q and from Fig. 73*b* it is observed that the moment of inertia is a maximum about the axis for which q is a minimum and vice versa.

The directions of the angular velocity vector $\boldsymbol{\omega}$ and the angular momentum vector $\mathbf{H}$ are generally not the same. To show this consider a rigid body rotating with an angular velocity $\boldsymbol{\omega} = \mathbf{i}\omega_x + \mathbf{j}\omega_y + \mathbf{k}\omega_z$ with coordinate axes chosen as principal axes so that $\mathbf{H} = \mathbf{i}I_1\omega_x + \mathbf{j}I_2\omega_y + \mathbf{k}I_3\omega_z$. These vectors can have the same direction only if $I_1 = I_2 = I_3$, in which case the ratios of

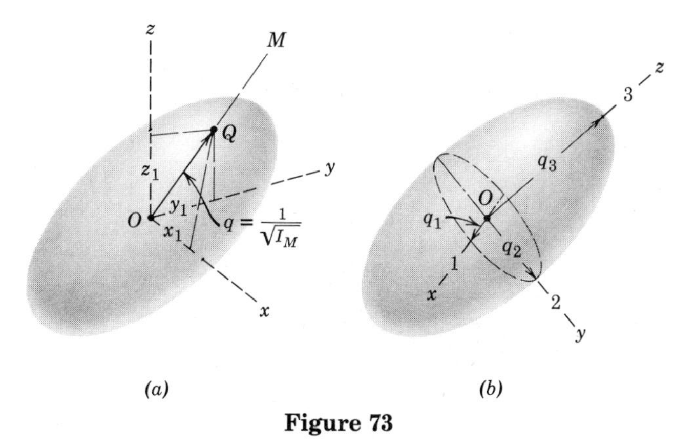

Figure 73

the components are the same and the vectors are parallel, or if the body is rotating about one of the principal axes which makes the remaining two of the components of each of the vectors zero.

In order to determine the principal moments of inertia from the moments and products of inertia corresponding to an arbitrary orientation of axes, consider a body which rotates with an angular velocity ω about one of its principal axes of inertia about which the moment of inertia is I. The ellipsoid of inertia for this case is illustrated in Fig. 74. By virtue of the discussion in the preceding paragraph, the angular momentum is along this same principal axis and is $\mathbf{H} = I\omega$. If the components of $\mathbf{H}$ are equated to the expressions in Eqs. 148 applied to this special case, there results, upon rearrangement of terms,

$$\begin{aligned} (I_{xx} - I)\omega_x - I_{xy}\omega_y - I_{xz}\omega_z &= 0 \\ -I_{yx}\omega_x + (I_{yy} - I)\omega_y - I_{yz}\omega_z &= 0 \\ -I_{zx}\omega_x - I_{zy}\omega_y + (I_{zz} - I)\omega_z &= 0 \end{aligned} \tag{152}$$

These equations will have a solution provided the determinant of the coefficients vanishes or

$$\begin{vmatrix} I_{xx} - I & -I_{xy} & -I_{xz} \\ -I_{yx} & I_{yy} - I & -I_{yz} \\ -I_{zx} & -I_{zy} & I_{zz} - I \end{vmatrix} = 0 \tag{153}$$

This equation is a cubic equation in I, and the three roots are the principal moments of inertia I_1, I_2, I_3. Although a proof may be cited, it is clear from the physical problem that all three roots are positive and real.

The direction cosines l, m, n of a principal inertia axis may be obtained directly from Eq. 152 since for these equations the directions of ω and the principal inertia axis coincide, Fig. 74. The substitutions $\omega_x = \omega l$, $\omega_y = \omega m$, $\omega_z = \omega n$ give

$$\begin{aligned} (I_{xx} - I)l - I_{xy}m - I_{xz}n &= 0 \\ -I_{yx}l + (I_{yy} - I)m - I_{yz}n &= 0 \\ -I_{zx}l - I_{zy}m + (I_{zz} - I)n &= 0 \end{aligned} \tag{154}$$

These equations along with $l^2 + m^2 + n^2 = 1$ will enable a solution for the direction cosines to be made for each of the three I's separately.

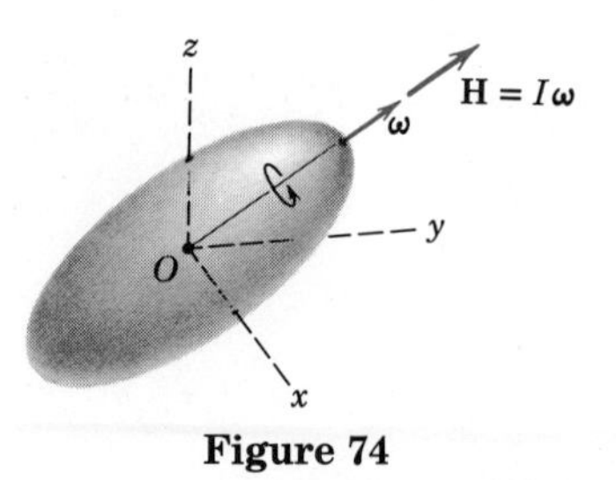

Figure 74

The transformation of inertia terms from a set of centroidal axes xyz to a *parallel* set of noncentroidal axes $x'y'z'$ is easily handled. If the components of displacement of the origin of $x'y'z'$ with respect to the origin of xyz are d_x, d_y, d_z, it is easily verified that

$$\begin{aligned} I_{x'x'} &= I_{xx} + m(d_y^2 + d_z^2) & I_{x'y'} &= I_{xy} + md_xd_y \\ I_{y'y'} &= I_{yy} + m(d_z^2 + d_x^2) & I_{x'z'} &= I_{xz} + md_xd_z \\ I_{z'z'} &= I_{zz} + m(d_x^2 + d_y^2) & I_{y'z'} &= I_{yz} + md_yd_z \end{aligned} \tag{155}$$

These transformations for the translation of axes constitute the *parallel-axis theorems* which can be used *only* if the origin of xyz is the mass center.

Sample Problem

8/1 The bent plate has a mass of 70 kg per square metre of surface area and revolves about the z-axis at the rate $\omega = 300$ rad/s. Determine (*a*) the moments and products of inertia with respect to the attached coordinate axes shown, (*b*) the principal moments of inertia and the direction cosines of the axis of minimum moment of inertia, and (*c*) the angular momentum **H** of the plate about O. Neglect the hub at the rotation axis and the thickness of the plates compared with their surface dimensions.

Solution. (*a*) The moments and products of inertia are written with the aid of Eqs. 155 by transfer from the parallel centroidal axes for each part.

First, the mass of each part is

$$m_A = (0.100)(0.125)(70) = 0.875 \text{ kg}$$

$$m_B = (0.075)(0.150)(70) = 0.788 \text{ kg}$$

For part A

$[I_{xx} = \bar{I}_{xx} + md^2]$

$$I_{xx} = \frac{0.875}{12}(\overline{0.100}^2 + \overline{0.125}^2) + 0.875(\overline{0.050}^2 + \overline{0.0625}^2)$$

$$= 0.007\,47 \text{ kg}\cdot\text{m}^2$$

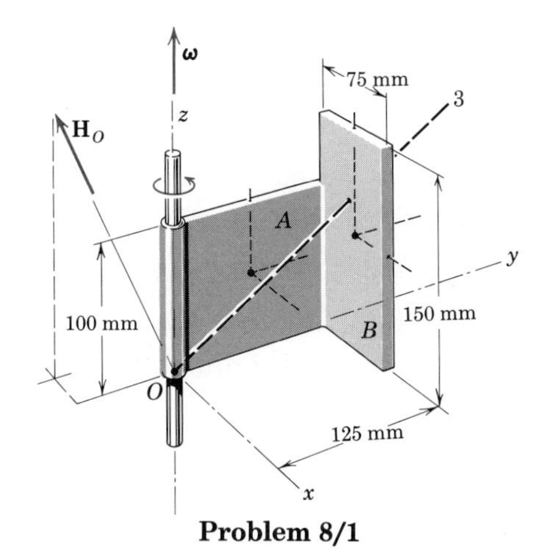

Problem 8/1

$[I_{yy} = \frac{1}{3}ml^2]$ $\quad I_{yy} = \frac{0.875}{3}(\overline{0.100^2}) = 0.002\,92 \text{ kg}\cdot\text{m}^2$

$[I_{zz} = \frac{1}{3}ml^2]$ $\quad I_{zz} = \frac{0.875}{3}(\overline{0.125^2}) = 0.004\,56 \text{ kg}\cdot\text{m}^2$

$[I_{xy} = \int xy\, dm, \quad I_{xz} = \int xz\, dm]$ $\quad I_{xy} = 0, \quad I_{xz} = 0$

$[I_{yz} = \bar{I}_{yz} + md_y d_z]$ $\quad I_{yz} = 0 + 0.875(0.0625)(0.050) = 0.002\,73 \text{ kg}\cdot\text{m}^2$

For part B

$[I_{xx} = \bar{I}_{xx} + md^2]$ $\quad I_{xx} = \frac{0.788}{12}(\overline{0.150^2}) + 0.788(\overline{0.125^2} + \overline{0.075^2}) = 0.018\,21 \text{ kg}\cdot\text{m}^2$

$[I_{yy} = \bar{I}_{yy} + md^2]$ $\quad I_{yy} = \frac{0.788}{12}(\overline{0.075^2} + \overline{0.150^2}) + 0.788(\overline{0.0375^2} + \overline{0.075^2})$

$= 0.007\,38 \text{ kg}\cdot\text{m}^2$

$[I_{zz} = \bar{I}_{zz} + md^2]$ $\quad I_{zz} = \frac{0.788}{12}(\overline{0.075^2}) + 0.788(\overline{0.125^2} + \overline{0.0375^2})$

$= 0.013\,81 \text{ kg}\cdot\text{m}^2$

$[I_{xy} = \bar{I}_{xy} + md_x d_y]$ $\quad I_{xy} = 0 + 0.788(0.0375)(0.125) = 0.003\,69 \text{ kg}\cdot\text{m}^2$

$[I_{xz} = \bar{I}_{xz} + md_x d_z]$ $\quad I_{xz} = 0 + 0.788(0.0375)(0.075) = 0.002\,21 \text{ kg}\cdot\text{m}^2$

$[I_{yz} = \bar{I}_{yz} + md_y d_z]$ $\quad I_{yz} = 0 + 0.788(0.125)(0.075) = 0.007\,38 \text{ kg}\cdot\text{m}^2$

The sum of the respective inertia terms gives for the two plates together

$$I_{xx} = 0.0257 \text{ kg}\cdot\text{m}^2 \qquad I_{xy} = 0.003\,69 \text{ kg}\cdot\text{m}^2$$
$$I_{yy} = 0.010\,30 \text{ kg}\cdot\text{m}^2 \qquad I_{xz} = 0.002\,21 \text{ kg}\cdot\text{m}^2 \qquad \textit{Ans.}$$
$$I_{zz} = 0.018\,36 \text{ kg}\cdot\text{m}^2 \qquad I_{yz} = 0.010\,12 \text{ kg}\cdot\text{m}^2$$

(b) The principal moments of inertia are given by the solution to Eq. 153. Substitution into the determinant, expansion, and simplification yield the cubic equation

$$I^3 - 0.0543I^2 + 0.000\,804I - 1.763(10^{-6}) = 0$$

Solution* gives the three roots

$$I_1 = 0.0265 \text{ kg}\cdot\text{m}^2 \text{ (maximum)}$$
$$I_2 = 0.0252 \text{ kg}\cdot\text{m}^2 \text{ (intermediate)} \qquad \textit{Ans.}$$
$$I_3 = 0.0026 \text{ kg}\cdot\text{m}^2 \text{ (minimum)}$$

The direction cosines l_3, m_3, n_3 for the axis of minimum moment of inertia are obtained by substitution into Eqs. 154 which gives

$$(0.0257 - 0.0026)l_3 - 0.003\,69m_3 - 0.002\,21n_3 = 0$$
$$-0.003\,69l_3 + (0.010\,30 - 0.0026)m_3 - 0.1012n_3 = 0$$
$$-0.002\,21l_3 - 0.010\,12m_3 + (0.018\,36 - 0.0026)n_3 = 0$$

Solution of these equations with $l_3^2 + m_3^2 + n_3^2 = 1$ yields

$$l_3 = 0.183 \qquad m_3 = 0.815 \qquad n_3 = 0.550 \qquad \textit{Ans.}$$

* Solution may be programmed for a digital computer, or an algebraic or graphical solution may be obtained.

It is observed that extreme accuracy would be required to obtain the direction cosines for the axes 1 and 2 in this particular problem since the maximum and the intermediate moments of inertia are almost equal and it would be difficult to distinguish between them.

(*c*) The angular momentum of the body is given by Eq. 147 where $\boldsymbol{\omega} = 300\mathbf{k}$ rad/sec. Thus

$$\mathbf{H}_O = 300(-0.002\,21\mathbf{i} - 0.010\,12\mathbf{j} + 0.018\,36\mathbf{k})\ \text{kg}\cdot\text{m}^2/\text{s} \qquad \textit{Ans.}$$

The principal axis O–3 of minimum moment of inertia and the vectors $\mathbf{H}_O$ and $\boldsymbol{\omega}$ are shown on the figure.

Problems

8/2 The slender rod of mass m and length l rotates about the y-axis with an angular velocity ω. By inspection write the expression for the angular momentum $\mathbf{H}$ of the rod about the origin of the x-y-z axes for the position shown. Verify the result by applying Eq. 147.

Ans. $\mathbf{H} = mb^2\omega\left(\mathbf{j} - \dfrac{l}{2b}\mathbf{k}\right)$

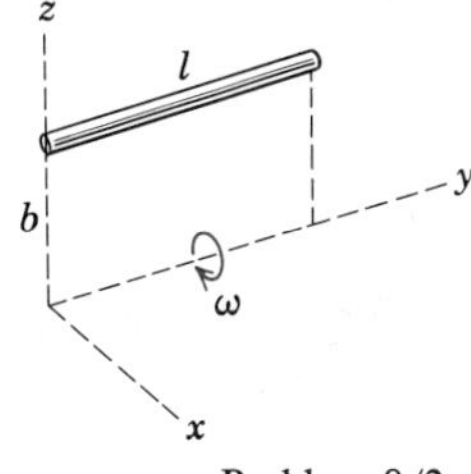

Problem 8/2

8/3 The bent rod has a mass m. Determine its moment of inertia about the diagonal axis OM.

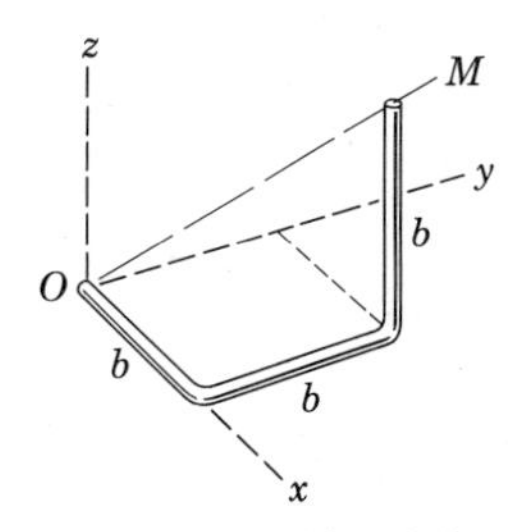

Problem 8/3

8/4 The slender rod of mass m and length l rotates about the y-axis as the element of a right circular cone. If the angular velocity about the y-axis is ω, determine the expression for the angular momentum of the rod with respect to the x-y-z axes for the particular position shown.

Ans. $\mathbf{H} = \frac{1}{3}ml^2\omega \sin\theta\ (\mathbf{j}\sin\theta - \mathbf{k}\cos\theta)$

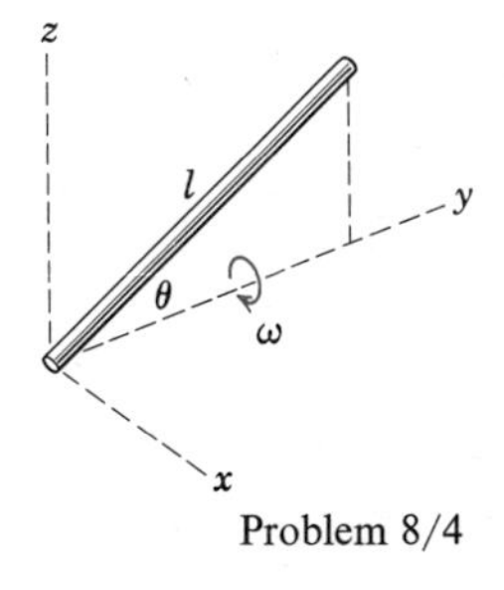

Problem 8/4

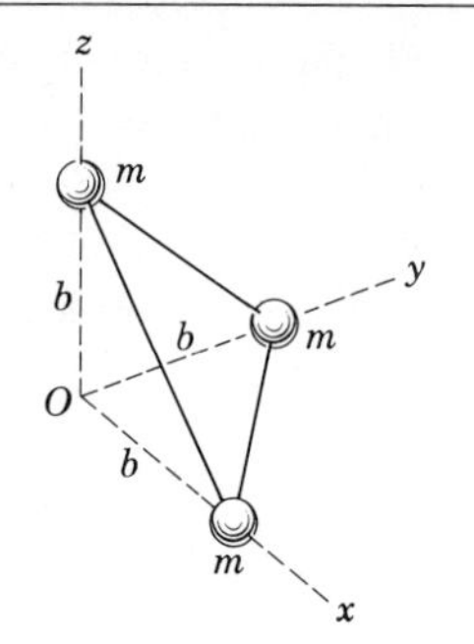

Problem 8/5

8/5 Prove that the moment of inertia of the three identical balls each of mass m and radius r has the same value for all axes through O.

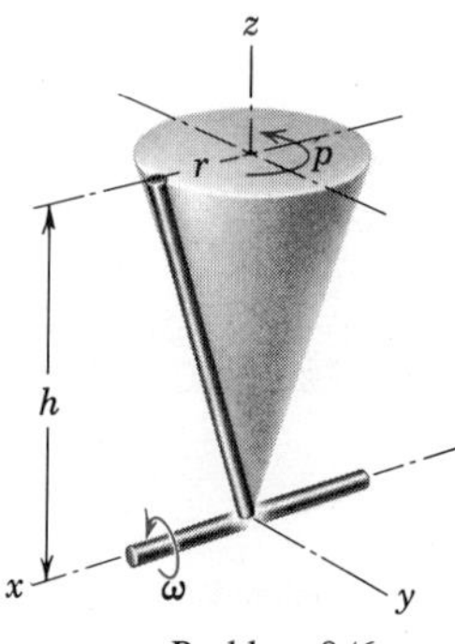

Problem 8/6

8/6 The slender bar of mass m and length $l = \sqrt{h^2 + r^2}$ is attached as an element of the cone which revolves about the z-axis with an angular rate p. Simultaneously the entire cone revolves about the x-axis with an angular rate ω. Determine the x-component of the angular momentum of the bar at the instant the bar crosses the x-z plane as shown.

$$\textit{Ans. } H_x = \frac{mh}{3}(h\omega - rp)$$

8/7 Determine the z-component of angular momentum for the bar of Prob. 8/6 for the position illustrated.

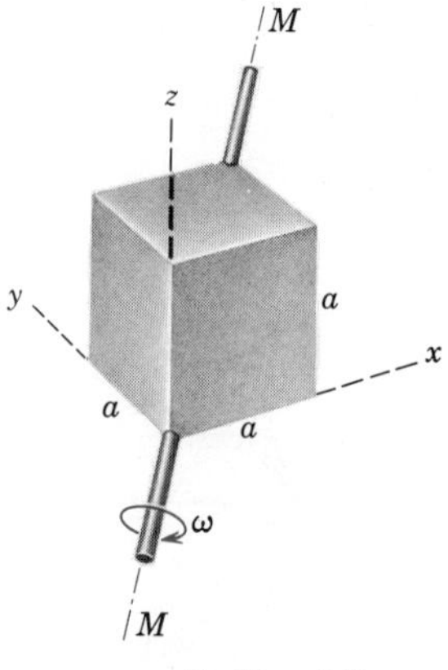

Problem 8/8

8/8 The solid cube of mass m and side a revolves about an axis M-M through a diagonal with an angular velocity ω. Write the expression for the angular momentum $\mathbf{H}$ of the cube with respect to the axes indicated.

$$\textit{Ans. } \mathbf{H} = \frac{ma^2\omega}{6\sqrt{3}}(\mathbf{i} + \mathbf{j} + \mathbf{k})$$

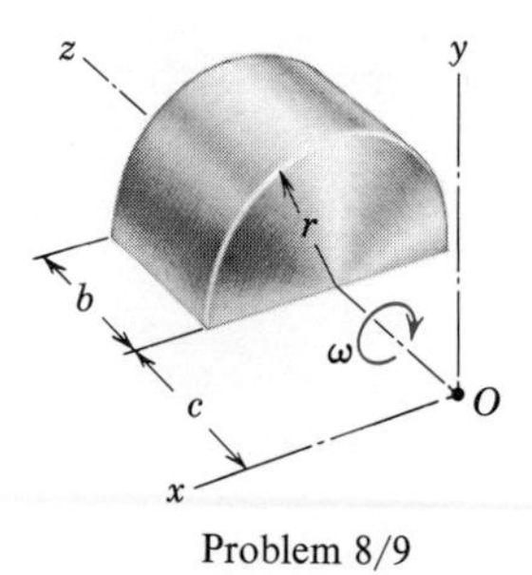

Problem 8/9

8/9 The half-circular cylinder revolves about the z-axis with an angular velocity ω as shown. Evaluate the angular momentum $\mathbf{H}$ with respect to the x-y-z axes. The mass of the cylinder is m.

8/10 The space capsule shown has a mass m with mass center G. Its radius of gyration about its axis of rotational symmetry z is k and that about either the x- or y-axis is k'. In space the capsule spins within its x-y-z reference frame at the rate $p = \dot{\phi}$. Simultaneously a point C on the z-axis moves in a circle with a frequency f about the z_0-axis, which has a constant direction in space. Determine the components of angular momentum of the capsule relative to the axes selected. Note that the x-axis always lies in the z-z_0 plane and that the y-axis is therefore normal to z_0.
Ans. $\mathbf{H} = 2\pi mf(-\mathbf{i}k'^2 \sin\theta + \mathbf{k}k^2 \cos\theta) + mk^2p\mathbf{k}$

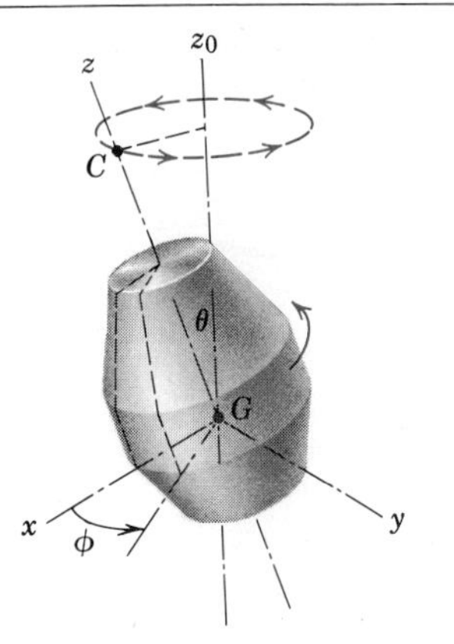

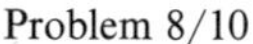
Problem 8/10

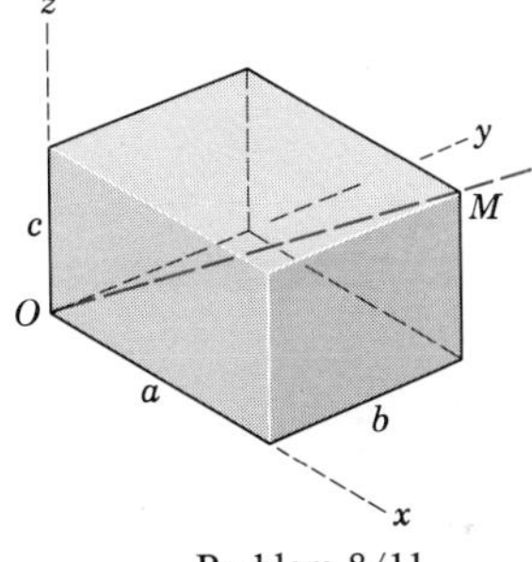

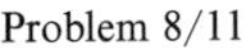
Problem 8/11

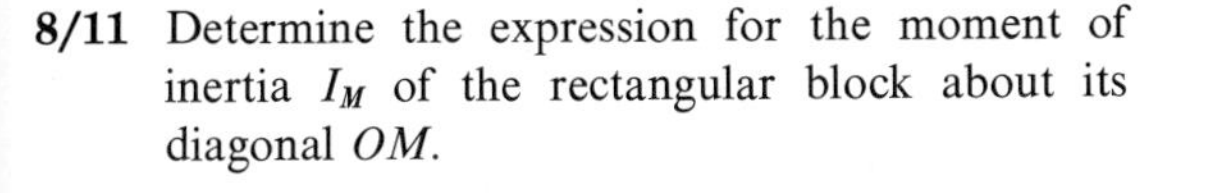
8/11 Determine the expression for the moment of inertia I_M of the rectangular block about its diagonal OM.

8/12 The solid circular cylinder of mass m, radius r, and length b revolves at an angular rate p about its geometric axis. Simultaneously the bracket and attached shaft axis revolve at the rate ω about the x-axis. Write the expressions for the x-, y-, z-components of angular momentum $\mathbf{H}$ of the cylinder about O.

$$Ans.\quad H_x = \left(\frac{b^2}{3} + \frac{r^2}{4} + h^2\right)m\omega$$

$$H_y = \tfrac{1}{2}mr^2p$$

$$H_z = 0$$

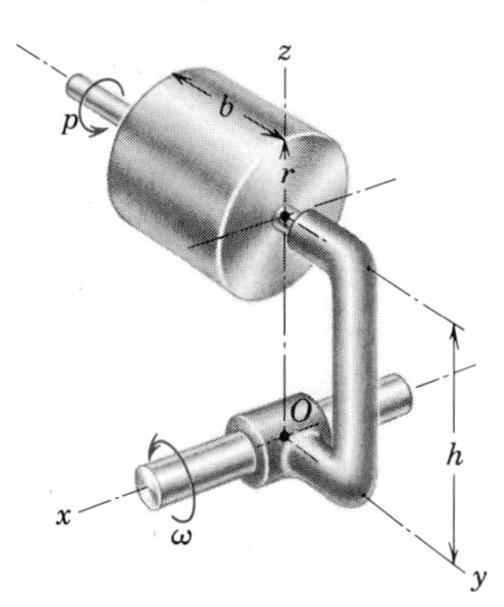

Problem 8/12

8/13 The solid circular disk of mass m and radius r rolls in a circle of radius b on the x-y plane without slipping. If the center line OC of the axle of the wheel rotates about the z-axis with an angular velocity ω, determine the expression for the angular momentum of the disk with respect to the fixed point O.

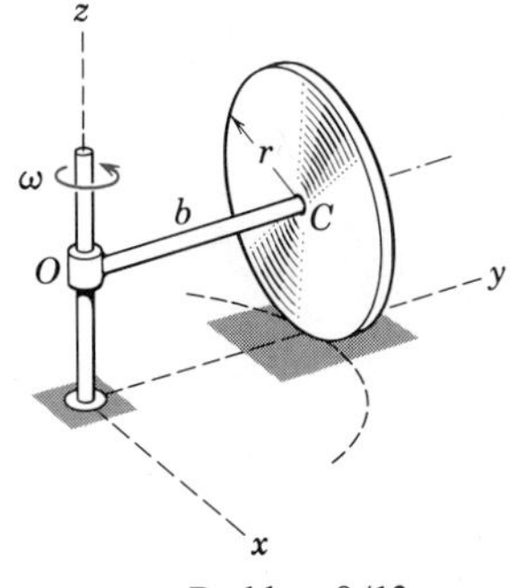

Problem 8/13

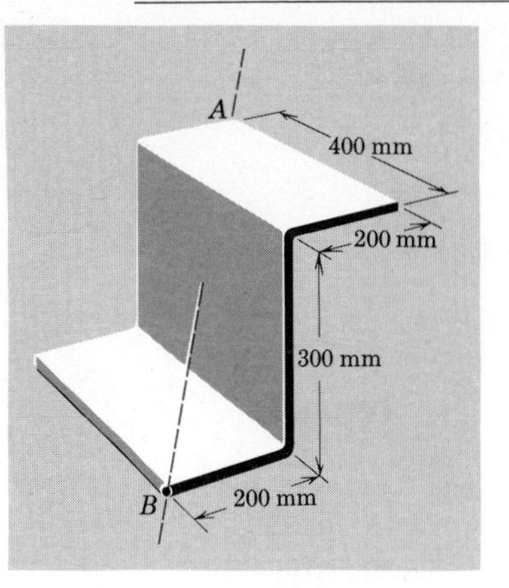

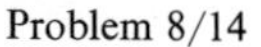

Problem 8/14

8/14 The steel plate with two right-angle bends has a thickness of approximately 15 mm and a total mass of exactly 33 kg. Calculate its moment of inertia about the diagonal axis through the corners A and B. *Ans.* $I = 0.670\ \text{kg}\cdot\text{m}^2$

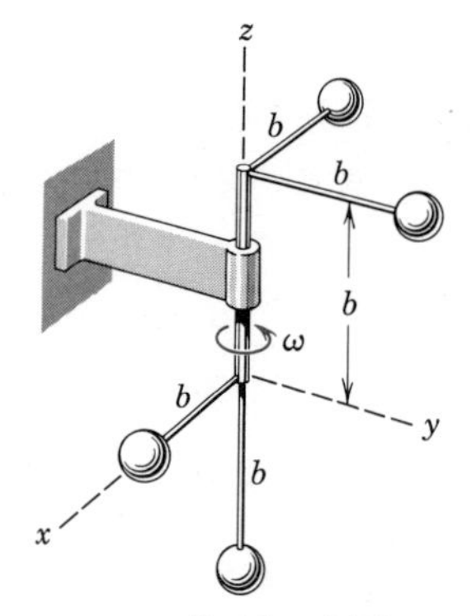

Problem 8/15

8/15 Each of the metal spheres has a mass m and a diameter which is small compared with the dimension b. Compute the values of the principal moments of inertia and determine the direction cosines for each of the principal axes of inertia. Draw the ellipsoid of inertia. If the assembly has an angular velocity ω about the z-axis, write the expression for the angular momentum $\mathbf{H}$ of the assembly.

Ans. $I_1 = 5.532mb^2$; $l_1 = -0.293$, $m_1 = 0.844$, $n_1 = -0.449$

$I_2 = 4.347mb^2$; $l_2 = 0.844$, $m_2 = 0.449$, $n_2 = 0.293$

$I_3 = 2.121mb^2$; $l_3 = -0.449$, $m_3 = 0.293$, $n_3 = 0.844$

$\mathbf{H} = mb^2\omega(\mathbf{i} - \mathbf{j} + 3\mathbf{k})$

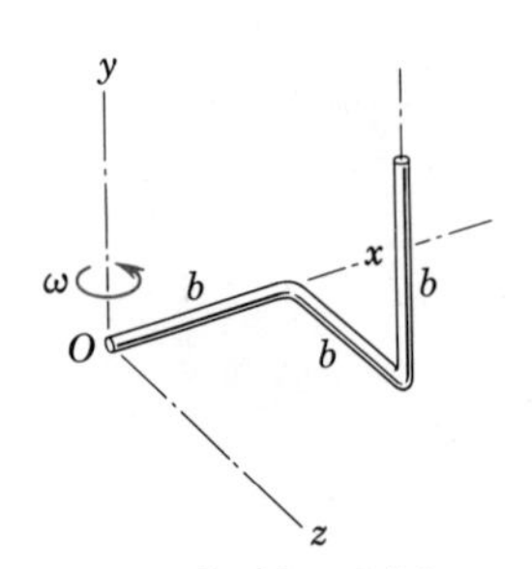

Problem 8/16

8/16 The rod with the right-angle bends has a mass m and rotates about the y-axis at the rate ω. Write the expressions for the moments and products of inertia for the axes selected and determine the principal moments of inertia.

Ans. $I_1 = 0.145mb^2$ minimum

$I_2 = 1.299mb^2$ maximum

$I_3 = 1.223mb^2$ intermediate

8/17 Use the results cited for the principal moments of inertia of the bent rod in Prob. 8/16 and specify the direction cosines of the axis of the intermediate value of principal moment of inertia. Construct the ellipsoid of inertia.

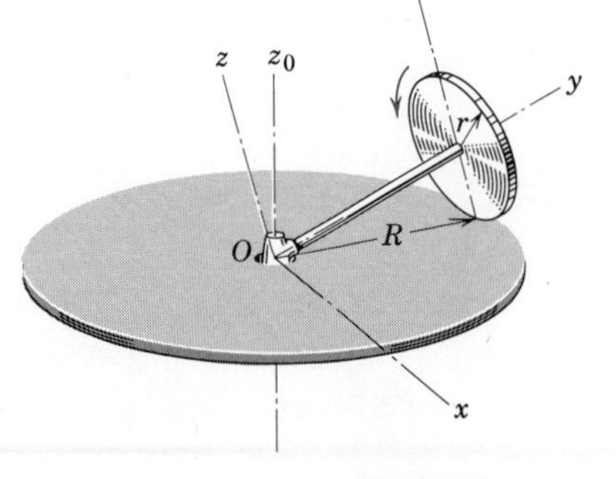

Problem 8/18

◀ **8/18** The uniform circular disk of mass m and radius r is mounted on its shaft which is pivoted at O about the vertical z_0-axis. If the disk rolls without slipping and makes one complete trip around the large fixed disk in time τ, write the expression for the angular momentum of the disk with respect to the x-y-z axes through O. (*Hint:* The spin relative to x-y-z is $R\omega/r$ where $\omega = 2\pi/\tau$.)

$$Ans.\ \mathbf{H}_O = \frac{2\pi mr^2}{\tau}\left[\frac{1}{2}\left(\frac{r}{R} - \frac{R}{r}\right)\mathbf{j} + \left(\frac{R^2}{r^2} - \frac{3}{4}\right)\sqrt{1 - \frac{r^2}{R^2}}\,\mathbf{k}\right]$$

◀ **8/19** In a test of the solar panels for a spacecraft the model shown is rotated about the vertical axis at the angular rate ω. If the mass per unit area of panel is ρ, write the expression for the angular momentum $\mathbf{H}_O$ of the assembly about the axes shown. Determine the maximum, minimum, and intermediate moments of inertia.

Ans.
$I_1 = 4\rho bc[(b^2 + c^2)/3 + a^2 + ac]$ maximum
$I_2 = 4\rho bc(c^2/3 + a^2 + ac)$ intermediate
$I_3 = \dfrac{4}{3}\rho b^3 c$ minimum

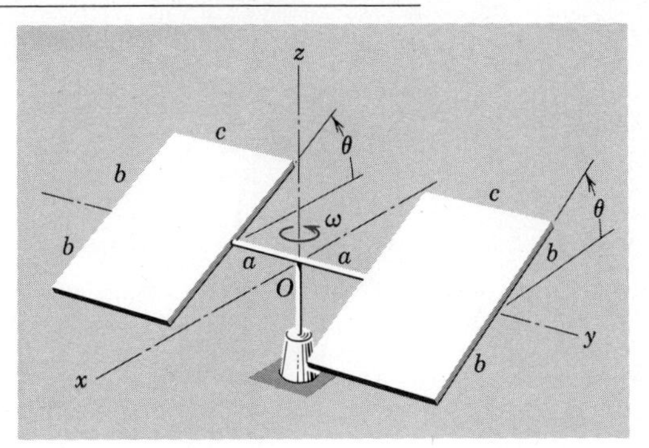

Problem 8/19

◀ **8/20** The thin quarter-circular plate of radius r and mass m is mounted on a shaft which rotates at the rate ω about the z-axis. The plane of the plate is tilted an angle ϕ with respect to the x-y plane normal to the shaft axis. Determine the expression for the angular momentum $\mathbf{H}$ of the plate with respect to the x-y-z axes.

$$Ans.\ \mathbf{H} = \frac{mr^2\omega}{2}\left(\frac{\sin 2\phi}{4}\mathbf{i} + \frac{\sin\phi}{\pi}\mathbf{j} + \frac{1 + \cos^2\phi}{2}\mathbf{k}\right)$$

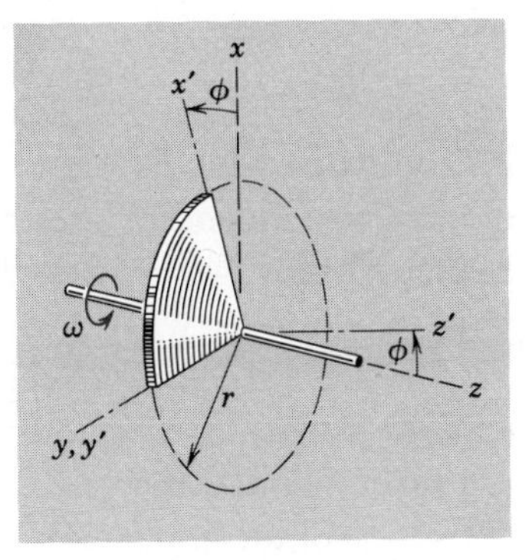

Problem 8/20

◀ **8/21** The half-cylindrical shell of radius r, length $2b$, and mass m revolves about the z-axis with an angular velocity ω as shown. Determine the expression for the angular momentum $\mathbf{H}$ of the shell with respect to the x-y-z axes.

$$Ans.\ \mathbf{H} = m\omega\left(\frac{2r^2}{\pi}\mathbf{j} + \left[\frac{r^2}{2} + \frac{b^2}{3}\right]\mathbf{k}\right)$$

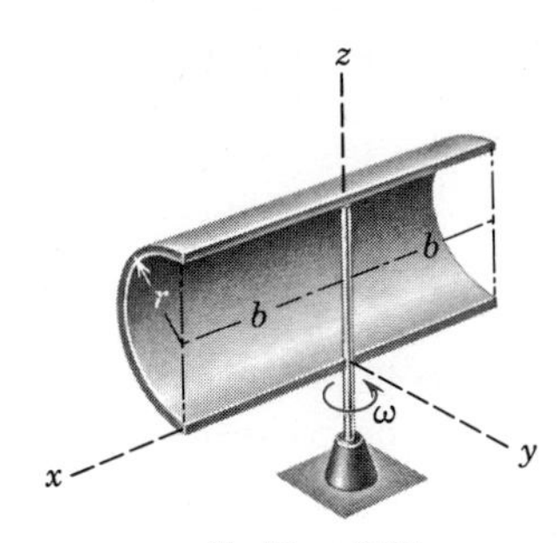

Problem 8/21

8/22 The half-cylindrical shell of mass m, radius r, and length b revolves about one edge along the z-axis with a constant rate ω as shown. Determine the angular momentum $\mathbf{H}$ of the shell with respect to the x-y-z axes.

$$Ans.\ \mathbf{H} = mr\omega\left(\frac{b}{2}\mathbf{i} + \frac{b}{\pi}\mathbf{j} + 2r\mathbf{k}\right)$$

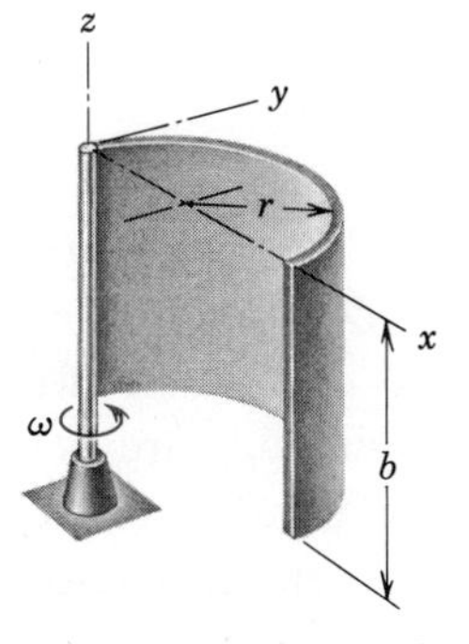

Problem 8/22

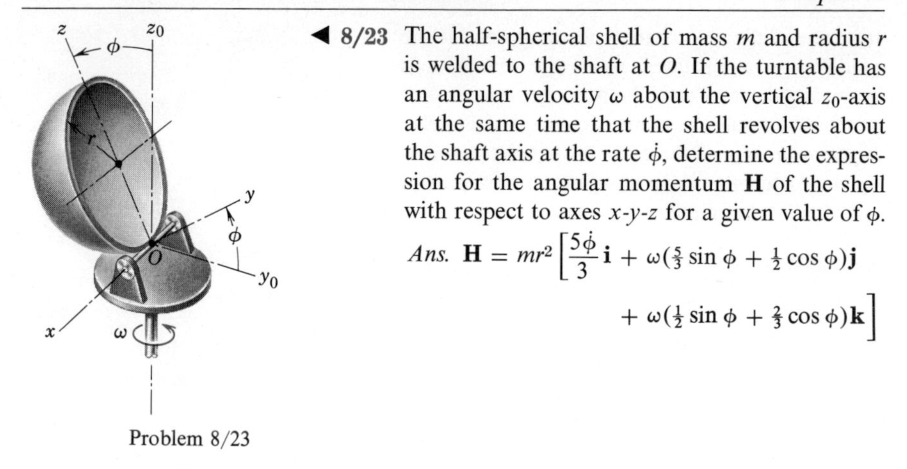

◀ **8/23** The half-spherical shell of mass m and radius r is welded to the shaft at O. If the turntable has an angular velocity ω about the vertical z_0-axis at the same time that the shell revolves about the shaft axis at the rate $\dot{\phi}$, determine the expression for the angular momentum $\mathbf{H}$ of the shell with respect to axes x-y-z for a given value of ϕ.

Ans. $\mathbf{H} = mr^2\left[\frac{5\dot{\phi}}{3}\mathbf{i} + \omega(\frac{5}{3}\sin\phi + \frac{1}{2}\cos\phi)\mathbf{j} + \omega(\frac{1}{2}\sin\phi + \frac{2}{3}\cos\phi)\mathbf{k}\right]$

Problem 8/23

---

42 Momentum and Energy Equations of Motion. With the description of angular momentum and the inertial properties of a rigid body established in the previous two articles, the general moment equations and energy equations may now be written for rigid-body motion.

The general moment relations for any system of constant mass, Eqs. 97 and 98, are expressed here by the single equation $\Sigma\mathbf{M} = \dot{\mathbf{H}}$ where the terms are taken about either a fixed point O or about the mass center G. In the derivation of the moment principle the derivative of $\mathbf{H}$ was taken with respect to an absolute coordinate system. When $\mathbf{H}$ is expressed in terms of components measured relative to a moving coordinate system x-y-z which has an angular velocity $\mathbf{\Omega}$, then by Eq. 142 the moment relation becomes

$$\Sigma\mathbf{M} = \left(\frac{d\mathbf{H}}{dt}\right)_{xyz} + \mathbf{\Omega} \times \mathbf{H}$$
$$= (\mathbf{i}\dot{H}_x + \mathbf{j}\dot{H}_y + \mathbf{k}\dot{H}_z) + \mathbf{\Omega} \times \mathbf{H}$$

The terms in parentheses represent that part of $\dot{\mathbf{H}}$ due to the change in magnitude of the components of $\mathbf{H}$, and the cross-product term represents that part due to the changes in direction of the components of $\mathbf{H}$. Expansion of the cross product and rearrangement of terms give

$$\begin{aligned} \Sigma\mathbf{M} = \ & \mathbf{i}(\dot{H}_x - H_y\Omega_z + H_z\Omega_y) \\ & +\mathbf{j}(\dot{H}_y - H_z\Omega_x + H_x\Omega_z) \\ & +\mathbf{k}(\dot{H}_z - H_x\Omega_y + H_y\Omega_x) \end{aligned} \qquad \textbf{(156)}$$

Equation 156 is the most general form of the moment equation about a fixed point O or about the mass center G. The Ω's are the angular velocity components of rotation of the reference axes, and the H-components in the case of a rigid body are as defined in Eq. 148 where the ω's are the components of the angular velocity of the body. Equation 156 is now applied to a rigid body where the coordinate axes are *attached* to the body. Under these conditions when expressed in the x-y-z coordinates, the *moments and products of inertia are invariant with time,* and $\mathbf{\Omega} = \boldsymbol{\omega}$. Thus for axes attached to the body, the three scalar components of Eq. 156 become

$$\begin{aligned}\Sigma M_x &= \dot{H}_x - H_y\omega_z + H_z\omega_y \\ \Sigma M_y &= \dot{H}_y - H_z\omega_x + H_x\omega_z \\ \Sigma M_z &= \dot{H}_z - H_x\omega_y + H_y\omega_x\end{aligned} \tag{157}$$

Equations 157 are the general moment equations for rigid-body motion with reference axes attached to the body, and they hold with respect to axes through a fixed point O or through the mass center G. If the mass center is selected, either the relative or absolute angular momentum may be used.

From the discussion following Eqs. 99 and 100 it may be observed that Eq. 156 and, hence, Eq. 157 are also valid about special points other than a fixed point or the mass center. For these special conditions the proper evaluation of the appropriate $\dot{\mathbf{H}}$ can be the cause of much difficulty, particularly when the special condition is fulfilled only at an instant of time. Consequently, $\Sigma\mathbf{M} = \dot{\mathbf{H}}$ will be applied only about the mass center or about a fixed point in the work which follows.

In Art. 41 it was shown that, in general, for any origin fixed to a rigid body, there are three principal axes of inertia with respect to which the products of inertia vanish. If the reference axes coincide with the principal axes of inertia with origin at the mass center G or at a point O fixed to the body and fixed in space, the factors I_{xy}, I_{yz}, I_{xz} will be zero, and Eqs. 157 become

$$\begin{aligned}\Sigma M_x &= I_{xx}\dot{\omega}_x - (I_{yy} - I_{zz})\omega_y\omega_z \\ \Sigma M_y &= I_{yy}\dot{\omega}_y - (I_{zz} - I_{xx})\omega_z\omega_x \\ \Sigma M_z &= I_{zz}\dot{\omega}_z - (I_{xx} - I_{yy})\omega_x\omega_y\end{aligned} \tag{158}$$

These relations are known as *Euler's equations** and are among the most useful of the motion equations of dynamics.

Euler's equations are frequently used to describe the motion of a rigid body which spins about an axis of symmetry. In this event it is convenient to permit the body to rotate about its axis of symmetry relative to the reference system. Figure 75 illustrates such a body which has rotational symmetry about the z-axis. The hypothetical plane rotates with the body, so that the angular velocity or spin of the body relative to x-y-z is measured by $\dot{\phi}$. Clearly the values of the moment of inertia integrals about any of the coordinate axes are not functions of ϕ or the time t. If the angular velocity of the coordinate axes is designated by $\mathbf{\Omega}$, then the momentum components from Eq. 148 are

$$\begin{aligned}H_x &= I_{xx}\omega_x = I_{xx}\Omega_x \\ H_y &= I_{yy}\omega_y = I_{yy}\Omega_y \\ H_z &= I_{zz}\omega_z = I_{zz}(\Omega_z + \dot{\phi})\end{aligned}$$

where ω_z is the total angular velocity of the body about the z-axis. With

* Named after Leonhard Euler (1707–1783), a Swiss mathematician.

these substitutions the components of Eq. 156 become

$$\begin{aligned} \Sigma M_x &= I_{xx}\dot{\Omega}_x - (I_{yy} - I_{zz})\Omega_y\Omega_z + I_{zz}\dot{\phi}\Omega_y \\ \Sigma M_y &= I_{yy}\dot{\Omega}_y - (I_{zz} - I_{xx})\Omega_z\Omega_x - I_{zz}\dot{\phi}\Omega_x \\ \Sigma M_z &= I_{zz}\dot{\Omega}_z + I_{zz}\ddot{\phi} \end{aligned} \tag{159}$$

As with Eqs. 156, 157, and 158, these *modified Euler's equations* apply to an origin through a fixed point O of rotation or through the mass center G. If a coordinate axis other than z is chosen as the spin axis, the equations must be rewritten.

Consider now the energy equation for rigid-body motion. This equation has been expressed by Eq. 94 and its application follows the procedure outlined previously for particle motion and for plane motion. The expression for the kinetic energy T of a rigid body with general motion, however, must be developed.

The expression for the kinetic energy T of a general mass system was derived in Art. 24 of Chapter 4 and is given by Eq. 93 which shows clearly the separate contributions to T due to the translational motion of the system and due to the motion relative to the mass center. The translational term is

$$\tfrac{1}{2}m\bar{v}^2 = \tfrac{1}{2}m\dot{\bar{\mathbf{r}}} \cdot \dot{\bar{\mathbf{r}}} = \tfrac{1}{2}\bar{\mathbf{v}} \cdot \mathbf{G}$$

where $\dot{\bar{\mathbf{r}}}$ is the velocity of the mass center and $\mathbf{G}$ is the linear momentum of the body.

The relative term of Eq. 93 is $\Sigma\frac{1}{2}m_i|\dot{\boldsymbol{\rho}}_i|^2$ where $\boldsymbol{\rho}_i$ is the position vector which locates the representative particle of mass m_i with respect to the mass center, Fig. 41. For a rigid body with an angular velocity $\boldsymbol{\omega}$, the velocity of m_i relative to the mass center is $\dot{\boldsymbol{\rho}}_i = \boldsymbol{\omega} \times \boldsymbol{\rho}_i$, and the kinetic energy term for the motion relative to the mass center becomes

$$\Sigma\tfrac{1}{2}m_i\dot{\boldsymbol{\rho}}_i \cdot \dot{\boldsymbol{\rho}}_i = \Sigma\tfrac{1}{2}m_i(\boldsymbol{\omega} \times \boldsymbol{\rho}_i) \cdot (\boldsymbol{\omega} \times \boldsymbol{\rho}_i)$$

This expression may be reduced by recalling that the dot and the cross may be interchanged for the triple scalar product, i.e., $\mathbf{P} \times \mathbf{Q} \cdot \mathbf{R} = \mathbf{P} \cdot \mathbf{Q} \times \mathbf{R}$.

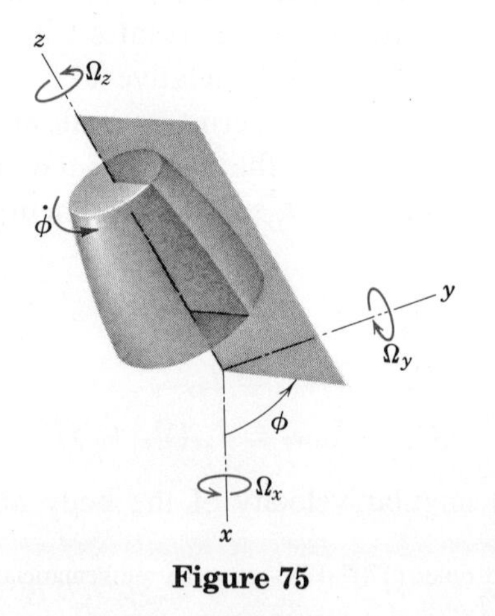

Figure 75

This transformation gives

$$\Sigma \tfrac{1}{2} m_i (\boldsymbol{\omega} \times \boldsymbol{\rho}_i) \cdot (\boldsymbol{\omega} \times \boldsymbol{\rho}_i) = \tfrac{1}{2}\boldsymbol{\omega} \cdot \Sigma(m_i \boldsymbol{\rho}_i \times [\boldsymbol{\omega} \times \boldsymbol{\rho}_i]) = \tfrac{1}{2}\boldsymbol{\omega} \cdot \overline{\mathbf{H}}$$

where $\overline{\mathbf{H}}$ is the integral expressed by Eq. 144 which represents both the relative and the absolute angular momentum about the mass center. Thus the general expression for the kinetic energy of a rigid body moving with mass-center velocity $\overline{\mathbf{v}}$ and angular velocity $\boldsymbol{\omega}$ is

$$T = \tfrac{1}{2}\overline{\mathbf{v}} \cdot \mathbf{G} + \tfrac{1}{2}\boldsymbol{\omega} \cdot \overline{\mathbf{H}} \tag{160}$$

Expansion of this vector equation by substitution of the expression for $\overline{\mathbf{H}}$ written from Eq. 147 yields

$$T = \tfrac{1}{2}m\bar{v}^2 + \tfrac{1}{2}(\bar{I}_{xx}\omega_x^{\,2} + \bar{I}_{yy}\omega_y^{\,2} + \bar{I}_{zz}\omega_z^{\,2}) - (\bar{I}_{xy}\omega_x\omega_y + \bar{I}_{xz}\omega_x\omega_z + \bar{I}_{yz}\omega_y\omega_z) \tag{161}$$

If the axes coincide with the principal axes of inertia, the kinetic energy is merely

$$T = \tfrac{1}{2}m\bar{v}^2 + \tfrac{1}{2}(\bar{I}_{xx}\omega_x^{\,2} + \bar{I}_{yy}\omega_y^{\,2} + \bar{I}_{zz}\omega_z^{\,2}) \tag{162}$$

When a rigid body is pivoted about a fixed point O or when there is a point O in the body which momentarily has zero velocity, the kinetic energy is $T = \Sigma\tfrac{1}{2}m_i\dot{\mathbf{r}}_i \cdot \dot{\mathbf{r}}_i$. This expression reduces to

$$T = \tfrac{1}{2}\boldsymbol{\omega} \cdot \mathbf{H}_O \tag{163}$$

where $\mathbf{H}_O$ is the angular momentum about O, as may be seen by replacing $\boldsymbol{\rho}_i$ in the previous derivation by $\mathbf{r}_i$, the position vector from O. Equations 160 and 163 are the three-dimensional counterparts of Eqs. 122 and 123 for plane motion.

The resultant of all external forces acting on a rigid body may be replaced by the resultant force $\Sigma\mathbf{F}$ acting through the mass center and a resultant couple $\Sigma\overline{\mathbf{M}}$ acting about the mass center. Work is done by the resultant force and the resultant couple at the respective rates $\Sigma\mathbf{F} \cdot \overline{\mathbf{v}}$ and $\Sigma\overline{\mathbf{M}} \cdot \boldsymbol{\omega}$ where $\overline{\mathbf{v}}$ is the linear velocity of the mass center and $\boldsymbol{\omega}$ is the angular velocity of the body. Integration over the time from condition 1 to condition 2 gives the total work done during the time interval. Thus

$$\begin{aligned} \int_{t_1}^{t_2} \Sigma\mathbf{F} \cdot \overline{\mathbf{v}}\, dt &= \tfrac{1}{2}\overline{\mathbf{v}} \cdot \mathbf{G}\Big]_1^2 \\ \int_{t_1}^{t_2} \Sigma\overline{\mathbf{M}} \cdot \boldsymbol{\omega}\, dt &= \tfrac{1}{2}\boldsymbol{\omega} \cdot \overline{\mathbf{H}}\Big]_1^2 \end{aligned} \tag{164}$$

These equations express the change in translational kinetic energy and the change in rotational kinetic energy, respectively, for the interval during which $\Sigma\mathbf{F}$ or $\Sigma\overline{\mathbf{M}}$ acts.

The relationships developed in this article will now be applied to three classes of space motion of rigid bodies in the articles which follow. As with all force-acceleration and force-momentum problems, a free-body diagram should be drawn of the body or system isolated to ensure the correct evaluation of the force and moment summations.

43 General Plane Motion. When all particles of a rigid body move in planes which are parallel to a fixed plane, the body is said to undergo general plane motion. Every line in such a body which is normal to the fixed plane remains parallel to itself at all times. In Fig. 76 a body with general plane motion is represented. If the x-y axes through G are chosen parallel to the fixed reference plane P of motion, then the components of the angular velocity of both the body and attached axes are $\omega_x = \omega_y = 0$, $\omega_z \neq 0$. For this case the angular momentum components from Eq. 148 become

$$H_x = -I_{xz}\omega_z \qquad H_y = -I_{yz}\omega_z \qquad H_z = I_{zz}\omega_z$$

and the moment relations of Eqs. 157 reduce to

$$\begin{aligned} \Sigma M_x &= -I_{xz}\dot{\omega}_z + I_{yz}\omega_z^2 \\ \Sigma M_y &= -I_{yz}\dot{\omega}_z - I_{xz}\omega_z^2 \\ \Sigma M_z &= I_{zz}\dot{\omega}_z \end{aligned} \qquad (165)$$

It is seen that the third moment equation is equivalent to the first of Eqs. 112 where the z-axis passes through the mass center or to the third of Eqs. 116 if the z-axis passes through a fixed point. Also, an equilibrium of moments $\Sigma M_x = 0$ and $\Sigma M_y = 0$ results if $I_{xz} = I_{yz} = 0$.

Equations 165 hold for an origin of coordinates at the mass center, as shown in Fig. 76, or for any origin on a fixed axis of rotation. There are other special choices of coordinates for which Eqs. 165 are valid, and these axes may be selected with the aid of Eq. 100*a*.

The three independent force equations of motion which also apply to general plane motion are clearly

$$\Sigma F_x = m\bar{a}_x \qquad \Sigma F_y = m\bar{a}_y \qquad \Sigma F_z = 0$$

Equations 165 find special use in describing the effect of dynamic unbalance in rotating shafts and in rolling assemblies. For such use it is frequently simpler to form the equivalence between the external forces and the $m\bar{a}$ and $\bar{I}\alpha$ resultants directly with the free-body diagram in place of using Eqs. 165.

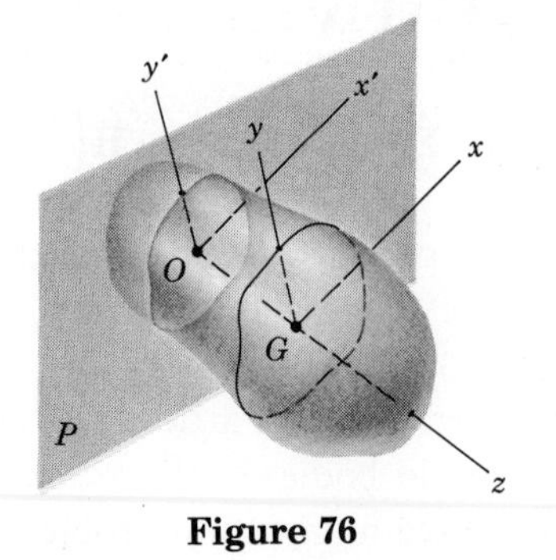

Figure 76

Sample Problem

8/24 The two circular disks each of mass m_1 are connected by the curved bar bent into quarter-circular arcs and welded to the disks. The bar has a mass m_2. The total mass of the assembly is $m = 2m_1 + m_2$. If the disks roll without slipping on a horizontal plane with a constant velocity v of the disk centers, determine the value of the friction force under each disk at the instant represented when the plane of the curved bar is horizontal.

Solution. The motion is identified as general plane motion since the planes of motion of all parts of the system are parallel. The free-body diagram shows the normal forces and friction forces at A and B and the total weight mg acting through the mass center G. The center O of the wheel has zero acceleration, and, by reference to Eq. 100*a*, it may be concluded that O is a valid origin of coordinates for the use of Eqs. 165. Point G would be an equally valid choice.

A moment equation about the y-axis eliminates all forces except F_A and requires the computation of the product of inertia I_{xz}. Thus

$$\left[I_{xz} = \int xz\, dm\right] \qquad I_{xz} = \int_0^{\pi/2} (r \sin\theta)(r\cos\theta)\rho r\, d\theta + \int_0^{\pi/2} (-r\sin\theta)(2r - r\cos\theta)\rho r\, d\theta$$

where ρ is the mass per unit length of rod. Integration gives

$$I_{xz} = -\rho r^3 = -\frac{m_2 r^2}{\pi}$$

The second of Eqs. 165 gives, with $\omega_z = v/r$ and $\dot{\omega}_z = 0$,

$$[\Sigma M_y = -I_{xz}\omega_z^2] \qquad F_A(2r) = -\left(-\frac{m_2 r^2}{\pi}\right)\frac{v^2}{r^2} \qquad F_A = \frac{m_2 v^2}{2\pi r} \qquad \textit{Ans.}$$

Since the acceleration of G is zero,

$$[\Sigma F_x = 0] \qquad F_B - F_A = 0 \qquad F_B = \frac{m_2 v^2}{2\pi r} \qquad \textit{Ans.}$$

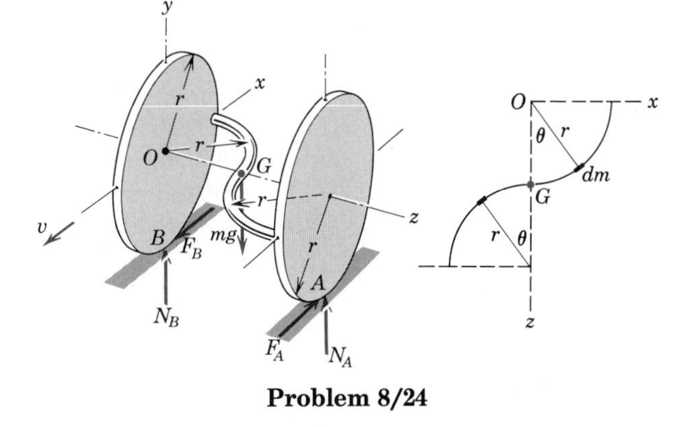

Problem 8/24

Problems

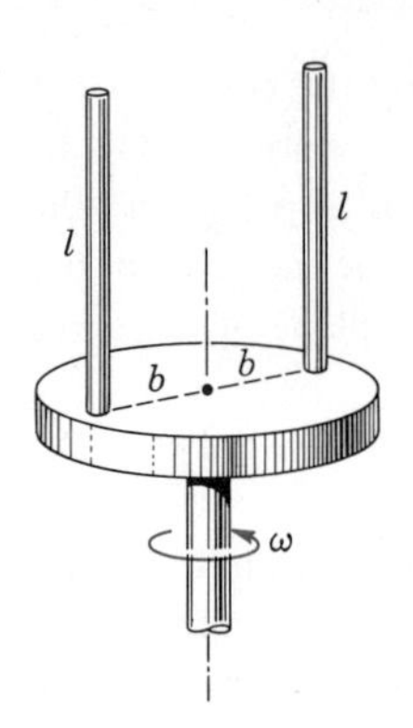

Problem 8/25

8/25 Each of the two rods of mass m is welded to the face of the disk which rotates about the vertical axis with a constant angular velocity ω. Determine the bending moment M acting on each rod at its base.

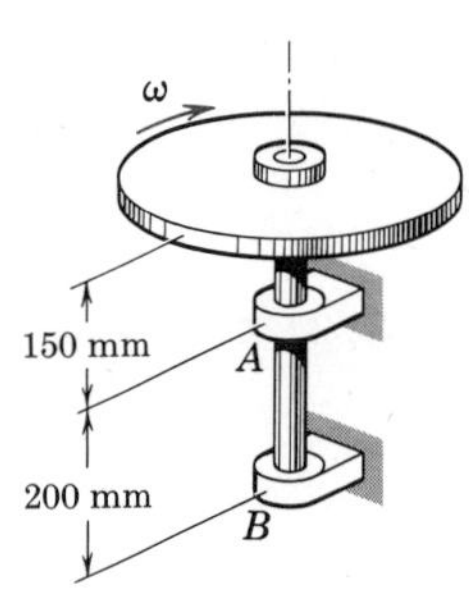

Problem 8/26

8/26 The 6-kg circular disk and attached shaft rotate at a constant speed $\omega = 10\,000$ rev/min. If the center of mass of the disk is 0.05 mm off center, determine the magnitudes of the horizontal forces A and B supported by the bearings because of the rotational unbalance.

Ans. $A = 576$ N, $B = 247$ N

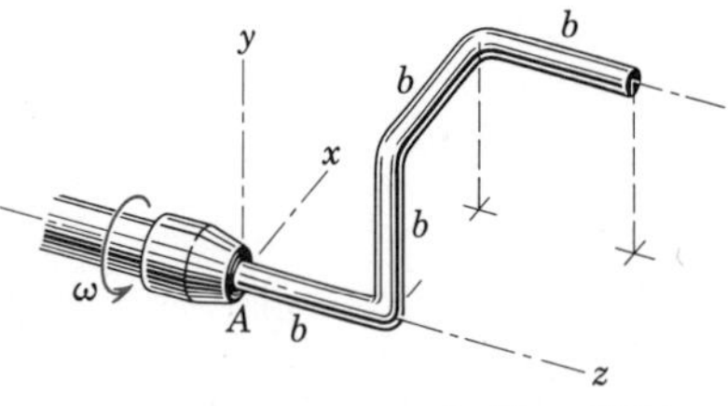

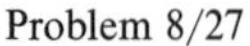

Problem 8/27

8/27 The irregular rod has a mass ρ per unit length and rotates about the shaft axis at the constant rate ω. Determine the expression for the bending moment M in the rod at A. Neglect the small moment caused by the weight of the rod.

Ans. $M = \sqrt{13}\rho b^3\omega^2$

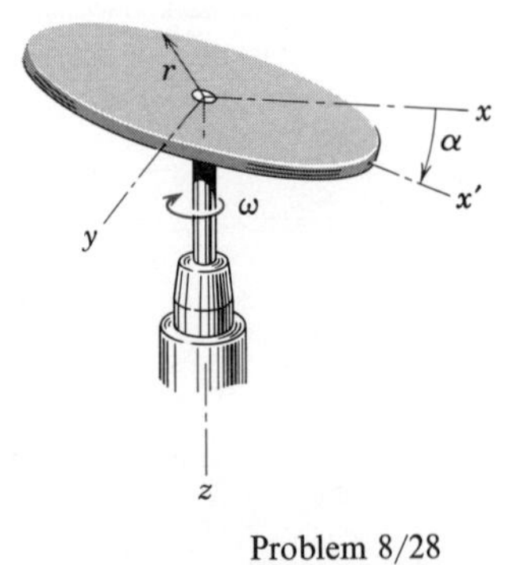

Problem 8/28

8/28 The circular disk of mass m and radius r is mounted on the vertical shaft with a small angle α between its plane and the plane of rotation of the shaft. Determine the expression for the bending moment M in the shaft due to the wobble of the disk at a shaft speed of ω rad/s.

8/29 The slender rod OA of mass m and length l is pivoted freely about a horizontal axis through O. The pivot at O and attached shaft rotate with a constant angular speed ω about the vertical z-axis. Write the expression for the angle θ assumed by the rod. What minimum value must ω attain before the rod will assume other than a vertical position?

$$Ans.\ \theta = \cos^{-1}\frac{3g}{2l\omega^2},\quad \omega_{\min} = \sqrt{\frac{3g}{2l}}$$

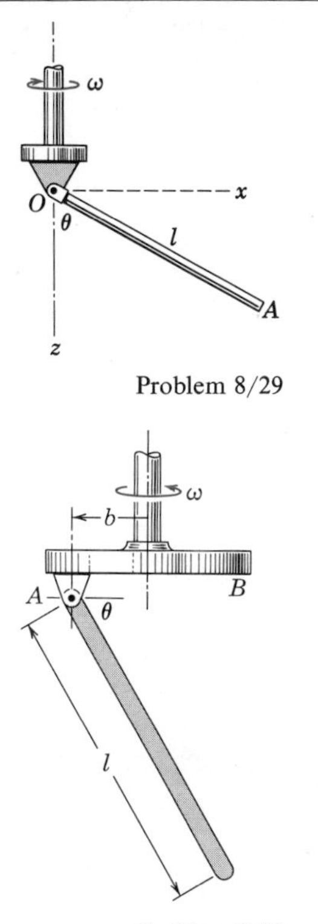

Problem 8/29

8/30 If the rod of Prob. 8/29 is welded to O at an angle θ with the vertical, determine the expression for the bending moment M in the rod at O due to the combined effect of the angular velocity ω of the shaft and the weight of the rod.

$$Ans.\ M_y = \frac{mgl}{2}\sin\theta\left(1 - \frac{2l}{3g}\omega^2\cos\theta\right)$$

8/31 The uniform slender rod of length l is freely hinged to the bracket A on the under side of the disk B. The disk rotates about a vertical axis with a constant angular velocity ω. Determine the value of ω which will permit the rod to maintain the position $\theta = 60°$ with $b = l/4$.

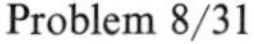

Problem 8/31

8/32 The partial ring has a mass ρ per unit length of rim and is welded to the shaft at A. Determine the bending moment M in the ring at A due to the effect of rotation of the ring for an angular velocity ω. *Ans.* $M = \frac{1}{2}\rho r^3\omega^2$

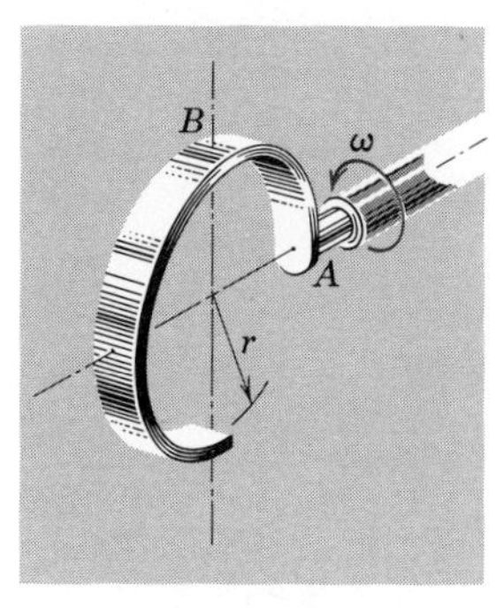

Problem 8/32

8/33 For the ring section of Prob. 8/32 determine the torsional moment T in the ring at B (moment about an axis tangent to the rim) as the ring starts from rest under the action of a torque M transmitted by the shaft to the ring at A. The mass per unit length of rim is ρ.

8/34 Determine the bending moment M at the tangency point A in the semi-circular rod of radius r and mass m as it rotates about the tangent axis with a constant angular velocity ω. Neglect the moment mgr produced by the weight compared with that caused by the rotation of the rod.

$$Ans.\ M = \frac{2mr^2\omega^2}{\pi}$$

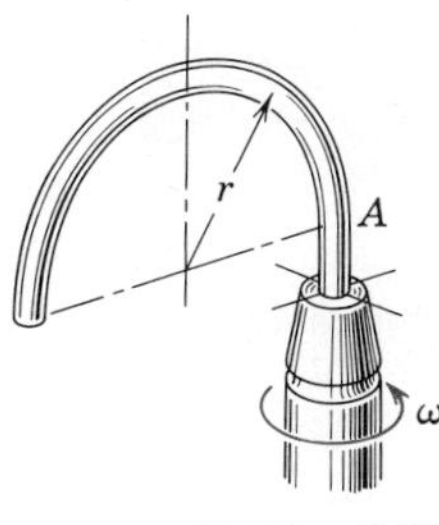

Problem 8/34

8/35 Work Sample Problem 8/24 using G as the origin of coordinates.

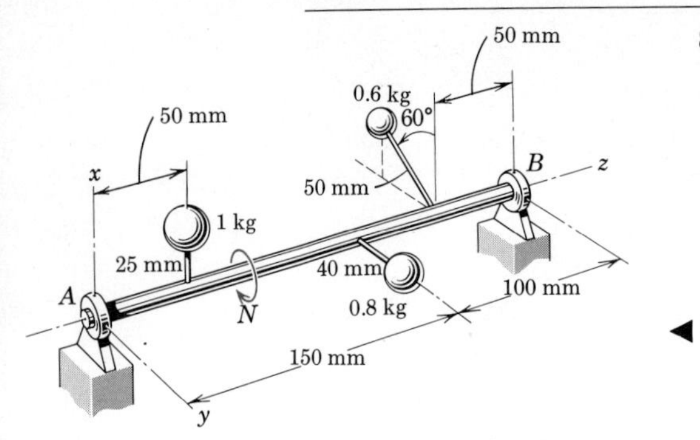

Problem 8/36

8/36 A shaft carries three unbalanced disk cams which are represented by the equivalent concentrated masses shown. Calculate the magnitudes of the reactions at the bearings A and B resulting from the dynamic unbalance at a shaft speed $N = 3600$ rev/min.

Ans. $A = 3440$ N, $B = 2430$ N

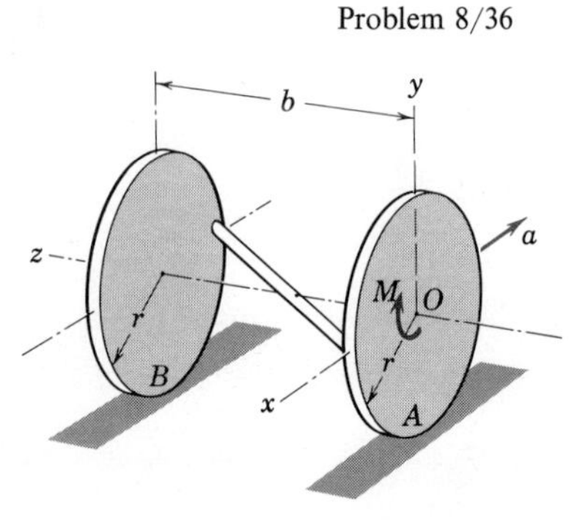

Problem 8/37

◀**8/37** Each of the two circular disks has a mass m and is welded to the end of the rigid rod of mass m_0 so that the disks have a common z-axis and are separated by a distance b. A couple M, applied to one of the disks with the assembly initially at rest, gives the centers of the disks an acceleration $\mathbf{a} = -a\mathbf{i}$. Friction is sufficient to prevent slipping. Derive expressions for the normal forces N_A and N_B exerted by the horizontal surface on the disks as they begin to roll. Express the results in terms of the acceleration a rather than the moment M.

$$Ans.\ N_A = mg + \frac{m_0 g}{2}\left(1 + \frac{a}{3g}\right)$$

$$N_B = mg + \frac{m_0 g}{2}\left(1 - \frac{a}{3g}\right)$$

◀**8/38** If a couple $\mathbf{M} = M\mathbf{k}$ is applied to one of the disks of Sample Problem 8/24 as they start from the rest position shown, determine the momentary values of the reactions N_A and N_B.

$$Ans.\ N_A = \frac{mg}{2} - \frac{Mm_2}{3\pi mr}$$

$$N_B = \frac{mg}{2} + \frac{Mm_2}{3\pi mr}$$

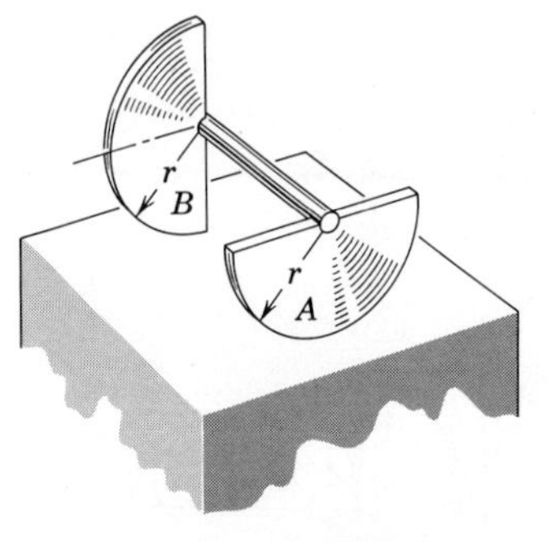

Problem 8/39

◀**8/39** The two half-circular disks A and B each of mass m are rigidly connected by a shaft of negligible mass. If they are released from rest on a horizontal surface from the position shown, determine the minimum coefficient of friction between the disks and the surface so that neither one will slip as the disks start to roll.

Ans. $f_B = 0.215$

◀**8/40** Determine the expressions for the normal and friction forces in Sample Problem 8/24 at an instant when the disks have revolved through an angle θ from the position shown.

$$Ans.\ N_A = \frac{mg}{2} + \frac{m_2 v^2}{2\pi r}\sin\theta$$

$$N_B = \frac{mg}{2} - \frac{m_2 v^2}{2\pi r}\sin\theta$$

$$F_A = F_B = \frac{m_2 v^2}{2\pi r}\cos\theta$$

44 Rotation about a Point. The rotation of a rigid body about either a fixed point O in the body or about its center of mass G may be described by the same basic equations. A spinning top, the rotor of a gyroscope, and a space capsule are examples of bodies whose motions can be described by the equations for rotation about a point. The moment equations for this general class of problems are fairly complex, and their complete solutions involve the use of elliptic integrals and somewhat lengthy computations. A large fraction of engineering problems where the motion is one of rotation about a point involves bodies of revolution about an axis of symmetry. This symmetry introduces simplifications and greatly facilitates solution of the equations.

Consider a body with complete axial symmetry, Fig. 77*a*, rotating about a fixed point O on its axis which is taken to be the z-direction. With O as origin, the x- and y-axes automatically become principal axes of inertia along with the z-axis. This same description may be used for the rotation of a similar symmetrical body about its center of mass G which is taken as the origin of coordinates, as shown with the gimbaled gyroscope rotor of Fig. 77*b*. Again the x- and y-axes are principal axes of inertia for point G. The same description may also be used to represent the rotation about its mass center of an axially symmetric body in space, such as the space capsule in Fig. 77*c*. In each case it is noted that, regardless of the rotation of the axes or of the rotation of the body relative to the axes (spin about the z-axis), the moments of inertia about the x- and y-axes remain constant with time. The principal moments of inertia will be designated $I_{zz} = I$ and $I_{xx} = I_{yy} = I_0$. The products of inertia are, of course, zero.

The modified Euler's equations, Eqs. 159, and Euler's angles, Fig. 68, are natural descriptions for the problem at hand since both permit rotation within a coordinate system which itself is rotating. It is highly desirable that a review of the kinematics of Euler's angles, Art. 37, as well as a review of the modified Euler's equations be conducted before proceeding further. Figure 78, which uses the same notation as Fig. 68, shows the case of rotation about the fixed point O along with the Euler's angles ψ, θ, ϕ. Plane A is the

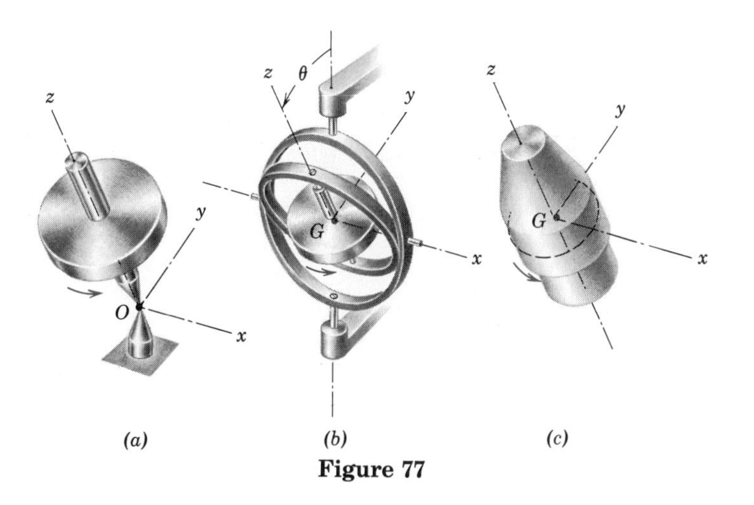

Figure 77

plane of the fixed axes X-Y, and plane B is normal to the spin axis z. The x-axis is the intersection of planes A and B, and ψ measures the angle of precession about the vertical. The rotation of the body relative to x-y-z is specified by the angle ϕ, and the inclination of the spin axis from the vertical is given by θ. The angular velocity of the x-y-z axes is $\mathbf{\Omega}$, given by Eq. 137, and the angular velocity of the body is $\boldsymbol{\omega}$, given by Eq. 138. The angular velocity components of $\mathbf{\Omega}$ and $\boldsymbol{\omega}$ as seen from the fixed X-Y-Z reference frame are

$$\begin{aligned} \Omega_x &= \dot{\theta} & \omega_x &= \dot{\theta} \\ \Omega_y &= \dot{\psi} \sin \theta & \omega_y &= \dot{\psi} \sin \theta \\ \Omega_z &= \dot{\psi} \cos \theta & \omega_z &= \dot{\psi} \cos \theta + \dot{\phi} \end{aligned}$$

It is important to note that the axes and the body have identical x- and y-components of angular velocity but that the z-components differ by the relative angular velocity $\dot{\phi}$.

Substitution of the angular velocity components and the new notation for the moments of inertia into Eqs. 159 yields upon combination

$$\begin{aligned} \Sigma M_x &= I_0(\ddot{\theta} - \dot{\psi}^2 \sin \theta \cos \theta) + I\dot{\psi}\omega_z \sin \theta \\ \Sigma M_y &= \frac{I_0}{\sin \theta} \frac{d}{dt} (\dot{\psi} \sin^2 \theta) - I\dot{\theta}\omega_z \\ \Sigma M_z &= I\dot{\omega}_z \end{aligned} \tag{166}$$

Equations 166 are the general equations of rotation of a symmetrical body about either a fixed point O or the mass center G. In a given problem the solution to the equations will depend on the moment sums applied to the body about the three coordinate axes. Three particular cases of rotation about a point are described in the sections which follow.

Case (a). Steady-state precession. The conditions under which the rotor of Fig. 78 spins at the constant rate $\dot{\phi}$ and precesses about the vertical with constant values of θ and $\dot{\psi}$ will now be investigated. Substitution of $\ddot{\phi} = 0$,

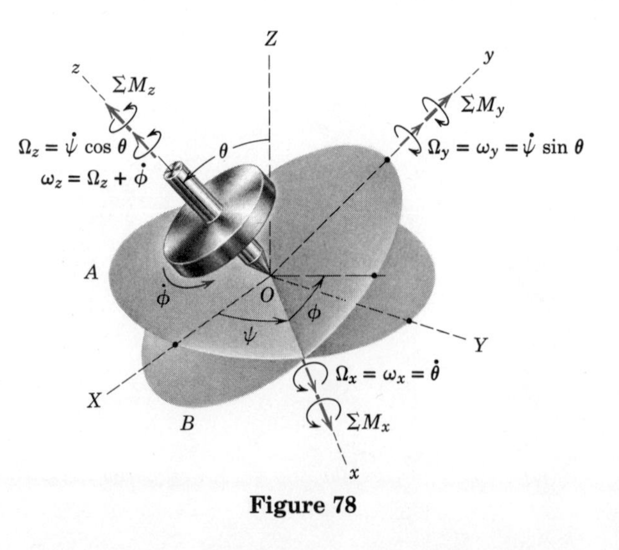

Figure 78

$\dot{\theta} = 0, \ddot{\psi} = 0$ into the moment relations, Eqs. 166, yields

$$\begin{aligned} \Sigma M_x &= \dot{\psi} \sin \theta [I\omega_z - I_0 \dot{\psi} \cos \theta] \\ \Sigma M_y &= 0 \\ \Sigma M_z &= 0 \end{aligned} \tag{167}$$

From these results it is seen that the required moment acting on the rotor about O (or about G) must be in the x-direction since the y- and z-components are zero. Furthermore, with the constant values of θ, $\dot{\psi}$, $\dot{\phi}$, the moment is seen to be of constant magnitude. It is also important to observe that the moment axis is perpendicular to the plane defined by the precession axis (Z-axis) and the spin axis (z-axis). For the rotor, Fig. 78, of mass m with center of mass G a distance $\bar{r}$ from O, the moment about O is a vector in the x-direction with the constant magnitude $\Sigma M_x = mg\bar{r} \sin \theta$.

By far the most common engineering examples of gyroscopic motion occur where precession takes place about an axis which is normal to the rotor axis. With $\theta = \pi/2$, and M standing for ΣM_x, the moment equation reduces to

► $$M = I\dot{\phi}\dot{\psi} \tag{168}$$

where ω_z becomes $\dot{\phi}$ with $\cos \theta = 0$. Equation 168 is the most commonly used of the relations for gyroscopic motion. It is observed that with $\theta = \pi/2$, the moment axis, the precession axis, and the spin axis form a right-handed orthogonal triad, as shown in Fig. 79 for rotation about the mass center. For this special condition of $\theta = \pi/2$, the gyroscopic moment or couple may be written by the vector equation

$$M\mathbf{i} = I\dot{\psi}\mathbf{j} \times \dot{\phi}\mathbf{k} \tag{168a}$$

which describes the correct relations among the three directions of moment, precession, and spin.

For a rotor with spin $\dot{\phi}$ subjected to a moment M applied about an axis through O or G normal to the spin axis, Eq. 168 gives the resulting rate of precession $\dot{\psi}$ about an axis normal to both the spin and moment axes. Or, for a rotor whose spin axis is forced to precess at the rate $\dot{\psi}$, Eq. 168 gives the resulting gyroscopic moment acting on the rotor about an axis through O or G

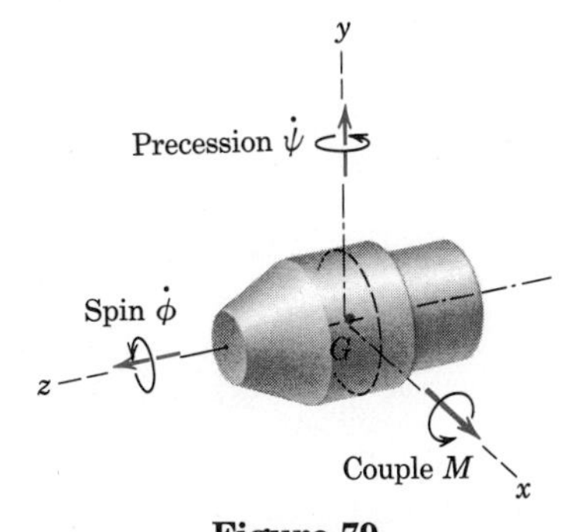

Figure 79

normal to both the spin axis and the precession axis. This gyroscopic moment would be the moment about O or G due to the reactive forces *on* the rotor as applied, for instance, by the bearings which support the rotor shaft. It is important to recognize that Eq. 168 is an equation of motion and relates the applied or resulting gyroscopic moment due to forces which act *on* the rotor, as disclosed by a free-body diagram, to the resulting motion of the rotor.

Physical insight into the behavior of a steadily precessing gyro with mutually perpendicular moment, spin, and precession axes is gained by recognizing from Fig. 80 that the change $d\mathbf{H}$ in angular momentum $\mathbf{H}$ is equal to the angular impulse $\mathbf{M}\,dt$. Thus in time dt the angular momentum vector swings from $\mathbf{H}$ to $\mathbf{H}'$, and the precessional rotation is $d\psi = |d\mathbf{H}|/H_z$. Substitution of $H_z = I\dot{\phi}$ and $|\mathbf{M}\,dt| = |d\mathbf{H}|$ from the fundamental moment equation gives $M\,dt = I\dot{\phi}\,d\psi$ or $M = I\dot{\phi}\dot{\psi}$ as before. Here, as in all cases, the vector change in angular momentum is in the direction of the applied moment. This fact is inherent in the basic moment-momentum equation and can always be used to establish the correct spatial relationships among the moment, precessional, and spin vectors. It is seen, therefore, from Fig. 80 that the spin axis always turns toward the moment axis. Just as the change in direction of the mass-center velocity is in the same direction as the resultant force, so does the change in angular momentum follow the direction of the applied moment.

For the steadily precessing gyro with θ, $\dot{\psi}$, and $\dot{\phi}$ constant and θ not limited to $\pi/2$, it is seen from Eqs. 167 with the substitution $\omega_z = \dot{\psi}\cos\theta + \dot{\phi}$ that the precessional velocity $\dot{\psi}$ must satisfy the quadratic equation

$$\dot{\psi}^2 + \frac{I\dot{\phi}}{(I - I_0)\cos\theta}\dot{\psi} - \frac{M}{(I - I_0)\sin\theta\cos\theta} = 0$$

where M stands for ΣM_x. The two roots of the equation give two precessional speeds

$$\dot{\psi} = \frac{I\dot{\phi}}{2(I_0 - I)\cos\theta}\left(1 \pm \sqrt{1 - \frac{4M(I_0 - I)\cos\theta}{I^2\dot{\phi}^2\sin\theta}}\right) \qquad (169)$$

which are real and distinct if

$$|\dot{\phi}| > \frac{2}{I}\sqrt{M(I_0 - I)\operatorname{ctn}\theta} \qquad (170)$$

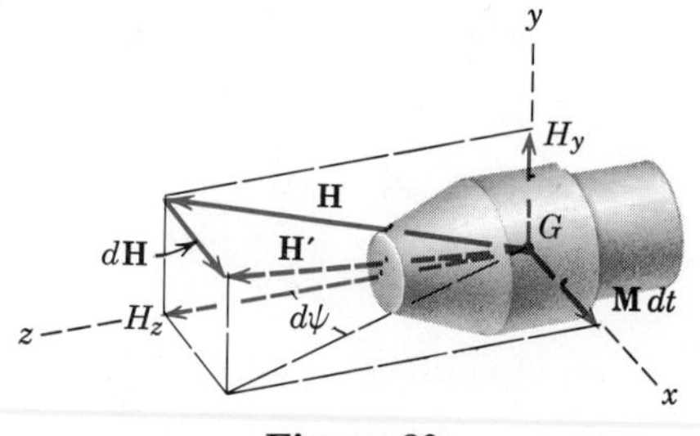

Figure 80

which ensures that the expression under the radical is positive. This relationship specifies the minimum value of $|\dot{\phi}|$ for which precession at the angle θ can occur. The binomial expansion of Eq. 169 gives

$$\dot{\psi} = \frac{I\dot{\phi}}{2(I_0 - I)\cos\theta}\left[1 \pm \left(1 - \frac{2M(I_0 - I)\cos\theta}{I^2\dot{\phi}^2 \sin\theta} + \cdots\right)\right]$$

where only the first two terms in the convergent series are shown. From this expression it is seen that for large values of $\dot{\phi}^2$ the two solutions may be approximated by

$$\dot{\psi}_1 = \frac{I\dot{\phi}}{(I_0 - I)\cos\theta} \quad \text{and} \quad \dot{\psi}_2 = \frac{M}{I\dot{\phi}\sin\theta} \tag{171}$$

where higher order terms have been neglected. The *fast precession* $\dot{\psi}_1$ is difficult to achieve experimentally by virtue of the large amount of energy required. The *slow precession* $\dot{\psi}_2$ is the one normally observed and corresponds to the lower energy level. For given values of M, I, and θ it is noted that the precessional velocity $\dot{\psi}_2$ increases as the spin velocity $\dot{\phi}$ decreases. This conclusion is good only when $\dot{\phi}$ is sufficiently large to make the approximation represented by Eq. 171 valid.

Case (b). Precession with zero moment. Consider, first, any rigid body rotating in space with zero moment about its mass center. With $\Sigma\mathbf{M} = 0$, Eq. 98 specifies that the angular momentum $\overline{\mathbf{H}}$ shall remain unchanged. Also, with zero moment, there is no change in rotational kinetic energy, so that Eq. 160 or the second of Eqs. 164 requires that

$$\omega \cdot \overline{\mathbf{H}} = \text{constant}$$

where ω is the absolute angular velocity of the rigid body. With constant $\overline{\mathbf{H}}$ this equation states that the projection of the vector ω on the vector $\overline{\mathbf{H}}$ remains constant. Thus the tip of the angular velocity vector must always lie in a fixed plane, called the *invariable plane,* which is normal to the line of $\overline{\mathbf{H}}$, called the *invariable line,* as illustrated in Fig. 81.

When axes x-y-z are rigidly attached to the body with origin at the mass center and oriented to coincide with the principal axes of inertia, the expression for the constant rotational part T of the kinetic energy from Eq. 162 may be written as

$$\omega \cdot \overline{\mathbf{H}} = 2T = I_{xx}\omega_x^2 + I_{yy}\omega_y^2 + I_{zz}\omega_z^2 = \text{constant}$$

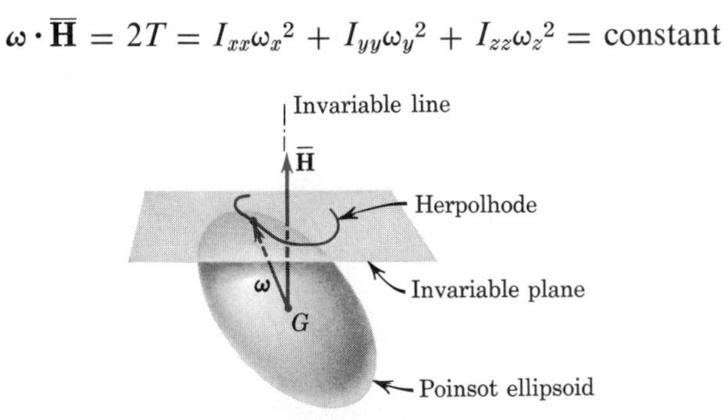

Figure 81

If (x, y, z) are the coordinates of the tip of ω, then the expression may be rewritten in the form

$$\frac{x^2}{2T/I_{xx}} + \frac{y^2}{2T/I_{yy}} + \frac{z^2}{2T/I_{zz}} = 1 \tag{172}$$

which is the equation of an ellipsoid, known as the *Poinsot ellipsoid.** By comparison with the expression for the ellipsoid of inertia in Art. 41, it is seen that the dimensions of the Poinsot ellipsoid and the ellipsoid of inertia differ only by a scale factor of $\sqrt{2T}$.

The constancy of the rotational kinetic energy may be expressed by

$$d(\omega \cdot \overline{\mathbf{H}}) = 0 \qquad \text{or} \qquad \overline{\mathbf{H}} \cdot d\omega = 0$$

since $\overline{\mathbf{H}}$ is constant. Thus the change $d\omega$ must be normal to $\overline{\mathbf{H}}$. Since the tip of ω lies on the surface of the Poinsot ellipsoid, it follows that the change in ω lies on the ellipsoidal surface which, consequently, must be tangent to the invariable plane. The path traced by the tip of ω on the invariable plane is known as the *herpolhode.* The curve traced on the Poinsot ellipsoid by the tip of ω is known as the *polhole.* As the motion proceeds, the ellipsoid rolls on the invariable plane with the tip of ω as the contact point.

The rolling-ellipsoid description for the moment-free motion of a rigid body is largely of classical and theoretical interest. Of more practical importance is the description of the rotation of a symmetrical rotor about its mass center. Such motion is encountered with spacecraft and projectiles which both spin and precess during flight. For this situation consider the symmetrical rotor in Fig. 82 for which the moment of external forces about its mass center G is zero. The Z-axis, which has a fixed direction in space, is chosen to coincide with the direction of the angular momentum $\overline{\mathbf{H}}$ which

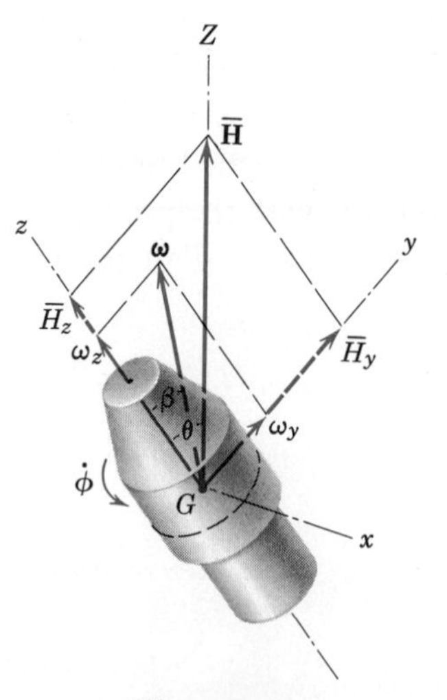

Figure 82

*Named after Louis Poinsot (1777–1859), a French engineer and mathematician who is noted for his work on the description of rotating bodies.

is constant since $\Sigma\overline{\mathbf{M}} = 0$. The *x-y-z* axes are attached in the manner described with Fig. 78. From Fig. 82 the three components of momentum are $\overline{H}_x = 0$, $\overline{H}_y = \overline{H}\sin\theta$, $\overline{H}_z = \overline{H}\cos\theta$. From the defining relations, Eqs. 148, with the notation of this article these components are also given by $\overline{H}_x = I_0\omega_x$, $\overline{H}_y = I_0\omega_y$, $\overline{H}_z = I\omega_z$. Thus $\omega_x = \Omega_x = 0$ so that θ is constant. This result means that the motion is one of steady precession about the constant $\overline{\mathbf{H}}$ vector. With no *x*-component the angular velocity $\boldsymbol{\omega}$ of the rotor lies in the *y-z* plane along with the *Z*-axis and makes an angle β with the *z*-axis. The relationship between β and θ is obtained from $\tan\theta = \overline{H}_y/\overline{H}_z = I_0\omega_y/(I\omega_z)$ which is

$$\tan\theta = \frac{I_0}{I}\tan\beta \tag{173}$$

Thus the angular velocity $\boldsymbol{\omega}$ makes a constant angle β with the spin axis.

The rate of precession is easily seen from Eq. 169 with $M = 0$ which gives for the nonzero value

$$\dot{\psi} = \frac{I\dot{\phi}}{(I_0 - I)\cos\theta} \tag{174}$$

It is clear from this relation that the direction of the precession depends on the relative magnitudes of the two moment of inertia terms. It is also noted that this precession is identical to the limiting expression for the high-speed precession for the steady-state motion described in case (*a*).

If $I_0 > I$, then $\beta < \theta$, as indicated in Fig. 83*a*, and the precession is said to be *direct*. Here the body cone rolls on the outside of the space cone as mentioned earlier in Art. 37 of Chapter 7.

If $I > I_0$, then $\theta < \beta$, as indicated in Fig. 83*b*, and the precession is said to be *retrograde*. In this instance the space cone is internal to the body cone, and $\dot{\psi}$ and $\dot{\phi}$ have opposite signs.

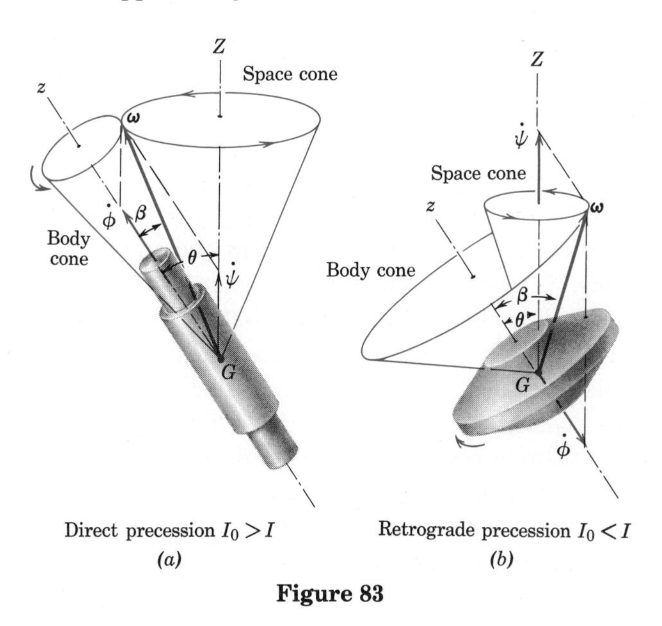

Direct precession $I_0 > I$
(*a*)

Retrograde precession $I_0 < I$
(*b*)

Figure 83

Case (c). Stability of rotation with zero moment. When a rigid body rotates about any one of its three principal axes of inertia through the mass center, the angular momentum $\overline{\mathbf{H}}$ and angular velocity $\boldsymbol{\omega}$ vectors coincide, and no precession would be indicated. On the other hand, not all of these motions are stable. Stability of motion is investigated by giving the body a slight perturbation away from the steady-motion position and then observing the behavior of the perturbation with time.

Consider a rigid body whose principal moments of inertia are I_1, I_2, I_3 and whose angular velocity components about the corresponding principal axes are ω_1, ω_2, ω_3. For the principal axes 1-2-3 rigidly attached to the body, Euler's equations, Eqs. 158, for zero moment become

$$
\begin{aligned}
I_1\dot{\omega}_1 &= (I_2 - I_3)\omega_2\omega_3 \\
I_2\dot{\omega}_2 &= (I_3 - I_1)\omega_3\omega_1 \\
I_3\dot{\omega}_3 &= (I_1 - I_2)\omega_1\omega_2
\end{aligned}
$$

For a rotation $\omega_1 = \omega_0$ about axis 1, the unperturbed condition gives $\omega_2 = \omega_3 = 0$. If a small perturbation is introduced, then the angular velocities about axes 1-2-3 may be written as $\omega_0 + e_1$, e_2, e_3, respectively, where the e's are very small angular-velocity increments. The second and third equations must be differentiated in order that e_3 may be eliminated from the one and e_2 eliminated from the other. In this process e_1 is neglected compared with ω_0, and terms involving the products of the e's are neglected. This step yields the linear differential equations

$$
\begin{aligned}
\ddot{e}_2 + \frac{(I_1 - I_2)(I_1 - I_3)}{I_2 I_3}\,\omega_0{}^2 e_2 &= 0 \\
\ddot{e}_3 + \frac{(I_1 - I_2)(I_1 - I_3)}{I_2 I_3}\,\omega_0{}^2 e_3 &= 0
\end{aligned}
$$

These equations are of the form $\ddot{x} + kx = 0$ which equation has a periodic solution for positive k and a nonlimiting hyperbolic solution for negative k. It is easily seen that both e_2 and e_3 will have a small periodic oscillation about their zero value if $I_1 > I_2$ and $I_1 > I_3$ which makes I_1 the maximum principal moment of inertia, or if $I_1 < I_2$ and $I_1 < I_3$ which makes I_1 the minimum principal moment of inertia. If I_1 is the intermediate principal moment of inertia, then it will be greater than one and less than the other of the remaining I's which would make k negative and mean that e_2 and e_3 would increase without limit. Thus, spin about either the maximum or the minimum inertia axis is a stable motion, whereas spin about the intermediate principal inertia axis is unstable since small values of e_2 and e_3 would tend to grow larger. A large value of k indicates a high frequency of perturbed oscillation and a strong restoring tendency. As k becomes smaller the perturbed motion approaches the critical condition $k = 0$ below which instability occurs. These conclusions have an important bearing on the design of spacecraft and missiles for stable motion.

Sample Problems

8/41 The cylindrical rotor B and attached shaft turn about the z-axis in a bearing which is a part of the clevis at O. The rotor and shaft have a mass of 12.5 kg with inertia properties closely approximated by those of the solid circular cylinder. The vertical shaft rotates at the constant rate $N_1 = 60$ rev/min. Rotor B rolls without slipping on the surface of the vertical cylinder A which rotates at the constant rate $N_2 = 240$ rev/min independently of the vertical shaft and clevis. Neglect any vertical friction force between B and A and determine the contact force N between the two rotors and the forces supported by the clevis pin at O.

Solution. The free-body diagram of the rotor and attached shaft discloses N and the force at O which is represented by its vertical and horizontal components. From the geometry of the figure, the angle θ is given by $75 + 100 \cos\theta = 200 \sin\theta$, from which a solution by successive approximations yields $\theta = 46.16°$.

The rotor precesses about the Z-axis at the rate $\dot{\psi} = 2\pi N_1/60 = 2\pi(60)/60 =$ 6.28 rad/s. The spin of the rotor relative to x-y-z due to N_1 alone (taking $N_2 = 0$) would be seen by fixing x-y-z and rotating the cylinder A in the direction opposite to N_1. This motion would produce a spin equal to $3\dot{\psi}/4 = 4.71$ rad/s in the negative z-direction. The spin for N_2 alone (taking $N_1 = 0$) would be $3N_2/4 = 3(2\pi)(240/60)/4 =$ 18.85 rad/s in the positive z-direction. Thus the total spin is $\dot{\phi} = 18.85 - 4.71 =$ 14.14 rad/s in the positive z-direction. The z-component of the total angular velocity $\boldsymbol{\omega}$ of the rotor becomes

$$\omega_z = \dot{\phi} - \dot{\psi}\cos\theta = 14.14 - 6.28\cos 46.16° = 9.79 \text{ rad/s}$$

Note that the angle θ in this problem is the supplement of the θ used in Art. 44. With this difference in mind the moment equation about the x-direction for steady precession, Eq. 167, becomes

$$12.5(9.81)(0.225)\sin\theta - N(0.2\cos\theta + 0.1\sin\theta) = (6.28\sin\theta)(9.79I + 6.28I_0\cos\theta)$$

Substitution of $\theta = 46.16°$ and the moments of inertia of

$$\left[I_{zz} = \tfrac{1}{2}mr^2\right] \qquad I = \tfrac{1}{2}(12.5)(0.1)^2 = 0.0625 \text{ kg}\cdot\text{m}^2$$

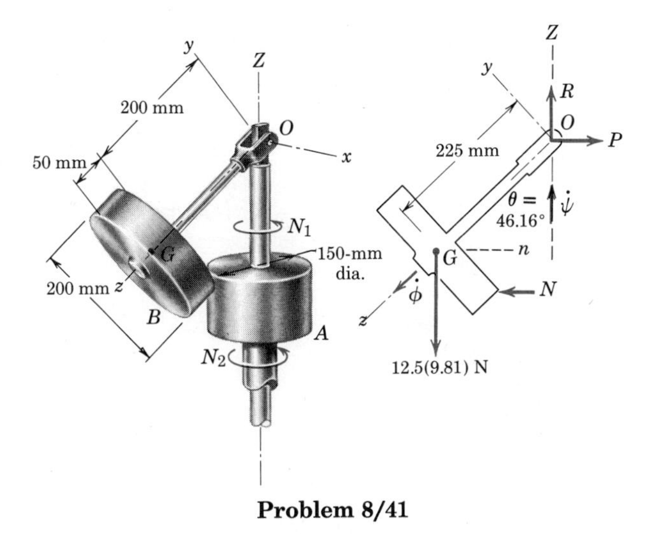

Problem 8/41

$[I_0 = \frac{1}{4}mr^2 + \frac{1}{12}ml^2 + md^2]$ $\qquad I_0 = 12.5\left(\frac{\overline{0.1^2}}{4} + \frac{\overline{0.05^2}}{12} + \overline{0.225^2}\right) = 0.667 \text{ kg}\cdot\text{m}^2$

gives

$$19.90 - 0.211N = 15.92 \qquad N = 18.90 \text{ N}$$ *Ans.*

The two force equations of motion are now easily applied to give

$[\Sigma F_n = m\bar{a}_n]$ $\quad P - 18.90 = 12.5(0.225)(\sin\theta)(6.28)^2 \qquad P = 99.0 \text{ N}$ *Ans.*

$[\Sigma F_z = 0]$ $\qquad R = 122.6 \text{ N}$ *Ans.*

8/42 A proposed space station is closely approximated by four uniform spherical shells each of mass m and radius r. The mass of the connecting structure and internal equipment may be neglected as a first approximation. If the station is designed to rotate about its z-axis at the rate of one revolution every 4 seconds, determine (a) the number n of complete cycles of precession for each revolution about the z-axis if the plane of rotation deviates only slightly from a fixed orientation, and (b) find the period τ of precession if the spin axis z makes an angle of 20° with respect to the axis of fixed orientation about which precession occurs. Draw the space and body cones for this latter condition.

Solution. (a) the number of precession cycles or wobbles for each revolution of the station about the z-axis would be the ratio of the precessional velocity $\dot{\psi}$ to the spin velocity $\dot{\phi}$ which, from Eq. 174, is

$$\frac{\dot{\psi}}{\dot{\phi}} = \frac{I}{(I_0 - I)\cos\theta}$$

The moments of inertia are

$$I_{zz} = I = 4[\tfrac{2}{3}mr^2 + m(2r)^2] = \tfrac{56}{3}\,mr^2$$

$$I_{xx} = I_0 = 2(\tfrac{2}{3})mr^2 + 2[\tfrac{2}{3}mr^2 + m(2r)^2] = \tfrac{32}{3}\,mr^2$$

It may be shown that the moment of inertia about any axis through O normal to z is the same.

With θ very small, $\cos\theta \approx 1$, and the ratio of angular rates becomes

$$n = \frac{\dot{\psi}}{\dot{\phi}} = \frac{\frac{56}{3}}{\frac{32}{3} - \frac{56}{3}} = -\frac{7}{3}$$ *Ans.*

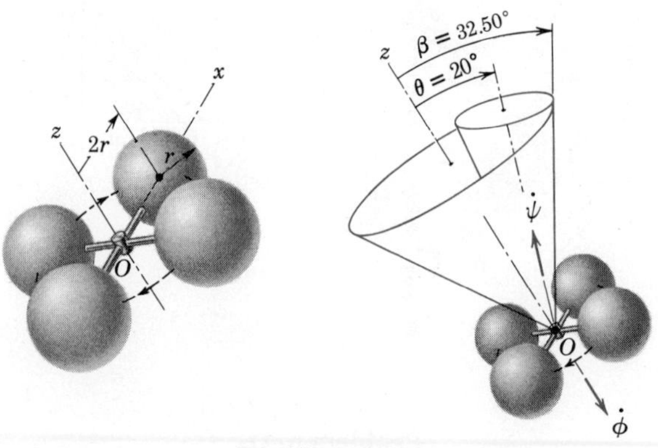

Problem 8/42

The minus sign indicates retrograde precession where, in the present case, $\dot{\psi}$ and $\dot{\phi}$ are essentially of opposite sense. Thus the station will make 7 wobbles for every 3 revolutions.

(*b*) For $\theta = 20°$ and $\dot{\phi} = 2\pi/4$ rad/s, the period of precession or wobble is $\tau = 2\pi/\dot{\psi}$, so that from Eq. 174

$$\tau = \frac{2\pi}{2\pi/4}\left|\frac{I_0 - I}{I}\cos\theta\right| = 4(\tfrac{3}{7})\cos 20° = 1.61 \text{ s} \qquad \textit{Ans.}$$

The precession is retrograde, and the body cone is external to the space cone as shown in the illustration where the body-cone angle, from Eq. 173, is

$$\tan\beta = \frac{I}{I_0}\tan\theta = \frac{\frac{56}{3}}{\frac{32}{3}}(0.3640) = 0.6370 \qquad \beta = 32.50°$$

Problems

8/43 The wheel spins at the rate p about its shaft OA, and the shaft in turn rotates about the vertical at the rate N. Determine whether the end A of the shaft deflects up or down as the shaft bends due to the gyroscopic effect during rotation.

Ans. End A deflects up

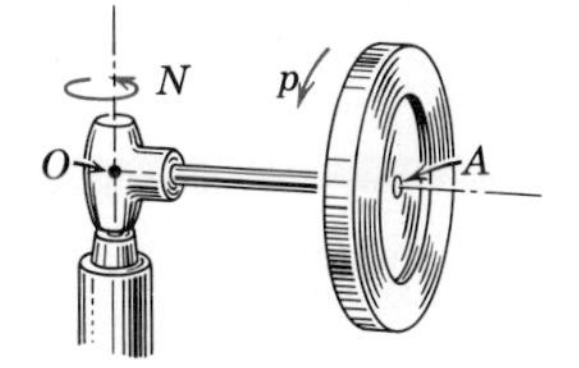

Problem 8/43

8/44 An airplane has a take-off speed of 225 km/h and retracts its main landing gear in the manner shown. Each of the two 800-mm-diameter wheels has a mass of 31 kg and a radius of gyration of 250 mm. If the retracting gear folds into the wing with an angular velocity of 0.5 rad/s, find the added bending moment M in the wheel bearing due to the gyroscopic action with the wheels still spinning at take-off speed.

Problem 8/44

8/45 The 225-kg rotor for a turbojet engine has a radius of gyration of 250 mm and rotates clockwise at 18 000 rev/min when viewed from the front of the airplane. If the airplane is traveling at 1000 km/h and making a turn to the right of 3-km radius, compute the gyroscopic moment M which the rotor bearings must support. Does the nose of the airplane tend to rise or fall as a result of the gyroscopic action?

Ans. $M = 2450$ N·m, Nose tends to rise

8/46 A car makes a turn to the right on a level road. Determine whether the normal reaction under the right rear wheel is increased or decreased as a result of the gyroscopic effect of the precessing wheels.

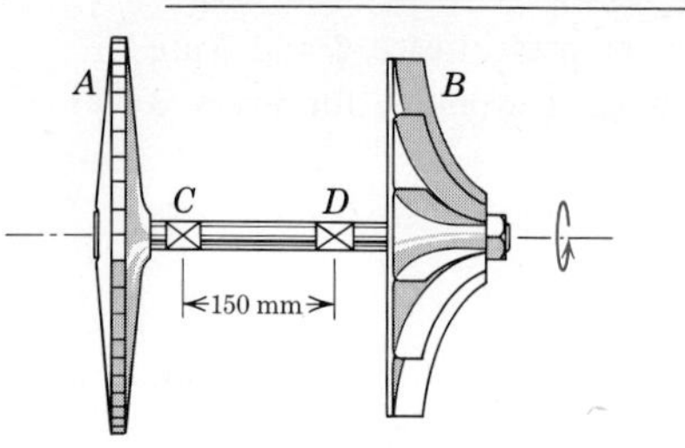

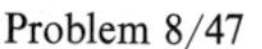

Problem 8/47

8/47 A small air compressor for an aircraft cabin consists of the 3.50-kg turbine A which drives the 2.40-kg blower B at a speed of 20 000 rev/min. The shaft of the assembly is mounted transversely to the direction of flight and is viewed from the rear of the aircraft in the figure. The radii of gyration of A and B are 79.0 mm and 71.0 mm, respectively. Calculate the radial forces exerted on the shaft by the bearings at C and D if the aircraft executes a clockwise roll (rotation about the longitudinal flight axis) of 2 rad/s viewed from the rear of the aircraft. Neglect the small moments caused by the weights of the rotors. *Ans.* $C = D = 948$ N

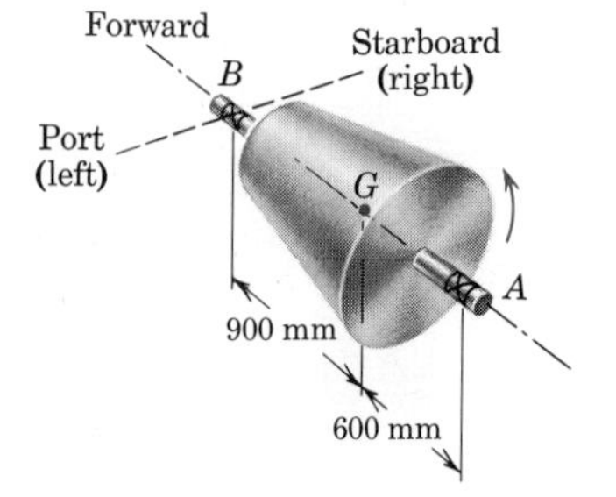

8/48 The turbine rotor in a ship's power plant has a mass of 1.5 t with center of mass at G and has a radius of gyration 225 mm. The rotor is mounted in bearings A and B with its axis in the horizontal fore-and-aft direction and turns counter-clockwise at 6000 rev/min when viewed from the stern. Determine the vertical components of the bearing reactions at A and B if the ship is making a turn to port (left) of 400-m radius at a speed of 22 knots (1 knot = 1.852 km/h). *Ans.* $A = 7.93$ kN, $B = 6.79$ kN

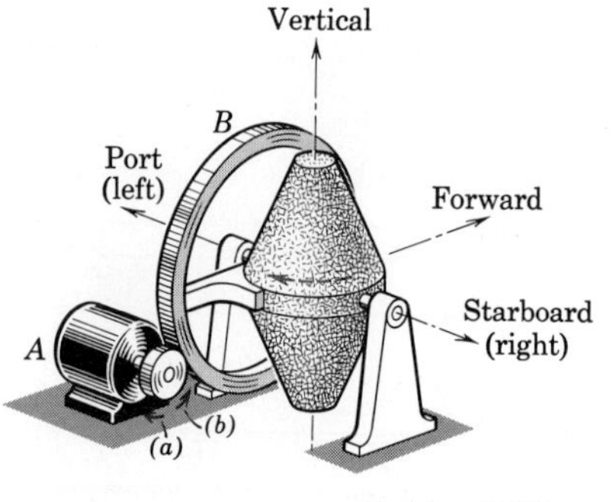

Problem 8/49

8/49 The figure shows a gyro mounted with vertical axis and used to stabilize a hospital ship against rolling. The motor A turns the pinion which precesses the gyro by rotating the large precession gear B and attached rotor assembly about a horizontal transverse axis in the ship. The rotor turns inside the housing at a clockwise speed of 960 rev/min when viewed from the top and has a mass of 80 t with radius of gyration of 1.45 m. Calculate the moment exerted on the hull structure by the gyro if the motor turns the precession gear B at the rate of 0.320 rad/s. In which of the two directions, (a) or (b), should the motor turn in order to counteract a roll of the ship to port?

8/50 An experimental car is equipped with a gyro stabilizer to counteract completely the tendency of the car to tip when rounding a curve (no change in normal force between tires and road). The rotor of the gyro has a mass m_0 and a radius of gyration k, and is mounted in fixed bearings on a shaft which is parallel to the rear axle of the car. The center of mass of the car is a distance h above the road, and the car is rounding an unbanked level turn at a speed v. At what speed p should the rotor turn and in what direction to counteract completely the tendency of the car to overturn for either a right or a left turn? The total mass of car and rotor is m.

$$Ans.\ p = \frac{mvh}{m_0k^2} \quad \text{opposite direction to car wheels}$$

8/51 The cylindrical shell is rotating in space about its geometric axis. If the axis has a slight wobble, for what ratios of l/r will the motion be direct or retrograde precession?

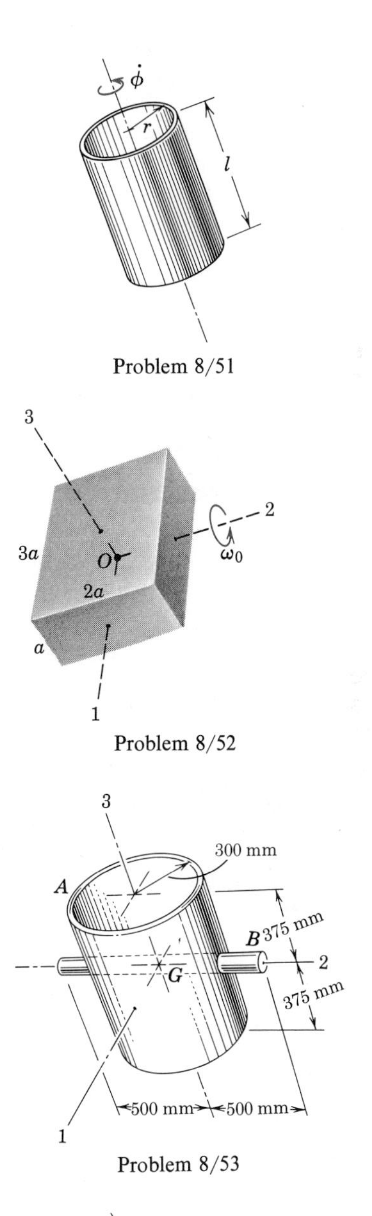

Problem 8/51

Problem 8/52

Problem 8/53

8/52 The rectangular block is released in space with an initial angular velocity ω_0 about the O-2 axis. Determine the angular velocity ω of the block if it were released instead about the O-1 axis with the same kinetic energy of rotation. In the first instance will the block continue to rotate about the O-2 axis?

8/53 The primary structure of a certain spacecraft is modeled by the cylindrical shell A which has a mass of 39 kg and the solid rod B which passes through the shell and has a mass of 12 kg. Indicate which of the principal axes 1, 2, 3 can be an axis about which stable rotation can occur in the absence of applied external moments.

Ans. Axes 1 and 2 stable
Axis 3 unstable

8/54 The thin ring is projected into the air with a spin velocity of 300 rev/min. If its geometric axis is observed to have a very slight precessional wobble, determine the frequency f of the wobble.

Ans. $f = 10$ Hz (cycles/second), retrograde precession

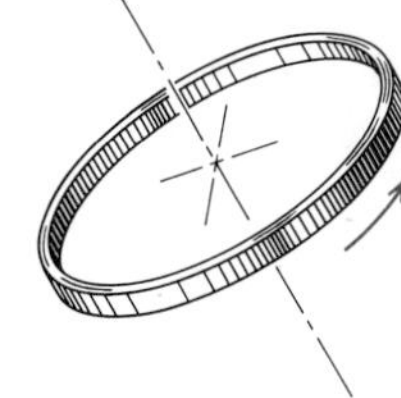

Problem 8/54

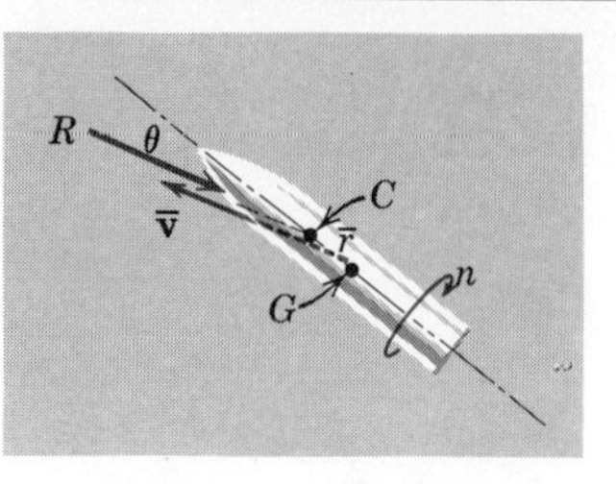

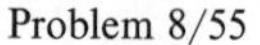
Problem 8/55

8/55 A projectile moving through the atmosphere with a velocity $\bar{\mathbf{v}}$ which makes a small angle θ with its geometric axis is subjected to a resultant aerodynamic force $\mathbf{R}$ essentially opposite in direction to $\bar{\mathbf{v}}$ as shown. If $\mathbf{R}$ passes through a point C slightly ahead of the mass center G, determine the expression for the minimum spin velocity n for which the projectile will be spin-stabilized with $\dot{\theta} = 0$. The moment of inertia about the spin axis is I and that about a transverse axis through G is I_0.

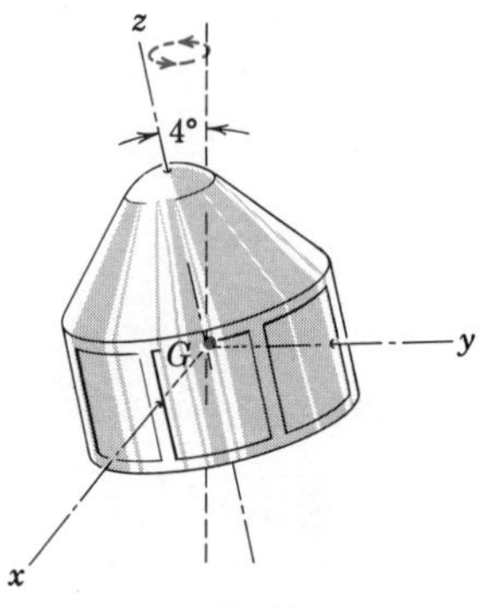

Problem 8/56

8/56 The spacecraft shown is symmetrical about its z-axis and has a radius of gyration of 720 mm about this axis. The radii of gyration about the x- and y-axes through the mass center are both equal to 540 mm. When moving in space the z-axis is observed to generate a cone with a total vertex angle of 4° as it precesses about the axis of total angular momentum. If the spacecraft has a spin velocity $\dot{\phi}$ about its z-axis of 1.5 rad/s, compute the period τ of each full precession. Is the spin vector in the positive or negative z-direction?

Ans. $\tau = 1.83$ s, spin vector in negative z-direction

8/57 Examine the motion of an axisymmetrical body in space under zero moment where I_0 and I approach equality.

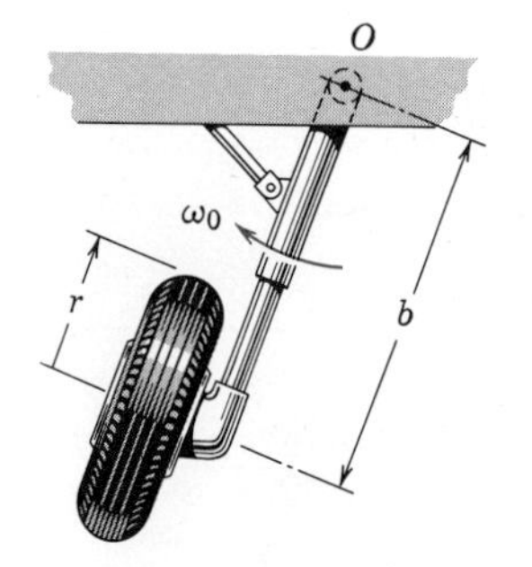

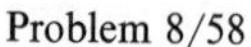
Problem 8/58

8/58 An airplane has just cleared the runway with a take-off speed v. Each of its freely spinning wheels has a mass m with a radius of gyration k about its axle and k_0 about a diametral axis through its mass center. The wheel is precessed at the angular rate ω_0 as the landing strut is folded into the wing about its pivot O. Write an expression for the kinetic energy T of the wheel relative to the airplane for the condition described.

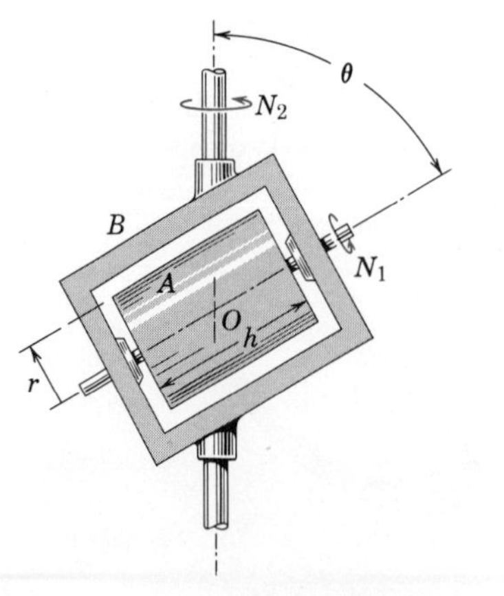

Problem 8/59

8/59 The solid circular cylinder A has a radius $r = 75$ mm, a length $h = 225$ mm, and a mass of 30 kg. The cylinder is given a rotational speed $N_1 = 300$ rev/min about its axis, which is inclined from the vertical axis of the frame B by the amount $\theta = 60°$. The frame, in turn, is given a rotational speed $N_2 = 150$ rev/min about the fixed vertical axis. Compute the kinetic energy T of the cylinder.

Ans. $T = 80.7$ J

8/60 If the geometric axis of the ring of Prob. 8/54 is observed to generate a cone with a total vertex angle $2\theta = 20°$, compute the rotational kinetic energy of the ring if it has a mass of 8 kg and a radius of 300 mm. Draw the space and body cones. *Ans.* $T = 377$ J

8/61 A solid right-circular cone of mass m, base radius $r = 100$ mm, and altitude $H = 3r = 300$ mm is suspended freely at its vertex O and is spinning about its geometric axis at the rate $\dot{\phi} = 2700$ rev/min. Determine the period for steady slow precession of the cone about the vertical. Does the result depend upon the angle θ made by the cone axis with the vertical?

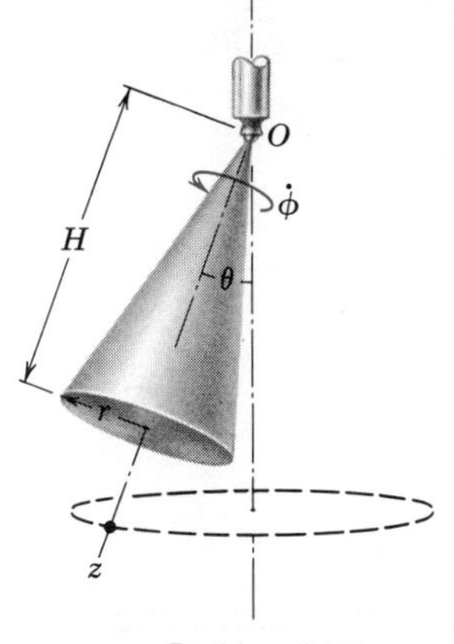

Problem 8/61

8/62 Calculate the rate $\dot{\psi}$ of fast precession for the solid cone of Prob. 8/61 if θ is very small. *Ans.* $\dot{\psi} = 154$ rev/min

8/63 Show why the axis of the symmetrical top of Fig. 78 will rise to the vertical and "sleep" if the top is spinning fast enough. Note that the pivot of the top would, in reality, be slightly rounded and friction would be present. What is the minimum spin rate n which a top can have in order to sleep? The moment of inertia about the spin axis is I, and that about a transverse axis through the pivot is I_0 which is greater than I. The top has a mass m with center of mass a distance $\bar{r}$ from O.

◀ **8/64** The housing for the electric motor is freely pivoted about the horizontal x-axis which passes through the mass center G of the rotor. If the motor is running at the constant rate $\dot{\phi} = p$, determine the angular acceleration $\ddot{\psi}$ which will result from the application of the moment M about the vertical pivot if $\dot{\gamma} = \dot{\psi} = 0$. The mass of the frame and housing is considered negligible compared with the mass m of the rotor. The radius of gyration of the rotor about the z-axis is k_z and that about the x-axis is k_x.

$$Ans.\ \ddot{\psi} = \frac{M/m}{k_x^2 \cos^2 \gamma + k_z^2 \sin^2 \gamma}$$

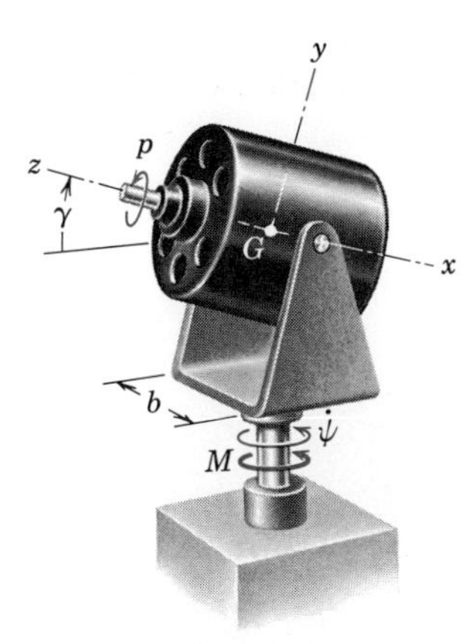

Problem 8/64

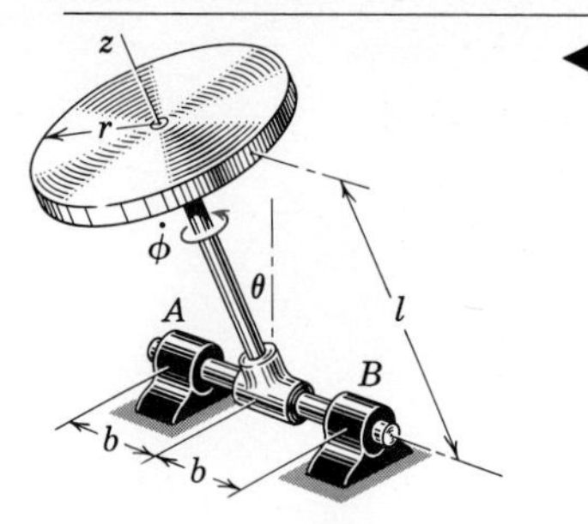

Problem 8/65

◀ **8/65** The solid circular disk of mass m and small thickness is spinning freely on its shaft at the rate $\dot{\phi}$. If the assembly is released in the vertical position at $\theta = 0$ with $\dot{\theta} = 0$, determine the horizontal components of the forces A and B exerted by the respective bearings on the horizontal shaft as the position $\theta = \pi/2$ is passed. Neglect the mass of the two shafts compared with m, and neglect all friction. Solve by using the appropriate moment equations.

Ans. $A_z = -\dfrac{m\dot{\theta}}{2}\left(\dfrac{r^2}{2b}\dot{\phi} + l\dot{\theta}\right)$

$B_z = \dfrac{m\dot{\theta}}{2}\left(\dfrac{r^2}{2b}\dot{\phi} - l\dot{\theta}\right)$

where $\dot{\theta} = 2\sqrt{\dfrac{2gl}{r^2 + 4l^2}}$

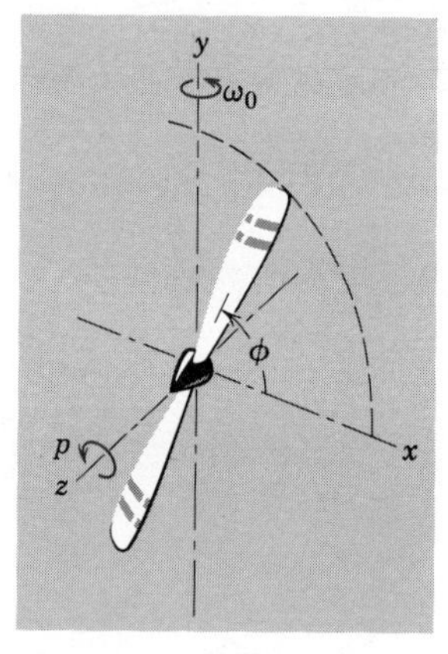

Problem 8/66

◀ **8/66** The two-bladed airplane propeller has a constant rotational speed p about its shaft axis z and a moment of inertia I about this same axis. The airplane is turning with a constant angular velocity ω_0 about the y-axis which remains vertical. A moment M_B acting on the propeller is supported by bending of the shaft, and the moment M_z is supported by torsion of the shaft. Determine each moment as a function of $\phi = pt$.

Ans. $M_B = 2Ip\omega_0 \sin \phi$ (Vector lies in the x-y plane perpendicular to blade)

$M_z = \frac{1}{2}I\omega_0^2 \sin 2\phi$

◀ **8/67** Solve Prob. 8/66 by direct use of Eqs. 156 for the axes shown which are not attached to the propeller.

◀ **8/68** The elements of a gyrocompass are shown in the figure where the rotor, at north latitude γ on the surface of the earth, is mounted in a single gimbal ring which is free to rotate about the fixed vertical y-axis. The rotor axis is, then, able to rotate in the horizontal x-z plane as measured by the angle β from the north direction. Assume that the gyro spins at the rate p and that it has mass moments of inertia about the spin axis and transverse axis through G of I and I_0, respectively. Show that the gyro axis oscillates about the north direction according to the equation $\ddot{\beta} + K^2\beta = 0$ where $K^2 = I\omega p \cos\gamma / I_0$, and that the period of oscillation about the north direction for small values of β is $\tau = 2\pi\sqrt{I_0/(I\omega p \cos\gamma)}$. (*Hint:* The components of the angular velocity Ω of the axes in terms of the angular velocity ω of the earth are

$$\Omega_x = -\omega \cos\gamma \sin\beta$$

$$\Omega_y = \omega \sin\gamma + \dot{\beta}$$

$$\Omega_z = \omega \cos\gamma \cos\beta$$

which may be used in the y-component of the moment equations, Eqs. 159, to determine β as a function of time. Note that the square of the angular velocity ω of the earth is small and may be neglected compared with the product ωp.)

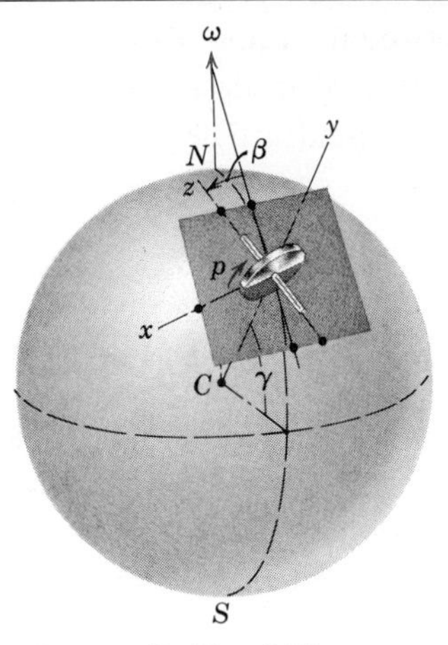

Problem 8/68

◀ **8/69** The two circular disks are rigidly connected by the light shaft and roll without slipping on the horizontal surface. If it requires a time $\tau = 4$ s for the disks to make one complete circle on the horizontal plane at constant speed, determine the normal forces N_1 and N_2 under the larger and smaller wheels, respectively. The disks have masses m and $4m$ as shown, and $r = 150$ mm.

Ans. $N_1 = 3.94mg$, $N_2 = 1.06mg$

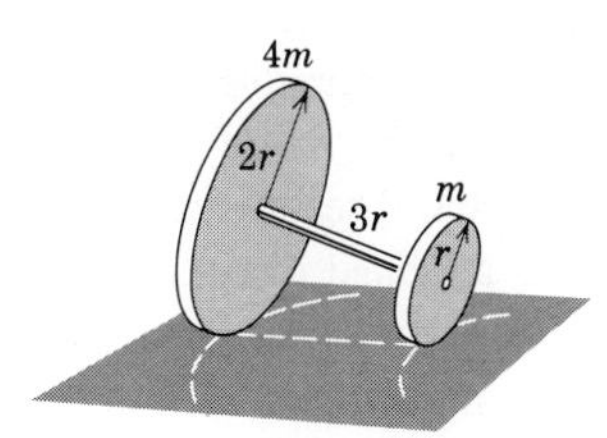

Problem 8/69

◀ **8/70** The homogeneous rectangular block of mass m is centrally mounted on the shaft A–A about which it rotates with a constant speed $\dot{\phi} = p$. Meanwhile the yoke is forced to rotate about the x-axis with a constant speed ω_0. Find the magnitude of the torque **M** as a function of ϕ. The center O of the block is the origin of the x-y-z coordinates. Principal axes 1-2-3 are attached to the block. *Ans.* $M = \frac{1}{12}m(c^2 - a^2)p\omega_0 \sin 2\phi$

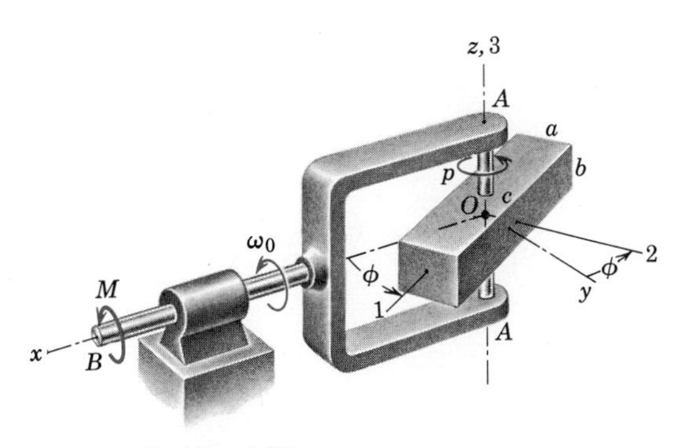

Problem 8/70

45 General Space Motion. In the previous two articles the motion equations were used to describe general plane motion and rotation about a point. In the present article these equations,

$$\Sigma\mathbf{F} = m\bar{\mathbf{a}} \quad \text{or} \quad \Sigma\mathbf{F} = \dot{\mathbf{G}} \qquad [92], [96]$$

$$\Sigma\overline{\mathbf{M}} = \dot{\overline{\mathbf{H}}} \qquad [98]$$

which were derived in Chapter 4, will be applied to the more general space motion of a rigid body. The moment equation was developed further in Eqs. 156 through 159 which are applicable to general space motion provided they are applied with reference to the mass center. It is clear that for the general case there will be six scalar motion equations, three force and three moment equations. Their application is illustrated in the Sample Problem which follows.

Sample Problem

8/71 Describe the motion of a thin circular disk which rolls on a horizontal surface without slipping. Determine the minimum velocity v of the disk which will ensure stable rolling in the vertical plane.

Solution. The figure shows the free-body diagram of the thin disk whose plane is tilted an angle θ from the vertical and which is rolling in the direction O–a at the instant shown. In addition to the weight mg and normal force N the friction force is represented by its two perpendicular components F_1 in the plane of the disk and F_2 which lies along the intersection of the x-z plane and the y_0-z_0 plane.

The y-axis remains horizontal and is parallel to the instantaneous direction of rolling O–a. The z-axis is the spin axis and intersects the vertical x_0-axis. The angle ψ measures the direction of the rolling path with respect to the fixed direction y_0. The x'-axis is attached to the disk, so that the angle ϕ and its time derivatives measure the angular motion of the disk relative to x-y-z. It should be noted that this reference system is very similar to that used in Fig. 78 for the spinning top.

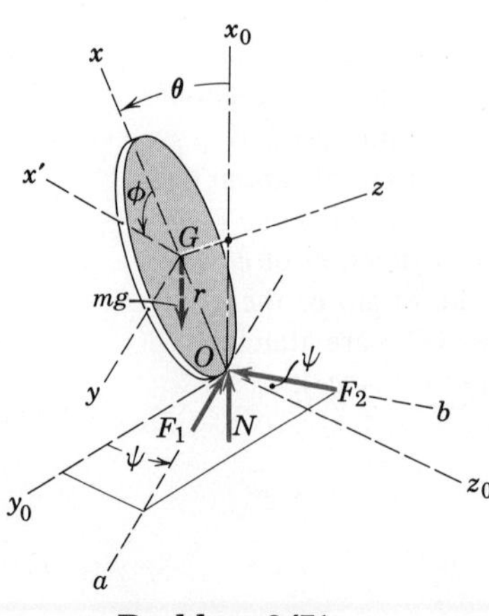

Problem 8/71

The components of angular velocity of both the axes and the disk are required for the moment equation. The *x-y-z* axes have angular velocity components which are

$$\Omega_x = \dot{\psi} \cos \theta \qquad \Omega_y = \dot{\theta} \qquad \Omega_z = \dot{\psi} \sin \theta$$

The disk has angular velocity components which are

$$\omega_x = \dot{\psi} \cos \theta \qquad \omega_y = \dot{\theta} \qquad \omega_z = \dot{\psi} \sin \theta + \dot{\phi}$$

Also, the angular momentum components of the disk about the principal axes are

$$\begin{aligned} \overline{H}_x &= \overline{I}_{xx}\omega_x = I_0\dot{\psi} \cos \theta \\ \overline{H}_y &= \overline{I}_{yy}\omega_y = I_0\dot{\theta} \\ \overline{H}_z &= \overline{I}_{zz}\omega_z = I(\dot{\psi} \sin \theta + \dot{\phi}) \end{aligned}$$

where $\overline{I}_{xx} = \overline{I}_{yy} = I_0$, $\overline{I}_{zz} = I$.

Direct substitution into Eqs. 156 written about the the mass center yields the component equations for the moment summation about the *x-y-z* axes as follows:

$$\begin{aligned} 0 &= I_0\ddot{\psi} \cos \theta - 2I_0\dot{\psi}\dot{\theta} \sin \theta + I\omega_z\dot{\theta} \\ Nr \sin \theta - F_2 r \cos \theta &= I_0\ddot{\theta} - I\omega_z\dot{\psi} \cos \theta + I_0\dot{\psi}^2 \sin \theta \cos \theta \qquad (a) \\ F_1 r &= I\dot{\omega}_z \end{aligned}$$

In order to obtain the three force equations the expressions for the components of acceleration $\overline{\mathbf{a}}$ of the mass center are required. If the disk does not slip, the velocity of the mass center is seen to be

$$\overline{\mathbf{v}} = 0\mathbf{i} + r\omega_z\mathbf{j} - r\dot{\theta}\mathbf{k}$$

where both $\dot{\psi}$ and $\dot{\phi}$ contribute to the *y*-component. With the principle of Eq. 142, the acceleration is

$$\begin{aligned} \overline{\mathbf{a}} = \dot{\overline{\mathbf{v}}} &= \left(\frac{d\overline{\mathbf{v}}}{dt}\right)_{xyz} + \mathbf{\Omega} \times \overline{\mathbf{v}} \\ &= (r\dot{\omega}_z\mathbf{j} - r\ddot{\theta}\mathbf{k}) + (\mathbf{i}\dot{\psi} \cos \theta + \mathbf{j}\dot{\theta} + \mathbf{k}\dot{\psi} \sin \theta) \times (r\omega_z\mathbf{j} - r\dot{\theta}\mathbf{k}) \\ &= \mathbf{i}r(-\dot{\theta}^2 - \omega_z\dot{\psi} \sin \theta) + \mathbf{j}r(\dot{\omega}_z + \dot{\theta}\dot{\psi} \cos \theta) + \mathbf{k}r(-\ddot{\theta} + \omega_z\dot{\psi} \cos \theta) \end{aligned}$$

With these acceleration components, the *x-y-z* components of Eq. 92, $\Sigma\mathbf{F} = m\overline{\mathbf{a}}$, give

$$\begin{aligned} (N - mg) \cos \theta + F_2 \sin \theta &= mr(-\dot{\theta}^2 - \omega_z\dot{\psi} \sin \theta) \\ -F_1 &= mr(\dot{\omega}_z + \dot{\theta}\dot{\psi} \cos \theta) \qquad (b) \\ (N - mg) \sin \theta - F_2 \cos \theta &= mr(-\ddot{\theta} + \omega_z\dot{\psi} \cos \theta) \end{aligned}$$

Equations (*a*) and (*b*) are the necessary and sufficient relations for the description of the rolling disk. Clearly their general solution would be difficult. It is possible, however, to examine the near-vertical motion of the disk from which the conditions of rolling stability may be deduced. With near-vertical rolling, the quantities $\dot{\theta}^2$, $\dot{\psi}^2$, $\dot{\theta}\dot{\psi}$, $\theta\dot{\psi}$ may be neglected since $\dot{\theta}$ and $\dot{\psi}$ are both small. Also $\sin \theta$ is replaced by θ and $\cos \theta$ by unity. With these simplifications $\omega_z = \dot{\phi}$, and Eqs. (*a*) and (*b*) become

$$0 = I_0\ddot{\psi} + I\omega_z\dot{\theta} \qquad (c)$$

$$Nr\theta - F_2 r = I_0\ddot{\theta} - I\omega_z\dot{\psi} \qquad (d)$$

$$F_1 r = I\dot{\omega}_z \tag{e}$$

$$N - mg + F_2\theta = mr(0) \tag{f}$$

$$-F_1 = mr\dot{\omega}_z \tag{g}$$

$$(N - mg)\theta - F_2 = mr(-\ddot{\theta} + \omega_z\dot{\psi}) \tag{h}$$

The elimination of F_1 between Eqs. (e) and (g) requires that $\dot{\omega}_z = 0$ and, hence,

$$\omega_z = n, \text{ constant}$$

To within the limits of the approximation, $\dot{\phi} = \omega_z = n$, so that $\ddot{\phi} = 0$. With ω_z constant, Eq. (c) may be integrated to give

$$I_0\dot{\psi} + In\theta = C, \text{ a constant} \tag{i}$$

Equations (d) and (h) are combined to eliminate F_2 and upon substitution of Eq. (i) give

$$(I_0 + mr^2)\ddot{\theta} + \left[(I + mr^2)\frac{I}{I_0}n^2 - mgr\right]\theta = (I + mr^2)\frac{Cn}{I_0} \tag{j}$$

Equation (j) is of the form $a\ddot{x} + bx = c$, where a, b, c are constants, and represents a stable harmonic oscillation if the ratio b/a is positive and an unstable hyperbolic solution if b/a is negative. Since $a > 0$ the requirement for rolling stability in the vertical plane is given by $b > 0$ so that

$$|n| > \sqrt{\frac{I_0 mgr}{I(I + mr^2)}} \qquad \textit{Ans.}$$

If the angular velocity n is any less than the value specified by the radical, the disk will become unstable.

For a thin uniform circular disk $I = \frac{1}{2}mr^2$, $I_0 = \frac{1}{4}mr^2$, and the minimum velocity of the disk for stable rolling in the upright position becomes

$$v = rn = \sqrt{gr/3}$$

Problems

8/72 Calculate the minimum velocity v of the center of a 600-mm-diameter hoop required to prevent wobbling as it rolls on a horizontal surface.

Ans. $v = 0.858$ m/s

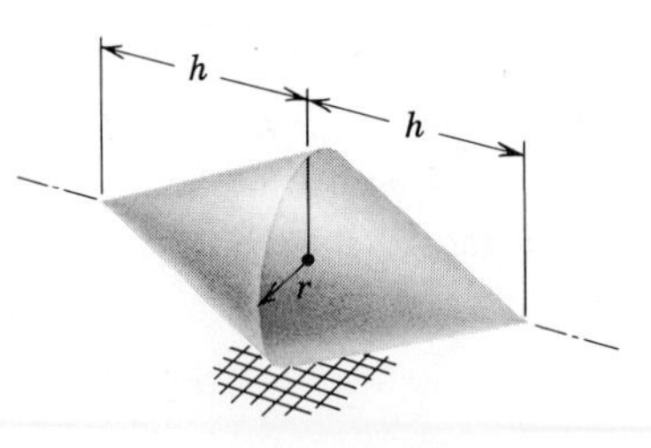

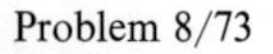
Problem 8/73

8/73 The two solid homogeneous cones roll on the periphery of their common base. If $h = 2r = 600$ mm, calculate the minimum velocity v of the center of the base in order that the cones will be stable when rolling on the level surface as shown.

Ans. $v = 2.04$ m/s

8/74 The rolling ring of negligible mass has a radius r and supports the slender rod of mass m and length L by means of light spokes. Prove that the rolling motion is always unstable for the near-vertical position of the ring.

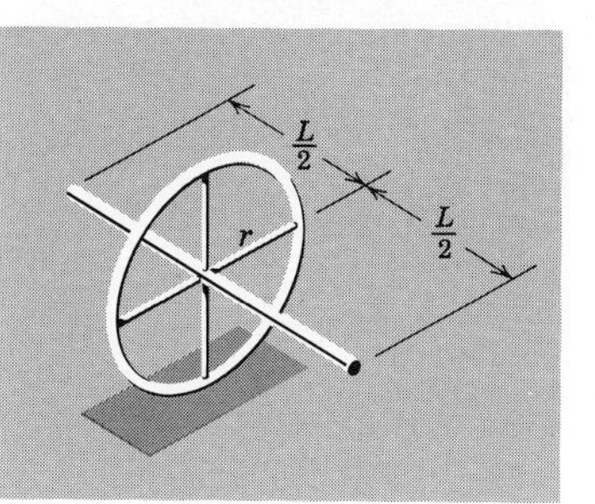

Problem 8/74

8/75 The circular disk of mass m and radius r rolls without slipping in a horizontal circle with a constant angle θ between the plane of the disk and the vertical. If the center of the disk moves in a circle of radius b with a velocity v, determine the kinetic energy T of the disk in terms of v and θ.

Ans. $T = \frac{1}{4}mv^2\left[3 - \frac{2r}{b}\sin\theta + \frac{r^2}{2b^2}(1 + \sin^2\theta)\right]$

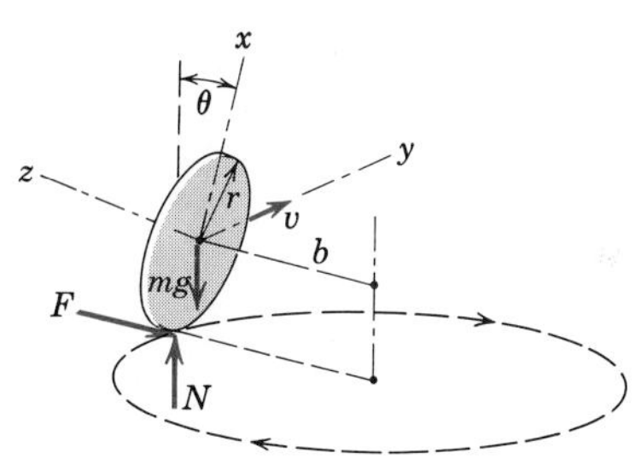

Problem 8/75

◀**8/76** With the aid of the free-body diagram for the rolling circular disk shown with Prob. 8/75, prove that b must satisfy the equation

$$\tan\theta = \frac{v^2}{2gb}\left(3 + \frac{r}{2b}\sin\theta\right).$$

◀**8/77** A coin held in the vertical plane is flipped so that it spins about a vertical diameter with an angular velocity $\dot{\psi}$ on a rough horizontal surface. If the coin is treated as a thin uniform disk of radius r, determine the expression for the value n of $\dot{\psi}$ below which the coin will not remain upright.

Ans. $n = 2\sqrt{\dfrac{g}{5r}}$

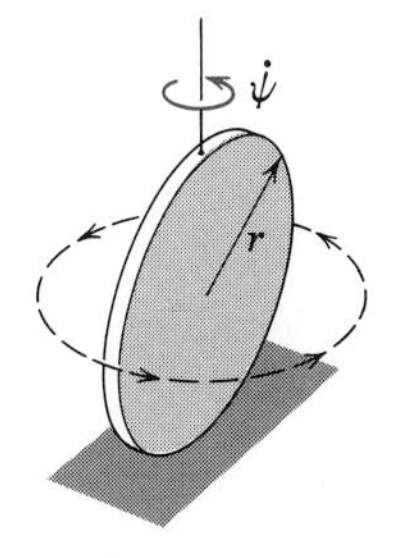

Problem 8/77

9 VIBRATION AND TIME RESPONSE

46 Introduction. An important and special class of problems in dynamics consists of the linear or angular motions of masses which oscillate or respond to applied disturbances in the presence of restoring forces. The response of a structure to earthquake or blast loading, the steady vibration of a rotating disk which is slightly out of balance, the oscillation of a pendulum, the roll of a ship, and the aeolian vibrations of power lines are but a few examples of this class of problems. In each case a mass or system of masses is initially or continuously disturbed in the presence of forces which tend to return the mass or system to its undisturbed position.

A useful engineering description of the time response of mass systems is accomplished by the solution of a mathematical model of an equivalent system which can be readily analyzed. The torsional vibrations of a ship's propeller shaft, for instance, can be described to within a close approximation by neglecting the mass of the shaft and replacing the propeller and the turbine by two disks, one concentrated on each end of the shaft. The transmission of force to a foundation from an unbalanced spring-mounted machine may usually be described by considering the machine to be a concentrated mass mounted on a single equivalent spring. When the body or system to be described is approximated by replacing the actual masses by concentrated masses connected by massless springs and subjected to concentrated retarding and disturbing forces, the system is often referred to as a *lumped-parameter* system. When the mass of a body is treated in its distributed state with a continuous variation of motion occurring throughout the mass, then the problem is referred to as a *distributed-parameter* system. The motions in the present chapter will be restricted to those which can be formulated by a lumped-parameter analysis.

The detailed study of vibrations and time response is a large subject for specialized study, and only a bare introduction to the topic will be given in this chapter. For a more complete treatment the student is directed to references on mechanical vibrations, linear systems, electric circuits, nonlinear oscillations, and pulse techniques.

47 Linear System Equation. By far the most useful equivalent system of the lumped-parameter variety is that of the concentrated mass mounted on an elastic spring and subjected to a retarding force and a disturbing force. Such a system is shown in Fig. 84*a* with the mass m in a general position which is displaced a distance x from the neutral or equilibrium position of the elastic spring whose stiffness is k. The mass is acted upon by an applied force $F = f(t)$ which is expressed as a function of the time t. Also, the mass is retarded by a force with a magnitude proportional to the velocity $\dot{x}$. This type of frictional retardation is termed *viscous damping* and is represented by the action of a dashpot or fluid damper under laminar flow conditions. Other types of damping forces may be encountered such as dry friction or Coulomb damping which is essentially independent of velocity, internal damping due to material hysteresis losses, turbulent-flow damping where the retarding force is more nearly proportional to the square of the velocity, and magnetic damping.

The free-body diagram discloses the applied force F, the restoring force $-kx$, and the retarding or damping force $-c\dot{x}$. The constant c is called the *viscous damping coefficient*. Application of Newton's second law for motion in the x-direction gives

$$F - kx - c\dot{x} = m\ddot{x}$$

which may be written as

$$m\ddot{x} + c\dot{x} + kx = F \tag{175a}$$

A system similar to that of Fig. 84*a* is shown in Fig. 84*b* where the applied force is transmitted through the spring attached to a foundation which has a displacement $\delta = \delta(t)$ from the initial position. If x is the absolute displacement of the mass measured from the equilibrium position when $\delta = 0$, then the spring has a tension $k(x - \delta)$, and the free-body diagram requires that

$$-c\dot{x} - k(x - \delta) = m\ddot{x}$$

or

$$m\ddot{x} + c\dot{x} + kx = k\delta \tag{175b}$$

It is seen, therefore, that Eq. 175*b* is equivalent to Eq. 175*a* when F is replaced by $k\delta$. The solutions of either of Eqs. 175 for various values of c, k, and F or δ cover a wide variety of oscillations and responses which can be used to describe the behavior of many engineering systems. Each of Eqs. 175 is seen to be a linear, second-order differential equation, and its solution can be obtained by several standard procedures.

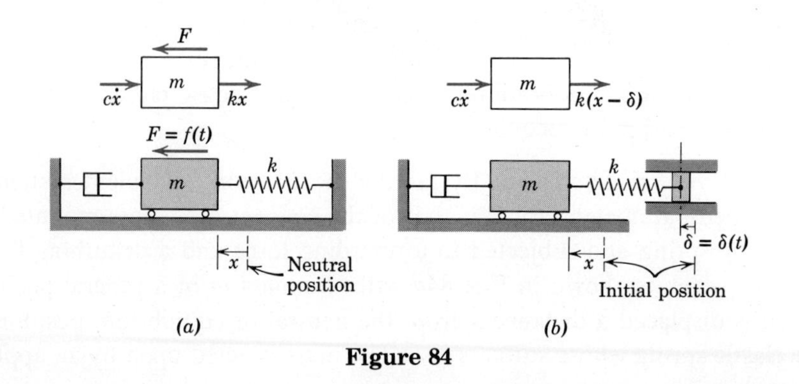

Figure 84

Before proceeding to the solution of the equation, it is important to observe the electric circuit analogy with its equivalent differential equation. Figure 85 shows a lumped series circuit consisting of a voltage E which is a function of the time, an inductance L, a capacitance C, and a resistance R. The voltage drop across each of the elements L, C, R in that order is $L\,\dfrac{di}{dt}$, $\dfrac{1}{C}\int i\,dt$, and Ri where i is the current. The sum of these drops must equal the applied voltage, so that

$$L\frac{di}{dt} + \frac{1}{C}\int i\,dt + Ri = E$$

But $i = \dot{q}$ where q is the electric charge, so that the equation for the circuit becomes

$$L\ddot{q} + R\dot{q} + \frac{1}{C}q = E \tag{176}$$

This equation has the same form as the equation for the mechanical "circuit." Thus by a simple interchange of symbols, the behavior of the electrical system may be used to predict the behavior of the mechanical system, or vice versa. The following table of mechanical and electrical equivalents will be found useful.

Mechanical-Electrical Equivalents

Mechanical			Electrical		
Quantity		Units	Quantity		Units
Mass	m	kg	Inductance	L	henry
Spring stiffness	k	N/m (= kg/s^2)	1/Capacitance	$1/C$	1/farad
Force	F	N (= kg·m/s^2)	Voltage	E	volt
Velocity	$\dot{x}$	m/s	Current	i	ampere
Displacement	x	m	Charge	q	coulomb
Viscous damping constant	c	N/(m/s) (= kg/s)	Resistance	R	ohm

The equivalence between the electrical quantities and their corresponding mechanical counterparts forms the basis for the development of electrical analog computers where the electrical response in various circuit com-

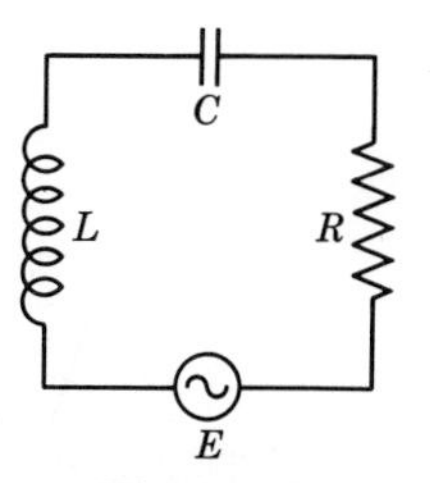

Figure 85

binations is used to predict the behavior of equivalent mechanical systems or other physical systems which obey analogous defining equations.

Solutions of Eqs. 175*a* and *b* are treated briefly in the following two articles.

48 Free Response. When the disturbing force F or the foundation deflection δ is zero, the linear differential equation becomes a homogeneous second-order equation. Its solution describes the oscillations and response of the equivalent mass when it is released from a displaced position.

Case (*a*) *Undamped response.* When the damping force is negligible or absent, the mass vibrates freely without energy loss, and its motion is described by the equation

$$m\ddot{x} + kx = 0 \tag{177}$$

The solution is a *simple harmonic* oscillation expressed by

$$x = C_1 \sin pt + C_2 \cos pt$$

where C_1 and C_2 are integration constants which depend on the manner in which the motion was begun. The solution is quickly verified by direct substitution upon which it is seen that $p = \sqrt{k/m}$. An alternative form of the solution may be written as

$$x = C \sin (pt + \phi)$$

where ϕ is a *phase angle*.

The displacement x may be described by the projection on the x-axis, Fig. 86, of a rotating vector whose length is the *amplitude* x_0 of the oscillation and whose angular velocity equals $p = \sqrt{k/m}$. From the figure it is seen that $x_0 = C = \sqrt{C_1{}^2 + C_2{}^2}$ and that $C_2/C_1 = \tan \phi$. If the time is counted from the position where the vector crosses the horizontal axis, then $\phi = 0$, and the solution is merely

$$x = x_0 \sin pt$$

The motion described is an oscillation where the time for each complete cycle is the *period* $\tau = 2\pi/p = 2\pi/\sqrt{k/m}$ and the number of cycles per unit time or *frequency* is $f = 1/\tau = \sqrt{k/m}/(2\pi)$. Frequency is expressed in *hertz* (Hz) where 1 Hz = 1 cycle/second. The angular velocity p of the rotating reference vector is known as the *circular frequency* and is expressed in radians per second (rad/s).

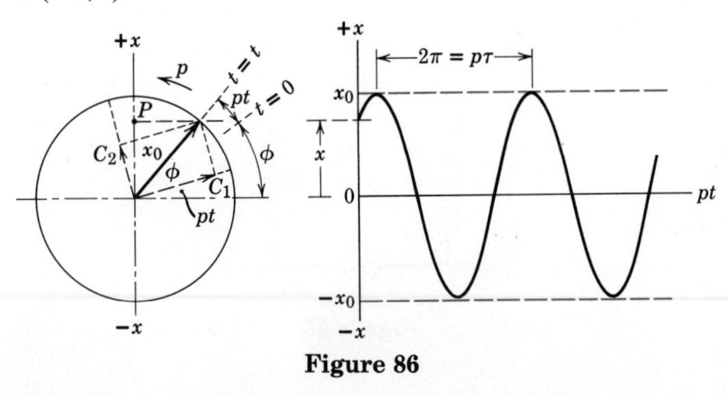

Figure 86

In the absence of a disturbing force F and a damping force $c\dot{x}$, the energy of the system is conserved, so that

$$T + V = \text{constant} \qquad \text{and} \qquad \frac{d}{dt}(T + V) = 0$$

Substitution of the expressions for T and V gives

$$\frac{d}{dt}(\tfrac{1}{2}m\dot{x}^2 + \tfrac{1}{2}kx^2) = m\dot{x}\ddot{x} + kx\dot{x} = 0$$

which is the same as Eq. 177 when $\dot{x}$ is cancelled. Derivation of the equation of motion by differentiation of the energy equation is of great advantage for systems with interacting members and reactive forces which do no work.

Energy considerations may be used to determine the period or frequency of oscillation for a linear conservative system without having to derive the differential equation of motion. Since energy is conserved, the maximum kinetic energy occurs at the position $x = 0$ and must equal the maximum potential energy at $x = x_0$ or

$$T_{\max} = V_{\max}$$

where V is taken to be zero when $T = T_{\max}$. From the solution for simple harmonic motion, the maximum velocity is $\dot{x}_{\max} = x_0 p$, so that

$$\tfrac{1}{2}m(x_0 p)^2 = \tfrac{1}{2}kx_0^2 \qquad \text{from which} \qquad p = \sqrt{k/m}$$

This direct determination of circular frequency may be used for any linear undamped oscillator.

Case (b) Damped response. When the damping force is not negligible, the free response from Eq. 175*a* or *b* is described by

$$\blacktriangleright \qquad m\ddot{x} + c\dot{x} + kx = 0 \tag{178}$$

Solution of the linear equation may be obtained with the substitution $x = Ce^{\alpha t}$ which gives the auxiliary algebraic equation

$$m\alpha^2 + c\alpha + k = 0$$

The two roots are

$$\alpha_1 = -\frac{c}{2m} + \sqrt{\left(\frac{c}{2m}\right)^2 - \frac{k}{m}} \qquad \alpha_2 = -\frac{c}{2m} - \sqrt{\left(\frac{c}{2m}\right)^2 - \frac{k}{m}}$$

If $c < 2\sqrt{km}$, the roots are complex. With the substitutions

$$b = \frac{c}{2m} \qquad q = \sqrt{\frac{k}{m} - \left(\frac{c}{2m}\right)^2}$$

the roots are $\alpha_1 = -b + iq$, $\alpha_2 = -b - iq$ where $i = \sqrt{-1}$. With the aid of the identity $e^{\pm iz} = \cos z \pm i \sin z$, the solution may be written as

$$x = e^{-bt}(C_1 \sin qt + C_2 \cos qt) \tag{179}$$

or

$$x = x_0 e^{-bt} \sin (qt + \phi)$$

where $x_0 = \sqrt{C_1^2 + C_2^2}$ and $\tan\phi = C_2/C_1$. This motion is clearly oscillatory with a decreasing amplitude bounded by the limiting curves $x = \pm x_0 e^{-bt}$ as shown in Fig. 87. The period $\tau = 2\pi/q$ is somewhat greater than that with no damping.

The damping coefficient c may be determined from an experimental record of the vibration by measuring two successive amplitudes such as x_1 and x_2 in the figure. With $t_2 = t_1 + \tau$, this ratio is

$$\frac{x_1}{x_2} = \frac{x_0 e^{-bt_1}}{x_0 e^{-b(t_1+\tau)}} = e^{b\tau}$$

The quantity $b\tau$ is known as the *logarithmic decrement* and is a direct measure of the damping coefficient. With the expressions for b and τ, and with the substitution of $c_{cr} = 2\sqrt{km}$ for the critical value of c, the logarithmic decrement may be written as

$$\log\frac{x_1}{x_2} = b\tau = \frac{2\pi\eta}{\sqrt{1-\eta^2}}$$

where

$$\eta = \frac{c}{c_{cr}}$$

If $c > 2\sqrt{km}$, both roots of the auxiliary equation are real, and a nonoscillatory motion ensues. The motion is said to be *overdamped,* and the mass, upon being released from a displaced condition, will creep back toward the neutral position but will not oscillate. The solution for the overdamped case may be written

$$x = e^{-bt}(A_1 e^{q't} + A_2 e^{-q't}) \tag{180}$$

where

$$q' = \sqrt{\left(\frac{c}{2m}\right)^2 - \frac{k}{m}} = q\sqrt{-1}$$

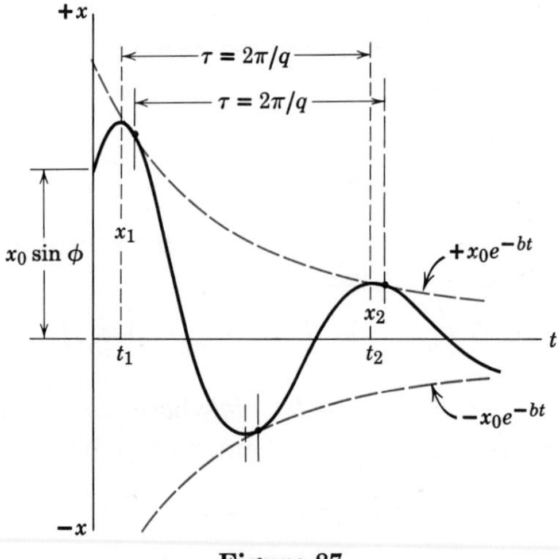

Figure 87

If $c = c_{cr} = 2\sqrt{km}$, the motion is said to be *critically damped,* and this condition represents the transition between a damped vibration and an over-damped creep. The solution for the critically damped case where the auxiliary equation has equal roots is

$$x = e^{-bt}(A_1 + A_2t)$$

Figure 88 indicates the three cases of viscous damping for a mass which is released from rest with an initial displacement x_0 from the undeformed position.

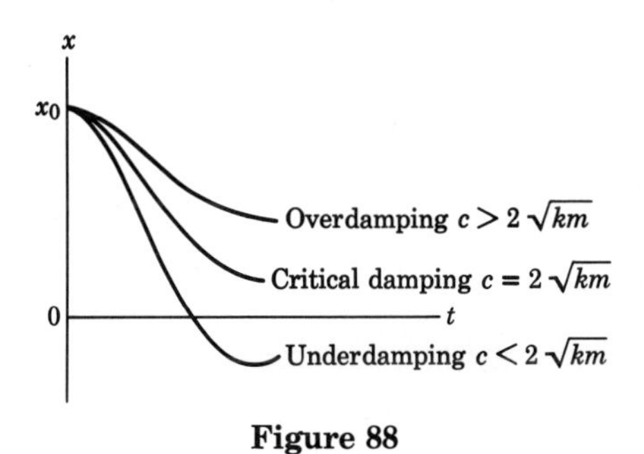

Figure 88

Sample Problem

9/1 The mechanism shown oscillates about the pivot O. If each spring is preset with an initial tension T_0 in the neutral position $\theta = 0$, determine the period for small oscillations by each of two methods. The assembly has a mass m with mass center at G and a moment of inertia I about O.

Solution: The free-body diagram of the mechanism is shown in any general displaced position θ. For small angles $\sin\theta$ is replaced by θ, and $\cos\theta$ becomes unity. The moment equation about O gives

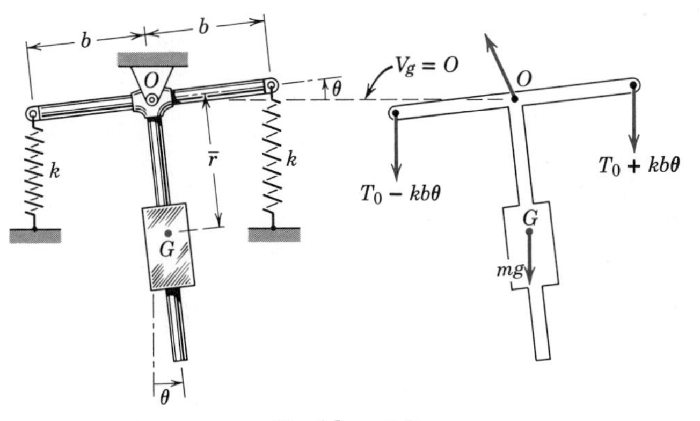

Problem 9/1

$[\Sigma M_O = I_O\alpha]$ $\qquad (T_0 - kb\theta)b - (T_0 + kb\theta)b - mg\bar{r}\theta = I\ddot{\theta}$

or

$$\ddot{\theta} + \frac{2kb^2 + mg\bar{r}}{I}\theta = 0$$

This equation is identical in form to Eq. 177, and the period is

$$\tau = \frac{2\pi}{p} = 2\pi\sqrt{\frac{I}{2kb^2 + mg\bar{r}}} \qquad \textit{Ans.}$$

An alternative solution is obtained by using the conservation-of-energy principle. The maximum kinetic energy occurs at $\theta = 0$ and is

$$T_{max} = \tfrac{1}{2}I\dot{\theta}^2{}_{max}$$

But for a simple harmonic oscillation, $\theta = \theta_0 \sin pt$, so that $\dot{\theta}_{max} = p\theta_0$. Thus

$$T_{max} = \tfrac{1}{2}Ip^2\theta_0{}^2$$

If the potential energy V is taken to be zero at $\theta = 0$, the potential energy at θ_0 becomes

$$\begin{aligned} V_{max} &= V_g + V_e \\ &= mg\bar{r}(1 - \cos\theta_0) + [\tfrac{1}{2}k(\delta + b\theta_0)^2 - \tfrac{1}{2}k\delta^2] + [\tfrac{1}{2}k(\delta - b\theta_0)^2 - \tfrac{1}{2}k\delta^2] \end{aligned}$$

For θ_0 small, the cosine is replaced by $1 - \dfrac{\theta_0{}^2}{2} + \cdots$ so that

$$V_g + V_e = mg\bar{r}\frac{\theta_0{}^2}{2} + kb^2\theta_0{}^2$$

or

$$V_{max} = \left(\frac{mg\bar{r}}{2} + kb^2\right)\theta_0{}^2$$

Now with $T_{max} = V_{max}$, there results

$$p^2 = \frac{1}{I}(mg\bar{r} + 2kb^2) \qquad \text{and} \qquad \tau = \frac{2\pi}{p} = 2\pi\sqrt{\frac{I}{mg\bar{r} + 2kb^2}} \qquad \textit{Ans.}$$

the same as before.

Problems

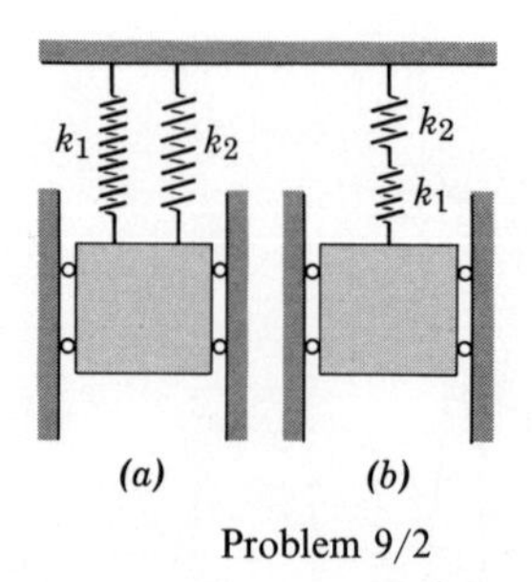

Problem 9/2

9/2 Replace the springs in each of the two cases shown by a single spring of stiffness k (equivalent spring constant) which will cause each weight to vibrate with its original frequency.

Ans. (*a*) $k = k_1 + k_2$

(*b*) $\dfrac{1}{k} = \dfrac{1}{k_1} + \dfrac{1}{k_2}$

9/3 The uniform rod of length l and mass m is suspended at its midpoint by a wire of length L. The resistance of the wire to torsion is proportional to its angle of twist θ and equals $(JG/L)\theta$ where J is the polar moment of inertia of the wire cross section and G is the shear modulus of elasticity. Derive the expression for the period τ of oscillation of the bar when it is set into rotation about the axis of the wire.

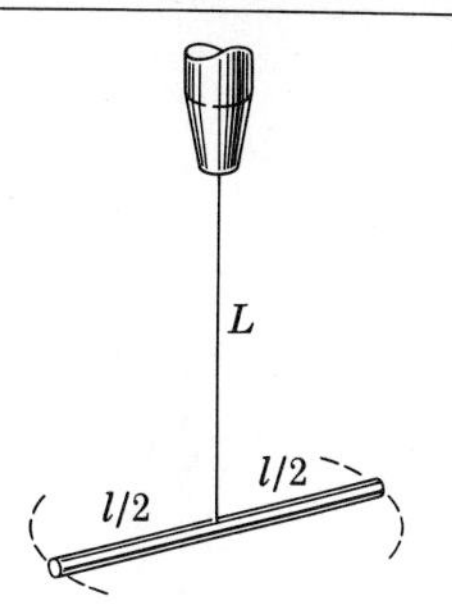

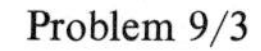
Problem 9/3

9/4 The flywheel is suspended from its center by a wire from a fixed support, and a period τ_1 is measured for torsional oscillation of the flywheel about the vertical axis. Two small weights, each of mass m, are next attached to the flywheel in opposite positions at a distance r from the center. This additional mass results in a slightly longer period τ_2. Write an expression for the moment of inertia I of the flywheel in terms of the measured quantities.

$$\textit{Ans. } I = \frac{2mr^2}{(\tau_2/\tau_1)^2 - 1}$$

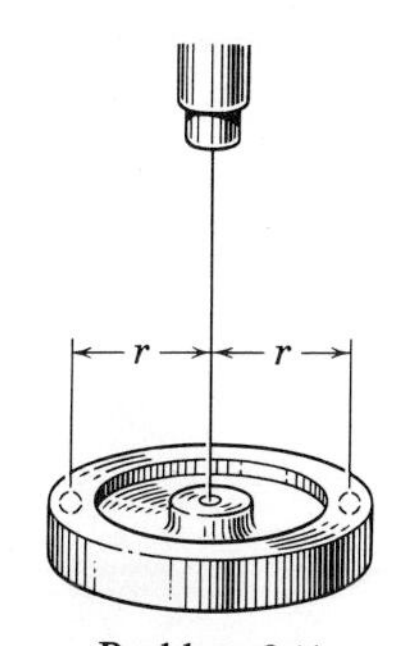

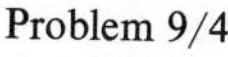
Problem 9/4

9/5 Determine the period τ for the uniform circular hoop of radius r as it oscillates with small amplitude about the horizontal knife edge.

$$\textit{Ans. } \tau = 2\pi\sqrt{\frac{2r}{g}}$$

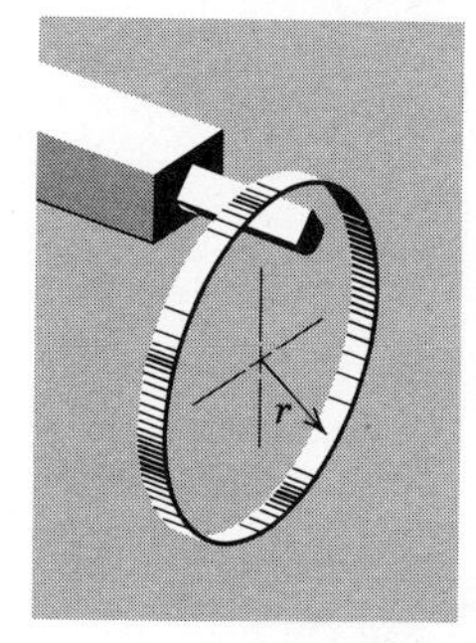

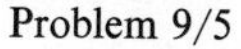
Problem 9/5

9/6 Determine the natural frequency f for small oscillations in the vertical plane about the bearing O for the semicircular disk of radius r.

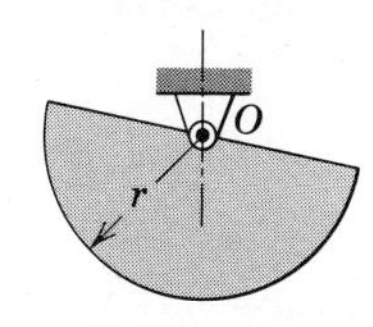

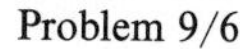
Problem 9/6

9/7 The center of mass G of the ship may be assumed to be at the center of the equivalent 15-m square section. The metacentric height h, determined by the intersection M of the force W acting through the center of buoyancy B with the center line of the ship, is 0.9 m. Determine the period τ of one complete roll of the ship if the amplitude is small and the resistance of the water is neglected. Neglect also the change in cross section of the ship at the bow and stern, and treat the ship as a uniform solid block of square cross section. *Ans.* $\tau = 12.95$ s

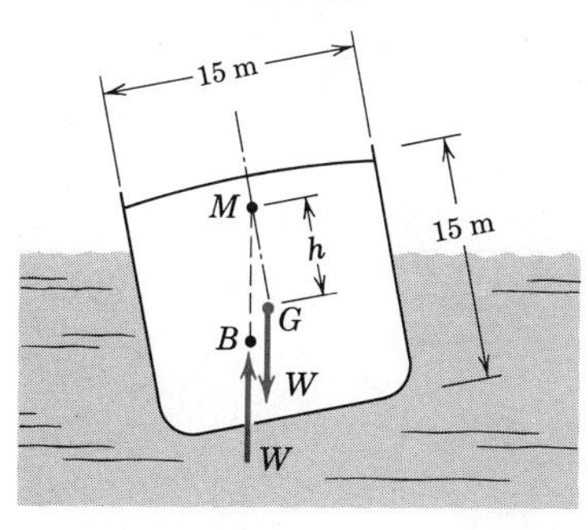

Problem 9/7

9/8 Turn the mechanism of Sample Problem 9/1 upside down and solve for the period τ for small oscillations. What can be said concerning stability?

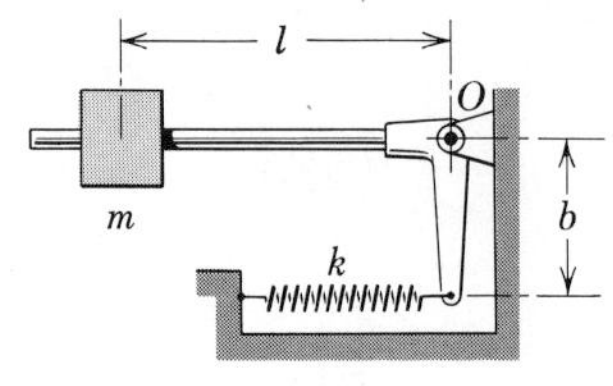

Problem 9/9

9/9 Determine the expression for the natural frequency f of small oscillations of the weighted arm about O. The stiffness of the spring is k and its length is adjusted so that the arm is in equilibrium in the horizontal position shown. Neglect the mass of the spring and arm compared with m.

$$Ans.\ f = \frac{1}{2\pi}\frac{b}{l}\sqrt{\frac{k}{m}}$$

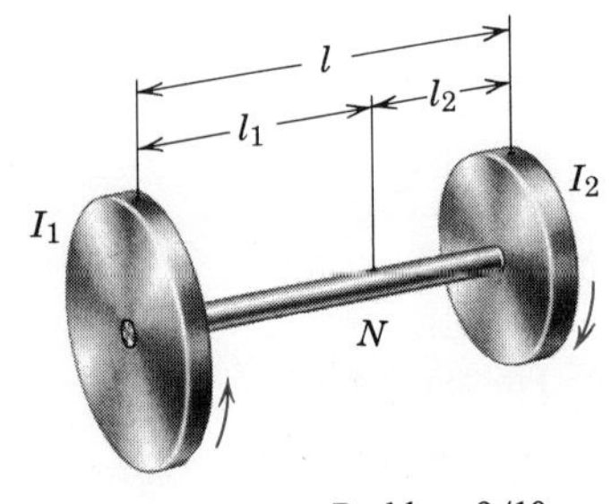

Problem 9/10

9/10 The torsional vibration of a shaft is frequently encountered when two rotors mounted on a shaft (such as a ship's propeller and its driving turbine) oscillate in opposite directions relative to one another. This oscillation represents a deviation from the steady rotation of the shaft and can be analyzed as an independent motion. For the simplified model shown, determine the location of section N where no oscillation of the shaft occurs (nodal section). Next find the period τ of the vibration. The rotors have moments of inertia I_1 and I_2, and the torsional resistance of the shaft to angular displacement θ in radians is $(JG/l)\theta$ where l is the length of the shaft in rotation (l_1 for I_1 and l_2 for I_2). The polar moment of inertia of the shaft cross section is J, and the shear modulus of the shaft is G. Neglect the mass of the shaft.

$$Ans.\ \tau = 2\pi\sqrt{\frac{I_1 I_2}{I_1 + I_2}\frac{l}{JG}}$$

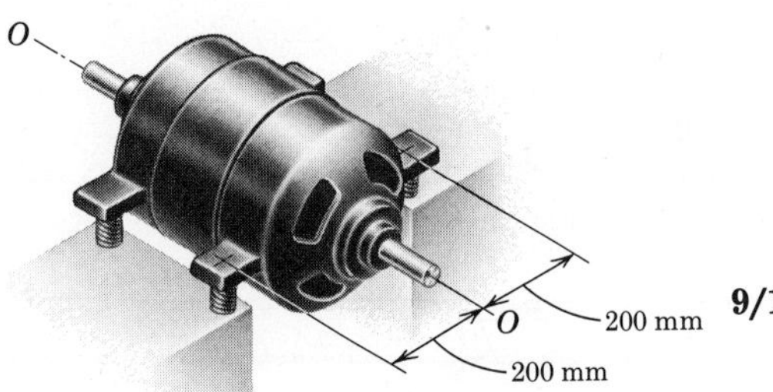

Problem 9/11

9/11 When the motor is slowly brought up to speed, a rather large vibratory oscillation of the entire motor about O-O occurs at a speed of 360 rev/min which shows that this speed corresponds to the natural frequency of free oscillation of the motor. If the motor has a mass of 43 kg and a radius of gyration of 100 mm about O-O, determine the stiffness k of each of the four identical spring mounts.

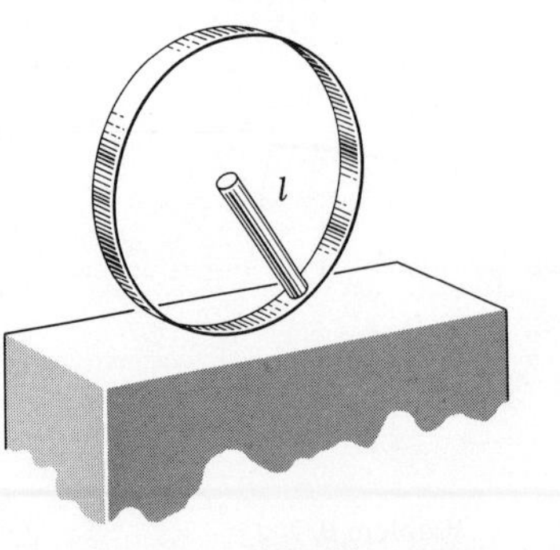

Problem 9/12

9/12 A uniform rod of mass m and length l is welded at one end to the rim of a light circular hoop of radius l. The other end lies at the center of the hoop. Determine the period τ for small oscillations about the vertical position of the bar if the hoop rolls on the horizontal surface without slipping.

$$Ans.\ \tau = 2\pi\sqrt{\frac{2l}{3g}}$$

9/13 The homogeneous solid cylinder of mass m and radius r rolls without slipping during its oscillation on the circular surface of radius R. If the motion is confined to small amplitudes, determine the period τ of oscillation. Solve by applying the equation of motion directly to the free-body diagram of the cylinder.

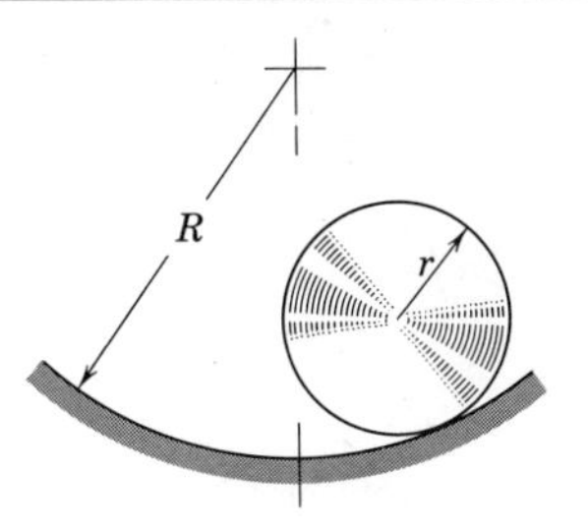

Problem 9/13

9/14 Derive the differential equation of motion for small oscillations of the rolling cylinder of Prob. 9/13 by direct differentiation of the energy equation. Write the expression for the frequency f of oscillation.

Ans. $f = \frac{1}{2\pi}\sqrt{\frac{2g}{3(R-r)}}$

9/15 An elastic spring of stiffness k is connected directly to a viscous fluid damper and released from rest with a displacement x_0 from the zero-force position. Neglect any mass of the system and express the displacement x as a function of the time t after release.

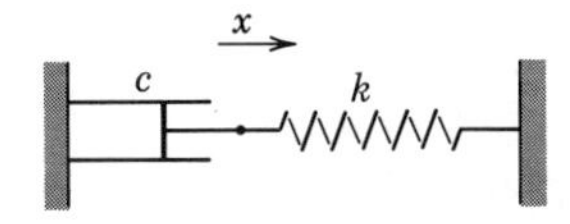

Problem 9/15

9/16 A linear harmonic oscillator having a mass of 1.10 kg is set into motion with viscous damping. If the frequency is 10 Hz and if two successive amplitudes a full cycle apart are measured to be 4.65 mm and 4.30 mm, compute the viscous damping coefficient c. *Ans.* $c = 1.722$ N·s/m

9/17 A mass m supported by an elastic spring of stiffness k is critically damped. Derive an expression for the displacement x_1 from the neutral position t_1 seconds after the mass is released from rest with a displacement x_0.

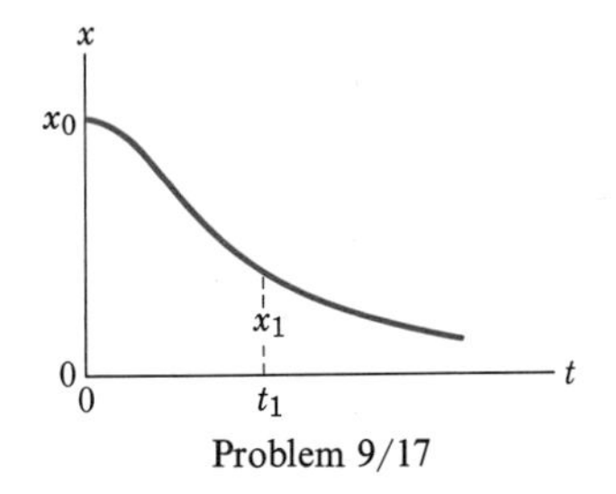

Problem 9/17

9/18 A linear oscillator with mass m, spring stiffness k, and damping coefficient c is set into motion when released from a displaced position. Derive an expression for the energy loss E in one cycle in terms of the amplitude x_1 at the start of the cycle. (See Fig. 87.)

Ans. $E = \frac{1}{2}kx_1^2(1 - e^{-2\pi c/(mq)})$

where $q = \sqrt{\frac{k}{m} - \left(\frac{c}{2m}\right)^2}$

9/19 If the amplitude of the seventh cycle of a damped linear oscillation is thirty times the amplitude of the nineteenth cycle, compute the damping ratio c/c_{cr}. *Ans.* $c/c_{cr} = 0.0451$

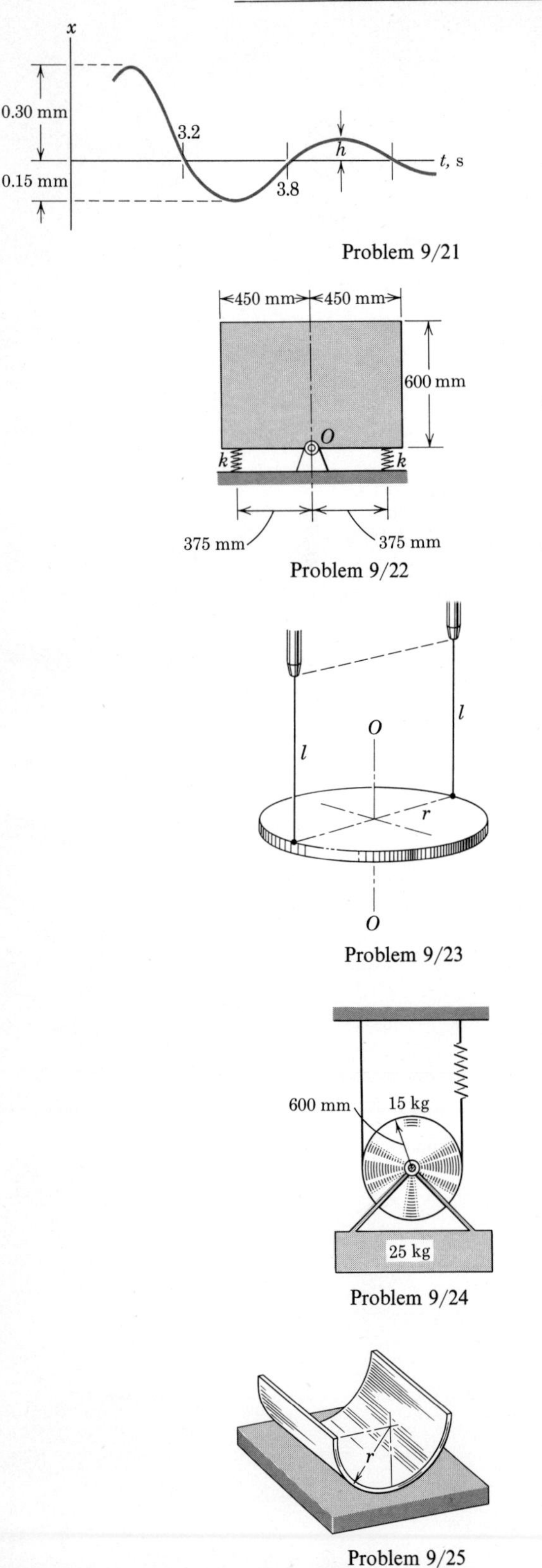

Problem 9/21

Problem 9/22

Problem 9/23

Problem 9/24

Problem 9/25

9/20 The period τ of damped linear oscillation for a certain 1-kg mass is 0.32 s. If the stiffness of the supporting linear spring is 850 N/m, calculate the damping coefficient c and the critical value c_{cr} for the motion.

Ans. $c = 43.1$ N·s/m, $c_{cr} = 58.3$ N·s/m

9/21 The signal from a vibration transducer records a decaying amplitude of a freely vibrating body with a mass of 4 kg. Calculate the indicated amplitude h, the damping coefficient c, and the stiffness k of the elastic support.

9/22 The homogeneous 250-kg rectangular block is pivoted about a horizontal axis through O and supported by two springs each of stiffness k. The base of the block is horizontal in the equilibrium position with each spring under a compressive force of 250 N. Determine the minimum stiffness k of the springs which will ensure vibration about the equilibrium position.

Ans. $k_{min} = 2.62$ kN/m

9/23 The uniform circular disk of mass m and radius r is suspended by the two wires of length l from two fixed points on the same horizontal line and a distance $2r$ apart. Determine the expression for the period τ for small oscillations about the central axis O-O. Neglect torsional resistance in the wires.

9/24 Determine the natural frequency f of vertical vibration of the 25-kg mass after it is released from a displaced position. The pulley has a mass of 15 kg with a centroidal radius of gyration of 450 mm, and the spring has a stiffness of 540 N/m.

Ans. $f = 1.063$ Hz

9/25 The semicircular cylindrical shell of radius r with small but uniform wall thickness is set into small rocking oscillation on the horizontal surface. If no slipping occurs, determine the expression for the period τ of each complete oscillation. Solve by the work-energy method.

Ans. $\tau = 2\pi\sqrt{\dfrac{r(\pi - 2)}{g}}$

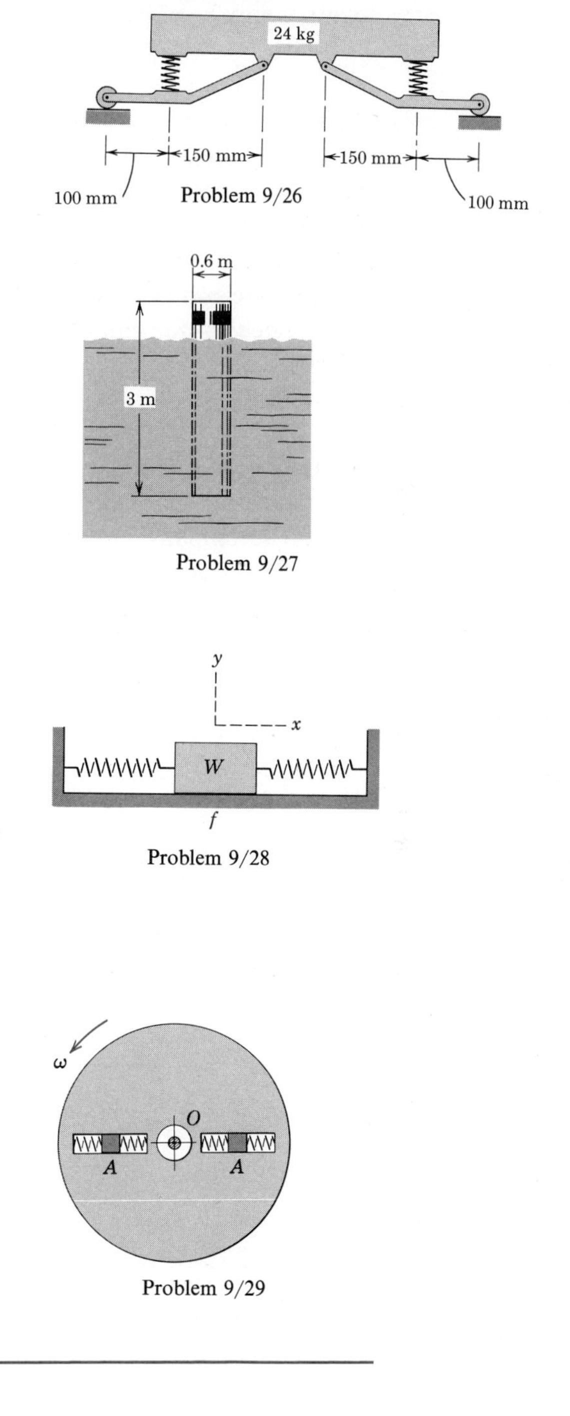

9/26 If the spring-loaded frame is given a slight vertical disturbance from its equilibrium position shown, determine its natural frequency f of vibration. The mass of the upper member is 24 kg, and that of the lower members is negligible. Each spring has a stiffness of 9 kN/m.

Ans. $f = 2.62$ Hz

◀ **9/27** The cylindrical buoy shown floats in salt water (1.03 t/m^3) and has a total mass of 0.8 t with a low center of mass so that it is stable in the upright position. When the buoy is raised 0.3 m above its floating position and released, it is observed to come to within 0.1 m of the release position upon rising the first time. If the frictional resistance is f newtons per square metre of submerged area of the vertical cylindrical surface for each metre per second of buoy velocity, calculate f assuming the submerged area subjected to damping forces is constant at its equilibrium value. *Ans.* $f = 37.6$ N·s/m^3

◀ **9/28** Investigate the case of Coulomb damping for the block shown, where the coefficient of kinetic friction is f and each spring has a stiffness $k/2$. The block is displaced a distance x_0 from the neutral position and released. Determine the differential equation of motion and solve. Plot the resulting vibration and indicate the rate of decay of amplitude with time.

Ans. $$x = \left(x_0 - \frac{fmg}{k}\right)\cos\sqrt{\frac{k}{m}}\,t + \frac{fmg}{k}$$

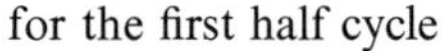

for the first half cycle

9/29 Each of the two slider blocks A has a mass m and is constrained to move in the smooth radial slots of the flywheel. Each of the four springs has a stiffness of $k/2$ and is in compression at all times. The blocks are both at a distance r_0 from O when the wheel is at rest. Determine the frequency f of vibration of the blocks for a constant speed ω of the flywheel. What is the significance of the condition when $\omega^2 \geq k/m$?

Ans. $$f = \frac{1}{2\pi}\sqrt{\frac{k}{m} - \omega^2}$$

---

49 Forced Response. When either the applied force $F = f(t)$ or the foundation movement $\delta = \delta(t)$ of the linear spring-mass system of Fig. 84 is not zero, a forced response results which may be described by a solution to the complete Eq. 175*a* or 175*b*. From the theory of linear differential equations it is recalled that the complete solution to the nonhomogeneous linear equation may be written as

$$x = x_c + x_p$$

where x_c is the general solution of the homogeneous equation, with $F = 0$ or $\delta = 0$, and where x_p is *any* particular solution to the entire equation. From the linearity of the differential equation it is noted that solutions to the complete equation can be added together to form new solutions. This property of the linear differential equation is known as the *principle of superposition.*

There are two classes of response which are of interest. The first class is that of a *steady-state* vibration which results from a regular and periodic forcing function $F(t)$ or $\delta(t)$ of constant amplitude. The second type of motion is a *transient response* where $F(t)$ or $\delta(t)$ does not have a steady-state periodic form.

(*a*) *Steady-state vibration.* The steady-state condition arises where the forcing function has a steady oscillation. If the oscillation is harmonic, such as would accompany a rotating device, the forcing function may be described by $F = F_0 \sin \omega t$ for the system of Fig. 84*a* or $\delta = \delta_0 \sin \omega t$ for the system of Fig. 84*b*. Either a sine or cosine expression may be used depending on how the time is counted. The motion equations for the respective systems become

$$\ddot{x} + \frac{c}{m}\dot{x} + \frac{k}{m}x = \frac{F_0}{m}\sin \omega t \tag{181a}$$

$$\ddot{x} + \frac{c}{m}\dot{x} + \frac{k}{m}x = \frac{k\delta_0}{m}\sin \omega t \tag{181b}$$

The first of these equations will now be solved. The complementary solution was obtained in Eq. 179 and is

$$x_c = e^{-bt}(C_1 \sin qt + C_2 \cos qt)$$

For the particular solution the expression

$$x_p = A \sin \omega t + B \cos \omega t$$

is tried where ω is the circular frequency of the applied force and where the coefficients A and B are to be determined from the differential equation. Direct substitution into Eq. 181*a* and rearrangement of terms give

$$\left[-A\omega^2 - \frac{c}{m}B\omega + \frac{k}{m}A - \frac{F_0}{m}\right]\sin \omega t + \left[-B\omega^2 + \frac{c}{m}A\omega + \frac{k}{m}B\right]\cos \omega t = 0$$

This equation must hold for all values of the time t, so that the coefficients of the $\sin \omega t$ and $\cos \omega t$ terms are identically zero. Thus

$$A\left(\frac{k}{m} - \omega^2\right) - B\left(\frac{c\omega}{m}\right) = \frac{F_0}{m}$$

$$A\left(\frac{c\omega}{m}\right) + B\left(\frac{k}{m} - \omega^2\right) = 0$$

which give

$$A = \frac{\dfrac{F_0}{m}\left(\dfrac{k}{m} - \omega^2\right)}{\left(\dfrac{k}{m} - \omega^2\right)^2 + \left(\dfrac{c\omega}{m}\right)^2} \qquad B = \frac{-\dfrac{F_0}{m}\dfrac{c\omega}{m}}{\left(\dfrac{k}{m} - \omega^2\right)^2 + \left(\dfrac{c\omega}{m}\right)^2}$$

With the substitution $p^2 = k/m$ and with the solution written in the form $x_p = x_0 \sin(\omega t - \phi)$, then

$$x_0 = \sqrt{A^2 + B^2} = \frac{F_0/m}{\sqrt{(p^2 - \omega^2)^2 + (c\omega/m)^2}}$$

$$\phi = \tan^{-1}\left(\frac{-B}{A}\right) = \tan^{-1}\left(\frac{c\omega/m}{p^2 - \omega^2}\right)$$

The angle ϕ is a phase angle between F_0 and x_0 considered as rotating vectors. The complete solution may now be written as

$$x = e^{-bt}(C_1 \sin qt + C_2 \cos qt) + \frac{F_0/m}{\sqrt{(p^2 - \omega^2)^2 + (c\omega/m)^2}} \sin(\omega t - \phi) \qquad (182)$$

The first term in Eq. 182 dies out with time since it contains the diminishing factor e^{-bt} and is known as the *transient solution.* The second term continues with the amplitude x_0 and has a circular frequency ω. Figure 89 is a schematic plot of Eq. 182. The period of the steady-state motion is seen to be $\tau = 2\pi/\omega$.

If the amplitude x_0 of the steady-state term is divided by the static displacement $\delta_0 = F_0/k$ which m would have under the action of a steady force F_0 only, the ratio x_0/δ_0, known as the *magnification factor,* becomes

$$\blacktriangleright \quad \frac{x_0}{\delta_0} = \frac{1}{\sqrt{(1 - \mu^2)^2 + (2\eta\mu)^2}} \quad \text{with} \quad \phi = \tan^{-1}\frac{2\eta\mu}{1 - \mu^2} \qquad \textbf{(183)}$$

In these expressions $\eta = c/c_{cr} = c/(2\sqrt{km})$ is known as the *damping ratio,* and $\mu = \omega/p$ is called the *frequency ratio.*

It is seen from Eq. 183 that the amplitude becomes extremely large as ω approaches p if the damping c is small. In the limit for $c = 0$, the amplitude

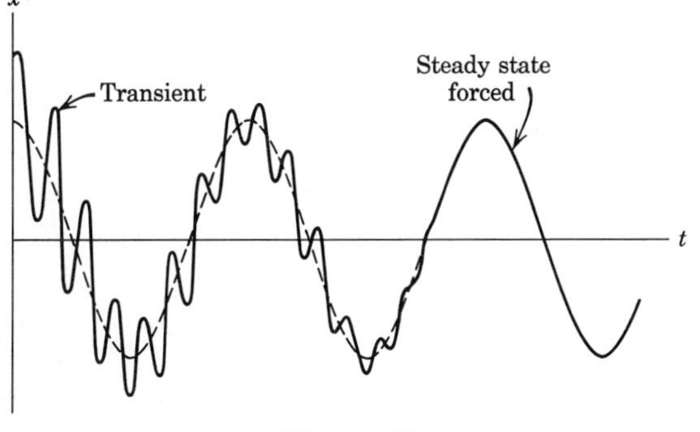

Figure 89

tends toward infinity as ω approaches the natural frequency p. This condition is called *resonance* and is a situation generally to be avoided. Thus, if rotating machinery is operated at or near the natural frequency for free vibrations of the system, abnormally large vibrations may result. Similarly, if the foundation to which the spring is attached in Fig. 84*b* vibrates at or near the natural frequency, large amplitudes of the vibrating mass will result. A plot of the magnitude of the magnification factor x_0/δ_0 as a function of the dimensionless frequency ratio μ for various damping ratios η is shown in Fig. 90*a* and gives what are called the *frequency response* curves. In Fig. 90*b* is shown a plot of the phase angle ϕ as a function of frequency ratio μ for various damping ratios η. The phase angle represents the time ϕ/ω by which the applied force F (or foundation motion δ) leads the resulting vibration x_p.

When the forcing function $F(t)$ or $\delta(t)$ of Eq. 175*a* or Eq. 175*b* is periodic and regular but nonharmonic, the response will be a steady-state motion, but it will no longer be represented by a single sine or cosine expression. In Fig. 91*a* are shown five examples of forcing functions or excitations which are periodic but nonharmonic. Three methods used to analyze the response of the mass to such forcing functions are the phase-plane method, harmonic analysis by expressing the input wave shape in a Fourier series, and the Laplace transform. These methods will not be developed here but are described in detail in a large number of references on vibrations, linear systems, and circuit analysis.

It is usually desirable to reduce as much as possible the forced vibrations which are generated in engineering structures and machines. Vibration reduction is normally accomplished in any of four ways: (1) reduction or elimination of the exciting force by balancing or other removal, (2) introduction of sufficient damping to limit the amplitude, (3) isolation of the body from the vibration source by providing elastic mountings of the proper stiffness, and (4) operation at a forced frequency sufficiently different from the natural frequency so as to avoid resonance.

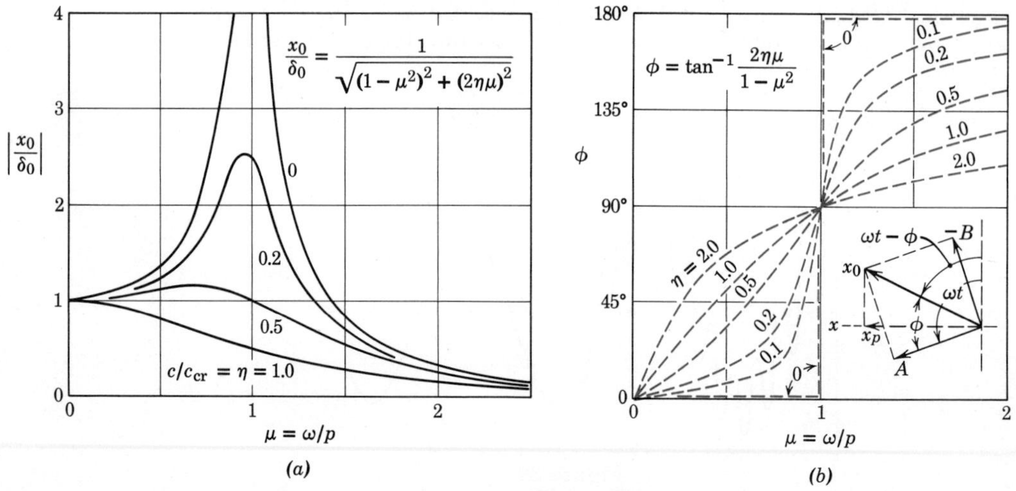

Figure 90

(*b*) *Transient response.* When the forcing function or excitation is nonperiodic, a nonsteady-state or transient response results. Examples of such excitation are shown in Fig. 91*b*. Included also is an example of a random input which has no repeated pattern. A few of the inputs which are expressible in relatively simple mathematical terms permit a direct integration of the differential equation. There are many others, however, for which one of the methods mentioned in the preceding paragraph may be used in determining the response. Again, the reader who wishes to learn more about the specialized subject of mechanical vibration is directed to references which cover the field of linear and nonlinear systems analysis.

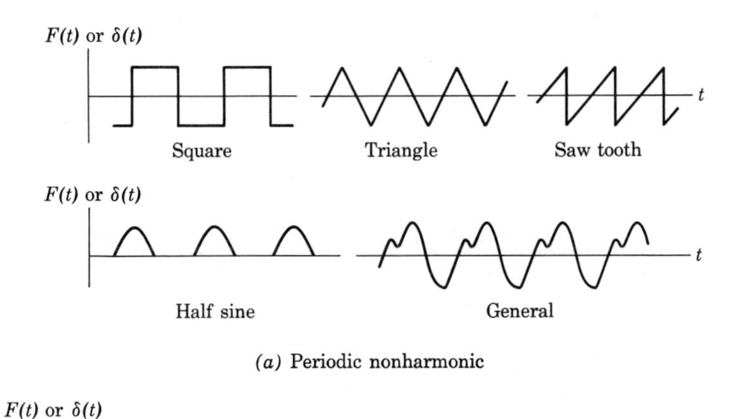

(*a*) Periodic nonharmonic

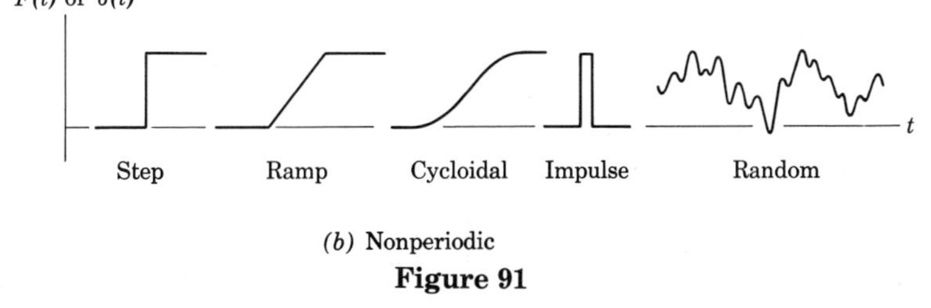

(*b*) Nonperiodic

Figure 91

Sample Problem

9/30 The elements of a *seismic* instrument are shown in the *a*-part of the figure where the mass m is supported by a spring of stiffness k and suppressed by internal viscous damping c. The entire assembly is subject to a motion δ of the frame. Examine (*a*) the characteristics of this instrument for measuring both the acceleration $\ddot{\delta}$ and the displacement δ of a vibrating structure to which it is attached, and (*b*) the response of the instrument to a constant acceleration a_1 starting from a rest position.

Solution: The response of the seismic mass is measured in terms of its displacement x from the equilibrium position relative to the case in which the mass moves. From the free-body diagram of m the equation of motion is

$$[\Sigma F = ma] \qquad -c\dot{x} - kx = m\frac{d^2}{dt^2}(x + \delta), \qquad \ddot{x} + \frac{c}{m}\dot{x} + \frac{k}{m}x = -\ddot{\delta}$$

where the damping force depends on the relative velocity $\dot{x}$ and the spring force depends on the relative displacement x.

(*a*) When the structure to which the instrument is attached has a harmonic oscillation $\delta = \delta_0 \sin \omega t$, the motion equation is

$$\ddot{x} + \frac{c}{m}\dot{x} + \frac{k}{m}x = \delta_0\omega^2 \sin \omega t$$

which is the same as Eq. 181*b* if ω^2 is substituted for k/m.

The steady-state solution is given by the second term of Eq. 182 where F_0/m is replaced by $\delta_0\omega^2$. Thus

$$x = \frac{\delta_0\omega^2}{\sqrt{(p^2 - \omega^2)^2 + (c\omega/m)^2}} \sin(\omega t - \phi)$$

If x_0 is the amplitude of the relative response x, the ratio x_0/δ_0 becomes

$$\frac{x_0}{\delta_0} = \frac{\mu^2}{\sqrt{(1-\mu^2)^2 + (2\eta\mu)^2}} \qquad \text{with} \qquad \phi = \tan^{-1}\frac{2\eta\mu}{1-\mu^2}$$

where, as before, $\mu = \omega/p$ and $\eta = c/(2\sqrt{km})$. Also, δ_0 is the actual amplitude of structure vibration whereas the δ_0 in Eq. 183 was defined as the static deflection of m under a force F_0. The plot of x_0/δ_0 as a function of $\mu = \omega/p$ is shown in the *b*-part of the figure. The similarity with the frequency response curves for absolute vibration, Fig. 90*a*, should be noted. The curves for no damping $\eta = 0$, critical damping $\eta = 1$, and an intermediate value with $\eta = 0.5$ are shown. The phase-angle relations are identical to those of Fig. 90*b*.

It is seen from the *b*-part of the present figure that when the natural frequency p of the seismic mass is high in comparison with ω, then μ is small and the denominator of

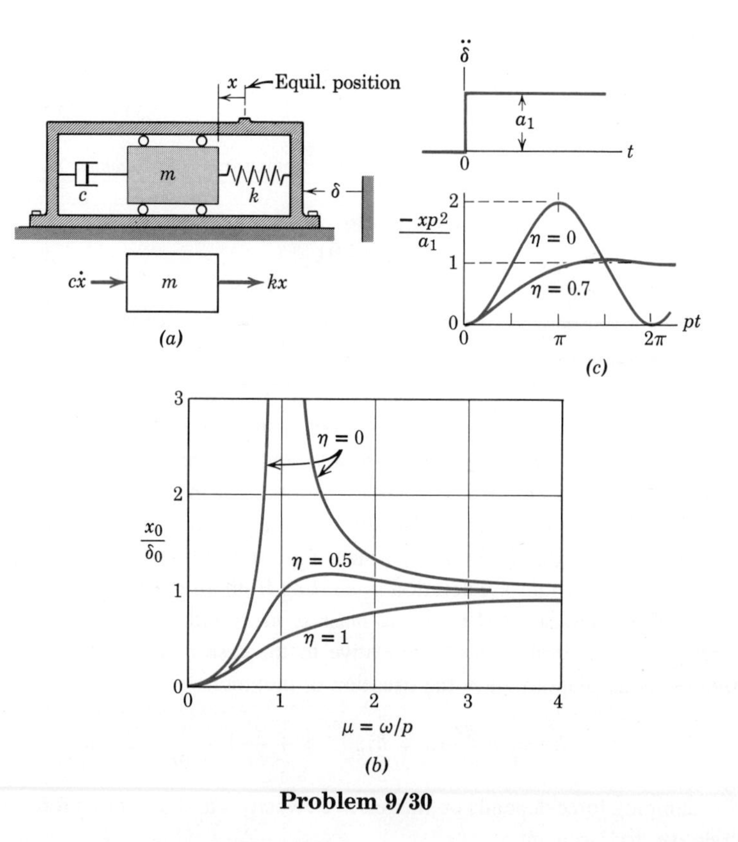

Problem 9/30

the expression for x_0/δ_0 approaches unity. With this approximation the relative displacement is

$$\frac{x_0}{\delta_0} = \mu^2 \qquad \text{or} \qquad \frac{x_0}{\delta_0} = \frac{\omega^2}{p^2}$$

But since $\delta = \delta_0 \sin \omega t$, the maximum acceleration is $|\ddot{\delta}|_{\max} = \delta_0\omega^2$ so that $|\ddot{\delta}|_{\max} = x_0 p^2$. Thus the maximum magnitude of the acceleration is proportional to the magnitude x_0 of the response, and the instrument may be used as an *accelerometer.*

It is also seen from the b-part of the figure that when the natural frequency p of the seismic mass is low in comparison with ω, say $\mu > 3$, then $x_0 \approx \delta_0$ regardless of the damping ratio η. Under these conditions, the instrument acts as a *displacement meter* and indicates the movement of the structure. For large values of μ, the major influence of damping is to control the phase shift ϕ.

(b) When there is a sudden acceleration $\ddot{\delta} = a_1$ starting from rest, as indicated in the c-part of the figure, the differential equation is

$$\ddot{x} + \frac{c}{m}\dot{x} + \frac{k}{m}x = -a_1$$

The particular solution is seen to be $x_p = -ma_1/k$, and the complete solution is

$$x = e^{-bt}(C_1 \sin qt + C_2 \cos qt) - ma_1/k$$

where the quantities q and b were defined with Eq. 179. Substitution of the initial conditions $x = 0$ and $\dot{x} = 0$ when $t = 0$ gives

$$x = \frac{ma_1}{k}\left[e^{-bt}\left(\frac{b}{q}\sin qt + \cos qt\right) - 1\right]$$

With the further substitution of $k/m = p^2$, $\eta = c/(2\sqrt{km})$, $b = \eta p$, $q = p\sqrt{1-\eta^2}$, the dimensionless displacement is

$$\frac{xp^2}{a_1} = e^{-\eta pt}\left(\frac{\eta}{\sqrt{1-\eta^2}}\sin p\sqrt{1-\eta^2}\,t + \cos p\sqrt{1-\eta^2}\,t\right) - 1 \qquad \textit{Ans.}$$

A plot of this response for the value $\eta = 0.7$ is shown in the c-part of the figure, and it is seen that only a very small overshoot is observed when pt goes from π to 2π. Also shown is the response for no damping, $\eta = 0$.

Problems

9/31 A spring-mounted oscillator with a natural undamped frequency of 4 Hz and a damping ratio $\eta = 0.2$ is mounted on a frame which vibrates with a frequency f which can be varied. If the amplitude of the absolute vibration of the oscillator is not to exceed twice that of the frame, specify by reference to Fig. 90 the permissible range of frequencies of the frame.

Ans. $f < 3.12$ Hz or $f > 4.44$ Hz

9/32 A seismic instrument has a 0.25-kg mass supported by a stiff spring with a stiffness of 3.6 kN/m. If the deflection x_0 of the mass relative to its case is measured to be 5.2 mm when the instrument is mounted on a machine which vibrates in the direction of x_0 with a frequency of 5 Hz, approximate the maximum acceleration a_0 of the machine. *Ans.* $a_0 = 7.6g$

9/33 The time rate of change of acceleration is called "jerk" J. Specify the maximum magnitude of the amplitude x_0, measured from the neutral position, for the undamped linear harmonic oscillator of mass m which will limit the jerk to a value J_0. The stiffness of the elastic supports is k.

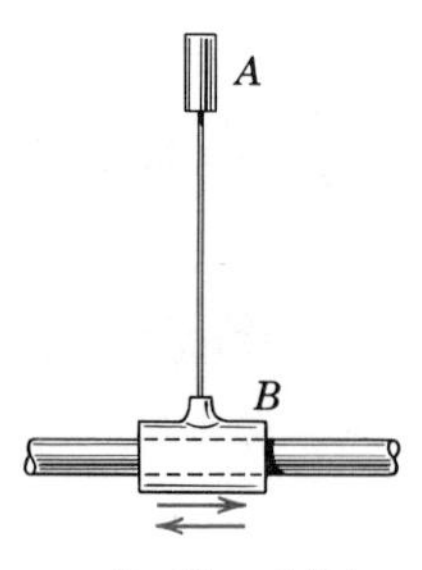

Problem 9/34

9/34 A static horizontal force of 10 N gives the 1.5-kg mass A a deflection of 12 mm against the elasticity of the light cantilever spring to which it is attached. If the base B of the spring is given a horizontal harmonic oscillation with a frequency of 2 Hz and an amplitude of 6 mm, calculate the amplitude x_0 of the resulting vibration of A. Assume negligible damping.
Ans. $x_0 = 8.38$ mm

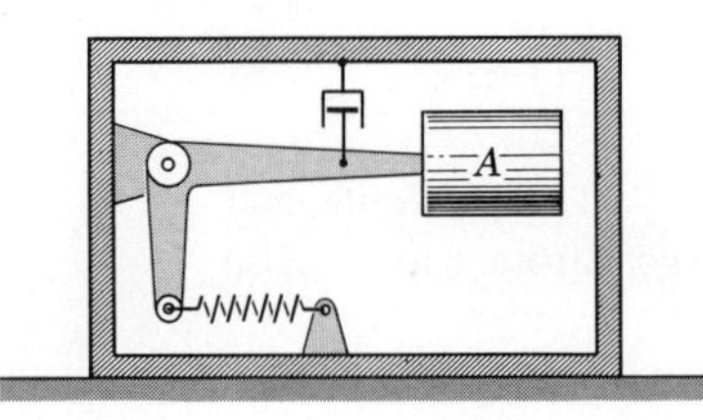

Problem 9/35

9/35 The seismic instrument shown is secured to a ship's deck near the stern where propeller-induced vibration is most pronounced. The ship has a single 3-bladed propeller which turns at 180 rev/min and operates partly out of water, thus causing a shock as each blade breaks the surface. The damping ratio of the instrument is $\eta = 0.5$, and its undamped natural frequency is 3 Hz. If the measured amplitude of A relative to its frame is 0.75 mm, compute the amplitude δ_0 of the vertical vibration of the deck.

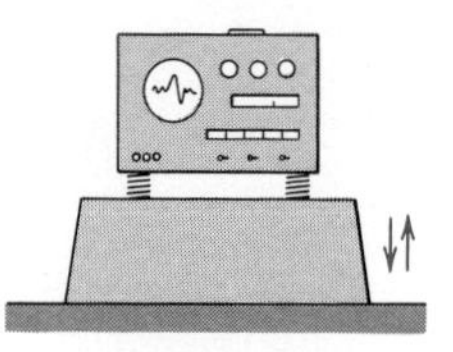
Problem 9/36

9/36 The instrument shown has a mass of 43 kg and is spring-mounted to the horizontal base. If the amplitude of vertical vibration of the base is 0.10 mm, calculate the range of frequencies f of the base vibration which must be prohibited if the amplitude of vertical vibration of the instrument is not to exceed 0.15 mm. Each of the four identical springs has a stiffness of 7.2 kN/m.
Ans. Prohibited range $2.38 < f < 5.32$ Hz

9/37 A spring-mounted machine with a mass of 24 kg is observed to vibrate harmonically in the vertical direction with an amplitude of 0.30 mm under the action of a vertical force which varies harmonically between F_0 and $-F_0$ with a frequency of 4 Hz. Damping is negligible. If a static force of magnitude F_0 causes a deflection of 0.60 mm, calculate the equivalent spring constant k for the springs which support the machine.

9/38 The circular disk of mass m is secured to an elastic shaft which is mounted in a rigid bearing at A. With the disk at rest a lateral force P applied to the disk produces a lateral deflection Δ, so that the equivalent spring constant is $k=P/\Delta$. If the center of mass of the disk is off center by a small amount e from the shaft center line, determine the expression for the lateral deflection δ of the shaft due to unbalance at a shaft speed ω in terms of the natural frequency $p = \sqrt{k/m}$ of lateral vibration of the shaft. At what critical speed ω_c would the deflection tend to become large? Neglect damping.

Ans. $\omega_c = p = \sqrt{k/m}$

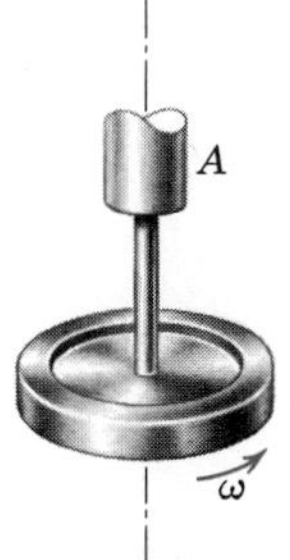

Problem 9/38

9/39 An electric motor has a rotor whose center of gravity is off center a small distance e. If the motor is mounted on spring pads and is constrained to vibrate in the vertical direction only, prove that Eq. 181a and its solution including Fig. 90 apply where the mass term m is the mass of the stator plus that of the rotor.

9/40 Show that the frequency response curves for the seismic instrument of Sample Prob. 9/30 can be used for the elastically mounted motor of Prob. 9/39 if the magnification factor is defined as the ratio $mx_0/(m'e)$ where m is the total mass of stator plus rotor, x_0 is the amplitude of the absolute vibration of the motor, m' is the mass of the rotor, and e is the eccentricity of the rotor.

9/41 A damped linear oscillator initially at rest in a neutral position is subjected to an impulse loading F as shown. If the duration of the pulse is small and if the total impulse is I, sketch the resulting response of the oscillator. Assume a square-wave form for the impulse.

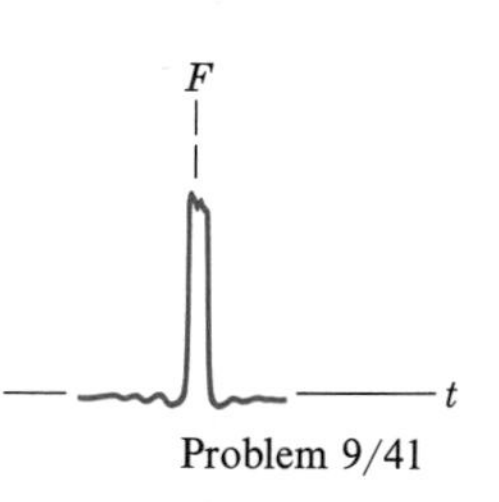

Problem 9/41

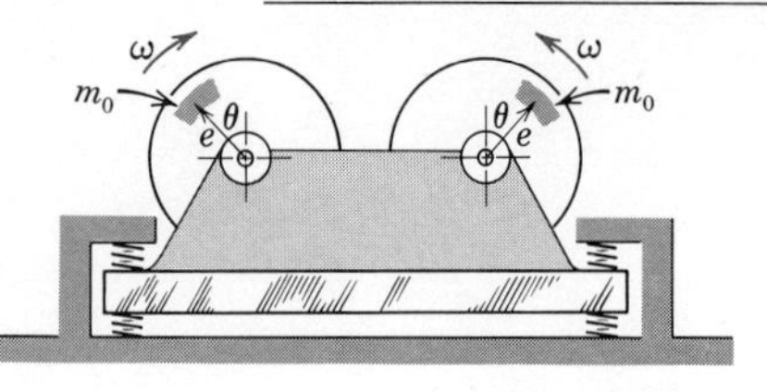

Problem 9/42

9/42 A device to produce vibrations consists of the two counter-rotating wheels each carrying an eccentric mass $m_0 = 1$ kg with a center of mass at a distance $e = 12$ mm from its axis of rotation. The wheels are synchronized so that the vertical positions of the unbalanced masses are always identical. The total mass of the device is 10 kg. Determine the two possible values of the equivalent spring constant k for the mounting which will permit the amplitude of the periodic force transmitted to the fixed mounting to be 1500 N due to the unbalance of the rotors at a speed of 1800 rev/min. Neglect damping.

Ans. $k = 823$ kN/m or 227 kN/m

9/43 An undamped spring-mass system is initially at rest in its equilibrium position at time $t = 0$. If the mass m is subjected to a force $F = F_0 e^{-bt}$ applied in the direction of the displacement x of m, determine the relationship between x and t which describes the subsequent motion. The constant b is a measure of the rate of decay of F with t.

Ans. $x = \dfrac{F_0}{mb^2 + k}\left[\dfrac{b}{p}\sin pt - \cos pt + e^{-bt}\right]$

where $p = \sqrt{k/m}$

9/44 Determine and plot the relation between the damping ratio η and the frequency ratio μ which will ensure that the relative displacement of a seismic mass to its frame will be proportional to the impressed acceleration. Show why $\eta = 0.7$ is a desired value for $\mu < 0.5$.

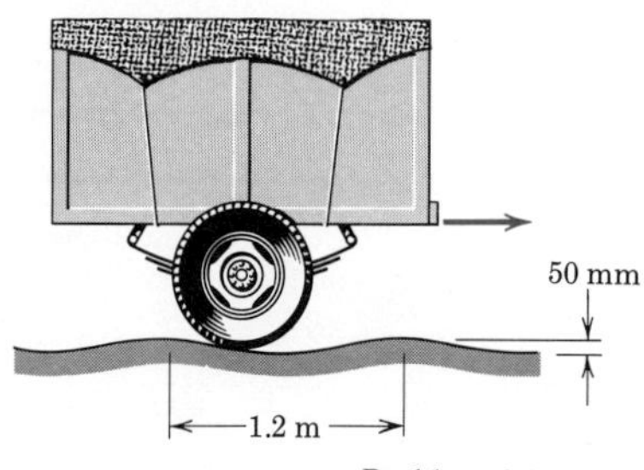

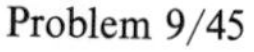

Problem 9/45

◀**9/45** Determine the amplitude of vertical vibration of the spring-mounted trailer as it travels at a velocity of 25 km/h over the corduroy road whose contour may be expressed by a sine or cosine term. The mass of the trailer is 500 kg and that of the wheels alone may be neglected. During the loading each 75 kg added to the load caused the trailer to sag 3 mm on its springs. Assume that the wheels are in contact with the road at all times and neglect damping. At what critical speed v_c is the vibration of the trailer greatest?

Ans. $x_0 = 14.75$ mm, $v_c = 15.23$ km/h

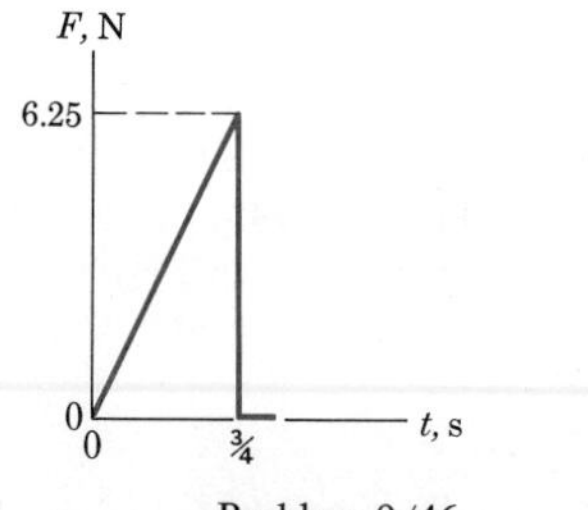

Problem 9/46

◀**9/46** Determine and plot the response $x = f(t)$ of a 0.75-kg undamped harmonic oscillator subjected to a force F which varies linearly with time for the first $\frac{3}{4}$ s as shown. The oscillator is initially at rest at time $t = 0$, and the stiffness of its elastic support is $k = 90$ N/m.

Ans. $x = 92.6\left(t - \dfrac{1}{10.95}\sin 10.95t\right)$ mm

◀ **9/47** Determine the expression for the power loss P averaged over a complete cycle due to the dissipation of frictional energy in a viscously damped linear oscillator. The forcing function is $F_0 \sin \omega t$. Use the notation of this article.

Ans. $P = \dfrac{F_0{}^2 c\omega^2/(2k^2)}{(1 - \mu^2)^2 + (2\eta\mu)^2}$

▼ ▼ ▼ ▼ ▼

50 Two-Degree-of-Freedom Systems. In the foregoing articles of this chapter the response of single-degree-of-freedom systems has been introduced. The response of systems having two or more degrees of freedom can be obtained by several available methods which involve somewhat increased complexity in the formulation of the defining equations and in their solution. Only a brief treatment of one two-degree-of-freedom system, shown in Fig. 92, will be presented here. For a detailed study of this subject the reader is referred to references in mechanical vibrations and in linear systems analysis.

The system of Fig. 92 consists of two masses m_1 and m_2 coupled by elastic springs of stiffness k_1, k_2, and k_3. The time-dependent force $F_1 = f_1(t)$ is applied externally to m_1. Damping will be neglected. The absolute displacements are x_1 and x_2 measured from the neutral positions. The equations of motion for m_1 and m_2 are

$$m_1\ddot{x}_1 + (k_1 + k_2)x_1 - k_2x_2 = F_1$$
$$m_2\ddot{x}_2 - k_2x_1 + (k_2 + k_3)x_2 = 0$$

If a harmonic forcing function $F_1 = F_0 \sin \omega t$ is applied, it would be expected that the resulting steady-state vibrations would also be harmonic with the same frequency. This assumption is easily verified by direct substitution of the trial solutions

$$x_1 = X_1 \sin \omega t \qquad \text{and} \qquad x_2 = X_2 \sin \omega t$$

into the motion equations. This substitution gives

$$X_1(k_1 + k_2 - m_1\omega^2) - k_2X_2 = F_0$$
$$X_2(k_2 + k_3 - m_2\omega^2) - k_2X_1 = 0$$

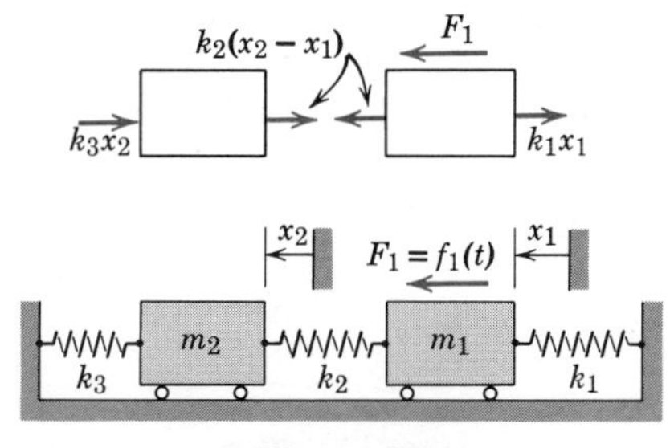

Figure 92

from which

$$X_1 = \frac{(k_2 + k_3 - m_2\omega^2)F_0}{(k_1 + k_2 - m_1\omega^2)(k_2 + k_3 - m_2\omega^2) - k_2^2}$$

$$X_2 = \frac{k_2 F_0}{(k_1 + k_2 - m_1\omega^2)(k_2 + k_3 - m_2\omega^2) - k_2^2}$$

With these values of X_1 and X_2 the assumed expressions for x_1 and x_2 satisfy the differential equations for all values of the time and are, hence, valid solutions. These solutions represent the steady-state motion and are particular integrals of the equations. It is assumed that the transient terms in the general solution have died out.

The critical frequencies for which the expressions for X_1 and X_2 become infinite are obtained by setting the common denominator equal to zero. The resulting relation is

$$\omega^4 - \left(\frac{k_1 + k_2}{m_1} + \frac{k_2 + k_3}{m_2}\right)\omega^2 + \frac{k_1k_2 + k_1k_3 + k_2k_3}{m_1m_2} = 0$$

which has the solutions

$$\omega^2 = \frac{k_1 + k_2}{2m_1} + \frac{k_2 + k_3}{2m_2} \pm \sqrt{\left(\frac{k_1 + k_2}{2m_1} + \frac{k_2 + k_3}{2m_2}\right)^2 - \frac{k_1k_2 + k_1k_3 + k_2k_3}{m_1m_2}} \tag{184}$$

The two values of ω^2 thus obtained give the critical frequencies at which resonance would occur, and these frequencies correspond to the two *natural modes* of free vibration of the system without F_1 acting.

The amplitude ratio for the two motions is

$$\frac{X_2}{X_1} = \frac{\dfrac{k_2}{m_2}}{\dfrac{k_2 + k_3}{m_2} - \omega^2} \tag{185}$$

When the lower value of ω^2 is substituted, it is found upon computation that X_2/X_1 is positive which indicates that the motions are in phase. For the higher value of ω^2 the ratio X_2/X_1 is found to be negative, and the two masses vibrate with a phase difference of 180 deg.

As an aid to the visualization of the resulting motions, the special case of $m = m_1 = m_2$, $k = k_1 = k_2$, and $k_3 = 0$ is treated. Direct substitution yields the two resonant frequencies

$$\omega_1^2 = \frac{3 - \sqrt{5}}{2}\frac{k}{m} \qquad \omega_2^2 = \frac{3 + \sqrt{5}}{2}\frac{k}{m}$$

The amplitude ratio becomes

$$\frac{X_2}{X_1} = \frac{\dfrac{k}{m}}{\dfrac{k}{m} - \omega^2}$$

Substitution of the lower value ω_1^2 gives $X_2/X_1 = +1.62$ which shows that the two motions are in phase. Substitution of the higher value ω_2^2 gives $X_2/X_1 = -0.62$ which shows that the two motions are 180 deg out of phase. With the further substitution $p^2 = k/m$, $\mu^2 = \omega^2/p^2$, and $\delta_0 = F_0/k$, the expressions for the steady-state displacements become

$$x_1 = \frac{(1 - \mu^2)\delta_0}{(2 - \mu^2)(1 - \mu^2) - 1} \sin \omega t$$

$$x_2 = \frac{\delta_0}{(2 - \mu^2)(1 - \mu^2) - 1} \sin \omega t \tag{186}$$

In dimensionless form the amplitude ratio for the motion of m_1 is

$$\frac{X_1}{\delta_0} = \frac{1 - \mu^2}{(2 - \mu^2)(1 - \mu^2) - 1} \tag{187}$$

and its magnitude is plotted in Fig. 93. It is noted that the amplitude X_1 of m_1 goes to zero when $\mu = 1$ which corresponds to the speed for resonance of the right-hand mass if the left-hand mass were absent. Thus, by coupling the added mass to the one to which F is applied, it is possible to eliminate the vibration of the right-hand mass entirely. This phenomenon forms the basis of the *dynamic vibration absorber* which may be designed for various ratios of the masses and corresponding values of the spring stiffnesses.

The foregoing problem is only one example of many multiple-degree-of-freedom systems which are encountered in practice. As long as the supporting springs are elastic and the damping is reasonably viscous, solutions may be obtained by any one of several methods of linear systems analysis. There are, however, situations where the coefficients of the restoring and

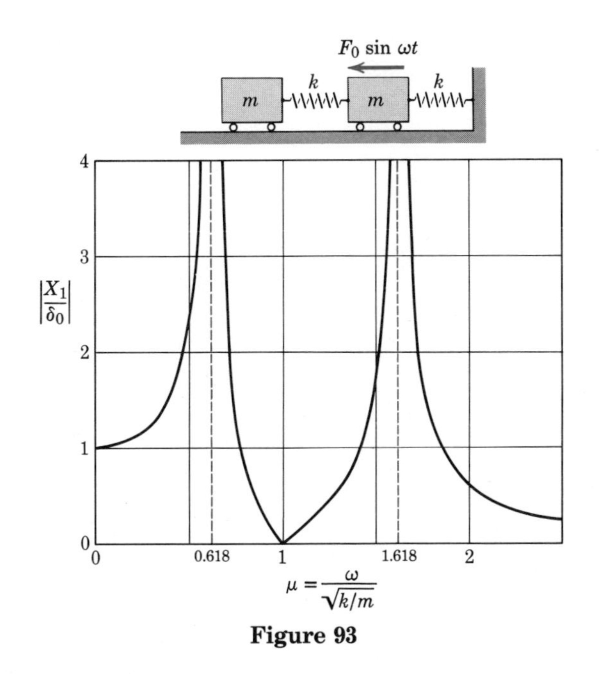

Figure 93

damping terms in the differential equation are functions of time, such as with a simple pendulum whose pivot is given a vertical harmonic oscillation. Problems of this type are described by the term *parametric excitation* and exhibit some rather unexpected phenomena. When the restoring and damping forces are nonlinear functions of the displacement and velocity, respectively, a nonlinear differential equation results, the solution to which is generally much more difficult than for the linear case. A simple oscillator whose spring becomes stiffer with displacement is one example of a nonlinear system. The subjects of parametric vibration and nonlinear response constitute important specialized studies, and the student is directed to more advanced references for their pursuit.

III DYNAMICS OF NONRIGID SYSTEMS

10 DYNAMICS OF NONRIGID SYSTEMS

51 Introduction. In Part I the dynamics of particle motion was developed, and in Part II the dynamics of rigid-body motion was treated. In the present chapter, Part III, the dynamics of nonrigid systems is introduced and applied to problems involving steady and unsteady mass flow, the propagation of waves in solids, and other selected problems. Also quite general formulations of the motion equations for any mass system are treated in the articles on Lagrange's equations and Hamilton's principle.

The dynamics of mass flow is of great importance in the description of fluid machinery of all types including turbines, pumps, nozzles, air-breathing jet engines, and rockets. The treatment of mass flow in the following two articles is not intended to take the place of a study of fluid mechanics but merely to present the basic principles and equations of momentum which find important usage in fluid mechanics and in the general flow of mass whatever be the form.

52 Steady Mass Flow. One of the most important cases of mass flow occurs during steady flow conditions where the rate at which mass enters a given volume equals the rate at which mass leaves the same volume. The volume in question may be a rigid container, fixed or moving, such as the volume in the nozzle of a jet aircraft or rocket, the space between blades in a gas turbine, the volume within the casing of a centrifugal pump, or the volume within the bend of a pipe through which a fluid is flowing at a steady rate. The design of such fluid machines is dependent on the analysis of the forces and moments which are developed through the corresponding momentum changes of the flowing mass.

Consider a rigid container, shown in section in Fig. 94*a*, into which mass flows in a steady stream at the rate m' through the entrance section of area A_1. Mass leaves the container through the exit section of area A_2 at the same rate, so that there is no accumulation or depletion of the total mass within the container during the period of observation. The velocity of the entering stream is $\mathbf{v}_1$ normal to A_1 and that of the leaving stream is $\mathbf{v}_2$ normal to A_2. If ρ_1 and ρ_2 are the respective densities of the two streams, the continuity of flow requires that

$$\rho_1 A_1 v_1 = \rho_2 A_2 v_2 \tag{188}$$

The forces which act are described by isolating either the mass of fluid within the container or the entire container and the fluid within it. The first approach would be used if the forces between the container and the fluid were to be described, and the second approach would be indicated when the forces external to the container are desired. The latter situation is the one usually desired in which case the *system isolated* consists of the fixed structure of the container and the fluid within it at a particular instant of time. This isolation is described by a free-body diagram of the mass within a closed space volume defined by the exterior surface of the container and the entrance and exit surfaces. All forces applied *externally* to this system must be accounted for, and in Fig. 94*a* the vector sum of this external force system is denoted by $\Sigma\mathbf{F}$. Included in $\Sigma\mathbf{F}$ are the forces exerted on the container at points of its attachment to other structures including attachments at A_1 and A_2, if present; the forces acting on the fluid within the container at A_1 and A_2 due to any static pressure which may exist in the fluid at these positions; and the weight of the fluid and structure if appreciable. The resultant $\Sigma\mathbf{F}$ of all of these external forces must equal $\dot{\mathbf{G}}$, the time rate of change of the linear momentum of the isolated system, according to Eq. 96 which was developed in Chapter 4 for any system of constant mass, rigid or nonrigid. The time rate of change of linear momentum is the vector difference between the rate at which linear momentum leaves the system and the rate at which linear momentum enters the system. If m' stands for the mass flow rate, $\dot{\mathbf{G}} = m'\mathbf{v}_2 - m'\mathbf{v}_1 = m'\,\Delta\mathbf{v}$. This result for $\dot{\mathbf{G}}$ may also be obtained by an incremental analysis. Figure 94*b* illustrates the system at time t when the system mass is that of the container, the fluid within it, and an increment Δm about to enter during time Δt. At time $t + \Delta t$ the same total mass is that of the container, the fluid within it, and an equal increment Δm which leaves the container in time Δt. The linear momentum of the container and fluid within it between the two sections A_1 and A_2 remains unchanged during Δt so that the change in momentum of the system in time Δt is

$$\Delta\mathbf{G} = (\Delta m)\mathbf{v}_2 - (\Delta m)\mathbf{v}_1 = \Delta m(\mathbf{v}_2 - \mathbf{v}_1)$$

Division by Δt and passage to the limit yield $\dot{\mathbf{G}} = m'\,\Delta\mathbf{v}$ where

$$m' = \lim_{\Delta t \to 0}\left(\frac{\Delta m}{\Delta t}\right) = \frac{dm}{dt}$$

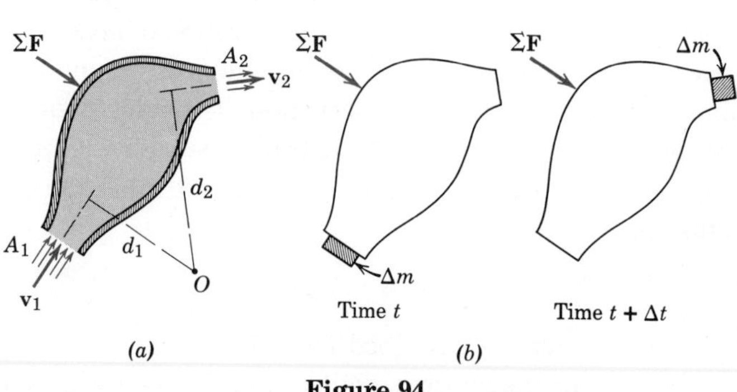

Figure 94

Thus by Eq. 96

$$\Sigma \mathbf{F} = m' \Delta \mathbf{v} \tag{189}$$

Equation 189 establishes the relation between the resultant force on a steady-flow system and the corresponding mass flow rate and vector velocity increment.

A similar formulation is obtained for the case of angular momentum in steady-flow systems. The resultant moment of all external forces about some fixed point O on or off the system, Fig. 94*a*, equals the time rate of change of angular momentum of the system about O. This fact was established in Eq. 97 which, for the case of steady flow in a single plane, becomes

$$\Sigma M_O = m'(v_2 d_2 - v_1 d_1) \tag{190}$$

When the velocities of the incoming and outgoing flows are not in the same plane, the equation may be written in vector form as

$$\Sigma \mathbf{M}_O = m'(\mathbf{d}_2 \times \mathbf{v}_2 - \mathbf{d}_1 \times \mathbf{v}_1) \tag{190a}$$

where $\mathbf{d}_1$ and $\mathbf{d}_2$ would be the position vectors to the centers of A_1 and A_2 from the fixed reference O. In both relations the mass center G may be used alternatively as a moment center by virtue of Eq. 98.

Equations 189 and 190 are very simple relations which find important use in describing relatively complex fluid actions. It should be noted that these equations relate *external* forces to the resultant changes in momentum and are independent of the flow path and momentum changes *internal* to the system.

The foregoing analysis may be applied to systems moving with constant speed by noting that the basic relations $\Sigma \mathbf{F} = \dot{\mathbf{G}}$ and $\Sigma \mathbf{M}_O = \dot{\mathbf{H}}_O$ or $\Sigma \overline{\mathbf{M}} = \dot{\overline{\mathbf{H}}}$ apply to systems moving with constant speed.* The only restriction is that the mass within the system remain constant with respect to time.

Three examples of the analysis of steady mass flow are given in the following sample problems which illustrate the application of the principles embodied in Eqs. 189 and 190.

Sample Problems

10/1 The smooth vane shown diverts the open stream of fluid of cross-sectional area A, mass density ρ, and velocity v. (*a*) Determine the force components R and F required to hold the vane in a fixed position. (*b*) Find the forces when the vane is given a constant velocity u less than v and in the direction of v. (*c*) Determine the optimum speed u for the generation of maximum power by the action of the fluid on the moving vane.

Solution. Part (*a*). The free-body diagram of the vane together with the fluid portion undergoing the momentum change is shown. The momentum equation may be applied to the isolated system for the change in motion in both the x- and y-directions. With

* The case of accelerating systems is treated in Art. 53 on variable mass.

the vane stationary the magnitude of the exit velocity v' equals that of the entering velocity v with fluid friction neglected. The changes in the velocity components are, then,

$$\Delta v_x = v' \cos\theta - v = -v(1 - \cos\theta) \quad \text{and} \quad \Delta v_y = v' \sin\theta - 0 = v \sin\theta$$

The mass rate of flow is $m' = \rho A v$, and substitution into Eq. 189 gives

$[\Sigma F_x = m' \Delta v_x]$ $\quad -F = \rho A v[-v(1 - \cos\theta)]$ $\quad F = \rho A v^2(1 - \cos\theta)$ *Ans.*

$[\Sigma F_y = m' \Delta v_y]$ $\quad R = \rho A v[v \sin\theta]$ $\quad R = \rho A v^2 \sin\theta$ *Ans.*

Part (*b*). In the case of the moving vane the final velocity of the fluid upon exit is the vector sum of the velocity u of the vane plus the velocity of the fluid relative to the vane. This combination is shown in the velocity diagram to the right of the figure for the exit conditions. The relative velocity is that which would be measured by an observer moving with the vane. This observer would measure $v - u$ metres of fluid passing over the vane per second, and the direction of this relative velocity is tangent to the vane at exit. The combination of these two velocity components gives the final absolute fluid velocity v' as shown. The x-component of v' is the sum of the components of its two parts, so $v'_x = (v - u)\cos\theta + u$. The change in x-velocity of the stream is

$$\Delta v_x = (v - u)\cos\theta + (u - v) = -(v - u)(1 - \cos\theta)$$

The y-component of v' is $(v - u)\sin\theta$, so that the change in the y-velocity of the stream is $\Delta v_y = (v - u)\sin\theta$.

The mass rate of flow m' is the mass undergoing momentum change per unit of time. This rate is the mass flowing over the vane per unit time and *not* the rate of issuance from the nozzle. Thus

$$m' = \rho A(v - u)$$

The impulse-momentum principle of Eq. 189 applied in the positive coordinate direction gives

$[\Sigma F_x = m' \Delta v_x]$ $\quad -F = \rho A(v - u)[-(v - u)(1 - \cos\theta)]$

$\quad F = \rho A(v - u)^2(1 - \cos\theta)$ *Ans.*

$[\Sigma F_y = m' \Delta v_y]$ $\quad R = \rho A(v - u)^2 \sin\theta$ *Ans.*

Part (*c*). Since R is normal to the velocity of the vane, it does no work. The work done by the force F shown is negative, but the power developed by the force (equal and opposite to F) exerted by the fluid on the moving vane is

$[P = Fu]$ $\quad P = \rho A(v - u)^2 u(1 - \cos\theta)$

The velocity of the vane for maximum power for the one blade in the stream is specified by

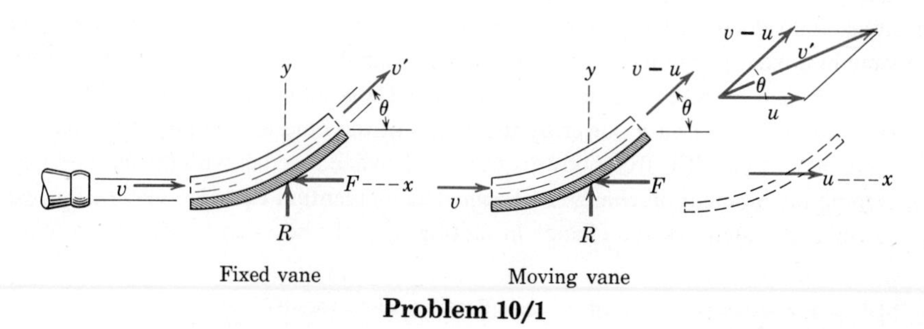

Fixed vane Moving vane

Problem 10/1

$\left[\frac{dP}{du} = 0\right]$ $\rho A(1 - \cos\theta)(v^2 - 4uv + 3u^2) = 0$

$$(v - 3u)(v - u) = 0 \qquad u = \frac{v}{3} \qquad \textit{Ans.}$$

The second solution $u = v$ gives a minimum condition of zero power. An angle $\theta = 180°$ completely reverses the direction of the fluid and clearly produces both maximum force and maximum power for any value of u.

10/2 The offset nozzle has a discharge area A at B and an inlet area A_0 at C. A liquid enters the nozzle at a static gage pressure p through the fixed pipe and issues from the nozzle with a velocity v in the direction shown. If the constant mass density of the liquid is ρ, write expressions for the tension T, shear Q, and bending moment M in the pipe at C.

Solution. The system isolated in the right-hand portion of the figure is the nozzle and the fluid within it. The free-body diagram of the system shows the tension T, shear Q, and bending moment M acting on the flange of the nozzle where it attaches to the fixed pipe. The force pA_0 on the fluid within the nozzle due to the static pressure is an additional external force.

Continuity of flow with constant density requires that

$$Av = A_0 v_0$$

where v_0 is the velocity of the fluid at entrance to the nozzle. The momentum principle of Eq. 189 applied to the system in the two coordinate directions gives

$[\Sigma F_x = m' \Delta v_x]$ $\qquad pA_0 - T = \rho Av(v\cos\theta - v_0)$

$$T = pA_0 + \rho Av^2\left(\frac{A}{A_0} - \cos\theta\right) \qquad \textit{Ans.}$$

$[\Sigma F_y = m' \Delta v_y]$ $\qquad -Q = \rho Av(-v\sin\theta - 0)$

$$Q = \rho Av^2 \sin\theta \qquad \textit{Ans.}$$

The moment principle of Eq. 190 applied in the clockwise sense gives

$[\Sigma M_O = m'(v_2 d_2 - v_1 d_1)]$ $\qquad M = \rho Av(va\cos\theta + vb\sin\theta - 0)$

$$M = \rho Av^2(a\cos\theta + b\sin\theta) \qquad \textit{Ans.}$$

The forces and moment acting on the pipe are equal and opposite to those shown acting on the nozzle.

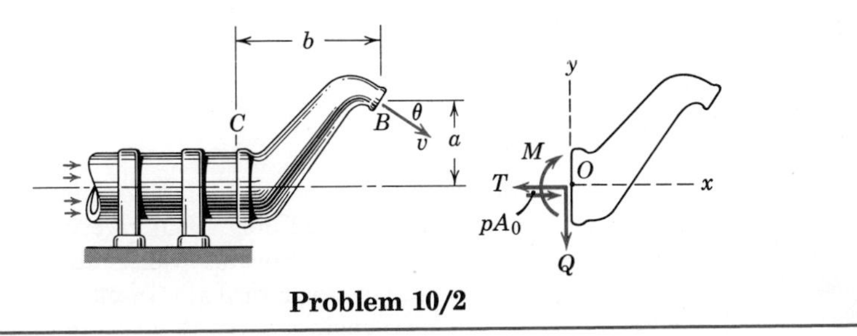

Problem 10/2

10/3 An air-breathing jet aircraft of total mass m consumes air at the mass rate m_a' and exhausts burned gas at the rate m_g' with a velocity u with respect to the aircraft. Fuel is consumed at the rate m_f'. The total aerodynamic forces of lift and drag, including the forces due to static pressure across the inlet and exhaust surfaces, are L and D, respectively. If the aircraft has a velocity v as shown, write the equation of motion for the aircraft in the direction of its flight.

Solution. The incoming air may be treated as a stream of fluid being diverted from an initial velocity of zero to a velocity v of the aircraft. By Eq. 189 this diversion is due to a force equal to $m_a'(v - 0) = m_a'v$, and the reaction to this force is shown as an external force on the free-body diagram. Similarly the exhaust gas may be treated as a steady stream of fluid being diverted from a forward velocity v to a rearward velocity $u - v$ with a net velocity change u. By Eq. 189 this change gives rise to a force equal to $m_g'u$ which is also shown as an external force. The system is now simulated by a mass m acted upon by the five external forces. The equation of motion is

$$[\Sigma F = ma] \qquad m_g'u - m_a'v - D - mg\sin\theta = m\dot{v}$$

If m_f' is the rate of burning fuel, $m_g' = m_f' + m_a'$. In practice m_f' is usually less than 2 per cent of m_a', so that the net thrust is

$$T = m_g'u - m_a'v \doteq m_g'(u - v)$$

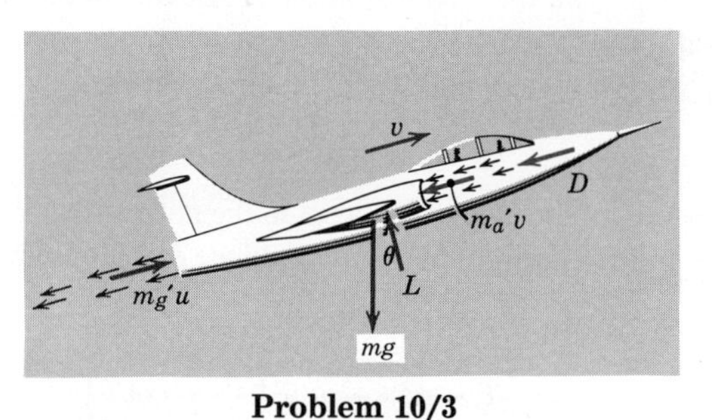

Problem 10/3

Problems

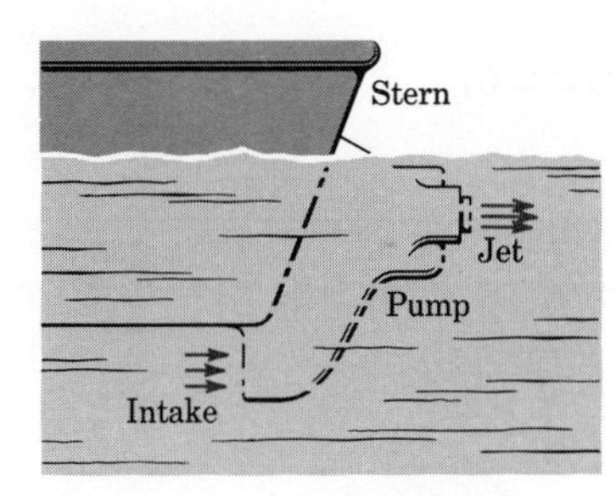

Problem 10/4

10/4 The small jet engine for marine use discharges salt water at the rate of 30 kg/s with a jet velocity of 40 m/s relative to the pump. Determine the hull resistance R at a constant hull speed of 10 knots. (1 knot equals 1.852 km/h.)

Ans. $R = 1046$ N

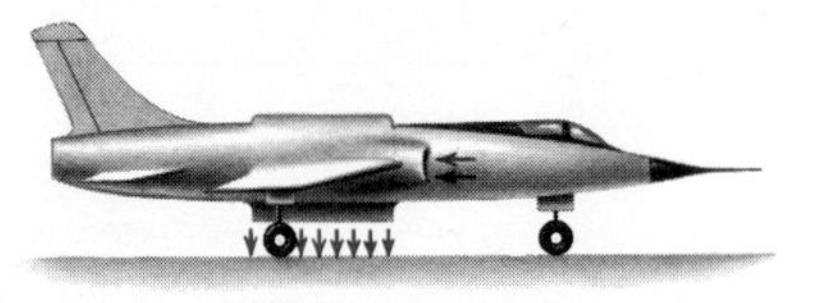

Problem 10/5

10/5 An experimental aircraft has a mass of 7 t and is designed for VTOL (vertical take-off and landing). Movable vanes deflect the exhaust stream downward at take-off. Each of the craft's two engines sucks in air at the rate of 60 kg/s at a density of 1.217 kg/m^3 and discharges the exhaust at a velocity of 600 m/s. Each engine uses fuel at the rate of 1 kg/s. Determine the initial vertical acceleration a.

Ans. $a = 0.647$ m/s^2

10/6 Shot peening is used to increase the fatigue strength of gear teeth and other machine parts. The device shown consists of a reservoir A which feeds the shot pellets into the center of a rotating impeller B that flings them at high speed and produces a spray of shot. The spray of high-velocity pellets impinges against the rotating gears C that are being treated. If m kg of shot are consumed per second and if the angular velocity is ω, determine the torque M required to drive the impeller.

10/7 A jet-engine noise suppressor consists of a movable duct which is secured directly behind the jet exhaust by cable A and deflects the blast directly upward. During ground test the engine sucks in air at the rate of 43 kg/s and burns fuel at the rate of 0.8 kg/s. The exhaust velocity is 720 m/s. Determine the tension T in the cable.

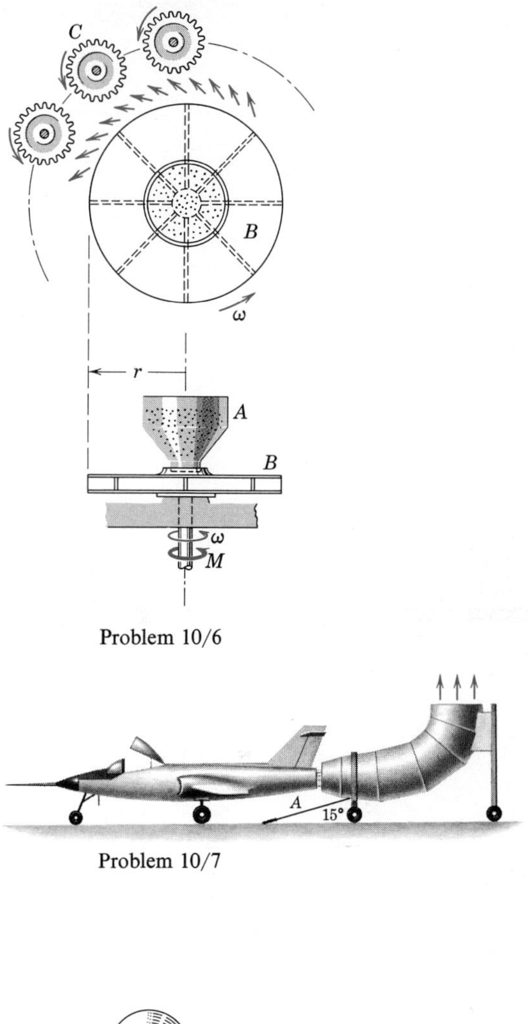

Problem 10/6

Problem 10/7

10/8 An air-breathing jet aircraft has a constant speed of 1500 km/h in horizontal flight at an altitude of 12 km. The turbojet engine consumes air at the rate of 110 kg/s at this speed and uses fuel at the rate of 0.97 kg/s. The gases are exhausted at a relative nozzle velocity of 780 m/s at the atmospheric pressure corresponding to the flight altitude. Determine the total drag D (air resistance) on the exterior surface of the aircraft and the useful power P (thrust horsepower) of the engine at this speed.

Ans. $D = 40.7$ kN, $P = 16.97$ MW

10/9 A 25-mm steel slab 1.2 m wide enters the rolls at the speed of 0.4 m/s and is reduced in thickness to 19 mm. Calculate the small horizontal thrust T on the bearings of each of the two rolls.

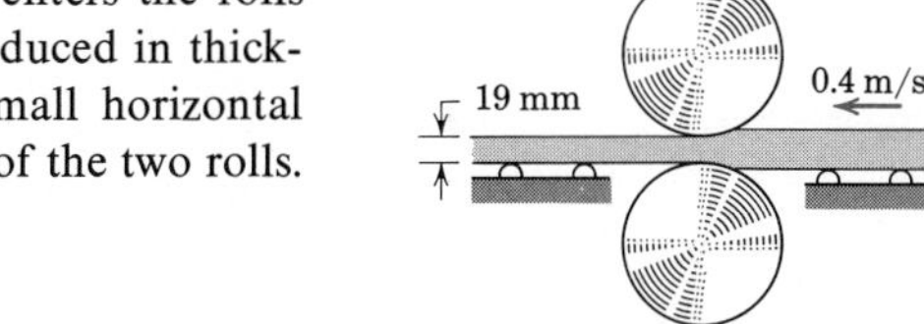

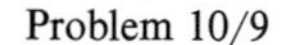

Problem 10/9

10/10 The jet aircraft has a mass of 4 t and has a drag (air resistance) of 24 kN at a speed of 800 km/h at a particular altitude. The aircraft consumes air at the rate of 90 kg/s through its intake scoop and uses fuel at the rate of 0.90 kg/s. If the exhaust has a rearward velocity of 600 m/s relative to the exhaust nozzle, determine the maximum angle of elevation α at which the jet can fly with a constant speed of 800 km/h at the particular altitude in question.

Ans. $\alpha = 15.58°$

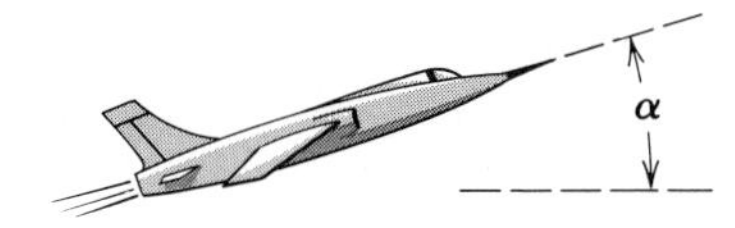

Problem 10/10

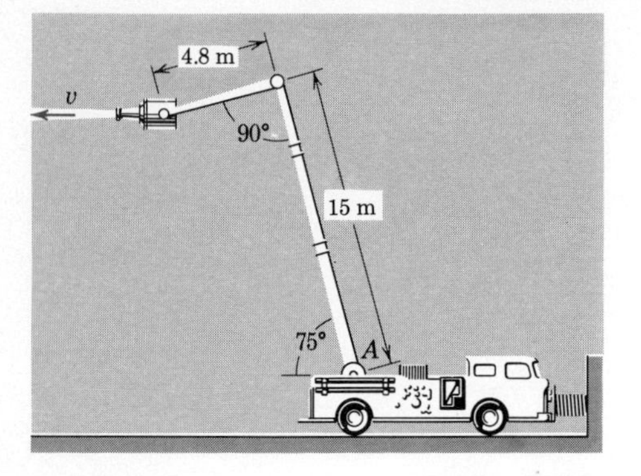

Problem 10/11

10/11 In a test of the operation of a "cherry picker" fire truck the equipment is free to roll with its brakes released. For the position shown the truck is observed to deflect the spring of stiffness $k = 15$ kN/m a distance of 150 mm because of the action of the horizontal stream of water issuing from the nozzle when the pump is activated. If the exit diameter of the nozzle is 30 mm, calculate the velocity v of the stream as it leaves the nozzle. Also determine the added moment M which the joint at A must resist when the pump is in operation with the nozzle in the position shown.

Problem 10/12

10/12 The snow plow eats its way through a 1.5-m drift and throws the snow off to the side at the rate of 30 t per minute. The snow has a velocity of 12 m/s relative to the plow in the 45° direction shown. Determine the lateral force R between the tires and the road and the tractive force F required to drive the plow at 15 km/h.

Ans. $R = 4.24$ kN, $F = 2.08$ kN

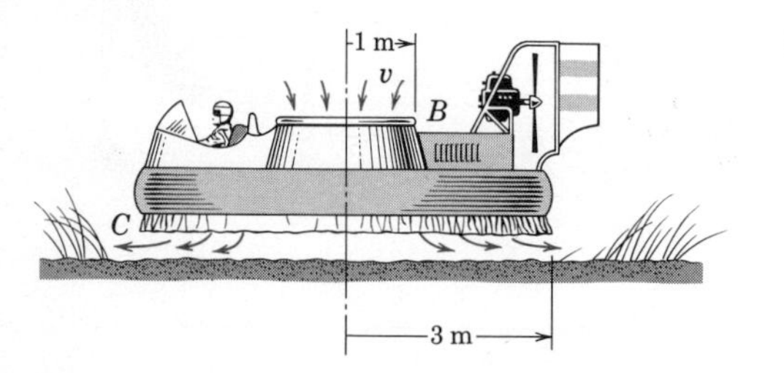

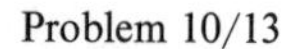

Problem 10/13

10/13 The experimental ground-effect machine has a total mass of 2.1 t. It hovers 0.3 to 0.5 m off the ground by pumping air at atmospheric pressure through the circular intake duct at B and discharging it horizontally under the periphery of the skirt C. For an intake velocity v of 45 m/s calculate the average air pressure p under the 6-m-diameter machine. The density of the air is 1.217 kg/m^3.

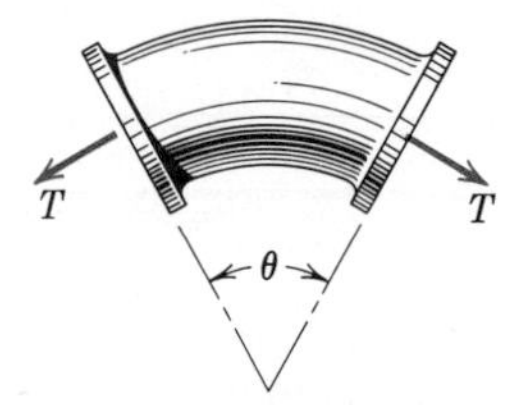

Problem 10/14

10/14 The pipe bend shown has a cross-sectional area A and is supported in its plane by the tension T applied to its flanges by the adjacent connecting pipes (not shown). If the velocity of the liquid is v, its mass density ρ, and its static pressure p, determine T and show that it is independent of the bend angle θ.

Ans. $T = A(p + \rho v^2)$

10/15 In the case of multiple vanes where each vane which enters the jet is followed immediately by another, as in a turbine or water wheel, determine the maximum power P which can be developed for a given blade angle and the corresponding optimum peripheral speed u of the vanes in terms of the jet velocity v for maximum power. Modify Sample Prob. 10/1 by assuming an infinite number of vanes so that the rate at which fluid leaves the nozzle equals the rate at which fluid passes over the vanes.

Ans. $P = \frac{1}{4}\rho A v^3(1 - \cos\theta)$, $u = v/2$

10/16 The 90° vane moves to the left with a constant velocity of 10 m/s against a stream of fresh water issuing from the 25-mm-diameter nozzle with a velocity of 20 m/s. Calculate the forces F_x and F_y on the vane required to support the motion.
Ans. $F_x = 442$ N, $F_y = 442$ N

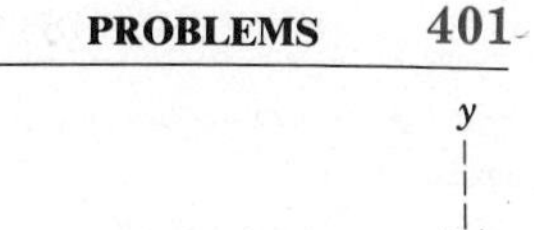
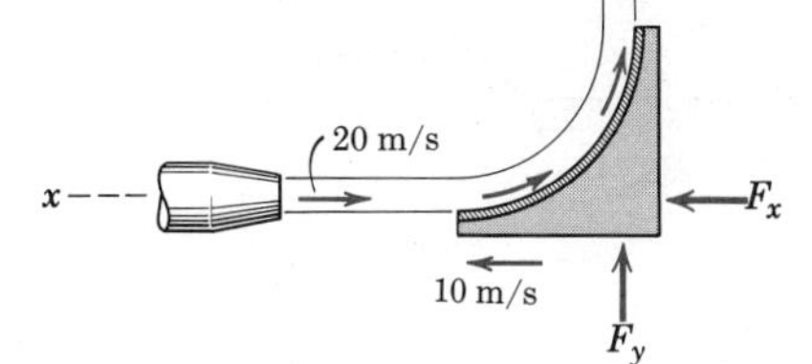

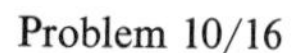
Problem 10/16

10/17 The vane shown is fastened to a vertical shaft at O and diverts an open stream of fresh water as shown. The water flows at the rate of 3 kg/s and has a velocity whose magnitude v remains constant at 300 m/s. Compute the x- and y-components of the force exerted on the vane by the pivot at O and find the couple M which must be applied to the vane by the shaft at O in order to hold the vane in the fixed position shown.

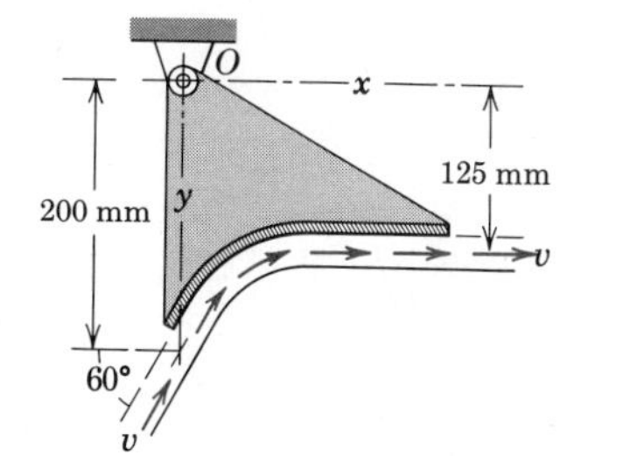

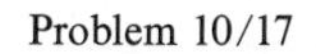
Problem 10/17

10/18 In the static test of a jet engine and exhaust nozzle assembly, air is sucked into the engine at the rate of 30 kg/s and fuel is burned at the rate of 0.6 kg/s. The flow area, static pressure, and axial-flow velocity for the three sections shown are as follows:

	Sec. *A*	Sec. *B*	Sec. *C*
Flow area, m^2	0.15	0.16	0.06
Static pressure, kPa	−14	140	14
Axial-flow velocity, m/s	120	315	600

Determine the tension T in the diagonal member of the supporting test stand and calculate the force F exerted on the nozzle flange at B by the bolts and gasket to hold the nozzle to the engine housing. *Ans.* $T = 25.0$ kN, $F = 12.84$ kN

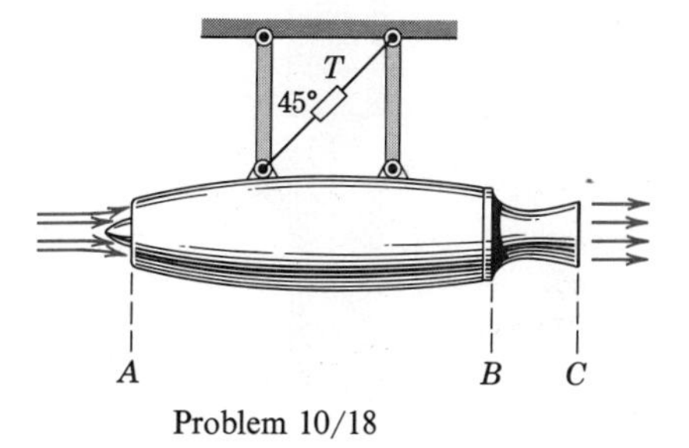

Problem 10/18

10/19 A ball of mass m is supported in a vertical jet of water with density ρ. If the stream of water issuing from the nozzle has a diameter d and velocity u, determine the height h above the nozzle at which the ball is supported. Assume that the jet remains intact and that there is no energy loss in the jet.

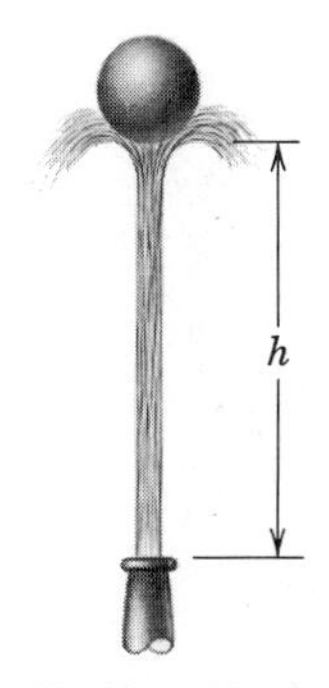

Problem 10/19

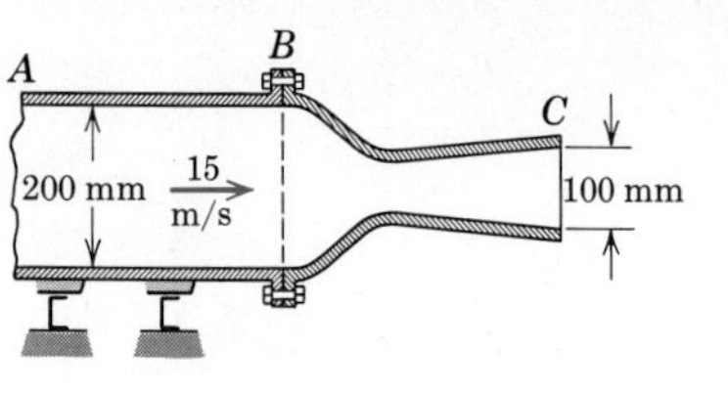

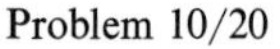
Problem 10/20

10/20 An air stream with a velocity of 15 m/s is pumped through the stationary duct A and exhausted through an experimental nozzle section BC. The average static pressure across section B is 1050 kPa gage, and the density of air at this pressure and at the temperature prevailing is 13.5 kg/m^3. The average static pressure across the exit section C is measured to be 14 kPa gage, and the corresponding density of air is 1.217 kg/m^3. Calculate the force T exerted on the nozzle flange at B by the bolts and the gasket to hold the nozzle in place.

Ans. $T = 28.7$ kN

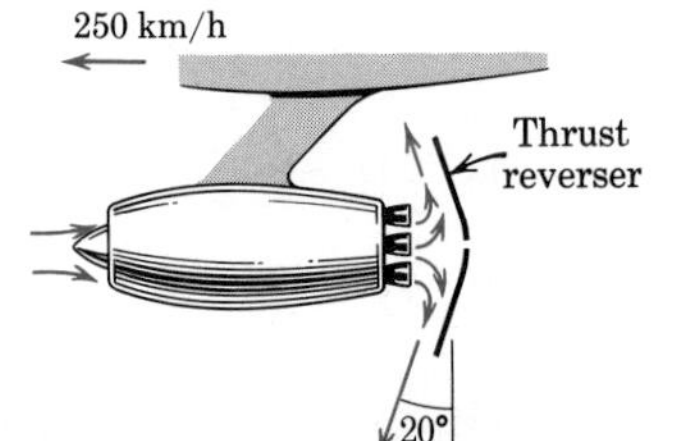

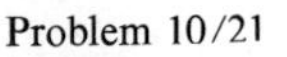
Problem 10/21

10/21 Reverse thrust to decelerate a jet aircraft during landing is obtained by swinging the exhaust deflectors into place as shown, thus causing a partial reversal of the exhaust stream. For a jet engine which consumes 40 kg of air per second at a ground speed of 250 km/h and which uses 0.6 kg of fuel per second, determine the reverse thrust as a fraction n of the forward thrust with deflectors removed. The exhaust velocity relative to the nozzle is 600 m/s. Assume air enters the engine with a relative velocity equal to the ground speed of the aircraft.

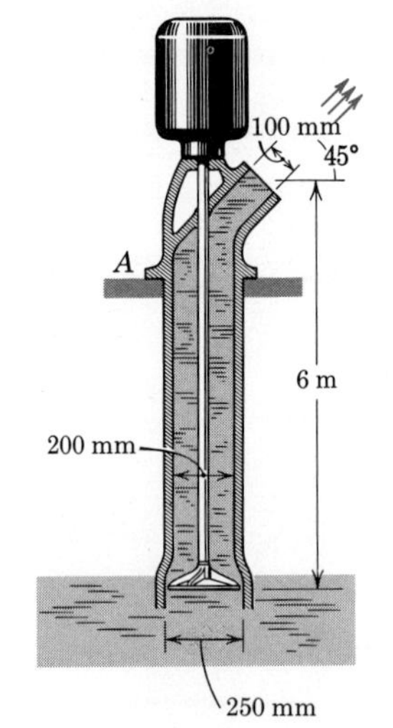

Problem 10/22

10/22 The sump pump has a net mass of 310 kg and pumps fresh water against a 6-m head at the rate of 0.125 m^3/s. Determine the vertical force R between the supporting base and the pump flange at A during operation. The mass of water in the pump may be taken to be the equivalent of a 200-mm-diameter column 6 m in height.

Ans. $R = 5980$ N

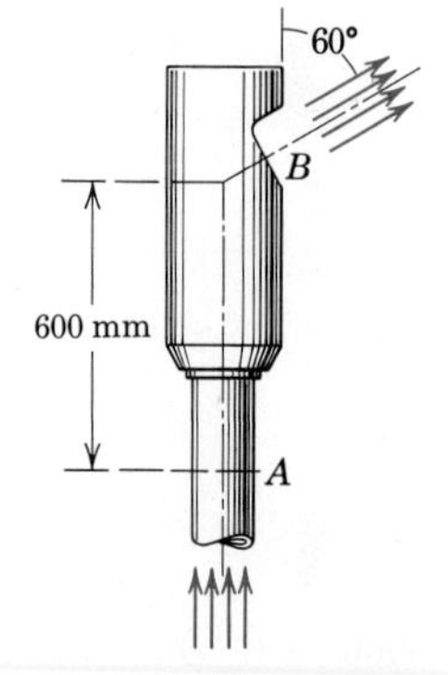

Problem 10/23

10/23 Air enters the pipe at A at the rate of 6 kg/s under a pressure of 1400 kPa gage and leaves the whistle at atmospheric pressure through the opening at B. The entering velocity of the air at A is 45 m/s, and the exhaust velocity at B is 360 m/s. Calculate the tension T, shear Q, and bending moment M in the pipe at A. The net flow area at A is 7500 mm^2.

10/24 The 180° pipe return discharges salt water (density 1030 kg/m^3) into the atmosphere at a constant rate of 0.05 m^3/s. The static pressure in the water at section A is 70 kPa above atmospheric pressure. The flow area of the pipe at A is 12 500 mm^2 and that at each of the two outlets is 2000 mm^2. If each of the six flange bolts is tightened with a torque wrench so that it is under a tension of 750 N, determine the average pressure p on the gasket between the two flanges. The flange area in contact with the gasket is 10 000 mm^2. Also determine the bending moment M in the pipe at section A if the left-hand side of the tee is blocked off and the flow rate is cut in half.

Ans. $p = 278$ kPa, $M = 64.4$ N·m

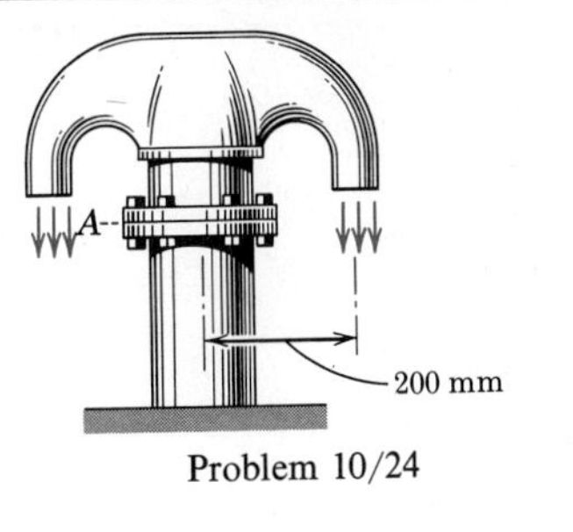

Problem 10/24

10/25 The helicopter shown has a mass m and hovers in position by imparting downward momentum to a column of air defined by the slip-stream boundary shown. Find the downward velocity v given to the air by the rotor at a section in the stream below the rotor where the pressure is atmospheric and the stream radius is r. Also find the power P required of the engine. Neglect the rotational energy of the air, any temperature rise due to air friction, and any change in air density ρ.

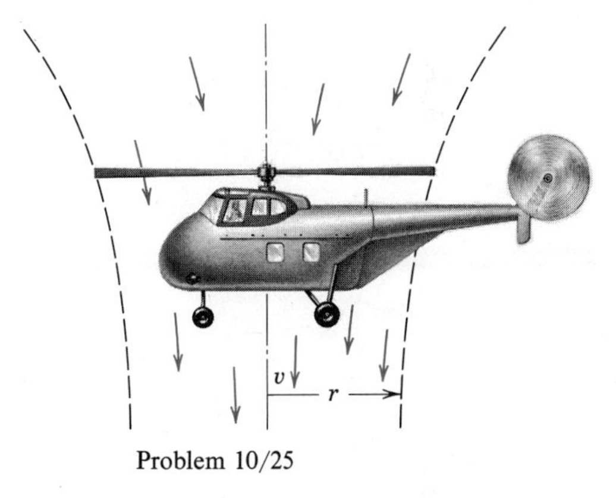

Problem 10/25

10/26 The centrifugal pump handles 20 m^3 of fresh water per minute with inlet and outlet velocities of 18 m/s. The impeller is turned clockwise through the shaft at O by a motor which delivers 40 kW at a pump speed of 900 rev/min. With the pump filled but not turning, the vertical reactions at C and D are each 250 N. Calculate the forces exerted by the foundation on the pump at C and D while the pump is running. The tensions in the connecting pipes at A and B are exactly balanced by the respective forces due to the static pressure in the water. (*Suggestion:* Isolate the entire pump and water within it between sections A and B and apply the momentum principle to the entire system.)

Ans. $C = 4340$ N up, $D = 3840$ N down

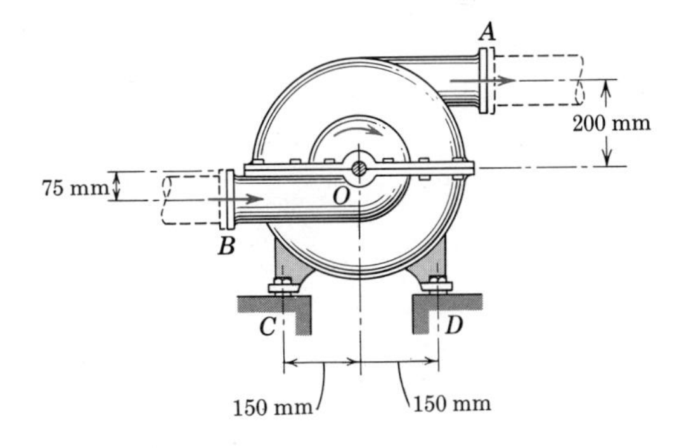

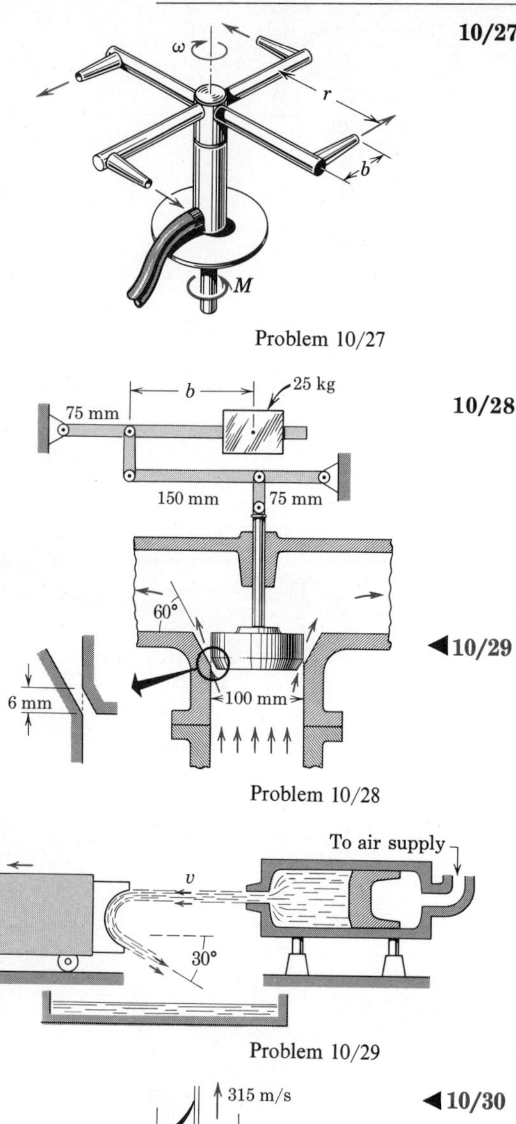

Problem 10/27

Problem 10/28

Problem 10/29

10/27 The sprinkler is made to rotate at the constant angular velocity ω and distributes water at the volume rate Q. Each of the four nozzles has an exit area A. Write an expression for the torque M on the shaft of the sprinkler necessary to maintain the given motion. For a given pressure and, hence, flow rate Q, at what speed ω_0 will the sprinkler operate with no applied torque? Let ρ be the density of the water.

Ans. $\omega_0 = \dfrac{Qr}{4A(r^2 + b^2)}$

10/28 Determine the setting b for the 25-kg mass required to limit the flow of fresh water through the valve to 1.2 m³/min. The conical plug valve discharges into the upper pipe at atmospheric pressure, and the static pressure in the inlet pipe is 280 kPa above atmospheric pressure.

Ans. $b = 116.9$ mm

◀**10/29** A test vehicle for impact studies has a mass of 1.6 t and is accelerated from rest by the action of a water jet which impinges against the deflector vane attached to the vehicle as shown. The fresh-water jet is generated by the action of a piston which is activated by air released from a high-pressure chamber. The jet is 150 mm in diameter and has a velocity which is essentially constant at 180 m/s for the short duration of the test. If frictional resistance is 10 per cent of the vehicle weight, determine the velocity u of the vehicle after the jet has acted for 4 s. Assume steady-flow conditions for the vane.

Ans. $u = 167.3$ m/s

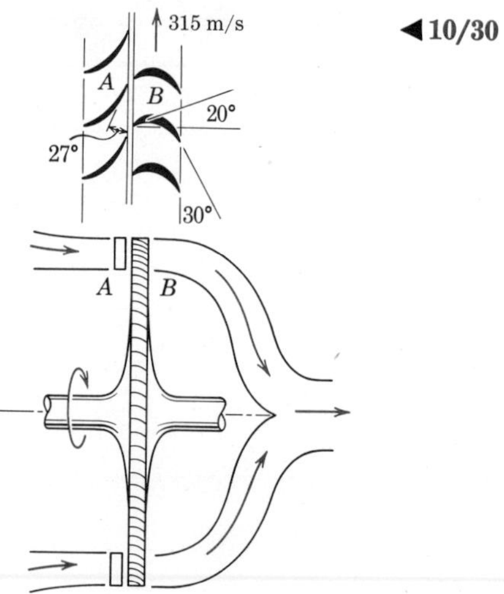

Problem 10/30

◀**10/30** In the figure is shown a detail of the stationary nozzle diaphragm A and the rotating blades B of a gas turbine. The products of combustion pass through the fixed diaphragm blades at the 27° angle and impinge on the moving rotor blades. The angles shown are selected so that the velocity of the gas relative to the moving blade at entrance is at the 20° angle for minimum turbulence, corresponding to a mean blade velocity of 315 m/s at a radius of 375 mm. If gas flows past the blades at the rate of 15 kg/s, determine the theoretical power output P of the turbine. Neglect fluid and mechanical friction with the resulting heat energy loss, and assume that all the gases are deflected along the surfaces of the blades with a velocity relative to the blade of constant magnitude.

Ans. $P = 1.197$ MW

◀ **10/31** In the figure is shown an impulse turbine wheel for a hydroelectric power plant which is to operate with a static head of water of 300 m at each of its 6 nozzles and is to rotate at the speed of 270 rev/min. Each wheel and generator unit is to develop an output power of 22 000 kW. The efficiency of the generator may be taken to be 0.90, and an efficiency of 0.85 for the conversion of the kinetic energy of the water jets to energy delivered by the turbine may be expected. The mean peripheral speed of such a wheel for greatest efficiency will be about 0.47 times the jet velocity. If each of the buckets is to have the shape shown, determine the necessary jet diameter d and wheel diameter D. Assume that the water acts on the bucket which is at the tangent point of each jet stream.

Ans. $d = 165.3$ mm, $D = 2.55$ m

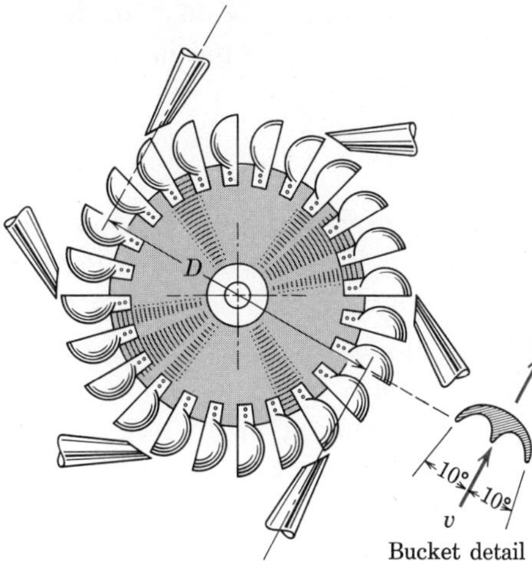

Problem 10/31

53 Variable Mass. The force-momentum relationship, $\Sigma \mathbf{F} = \dot{\mathbf{G}}$, and the moment-momentum relationships, $\Sigma \mathbf{M}_O = \dot{\mathbf{H}}_O$ and $\Sigma \overline{\mathbf{M}} = \dot{\overline{\mathbf{H}}}$, were developed in Chapter 4 as Eqs. 96, 97, and 98, respectively. In the derivation of these equations the summations were taken over a fixed number of particles, so that the mass of the system to be analyzed was constant with respect to time. In Art. 52 these momentum principles were extended in Eqs. 189 and 190 to describe the action of forces on a system defined by a geometric volume through which passes a steady flow of mass. The amount of mass within the control volume was, therefore, constant with respect to time. When the mass within the boundary of a system under consideration is not constant with respect to time, the foregoing relationships are no longer valid.*

The equations of Newtonian mechanics which describe the motion of a system whose mass is time-dependent are developed in the present article. A simplified one-dimensional case is treated first where the mass is "lumped" or concentrated. Next, an exact formulation is developed where the mass of the system is distributed over a finite volume and the system may have any general motion.

Part A: Simplified Case

For this case consider first a body, Fig. 95*a*, which gains mass by virtue of its overtaking and swallowing a stream of matter. The mass of the body and its velocity at any instant are m and v, respectively. The stream of matter is assumed to be moving in the same direction as m with a constant velocity v_0 less than v. The force exerted by m on the particles of the stream to accelerate them from a velocity v_0 to a greater velocity v is $R = m'(v - v_0) = \dot{m}u$ where the time rate of increase of m is $m' = \dot{m}$ and where u is the relative velocity with which the particles approach m. In addition to R, all other

* In relativistic mechanics the mass is found to be a function of velocity, and its time derivative has a meaning different from that in Newtonian mechanics.

forces acting on m in the direction of its motion are designated by ΣF. The equation of motion of m from Newton's second law is, therefore, $\Sigma F - R = m\dot{v}$ or

$$\Sigma F = m\dot{v} + \dot{m}u \tag{191}$$

Similarly, if the body loses mass by expelling it at a velocity v_0 less than v, Fig. 95*b*, the force R required to decelerate the particles from a velocity v to a lesser velocity v_0 is $R = m'(-v_0 - [-v]) = m'(v - v_0)$. But $m' = -\dot{m}$ since m is decreasing. Also the relative velocity with which the particles leave m is $u = v - v_0$. Thus the force R becomes $R = -\dot{m}u$. If ΣF denotes the resultant of all other forces acting on m in the direction of its motion, Newton's second law requires $\Sigma F + R = m\dot{v}$ or

$$\Sigma F = m\dot{v} + \dot{m}u$$

which is the same relationship as in the case where m is gaining mass. Equation 191, therefore, may be used as the equation of motion of m whether it is gaining or losing mass. A frequent error in the use of the force-momentum equation is to express the partial force sum ΣF as

$$\Sigma F = \frac{d}{dt}(mv) = m\dot{v} + \dot{m}v$$

From the foregoing analyses it is seen that this direct differentiation of the linear momentum gives the correct force ΣF *only* when the body picks up mass initially at rest or when it expels mass at a zero absolute velocity. In both instances $v_0 = 0$ and $u = v$.

Equation 191 may also be obtained by a direct differentiation of the momentum from the basic relation $\Sigma F = \dot{G}$ provided a proper system of constant total mass is chosen. To illustrate this approach, the case where m is losing mass is taken, and Fig. 95*c* shows the system of m and an arbitrary portion m_0 of the stream of ejected mass. The mass of this system is $m + m_0$ and is constant. The ejected stream of mass is assumed to move

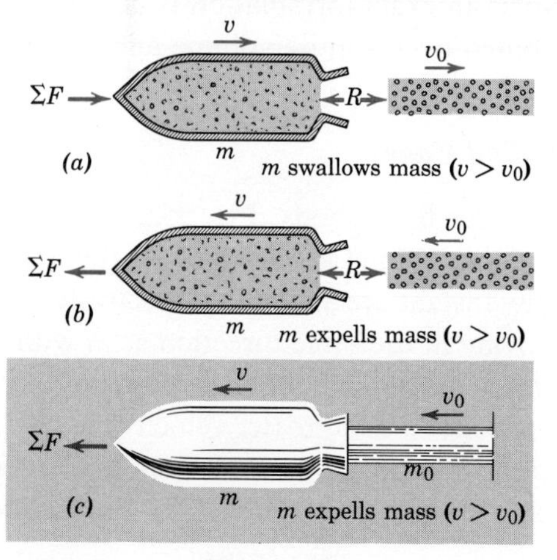

Figure 95

undisturbed once separated from m, and the only force external to the entire system is ΣF which is applied directly to m as before. The reaction $R = -\dot{m}u$ is internal to the system and is not disclosed. With constant total mass the momentum principle $\Sigma F = \dot{G}$ is applicable and becomes

$$\Sigma F = \frac{d}{dt}(mv + m_0 v_0) = m\dot{v} + \dot{m}v + \dot{m}_0 v_0 + m_0 \dot{v}_0.$$

Clearly $\dot{m}_0 = -\dot{m}$, and the velocity of the ejected mass with respect to m is $u = v - v_0$. Also $\dot{v}_0 = 0$ since m_0 moves undisturbed with no acceleration once free of m. Thus the relation becomes

$$\Sigma F = m\dot{v} + \dot{m}u$$

which is identical with the result of the previous formulation.

The case of m losing mass is clearly descriptive of rocket propulsion. Figure 96*a* shows a vertically ascending rocket, the system for which is the mass within the volume defined by the exterior surface of the rocket and the exit plane across the nozzle. External to this system the free-body diagram discloses the instantaneous values of gravitational attraction mg, aerodynamic resistance R, and the force pA due to the average static pressure across the nozzle exit plane of area A. Also the rate of mass flow is $m' = -\dot{m}$. Thus the equation of motion of the rocket $\Sigma F = m\dot{v} + \dot{m}u$ becomes

$$m'u + pA - mg - R = m\dot{v} \tag{192}$$

Equation 192 is of the form "$\Sigma F = ma$" where the first term in "ΣF" is the thrust $T = m'u$. Thus the rocket may be simulated as a body to which an external thrust T is applied, Fig. 96*b*, and the problem may then be analyzed like any other "$F = ma$" problem except that m is a function of time.

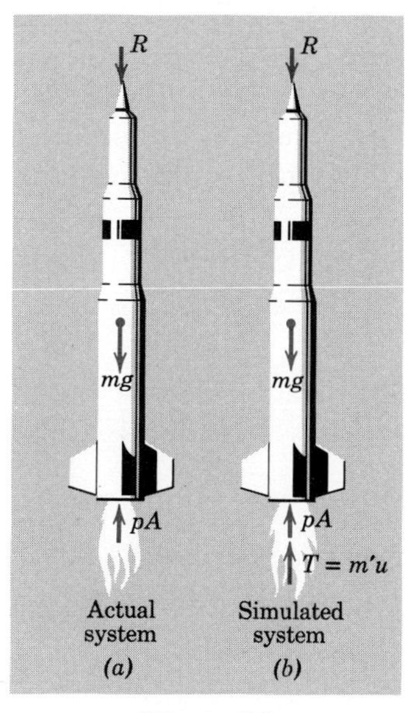

Figure 96

It may be observed that, during the initial stages of motion when the magnitude of the velocity v of the rocket is less than the relative exhaust velocity u, the absolute velocity v_0 of the exhaust gases will be directed rearward. On the other hand, when the rocket reaches a velocity v whose magnitude is greater than u, the absolute velocity v_0 of the exhaust gases will be directed forward. For a given mass rate of flow the rocket thrust T depends only on the relative exhaust velocity u and not on the magnitude or direction of the absolute velocity v_0 of the exhaust gases.

Sample Problems

10/32 The end of a chain of length l and mass ρ per unit length which is piled on a platform is lifted vertically with a constant velocity v by a variable force P. Find P as a function of the height x of the end above the platform.

Solution. It will be assumed that the chain is of the open-link type so that each link acquires its velocity v abruptly from its rest condition.

The principle of impulse and momentum for a system of particles expressed by Eq. 96 will be applied to the entire chain considered as the system of constant mass. The free-body diagram of the system shows the unknown force P, the total weight of all links ρgl, and the force $\rho g(l - x)$ exerted by the platform on those links which are at rest upon it. The momentum of the system at any position is

$$G_x = \rho x v$$

and the momentum equation gives

$$\left[\Sigma F_x = \frac{dG_x}{dt}\right] \qquad P + \rho g(l - x) - \rho gl = \frac{d}{dt}(\rho x v) \qquad P = \rho(gx + v^2) \qquad \textit{Ans.}$$

The force P is seen to be equal to the weight of the portion of the chain which is off the platform plus the added term which accounts for the time rate of increase of momentum of the chain.

Solution by Eq. 191 may also be made by considering the moving part x of the chain as a body which gains mass. The force ΣF is the resultant of all forces on the moving

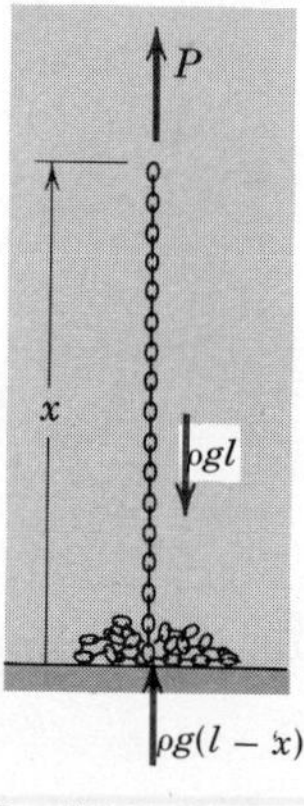

Problem 10/32

mass ρx of the chain except that exerted by the particles which are accumulated and is $\Sigma F = P - \rho g x$. The velocity is constant so $\dot{v} = 0$. The rate of increase of mass is $\dot{m} = \rho v$. The relative velocity u of the attaching particles is v. Thus

$$[\Sigma F = m\dot{v} + \dot{m}u] \qquad P - \rho g x = 0 + \rho v(v) \qquad P = \rho(gx + v^2)$$

10/33 A rocket of initial total mass m_0 is fired vertically up from the north pole and accelerates until the fuel, which burns at a constant rate, is exhausted. The relative nozzle velocity of the exhaust gas has a constant value u, and the nozzle exhausts at atmospheric pressure throughout the flight. If the residual mass of the rocket structure and machinery is m_b when burn-out occurs, determine the expression for the maximum velocity reached by the rocket. Neglect atmospheric resistance and the variation of gravity with altitude.*

Solution. The rocket thrust is $T = m'u = -\dot{m}u$ and is shown as a force on the simulated rocket system. With the neglect of p and R, Eq. 192, or Newton's second law, gives

$$-\dot{m}u - mg = m\dot{v}$$

Multiplication by dt, division by m, and rearrangement give

$$dv = -u\frac{dm}{m} - g\,dt$$

The velocity v corresponding to the time t is given by the integration

$$\int_0^v dv = -u\int_{m_0}^{m}\frac{dm}{m} - g\int_0^t dt$$

or

$$v = u\ln\frac{m_0}{m} - gt$$

Since the fuel is burned at the constant rate $m' = -\dot{m}$, the mass at any time t is $m = m_0 + \dot{m}t$. The time at which burnout occurs is, therefore, $t_b = (m_0 - m_b)/(-\dot{m})$ and gives the condition for maximum velocity. Thus

$$v_{max} = u\ln\frac{m_0}{m_b} + \frac{g}{\dot{m}}(m_0 - m_b) \qquad \textit{Ans.}$$

The quantity $\dot{m}$ is a negative number since the mass decreases with time.

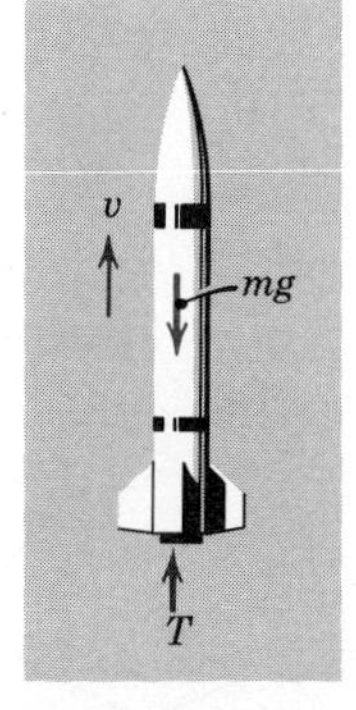

Problem 10/33

*The neglect of atmospheric resistance is not a bad assumption for a first approximation inasmuch as the velocity of the ascending rocket is smallest in the dense part of the atmosphere and greatest in the rarefied region. Also for an altitude of 320 km the acceleration due to gravity is 91 per cent of the value at the surface of the earth.

Part B: Exact Case

In formulating Eqs. 191 and 192 it was assumed that all mass particles within m have identical motions and that the velocity transition between m and m_0 occurs abruptly at the boundary between the two masses. Although these assumptions give close approximations to the physical problem, a more general treatment is needed for an exact description of the motion of a time-dependent mass where the assumptions cited are not strictly true.

For the general case of variable mass consider the system in Fig. 97 defined by a closed space surface whose shape and position may vary with time. The quantity and distribution of mass within the system are also time-dependent. Mass is assumed to enter the system through section E in a continuous stream. The total mass of the system at time t, shown in Fig. 97, is m. The mass increment Δm is the mass which enters the system in time Δt. The position of Δm from a fixed reference point O is given by the vector $\mathbf{r}_0$ and from the center of mass G of m by the vector $\boldsymbol{\rho}_0$. The position of the entrance section from these same two reference points is given by $\mathbf{r}_e$ and $\boldsymbol{\rho}_e$, respectively. It is noted at this point that whereas $\mathbf{r}_e = \mathbf{r}_0$ and $\boldsymbol{\rho}_e = \boldsymbol{\rho}_0$ in the limit as $\Delta t \to 0$, their derivatives represent different velocities so that $\dot{\mathbf{r}}_e \neq \dot{\mathbf{r}}_0$ and $\dot{\boldsymbol{\rho}}_e \neq \dot{\boldsymbol{\rho}}_0$. The instantaneous position of the center of mass G of the varying mass is located by the position vector $\bar{\mathbf{r}}$, and a representative particle of mass m_i of the system is located from O and G by the respective vectors $\mathbf{r}_i$ and $\boldsymbol{\rho}_i$.

The linear momentum of the system at time t is the sum of the linear momenta of all particles within the boundary at time t and is

$$\mathbf{G} = \Sigma m_i \dot{\mathbf{r}}_i$$

At time $t + \Delta t$, the linear momentum is

$$\mathbf{G} + \Delta\mathbf{G} = \Sigma m_i(\dot{\mathbf{r}}_i + \Delta\dot{\mathbf{r}}_i) + \Delta m(\dot{\mathbf{r}}_0 + \Delta\dot{\mathbf{r}}_0)$$

where the momentum of the increment Δm must be included since Δm is now a part of the system. Subtraction of the first expression from the second expression, division by Δt, and passage to the limit yield

$$\dot{\mathbf{G}} = \Sigma m_i \ddot{\mathbf{r}}_i + \dot{m}\dot{\mathbf{r}}_0$$

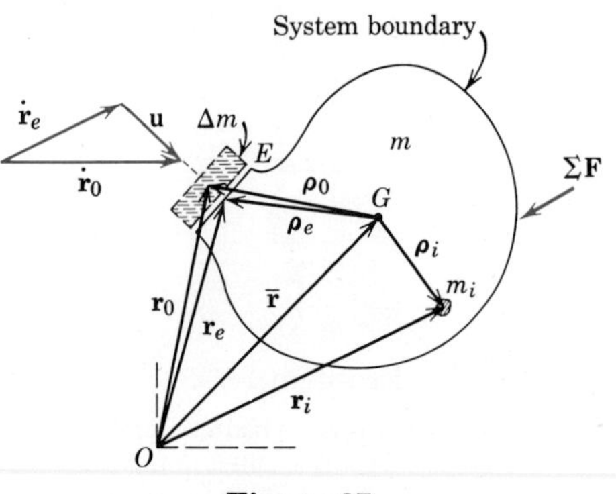

Figure 97

where the higher order term $\Delta m\, \Delta\dot{\mathbf{r}}_0$ disappears in the limit. The resultant force external to the system (including any static pressure acting over the section boundary at E) is $\Sigma\mathbf{F} = \Sigma m_i \ddot{\mathbf{r}}_i$. Combination with the expression for $\dot{\mathbf{G}}$ yields

$$\Sigma\mathbf{F} = \dot{\mathbf{G}} - \dot{m}\dot{\mathbf{r}}_0 \quad \textbf{(193)}$$

Equation 193 is the force-momentum relation for a system with time-dependent mass and is a generalization of Eq. 96 which has already been used extensively. The new equation states that the resultant of external forces which act on a system of time-dependent mass equals the time rate of change of the linear momentum of the varying mass minus the rate at which linear momentum is being added to the system by incoming mass.

In applying Eq. 193 it is convenient to express $\mathbf{G}$ in terms of measured quantities. First, the mass center is located by the moment principle at time t and at time $t + \Delta t$ by the respective expressions

$$m\bar{\mathbf{r}} = \Sigma m_i \mathbf{r}_i$$

and

$$m\bar{\mathbf{r}} + \Delta(m\bar{\mathbf{r}}) = \Sigma m_i(\mathbf{r}_i + \Delta\mathbf{r}_i) + \Delta m(\mathbf{r}_0 + \Delta\mathbf{r}_0)$$

Subtraction of the first from the second expression, division by Δt, passage to the limit, and neglect of the second order term give

$$\frac{d}{dt}(m\bar{\mathbf{r}}) = \Sigma m_i \dot{\mathbf{r}}_i + \dot{m}\mathbf{r}_0$$

so that

$$\mathbf{G} = \frac{d}{dt}(m\bar{\mathbf{r}}) - \dot{m}\mathbf{r}_0$$

Substitution of $\mathbf{r}_0 - \bar{\mathbf{r}} = \boldsymbol{\rho}_0 = \boldsymbol{\rho}_e$ gives

$$\mathbf{G} = m\dot{\bar{\mathbf{r}}} - \dot{m}\boldsymbol{\rho}_e \quad (194)$$

From Eq. 194 it may be concluded that the linear momentum of a time-varying mass depends on the location with respect to the mass center of the position where mass is added.

With the substitutions $\dot{\bar{\mathbf{r}}} = \dot{\mathbf{r}}_e - \dot{\boldsymbol{\rho}}_e$ and $\mathbf{u} = \dot{\mathbf{r}}_0 - \dot{\mathbf{r}}_e$ into the differentiated expression for $\mathbf{G}$, Eq. 193 becomes, upon collection of terms,

$$\Sigma\mathbf{F} = m\ddot{\mathbf{r}}_e - \dot{m}\mathbf{u} - \frac{d^2}{dt^2}(m\boldsymbol{\rho}_e) \quad (195)$$

Expression of the equation of motion in this form, which includes the acceleration $\ddot{\mathbf{r}}_e$ of the position E on the system boundary rather than the acceleration of the mass center, will be found convenient since the motion of the structure bounding the mass is usually the primary concern. Such is clearly the case with a rocket.

If mass is leaving the system rather than entering the system, $\dot{m}$ will be of opposite sign and $\mathbf{u}$ will be reversed in direction. If mass is entering or leaving the system through more than one opening, the last term in Eq. 193

may be replaced by appropriate terms one for each entering or leaving stream.

The relation between moment and angular momentum for the varying-mass system is obtained by starting with the expression for the angular momentum $\overline{\mathbf{H}}$ of the system about the mass center G. At time t this angular momentum is

$$\overline{\mathbf{H}} = \Sigma(\boldsymbol{\rho}_i \times m_i\dot{\mathbf{r}}_i)$$

and at time $t + \Delta t$, the angular momentum becomes

$$\overline{\mathbf{H}} + \Delta\overline{\mathbf{H}} = \Sigma[(\boldsymbol{\rho}_i + \Delta\boldsymbol{\rho}_i) \times m_i(\dot{\mathbf{r}}_i + \Delta\dot{\mathbf{r}}_i)] + (\boldsymbol{\rho}_0 + \Delta\boldsymbol{\rho}_0) \times \Delta m(\dot{\mathbf{r}}_0 + \Delta\dot{\mathbf{r}}_0)$$

Subtraction of the expression for $\overline{\mathbf{H}}$, division by Δt, passage to the limit, and the neglect of higher order terms yield

$$\dot{\overline{\mathbf{H}}} = \Sigma(\boldsymbol{\rho}_i \times m_i\ddot{\mathbf{r}}_i) + \Sigma(\dot{\boldsymbol{\rho}}_i \times m_i\dot{\mathbf{r}}_i) + \boldsymbol{\rho}_e \times \dot{m}\dot{\mathbf{r}}_0$$

where $\boldsymbol{\rho}_e$ has been substituted for $\boldsymbol{\rho}_0$. The first summation is merely the moment sum about G of the resultant forces on all particles of the system, which sum equals the resultant moment $\Sigma\overline{\mathbf{M}}$ about G of all external forces acting on the system since the moments of internal forces cancel.

When the substitution $\dot{\boldsymbol{\rho}}_i = \dot{\mathbf{r}}_i - \dot{\bar{\mathbf{r}}}$ is made, the second summation becomes

$$\Sigma(\dot{\boldsymbol{\rho}}_i \times m_i\dot{\mathbf{r}}_i) = \Sigma(\dot{\mathbf{r}}_i - \dot{\bar{\mathbf{r}}}) \times m_i\dot{\mathbf{r}}_i = \mathbf{0} - \dot{\bar{\mathbf{r}}} \times \Sigma m_i\dot{\mathbf{r}}_i = -\dot{\bar{\mathbf{r}}} \times \mathbf{G}$$

Further simplification occurs upon substitution of Eq. 194, so that

$$\Sigma(\dot{\boldsymbol{\rho}}_i \times m_i\dot{\mathbf{r}}_i) = -\dot{\bar{\mathbf{r}}} \times (m\dot{\bar{\mathbf{r}}} - \dot{m}\boldsymbol{\rho}_e) = \dot{\bar{\mathbf{r}}} \times \dot{m}\boldsymbol{\rho}_e$$

The expression for $\dot{\overline{\mathbf{H}}}$ becomes

$$\begin{aligned} \dot{\overline{\mathbf{H}}} &= \Sigma\overline{\mathbf{M}} + \dot{\bar{\mathbf{r}}} \times \dot{m}\boldsymbol{\rho}_e + \boldsymbol{\rho}_e \times \dot{m}\dot{\mathbf{r}}_0 \\ &= \Sigma\overline{\mathbf{M}} + \boldsymbol{\rho}_e \times \dot{m}(\dot{\mathbf{r}}_0 - \dot{\bar{\mathbf{r}}}) \end{aligned}$$

upon reordering the cross-product term. With the further substitution of $\dot{\mathbf{r}}_0 - \dot{\bar{\mathbf{r}}} = \dot{\boldsymbol{\rho}}_0$, the moment equation becomes

$$\blacktriangleright \qquad \Sigma\overline{\mathbf{M}} = \dot{\overline{\mathbf{H}}} - \boldsymbol{\rho}_e \times \dot{m}\dot{\boldsymbol{\rho}}_0 \qquad \textbf{(196)}$$

Equation 196 is the moment equation for a system with time-dependent mass and is a generalization of Eq. 98 which was developed in Chapter 4 for a system of constant mass and used extensively in the description of rigid-body motion. The new equation states that the resultant moment about the mass center of all external forces which act on a system of time-dependent mass equals the time rate of change of angular momentum about the mass center of the varying mass minus the rate at which the added mass increases the angular momentum relative to G. The term $\dot{\boldsymbol{\rho}}_0$ is the velocity of the entering mass relative to G and equals the relative velocity $\mathbf{u}$ only if $\dot{\boldsymbol{\rho}}_e$ is zero.

A similar moment equation may be written about the fixed point O in which case $\mathbf{r}_e$ replaces $\boldsymbol{\rho}_e$, $\dot{\mathbf{r}}_0$ replaces $\dot{\boldsymbol{\rho}}_0$, and the angular momentum is taken about O. The resulting equation is

$$\Sigma \mathbf{M}_O = \dot{\mathbf{H}}_O - \mathbf{r}_e \times \dot{m}\dot{\mathbf{r}}_0 \qquad \textbf{(197)}$$

Equation 197 is a generalization of Eq. 97 for a constant mass system.

In the foregoing derivations of the force and moment equations for a system of time-dependent mass, any variation in the magnitude or direction of the relative velocity $\mathbf{u}$ of entering mass over the entrance section was assumed to be absent. If the mass enters the system in a continuous but varying manner over an appreciable area of the system boundary, the last term on the right-hand side of Eqs. 193, 196, and 197 must be replaced by an appropriate integral. If a differential element of the boundary area through which mass is passing is represented by the vector $d\mathbf{A}$ normal to the surface with the positive sense taken outward, the rate at which mass enters dA becomes $-\rho\mathbf{u} \cdot d\mathbf{A}$ where ρ stands for the density of the incoming mass. The minus sign accounts for the fact that $\mathbf{u}$ is positive inward in the sense opposite to $d\mathbf{A}$. The rate at which linear momentum is added to the system through dA is, then, $\dot{\mathbf{r}}_0(-\rho\mathbf{u} \cdot d\mathbf{A})$, and the total rate for the system is the integral of this quantity over the flow area of the boundary. Equation 193 is thus replaced by

$$\Sigma \mathbf{F} = \dot{\mathbf{G}} + \int \dot{\mathbf{r}}_0(\rho\mathbf{u} \cdot d\mathbf{A}) \qquad (198)$$

The rate at which angular momentum relative to the mass center G is added to the system through dA is $\boldsymbol{\rho}_e \times \dot{\boldsymbol{\rho}}_0(-\rho\mathbf{u} \cdot d\mathbf{A})$, and the total rate of adding angular momentum relative to G is the integral of this expression over the flow area. Thus Eq. 196 becomes

$$\Sigma \overline{\mathbf{M}} = \dot{\overline{\mathbf{H}}} + \int \boldsymbol{\rho}_e \times \dot{\boldsymbol{\rho}}_0(\rho\mathbf{u} \cdot d\mathbf{A}) \qquad (199)$$

Similarly, when the moment equation is written about a fixed point O, Eq. 197 becomes

$$\Sigma \mathbf{M}_O = \dot{\mathbf{H}}_O + \int \mathbf{r}_e \times \dot{\mathbf{r}}_0(\rho\mathbf{u} \cdot d\mathbf{A}) \qquad (200)$$

Equations 193, 196, and 197 and Eqs. 198, 199, and 200 are generalizations of the classical momentum equations presented in Chapter 4 and permit the proper motion equations to be written directly for the linear and angular motions of time-dependent mass systems. The equations for variable mass presented here are uncommon in the literature inasmuch as variable-mass problems are of relatively recent interest.

Sample Problem

10/34 Write the exact equation of motion for the vertical ascent of a rocket of total mass m. The relative velocity of the expelled gases is U. The center of mass of the rocket is a distance h from the exhaust nozzle, the average pressure across the nozzle of exit area A is p, and the aerodynamic resistance is R.

Solution. Let the vertical coordinate of the rocket nozzle be x which replaces $\mathbf{r}_e$ of Eq. 195. Also $-h$ replaces $\boldsymbol{\rho}_e$ and $-U$ replaces $\mathbf{u}$. With these substitutions, Eq. 195 becomes

$$-R - mg + pA = m\ddot{x} + \dot{m}U + \frac{d^2}{dt^2}(mh)$$

or

$$-R - mg + pA + m'U = m\ddot{x} + \frac{d^2}{dt^2}(mh) \qquad \textit{Ans.}$$

The term $m'U$ where $m' = -\dot{m}$ represents the thrust. This equation is of the form "$\Sigma F_x = ma_x$" for the simulated rocket except that it contains the correction term $d^2(mh)/dt^2$. It is noted that this correction term exists even when h is constant (radially-burning solid fuel) provided $\ddot{m}$ is not zero.

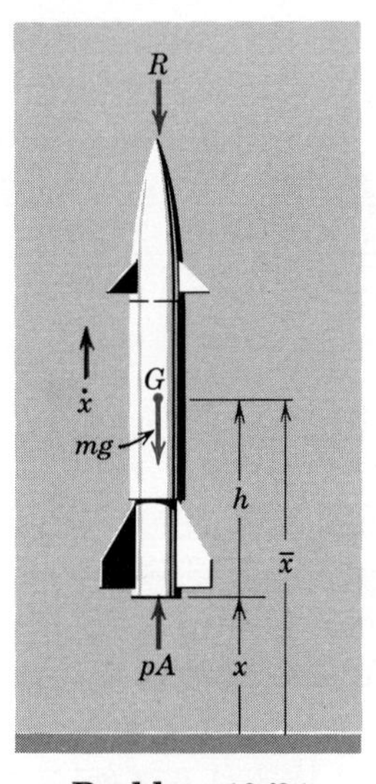

Problem 10/34

Problems

10/35 The Saturn V rocket has a total launch mass of 3160 t, and each of its five F-1 engines develops a thrust of 7500 kN. A total of 2280 t of fuel is burned by the first-stage F-1 engines in 160 s at a constant rate. Calculate the initial vertical lift-off acceleration a of the rocket and the exhaust velocity u of the burned gases relative to the nozzle.

Ans. $a = 2.06$ m/s^2, $u = 2630$ m/s

10/36 The mass m of a raindrop increases as it picks up moisture during its vertical descent through still air. If the air resistance to motion of the drop is R and its downward velocity is v, write the equation of motion for the drop and show that the relation $\Sigma F = d(mv)/dt$ is obeyed as a special case of the variable-mass equation.

10/37 The tank of water is at rest on a horizontal surface when a 250-N force is applied to it as shown. If water issues from the rear discharge pipe at the rate of 20 kg/s with a velocity of 2.4 m/s relative to the opening in the direction shown, calculate the initial acceleration a of the tank if its total mass at the start is 300 kg. Neglect the rotational inertia of the wheels.

Ans. $a = 0.913 \text{ m/s}^2$

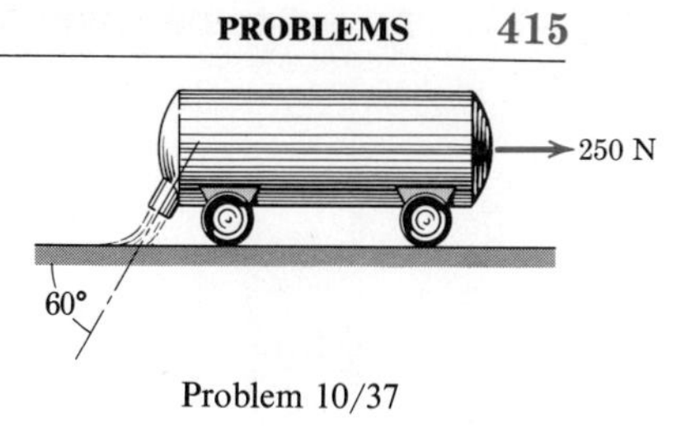

Problem 10/37

10/38 A small rocket of initial mass m_0 is fired vertically upward near the surface of the earth (g constant). If air resistance is neglected, determine the manner in which the mass m of the rocket must vary as a function of the time t after launching in order that the rocket may have a constant vertical acceleration a with a constant relative velocity u of the escaping gases with respect to the nozzle.

10/39 The end of a pile of loose-link chain of mass ρ per unit length is being pulled horizontally along the surface by a constant force P. If the coefficient of friction between the chain and the surface is f, determine the acceleration a of the chain in terms of x and $\dot{x}$.

Ans. $a = \dfrac{P}{\rho x} - fg - \dfrac{\dot{x}^2}{x}$

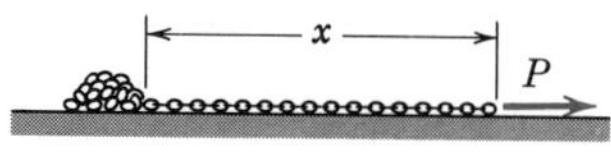

Problem 10/39

10/40 The upper end of the open-link chain of length L and mass ρ per unit length is released from rest with the lower end just touching the platform of the scale. Determine the expression for the force F read on the scale as a function of the distance x through which the upper end has fallen. (*Comment:* The chain acquires a free-fall velocity of $\sqrt{2gx}$ since the links on the scale exert no force on those above, which are still falling freely. Work the problem in two ways, first, by evaluating the time rate of change of momentum for the entire chain and, second, by considering the force F to be composed of the weight of the links at rest on the scale plus the force necessary to divert an equivalent stream of fluid.)

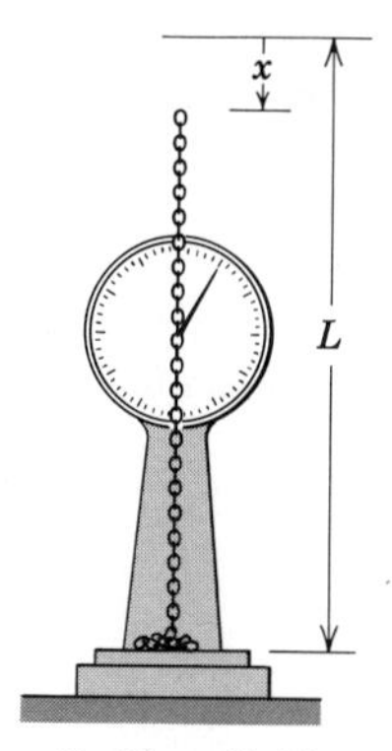

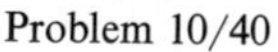

Problem 10/40

10/41 The left end of the open-link chain of length l and mass ρ per unit length is released from rest at $x = 0$. Determine the expression for the tension T in the chain at its support at A in terms of x. Also determine the energy loss ΔE during the entire motion from $x = 0$ to $x = 2l$.

Ans. $T = \dfrac{3\rho g x}{2}$, $\Delta E = \rho g l^2$

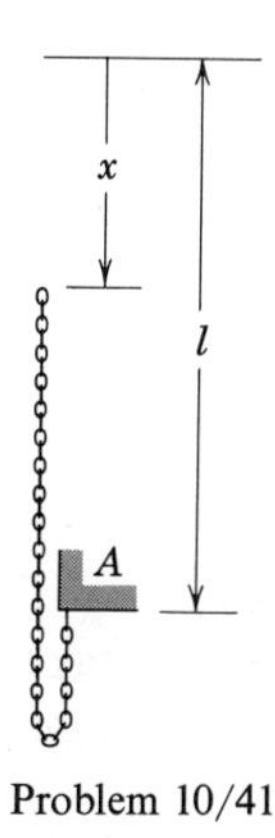

Problem 10/41

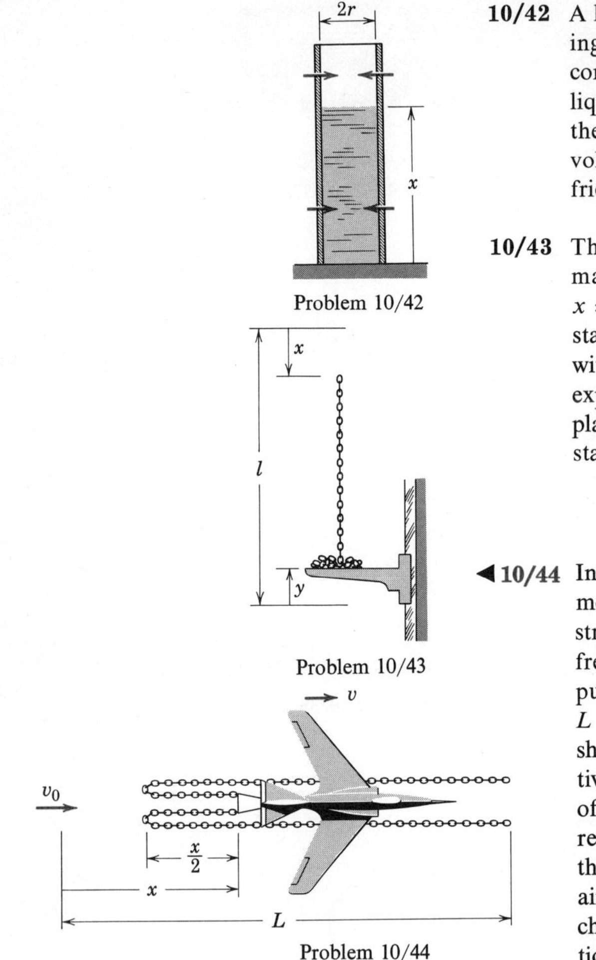

Problem 10/42

Problem 10/43

Problem 10/44

10/42 A liquid of density ρ is contained in the collapsing vertical tube whose radius r varies at the constant rate $\dot{r} = -c$. Find the pressure p in the liquid at the base of the tube as a function of the height x of the liquid column. The constant volume of the liquid is V. Neglect any fluid friction.

10/43 The open-link chain of total length l and of mass ρ per unit length is released from rest at $x = 0$ at the same instant that the platform starts from rest at $y = 0$ and moves vertically up with a constant acceleration a. Determine the expression for the total force R exerted on the platform by the chain t seconds after the motion starts.

Ans. $R = \frac{3}{2}\rho(a + g)^2t^2$

◀**10/44** In the figure is shown a system used to arrest the motion of an airplane landing on a field of restricted length. The plane of mass m rolling freely with a velocity v_0 engages a hook which pulls the ends of two heavy chains each of length L and mass ρ per unit length in the manner shown. A conservative calculation of the effectiveness of the device neglects the retardation of chain friction on the ground and any other resistance to the motion of the airplane. With these assumptions compute the velocity v of the airplane at the instant that the last link of each chain is put in motion. Also determine the relation between displacement x and the time t after contact with the chain. Assume each link of the chain acquires its velocity v suddenly upon contact with the moving links.

Ans. $v = \dfrac{v_0}{1 + 2\rho L/m}$

$$x = \frac{m}{\rho}\left[\sqrt{1 + \frac{2v_0 t\rho}{m}} - 1\right]$$

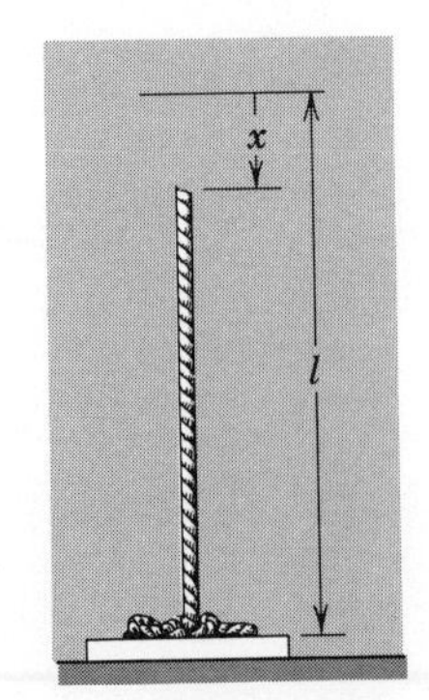

Problem 10/45

◀**10/45** A rope or hinged-link bicycle-type chain of length l and mass ρ per unit length is released from rest with $x = 0$. Determine the expression for the total force R exerted on the fixed platform by the chain as a function of x. Note that the hinged-link chain is a conservative system during all but the last increment of motion. Compare the result with that of Prob. 10/43 if the upward motion of the platform in that problem is taken to be zero.

Ans. $R = \rho g x \dfrac{4l - 3x}{2(l - x)}$

10/46 One end of the pile of chain falls through a hole in its support and pulls the remaining links after it in a steady flow. If the links which are initially at rest acquire the velocity of the chain suddenly and without frictional resistance or interference from the support or from adjacent links, find the velocity v of the chain as a function of x if $v = 0$ when $x = 0$. Also find the acceleration a of the falling chain and the energy ΔE lost from the system as the last link leaves the platform. (*Hint:* Apply Eq. 191 and treat the product xv as the variable when solving the differential equation. Also note at the appropriate step that $dx = v\,dt$.) The total length of the chain is L, and its mass per unit length is ρ.

$$\textit{Ans. } v = \sqrt{\frac{2gx}{3}}, \quad a = \frac{g}{3}, \quad \Delta E = \frac{\rho g L^2}{6}$$

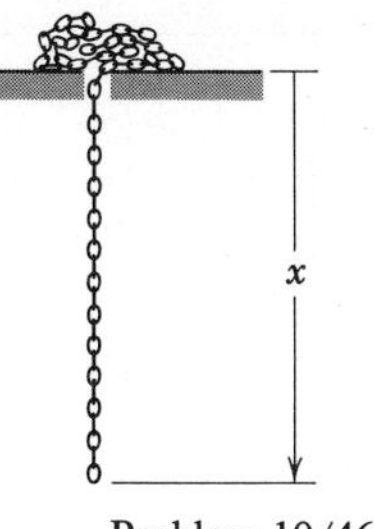

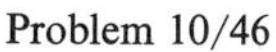
Problem 10/46

10/47 Replace the pile of chain in Prob. 10/46 by a coil of rope of mass ρ per unit length and total length L as shown, and determine the velocity of the falling section in terms of x if it starts from rest at $x = 0$. Show that the acceleration is constant at $g/2$. The rope is considered to be perfectly flexible in bending but inextensible and constitutes a conservative system (no energy loss). Rope elements acquire their velocity in a continuous manner from zero to v in a small transition section of the rope at the top of the coil. For comparison with the chain of Prob. 10/46 this transition section may be considered to have negligible length without violating the requirement that there be no energy loss in the present problem. Also determine the force R exerted by the platform on the coil in terms of x and explain why R becomes zero when $x = 2L/3$. Neglect the dimensions of the coil compared with x.

$$\textit{Ans. } v = \sqrt{gx}, \quad R = \rho g\left(L - \frac{3x}{2}\right)$$

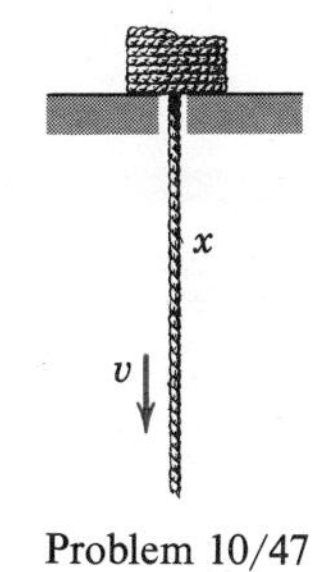

Problem 10/47

Problem 10/48

◀**10/48** A drop test is conducted on an instrument mounted in the carriage A which has a mass m including the instrument. Two flexible electric cables, each of total length l and mass ρ per unit length, transmit signals from the accelerometers located in A and are attached as shown. The carriage A is dropped from rest at $x = 0$ and strikes the base at $x = l/2$. Find the acceleration $\ddot{x}$ of A as a function of x and show that it is greater than g. Also find the force of support R at the fixed end of each cable as a function of x. It may be assumed that each cable hangs in two vertical lengths with a connecting loop of negligible dimensions.

$$\textit{Ans. } \ddot{x} = g\left\{1 + \rho x \frac{m + \rho\left(l - \dfrac{x}{2}\right)}{[m + \rho(l - x)]^2}\right\}$$

$$R = \frac{\rho g}{2}\left\{l + x + x\frac{m + \rho\left(l - \dfrac{x}{2}\right)}{m + \rho(l - x)}\right\}$$

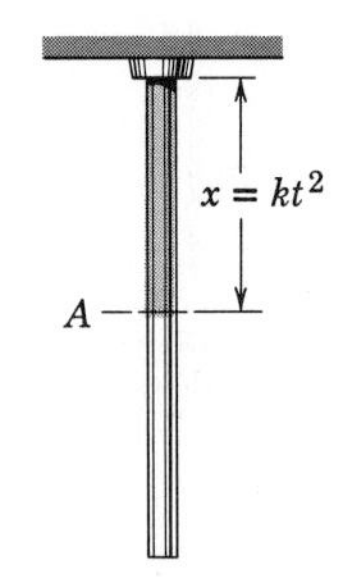

Problem 10/49

10/49 As an exercise to test the applicability of Eq. 195 consider a varying section of a vertical bar hanging under the action of its own weight. The system to be analyzed is the mass within the bar above section A. Section A descends according to $x = kt^2$ where k is a constant and t is time. Apply Eq. 195 and verify the known fact that the resultant force on any portion of the static bar is zero. The mass of the bar per unit length is ρ.

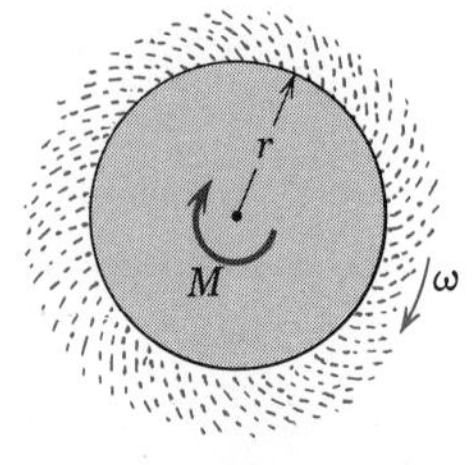

Problem 10/50

10/50 The homogeneous solid circular disk of unit thickness has a density ρ. The disk rotates about its center under the action of a torque M and loses mass by a uniform disintegration of its rim at the constant rate $c = -\dot{r}$. Determine the angular acceleration $\dot{\omega}$ of the disk as a function of its radius r. Compare the result with the erroneous answer obtained by assuming $M = \dot{H}$.

$$\textit{Ans. } \dot{\omega} = \frac{2M}{\rho\pi r^4}$$

10/51 Use the results of Sample Problem 10/34 to determine the expression for the magnitude of the ratio n of the correction term in the force equation of motion to the rocket thrust $\dot{m}U$ for the solid-propellant rocket in terms of y, $\dot{y}$, and $\ddot{y}$ where the grain is a solid cylinder of radius r_0 encased in an inhibitor so that it burns from the end only. Compute n for the start of the burning where $y = 1.2$ m, $\dot{y} = 50$ mm/s, and $\ddot{y} = 150$ mm/s^2. The velocity U of the exhaust gases relative to the nozzle is constant at 1200 m/s.

$$Ans.\ n = \frac{y}{U}\left(\frac{\dot{y}}{y} + \frac{\ddot{y}}{\dot{y}}\right) = 0.00304$$

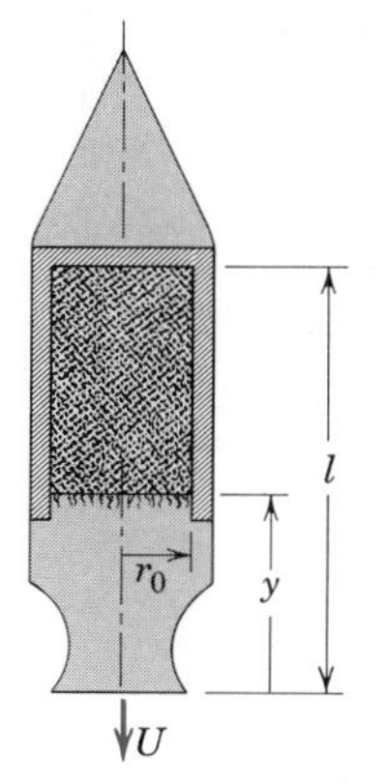

Problem 10/51

10/52 Use the results of Sample Problem 10/34 to determine the expression for the magnitude of the ratio n of the correction term in the force equation of motion to the rocket thrust $\dot{m}U$ for the solid-propellant rocket in terms of r, $\dot{r}$, and $\ddot{r}$ where the hollow grain burns from the inside out. Compute n at the start of burning when $r = 25$ mm, $\dot{r} = 50$ mm/s, and $\ddot{r} = 150$ mm/s^2. The velocity U of the exhaust gases relative to the nozzle is constant at 1200 m/s, and the center of mass of the grain is 2.4 m from the nozzle exit plane. *Ans.* $n = 0.01$

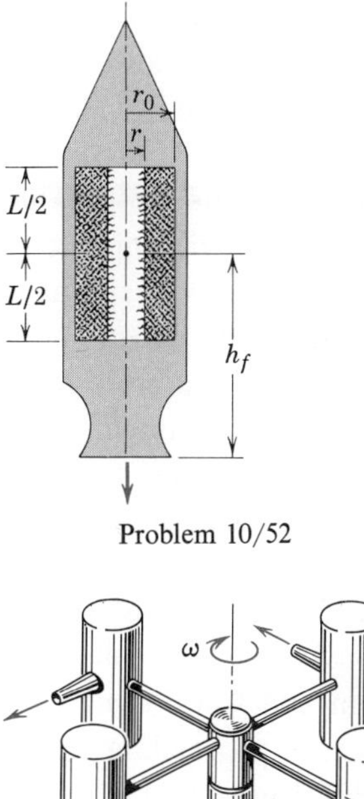

Problem 10/52

10/53 A rotating sprinkler contains all of its water supply in four cylindrical tanks at a distance r from the axis of rotation. Water issues through the nozzles, each of exit area A, with a constant total volume rate Q under the action of air pressure in the tanks. Determine the retarding moment M which must be applied to the sprinkler in order to maintain a constant angular velocity ω of the sprinkler in the direction shown. At what speed ω_0 will the sprinkler operate with no applied torque? Let ρ be the density of the water. Compare the result with that of Prob. 10/27.

$$Ans.\ M = \rho Q\left(\frac{Qr}{4A} - b^2\omega\right), \quad \omega_0 = \frac{Qr}{4Ab^2}$$

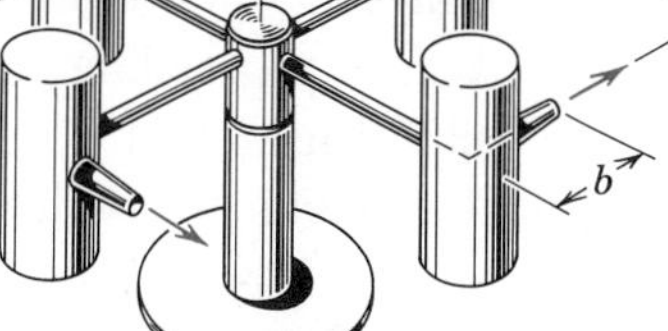

Problem 10/53

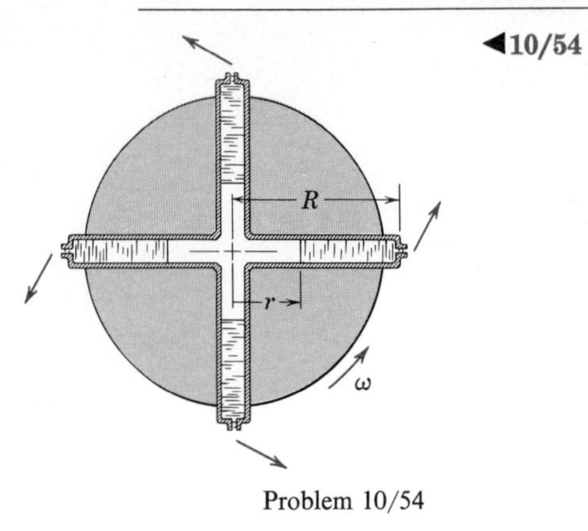

Problem 10/54

◀**10/54** A disk with four identical tubes filled with liquid is spinning freely with an angular velocity ω_0 about its vertical axis when orifices, which point in the radial direction, are opened in the ends of the tubes. For the limiting case where the mass of the disk and tubes is negligible compared with the mass of the liquid, determine the angular velocity ω of the disk when all of the liquid has left the tubes. Does ω depend on the size of the orifices, the time required to drain the tubes, or the viscosity of the liquid? Neglect mechanical friction. *Ans.* $\omega = 0.0777\omega_0$

54 Wave Propagation. The propagation of waves of mechanical force and displacement through deformable solids is a phenomenon which involves dynamics and material properties. A knowledge of wave propagation is important in the design of cables subjected to transverse disturbances, bars which withstand longitudinal impact, and other machine and structural components which must be designed to withstand high loading rates.

In Chapter 9 the time response of masses in both linear and angular motion was discussed under the assumption that each vibrating mass was concentrated at a point for linear motion or was rotating as a rigid unit in a single plane for angular motion. Such an approach was termed a lumped-parameter analysis. In discussing the propagation of waves in continuous media, it is necessary to consider the disturbance as a continuous function of both time and position within the body, and this type of description is referred to as a *distributed-parameter* analysis. The description of wave propagation which follows constitutes only a brief introduction to the formulation and solution of the equations which govern this phenomenon.

Before proceeding to the mathematical formulation of wave motion, it is helpful to obtain physical insight into the phenomenon. The mechanical model of Fig. 98, which shows a series of small balls mounted on elastic rods, will serve to illustrate the transfer of force and displacement in wave motion. If the ball on the extreme left is pulled to the left and released, a succession of collisions between balls will proceed from left to right as the disturbance "wave" travels along the line of balls. The greater the number of balls and the closer their spacing, the more nearly does the experiment represent a continuous function of the horizontal coordinate.

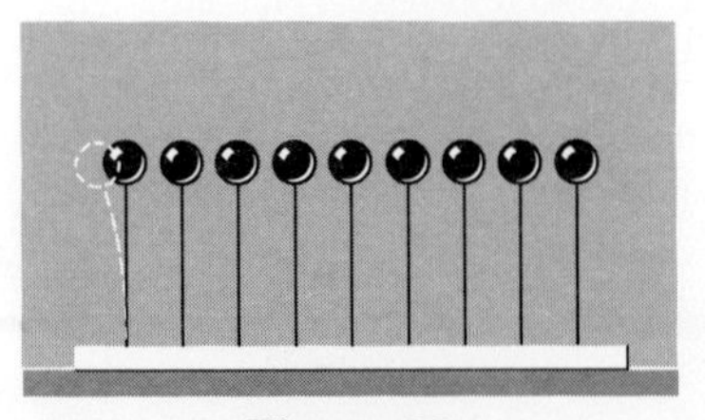

Figure 98

Three examples of one-dimensional wave motion in elastic bodies involving the solutions of equations of identical mathematical form will be given. The governing equations are formulated first, and a single description of the solutions of the characteristic wave equation is then given.

Longitudinal waves. Consider now the transmission of a longitudinal wave of stress along an elastic rod, Fig. 99, of uniform cross-sectional area A and elastic modulus E. The elastic displacement of a section whose coordinate is x is u, and the simultaneous displacement of a section whose coordinate is $x + dx$ is $u + (\partial u/\partial x)\, dx$. The partial derivative is used since u is also a function of the time t, and change with respect to x alone is required for the same instant of time. It is assumed that the horizontal displacements u of all points on a given cross section are identical so that each plane section remains plane. The displacement u is finite but small. The linear elastic strains at the two sections are

$$e_x = \frac{\partial u}{\partial x} \qquad \text{and} \qquad e_x + \frac{\partial e_x}{\partial x}\, dx = \frac{\partial u}{\partial x} + \frac{\partial^2 u}{\partial x^2}\, dx$$

The free-body diagram of the element of the rod shows the force on each section which equals the stress times the area. If ρ represents the density of the elastic body, the equation of motion in the x-direction becomes

$$EA\, \frac{\partial^2 u}{\partial x^2}\, dx = \rho A\, \frac{\partial^2 u}{\partial t^2}\, dx$$

where the acceleration of the center of the element is $\partial^2 u/\partial t^2$ plus a higher order term which vanishes. Thus

$$\frac{\partial^2 u}{\partial t^2} = \frac{E}{\rho}\, \frac{\partial^2 u}{\partial x^2} \tag{201}$$

Equation 201 is known as the *wave equation* in one dimension, and its solution provides a description of the propagation of longitudinal stress waves in the rod.

Torsional waves. Consider next an elastic circular shaft, Fig. 100, of density ρ and with radius r and shear modulus G, in which angular oscillations occur along its length. A longitudinal surface element in the untwisted state

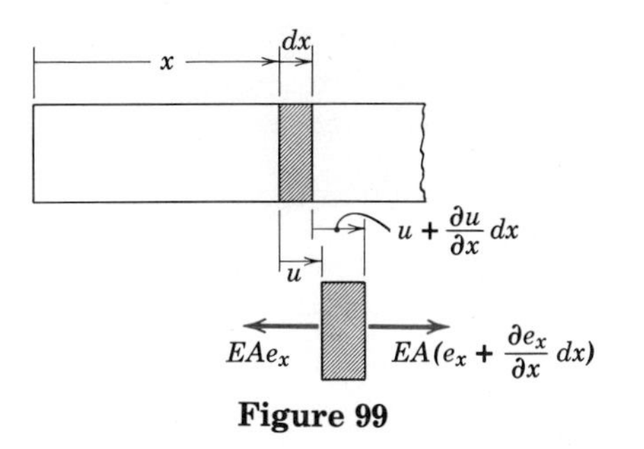

Figure 99

is shown as *a-a* and in the twisted state as *a′-a′*. The angular displacement of a section whose coordinate is x is θ measured from the untwisted condition. Simultaneously the angular displacement of an adjacent section whose coordinate is $x + dx$ is $\theta + (\partial\theta/\partial x)\, dx$. Again the partial derivative is used since θ will be a function of time t as well as x. It is assumed that radial lines in each section remain radial and that the angular displacements are small but finite. The angular strain at the first section is γ, and for small strains it may be replaced by its tangent. Thus

$$\gamma = \frac{r \dfrac{\partial\theta}{\partial x}\, dx}{dx} = r\frac{\partial\theta}{\partial x}$$

The angular strain at the section whose coordinate is $x + dx$ is, then,

$$r\frac{\partial\theta}{\partial x} + \frac{\partial}{\partial x}\left(r\frac{\partial\theta}{\partial x}\right) dx$$

The free-body diagram of the elemental slice in the lower part of the figure shows the torsional moments on both sections. Since the torsional moment for a circular shaft is $M = JG\gamma/r$ where J is the polar moment of inertia of the section, the moment on the left end of the elemental disk is $M = JG(\partial\theta/\partial x)$, and the moment on the right end is

$$M + \frac{\partial M}{\partial x}\, dx = JG\frac{\partial\theta}{\partial x} + JG\frac{\partial^2\theta}{\partial x^2}\, dx$$

The equation for rotational motion of the element is obtained by equating the net moment on the element to the product of its mass moment of inertia $dI = \rho J\, dx$ and its angular acceleration $\partial^2\theta/\partial t^2$. Thus

$$JG\frac{\partial^2\theta}{\partial x^2}\, dx = \rho J\frac{\partial^2\theta}{\partial t^2}\, dx$$

or

$$\frac{\partial^2\theta}{\partial t^2} = \frac{G}{\rho}\frac{\partial^2\theta}{\partial x^2} \tag{202}$$

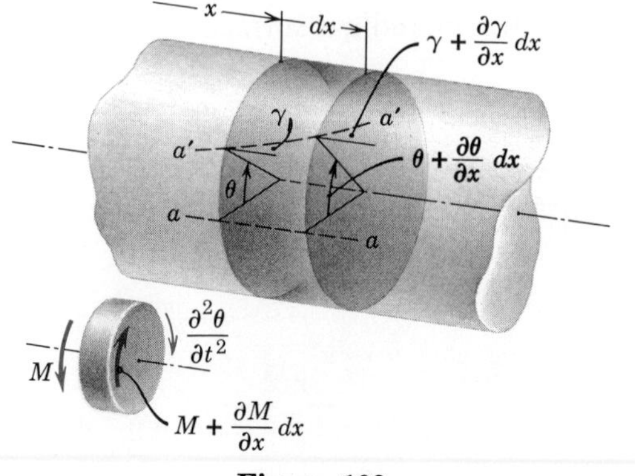

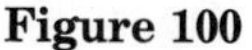

Figure 100

which is the wave equation for the transmission of torsional waves along the shaft. Equation 202 is identical in form to Eq. 201.

Transverse waves. Consider a taut wire or string of mass ρ' per unit length stretched under a tension T between two fixed points. The upper half of Fig. 101 shows a portion of this wire at an instant of time with the origin of coordinates at the fixed left end of the wire. An element of horizontal length dx is displaced in the transverse y-direction. The displacements shown are greatly exaggerated in order to reveal the mathematical relations. The lower half of the figure shows the free-body diagram of the element under the action of the two tensions which are assumed to be of the same magnitude but which differ in direction because of the curvature of the wire. It is further assumed that all motions occur in the y-direction.

For small deflections the substitution $\sin\theta = \theta = \tan\theta = \partial y/\partial x$ may be made. Again partial derivatives are used inasmuch as y is a function of both x and t. The downward component of force acting on the left end of the element is, therefore, $T\sin\theta = T(\partial y/\partial x)$, and the upward component of the tension on the right end of the element is

$$T\sin\left(\theta + \frac{\partial\theta}{\partial x}\,dx\right) = T\left(\frac{\partial y}{\partial x} + \frac{\partial^2 y}{\partial x^2}\,dx\right)$$

The vertical acceleration of the element is $\partial^2 y/\partial t^2$, so that the equation of motion in the transverse direction is

$$T\frac{\partial^2 y}{\partial x^2}\,dx = \rho'\,dx\,\frac{\partial^2 y}{\partial t^2}$$

or

$$\frac{\partial^2 y}{\partial t^2} = \frac{T}{\rho'}\frac{\partial^2 y}{\partial x^2} \tag{203}$$

which is the wave equation for transverse vibrations of a taut string or wire. It is noted that this equation has the same form as that of Eqs. 201 and 202.

Consider now the solutions to all three wave equations. The equations will be written as

$$\blacktriangleright \qquad \frac{\partial^2 \phi}{\partial t^2} = c^2\frac{\partial^2 \phi}{\partial x^2} \tag{204}$$

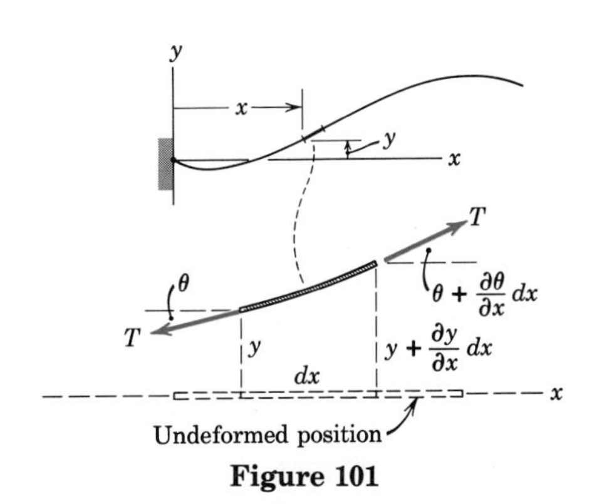

Figure 101

where ϕ represents the magnitude of the disturbance, whether it is the longitudinal displacement u, the angular displacement θ, the transverse displacement y, or some other disturbance described by the same equation. The characteristic scalar constant of the particular system is

$$c^2 = \frac{E}{\rho} \text{ for longitudinal elastic waves}$$

$$c^2 = \frac{G}{\rho} \text{ for torsional elastic waves}$$

$$c^2 = \frac{T}{\rho'} \text{ for transverse waves in a stretched wire}$$

Solutions of the wave equation. The disturbance ϕ will be some function F of both x and t, so that the functional relation $\phi = F(x, t)$ may be written. If the disturbance travels in the x-direction at a constant velocity v without changing shape, then for a ϕ-x_1 coordinate system moving with the disturbance, Fig. 102*a*, the functional relation in the moving system is merely $\phi = f(x_1)$. In the fixed ϕ-x coordinate system, Fig. 102*b*, the horizontal coordinate is given by $x = x_1 + vt$, so that the same disturbance takes the form

$$\phi = f(x - vt) \tag{205}$$

To show that the expression of Eq. 205 is a solution to Eq. 204, let $w = x - vt$. It is seen that

$$\frac{\partial\phi}{\partial x} = \frac{\partial\phi}{\partial w}\frac{\partial w}{\partial x} = \frac{\partial\phi}{\partial w} \quad \text{and} \quad \frac{\partial^2\phi}{\partial x^2} = \frac{\partial}{\partial w}\left(\frac{\partial\phi}{\partial w}\right)\frac{\partial w}{\partial x} = \frac{\partial^2\phi}{\partial w^2}$$

$$\frac{\partial\phi}{\partial t} = \frac{\partial\phi}{\partial w}\frac{\partial w}{\partial t} = -v\frac{\partial\phi}{\partial w} \quad \text{and} \quad \frac{\partial^2\phi}{\partial t^2} = \frac{\partial}{\partial w}\left(-v\frac{\partial\phi}{\partial w}\right)\frac{\partial w}{\partial t} = v^2\frac{\partial^2\phi}{\partial w^2}$$

Substitution into Eq. 204 gives

$$v^2\frac{\partial^2\phi}{\partial w^2} = c^2\frac{\partial^2\phi}{\partial w^2}$$

which holds true if $c = v$, the velocity of propagation of the disturbance.

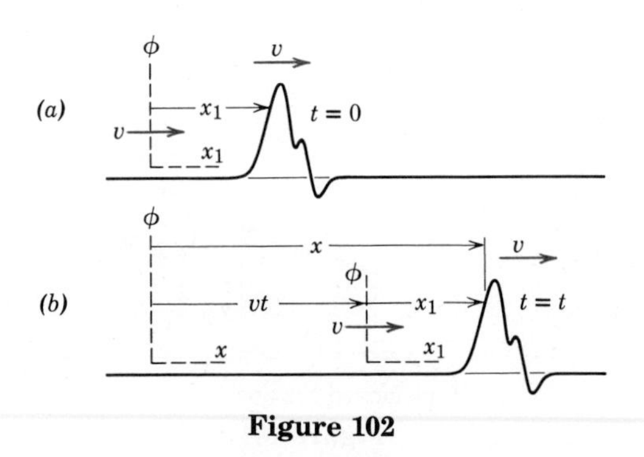

Figure 102

Consequently

$$\phi = f(x - ct) \tag{206}$$

is a solution of the wave equation and describes a *traveling wave* having constant shape and moving at a constant rate c in the plus x-direction. The velocity c is known as the *phase velocity* and is a constant for all traveling waves in the particular system involved. By similar reasoning it follows that a function of $x + ct$ would represent the solution for a disturbance traveling in the negative x-direction with a velocity c.

Since the wave equation is a linear differential equation, valid solutions may be formed by the superposition of solutions. Thus the disturbance

$$\phi = f_1(x - ct) + f_2(x + ct)$$

is also a solution and represents the simultaneous action of two waves traveling in opposite directions.

If the traveling wave is harmonic and has the form $\phi = \phi_0 \sin 2\pi(x/\lambda)$ at time $t = 0$ where λ is the wavelength, then at time t it will be given by

$$\phi = \phi_0 \sin \frac{2\pi}{\lambda}(x - ct)$$

as may be seen from Fig. 103. If the reference circular frequency p is introduced, then $\lambda = 2\pi c/p$, and the expression may be written

$$\phi = \phi_0 \sin p\left(\frac{x}{c} - t\right)$$

which is a valid solution to Eq. 204. Other solutions may be obtained by using a cosine expression or by reversing the sign of t for waves traveling in the minus x-direction. Nonsinusoidal wave shapes may be described by superposing solutions in the form of a Fourier series

$$\phi = \sum_n A_n \sin p_n\left(\frac{x}{c} - t\right) + \sum_n B_n \cos p_n\left(\frac{x}{c} - t\right) \tag{207}$$

for waves traveling in the positive x-direction, for example, where $p_n = np$.

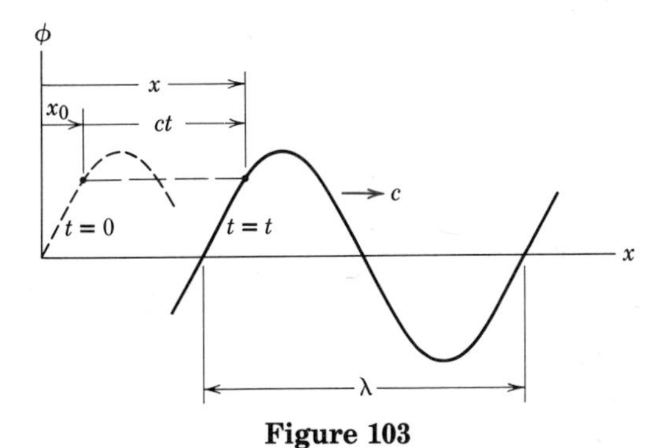

Figure 103

If two sinusoidal waves of equal amplitude and wave length are moving in opposite directions, the disturbances may be added to obtain

$$\phi = \phi_1 \sin p\left(\frac{x}{c} - t\right) + \phi_1 \sin p\left(\frac{x}{c} + t\right)$$

which is easily transformed to the expression

$$\phi = 2\phi_1 \cos pt \sin \frac{px}{c} \tag{208}$$

by the trigonometric identity $\sin (A - B) + \sin (A + B) = 2 \sin A \cos B$. It is observed that for any given value of x the disturbance is a simple harmonic oscillation except at the *nodal points* $x = \lambda/2, \lambda, 3\lambda/2, \ldots$, where the disturbance is zero for all values of t. Also, at a given value of t, the wave shape is that of a sine curve. This solution, shown in Fig. 104 for one value of p, gives what are called *standing waves* which are readily observed in a vibrating tight string. For standing waves in a string stretched between two fixed points a distance l apart at $x = 0$ and $x = l$, the disturbance ϕ must be zero at both points, which are nodal points. Thus $\sin (px/c) = 0$ at each of these positions. There are various values of p that will satisfy the condition at $x = l$ given by

$$\sin \frac{p_n l}{c} = 0 \qquad \text{where} \qquad p_n = n\pi \frac{c}{l}$$

The integers $n = 1, 2, 3, \ldots$ give the frequencies $p_1, p_2, p_3, \ldots p_n$ which correspond to the *normal modes* of oscillation. There are an infinite number of modes or degrees of freedom possible. The *fundamental frequency* (*first harmonic*) is given by $n = 1$, and higher frequencies (harmonics) are given by higher values of n. This type of problem where the boundary conditions are satisfied only for certain discrete values of a parameter is called a *characteristic-value problem,* and the various values of the parameter p_n are known as characteristic values or *eigenvalues.*

A second general approach to the solution of the wave equation lies in separating the variables by taking solutions of the form

$$\phi = f_1(t) f_2(x)$$

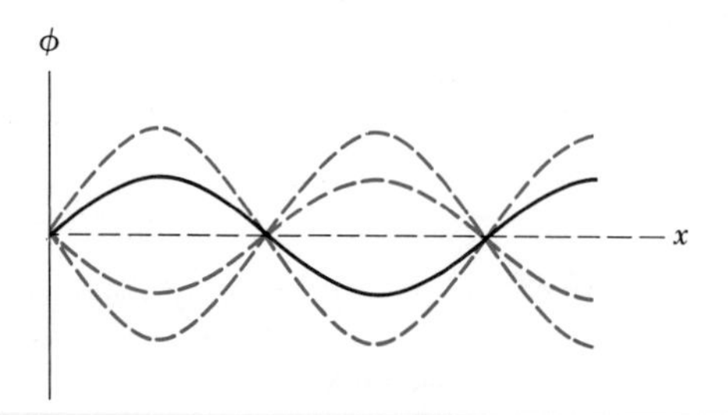

Figure 104

where f_1 is a function of t only and f_2 is a function of x only. Substitution into Eq. 204 and division by $f_1(x)f_2(t)$ give

$$\frac{1}{f_1}\frac{d^2f_1}{dt^2} = \frac{c^2}{f_2}\frac{d^2f_2}{dx^2}$$

Ordinary derivatives are used since each function contains one variable only. The equals sign separates two quantities each of which is a function of a different variable. If these quantities are equal for all values of the variables, they must both be equal to the same constant which is taken here as $-p^2$. With this reasoning the two ordinary and linear differential equations

$$\frac{d^2f_1}{dt^2} + p^2f_1 = 0 \qquad \text{and} \qquad \frac{d^2f_2}{dx^2} + \frac{p^2}{c^2}f_2 = 0$$

are formed. These familiar equations were treated in Art. 48 in the discussion of undamped free vibrations, and the following solutions may be written directly in terms of sin pt and cos pt for f_1 and sin (px/c) and cos (px/c) for f_2. The four combinations of the products of the functions may be added to obtain the expression

$$\phi = C_1 \sin pt \sin \frac{px}{c} + C_2 \sin pt \cos \frac{px}{c} + C_3 \cos pt \sin \frac{px}{c} + C_4 \cos pt \cos \frac{px}{c} \qquad (209)$$

With the various values of p designated by $p_1, p_2, p_3 \ldots p_n$, a sum of the solutions may be written in the form

$$\phi = \sum_n \left(C_{1_n} \sin p_nt \sin \frac{p_nx}{c} + C_{2_n} \sin p_nt \cos \frac{p_nx}{c} + C_{3_n} \cos p_nt \sin \frac{p_nx}{c} + C_{4_n} \cos p_nt \cos \frac{p_nx}{c} \right) \qquad (210)$$

If the sum is taken from $n = 1$ to $n = \infty$, a general solution for a finite interval may be formed which can describe the propagation of any disturbance ϕ. If Eq. 210 is used to describe a wave form traveling in the plus x-direction, for example, it may be made equivalent to Eq. 207 by the substitutions $C_{3_n} = -C_{2_n} = A_n$ and $C_{1_n} = C_{4_n} = B_n$.

As noted in the foregoing solutions of the wave equation it is necessary to satisfy the boundary conditions. Thus for a taut wire secured at one end, the lateral displacement y at this position must be zero, and this end becomes a node. For a bar with one end fixed, the displacement u must be zero at this end. If one end is free, the strain $\partial u/\partial x$ is zero. Additionally, it may be shown that a pulse which travels longitudinally along a bar will be reflected at the end of the bar as a mirror image of the pulse with the opposite sign if the end is rigidly fixed and as a mirror image with the same sign if the end of the bar is free.

The three physical problems cited all lead to a one-dimensional wave equation. When waves are propagated in two and three dimensions, additional terms are needed. The wave equation in three dimensions has the form

$$\nabla^2\phi = \frac{1}{c^2}\frac{\partial^2\phi}{\partial t^2} \tag{211}$$

where the symbol ∇^2 stands for the Laplacian operator

$$\nabla^2 = \frac{\partial^2}{\partial x^2} + \frac{\partial^2}{\partial y^2} + \frac{\partial^2}{\partial z^2}$$

The Laplacian operator, and hence the wave equation, may be expressed in cylindrical or spherical coordinates if the geometry of the problem so indicates.

The wave equation is used to describe the propagation of other wave phenomena including sound, light, and electromagnetic energy. Much of these analyses parallels that introduced in this article.

Sample Problem

10/55 A circular shaft of length l is rigidly fixed at one end ($x = l$) and free at the other end ($x = 0$). Determine the expressions for the frequencies of the first four modes of torsional vibration of the shaft. The shear modulus of the shaft is G, and the mass density is ρ.

Solution: A solution to the wave equation will be found which will satisfy the following conditions. At $x = 0$, the external moment is zero so that $\partial\theta/\partial x = 0$. At $x = l$, there can be no displacement and $\theta = 0$. In Eq. 210 only the terms in $\cos p_n x/c$ are retained since terms in $\sin p_n x/c$ would not yield a zero moment on the end $x = 0$ of the shaft. Thus the function

$$\phi = \theta = \sum_n (C_{2_n}\sin p_n t + C_{4_n}\cos p_n t)\cos\frac{p_n x}{c}$$

will be used for the angular displacement solution.

The condition $\partial\theta/\partial x = 0$ for $x = 0$ is satisfied, and the further requirement that $\theta = 0$ for $x = l$ regardless of t gives

$$0 = \sum_n (C_{2_n}\sin p_n t + C_{4_n}\cos p_n t)\cos\frac{p_n l}{c}$$

This statement should hold for all values of t which requires that the argument of the cosine term must satisfy the following condition:

$$\frac{p_n l}{c} = \frac{2n-1}{2}\pi \quad \text{or} \quad p_n = \frac{2n-1}{2}\frac{c}{l}\pi$$

The frequency f of the torsional waves is

$$f_n = \frac{p_n}{2\pi} = \frac{2n-1}{4}\frac{c}{l}$$

and with $c = \sqrt{G/l}$ has the successive values

$$f_1 = \frac{\sqrt{G/l}}{4l}, \quad f_2 = \frac{3\sqrt{G/l}}{4l}, \quad f_3 = \frac{5\sqrt{G/l}}{4l}, \quad f_4 = \frac{7\sqrt{G/l}}{4l}, \qquad \textit{Ans.}$$

for the first four modes of torsional oscillation.

Problems

10/56 If a sudden impact occurs on the end of an unsupported steel bar 2 m in length, compute the time required for the initial longitudinal pulse of stress to return to the point of impact following reflection at the other end. The elastic modulus for steel is $210(10^6)$ kPa.

Ans. $t = 0.772$ ms

10/57 A 6-mm-diameter steel wire is stretched between two fixed points 30 m apart and is plucked at one end. Determine the tension T to which the wire must be preset in order that the disturbance will reach the other end in 0.3 s.

Ans. $T = 2210$ N

10/58 Derive the differential equation for the free lateral vibrations of an elastic beam. For small lateral deflections y, the equation relating the bending moment M, the elastic modulus E, and the moment of inertia I of the cross-sectional area is

$$EI\frac{d^2y}{dx^2} = M$$

Neglect any longitudinal motion in the direction x of the length of the beam and neglect any rotational acceleration of elements of the beam. The mass per unit length of the beam is ρ.

$$\textit{Ans.} \quad \frac{\partial^4 y}{\partial x^4} + \frac{\rho}{EI}\frac{\partial^2 y}{\partial t^2} = 0$$

10/59 For a traveling longitudinal stress wave in a long elastic bar, show that the kinetic energy T per unit volume equals the elastic potential energy V_e per unit volume at any specified position in the bar.

◀ **10/60** Determine the time average $\dot{\bar{U}}$ of the total mechanical energy which is propagated with a sinusoidal longitudinal stress wave per unit time across a given section (x = constant) in an elastic rod if the magnitude of the compressive stress is σ_0. The cross-sectional area of the rod is A, its elastic modulus is E, and its density is ρ.

$$\textit{Ans.} \quad \dot{\bar{U}} = \frac{\sigma_0{}^2 A}{2\sqrt{\rho E}}$$

55 Generalized Coordinates and Lagrange's Equations. For systems with multiple degrees of freedom it is generally convenient to choose coordinates, called *generalized coordinates,* which are independent of one another. A very powerful approach to the determination of the equations of motion of a dynamical system using generalized coordinates is due to Lagrange* and will be presented in this article.

Consider, first, the simple pendulum of mass m_1 with fixed length r_1 shown in Fig. 105*a*. The single coordinate θ_1 will determine uniquely the position of m_1, since the simple pendulum is a system of one degree of freedom. The two coordinates x_1 and y_1 could also be used to locate m_1 but would require the inclusion of the *equation of constraint*

$$x_1^2 + y_1^2 = r_1^2$$

since they are not independent and hence are not generalized coordinates. The double pendulum of Fig. 105*b* is a system of two degrees of freedom. The positions of the two masses may be described uniquely by the two coordinates θ_1 and θ_2 which are generalized coordinates since one can be changed physically without changing the other. If rectangular coordinates were used, the two equations of constraint would be

$$x_1^2 + y_1^2 = r_1^2$$
$$(x_2 - x_1)^2 + (y_2 - y_1)^2 = r_2^2$$

It is seen that the number of defining coordinates minus the number of equations of constraint equals the number of degrees of freedom and, hence, the number of generalized coordinates. A system with k degrees of freedom generally requires k generalized coordinates‡ which are designated as

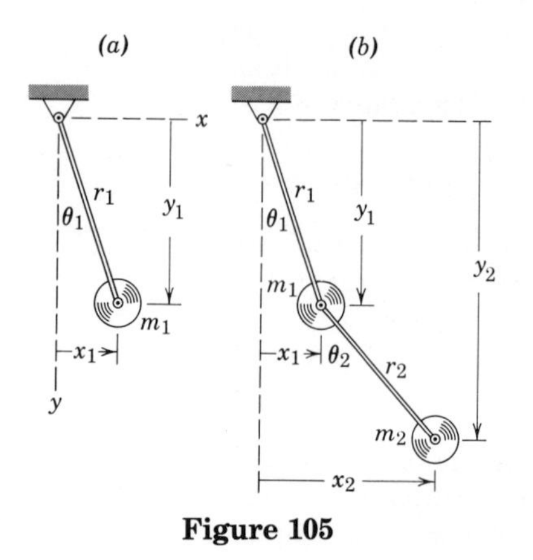

Figure 105

* First presented by J. L. Lagrange (1736–1813) in his *Mécanique Analytique* (1788). Lagrange was a French mathematician known for his early work in the calculus of variations.

‡ The number of generalized coordinates equals the number of degrees of freedom for *holonomic systems* where the equations of constraint can be expressed in terms of the coordinates alone or of the coordinates and the time. Certain nonholonomic systems exist where the constraints are specified by nonintegrable expressions involving the velocities, and more coordinates are required than there are degrees of freedom. Nonholonomic systems are not encountered frequently and are not considered here.

$q_1, q_2, q_3, \ldots q_k$. Generalized coordinates may be lengths or angles or any other set of independent quantities which define the position of the system.

The equations of Lagrange will first be developed for the motion of a single particle of mass m acted upon by a resultant force with components F_x, F_y, F_z. Extension to cover a general system of particles will then be made. The scalar components of the motion equation for m are each multiplied by the respective virtual displacement and added together to give the virtual work equation

$$F_x\,\delta x + F_y\,\delta y + F_z\,\delta z = m(\ddot{x}\,\delta x + \ddot{y}\,\delta y + \ddot{z}\,\delta z) \tag{212}$$

which will next be rewritten in terms of the generalized coordinates of m.

The functional relationships between the rectangular coordinates x, y, z of m and the generalized coordinates $q_1, q_2, q_3, \ldots q_k$ of the system of which m is a part may be expressed as*

$$\begin{aligned} x &= f_1(q_1, q_2, q_3, \ldots q_k) \\ y &= f_2(q_1, q_2, q_3, \ldots q_k) \\ z &= f_3(q_1, q_2, q_3, \ldots q_k) \end{aligned}$$

The variations in x, y, and z in terms of the corresponding variations in $q_1, q_2, q_3, \ldots q_k$ are

$$\begin{aligned} \delta x &= \frac{\partial x}{\partial q_1}\delta q_1 + \frac{\partial x}{\partial q_2}\delta q_2 + \frac{\partial x}{\partial q_3}\delta q_3 + \cdots + \frac{\partial x}{\partial q_k}\delta q_k \\ \delta y &= \frac{\partial y}{\partial q_1}\delta q_1 + \frac{\partial y}{\partial q_2}\delta q_2 + \frac{\partial y}{\partial q_3}\delta q_3 + \cdots + \frac{\partial y}{\partial q_k}\delta q_k \\ \delta z &= \frac{\partial z}{\partial q_1}\delta q_1 + \frac{\partial z}{\partial q_2}\delta q_2 + \frac{\partial z}{\partial q_3}\delta q_3 + \cdots + \frac{\partial z}{\partial q_k}\delta q_k \end{aligned}$$

where the δ's refer to infinitesimal virtual changes in the coordinates. Inasmuch as the q's are independent, the motion may be examined for the effect of a variation in one of them while holding the remaining ones constant. Thus, take $\delta q_1 \neq 0$, $\delta q_2 = \delta q_3 = \cdots = \delta q_k = 0$ which give

$$\delta x = \frac{\partial x}{\partial q_1}\delta q_1 \qquad \delta y = \frac{\partial y}{\partial q_1}\delta q_1 \qquad \delta z = \frac{\partial z}{\partial q_1}\delta q_1$$

Substitution into Eq. 212 gives

$$\left(F_x\frac{\partial x}{\partial q_1} + F_y\frac{\partial y}{\partial q_1} + F_z\frac{\partial z}{\partial q_1}\right)\delta q_1 = m\left(\ddot{x}\frac{\partial x}{\partial q_1} + \ddot{y}\frac{\partial y}{\partial q_1} + \ddot{z}\frac{\partial z}{\partial q_1}\right)\partial q_1 \tag{213}$$

The left-hand side of the equation is the work done by the F's during the virtual displacement δq_1. If the quantity

$$Q_1 = F_x\frac{\partial x}{\partial q_1} + F_y\frac{\partial y}{\partial q_1} + F_z\frac{\partial z}{\partial q_1}$$

* The case where the rectangular coordinates are functions of the time t in addition to the generalized coordinates will not be discussed here.

is introduced and defined as a *generalized force,* $Q_1\,\delta q_1$ becomes the virtual work done. The generalized force, then, is the quantity which when multiplied by the generalized virtual displacement gives the correct expression for the virtual work done by the actual forces acting on the system. A generalized force may have the dimensions of moment rather than force, for example, if the corresponding generalized coordinate is a dimensionless angle.

To simplify the right-hand side of Eq. 213, it is seen that

$$\frac{d}{dt}\left(\dot{x}\frac{\partial x}{\partial q_1}\right) = \ddot{x}\frac{\partial x}{\partial q_1} + \dot{x}\frac{d}{dt}\left(\frac{\partial x}{\partial q_1}\right)$$

or

$$\ddot{x}\frac{\partial x}{\partial q_1} = \frac{d}{dt}\left(\dot{x}\frac{\partial x}{\partial q_1}\right) - \dot{x}\frac{d}{dt}\left(\frac{\partial x}{\partial q_1}\right)$$

Also

$$\dot{x} = \frac{dx}{dt} = \frac{\partial x}{\partial q_1}\dot{q}_1 + \frac{\partial x}{\partial q_2}\dot{q}_2 + \frac{\partial x}{\partial q_3}\dot{q}_3$$

from which partial differentiation with respect to each $\dot{q}$ term gives

$$\frac{\partial \dot{x}}{\partial \dot{q}_1} = \frac{\partial x}{\partial q_1} \qquad \frac{\partial \dot{x}}{\partial \dot{q}_2} = \frac{\partial x}{\partial q_2} \qquad \frac{\partial \dot{x}}{\partial \dot{q}_3} = \frac{\partial x}{\partial q_3}$$

This step is referred to as the "cancellation of the dots." Thus

$$\ddot{x}\frac{\partial x}{\partial q_1} = \frac{d}{dt}\left(\dot{x}\frac{\partial \dot{x}}{\partial \dot{q}_1}\right) - \dot{x}\frac{\partial \dot{x}}{\partial q_1}$$

where the order of differentiation in the last term has been interchanged. With the further simplifications

$$\dot{x}\frac{\partial \dot{x}}{\partial \dot{q}_1} = \frac{\partial}{\partial \dot{q}_1}\left(\frac{\dot{x}^2}{2}\right) \qquad \text{and} \qquad \dot{x}\frac{\partial \dot{x}}{\partial q_1} = \frac{\partial}{\partial q_1}\left(\frac{\dot{x}^2}{2}\right)$$

the expression containing the acceleration term may be written as

$$\ddot{x}\frac{\partial x}{\partial q_1} = \frac{d}{dt}\left[\frac{\partial}{\partial \dot{q}_1}\left(\frac{\dot{x}^2}{2}\right)\right] - \frac{\partial}{\partial q_1}\left(\frac{\dot{x}^2}{2}\right)$$

Equivalent expressions exist for the y- and z-coordinates. Substitution into Eq. 213 now gives

$$Q_1\,\delta q_1 = m\left\{\frac{d}{dt}\left[\frac{1}{2}\frac{\partial}{\partial \dot{q}_1}(\dot{x}^2 + \dot{y}^2 + \dot{z}^2)\right] - \frac{1}{2}\frac{\partial}{\partial q_1}(\dot{x}^2 + \dot{y}^2 + \dot{z}^2)\right\}\delta q_1$$

But the kinetic energy of m is $T = \frac{1}{2}m(\dot{x}^2 + \dot{y}^2 + \dot{z}^2)$, so that the virtual work equation upon cancellation of δq_1 may be written

$$\frac{d}{dt}\left(\frac{\partial T}{\partial \dot{q}_1}\right) - \frac{\partial T}{\partial q_1} = Q_1$$

Similar equations hold for the variations $\delta q_2, \delta q_3, \ldots \delta q_k$, each taken separately. Thus the relations may be expressed by the equation

$$\frac{d}{dt}\left(\frac{\partial T}{\partial \dot{q}_j}\right) - \frac{\partial T}{\partial q_j} = Q_j \tag{214}$$

where $j = 1, 2, 3, \ldots k$.

Equations 214 are Lagrange's equations written for a single particle. Successive application, one for each degree of freedom j, will yield the motion equation of the particle corresponding to each degree of freedom.

If the forces acting on m are conservative, they can be derived from a potential energy function V, and the generalized force may be written $Q_j = -\partial V/\partial q_j$. With this substitution and the introduction of the quantity $L = T - V$, known as the *Lagrangian function* or the *kinetic potential,* Lagrange's equations for a particle may be written

$$\frac{d}{dt}\left(\frac{\partial L}{\partial \dot{q}_j}\right) - \frac{\partial L}{\partial q_j} = 0 \qquad j = 1, 2, 3, \ldots k \tag{215}$$

It is noted that the potential energy V is a function of the q's (position) but not the $\dot{q}$'s (velocity) so that $\partial V/\partial \dot{q}_j = 0$. Attention is called to the fact that Eqs. 214 are valid for both conservative and nonconservative forces, whereas Eq. 215 is valid for conservative forces only.

Lagrange's equations are of little use when restricted to the motion of a single particle. Their main use comes in the analysis of systems of particles, particularly those with multiple degrees of freedom. To extend the equations to a system of particles let i denote the ith particle of a system of n particles. Equation 214 may be written for this particle and summed over the n particles of the system. Thus

$$\sum_{i=1}^{n}\left\{\frac{d}{dt}\left(\frac{\partial T_i}{\partial \dot{q}_j}\right) - \frac{\partial T_i}{\partial q_j}\right\} = \sum_{i=1}^{n} Q_{ji}$$

where $j = 1, 2, 3, \ldots k$, with one summation equation written for each j. But the kinetic energy of the system is $T = \sum_{i=1}^{n} T_i$, and the resultant generalized force corresponding to q_j is the sum $\sum_{i=1}^{n} Q_{ji} = Q_j$ which eliminates internal forces. Thus Lagrange's equations for a general system of particles, conservative or nonconservative, becomes

$$\blacktriangleright \qquad \frac{d}{dt}\left(\frac{\partial T}{\partial \dot{q}_j}\right) - \frac{\partial T}{\partial q_j} = Q_j \qquad j = 1, 2, 3, \ldots k, \tag{214a}$$

and for a conservative system they become

$$\blacktriangleright \qquad \frac{d}{dt}\left(\frac{\partial L}{\partial \dot{q}_j}\right) - \frac{\partial L}{\partial q_j} = 0 \qquad j = 1, 2, 3, \ldots k \tag{215a}$$

where T and $L = T - V$ are the energy terms for the entire system and where, as before, there are as many equations, k, as there are degrees of freedom.

Lagrange's equations provide a powerful means of obtaining the differential equations of motion expressed in the independent coordinates of the problem merely by a differentiation of the energy written in these same generalized coordinates. This approach is particularly useful for systems with multiple degrees of freedom. It has the advantage of not involving the forces of constraint which do no work and which often complicate the formulation of the equations of motion from the force-mass-acceleration equations. Furthermore, the Lagrangian approach requires an expression for the velocities rather than for the accelerations which, in turn, frequently require considerable kinematics to determine. Lagrange's equations are among the most useful of the advanced methods of mechanics and have also found extensive use in the analysis of electrical and electro-mechanical systems.

Sample Problem

10/61 The small cylindrical mass m is confined to slide on the slender rod of length l and mass m_0 which is free to rotate in the vertical plane about O. The spring which supports the mass has a stiffness k, and $r = r_0$ when it is uncompressed. Write the differential equations of motion corresponding to the two degrees of freedom of the system. Neglect friction and the mass of the spring.

Solution: If the horizontal line through O is taken as the datum plane for zero potential energy, then the potential energy for the system in the general position is

$$V = V_e + V_g = \tfrac{1}{2}k(r - r_0)^2 - mgr\cos\theta - m_0 g\frac{l}{2}\cos\theta$$

The kinetic energy of the system is

$$T = \tfrac{1}{2}m(\dot{r}^2 + r^2\dot{\theta}^2) + \tfrac{1}{2}\tfrac{1}{3}m_0 l^2\dot{\theta}^2$$

Since the system is conservative, Eq. 215*a* may be used where $L = T - V$, $q_1 = r$, and $q_2 = \theta$. For the r-coordinate the derivatives are

$$\frac{\partial L}{\partial q_1} = \frac{\partial L}{\partial r} = mr\dot{\theta}^2 - k(r - r_0) + mg\cos\theta$$

$$\frac{d}{dt}\left(\frac{\partial L}{\partial \dot{q}_1}\right) = \frac{d}{dt}\left(\frac{\partial L}{\partial \dot{r}}\right) = \frac{d}{dt}(m\dot{r}) = m\ddot{r}$$

Substitution into Eq. 215*a* gives

$$m(\ddot{r} - r\dot{\theta}^2) + k(r - r_0) - mg\cos\theta = 0 \qquad \textit{Ans.}$$

The derivatives for the θ-coordinate are

$$\frac{\partial L}{\partial q_2} = \frac{\partial L}{\partial \theta} = -mgr\sin\theta - m_0 g\frac{l}{2}\sin\theta$$

$$\frac{d}{dt}\left(\frac{\partial L}{\partial \dot{q}_2}\right) = \frac{d}{dt}\left(\frac{\partial L}{\partial \dot{\theta}}\right) = \frac{d}{dt}\left(mr^2\dot{\theta} + \frac{1}{3}m_0 l^2\dot{\theta}\right) = \left(mr^2 + \frac{m_0 l^2}{3}\right)\ddot{\theta} + 2mr\dot{r}\dot{\theta}$$

Substitution into Eq. 215*a* gives

$$\left(mr^2 + \frac{m_0 l^2}{3}\right)\ddot{\theta} + 2mr\dot{r}\dot{\theta} + \left(mr + \frac{m_0 l}{2}\right)g\sin\theta = 0 \qquad \textit{Ans.}$$

It is noted that the two motion equations were determined by a straightforward process of differentiation as called for by Lagrange's equations. The necessity for introducing the normal force N between the slider and the rod or the reaction R on the bearing at O is eliminated by the method, as only those forces external to the system which do work are involved. Lagrange's method establishes the differential equations of motion but it does not solve them.

If Eq. 214*a* is used in place of Eq. 215*a*, it is necessary to remove the spring as an energy source and compute the generalized forces acting externally to the system composed of the rod and mass without the spring. The external forces which do work on the system are the weights of the two members and the spring force. For a virtual displacement δr the work done is

$$Q_r\,\delta r = mg\cos\theta\,(\delta r) - k(r - r_0)(\delta r) \qquad Q_r = mg\cos\theta - k(r - r_0)$$

For a virtual displacement $\delta\theta$ the work done is

$$Q_\theta\,\delta\theta = -mg\sin\theta\,(r\,\delta\theta) - m_0 g\sin\theta\left(\frac{l}{2}\delta\theta\right)$$

or

$$Q_\theta = -\left(mr + m_0\frac{l}{2}\right)g\sin\theta$$

The reader should use these generalized forces together with the derivatives of the kinetic energy T in Eqs. 214*a* so that the equivalence with Eqs. 215*a* may be observed.

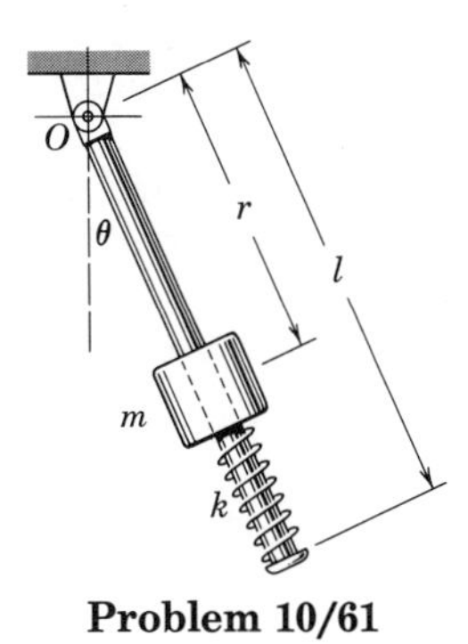

Problem 10/61

Problems

10/62 Derive the characteristic differential equation of motion for the simple linear spring-mass oscillator by using Lagrange's equations. The mass is m and the spring stiffness is k.

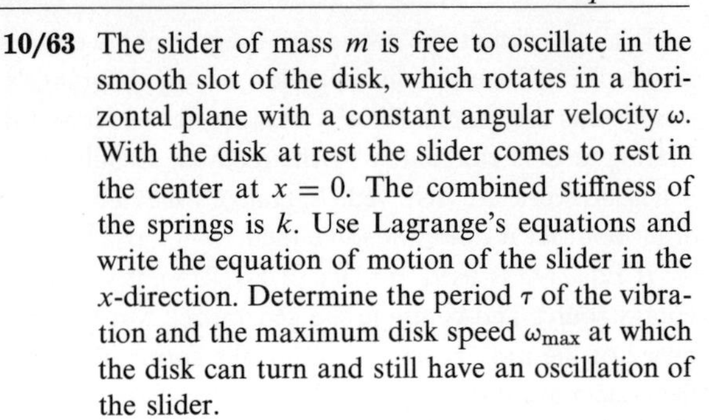

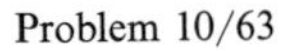

Problem 10/63

10/63 The slider of mass m is free to oscillate in the smooth slot of the disk, which rotates in a horizontal plane with a constant angular velocity ω. With the disk at rest the slider comes to rest in the center at $x = 0$. The combined stiffness of the springs is k. Use Lagrange's equations and write the equation of motion of the slider in the x-direction. Determine the period τ of the vibration and the maximum disk speed ω_{max} at which the disk can turn and still have an oscillation of the slider.

$$\textit{Ans. } \tau = \frac{2\pi}{\sqrt{\dfrac{k}{m} - \omega^2}}, \quad \omega_{max} = \sqrt{k/m}$$

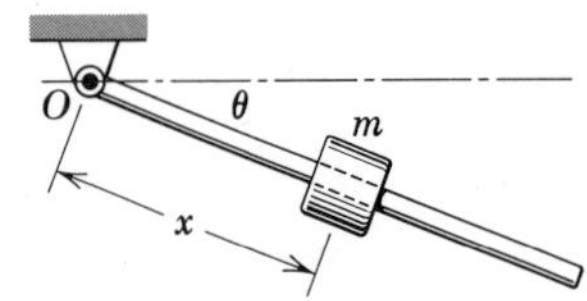

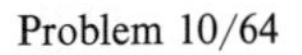

Problem 10/64

10/64 The collar of mass m slides freely along the rod, which is pivoted about a horizontal axis through O. The rod has a length L, a mass m_0, and a radius of gyration k about O. If the rod and collar are released from rest in the positions $\theta = 0$ and $x = x_0$, write the differential equations which describe the subsequent motion.

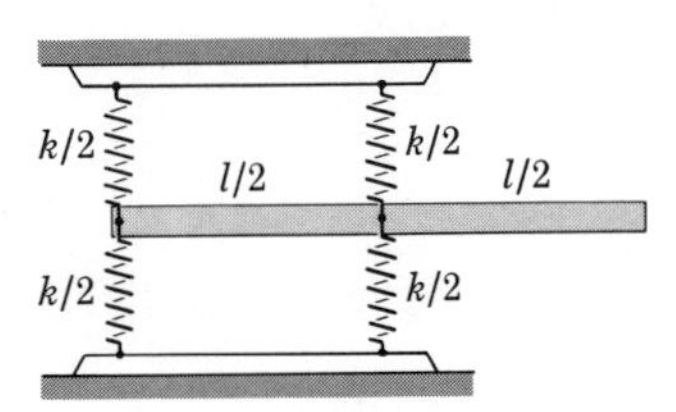

Problem 10/65

10/65 The uniform beam of length l and mass m is attached to the four identical springs each of stiffness $k/2$ and oscillates in the horizontal plane of the figure. Derive the two differential equations of motion for small angular oscillations and small linear vibrations in the direction transverse to the beam.

$$\textit{Ans. } m\ddot{x} + 2kx - k\frac{l\theta}{2} = 0$$

$$m\ddot{\theta} - 6k\left(\frac{x}{l} - \frac{\theta}{2}\right) = 0$$

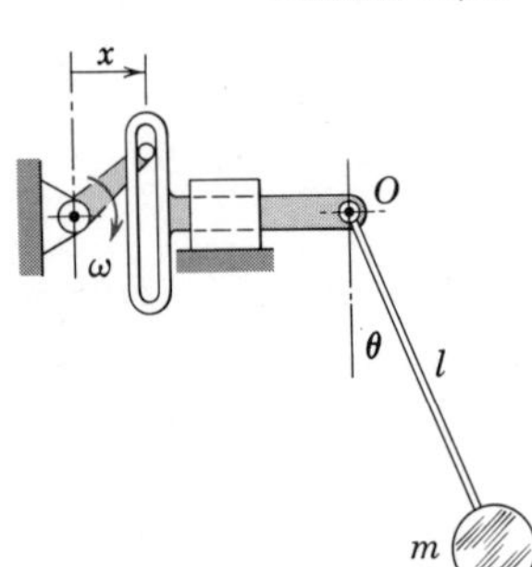

Problem 10/66

10/66 The support O for the simple pendulum of mass m and length l is given a horizontal oscillation which is described by $x = x_0 \sin \omega t$. Derive the equation of motion for the pendulum.

$$\textit{Ans. } \ddot{\theta} + \frac{g}{l}\sin\theta = \frac{x_0}{l}\omega^2 \sin \omega t \cos\theta$$

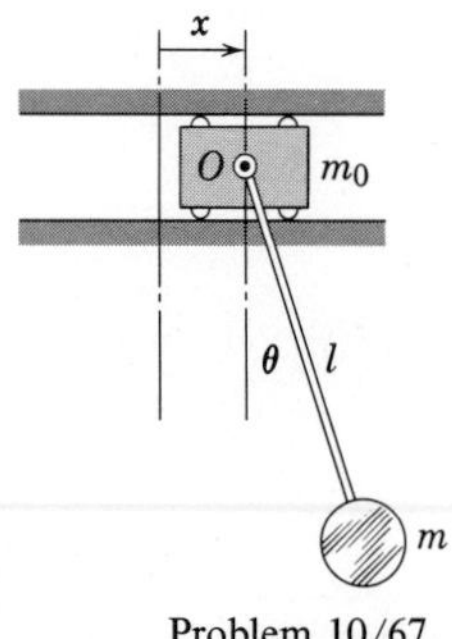

Problem 10/67

10/67 The simple pendulum of mass m and length l is suspended from the mass m_0 which is free to move in the horizontal direction. The pendulum is released from a displaced position, and a coupled motion of the two masses takes place. Derive the differential equations of motion for the system. Determine the period τ of the pendulum for small oscillations by taking $\sin\theta \approx 0$, $\cos\theta \approx 1$, and $\dot{\theta}^2 \approx 0$.

$$\textit{Ans. } \tau = 2\pi\sqrt{\frac{m_0}{m + m_0}\frac{l}{g}}$$

10/68 Determine the moment M on the shaft of the mechanism of Prob. 8/70 repeated here which will produce a constant speed ω_0 about the x-axis. The speed $\dot{\phi} = p$ is constant. Work by Lagrange's equations and check the result with the given answer.

Ans. $M = \frac{1}{12}m(c^2 - a^2)p\omega_0 \sin 2\phi$

10/69 By the direct use of Lagrange's equations derive the first of Eqs. 166 for the rotation of a symmetrical body about a fixed point O on its axis of symmetry. Use the notation of Art. 44. Note that the Eulerian angles are generalized coordinates.

10/70 The semicircular disk of radius r and mass m is pivoted freely as a pendulum about its x-axis. The supporting disk and brackets have a moment of inertia I_0 about the vertical z-axis around which they rotate freely without driving or retarding torque. Prove that the motion of the half-disk pendulum is unaffected by the rotation of the supporting disk.

10/71 The slender bar of mass m and length l swings as a pendulum about its upper end which is spring-mounted in the vertical guide. The spring has a stiffness k. Derive the two differential equations which describe the resulting motion of the system.

10/72 The pivot for a simple pendulum of length l and mass m_1 is attached to a mass m_2 which is spring-mounted so that it may oscillate in the vertical direction. If the spring has a stiffness k, determine the two equations of motion and simplify them for the case of small displacements and small velocities. Measure the downward displacement x of m_2 from the equilibrium position.

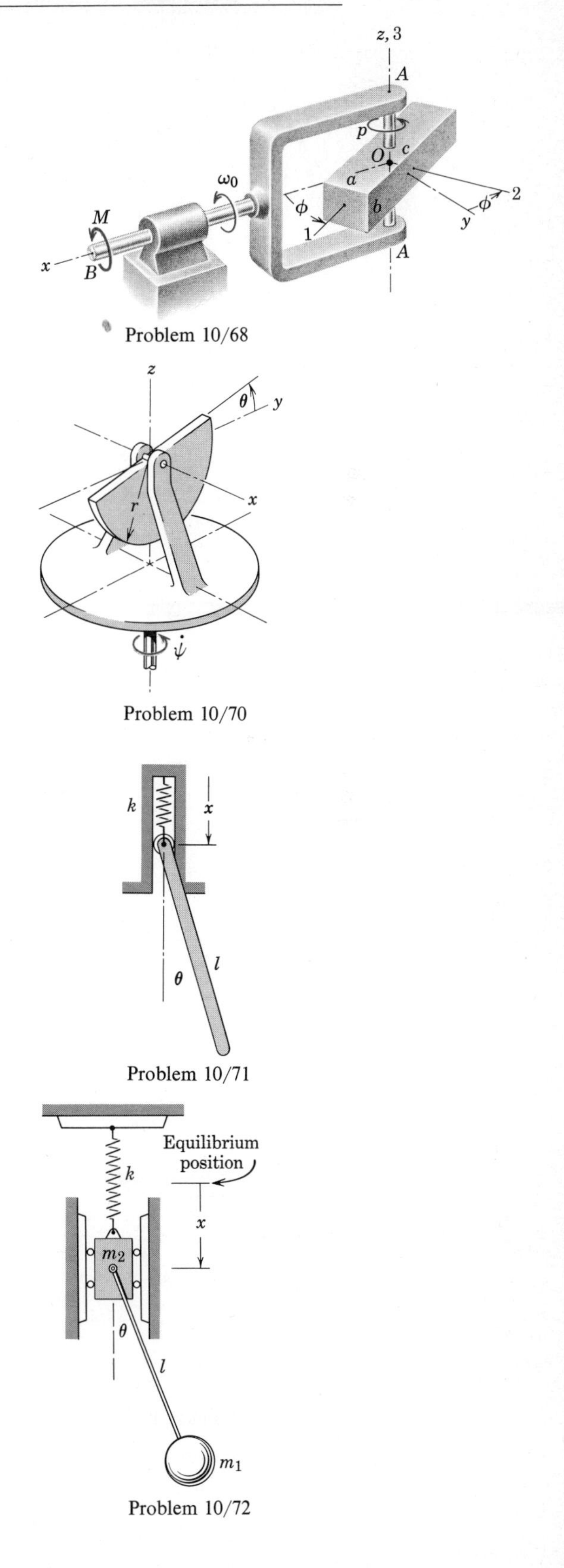

Problem 10/68

Problem 10/70

Problem 10/71

Problem 10/72

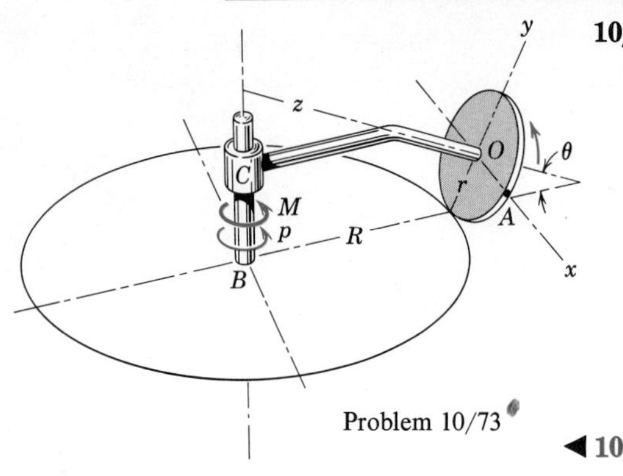

Problem 10/73

10/73 The mechanism of Prob. 7/31 is repeated here. The circular disk of mass m and radius r is mounted on the bent axle CO and rolls without slipping around the horizontal circle of radius R. Determine the moment M applied to the axle at C necessary to give the assembly an angular acceleration $\dot{p}$. The moment of inertia of the bent axle about the vertical axis is I_1.

Ans.
$$M = \{I_1 + \tfrac{3}{2}m(R + r\sin\theta)^2 + \tfrac{1}{4}mr^2\cos^2\theta\}\dot{p}$$

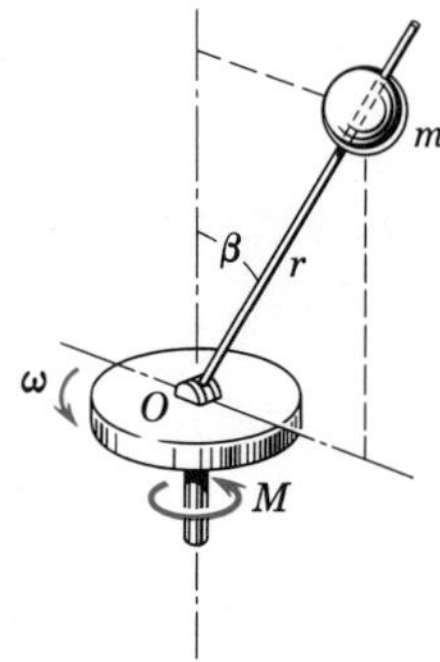

Problem 10/74

◀ **10/74** The small mass m is free to slide on the light rod pivoted at O on the disk which rotates about its vertical axis at the constant speed ω. An internal mechanism lowers the rod at the constant rate $\dot{\beta} = k$ while the mass is sliding on the rod. Write the three equations of motion of m in terms of its generalized coordinates and determine an expression for the moment M necessary to maintain the constant speeds ω and k.

$$\textit{Ans. } M = (2mr\omega\sin\beta)\frac{d}{dt}(r\sin\beta)$$

◀ **10/75** Replace the half-disk of Prob. 10/70 by a slender rod of mass m and length l pivoted at its end about the x-axis and free to swing in the y-z plane. The supporting disk and brackets have a large moment of inertia I_0 about the vertical z-axis. When the rod swings through the vertical position, its angular velocity is $\dot{\theta} = \Omega_0$ and $\dot{\psi} = \omega_0$. Show that the motion of the rod is affected by the rotation of the supporting disk and determine the period τ of the rod for small amplitudes of oscillation. What limitation exists for ω_0?

$$\textit{Ans. } \tau = \frac{2\pi}{\sqrt{\dfrac{3g}{2l} - \omega_0^2}}$$

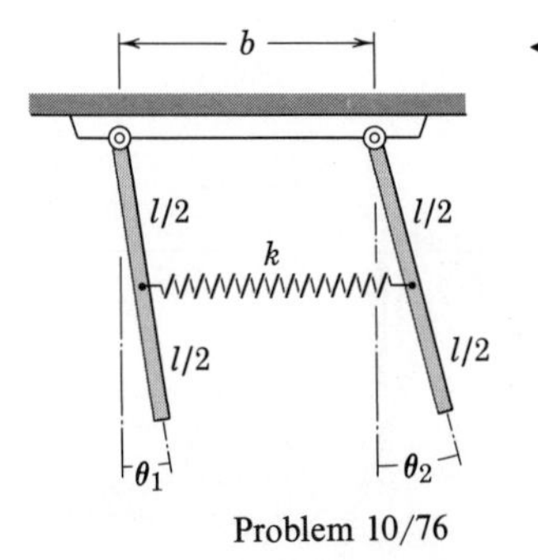

Problem 10/76

◀ **10/76** The two slender bars each of mass m and length l are connected at their midpoints by a spring of stiffness k and unstretched length b. If the bars swing in the vertical plane and are released from rest with $\theta_1 = 0$ and $\theta_2 = \theta_0$, determine the equations of motion and solve for θ_1 as a function of the time t after release if θ_0 is a very small angle.

$$\textit{Ans. } \theta_1 = \frac{\theta_0}{2}(\cos pt - \cos \mu t)$$

$$\text{where } p = \sqrt{\frac{3g}{2l}},\quad \mu = \sqrt{\frac{3}{2}\left(\frac{k}{m} + \frac{g}{l}\right)}$$

56 Hamilton's Principle. One of the most general of the variational principles of mechanics is Hamilton's principle* which provides a means for finding the equations of motion of a dynamical system by determining the stationary value of a scalar integral. The method is based on a comparison of possible paths (motions) which could occur from time t_1 to time t_2 when the actual motion is known at the two times but not during the interval between them.

In Fig. 106 the true path of a particle of mass m_i of any mass system is represented schematically by the full line between the two points A and B during the time from t_1 to t_2. At time t the particle occupies the intermediate position shown. A varied path between the same two points which could be taken by the particle during the same time interval is shown by the dotted line, and at the same time t the particle would occupy the position C if it were following the varied path. At time t the position vector of m_i on the true path is $\mathbf{r}_i$, and on the varied path at the same time t the position vector is $\mathbf{r}_i + \delta\mathbf{r}_i$. The notation $\delta\mathbf{r}_i$ is used to denote an infinitesimal virtual change in the position vector with no change in t, whereas $d\mathbf{r}_i$ refers to an infinitesimal change along the actual path during which the time changes an amount dt.

The equation of virtual work is formed by taking the scalar product of $\delta\mathbf{r}_i$ with both sides of the motion equation for m_i and summing over all particles of the entire system. If $\mathbf{F}_i$ is the resultant force on m_i, this sum gives

$$\sum_i \mathbf{F}_i \cdot \delta\mathbf{r}_i = \sum_i m_i\ddot{\mathbf{r}}_i \cdot \delta\mathbf{r}_i$$

where the force-summation term represents the net virtual work done by the external forces on the system. The right-hand side of the equation can be transformed to include the kinetic energy term by the substitution

$$\frac{d}{dt}(\dot{\mathbf{r}}_i \cdot \delta\mathbf{r}_i) = \ddot{\mathbf{r}}_i \cdot \delta\mathbf{r}_i + \dot{\mathbf{r}}_i \cdot \delta\dot{\mathbf{r}}_i$$

In the limit the velocity $\dot{\mathbf{r}}_i$ along the true path is parallel to the velocity $\dot{\mathbf{r}}_i + \delta\dot{\mathbf{r}}_i$ on the varied path, so that $\delta\dot{\mathbf{r}}_i$ and $\dot{\mathbf{r}}_i$ are parallel. Thus, their dot product is $\dot{\mathbf{r}}_i \cdot \delta\dot{\mathbf{r}}_i = \delta(\frac{1}{2}|\dot{\mathbf{r}}_i|^2)$, and

$$m_i\ddot{\mathbf{r}}_i \cdot \delta\mathbf{r}_i = m_i\frac{d}{dt}(\dot{\mathbf{r}}_i \cdot \delta\mathbf{r}_i) - m_i\delta(\tfrac{1}{2}|\dot{\mathbf{r}}_i|^2)$$

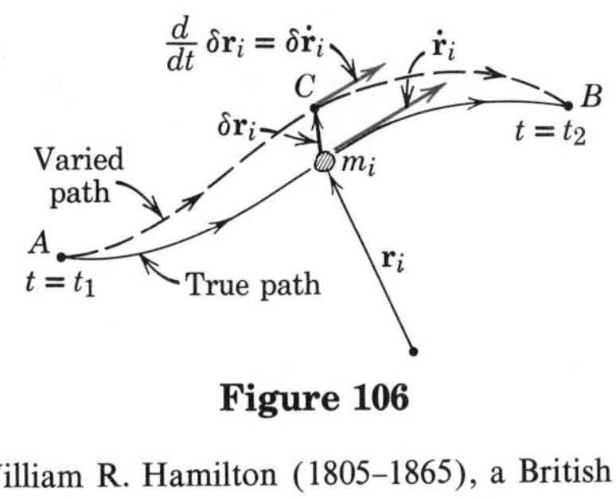

Figure 106

* Formulated by Sir William R. Hamilton (1805–1865), a British mathematician.

or
$$m_i\ddot{\mathbf{r}}_i \cdot \delta\mathbf{r}_i = \frac{d}{dt}[m_i(\dot{\mathbf{r}}_i \cdot \delta\mathbf{r}_i)] - \delta T_i$$

where the kinetic energy of the particle is $T_i = \frac{1}{2}m_i|\dot{\mathbf{r}}_i|^2$. Substitution into the virtual-work equation along with the expressions for total work $\delta U = \sum_i \mathbf{F}_i \cdot \delta\mathbf{r}_i$ and the virtual change in the total kinetic energy $\delta T = \delta \sum_i T_i = \sum_i \delta T_i$ gives

$$\delta U = \sum_i \frac{d}{dt}[m_i(\dot{\mathbf{r}}_i \cdot \delta\mathbf{r}_i)] - \delta T$$

For a conservative system the forces are derivable from the potential energy V, and $\delta U = -\delta V$. Therefore

$$\delta(T - V) = \frac{d}{dt}\sum_i m_i(\dot{\mathbf{r}}_i \cdot \delta\mathbf{r}_i)$$

Multiplication by dt and formation of the integral over the time interval give

$$\int_{t_1}^{t_2} \delta(T - V)\, dt = \int_{t=t_1}^{t=t_2} d\sum_i m_i(\dot{\mathbf{r}}_i \cdot \delta\mathbf{r}_i)$$

or

$$\delta\int_{t_1}^{t_2} (T - V)\, dt = \sum_i m_i(\dot{\mathbf{r}}_i \cdot \delta\mathbf{r}_i)\Big]_{t=t_1}^{t=t_2}$$

But at both limits $\delta\mathbf{r}_i = 0$, so that the right side of the equation is zero which leaves

$$\delta\int_{t_1}^{t_2} (T - V)\, dt = 0 \qquad \text{or} \qquad \delta\int_{t_1}^{t_2} L\, dt = 0 \tag{216}$$

Equation 216 is a statement of *Hamilton's principle* for a conservative dynamical system and shows that the correct expression for the *Lagrangian function* $L = T - V$ for the system is that expression which produces a stationary value of the integral taken between two different times for which the configuration of the system is known. Hamilton's principle selects the correct dynamical path from other possible paths. By "dynamical path" is meant the correct functional relationship between the coordinates and the time which correctly describes the motion during the interval selected. Hamilton's principle, like Lagrange's equations, will determine the differential equations of motion, but it says nothing concerning their solution. The principle may be used, for instance, to determine the defining equation of motion for a distributed-parameter system with an infinite number of degrees of freedom.

An alternative form of Eq. 216 may be written for a conservative system where the total dynamical energy $E = T + V$ is constant. Thus the Lagranian function becomes $T - V = 2T - E$, and since E is constant and the variation of its integral between any two points vanishes, Eq. 216 becomes

$$\delta \int_{t_1}^{t_2} 2T\,dt = 0 \tag{217}$$

Equation 217 is called the *principle of least action* and requires that the time integral of twice the kinetic energy must have a stationary value when compared with the same integral for varied paths.

Hamilton's principle and the principle of least action have found important but limited application in engineering problems. Their use will undoubtedly grow with time as the complexity and generality of design situations increase.

A REVIEW PROBLEMS

In the preceding chapters the problems which are included with the various articles illustrate application of the particular topics involved. Thus the problem category and method of solution for these problems are to a great extent indicated automatically by their association with the article. The student of mechanics should develop ability to classify a new problem by recognizing the topic or topics involved and by selecting the appropriate method or methods of solution. The following review problems in Appendix A are included to help the student develop this ability. Problems associated with the more advanced sections of the book are similarly identified by the gray band along the outer margins of the page. Problems are arranged in approximate order of increasing difficulty in each of the sections. Some problems include more than one topic or may be worked by more than one method. It is suggested that the student use his time to outline the solution to as many of the problems as possible rather than to concentrate on the complete solution of only a limited few of them. In this way a more comprehensive review can be completed. The answers to all of the problems are included.

A1 A jet airliner with a mass of 115 t develops a constant jet thrust of 90 kN for each of its four engines. Determine the length s of runway required for take-off at a speed of 270 km/h. Neglect air resistance. *Ans.* $s = 898$ m

A2 The time derivative of a vector $\mathbf{V}$ makes an angle of 120° with $\mathbf{V}$ at an instant when $\mathbf{V}$ has a magnitude of 10 units. If the magnitude of the time derivative of the vector is 6 units/s at this instant, determine the corresponding derivative of the magnitude and the angular velocity ω of the vector.

Ans. $\frac{d}{dt}|\mathbf{V}| = -3$ units/s

$\omega = 0.520$ rad/s

A3 The jet transport has a mass of 90 t and its four engines develop a total thrust T of 240 kN at its take-off speed of 280 km/h. Immediately after take-off the direction of its actual velocity v and acceleration a makes an angle of 7° with the horizontal, whereas the aircraft axis and thrust are directed 10° above the horizontal. Neglect air resistance (drag) in the direction opposite to the velocity and compute the acceleration $a = \dot{v}$. *Ans.* $a = 1.467\ \text{m/s}^2$

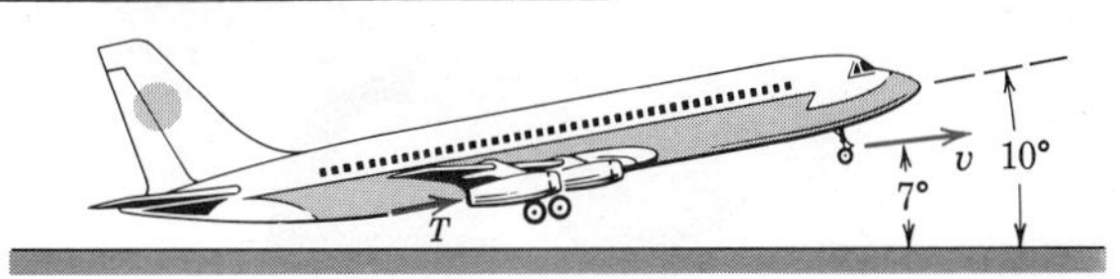

Problem A3

A4 The end of the coil of rope of total length L is released from rest at $x = 0$ and falls with increasing vertical velocity as the rope peels off the coils and passes through the center of the coil in a smooth and continuous flow. Neglect all friction and derive the expression for the velocity v of the rope as the upper end passes through the opening at $x = 0$. The mass of the rope per unit length is ρ, and the initial height of the coil with $x = 0$ is h. *Ans.* $v = \sqrt{g(L + h)}$

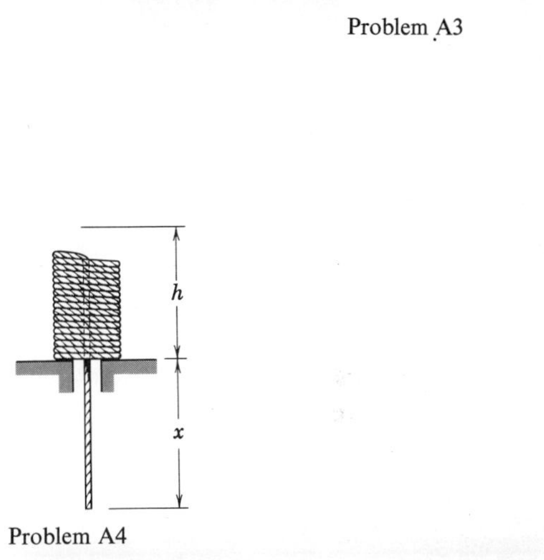

Problem A4

A5 The two spacecraft A and B are to perform a docking operation in space. Spacecraft A moves with a constant velocity of 27 000 km/h relative to the earth. When the separation distance is 15 km, spacecraft B, which has a mass of 800 kg, has a velocity of 27 150 km/h. If the retro rockets on B are fired at this point, determine the constant thrust F which is required to bring A and B together with zero relative velocity. What total impulse I is required? Assume collinear straight-line motion during the interval.
Ans. $F = 46.3$ N, $I = 33.3$ kN·s

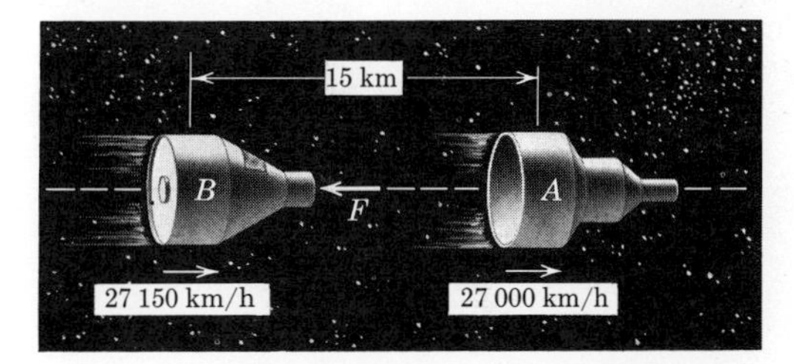

Problem A5

A6 The path of a nonrotating spacecraft A as it nears a planet P lies in the plane of rotation of the planet about the sun. The planet has a velocity $v_p = 126(10^3)$ km/h in its near-circular orbit around the sun. If the planet appears to have a velocity $v_r = 22.2(10^3)$ km/h away from the sun as seen from the spacecraft as it passes from the shaded to the lighted side of the planet, calculate the magnitude of the velocity $\mathbf{v}_A$ of the spacecraft as measured in the solar reference system. *Ans.* $v_A = 127.9(10^3)$ km/h

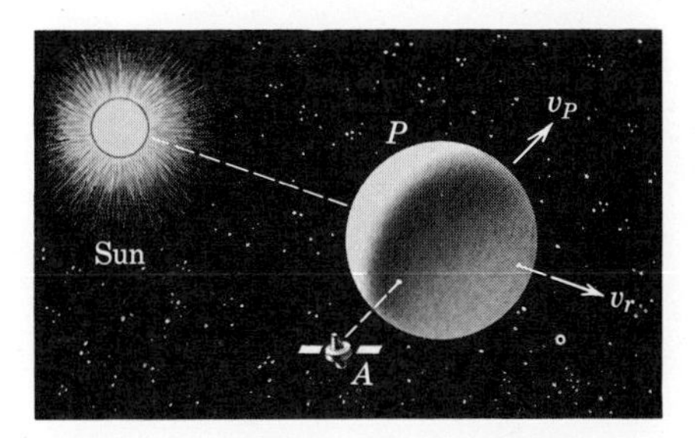

Problem A6

A7 A flexible rope secured at end B wraps around the fixed cylinder as it falls from the dotted position of release. Discuss the energy transformation as the rope falls and, in particular, consider the velocity of the end A of the rope as it hits the surface of the cylinder.

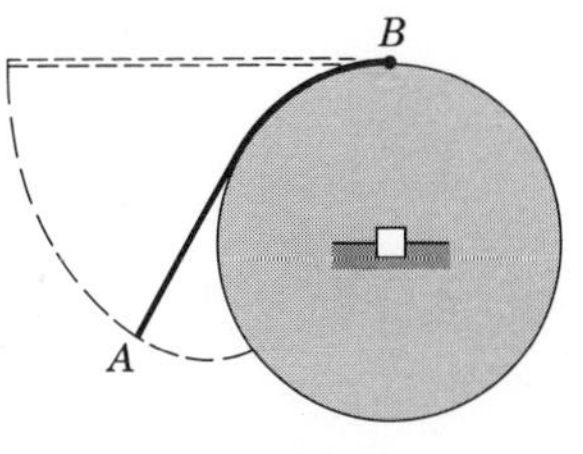

Problem A7

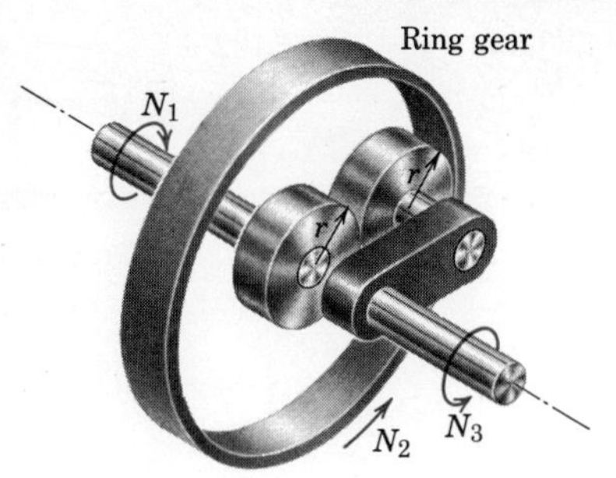

Problem A8

A8 The planetary-gear system shown acts as a speed reducer. If the input speed is N_1 and if the ring gear turns in the opposite direction with a speed $N_2 > N_1/2$, write the expression for the speed N_3 of the output shaft.

Ans. $N_3 = \frac{1}{4}(3N_2 - N_1)$

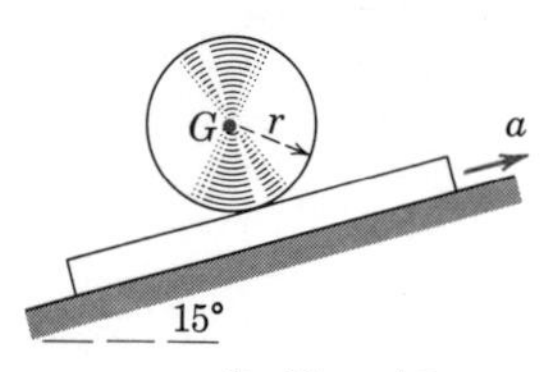

Problem A9

A9 Determine the acceleration a of the supporting surface required to keep the center G of the circular cylinder in a fixed position during the motion. Friction is sufficient to prevent any slipping between the cylinder and its support.

Ans. $a = 5.08 \text{ m/s}^2$

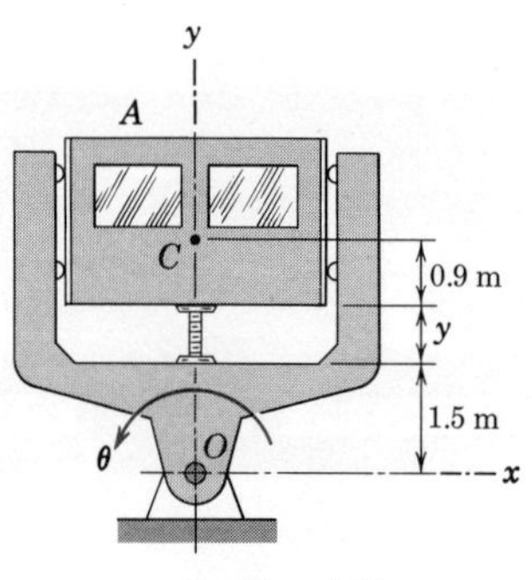

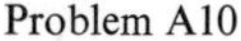
Problem A10

A10 A test chamber A is used for studying motion sickness. It is capable of oscillation about a horizontal axis through O according to $\theta = \theta_0 \sin 2\pi f_1 t$, and at the same time the chamber has a linear motion $y = y_0 \sin 2\pi f_2 t$ relative to the frame. For a certain series of tests the amplitudes are set at $\theta_0 = \pi/4$ rad and $y_0 = 150$ mm, while the frequencies are $f_1 = \frac{1}{4}$ Hz and $f_2 = \frac{1}{2}$ Hz. Determine the vector expression for the acceleration of point C in the chamber at the instant when $t = 2$ s.

Ans. $\mathbf{a} = \dfrac{3\pi^3}{80}(\mathbf{i} - \pi\mathbf{j}) \text{ m/s}^2$

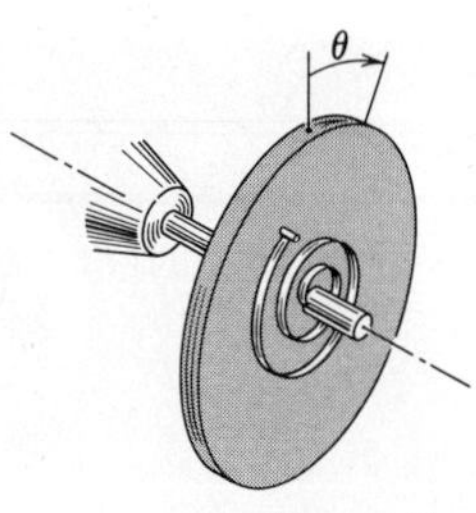

Problem A11

A11 The circular disk of mass m and radius of gyration k is is attached to the torsion spring, which is secured to the fixed shaft. The disk is otherwise free to rotate on the shaft. The spring exerts a torque $M = C\theta$ on the disk, where C is a constant and where θ is the angle of rotation of the disk in radians from the neutral position of the spring. If the disk is released from rest with an angular displacement θ_0, determine the maximum power output P of the spring and the angle θ_m at which this condition occurs.

Ans. $P = \dfrac{C}{2k}\sqrt{\dfrac{C}{m}}\theta_0^2, \quad \theta_m = \theta_0/\sqrt{2}$

A12 Determine the tension T in the wire which supports a simple pendulum of mass m in terms of the angle of swing θ from the vertical. The pendulum is released from rest at an angle θ_0 from the vertical. The angle θ is not a small angle.

Ans. $T = mg(3\cos\theta - 2\cos\theta_0)$

A13 The upper pair of connected spheres each of mass m moves initially with a velocity v before colliding with the lower pair of identical spheres initially at rest. The coefficient of restitution between the two colliding spheres is e. Prove that the angular velocities of the two links which connect the spheres have the same magnitude immediately after impact.

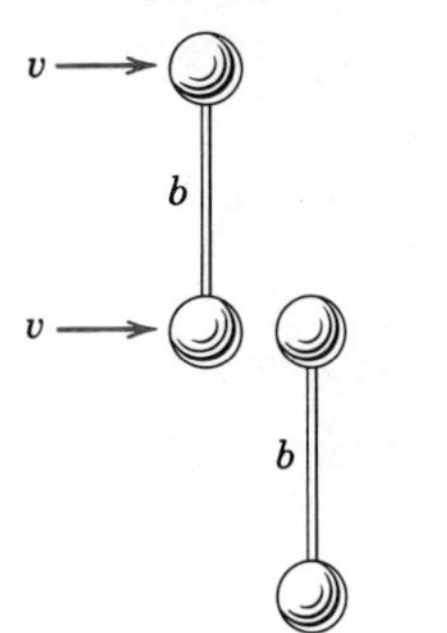

Problem A13

A14 The tractor is used to hoist a steel casing of mass m out of a dry well with the cable and pulley arrangement shown. When $x = 0$, end A and pulley B coincide at C. Write the expression for the tension T in the cable below B as a function of x if the tractor moves forward with a constant velocity v. The radius of the pulleys may be neglected, and the casing is free from the sides of the well.

$$Ans.\ T = mg\left(1 + \frac{h^2v^2}{2g\sqrt{(x^2 + h^2)^3}}\right)$$

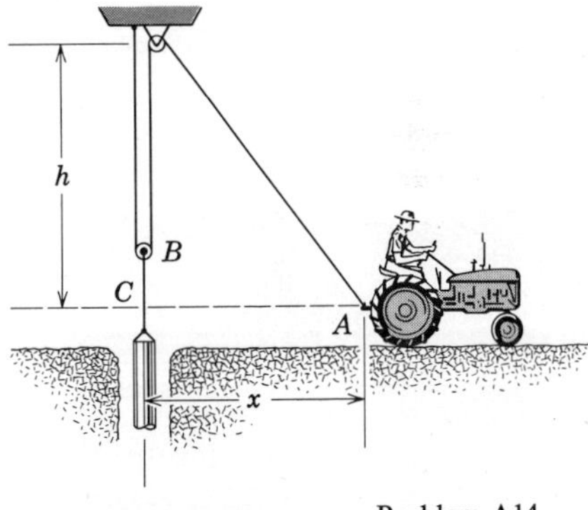

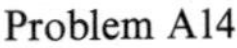

Problem A14

A15 The slotted disk is rotating about a vertical axis through O and starts from rest with a constant angular acceleration $\dot{\omega}$. The small slider A has a mass m and is free to move a slight amount before one or the other of the wires becomes taut.

(a) Find the initial acceleration a_r of A relative to the slot before being restrained by the wire.

(b) Determine the tension T in the tight wire prior to the acquisition of any appreciable angular velocity.

(c) Determine the angular velocity during the acceleration period when tension is transferred from one wire to the other.

$$Ans.\ a_r = r\dot{\omega}/\sqrt{2}$$
$$T = mr\dot{\omega}/\sqrt{2}$$
$$\omega = \sqrt{\dot{\omega}}$$

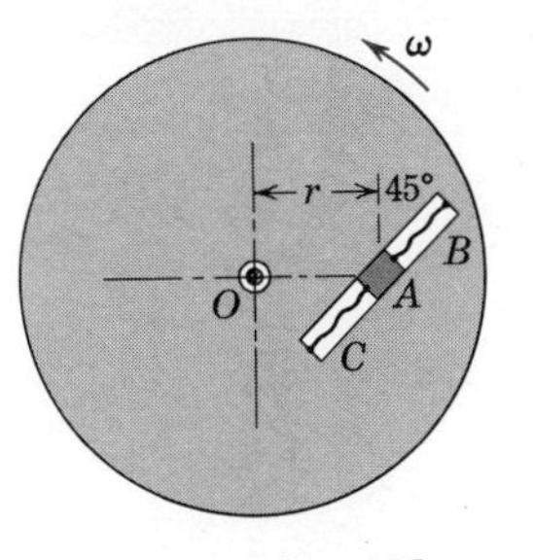

Problem A15

A16 The split ring rotates about the axis O-O, which is normal to the plane of the ring, and starts from rest with an initial angular acceleration α. For this condition determine the bending moment M and shear V in the ring at section A for θ essentially zero. The mass per unit length of rod from which the ring is made is ρ. Neglect gravitational forces. *Ans.* $M = 2\pi\rho r^3\alpha$, $V = 2\rho r^2\alpha$

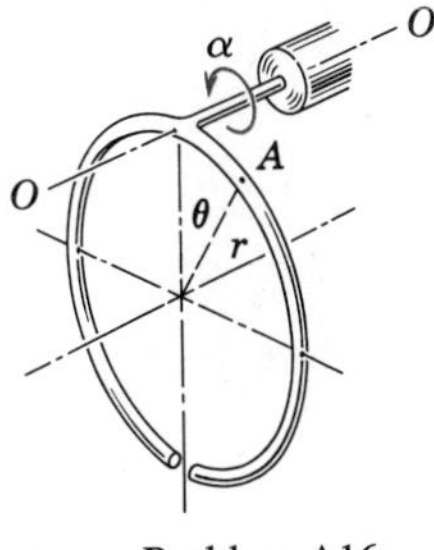

Problem A16

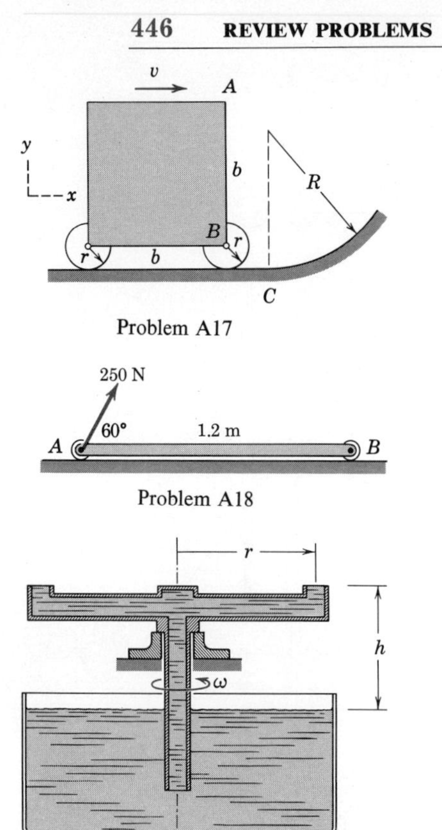

Problem A17

A17 The square frame mounted on corner rollers of radius r is moving in a straight line with a velocity v before it contacts the circular path of radius R at C. Determine the acceleration of corner A at the instant after the rollers at B pass C.

Ans. $\mathbf{a}_A = \dfrac{v^2}{R - r}(-\tfrac{1}{2}\mathbf{i} + \mathbf{j})$

Problem A18

A18 The uniform 1.2-m bar, initially at rest on the horizontal surface, has a mass of 15 kg and is supported at its ends by small rollers. Compute the initial angular acceleration α of the bar and the initial reaction under roller B resulting from the application of the 250-N force.

Ans. $\alpha = 23.8\text{ rad/s}^2$, $B = 145.0\text{ N}$

A19 The rotating tubes will pump water through the height h and discharge it through their open ends if the rotational speed ω is sufficiently great. Determine the minimum speed ω at which pumping will begin, assuming the pump is primed.

Ans. $\omega = \sqrt{2gh}/r$

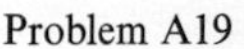
Problem A19

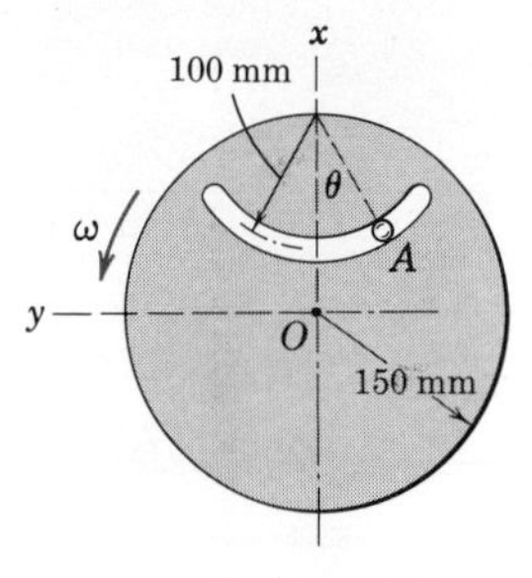

Problem A20

A20 The disk with the circular slot of 100-mm radius rotates about O with an angular velocity ω. Simultaneously, the pin A oscillates in the slot. Determine the acceleration a_A of A if, at the instant considered, $\omega = 10\text{ rad/s}$, $\dot{\omega} = 25\text{ rad/s}^2$, $\theta = 0$, $\dot{\theta} = 3\text{ rad/s}$, and $\ddot{\theta} = 5\text{ rad/s}^2$.

Ans. $\mathbf{a}_A = 1900\mathbf{i} + 750\mathbf{j}\text{ mm/s}^2$

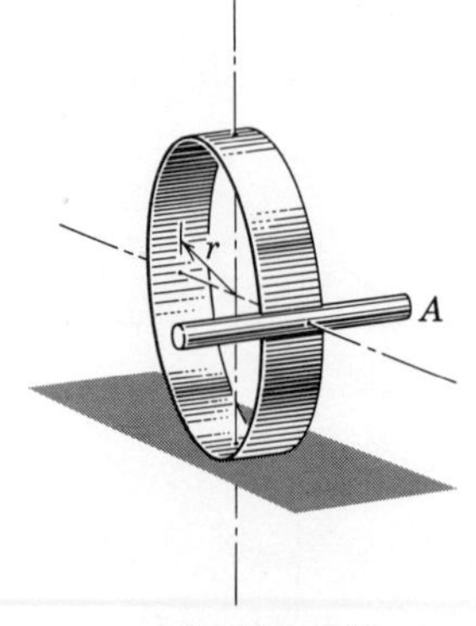

Problem A21

A21 The pipe A of mass m is welded to the rim of a circular hoop of radius r and negligible mass which is free to roll on a horizontal plane. If the hoop is released from rest with the pipe initially in a horizontal plane through the center of the hoop, determine the initial angular acceleration α of the hoop if the coefficient of friction between the hoop and the plane is (a) greater than unity and (b) less than unity. (*Hint:* The resultant of all forces acting on the hoop and pipe taken together passes through the center of the pipe which may be considered a particle.)

Ans. (a) $\alpha = \dfrac{g}{2r}$, (b) $\alpha = \dfrac{g}{r}$

A22 The motor M supplies 240 N·m of starting torque to its 100-mm-radius pinion in order to elevate the assembly in the smooth vertical guides. The large gear A meshes with the pinion and carries the cable drum of 200-mm radius. If the gear and attached drum together have a mass of 100 kg and a combined radius of gyration of 150 mm, and if the mass of the entire assembly is 450 kg, determine the initial upward acceleration of the carriage. Neglect the moments of inertia of the pinion and two small sheaves. *Ans.* $\ddot{x} = 4.13$ m/s

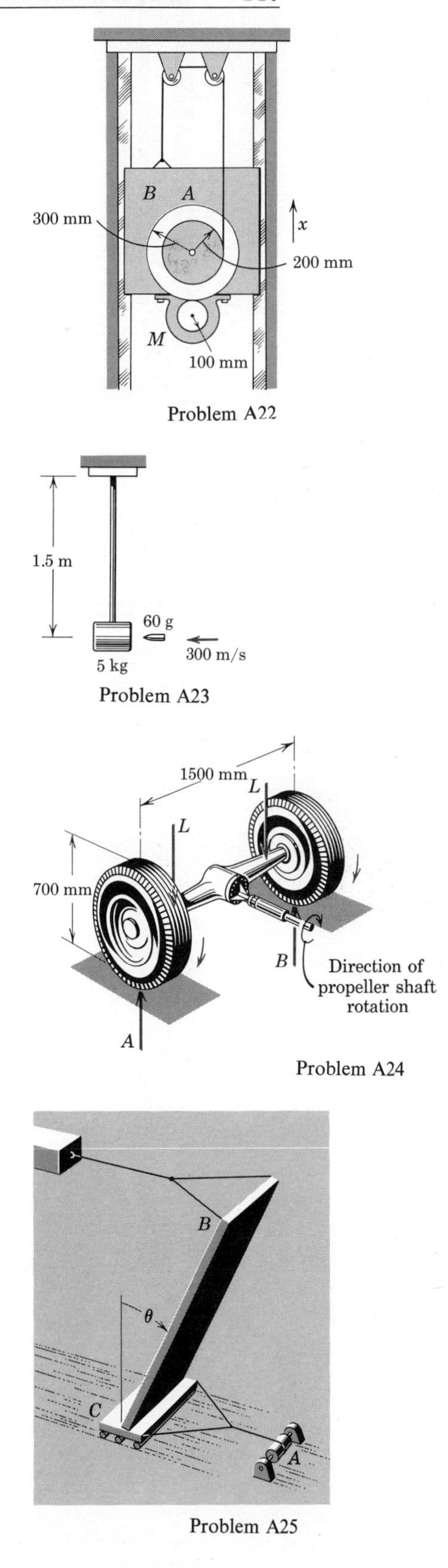

Problem A22

Problem A23

Problem A24

Problem A25

A23 A 60-g bullet is fired with a velocity of 300 m/s at the 5-kg block mounted on a stiff but light cantilever beam. The bullet is embedded in the block which is then observed to vibrate with a frequency of 4 Hz. Compute the amplitude A of the vibration and find the damping constant c in N·s/m if the amplitude at the end of the 10th complete oscillation is 0.6 that of the initial displacement. *Ans.* $c = 2.07$ N·s/m

A24 The rear axle of an automobile is shown in the sketch. The gear ratio in the differential is 3.7:1 (propeller-shaft speed is 3.7 times the rear-wheel speed). The car has a total mass of 1.8 t with center of mass midway between the front and rear axles and 600 mm above the road. The wheelbase is 3000 mm. During acceleration on a level road the engine delivers 75 kW to the rear wheels at a speed of 80 km/h. Determine the normal forces A and B under the rear wheels at this speed and corresponding acceleration. Assume that the rear wheels do not slip and that there are no vertical forces on the rear-wheel assembly other than those indicated. Neglect the mass of the wheels compared with the total mass of the car. *Ans.* $A = 4.54$ kN, $B = 4.96$ kN

A25 The large concrete slab shown is being tilted slowly into position by the winch at A. In the position shown the cable at B is horizontal, and θ is 30°. If the horizontal lower cable breaks at the winch in this position, determine the initial acceleration a of the bottom C of the slab which is free to move on rollers on the horizontal surface. *Ans.* $a = 3\sqrt{3}g/8$

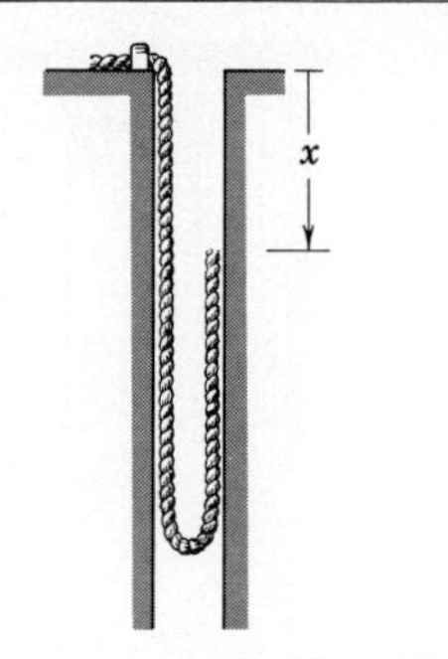

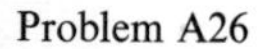
Problem A26

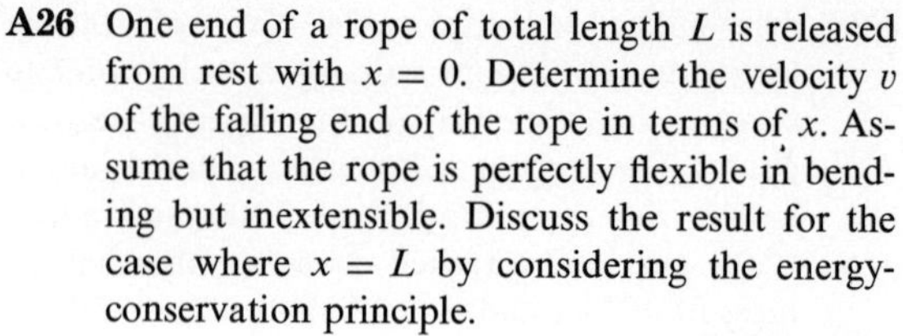
A26 One end of a rope of total length L is released from rest with $x = 0$. Determine the velocity v of the falling end of the rope in terms of x. Assume that the rope is perfectly flexible in bending but inextensible. Discuss the result for the case where $x = L$ by considering the energy-conservation principle.

$$Ans.\ v = \sqrt{\frac{gx(2L - x)}{L - x}}$$

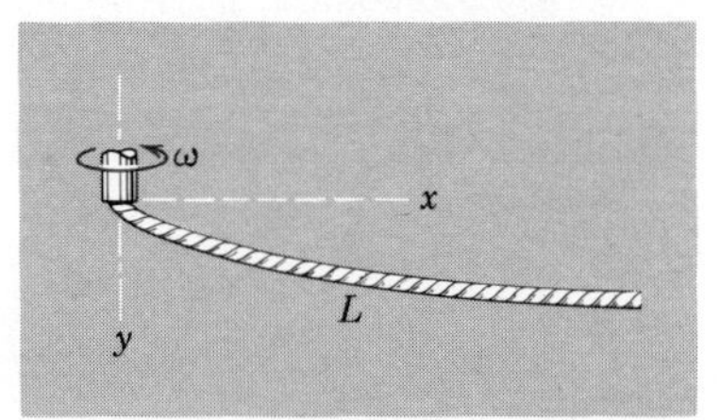

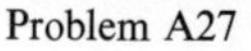
Problem A27

A27 A rope of length L is whirled around a fixed vertical axis with an angular velocity ω. Determine the shape assumed by the rope if ω is large so that the distance along the rope to any element may be approximated by the horizontal radius to the element.

$$Ans.\ y = \frac{2g}{\omega^2}\ln\left(1 + \frac{x}{L}\right)$$

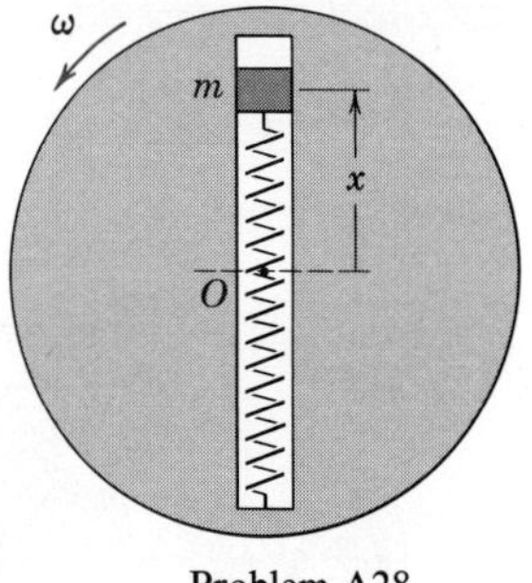

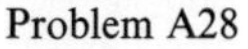
Problem A28

A28 The slotted disk rotates in the horizontal plane about its center O at a constant angular speed ω, and the slider of mass m with attached spring slides in the slot with negligible friction. The spring has a stiffness k and is undeformed when the slider is at the center of the disk. During rotation of the disk the slider is released from rest relative to the disk at the position for which $x = x_0$ where it was initially latched to the disk. If $\omega^2 < k/m$, determine (a) the initial value of $\ddot{x}$ just after the slider is released, and (b) the horizontal force F on the slider as it first passes the center of the disk.

$$Ans.\ (a)\ \ddot{x} = -\left(\frac{k}{m} - \omega^2\right)x_0$$

$$(b)\ F = 2mx_0\omega\sqrt{\frac{k}{m} - \omega^2}$$

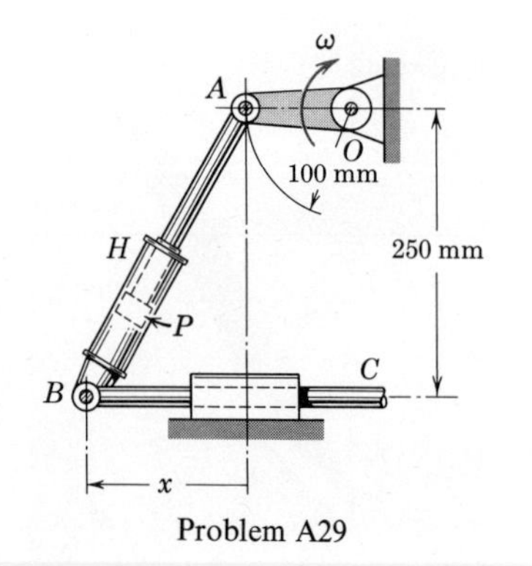

Problem A29

A29 The angular velocity of the crank OA is $\omega = \frac{3}{2}$ rad/s and $\dot{\omega} = 0$ in the horizontal position shown. Simultaneously, for the control rod BC, $x = 150$ mm, $\dot{x} = -100$ mm/s, and $\ddot{x} = 0$. For these conditions determine the velocity u and acceleration $\dot{u}$ of the piston P with respect to the hydraulic cylinder H and calculate the angular acceleration α of the cylinder axis.

Ans. $u = 0.0772$ m/s
$\dot{u} = 0.207$ m/s
$\alpha = 0.958$ rad/s^2

A30 At the instant represented $\theta = \pi/4$, $\dot{\theta} = 8$ rad/s, and $\ddot{\theta} = 0$ for the crank OA. For this position determine the velocity v_P and acceleration a_P of the point P on the shaft of the slotted link.

Ans. $v_P = 0.239$ m/s up
$a_P = 7.14$ m/s^2 down

Problem A30

A31 The hydraulic linkage AC is elongating at the constant rate $\dot{x} = 0.25$ m/s. For the instant when $\theta = \beta = 60°$, determine the velocity and acceleration of end B.

Ans. $v_B = 0.577$ m/s
$a_B = 0.0867$ m/s^2

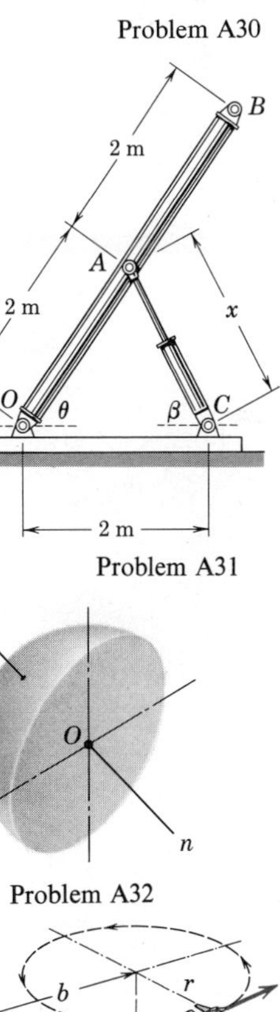

Problem A31

A32 Prove that the moment of inertia of the homogeneous solid hemisphere about any axis n-n through its center O is the same for all axes through O. Does the same conclusion hold for a hemispherical shell?

Problem A32

A33 An aircraft A is flying with a constant speed v in a horizontal circle of radius r at an altitude h. Determine the components of acceleration in the spherical coordinate system of the radar tracking antenna located at O.

$$\textit{Ans.}\quad a_R = -\frac{v^2}{r}\cos\phi\cos\theta$$

$$a_\theta = \frac{v^2}{r}\sin\theta$$

$$a_\phi = \frac{v^2}{r}\sin\phi\cos\theta$$

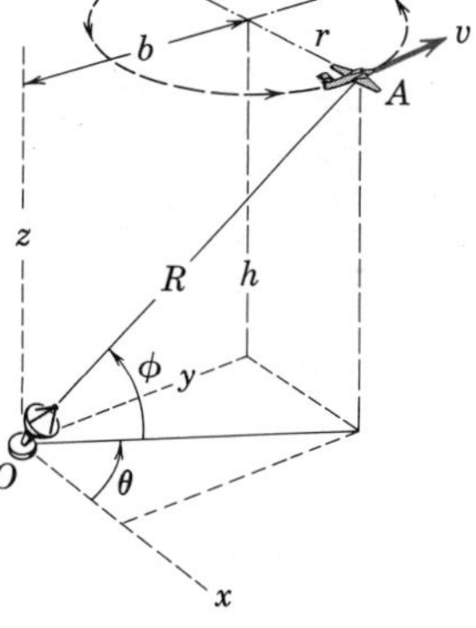

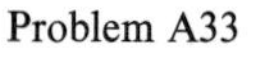
Problem A33

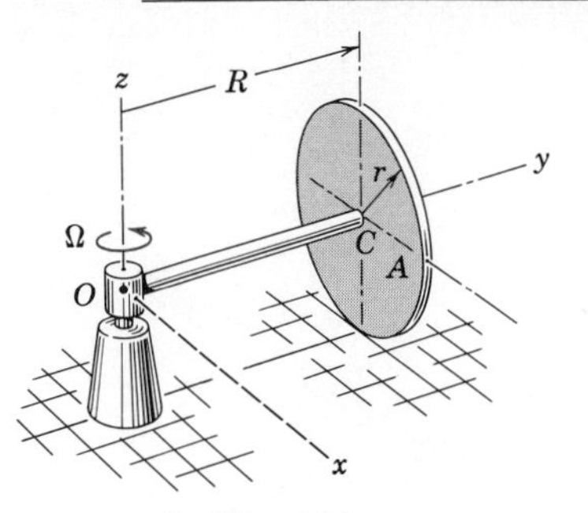

Problem A34

A34 The shaft at O rotates about the vertical z-axis with a constant angular velocity Ω and causes the wheel to roll in a horizontal circle of radius R. Determine the acceleration of point A on the rim at the instant represented. The wheel rolls without slipping.

$$\textit{Ans.}\ \mathbf{a}_A = -\Omega^2\left(\frac{r^2 + R^2}{r}\mathbf{i} + R\mathbf{j}\right)$$

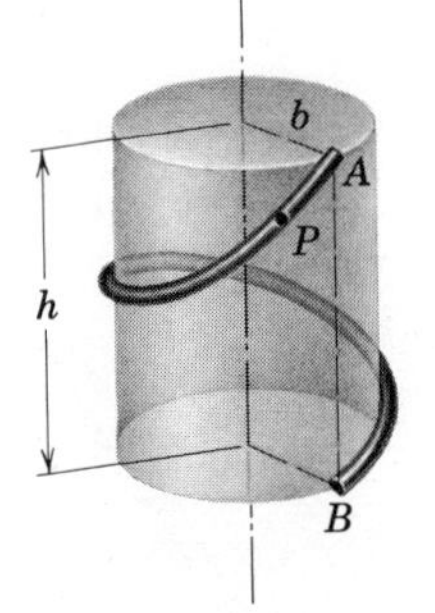

Problem A35

A35 A particle of mass m is released from rest at A and slides down the smooth spiral tube of constant helix angle. Determine the force N exerted by the tube on the particle as it completes one full turn at B.

$$\textit{Ans.}\ N = \frac{mg}{1 + \left(\dfrac{h}{2\pi b}\right)^2}\sqrt{1 + \frac{h^2}{b^2}\left(4 + \frac{1}{4\pi^2}\right)}$$

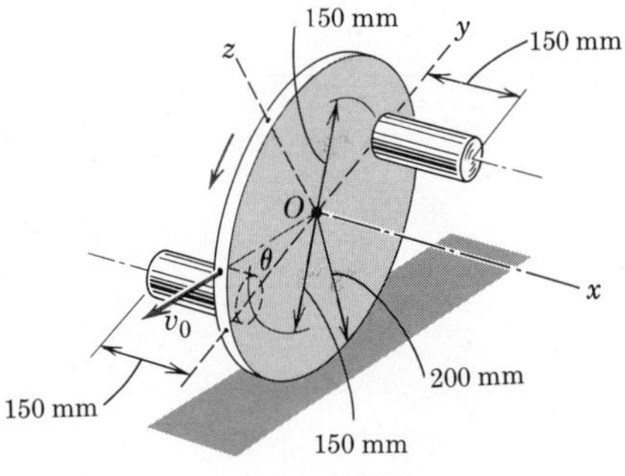

Problem A36

A36 The circular has a mass of 24 kg and rolls in the vertical plane without slipping with a velocity $v_O = 0.9$ m/s of its center O in the direction shown. Two stub shafts each having a mass of 4 kg are welded to the disk in the positions shown. Neglect the thickness of the disk and the diameter of the shafts compared with the other dimensions and compute the angular momentum about the mass center O for the orientation of axes shown.

$$\textit{Ans.}\ \mathbf{H}_O = 2.97\mathbf{i} - 0.405\mathbf{j}\ \text{kg}\cdot\text{m}^2/\text{s}$$

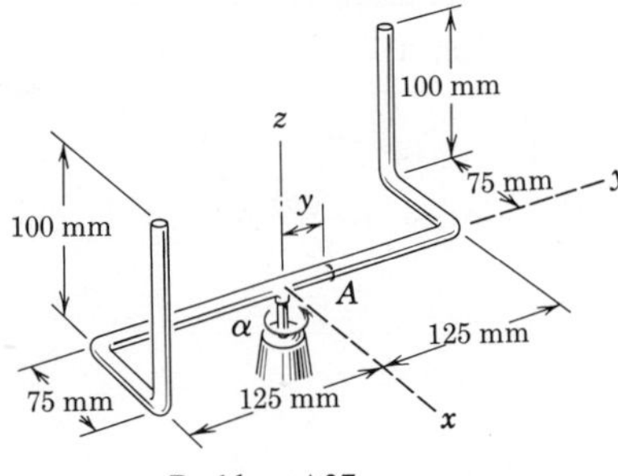

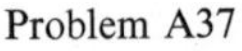

Problem A37

A37 The bent rod shown rotates about the z-axis and starts from rest with an angular acceleration $\alpha = 10^3$ rad/s^2 resulting from an applied torque. The material of the rod has a mass of 1.2 kg per metre of length. Neglect the small effect of gravity and determine the bending moment M in the rod at section A for y negligibly small.

$$\textit{Ans.}\ M = 4.93\ \text{N}\cdot\text{m}$$

A38 Assume that the slotted disk in Prob. A28 has a moment of inertia I about its center O and that it is rotating freely with no externally applied torque. Derive the differential equation for the oscillation of m in the slot after having been released from rest relative to the disk at the position $x = x_0$ at which instant the angular velocity of the disk was ω_0.

$$\textit{Ans.}\ \ddot{x} + \left\{\frac{k}{m} - \left[\frac{m x_0^2 + I}{m x^2 + I}\right]^2 \omega_0^2\right\} x = 0$$

A39 The disk rotates in the fixed x-y plane about the z-axis. The link AB is connected to the rim of the disk and to the slider at A by ball-and-socket joints. Point A moves in the y-z plane and is a distance r from the z-axis. Determine the velocity $\dot{z}$ of A in terms of θ if the disk turns with a constant speed $p = \dot{\theta}$. Also find the components of the angular velocity $\boldsymbol{\omega}$ of the center line of the link AB at the instant when $\theta = \pi/2$.

$$\textit{Ans. } \dot{z} = \frac{r^2 p \sin\theta}{\sqrt{l^2 - 2r^2(1 + \cos\theta)}}$$

$$\omega_x = \frac{r^2 - l^2}{l^2\sqrt{l^2 - 2r^2}}\, rp$$

$$\omega_y = \frac{r^3 p}{l^2\sqrt{l^2 - 2r^2}}$$

$$\omega_z = -\frac{r^2 p}{l^2}$$

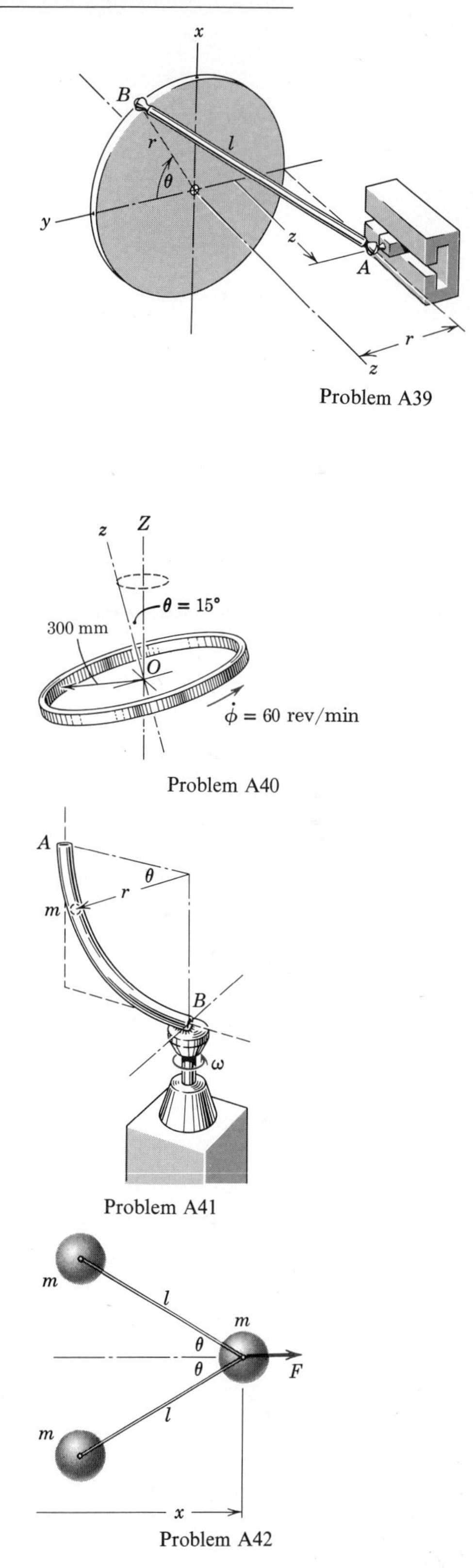

Problem A39

Problem A40

Problem A41

Problem A42

A40 The uniform circular ring has a mass of 8 kg and a radius of 300 mm. The ring is released in space with a spin velocity $\dot{\phi}$ of 60 rev/min about its z-axis and with a precessional wobble about the Z-axis corresponding to $\theta = 15°$. Calculate the kinetic energy of rotation T_{rot} of the ring about its center O. *Ans.* $T_{rot} = 16.25$ J

A41 The smooth tube is caused to rotate about the vertical axis with a constant angular speed ω while a small ball of mass m is released from rest relative to the tube at end A. Determine the value of ω which will permit the ball to reach the bottom B of the tube with zero velocity. For this value of ω determine the components of the force exerted by the tube on the ball in terms of θ. The component in the plane of the tube is N_1, and the component normal to the plane of the tube is N_2.

$$\textit{Ans. } \omega = \sqrt{2g/r}$$

$$N_1 = mg(3\sin\theta + 2\cos 2\theta)$$

$$N_2 = 4mg\sin\theta\sqrt{\sin\theta(1 - \sin\theta)}$$

A42 The three identical bodies each of mass m are connected by the two hinged links of negligible mass and equal length l. If a force F is applied as shown to the central body of the system at rest in space, determine the initial acceleration $\ddot{x}$ of this body.

$$\textit{Ans. } \ddot{x} = \frac{F/m}{3 - 2\sin^2\theta}$$

B VECTORS AND MATRICES

In the sections which follow a brief outline is presented of vector analysis and selected topics of matrix analysis which are useful in dynamics. Detailed explanations and proofs are not cited, and the reader is referred to mathematical references covering these subjects for a more complete development.

PART 1 VECTORS

B1 Notation. Vector quantities may be described by any notation which accounts properly for both their magnitudes and directions. Scalar algebra may be used to handle the magnitude relationships, and geometric diagrams with appropriate trigonometry may be used to account for the directional properties. Such analysis is generally quite adequate or preferable for problems with one or two geometric dimensions. For much of the analysis in three geometric dimensions, the vector notation invented by Gibbs* is of great advantage and is used widely. The following development is designed to serve as both an introduction to and a concise summary of the relationships of the algebra and calculus of vectors as they are employed in basic mechanics using Gibbs' notation.

It is absolutely essential to adopt a consistent notation that will always distinguish vector quantities from scalar quantities. In the printed page, boldface type is used for all vector quantities, and lightface italic type is used

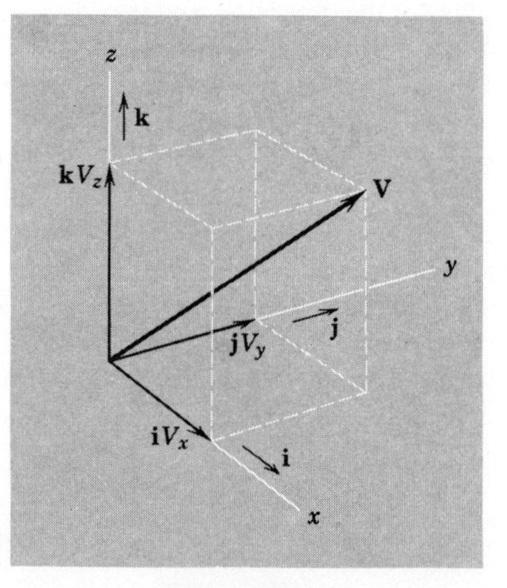

Figure B1

[1] Josiah Willard Gibbs (1839–1903), a professor of mathematical physics at Yale University.

for scalar quantities. In handwritten work a distinguishing symbol, such as an underline, should be used to designate vector quantities.

A vector **V** is represented in three-dimensional space, Fig. B1, in terms of the vector sum of its three mutually perpendicular components

$$\mathbf{V} = \mathbf{i}V_x + \mathbf{j}V_y + \mathbf{k}V_z \tag{B1}$$

where **i**, **j**, **k** are the unit vectors in the *x*-, *y*-, and *z*-directions, respectively. A unit vector is a vector with a specified direction and a magnitude of unity. Thus the vector quantity $\mathbf{i}V_x$ has a direction specified by the unit vector **i** in the *x*-direction and a magnitude equal to V_x, the *x*-component of the vector **V**.

The scalar magnitude of **V** is

$$|\mathbf{V}| = V = \sqrt{V_x^2 + V_y^2 + V_z^2} \tag{B2}$$

Any vector **V** may be multiplied by a scalar a to give the vector $\mathbf{V}a$ or $a\mathbf{V}$, which product has the magnitude aV and the direction of **V**.

B2 Addition. Two vectors **P** and **Q**, Fig. B2*a*, may be added to obtain their resultant or sum **P** + **Q**, as shown in Fig. B2*b* where the two vectors are the two legs of the parallelogram. For any convenient orientation of reference axes, the sum may be written in component form as

$$\begin{aligned}\mathbf{P} + \mathbf{Q} &= (\mathbf{i}P_x + \mathbf{j}P_y + \mathbf{k}P_z) + (\mathbf{i}Q_x + \mathbf{j}Q_y + \mathbf{k}Q_z) \\ &= \mathbf{i}(P_x + Q_x) + \mathbf{j}(P_y + Q_y) + \mathbf{k}(P_z + Q_z)\end{aligned} \tag{B3}$$

As seen from Fig. B2*b*, the sum may be obtained by combining the vectors head-to-tail in either order in the triangle form of vector addition to obtain their sum. Thus

$$\mathbf{P} + \mathbf{Q} = \mathbf{Q} + \mathbf{P} \tag{B4}$$

and vector addition is seen to be *commutative.*

Vector addition is also *associative* as observed in Fig. B3, where it is clear

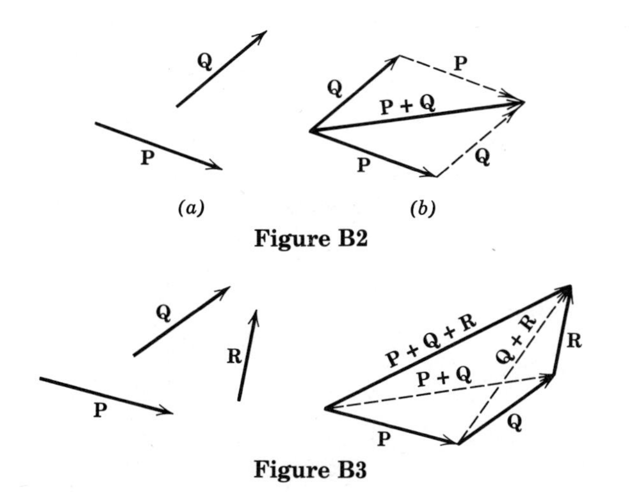

Figure B2

Figure B3

that the sum of the three vectors **P**, **Q**, **R** may be arrived at by adding **P** to the sum of **Q** and **R** or by adding the sum of **P** and **Q** to **R**. Thus

$$\mathbf{P} + (\mathbf{Q} + \mathbf{R}) = (\mathbf{P} + \mathbf{Q}) + \mathbf{R} \tag{B5}$$

The subtraction of a vector is the same as the addition of a negative vector. Thus the vector difference between **P** and **Q** of Fig. B4 is $\mathbf{P} - \mathbf{Q} = \mathbf{P} + (-\mathbf{Q})$.

B3 Dot or Scalar Product. There are two distinctly different ways in which vectors may be multiplied. In the first way the *dot* or *scalar product* of two vectors **P** and **Q**, Fig. B5*a*, is defined as

$$\mathbf{P} \cdot \mathbf{Q} = PQ \cos \theta \tag{B6}$$

where θ is the angle between them. This product may be viewed as the magnitude of **P** multiplied by the component $Q \cos \theta$ of **Q** in the direction of **P**, Fig. B5*b*, or as the magnitude of **Q** multiplied by the component $P \cos \theta$ of **P** in the direction of **Q**, Fig. B5*c*. The commutative law

$$\mathbf{P} \cdot \mathbf{Q} = \mathbf{Q} \cdot \mathbf{P} \tag{B7}$$

holds for the dot product, since the order of the scalar terms in the scalar multiplication may be interchanged without affecting the product.

From the definition of the dot product, it follows that

$$\begin{aligned} \mathbf{i} \cdot \mathbf{i} &= \mathbf{j} \cdot \mathbf{j} = \mathbf{k} \cdot \mathbf{k} = 1 \\ \mathbf{i} \cdot \mathbf{j} &= \mathbf{j} \cdot \mathbf{i} = \mathbf{i} \cdot \mathbf{k} = \mathbf{k} \cdot \mathbf{i} = \mathbf{j} \cdot \mathbf{k} = \mathbf{k} \cdot \mathbf{j} = 0 \end{aligned}$$

Thus

$$\begin{aligned} \mathbf{P} \cdot \mathbf{Q} &= (\mathbf{i}P_x + \mathbf{j}P_y + \mathbf{k}P_z) \cdot (\mathbf{i}Q_x + \mathbf{j}Q_y + \mathbf{k}Q_z) \\ &= P_xQ_x + P_yQ_y + P_zQ_z \end{aligned} \tag{B8}$$

and

$$\mathbf{P} \cdot \mathbf{P} = P_x^2 + P_y^2 + P_z^2$$

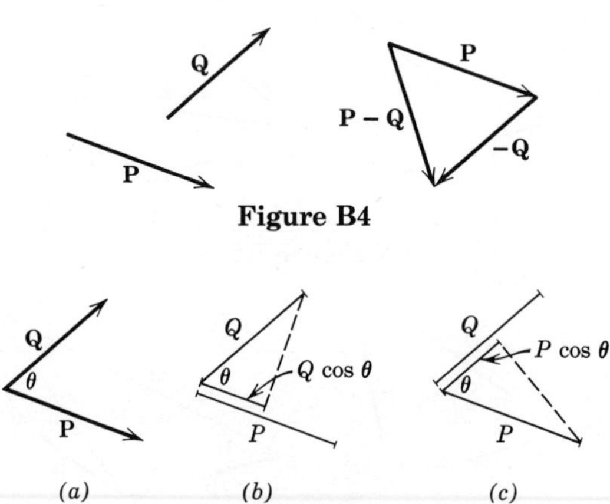

Figure B4

(*a*) (*b*) (*c*)

Figure B5

It follows from the definition of the dot product that two vectors **P** and **Q** are perpendicular when their dot product vanishes, $\mathbf{P} \cdot \mathbf{Q} = 0$.

The angle θ between two vectors $\mathbf{P}_1$ and $\mathbf{P}_2$ may be found from their dot product expression $\mathbf{P}_1 \cdot \mathbf{P}_2 = P_1P_2 \cos\theta$, which gives

$$\cos\theta = \frac{\mathbf{P}_1 \cdot \mathbf{P}_2}{P_1P_2} = \frac{P_{1_x}P_{2_x} + P_{1_y}P_{2_y} + P_{1_z}P_{2_z}}{P_1P_2} = l_1l_2 + m_1m_2 + n_1n_2$$

where l, m, n stand for the respective direction cosines of the vectors. It is also observed that two vectors are perpendicular when their direction cosines obey the relation $l_1l_2 + m_1m_2 + n_1n_2 = 0$.

The distributive law holds for the dot product, as is easily seen from the following expansion.

$$\begin{aligned}\mathbf{P} \cdot (\mathbf{Q}+\mathbf{R}) &= (\mathbf{i}P_x+\mathbf{j}P_y+\mathbf{k}P_z) \cdot (\mathbf{i}[Q_x+R_x]+\mathbf{j}[Q_y+R_y]+\mathbf{k}[Q_z+R_z]) \\ &= P_x(Q_x + R_x) + P_y(Q_y + R_y) + P_z(Q_z + R_z) \\ &= (P_xQ_x + P_yQ_y + P_zQ_z) + (P_xR_x + P_yR_y + P_zR_z)\end{aligned}$$

so that

$$\mathbf{P} \cdot (\mathbf{Q} + \mathbf{R}) = \mathbf{P} \cdot \mathbf{Q} + \mathbf{P} \cdot \mathbf{R} \tag{B9}$$

B4 Cross or Vector Product. The second form of multiplication of vectors is known as the *cross* or *vector product.* For the vectors **P** and **Q** of Fig. B6 this product is written as $\mathbf{P} \times \mathbf{Q}$ and is defined as a vector whose magnitude equals the product of the magnitudes of **P** and **Q** multiplied by the sine of the angle θ (less than 180 deg) between them. The direction of $\mathbf{P} \times \mathbf{Q}$ is normal to the plane defined by **P** and **Q**, and the sense of $\mathbf{P} \times \mathbf{Q}$ is in the direction of the advancement of a right-hand screw when revolved from **P** to **Q** through the smaller of the two angles between them. If **n** is a unit vector with the direction and sense of $\mathbf{P} \times \mathbf{Q}$, the cross product may be written

$$\mathbf{P} \times \mathbf{Q} = \mathbf{n}PQ \sin\theta \tag{B10}$$

By using the right-hand rule and reversing the order of vector multiplica-

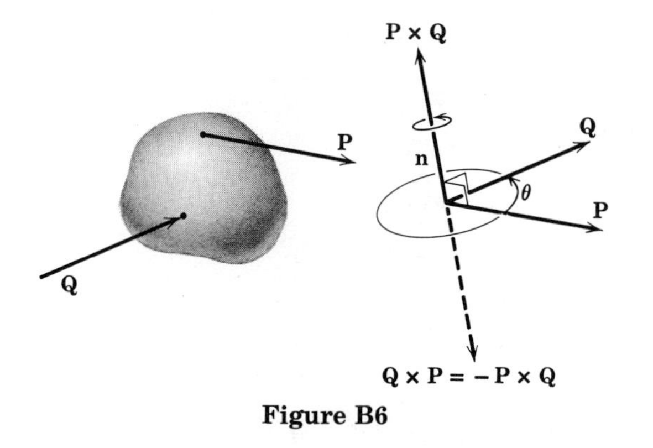

Figure B6

tion, it is seen from Fig. B6 that $\mathbf{P} \times \mathbf{Q} = -\mathbf{Q} \times \mathbf{P}$. Thus the commutative law does not hold for the cross product.

The distributive law does hold for the cross product and may be shown geometrically. In Fig. B7 a triangular prism 1-2-3-4-5-6 is formed from the three arbitrary vectors **P**, **Q**, **R** as shown. The magnitude of the area of the parallelogram face 1-4-5-2 is the base $|\mathbf{Q}|$ times the altitude, which is $|\mathbf{P}|$ times the sine of the angle between **P** and **Q**. Thus this area is represented by the magnitude of the cross product $\mathbf{P} \times \mathbf{Q}$. The area of this face has a direction normal to its plane, and with its positive sense out from the figure, it may be represented by the vector $\mathbf{P} \times \mathbf{Q}$. In similar fashion the area 2-5-6-3 is represented by $\mathbf{P} \times \mathbf{R}$, and the area 1-3-6-4 by $(\mathbf{Q} + \mathbf{R}) \times \mathbf{P}$. The parallel triangular faces are $\frac{1}{2}(\mathbf{Q} \times \mathbf{R})$ and $-\frac{1}{2}(\mathbf{Q} \times \mathbf{R})$. Since the prism is a closed surface, the sum of its surfaces represented as vectors must vanish. Therefore

$$(\mathbf{Q} + \mathbf{R}) \times \mathbf{P} + \mathbf{P} \times \mathbf{Q} + \mathbf{P} \times \mathbf{R} + \tfrac{1}{2}(\mathbf{Q} \times \mathbf{R}) - \tfrac{1}{2}(\mathbf{Q} \times \mathbf{R}) = \mathbf{0}$$

Reversing the order and the sign of the first term and dropping the last two terms give

$$\mathbf{P} \times (\mathbf{Q} + \mathbf{R}) = \mathbf{P} \times \mathbf{Q} + \mathbf{P} \times \mathbf{R} \tag{B11}$$

which proves the distributive law for the cross product.

From the definition of the cross product the following relations between the unit vectors are apparent:

$$\mathbf{i} \times \mathbf{j} = \mathbf{k} \qquad \mathbf{j} \times \mathbf{k} = \mathbf{i} \qquad \mathbf{k} \times \mathbf{i} = \mathbf{j}$$
$$\mathbf{j} \times \mathbf{i} = -\mathbf{k} \qquad \mathbf{k} \times \mathbf{j} = -\mathbf{i} \qquad \mathbf{i} \times \mathbf{k} = -\mathbf{j}$$
$$\mathbf{i} \times \mathbf{i} = \mathbf{j} \times \mathbf{j} = \mathbf{k} \times \mathbf{k} = \mathbf{0}$$

With the aid of these identities and the distributive law the vector product may be written

$$\begin{aligned} \mathbf{P} \times \mathbf{Q} &= (\mathbf{i}P_x + \mathbf{j}P_y + \mathbf{k}P_z) \times (\mathbf{i}Q_x + \mathbf{j}Q_y + \mathbf{k}Q_z) \\ &= \mathbf{i}(P_yQ_z - P_zQ_y) + \mathbf{j}(P_zQ_x - P_xQ_z) + \mathbf{k}(P_xQ_y - P_yQ_x) \end{aligned} \tag{B12}$$

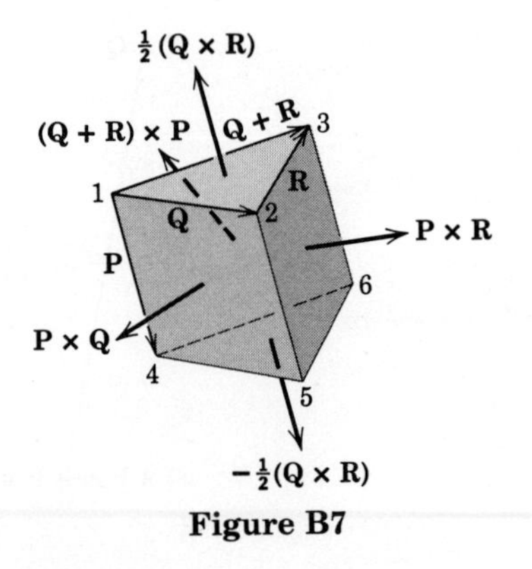

Figure B7

upon rearrangement of terms. This expression may be written compactly as the determinant

$$\mathbf{P} \times \mathbf{Q} = \begin{vmatrix} \mathbf{i} & \mathbf{j} & \mathbf{k} \\ P_x & P_y & P_z \\ Q_x & Q_y & Q_z \end{vmatrix} \tag{B12a}$$

which is easily verified by carrying out the expansion.

B5 Additional Relations. Two additional relations of vector algebra will be stated without proof, although their validity is not difficult to show geometrically.

The *triple scalar product* is the dot product of two vectors where one of them is specified as a cross product of two additional vectors. This product is a scalar and is given by any one of the equivalent expressions

$$(\mathbf{P} \times \mathbf{Q}) \cdot \mathbf{R} = \mathbf{R} \cdot (\mathbf{P} \times \mathbf{Q}) = -\mathbf{R} \cdot (\mathbf{Q} \times \mathbf{P})$$

Actually the parentheses are not needed since it would be meaningless to write $\mathbf{P} \times (\mathbf{Q} \cdot \mathbf{R})$. It may be shown that

$$\mathbf{P} \times \mathbf{Q} \cdot \mathbf{R} = \mathbf{P} \cdot \mathbf{Q} \times \mathbf{R} \tag{B13}$$

which establishes the rule that the dot and the cross may be interchanged without changing the value of the triple scalar product. Furthermore, it may be seen upon expansion that

$$\mathbf{P} \times \mathbf{Q} \cdot \mathbf{R} = \begin{vmatrix} P_x & P_y & P_z \\ Q_x & Q_y & Q_z \\ R_x & R_y & R_z \end{vmatrix} \tag{B14}$$

The *triple vector product* is the cross product of two vectors where one of them is specified as a cross product of two additional vectors. This product is a vector and is given by any one of the equivalent expressions

$$(\mathbf{P} \times \mathbf{Q}) \times \mathbf{R} = -\mathbf{R} \times (\mathbf{P} \times \mathbf{Q}) = \mathbf{R} \times (\mathbf{Q} \times \mathbf{P})$$

Here the parentheses must be used since an expression $\mathbf{P} \times \mathbf{Q} \times \mathbf{R}$ would be ambiguous because it would not identify the vector to be crossed. It may be shown that the triple vector product is equivalent to

$$(\mathbf{P} \times \mathbf{Q}) \times \mathbf{R} = \mathbf{R} \cdot \mathbf{P}\mathbf{Q} - \mathbf{R} \cdot \mathbf{Q}\mathbf{P}$$

or (B15)

$$\mathbf{P} \times (\mathbf{Q} \times \mathbf{R}) = \mathbf{P} \cdot \mathbf{R}\mathbf{Q} - \mathbf{P} \cdot \mathbf{Q}\mathbf{R}$$

The first term in the first expression, for example, is the dot product $\mathbf{R} \cdot \mathbf{P}$, a scalar, multiplied by the vector $\mathbf{Q}$. The validity of Eqs. B13 and B15 may be checked easily by carrying out the indicated operations with three arbitrary vectors with numerical or algebraic coefficients.

B6 Derivatives of Vectors. The derivative of a vector $\mathbf{P}$ with respect to a scalar, such as the time t, is the limit of the ratio of the change $\Delta\mathbf{P}$ in $\mathbf{P}$ to the corresponding change Δt in t as Δt approaches zero. Thus

$$\frac{d\mathbf{P}}{dt} = \lim_{\Delta t \to 0} \frac{\Delta \mathbf{P}}{\Delta t}$$

$$= \lim_{\Delta t \to 0} \left(\mathbf{i}\frac{\Delta P_x}{\Delta t} + \mathbf{j}\frac{\Delta P_y}{\Delta t} + \mathbf{k}\frac{\Delta P_z}{\Delta t} \right)$$

where $\Delta \mathbf{P}$ has been expressed in terms of its components. It follows that

$$\frac{d\mathbf{P}}{dt} = \mathbf{i}\frac{dP_x}{dt} + \mathbf{j}\frac{dP_y}{dt} + \mathbf{k}\frac{dP_z}{dt}$$

and

$$\frac{d^n\mathbf{P}}{dt^n} = \mathbf{i}\frac{d^nP_x}{dt^n} + \mathbf{j}\frac{d^nP_y}{dt^n} + \mathbf{k}\frac{d^nP_z}{dt^n} \tag{B16}$$

The derivative of the sum of two vectors is simply

$$\frac{d(\mathbf{P} + \mathbf{Q})}{dt} = \lim_{\Delta t \to 0} \frac{\Delta(\mathbf{P} + \mathbf{Q})}{\Delta t} = \lim_{\Delta t \to 0} \left(\frac{\Delta \mathbf{P}}{\Delta t} + \frac{\Delta \mathbf{Q}}{\Delta t} \right)$$

$$= \frac{d\mathbf{P}}{dt} + \frac{d\mathbf{Q}}{dt} \tag{B17}$$

since the limit of the sum is the same as the sum of the limits of the terms.

The derivative of the product of a vector $\mathbf{P}$ and a scalar u obeys the same rule as for the product of two scalar quantities and is

$$\frac{d(\mathbf{P}u)}{dt} = \lim_{\Delta t \to 0} \frac{(\mathbf{P} + \Delta\mathbf{P})(u + \Delta u) - \mathbf{P}u}{\Delta t}$$

$$= \lim_{\Delta t \to 0} \frac{\mathbf{P}\Delta u + u\Delta\mathbf{P}}{\Delta t} = \lim_{\Delta t \to 0} \left(\mathbf{P}\frac{\Delta u}{\Delta t} + u\frac{\Delta \mathbf{P}}{\Delta t} \right)$$

$$= \mathbf{P}\frac{du}{dt} + u\frac{d\mathbf{P}}{dt} \tag{B18}$$

The derivatives of the scalar (dot) product and the vector (cross) product of two vectors also obey the same rules as for the product of two scalar quantities. Thus for the dot product

$$\frac{d(\mathbf{P}\cdot\mathbf{Q})}{dt} = \lim_{\Delta t \to 0} \frac{(\mathbf{P} + \Delta\mathbf{P})\cdot(\mathbf{Q} + \Delta\mathbf{Q}) - \mathbf{P}\cdot\mathbf{Q}}{\Delta t}$$

$$= \lim_{\Delta t \to 0} \frac{\mathbf{P}\cdot\Delta\mathbf{Q} + \mathbf{Q}\cdot\Delta\mathbf{P} + \Delta\mathbf{P}\cdot\Delta\mathbf{Q}}{\Delta t}$$

$$= \lim_{\Delta t \to 0} \left(\mathbf{P}\cdot\frac{\Delta\mathbf{Q}}{\Delta t} + \frac{\Delta\mathbf{P}}{\Delta t}\cdot\mathbf{Q} + \frac{\Delta\mathbf{P}\cdot\Delta\mathbf{Q}}{\Delta t} \right)$$

$$= \mathbf{P}\cdot\frac{d\mathbf{Q}}{dt} + \frac{d\mathbf{P}}{dt}\cdot\mathbf{Q} \tag{B19}$$

The third term drops out in the limit, since it is of a higher order than the ones that remain.

Similarly, for the cross product the derivative is

$$\frac{d(\mathbf{P}\times\mathbf{Q})}{dt} = \lim_{\Delta t\to 0} \frac{(\mathbf{P}+\Delta\mathbf{P})\times(\mathbf{Q}+\Delta\mathbf{Q}) - \mathbf{P}\times\mathbf{Q}}{\Delta t}$$

$$= \lim_{\Delta t\to 0} \frac{\mathbf{P}\times\Delta\mathbf{Q} + \Delta\mathbf{P}\times\mathbf{Q} + \Delta\mathbf{P}\times\Delta\mathbf{Q}}{\Delta t}$$

$$= \lim_{\Delta t\to 0} \left(\mathbf{P}\times\frac{\Delta\mathbf{Q}}{\Delta t} + \frac{\Delta\mathbf{P}}{\Delta t}\times\mathbf{Q} + \frac{\Delta\mathbf{P}\times\Delta\mathbf{Q}}{\Delta t}\right)$$

$$= \mathbf{P}\times\frac{d\mathbf{Q}}{dt} + \frac{d\mathbf{P}}{dt}\times\mathbf{Q} \tag{B20}$$

Again the third term disappears in the limit, since it is of a higher order than those that remain. The only special caution to be observed in the differentiation of cross products is that the order of the quantities on the two sides of the cross sign must be preserved, since the quantities in cross-product multiplication are not commutative.

B7 Integration of Vectors. Integration of vectors poses no special problem. If **V** is a function of x, y, and z and an element of volume is $d\tau = dx\,dy\,dz$, the integral of **V** over the volume may be written as the vector sum of the three integrals of its components. Thus

$$\int \mathbf{V}\,d\tau = \mathbf{i}\int V_x\,d\tau + \mathbf{j}\int V_y\,d\tau + \mathbf{k}\int V_z\,d\tau \tag{B21}$$

B8 Gradient. Consider a scalar function defined by $\phi = f(x, y, z)$. Successive constant values of ϕ such as ϕ_1, ϕ_2, ϕ_3, . . . define adjoining space surfaces, Fig. B8. From any point A on one of the surfaces, there is one path n which goes most directly from surface to surface. The quantity

$$\mathbf{F} = \mathbf{i}\frac{\partial\phi}{\partial x} + \mathbf{j}\frac{\partial\phi}{\partial y} + \mathbf{k}\frac{\partial\phi}{\partial z}$$

called the *gradient* of ϕ, is a vector in the n-direction and represents the maximum space rate of change of ϕ.

The gradient of ϕ may be written in compact notation using the vector

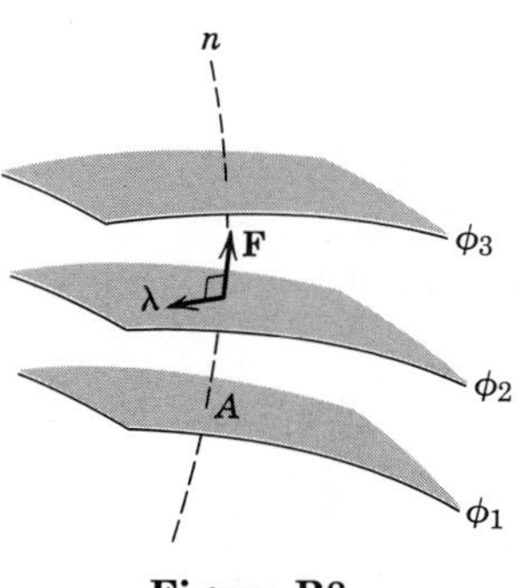

Figure B8

operator ∇ ("del"), so that

$$\mathbf{F} = \nabla\phi$$

where (B22)

$$\nabla = \mathbf{i}\frac{\partial}{\partial x} + \mathbf{j}\frac{\partial}{\partial y} + \mathbf{k}\frac{\partial}{\partial z}$$

The fact that $\nabla\phi$ is perpendicular to the surface ϕ = constant is shown by taking a unit vector $\boldsymbol{\lambda} = \mathbf{i}\,dx + \mathbf{j}\,dy + \mathbf{k}\,dz$ which lies in that surface and noting that $\boldsymbol{\lambda}$ and $\nabla\phi$ satisfy the condition for perpendicularity, namely,

$$\boldsymbol{\lambda} \cdot \nabla\phi = 0$$

Thus

$$dx\frac{\partial\phi}{\partial x} + dy\frac{\partial\phi}{\partial y} + dz\frac{\partial\phi}{\partial z} = d\phi = 0$$

where dx, dy, dz lie in the surface ϕ = constant for which $d\phi = 0$.

The dot product of the vector ∇ into itself gives the scalar operator

$$\nabla \cdot \nabla = \frac{\partial^2}{\partial x^2} + \frac{\partial^2}{\partial y^2} + \frac{\partial^2}{\partial z^2} \tag{B23}$$

which is called the *Laplacian*. The expression $\nabla^2\phi = 0$ is known as Laplace's Equation.

B9 Divergence. When the vector operator ∇ is dotted into a vector **V**, the result is

$$\nabla \cdot \mathbf{V} = \frac{\partial V_x}{\partial x} + \frac{\partial V_y}{\partial y} + \frac{\partial V_z}{\partial z} \tag{B24}$$

which is known as the *divergence* of **V**.

B10 Curl. When the vector operator ∇ is crossed into a vector **V**, the result is

$$\nabla \times \mathbf{V} = \mathbf{i}\left(\frac{\partial V_z}{\partial y} - \frac{\partial V_y}{\partial z}\right) + \mathbf{j}\left(\frac{\partial V_x}{\partial z} - \frac{\partial V_z}{\partial x}\right) + \mathbf{k}\left(\frac{\partial V_y}{\partial x} - \frac{\partial V_x}{\partial y}\right)$$

$$= \begin{vmatrix} \mathbf{i} & \mathbf{j} & \mathbf{k} \\ \dfrac{\partial}{\partial x} & \dfrac{\partial}{\partial y} & \dfrac{\partial}{\partial z} \\ V_x & V_y & V_z \end{vmatrix} \tag{B22}$$

This expression is known as the *curl* of **V**.

B11 Other Operations. It may be shown that the curl of the gradient is identically zero

$$\nabla \times \nabla\phi = \mathbf{0}$$

and that the divergence of the curl is also identically zero

$$\nabla \cdot \nabla \times \mathbf{V} = 0$$

Also, it may be proved that

$$\nabla \times (\nabla \times \mathbf{V}) = \nabla(\nabla \cdot \mathbf{V}) - (\nabla \cdot \nabla)\mathbf{V}$$

which may be written merely as

$$\nabla \times \nabla \times \mathbf{V} = \nabla\nabla \cdot \mathbf{V} - \nabla \cdot \nabla\mathbf{V} \tag{B23}$$

PART 2 MATRICES

B12 Notation. A convenient shorthand which offers considerable advantage when working with systems of linear equations is contained in matrix notation. The notation permits a systematic handling of the large array of quantities involved in the solutions to such equations and, consequently, provides a convenient form for use in computer programming.

Consider the three quantities y_1, y_2, y_3 related to the three quantities x_1, x_2, x_3 with the a's as coefficients by the linear equations

$$\begin{aligned} y_1 &= a_{11}x_1 + a_{12}x_2 + a_{13}x_3 \\ y_2 &= a_{21}x_1 + a_{22}x_2 + a_{23}x_3 \\ y_3 &= a_{31}x_1 + a_{32}x_2 + a_{33}x_3 \end{aligned} \tag{B24}$$

In matrix notation these equations may be represented by the array

$$\begin{Bmatrix} y_1 \\ y_2 \\ y_3 \end{Bmatrix} = \begin{bmatrix} a_{11} & a_{12} & a_{13} \\ a_{21} & a_{22} & a_{23} \\ a_{31} & a_{32} & a_{33} \end{bmatrix} \begin{Bmatrix} x_1 \\ x_2 \\ x_3 \end{Bmatrix} \tag{B25}$$

where the terms in the given equations are formed by multiplying successive elements in the matrix of the coefficients by the corresponding elements in the column matrix of the x's. The set of equations may be represented by the further simplified matrix notation

$$\{y\} = [a]\{x\} \tag{B26}$$

where $\{x\}$ and $\{y\}$ are the *column matrices* and $[a]$ is a *square matrix,* which is formed when the number of rows equals the number of columns. A square matrix with n rows and n columns is said to be of *order n*.

Equation B26 may also be written in terms of the summation

$$y_i = \sum_{j=1}^{3} a_{ij}x_j \qquad \text{where } i = 1, 2, 3 \tag{B27}$$

The terms a_{ij} are known as *elements* or *components* of the matrix, where the first index in the sequence identifies the number of the row and the second index identifies the number of the column, regardless of what symbols appear. Thus in a_{ij} the row is denoted by i, whereas in a_{ki}, the row is denoted by k and the column by i. With this index notation the matrix $[a]$ may also be written as $[a_{ij}]$.

A square matrix which contains only the main diagonal terms

$$\begin{bmatrix} a_{11} & 0 & 0 \\ 0 & a_{22} & 0 \\ 0 & 0 & a_{33} \end{bmatrix}$$

is known as a *diagonal matrix.* The matrix

$$\begin{bmatrix} a_{11} & b & c \\ b & a_{22} & d \\ c & d & a_{33} \end{bmatrix}$$

for which $a_{ij} = a_{ji}$ is known as a *symmetric matrix,* and the matrix

$$\begin{bmatrix} 0 & -b & -c \\ b & 0 & d \\ c & -d & 0 \end{bmatrix}$$

is an example of an *anti-symmetric matrix* where $a_{ij} = -a_{ji}$. In this case the main diagonal terms must be zero.

B13 Matrix Algebra. The sum of two matrices of equal numbers of rows and columns is the matrix formed by adding corresponding elements. Thus

$$[a] + [b] = [a_{ij} + b_{ij}] \tag{B28}$$

The product of a scalar s and a matrix is the matrix formed by multiplying each element by s. Thus

$$s[a] = [sa_{ij}] \tag{B29}$$

The product of two matrices may be defined only where the number of columns in the first equals the number of rows in the second. Thus

$$[c] = [a][b] \tag{B30}$$

where

$$c_{ij} = \sum_k a_{ik} b_{kj}$$

In general, the commutative law for matrix multiplication does not hold, so that

$$[a][b] \neq [b][a]$$

The associative law for matrix multiplication does hold, so that

$$[a]([b][c]) = ([a][b])[c]$$

The distributive law also holds, and

$$[a]([b] + [c]) = [a][b] + [a][c]$$

B14 Inverse and Multiple Transformations. The set of Eqs. B26 may be rewritten to express the x's in terms of the y's. In matrix notation this new set of independent equations is represented by

$$\{x\} = [a]^{-1}\{y\} \tag{B31}$$

where the matrix $[a]^{-1}$ is known as the *inverse matrix* of $[a]$. The inverse matrix may be determined from the original matrix $[a]$ with the aid of Cramer's rule for the solution of linear equations. The resulting expression for the inverse matrix is

$$[a]^{-1} = \frac{[A_{ij}]^t}{|a|} = \frac{[A_{ji}]}{|a|} \tag{B32}$$

where the matrix $[A_{ij}]^t = [A_{ji}]$ is the *transpose* of the cofactor matrix $[A_{ij}]$ and is formed by merely interchanging rows and columns in $[A_{ij}]$. The superscript t refers to the transpose of the matrix and is not an exponent. For a square matrix this transposition involves simply the interchange of terms across the main diagonal. As an example of this operation,

$$\begin{bmatrix} a & b & c \\ d & e & f \\ g & h & i \end{bmatrix}^t = \begin{bmatrix} a & d & g \\ b & e & h \\ c & f & i \end{bmatrix}$$

Each element A_{ij} in $[A_{ij}]$ is the cofactor of a_{ij} in $|a|$ and equals $(-1)^{i+j}$ times the minor determinant M_{ij} formed from the elements of $|a|$ with the i-th row and the j-th column removed. Thus $A_{ij} = +M_{ij}$ if $i + j$ is even and $A_{ij} = -M_{ij}$ if $i + j$ is odd. The element A_{12} for the array of Eq. B25, for example, is $A_{12} = -(a_{21}a_{33} - a_{23}a_{31})$. The scalar quantity $|a|$ is the determinant of the elements in $[a]$ and cannot be zero if the governing equations in x and y are independent.

The procedure, then, for writing the inverse transform matrix $[a]^{-1}$ is to form the matrix $[A_{ij}]$ of the cofactors and then to exchange terms across the diagonal to get $[A_{ji}]$ which, when multiplied by $1/|a|$, gives $[a]^{-1}$.

It may be demonstrated that a matrix and its inverse obey the relation

$$[a][a]^{-1} = [1] \qquad \text{or} \qquad [a]^{-1}[a] = [1]$$

where $[1]$ is called the *unit matrix* and is

$$[1] = \begin{bmatrix} 1 & 0 & 0 \\ 0 & 1 & 0 \\ 0 & 0 & 1 \end{bmatrix}$$

If the x's of Eq. B26 are expressed in terms of other quantities $\{z\}$ such that

$$\{x\} = [b]\{z\}$$

then substitution into Eq. B26 yields

$$\{y\} = [a][b]\{z\} = [c]\{z\} \tag{B33}$$

The elements in the new matrix $[c]$ are formed from the elements of $[a]$ and $[b]$ by the rule given in Eq. B30.

The inverse multiple transformation may also occur where

$$\{z\} = [b]^{-1}\{x\}$$

which, upon substitution of Eq. B31, becomes

$$\{z\} = [b]^{-1}[a]^{-1}\{y\} = [c]^{-1}\{y\} \tag{B34}$$

where $[c]^{-1}$ may be formed directly from $[b]^{-1}[a]^{-1}$ by the rule of Eq. B30 with the appropriate change in symbols.

B15 Orthogonal Coordinate Transformations. When matrices are used for transforming quantities between orthogonal coordinate systems which are rotated with respect to one another, certain important simplifications occur. Under these conditions it may be shown that the inverse and the transpose of the transformation matrix are the same. Thus

$$[a]^{-1} = [a]^t \tag{B35}$$

Also the determinant of the elements of the transformation matrix equals unity, or

$$|a| = 1 \tag{B36}$$

and the cofactor matrices become

$$[A_{ij}] = [a] \qquad \text{and} \qquad [A_{ji}] = [a]^{-1} \tag{B37}$$

Transformations of this type were introduced in Art. 14 in the discussion of space coordinate systems.

B16 Symmetric Matrices and Tensors. The matrix $[a]$ in Eq. B25 is *symmetric* if $a_{ij} = a_{ji}$. The inertia matrix $[I]$, introduced in Art. 41, and the tensor which expresses the components of normal and shear stress at a point in a body are examples of symmetric matrices or tensors of the second rank. With a symmetrical array of elements it is seen that

$$\begin{aligned} [a]^t &= [a] \\ [A_{ij}] &= [A_{ji}] \\ [A_{ij}]^t &= [A_{ij}] \\ [a]^{-1} &= \frac{[A_{ij}]}{|a|} \end{aligned} \tag{B38}$$

These relations for symmetric matrices or tensors greatly simplify their transformations.

The transformation of a symmetric tensor of second rank, such as the inertia tensor $[I]$ of Eq. 149, from an orthogonal coordinate system α to a rotated orthogonal coordinate system β may be expressed in the following way. Equation 149 written for each of the two systems gives

$$\{H\}_\alpha = [I]_\alpha\{\omega\}_\alpha \qquad \text{and} \qquad \{H\}_\beta = [I]_\beta\{\omega\}_\beta$$

where the vectors **H** and ω in the two systems are related by the orthogonal transform $[T]$ for rotation of axes from α to β by the expressions

$$\{H\}_\beta = [T]\{H\}_\alpha \qquad \text{and} \qquad \{\omega\}_\beta = [T]\{\omega\}_\alpha$$

Examples of orthogonal transform matrices $[T]$ were worked out in Art. 14 and expressed in Eqs. 30, 33, and 35 for rotations between rectangular, cylindrical, and spherical coordinates. Elimination of $\{H\}_\alpha$ between the first and third relations and substitution of the resulting combination into the second relation with the sides of the equation reversed give

$$[I]_\beta\{\omega\}_\beta = [T][I]_\alpha\{\omega\}_\alpha$$

The fourth relation may be written $\{\omega\}_\alpha = [T]^{-1}\{\omega\}_\beta$ which, when substituted into the new combination, requires that

$$[I]_\beta = [T][I]_\alpha[T]^{-1}$$

or

$$[I]_\beta = [T][I]_\alpha[T]^t \tag{B39}$$

since the transpose and the inverse matrices are the same under orthogonal conditions. Equations B39 are the resulting tensor transformation relations. If i, j are the respective row and column indices for the inertia tensor in system β and k, l are the respective row and column indices for the inertia tensor in system α, and $[T]$ is the rotation transformation matrix from α to β, then the elements of I_β of Eq. B39 may be expressed in index notation by the sums

$$I_{ij} = \sum_k \sum_l T_{ik} T_{jl} I_{kl} \tag{B40}$$

where the indices have the values 1, 2, 3.

C USEFUL TABLES

Table C1 Properties

A. Density, ρ

	t/m^3	lbm/ft^3		t/m^3	lbm/ft^3
Aluminum	2.69	168	Iron (cast)	7.21	450
Concrete (av.)	2.40	150	Lead	11.37	710
Copper	8.91	556	Mercury	13.57	847
Earth (wet, av.)	1.76	110	Oil (av.)	0.90	56
(dry, av.)	1.28	80	Steel	7.83	489
Glass	2.59	162	Titanium	3.08	192
Gold	19.30	1205	Water (fresh)	1.00	62.4
Ice	0.90	56	(salt)	1.03	64
			Wood (soft pine)	0.48	30
			(hard, oak)	0.80	50

B. Coefficients of Friction

(The coefficients in the following table represent typical values under normal working conditions. Actual coefficients for a given situation will depend on the exact nature of the contacting surfaces. A variation of 25 to 100 per cent or more from these values could be expected in an actual application, depending on prevailing conditions of cleanliness, surface finish, pressure, lubrication, and velocity.)

Contacting Surface	Typical Values of Coefficient of Friction, f	
	Static	Kinetic
Steel on steel (dry)	0.6	0.4
Steel on steel (greasy)	0.1	0.05
Teflon on steel	0.04	0.04
Steel on babbitt (dry)	0.4	0.3
Steel on babbitt (greasy)	0.1	0.07
Brass on steel (dry)	0.5	0.4
Brake lining on cast iron	0.4	0.3
Rubber tires on smooth pavement (dry)	0.9	0.8
Wire rope on iron pulley (dry)	0.2	0.15
Hemp rope on metal	0.3	0.2
Metal on ice	—	0.02

	Coefficient of Rolling Friction, f_r
Pneumatic tires on smooth pavement	0.02
Steel tires on steel rails	0.006

Table C2 Solar System Constants

Universal gravitational constant $K = 6.673(10^{-11})\ \text{m}^3/(\text{kg}\cdot\text{s}^2)$
$= 3.439(10^{-8})\ \text{ft}^4/(\text{lbf-s}^4)$

Mass of Earth $m = 5.976(10^{24})$ kg
$= 4.095(10^{23})\ \text{lbf-s}^2/\text{ft}$

Period of Earth's rotation (1 sidereal day) = 23 h 56 min 4 s
= 23.9344 h

Angular velocity of Earth $\omega = 0.7292(10^{-4})$ rad/s

Angular velocity of Earth-Sun line $\omega' = 0.1991(10^{-6})$ rad/s

Mean velocity of Earth's center about Sun = 107 200 km/h

Body	Mean Distance to Sun km (mi)	Eccentricity of Orbit e	Period of Orbit solar days	Mean Diameter km (mi)	Mass Relative to Earth	Surface Gravitational Acceleration m/s^2 (ft/s^2)	Escape Velocity km/s (mi/s)
Sun	—	—	—	1 392 000 (865 000)	333 000	274 (898)	616 (383)
Moon	384 398* (238 854)*	0.055	27.32	3 476 (2 160)	0.0123	1.62 (5.32)	2.37 (1.47)
Mercury	57.3×10^6 (35.6×10^6)	0.206	87.97	5 000 (3 100)	0.054	3.47 (11.4)	4.17 (2.59)
Venus	108×10^6 (67.2×10^6)	0.0068	224.70	12 400 (7 700)	0.815	8.44 (27.7)	10.24 (6.36)
Earth	149.6×10^6 (92.96×10^6)	0.0167	365.26	12 742† (7 917)†	1.000	9.821‡ (32.22)‡	11.18 (6.95)
Mars	227.9×10^6 (141.6×10^6)	0.093	686.98	6 775 (4 210)	0.107	3.73 (12.3)	5.03 (3.13)

* Mean distance to Earth (center-to-center)

† Diameter of sphere of equal volume
polar diameter = 12 713 km
= 7 900 mi
equatorial diameter = 12 755 km
= 7 926 mi

‡ For nonrotating spherical Earth, equivalent to absolute value at sea level and latitude 37.5 deg

Table C3. Mathematical Relations

A. Series (expression in bracket following series indicates range of convergence)

$$(1 \pm x)^n = 1 \pm nx + \frac{n(n-1)}{2!}x^2 \pm \frac{n(n-1)(n-2)}{3!}x^3 + \cdots \qquad [x^2 < 1]$$

$$\sin x = x - \frac{x^3}{3!} + \frac{x^5}{5!} - \frac{x^7}{7!} + \cdots \qquad [x^2 < \infty]$$

$$\cos x = 1 - \frac{x^2}{2!} + \frac{x^4}{4!} - \frac{x^6}{6!} + \cdots \qquad [x^2 < \infty]$$

$$\sinh x = \frac{e^x - e^{-x}}{2} = x + \frac{x^3}{3!} + \frac{x^5}{5!} + \frac{x^7}{7!} + \cdots \qquad [x^2 < \infty]$$

$$\cosh x = \frac{e^x + e^{-x}}{2} = 1 + \frac{x^2}{2!} + \frac{x^4}{4!} + \frac{x^6}{6!} + \cdots \qquad [x^2 < \infty]$$

$$f(x) = \frac{a_0}{2} + \sum_{n=1}^{\infty} a_n \cos\frac{n\pi x}{l} + \sum_{n=1}^{\infty} b_n \sin\frac{n\pi x}{l}$$

$$\text{where } a_n = \frac{1}{l}\int_{-l}^{l} f(x)\cos\frac{n\pi x}{l}\,dx, \quad b_n = \frac{1}{l}\int_{-l}^{l} f(x)\sin\frac{n\pi x}{l}\,dx$$

[Fourier expansion for $-l < x < l$]

B. Derivatives

$$\frac{dx^n}{dx} = nx^{n-1}, \qquad \frac{d(uv)}{dx} = u\frac{dv}{dx} + v\frac{du}{dx}, \qquad \frac{d\left(\frac{u}{v}\right)}{dx} = \frac{v\frac{du}{dx} - u\frac{dv}{dx}}{v^2}$$

$$\lim_{\Delta x \to 0} \sin \Delta x = \sin dx = \tan dx = dx$$

$$\lim_{\Delta x \to 0} \cos \Delta x = \cos dx = 1$$

$$\frac{d \sin x}{dx} = \cos x, \qquad \frac{d \cos x}{dx} = -\sin x, \qquad \frac{d \tan x}{dx} = \sec^2 x$$

$$\frac{d \sinh x}{dx} = \cosh x, \qquad \frac{d \cosh x}{dx} = \sinh x, \qquad \frac{d \tanh x}{dx} = \text{sech}^2\, x$$

C. Integrals

$$\int x^n\, dx = \frac{x^{n+1}}{n+1}$$

$$\int \frac{dx}{x} = \ln x$$

$$\int \sqrt{a + bx}\, dx = \frac{2}{3b}\sqrt{(a + bx)^3}$$

$$\int x\sqrt{a + bx}\, dx = \frac{2}{15b^2}(3bx - 2a)\sqrt{(a + bx)^3}$$

$$\int \frac{dx}{\sqrt{a + bx}} = \frac{2\sqrt{a + bx}}{b}$$

$$\int \frac{x\, dx}{a + bx} = \frac{1}{b^2}[a + bx - a \ln (a + bx)]$$

$$\int \frac{x\, dx}{(a + bx)^n} = \frac{(a + bx)^{1-n}}{b^2}\left(\frac{a + bx}{2 - n} - \frac{a}{1 - n}\right)$$

$$\int \frac{dx}{a + bx^2} = \frac{1}{\sqrt{ab}}\tan^{-1}\frac{x\sqrt{ab}}{a} \quad \text{or} \quad \frac{1}{\sqrt{-ab}}\tanh^{-1}\frac{x\sqrt{-ab}}{a}$$

$$\int \sqrt{x^2 \pm a^2}\, dx = \tfrac{1}{2}[x\sqrt{x^2 \pm a^2} \pm a^2 \ln (x + \sqrt{x^2 \pm a^2})]$$

$$\int \sqrt{a^2 - x^2}\, dx = \tfrac{1}{2}\left(x\sqrt{a^2 - x^2} + a^2 \sin^{-1}\frac{x}{a}\right)$$

$$\int x\sqrt{a^2 - x^2}\, dx = -\tfrac{1}{3}\sqrt{(a^2 - x^2)^3}$$

$$\int x^2\sqrt{a^2 - x^2}\, dx = -\frac{x}{4}\sqrt{(a^2 - x^2)^3} + \frac{a^2}{8}\left(x\sqrt{a^2 - x^2} + a^2 \sin^{-1}\frac{x}{a}\right)$$

$$\int x^3\sqrt{a^2 - x^2}\, dx = -\tfrac{1}{5}(x^2 + \tfrac{2}{3}a^2)\sqrt{(a^2 - x^2)^3}$$

$$\int \frac{dx}{\sqrt{a + bx + cx^2}} = \frac{1}{\sqrt{c}} \ln\left(\sqrt{a + bx + cx^2} + x\sqrt{c} + \frac{b}{2\sqrt{c}}\right) \text{or} = \frac{-1}{\sqrt{-c}}\sin^{-1}\left(\frac{b + 2cx}{\sqrt{b^2 - 4ac}}\right)$$

$$\int \frac{dx}{\sqrt{x^2 \pm a^2}} = \ln (x + \sqrt{x^2 \pm a^2})$$

$$\int \frac{dx}{\sqrt{a^2 - x^2}} = \sin^{-1}\frac{x}{a}$$

$$\int x\sqrt{x^2 \pm a^2}\, dx = \tfrac{1}{3}\sqrt{(x^2 \pm a^2)^3}$$

$$\int x^2\sqrt{x^2 \pm a^2}\, dx = \frac{x}{4}\sqrt{(x^2 \pm a^2)^3} \mp \frac{a^2}{8}x\sqrt{x^2 \pm a^2} - \frac{a^4}{8}\ln (x + \sqrt{x^2 \pm a^2})$$

$$\int \frac{x\, dx}{\sqrt{x^2 - a^2}} = \sqrt{x^2 - a^2}$$

$$\int \frac{x\,dx}{\sqrt{a^2 \pm x^2}} = \pm\sqrt{a^2 \pm x^2}$$

$$\int \sin x\,dx = -\cos x$$

$$\int \cos x\,dx = \sin x$$

$$\int \sec x\,dx = \frac{1}{2}\ln\frac{1+\sin x}{1-\sin x}$$

$$\int \sin^2 x\,dx = \frac{x}{2} - \frac{\sin 2x}{4}$$

$$\int \cos^2 x\,dx = \frac{x}{2} + \frac{\sin 2x}{4}$$

$$\int \sin x \cos x\,dx = \frac{\sin^2 x}{2}$$

$$\int \sin^3 x\,dx = -\frac{\cos x}{3}(2 + \sin^2 x)$$

$$\int \cos^3 x\,dx = \frac{\sin x}{3}(2 + \cos^2 x)$$

$$\int x \sin x\,dx = \sin x - x\cos x$$

$$\int x \cos x\,dx = \cos x + x\sin x$$

$$\int x^2 \sin x\,dx = 2x\sin x - (x^2 - 2)\cos x$$

$$\int x^2 \cos x\,dx = 2x\cos x + (x^2 - 2)\sin x$$

$$\int \sinh x\,dx = \cosh x$$

$$\int \cosh x\,dx = \sinh x$$

$$\int \tanh x\,dx = \ln\cosh x$$

$$\int \ln x\,dx = x\ln x - x$$

$$\int e^{ax}\,dx = \frac{e^{ax}}{a}$$

Radius of curvature

$$\rho_{xy} = \frac{\left[1 + \left(\frac{dy}{dx}\right)^2\right]^{3/2}}{\frac{d^2y}{dx^2}}$$

$$\rho_{r\theta} = \frac{\left[r^2 + \left(\frac{dr}{d\theta}\right)^2\right]^{3/2}}{r^2 + 2\left(\frac{dr}{d\theta}\right)^2 - r\frac{d^2r}{d\theta^2}}$$

$$\int xe^{ax}\,dx = \frac{e^{ax}}{a^2}(ax - 1)$$

$$\int e^{ax}\sin px\,dx = \frac{e^{ax}(a\sin px - p\cos px)}{a^2 + p^2}$$

$$\int e^{ax}\cos px\,dx = \frac{e^{ax}(a\cos px + p\sin px)}{a^2 + p^2}$$

$$\int e^{ax}\sin^2 x\,dx = \frac{e^{ax}}{4 + a^2}\left(a\sin^2 x - \sin 2x + \frac{2}{a}\right)$$

$$\int e^{ax}\cos^2 x\,dx = \frac{e^{ax}}{4 + a^2}\left(a\cos^2 x + \sin 2x + \frac{2}{a}\right)$$

$$\int e^{ax}\sin x\cos x\,dx = \frac{e^{ax}}{4 + a^2}\left(\frac{a}{2}\sin 2x - \cos 2x\right)$$

Table C4. Properties of Plane Figures

Figure	Centroid	Area Moments of Inertia
Arc Segment	$\bar{r} = \dfrac{r \sin \alpha}{\alpha}$	—
Quarter and Semicircular Arcs	$\bar{y} = \dfrac{2r}{\pi}$	—
Triangular Area	$\bar{x} = \dfrac{a+b}{3}$ $\bar{y} = \dfrac{h}{3}$	$I_x = \dfrac{bh^3}{12}$ $\bar{I}_x = \dfrac{bh^3}{36}$ $I_{x_1} = \dfrac{bh^3}{4}$
Rectangular Area	—	$I_x = \dfrac{bh^3}{3}$ $\bar{I}_x = \dfrac{bh^3}{12}$ $\bar{J} = \dfrac{bh}{12}(b^2 + h^2)$
Area of Circular Sector	$\bar{x} = \dfrac{2}{3}\dfrac{r \sin \alpha}{\alpha}$	$I_x = \dfrac{r^4}{4}(\alpha - \frac{1}{2}\sin 2\alpha)$ $I_y = \dfrac{r^4}{4}(\alpha + \frac{1}{2}\sin 2\alpha)$ $J = \frac{1}{2}r^4\alpha$
Quarter Circular Area	$\bar{x} = \bar{y} = \dfrac{4r}{3\pi}$	$I_x = I_y = \dfrac{\pi r^4}{16}$ $J = \dfrac{\pi r^4}{8}$
Area of Elliptical Quadrant Area $A = \dfrac{\pi ab}{4}$	$\bar{x} = \dfrac{4a}{3\pi}$ $\bar{y} = \dfrac{4b}{3\pi}$	$I_x = \dfrac{\pi ab^3}{16}$ $I_y = \dfrac{\pi a^3 b}{16}$ $J = \dfrac{\pi ab}{16}(a^2 + b^2)$

Table C5. Properties of Homogeneous Solids
(m = mass of body shown)

Body	Mass Center	Moments of Inertia
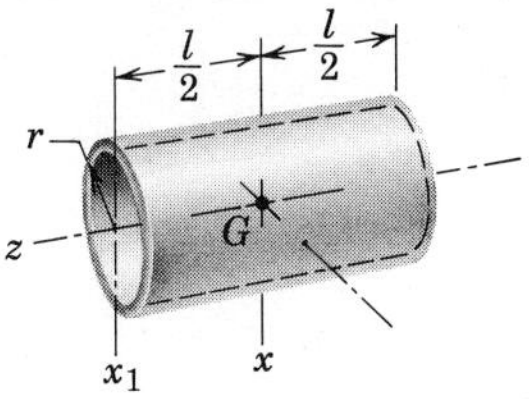 Circular Cylindrical Shell	—	$I_{xx} = \frac{1}{2}mr^2 + \frac{1}{12}ml^2$ $I_{x_1x_1} = \frac{1}{2}mr^2 + \frac{1}{3}ml^2$ $I_{zz} = mr^2$
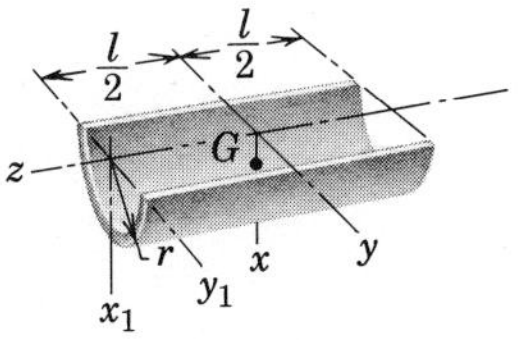 Half Cylindrical Shell	$\bar{x} = \frac{2r}{\pi}$	$I_{xx} = I_{yy}$ $= \frac{1}{2}mr^2 + \frac{1}{12}ml^2$ $I_{x_1x_1} = I_{y_1y_1}$ $= \frac{1}{2}mr^2 + \frac{1}{3}ml^2$ $I_{zz} = mr^2$
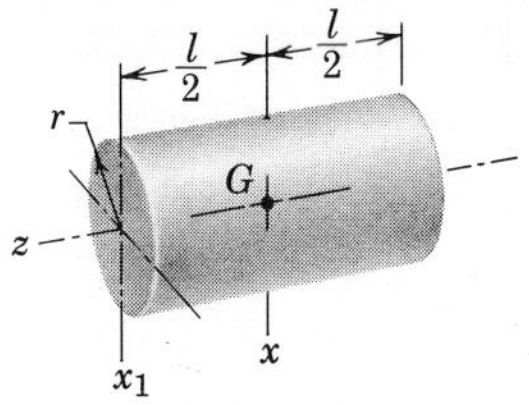 Circular Cylinder	—	$I_{xx} = \frac{1}{4}mr^2 + \frac{1}{12}ml^2$ $I_{x_1x_1} = \frac{1}{4}mr^2 + \frac{1}{3}ml^2$ $I_{zz} = \frac{1}{2}mr^2$
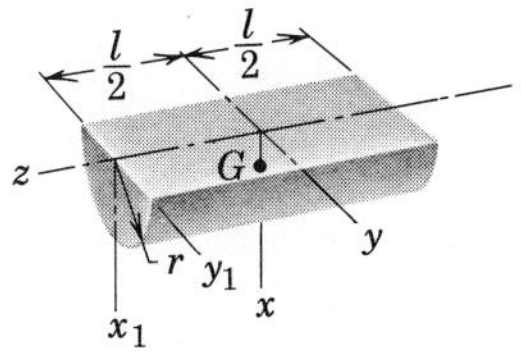 Semicylinder	$\bar{x} = \frac{4r}{3\pi}$	$I_{xx} = I_{yy}$ $= \frac{1}{4}mr^2 + \frac{1}{12}ml^2$ $I_{x_1x_1} = I_{y_1y_1}$ $= \frac{1}{4}mr^2 + \frac{1}{3}ml^2$ $I_{zz} = \frac{1}{2}mr^2$
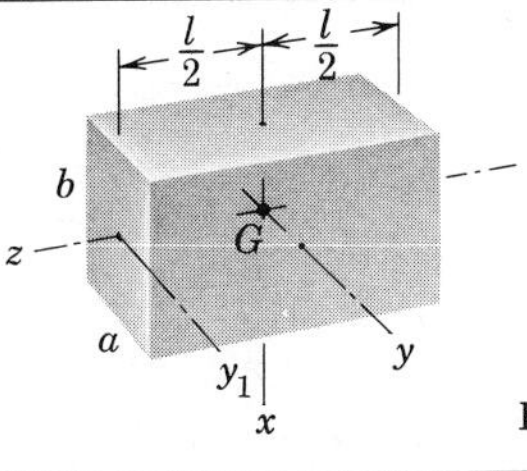 Rectangular Parallelepiped	—	$I_{xx} = \frac{1}{12}m(a^2 + l^2)$ $I_{yy} = \frac{1}{12}m(b^2 + l^2)$ $I_{zz} = \frac{1}{12}m(a^2 + b^2)$ $I_{y_1y_1} = \frac{1}{12}mb^2 + \frac{1}{3}ml^2$

Table C5. *Continued*
(m = mass of body shown)

Body	Mass Center	Moments of Inertia
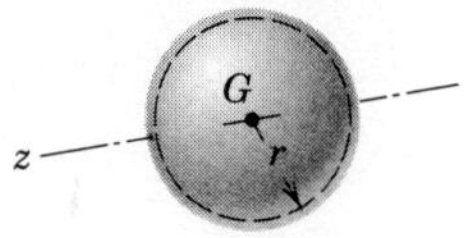 Spherical Shell	—	$I_{zz} = \frac{2}{3}mr^2$
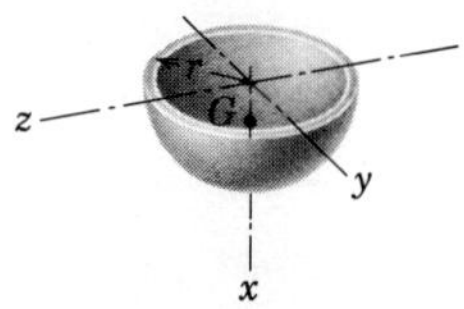 Hemispherical Shell	$\bar{x} = \frac{r}{2}$	$I_{xx} = I_{yy} = I_{zz} = \frac{2}{3}mr^2$
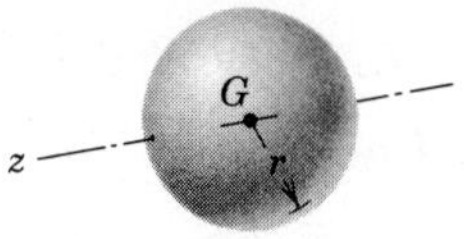 Sphere	—	$I_{zz} = \frac{2}{5}mr^2$
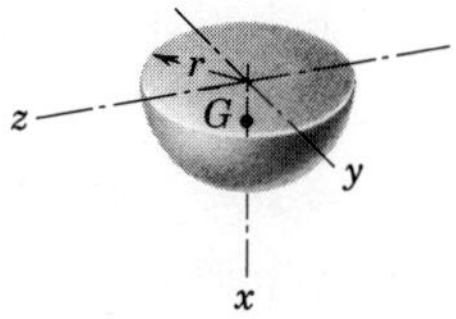 Hemisphere	$\bar{x} = \frac{3r}{8}$	$I_{xx} = I_{yy} = I_{zz} = \frac{2}{5}mr^2$
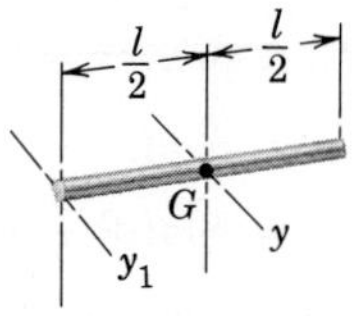 Uniform Slender Rod	—	$I_{yy} = \frac{1}{12}ml^2$ $I_{y_1y_1} = \frac{1}{3}ml^2$

Table C5. *Continued*
(m = mass of body shown)

Body	Mass Center	Moments of Inertia
Quarter Circular Rod	$\bar{x} = \bar{y} = \dfrac{2r}{\pi}$	$I_{xx} = I_{yy} = \frac{1}{2}mr^2$ $I_{zz} = mr^2$
Elliptical Cylinder	—	$I_{xx} = \frac{1}{4}ma^2 + \frac{1}{12}ml^2$ $I_{yy} = \frac{1}{4}mb^2 + \frac{1}{12}ml^2$ $I_{zz} = \frac{1}{4}m(a^2 + b^2)$ $I_{y_1y_1} = \frac{1}{4}mb^2 + \frac{1}{3}ml^2$
Conical Shell	$\bar{z} = \dfrac{2h}{3}$	$I_{yy} = \frac{1}{4}mr^2 + \frac{1}{2}mh^2$ $I_{y_1y_1} = \frac{1}{4}mr^2 + \frac{1}{6}mh^2$ $I_{zz} = \frac{1}{2}mr^2$
Half Conical Shell	$\bar{x} = \dfrac{4r}{3\pi}$ $\bar{z} = \dfrac{2h}{3}$	$I_{xx} = I_{yy} = \frac{1}{4}mr^2 + \frac{1}{2}mh^2$ $I_{x_1x_1} = I_{y_1y_1} = \frac{1}{4}mr^2 + \frac{1}{6}mh^2$ $I_{zz} = \frac{1}{2}mr^2$
Right Circular Cone	$\bar{z} = \dfrac{3h}{4}$	$I_{yy} = \frac{3}{20}mr^2 + \frac{3}{5}mh^2$ $I_{y_1y_1} = \frac{3}{20}mr^2 + \frac{1}{10}mh^2$ $I_{zz} = \frac{3}{10}mr^2$

Table C5. *Continued*
(m = mass of body shown)

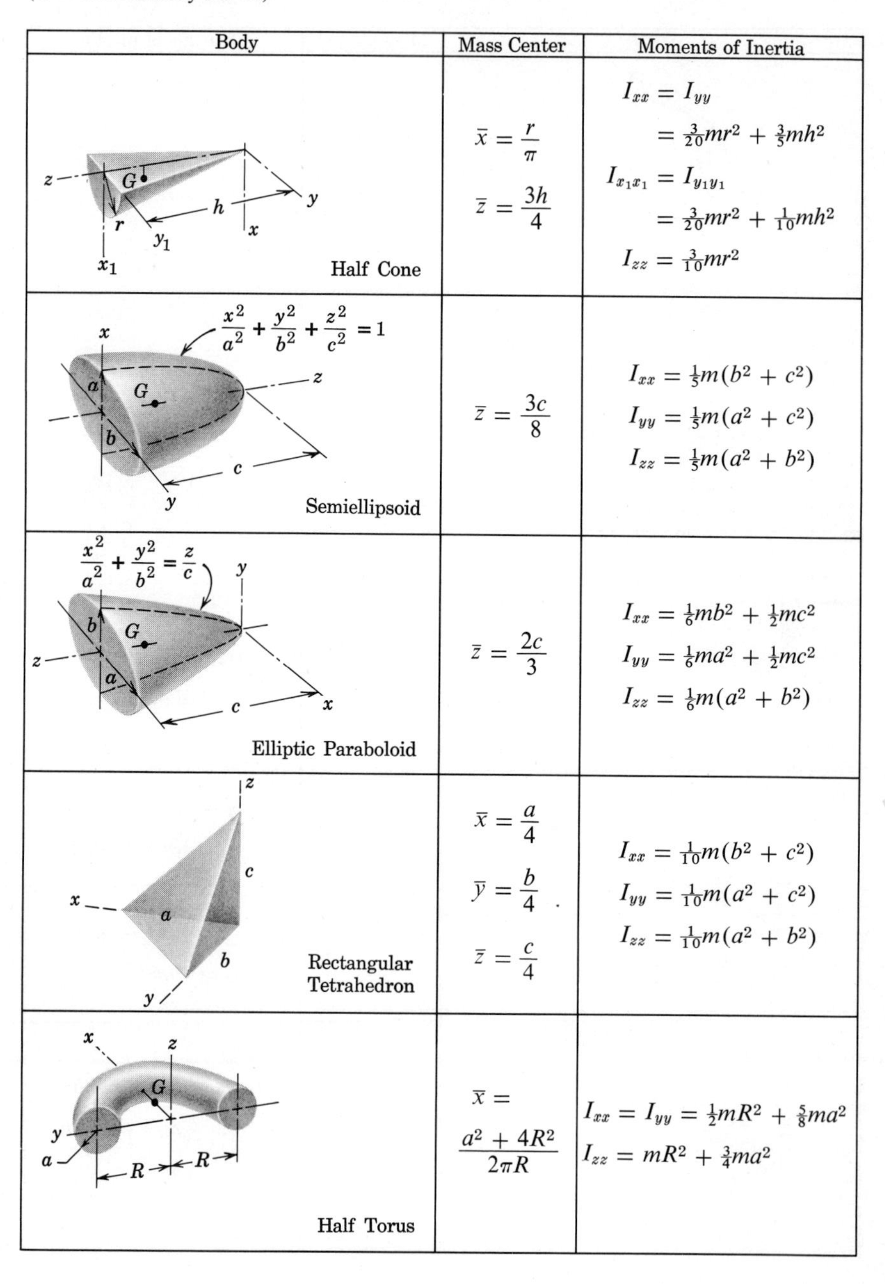

Body	Mass Center	Moments of Inertia
Half Cone	$\bar{x} = \frac{r}{\pi}$ $\bar{z} = \frac{3h}{4}$	$I_{xx} = I_{yy} = \frac{3}{20}mr^2 + \frac{3}{5}mh^2$ $I_{x_1x_1} = I_{y_1y_1} = \frac{3}{20}mr^2 + \frac{1}{10}mh^2$ $I_{zz} = \frac{3}{10}mr^2$
Semiellipsoid $\frac{x^2}{a^2} + \frac{y^2}{b^2} + \frac{z^2}{c^2} = 1$	$\bar{z} = \frac{3c}{8}$	$I_{xx} = \frac{1}{5}m(b^2 + c^2)$ $I_{yy} = \frac{1}{5}m(a^2 + c^2)$ $I_{zz} = \frac{1}{5}m(a^2 + b^2)$
Elliptic Paraboloid $\frac{x^2}{a^2} + \frac{y^2}{b^2} = \frac{z}{c}$	$\bar{z} = \frac{2c}{3}$	$I_{xx} = \frac{1}{6}mb^2 + \frac{1}{2}mc^2$ $I_{yy} = \frac{1}{6}ma^2 + \frac{1}{2}mc^2$ $I_{zz} = \frac{1}{6}m(a^2 + b^2)$
Rectangular Tetrahedron	$\bar{x} = \frac{a}{4}$ $\bar{y} = \frac{b}{4}$ $\bar{z} = \frac{c}{4}$	$I_{xx} = \frac{1}{10}m(b^2 + c^2)$ $I_{yy} = \frac{1}{10}m(a^2 + c^2)$ $I_{zz} = \frac{1}{10}m(a^2 + b^2)$
Half Torus	$\bar{x} = \frac{a^2 + 4R^2}{2\pi R}$	$I_{xx} = I_{yy} = \frac{1}{2}mR^2 + \frac{5}{8}ma^2$ $I_{zz} = mR^2 + \frac{3}{4}ma^2$

INDEX